C0-ASZ-315

ELECTRIC POWER DISTRIBUTION HANDBOOK

The ELECTRIC POWER ENGINEERING Series
Series Editor Leo L. Grigsby

Published Titles

Computational Methods for Electric Power Systems
Mariesa Crow

Distribution System Modeling and Analysis
William H. Kersting

Electric Drives
Ion Boldea and Syed Nasar

Electrical Energy Systems
Mohamed E. El-Hawary

Electric Power Substations Engineering
John D. McDonald

Electric Power Transformer Engineering
James H. Harlow

Electromechanical Systems, Electric Machines, and Applied Mechatronics
Sergey E. Lyshevski

The Induction Machine Handbook
Ion Boldea and Syed Nasar

Linear Synchronous Motors: Transportation and Automation Systems
Jacek Gieras and Jerry Piech

Power Quality
C. Sankaran

Power System Operations and Electricity Markets
Fred I. Denny and David E. Dismukes

Electric Power Distribution Handbook
Tom Short

Forthcoming Titles

The Electric Generators Handbook
Ion Boldea

ELECTRIC POWER DISTRIBUTION HANDBOOK

T.A. SHORT

CRC PRESS

Boca Raton London New York Washington, D.C.

Library of Congress Cataloging-in-Publication Data

Short, Tom, 1966–
Electric power distribution handbook / Tom Short.
p. cm. — (Electric power engineering series)
Includes bibliographical references and index.
ISBN 0-8493-1791-6 (alk. paper)
1. Electric power distribution—Handbooks, manuals, etc. I. Title. II. Series.

TK3001.S47 2003
621.319—dc21 2003055214

Direct all inquiries to CRC Press LLC, 2000 N.W. Corporate Blvd., Boca Raton, Florida 33431.

International Standard Book Number 0-8493-1791-6
Library of Congress Card Number 2003055214
Printed in the United States of America 1 2 3 4 5 6 7 8 9 0
Printed on acid-free paper

Dedication

To the future. To Jared. To Logan.

Preface

In industrialized countries, distribution systems deliver electricity literally everywhere, taking power generated at many locations and delivering it to end users. Generation, transmission, and distribution — of these big three components of the electricity infrastructure, the distribution system gets the least attention. Yet, it is often the most critical component in terms of its effect on reliability and quality of service, cost of electricity, and aesthetic (mainly visual) impacts on society.

Like much of the electric utility industry, several political, economic, and technical changes are pressuring the way distribution systems are built and operated. Deregulation has increased pressures on electric power utilities to cut costs and has focused emphasis on reliability and quality of electric service. The great fear of deregulation is that service will suffer because of cost cutting. Regulators and utility consumers are paying considerable attention to reliability and quality. Another change that is brewing is the introduction of distributed generation on the distribution system. Generators at the distribution level can cause problems (and have benefits if properly applied). Customers are pressing for lower costs, better reliability, and less visual impact from utility distribution systems.

Deregulation and technical changes increase the need by utility engineers for better information. This book helps fill some of those needs in the area of electric distribution systems. The first few chapters of the book focus on equipment-oriented information and applications such as choosing transformer connections, sizing and placing capacitors, and setting regulators.

The middle portion of this handbook contains many sections targeting reliability and power quality. The performance of the distribution system determines greater than 90% of the reliability of service to customers (the high-voltage transmission and generation system determines the rest). If performance is increased, it will have to be done on the distribution system.

Near the end, we tackle lightning protection, grounding, and safety. Safety is a very important consideration in the design, operation, and maintenance of distribution facilities. The last chapter on distributed generation provides information to help utilities avoid problems caused by the introduction of distributed generation.

I hope you find useful information in this book. If it is not in here, hopefully one of the many bibliographic references will lead you to what you are looking for. Please feel free to email me feedback on this book including errors, comments, opinions, or new sources of information. I would like to hear from you. Also, if you need my help with any interesting consulting or research opportunities, I would love to hear from you.

Tom Short
EPRI PEAC
Schenectady, New York
t.short@ieee.org

Foreword

The public profile of electricity and those who provide it is likely to dramatically increase. Unlike the industrial analog economy of the 20th century, the networked digital economy only runs on electricity — that is, a perfect stream of electrons seamlessly linked with a real-time flow of information. Productivity and competitive advantage depend on rapidly facilitating this transformation.

The challenges of the digital economy come at a time of rapid change for the electric power industry and especially for its power delivery system. Meanwhile, many areas are now considering ways to increase competition in retail markets and provide customers with greater choice among electricity providers. The capability of the power delivery system, however, has not kept up with the new demands brought by deregulation. Most transmission and distribution systems were designed more than a half-century ago, when long-distance power transfer was used mainly for economic exchange among a few utilities and when the reliability requirements of distribution systems were much less severe than in a digital economy. So far, the needed improvements in both capacity and reliability have not been made.

While utility loads have grown and users are more demanding, the distribution infrastructure has not kept pace. Consider the period from 1989 to 1999 when total electricity demand in the U.S. rose by nearly 30% — over the same period, expenditures by investor-owned utilities for distribution system construction fell by about 10% in real terms (based on data from EEI's *Statistical Yearbook*, 1999).

Mr. Short's handbook provides many tools to help address the challenges of providing a more reliable distribution system given significant cost constraints. In addition to a wealth of classic information on distribution practices, his handbook provides new insights based on recent research by EPRI, the Consortium for Electric Infrastructure to Support a Digital Society (CEIDS), IEEE, and others.

As someone who started his utility career as an electrician, I appreciate the practical advice provided by this handbook. While the book has leading-edge research and a great deal of in-depth technical information, it also manages to offer considerable practical information. As a former practitioner, I also appreciate the handbook's consideration of issues affecting safety.

This handbook is a fine addition to CRC Press's Electric Power Engineering Series, which distribution engineers should find useful for many years.

Clark W. Gellings, PE
Vice President — Power Delivery and Markets
EPRI
Palo Alto, California

Acknowledgments

First and foremost, I would like to thank my wife Kristin — thank you for your strength, thank you for your help, thank you for your patience, and thank you for your love. My play buddies, Logan and Jared, energized me and made me laugh. My family was a source of inspiration. I would like to thank my parents, Bob and Sandy, for their influence and education over the years.

EPRI PEAC provided a great deal of support on this project. I would like to recognize the reviews, ideas, and support of Phil Barker and Dave Crudele here in Schenectady, New York, and also Arshad Mansoor, Mike Howard, Charles Perry, Arindam Maitra, and the rest of the energetic crew in Knoxville, TN.

Many other people reviewed portions of the draft and provided input and suggestions, including Dave Smith (Power Technologies, Inc.), Dan Ward (Dominion Virginia Power), Jim Stewart (Consultant, Scotia, NY), Conrad St. Pierre (Electric Power Consultants), Karl Fender (Cooper Power Systems), John Leach (Hi-Tech Fuses, Inc.), and Rusty Bascom (Power Delivery Consultants).

Thanks to Power Technologies, Inc., for opportunities and mentoring during my early career with the help of several talented, helpful engineers, including Jim Burke, Phil Barker, Dave Smith, Jim Stewart, and John Anderson. Over the years, several clients have also educated me in many ways; two that stand out are Ron Ammon (Keyspan, retired) and Clay Burns (National Grid).

EPRI has been supportive of this project, including a review by Luther Dow. EPRI has also sponsored a number of interesting distribution research projects that I have been fortunate enough to be involved with, and EPRI has allowed me to share some of those efforts here.

As a side-note, I would like to recognize the efforts of linemen in the electric power industry. These folks do the real work of building the lines and keeping the power on. As a tribute to them, a trailer at the end of each chapter reveals a bit of the lineman's character and point of view.

Credits

Tables 4.3 to 4.7 and 4.13 are reprinted with permission from IEEE Std. C57.12.00-2000. *IEEE Standard General Requirements for Liquid-Immersed Distribution, Power, and Regulating Transformers.* Copyright 2000 by IEEE.

Figure 4.17 is reprinted with permission from ANSI/IEEE Std. C57.105-1978. *IEEE Guide for Application of Transformer Connections in Three-Phase Distribution Systems.* Copyright 1978 by IEEE.

Tables 6.2, 6.4, and 6.5 are reprinted with permission from IEEE Std. 18-2002. *IEEE Standard for Shunt Power Capacitors.* Copyright 2002 by IEEE.

Table 6.3 is reprinted with permission from ANSI/IEEE Std. 18-1992. *IEEE Standard for Shunt Power Capacitors.* Copyright 1993 by IEEE.

Table 7.9 is reprinted with permission from IEEE Std. 493-1997. *IEEE Recommended Practice for the Design of Reliable Industrial and Commercial Power Systems (Gold Book).* Copyright 1998 by IEEE.

Table 8.3 is reprinted with permission from IEEE Std. C37.112-1996. *IEEE Standard Inverse-Time Characteristic Equations for Overcurrent Relays.* Copyright 1997 by IEEE.

Tables 11.1 to 11.3 are reprinted with permission from IEEE Std. 519-1992. *IEEE Recommended Practices and Requirements for Harmonic Control in Electrical Power Systems.* Copyright 1993 by IEEE.

Figure 11.19 is reprinted with permission from IEEE Std. 141-1993. *IEEE Recommended Practice for Electric Power Distribution for Industrial Plants.* Copyright 1994 by IEEE.

Tables 12.4 and 12.5 are reprinted with permission from IEEE Std. C62.22-1997. *IEEE Guide for the Application of Metal-Oxide Surge Arresters for Alternating-Current Systems.* Copyright 1998 by IEEE.

Figure 12.16 is reprinted with permission from IEEE Std. 4-1995. *IEEE Standard Techniques for High-Voltage Testing.* Copyright 1995 by IEEE.

Figure 12.19 is reprinted with permission from IEEE Std. 1299/C62.22.1-1996. IEEE Guide for the Connection of Surge Arresters to Protect Insulated, Shielded Electric Power Cable Systems. Copyright 1997 by IEEE.

Section 12.7.2.1 is reprinted with permission from IEEE Std. 1410-1997. *IEEE Guide for Improving the Lightning Performance of Electric Power Overhead Distribution Lines.* Copyright 1997 by IEEE.

Tables 13.3 and 13.4 are reprinted with permission from IEEE Std. 142-1991. *IEEE Recommended Practice for Grounding of Industrial and Commercial Power Systems.* Copyright 1992 by IEEE.

Figure 13.15 is reprinted with permission from IEEE Std. 80-2000. *IEEE Guide for Safety in AC Substation Grounding.* Copyright 2000 by IEEE.

The IEEE disclaims any responsibility or liability resulting from the placement and use in the described manner.

Contents

1

Fundamentals of Distribution Systems

Electrification in the early 20th century dramatically improved productivity and increased the well-being of the industrialized world. No longer a luxury — now a necessity — electricity powers the machinery, the computers, the health-care systems, and the entertainment of modern society. Given its benefits, electricity is inexpensive, and its price continues to slowly decline (after adjusting for inflation — see Figure 1.1).

Electric power distribution is the portion of the power delivery infrastructure that takes the electricity from the highly meshed, high-voltage transmission circuits and delivers it to customers. Primary distribution lines are "medium-voltage" circuits, normally thought of as 600 V to 35 kV. At a distribution substation, a substation transformer takes the incoming transmission-level voltage (35 to 230 kV) and steps it down to several distribution primary circuits, which fan out from the substation. Close to each end user, a distribution transformer takes the primary-distribution voltage and steps it down to a low-voltage secondary circuit (commonly 120/240 V; other utilization voltages are used as well). From the distribution transformer, the secondary distribution circuits connect to the end user where the connection is made at the service entrance. Figure 1.2 shows an overview of the power generation and delivery infrastructure and where distribution fits in. Functionally, distribution circuits are those that feed customers (this is how the term is used in this book, regardless of voltage or configuration). Some also think of distribution as anything that is radial or anything that is below 35 kV.

The distribution infrastructure is extensive; after all, electricity has to be delivered to customers concentrated in cities, customers in the suburbs, and customers in very remote regions; few places in the industrialized world do not have electricity from a distribution system readily available. Distribution circuits are found along most secondary roads and streets. Urban construction is mainly underground; rural construction is mainly overhead. Suburban structures are a mix, with a good deal of new construction going underground.

A mainly urban utility may have less than 50 ft of distribution circuit for each customer. A rural utility can have over 300 ft of primary circuit per customer.

Several entities may own distribution systems: municipal governments, state agencies, federal agencies, rural cooperatives, or investor-owned utili-

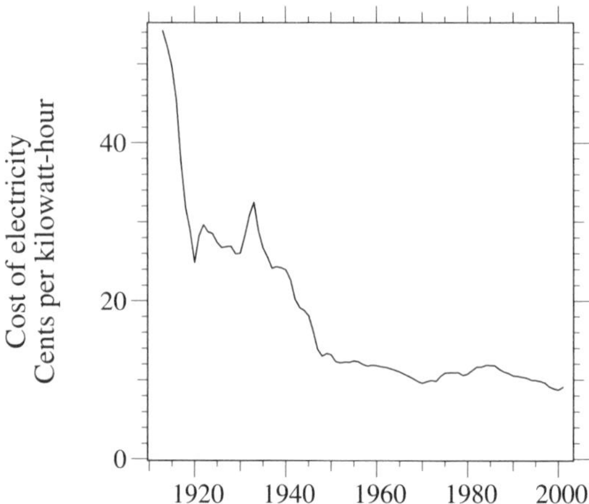

FIGURE 1.1
Cost of U.S. electricity adjusted for inflation to year 2000 U.S. dollars. (Data from U.S. city average electricity costs from the U.S. Bureau of Labor Statistics.)

ties. In addition, large industrial facilities often need their own distribution systems. While there are some differences in approaches by each of these types of entities, the engineering issues are similar for all.

For all of the action regarding deregulation, the distribution infrastructure remains a natural monopoly. As with water delivery or sewers or other utilities, it is difficult to imagine duplicating systems to provide true competition, so it will likely remain highly regulated.

Because of the extensive infrastructure, distribution systems are capital-intensive businesses. An Electric Power Research Institute (EPRI) survey found that the distribution plant asset carrying cost averages 49.5% of the total distribution resource (EPRI TR-109178, 1998). The next largest component is labor at 21.8%, followed by materials at 12.9%. Utility annual distribution budgets average about 10% of the capital investment in the distribution system. On a kilowatt-hour basis, utility distribution budgets average 0.89 cents per kilowatt-hour (see Table 1.1 for budgets shown relative to other benchmarks).

Low cost, simplification, and standardization are all important design characteristics of distribution systems. Few components and/or installations are individually engineered on a distribution circuit. Standardized equipment and standardized designs are used wherever possible. "Cookbook" engineering methods are used for much of distribution planning, design, and operations.

Distribution planning is the study of future power delivery needs. Planning goals are to provide service at low cost and high reliability. Planning requires a mix of geographic, engineering, and economic analysis skills. New circuits (or other solutions) must be integrated into the existing distribution system within a variety of economic, political, environmental, electrical, and geographic constraints. The planner needs estimates of load

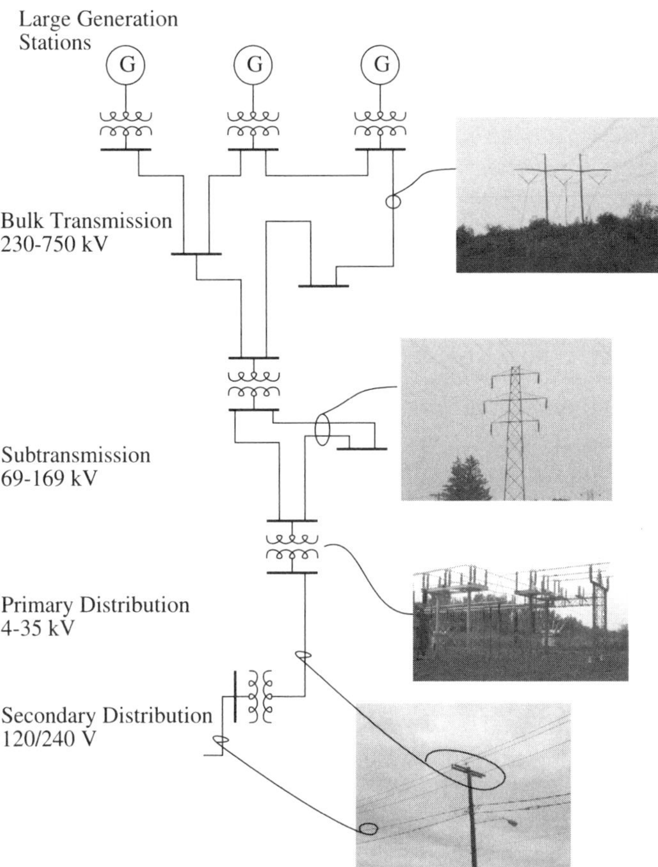

FIGURE 1.2
Overview of the electricity infrastructure.

TABLE 1.1

Surveyed Annual Utility Distribution Budgets in U.S. Dollars

	Average	Range
Per dollar of distribution asset	0.098	0.0916–0.15
Per customer	195	147–237
Per thousand kWH	8.9	3.9–14.1
Per mile of circuit	9,400	4,800–15,200
Per substation	880,000	620,000–1,250,000

Source: EPRI TR-109178, *Distribution Cost Structure — Methodology and Generic Data,* Electric Power Research Institute, Palo Alto, CA, 1998.

growth, knowledge of when and where development is occurring, and local development regulations and procedures. While this book has some material that should help distribution planners, many of the tasks of a planner, like load forecasting, are not discussed. For more information on distribution planning, see Willis's *Power Distribution Planning Reference Book* (1997), IEEE's *Power Distribution Planning* tutorial (1992), and the *CEA Distribution Planner's Manual* (1982).

1.1 Primary Distribution Configurations

Distribution circuits come in many different configurations and circuit lengths. Most share many common characteristics. Figure 1.3 shows a "typical" distribution circuit, and Table 1.2 shows typical parameters of a distribution circuit. A *feeder* is one of the circuits out of the substation. The main feeder is the three-phase backbone of the circuit, which is often called the *mains* or *mainline*. The mainline is normally a modestly large conductor such as a 500- or 750-kcmil aluminum conductor. Utilities often design the main feeder for 400 A and often allow an emergency rating of 600 A. Branching from the mains are one or more *laterals*, which are also called taps, lateral taps, branches, or branch lines. These laterals may be single-phase, two-phase, or three-phase. The laterals normally have fuses to separate them from the mainline if they are faulted.

The most common distribution primaries are four-wire, multigrounded systems: three-phase conductors plus a multigrounded neutral. Single-phase loads are served by transformers connected between one phase and the neutral. The neutral acts as a return conductor and as an equipment safety ground (it is grounded periodically and at all equipment). A single-phase line has one phase conductor and the neutral, and a two-phase line has two phases and the neutral. Some distribution primaries are three-wire systems (with no neutral). On these, single-phase loads are connected phase to phase, and single-phase lines have two of the three phases.

There are several configurations of distribution systems. Most distribution circuits are radial (both primary and secondary). Radial circuits have many advantages over networked circuits including

- Easier fault current protection
- Lower fault currents over most of the circuit
- Easier voltage control
- Easier prediction and control of power flows
- Lower cost

Distribution primary systems come in a variety of shapes and sizes (Figure 1.4). Arrangements depend on street layouts, the shape of the area covered

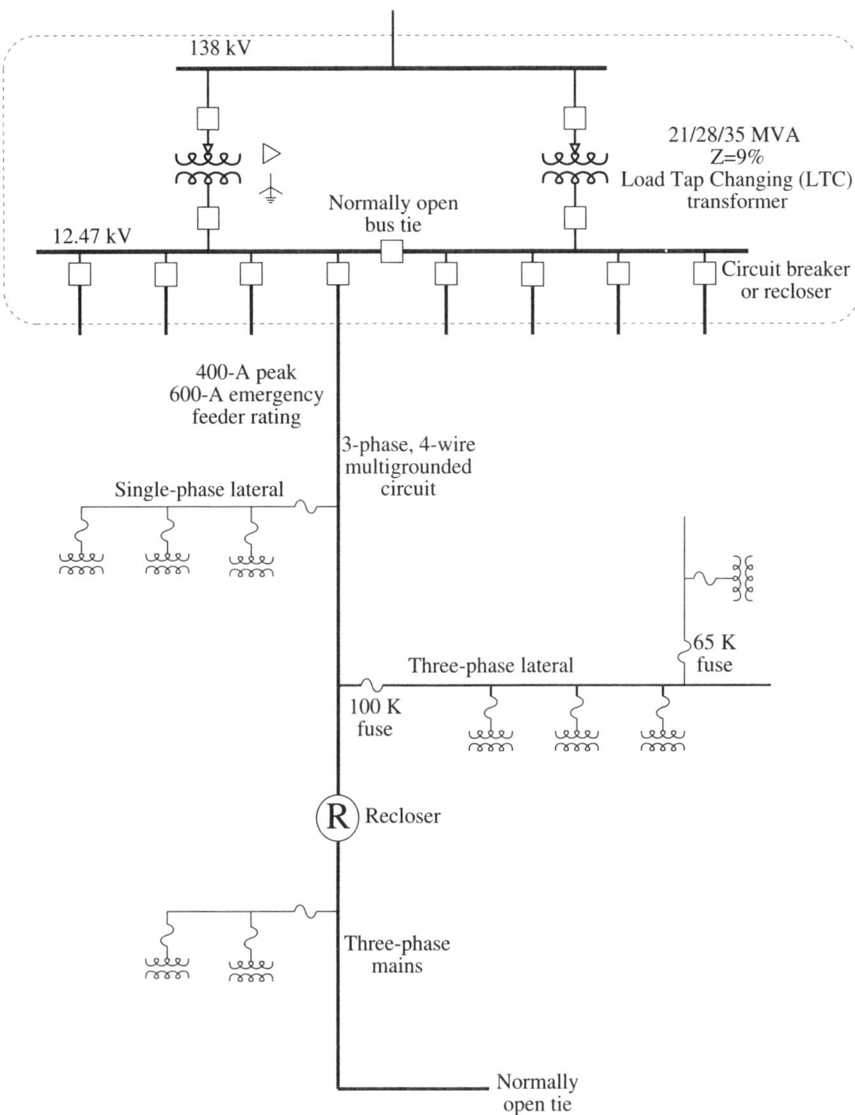

FIGURE 1.3
Typical distribution substation with one of several feeders shown (many lateral taps are left off). (Copyright © 2000. Electric Power Research Institute. 1000419. *Engineering Guide for Integration of Distributed Generation and Storage Into Power Distribution Systems.* Reprinted with permission.)

TABLE 1.2

Typical Distribution Circuit Parameters

	Most Common Value	Other Common Values
Substation characteristics		
Voltage	12.47 kV	4.16, 4.8, 13.2, 13.8, 24.94, 34.5 kV
Number of station transformers	2	1–6
Substation transformer size	21 MVA	5–60 MVA
Number of feeders per bus	4	1–8
Feeder characteristics		
Peak current	400 A	100–600 A
Peak load	7 MVA	1–15 MVA
Power factor	0.98 lagging	0.8 lagging–0.95 leading
Number of customers	400	50–5000
Length of feeder mains	4 mi	2–15 mi
Length including laterals	8 mi	4–25 mi
Area covered	25 mi^2	0.5–500 mi^2
Mains wire size	500 kcmil	4/0–795 kcmil
Lateral tap wire size	1/0	#4–2/0
Lateral tap peak current	25 A	5–50 A
Lateral tap length	0.5 mi	0.2–5 mi
Distribution transformer size (1 ph)	25 kVA	10–150 kVA

by the circuit, obstacles (like lakes), and where the big loads are. A common suburban layout has the main feeder along a street with laterals tapped down side streets or into developments. Radial distribution feeders may also have extensive branching — whatever it takes to get to the loads. An *express feeder* serves load concentrations some distance from the substation. A three-phase mainline runs a distance before tapping loads off to customers. With many circuits coming from one substation, a number of the circuits may have express feeders; some feeders cover areas close to the substation, and express feeders serve areas farther from the substation.

For improved reliability, radial circuits are often provided with normally open tie points to other circuits as shown in Figure 1.5. The circuits are still operated radially, but if a fault occurs on one of the circuits, the tie switches allow some portion of the faulted circuit to be restored quickly. Normally, these switches are manually operated, but some utilities use automated switches or reclosers to perform these operations automatically.

A primary-loop scheme is an even more reliable service that is sometimes offered for critical loads such as hospitals. Figure 1.6 shows an example of a primary loop. The key feature is that the circuit is "routed through" each

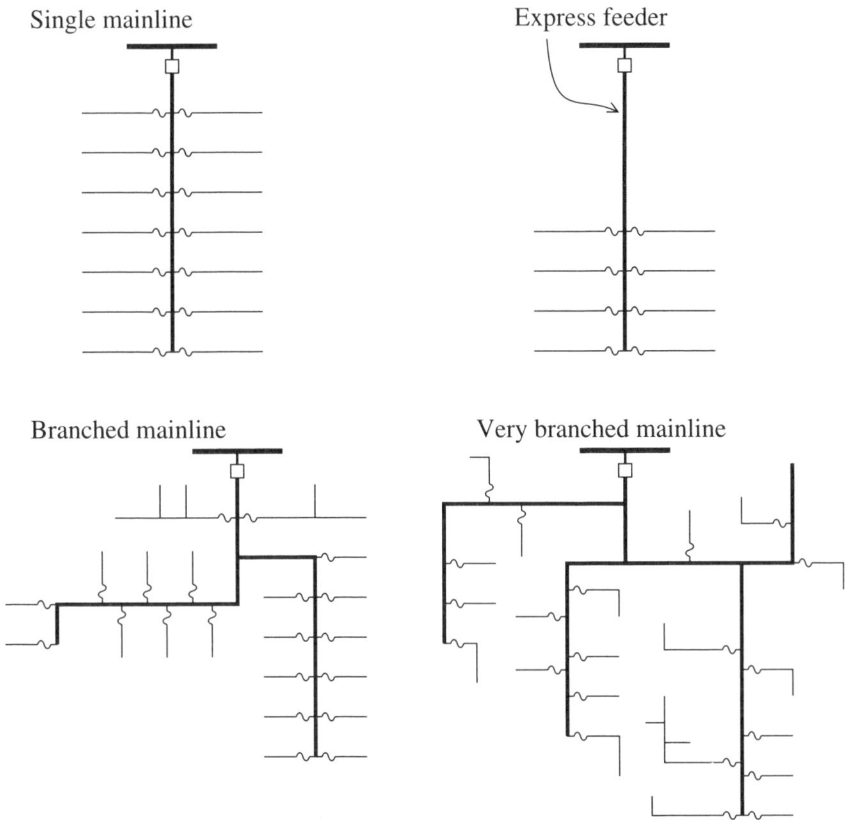

FIGURE 1.4
Common distribution primary arrangements.

critical customer transformer. If any part of the primary circuit is faulted, all critical customers can still be fed by reconfiguring the transformer switches.

Primary-loop systems are sometimes used on distribution systems for areas needing high reliability (meaning limited long-duration interruptions). In the open-loop design where the loop is left normally open at some point, primary-loop systems have almost no benefits for momentary interruptions or voltage sags. They are rarely operated in a closed loop. A widely reported installation of a sophisticated *closed* system has been installed in Orlando, FL, by Florida Power Corporation (Pagel, 2000). An example of this type of closed-loop primary system is shown in Figure 1.7. Faults on any of the cables in the loop are cleared in less than six cycles, which reduces the duration of the voltage sag during the fault (enough to help many computers). Advanced relaying similar to transmission-line protection is necessary to coordinate the protection and operation of the switchgear in the looped system. The relaying scheme uses a transfer trip with permissive over-reaching (the relays at each end of the cable must agree there is a fault

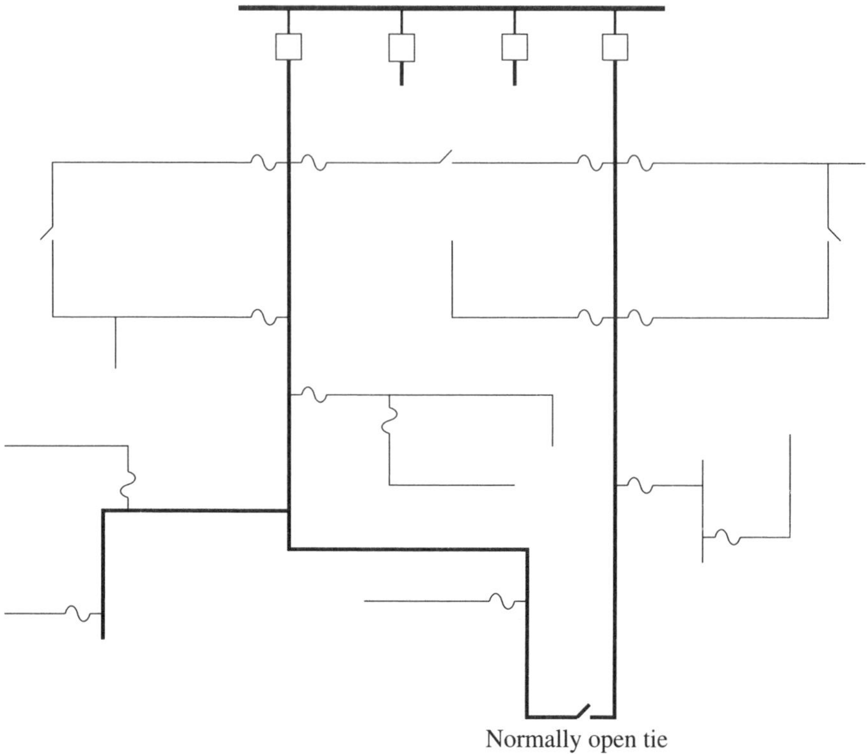

FIGURE 1.5
Two radial circuits with normally open ties to each other. (Copyright © 2000. Electric Power Research Institute. 1000419. *Engineering Guide for Integration of Distributed Generation and Storage Into Power Distribution Systems.* Reprinted with permission.)

between them with communications done on fiberoptic lines). A backup scheme uses directional relays, which will trip for a fault in a certain direction unless a blocking signal is received from the remote end (again over the fiberoptic lines).

Critical customers have two more choices for more reliable service where two primary feeds are available. Primary selective and secondary selective schemes both are normally fed from one circuit (see Figure 1.8). So, the circuits are still radial. In the event of a fault on the primary circuit, the service is switched to the backup circuit. In the primary selective scheme, the switching occurs on the primary, and in the secondary selective scheme, the switching occurs on the secondary. The switching can be done manually or automatically, and there are even static transfer switches that can switch in less than a half cycle to reduce momentary interruptions and voltage sags.

Today, the primary selective scheme is preferred mainly because of the cost associated with the extra transformer in a secondary selective scheme. The normally closed switch on the primary-side transfer switch opens after

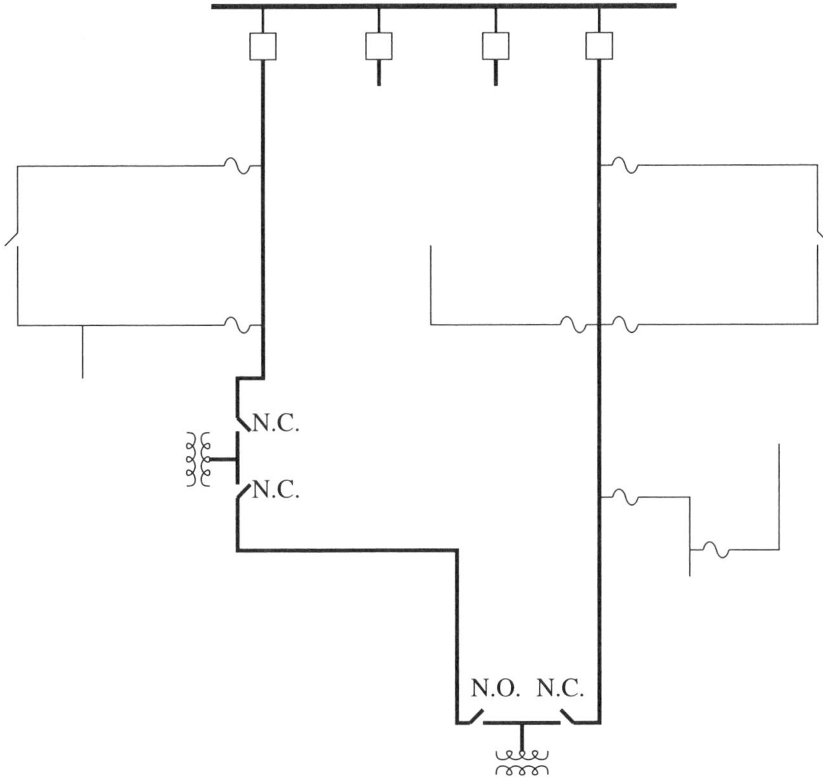

FIGURE 1.6
Primary loop distribution arrangement. (Copyright © 2000. Electric Power Research Institute. 1000419. *Engineering Guide for Integration of Distributed Generation and Storage Into Power Distribution Systems.* Reprinted with permission.)

sensing a loss of voltage. It normally has a time delay on the order of seconds — enough to ride through the distribution circuit's normal reclosing cycle. The opening of the switch is blocked if there is an overcurrent in the switch (the switch doesn't have fault interrupting capability). Transfer is also disabled if the alternate feed does not have proper voltage. The switch can return to normal through either an open or a closed transition; in a closed transition, both distribution circuits are temporarily paralleled.

1.2 Urban Networks

Some distribution circuits are not radial. The most common are the grid and spot secondary networks. In these systems, the secondary is networked together and has feeds from several primary distribution circuits. The spot

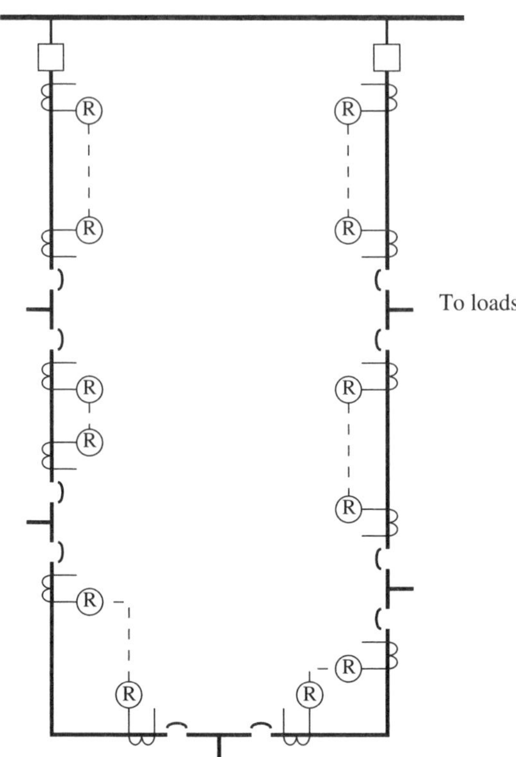

FIGURE 1.7
Example of a closed-loop distribution system.

network feeds one load such as a high-rise building. The grid network feeds several loads at different points in an area. Secondary networks are very reliable; if any of the primary distribution circuits fail, the others will carry the load without causing an outage for any customers.

The spot network generally is fed by three to five primary feeders (see Figure 1.9). The circuits are generally sized to be able to carry all of the load with the loss of either one or two of the primary circuits. Secondary networks have network protectors between the primary and the secondary network. A network protector is a low-voltage circuit breaker that will open when there is reverse power through it. When a fault occurs on a primary circuit, fault current backfeeds from the secondary network(s) to the fault. When this occurs, the network protectors will trip on reverse power. A spot network operates at 480Y/277 V or 208Y/120 V in the U.S.

Secondary grid networks are distribution systems that are used in most major cities. The secondary network is usually 208Y/120 V in the U.S. Five to ten primary distribution circuits (e.g., 12.47-kV circuits) feed the secondary network at multiple locations. Figure 1.10 shows a small part of a secondary network. As with a spot network, network protectors provide protection for faults on the primary circuits. Secondary grid networks can have peak loads

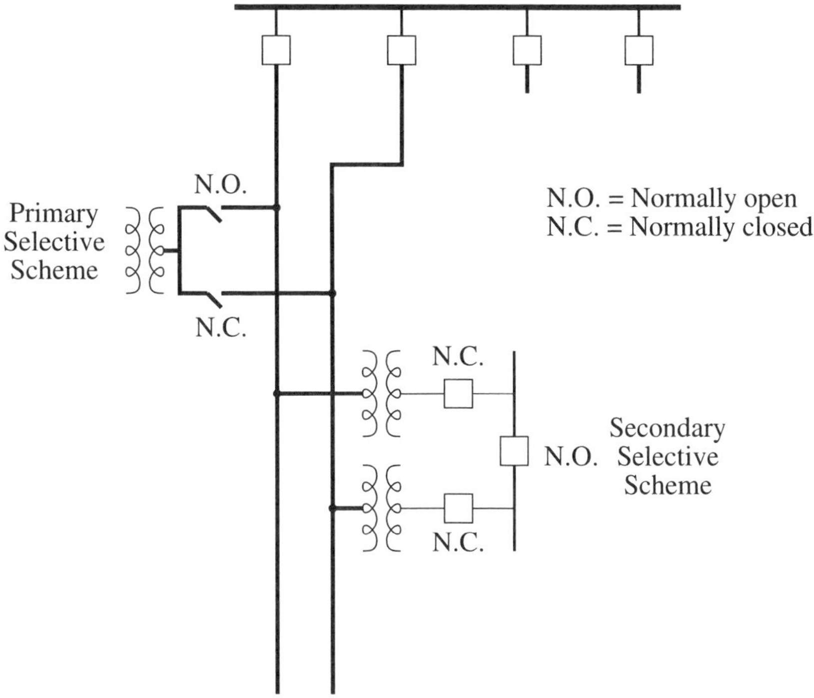

FIGURE 1.8
Primary and secondary selective schemes. (Copyright © 2000. Electric Power Research Institute. 1000419. *Engineering Guide for Integration of Distributed Generation and Storage Into Power Distribution Systems.* Reprinted with permission.)

of 5 to 50 MVA. Most utilities limit networks to about 50 MVA, but some networks are over 250 MVA. Loads are fed by tapping into the secondary networks at various points. Grid networks (also called street networks) can supply residential or commercial loads, either single or three phase. For single-phase loads, three-wire service is provided to give 120 V and 208 V (rather than the standard three-wire residential service, which supplies 120 V and 240 V).

Networks are normally fed by feeders originating from one substation bus. Having one source reduces circulating current and gives better load division and distribution among circuits. It also reduces the chance that network protectors stay open under light load (circulating current can trip the protectors). Given these difficulties, it is still possible to feed grid or spot networks from different substations or electrically separate buses.

The network protector is the key to automatic isolation and continued operation. The network protector is a three-phase low-voltage air circuit breaker with controls and relaying. The network protector is mounted on the network transformer or on a vault wall. Standard units are available with continuous ratings from 800 to 5000 A. Smaller units can interrupt 30 kA symmetrical, and larger units have interrupt ratings of 60 kA (IEEE Std.

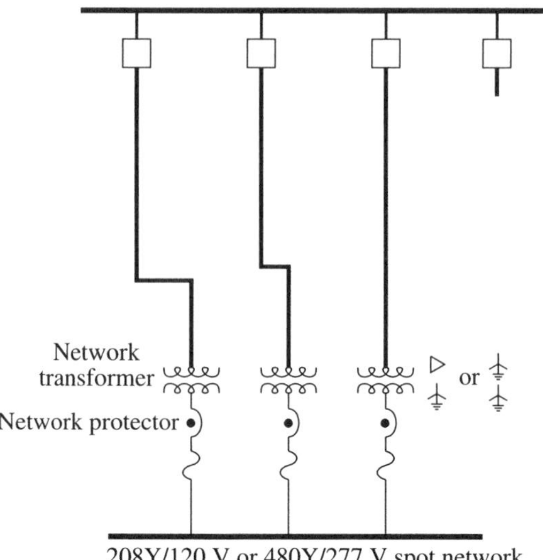

FIGURE 1.9
Spot network. (Copyright © 2000. Electric Power Research Institute. 1000419. *Engineering Guide for Integration of Distributed Generation and Storage Into Power Distribution Systems.* Reprinted with permission.)

C57.12.44–2000). A network protector senses and operates for reverse power flow (it does not have forward-looking protection). Protectors are available for either 480Y/277 V or 216Y/125 V.

The tripping current on network protectors can be changed, with low, nominal, and high settings, which are normally 0.05 to 0.1%, 0.15 to 0.20%, and 3 to 5% of the network protector rating. For example, a 2000-A network protector has a low setting of 1 A, a nominal setting of 4 A, and a high setting of 100 A (IEEE Std. C57.12.44–2000). Network protectors also have fuses that provide backup in case the network protector fails to operate, and as a secondary benefit, provide protection to the network protector and transformer against faults in the secondary network that are close.

The closing voltages are also adjustable: a 216Y/125-V protector has low, medium, and high closing voltages of 1 V, 1.5 V, and 2 V, respectively; a 480Y/277-V protector has low, medium, and high closing voltages of 2.2 V, 3.3 V, and 4.4 V, respectively.

1.3 Primary Voltage Levels

Most distribution voltages are between 4 and 35 kV. In this book, unless otherwise specified, voltages are given as line-to-line voltages; this follows normal industry practice, but it is sometimes a source of confusion. The four

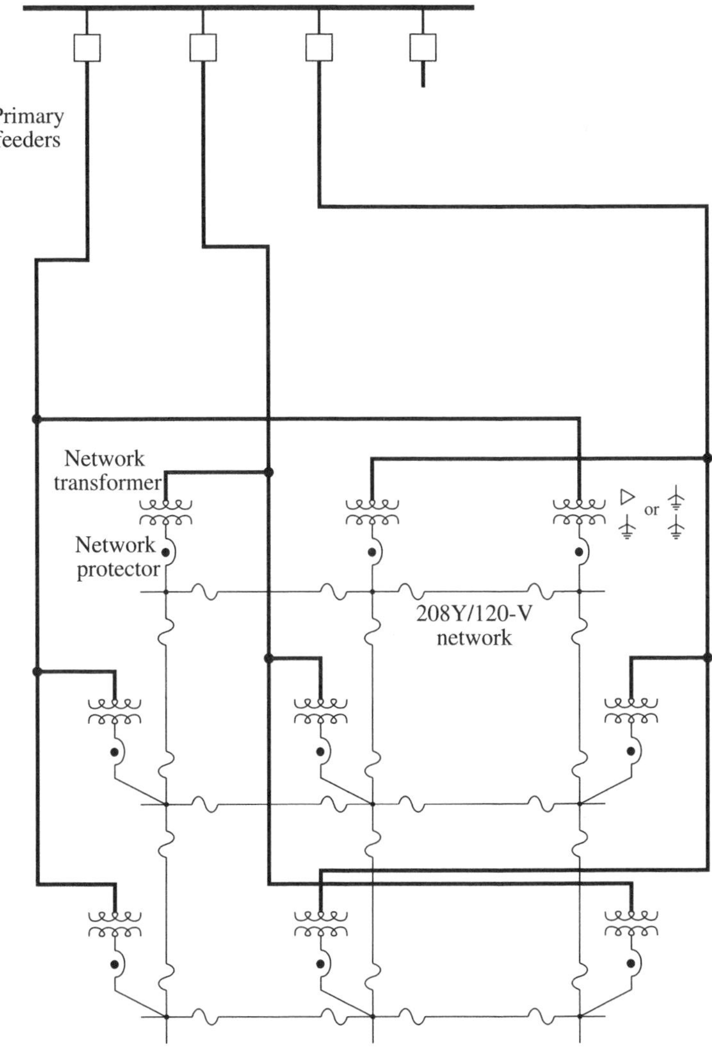

FIGURE 1.10
Portion of a grid network. (Copyright © 2000. Electric Power Research Institute. 1000419. *Engineering Guide for Integration of Distributed Generation and Storage Into Power Distribution Systems.* Reprinted with permission.)

major voltage classes are 5, 15, 25, and 35 kV. A voltage class is a term applied to a set of distribution voltages and the equipment common to them; it is not the actual system voltage. For example, a 15-kV insulator is suitable for application on any 15-kV class voltage, including 12.47 kV, 13.2 kV, and 13.8 kV. Cables, terminations, insulators, bushings, reclosers, and cutouts all have a voltage class rating. Only voltage-sensitive equipment like surge arresters, capacitors, and transformers have voltage ratings dependent on the actual system voltage.

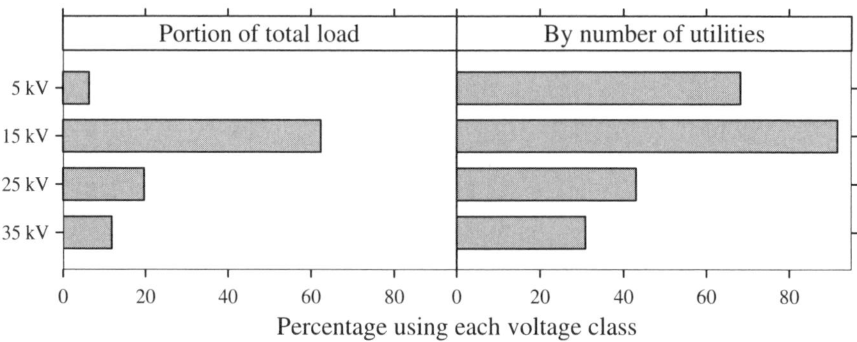

FIGURE 1.11
Usage of different distribution voltage classes (n = 107). (Data from [IEEE Working Group on Distribution Protection, 1995].)

Utilities most widely use the 15-kV voltages as shown by the survey results of North American utilities in Figure 1.11. The most common 15-kV voltage is 12.47 kV, which has a line-to-ground voltage of 7.2 kV.

The dividing line between distribution and subtransmission is often gray. Some lines act as both subtransmission and distribution circuits. A 34.5-kV circuit may feed a few 12.5-kV distribution substations, but it may also serve some load directly. Some utilities would refer to this as subtransmission, others as distribution.

The last half of the 20th century saw a move to higher voltage primary distribution systems. Higher-voltage distribution systems have advantages and disadvantages (see Table 1.3 for a summary). The great advantage of higher voltage systems is that they carry more power for a given current (Table 1.4 shows maximum power levels typically supplied by various distribution voltages). Less current means lower voltage drop, fewer losses, and more power-carrying capability. Higher voltage systems need fewer voltage

TABLE 1.3
Advantages and Disadvantages of Higher Voltage Distribution

Advantages	Disadvantages
Voltage drop — A higher-voltage circuit has less voltage drop for a given power flow.	*Reliability* — An important disadvantage of higher voltages: longer circuits mean more customer interruptions.
Capacity — A higher-voltage system can carry more power for a given ampacity.	*Crew safety and acceptance* — Crews do not like working on higher-voltage distribution systems.
Losses — For a given level of power flow, a higher-voltage system has fewer line losses.	*Equipment cost* — From transformers to cable to insulators, higher-voltage equipment costs more.
Reach — With less voltage drop and more capacity, higher voltage circuits can cover a much wider area.	
Fewer substations — Because of longer reach, higher-voltage distribution systems need fewer substations.	

TABLE 1.4

Power Supplied by Each Distribution Voltage for a Current of 400 A

System Voltage (kV)	Total Power (MVA)
4.8	3.3
12.47	8.6
22.9	15.9
34.5	23.9

regulators and capacitors for voltage support. Utilities can use smaller conductors on a higher voltage system or carry more power on the same size conductor. Utilities can run much longer distribution circuits at a higher primary voltage, which means fewer distribution substations. Some fundamental relationships are:

- *Power* — For the same current, power changes linearly with voltage.

$$P_2 = \frac{V_1}{V_1} P_1$$

  when $I_2 = I_1$

- *Current* — For the same power, increasing the voltage decreases current linearly.

$$I_2 = \frac{V_1}{V_2} I_1$$

  when $P_2 = P_1$

- *Voltage drop* — For the same power delivered, the percentage voltage drop changes as the ratio of voltages squared. A 12.47-kV circuit has four times the percentage voltage drop as a 24.94-kV circuit carrying the same load.

$$V_{\%2} = \left(\frac{V_1}{V_2}\right)^2 V_{\%1}$$

  when $P_2 = P_1$

- *Area coverage* — For the same load density, the area covered increases linearly with voltage: A 24.94-kV system can cover twice the area of a 12.47-kV system; a 34.5-kV system can cover 2.8 times the area of a 12.47-kV system.

$$A_2 = \frac{V_2}{V_1} A_1$$

where

V_1, V_2 = voltage on circuits 1 and 2

P_1, P_2 = power on circuits 1 and 2

I_1, I_2 = current on circuits 1 and 2

$V_{\%1}$, $V_{\%2}$ = voltage drop per unit length in percent on circuits 1 and 2

A_1, A_2 = area covered by circuits 1 and 2

The squaring effect on voltage drop is significant. It means that doubling the system voltage quadruples the load that can be supplied over the same distance (with equal percentage voltage drop); or, twice the load can be supplied over twice the distance; or, the same load can be supplied over four times the distance.

Resistive line losses are also lower on higher-voltage systems, especially in a voltage-limited circuit. Thermally limited systems have more equal losses, but even in this case higher voltage systems have fewer losses.

Line crews do not like higher voltage distribution systems as much. In addition to the widespread perception that they are not as safe, gloves are thicker, and procedures are generally more stringent. Some utilities will not glove 25- or 35-kV voltages and only use hotsticks.

The main disadvantage of higher-voltage systems is reduced reliability. Higher voltages mean longer lines and more exposure to lightning, wind, dig-ins, car crashes, and other fault causes. A 34.5-kV, 30-mi mainline is going to have many more interruptions than a 12.5-kV system with an 8-mi mainline. To maintain the same reliability as a lower-voltage distribution system, a higher-voltage primary must have more switches, more automation, more tree trimming, or other reliability improvements. Higher voltage systems also have more voltage sags and momentary interruptions. More exposure causes more momentary interruptions. Higher voltage systems have more voltage sags because faults further from the substation can pull down the station's voltage (on a higher voltage system the line impedance is lower relative to the source impedance).

Cost comparison between circuits is difficult (see Table 1.5 for one utility's cost comparison). Higher voltage equipment costs more — cables, insulators, transformers, arresters, cutouts, and so on. But higher voltage circuits can use smaller conductors. The main savings of higher-voltage distribution is fewer substations. Higher voltage systems also have lower annual costs from losses. As far as ongoing maintenance, higher voltage systems require less substation maintenance, but higher voltage systems should have more tree trimming and inspections to maintain reliability.

Conversion to a higher voltage is an option for providing additional capacity in an area. Conversion to higher voltages is most beneficial when substation

TABLE 1.5

Costs of 34.5 kV Relative to 12.5 kV

Item	Underground	Overhead
Subdivision without bulk feeders	1.25	1.13
Subdivision with bulk feeders	1.00	0.85
Bulk feeders	0.55	0.55
Commercial areas	1.05–1.25	1.05–1.25

Source: Jones, A.I., Smith, B.E., and Ward, D.J., "Considerations for Higher Voltage Distribution," *IEEE Transactions on Power Delivery*, vol. 7, no. 2, pp. 782–8, April 1992.

space is hard to find and load growth is high. If the existing subtransmission voltage is 34.5 kV, then using that voltage for distribution is attractive; additional capacity can be met by adding customers to existing 34.5-kV lines (a neutral may need to be added to the 34.5-kV subtransmission line).

Higher voltage systems are also more prone to ferroresonance. Radio interference is also more common at higher voltages.

Overall, the 15-kV class voltages provide a good balance between cost, reliability, safety, and reach. Although a 15-kV circuit does not naturally provide long reach, with voltage regulators and feeder capacitors it can be stretched to reach 20 mi or more. That said, higher voltages have advantages, especially for rural lines and for high-load areas, particularly where substation space is expensive.

Many utilities have multiple voltages (as shown by the survey data in Figure 1.11). Even one circuit may have multiple voltages. For example, a utility may install a 12.47-kV circuit in an area presently served by 4.16 kV. Some of the circuit may be converted to 12.47 kV, but much of it can be left as is and coupled through 12.47/4.16-kV step-down transformer banks.

1.4 Distribution Substations

Distribution substations come in many sizes and configurations. A small rural substation may have a nominal rating of 5 MVA while an urban station may be over 200 MVA. Figure 1.12 through Figure 1.14 show examples of small, medium, and large substations. As much as possible, many utilities have standardized substation layouts, transformer sizes, relaying systems, and automation and SCADA (supervisory control and data acquisition) facilities. Most distribution substation bus configurations are simple with limited redundancy.

Transformers smaller than 10 MVA are normally protected with fuses, but fuses are also used for transformers to 20 or 30 MVA. Fuses are inexpensive and simple; they don't need control power and take up little space. Fuses are not particularly sensitive, especially for evolving internal faults. Larger transformers normally have relay protection that operates a circuit switcher

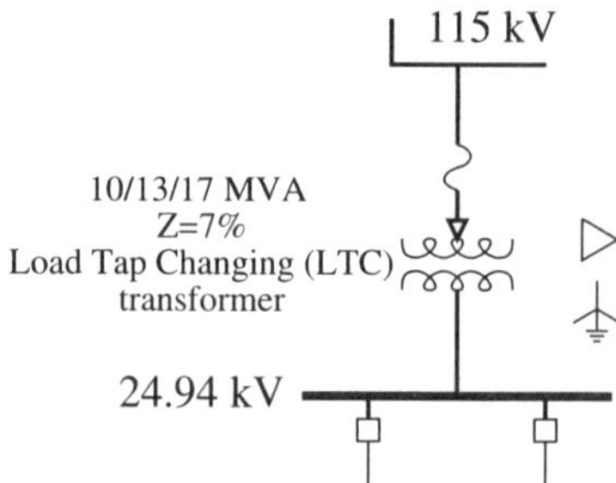

FIGURE 1.12
Example rural distribution substation.

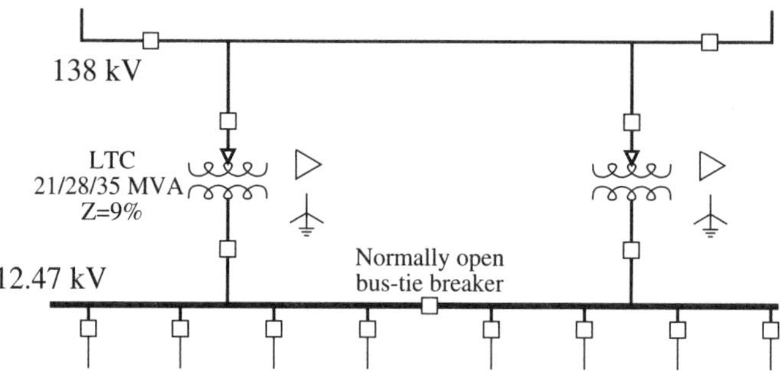

FIGURE 1.13
Example suburban distribution substation.

or a circuit breaker. Relays often include differential protection, sudden-pressure relays, and overcurrent relays. Both the differential protection and the sudden-pressure relays are sensitive enough to detect internal failures and clear the circuit to limit additional damage to the transformer. Occasionally, relays operate a high-side grounding switch instead of an interrupter. When the grounding switch engages, it creates a bolted fault that is cleared by an upstream device or devices.

The feeder interrupting devices are normally relayed circuit breakers, either free-standing units or metal-enclosed switchgear. Many utilities also use reclosers instead of breakers, especially at smaller substations.

Station transformers are normally protected by differential relays which trip if the current into the transformer is not very close to the current out of the transformer. Relaying may also include pressure sensors. The high-side protective device is often a circuit switcher but may also be fuses or a circuit breaker.

Two-bank stations are very common (Figure 1.13); these are the standard design for many utilities. Normally, utilities size the transformers so that if either transformer fails, the remaining unit can carry the entire substation's load. Utility practices vary on how much safety margin is built into this calculation, and load growth can eat into the redundancy.

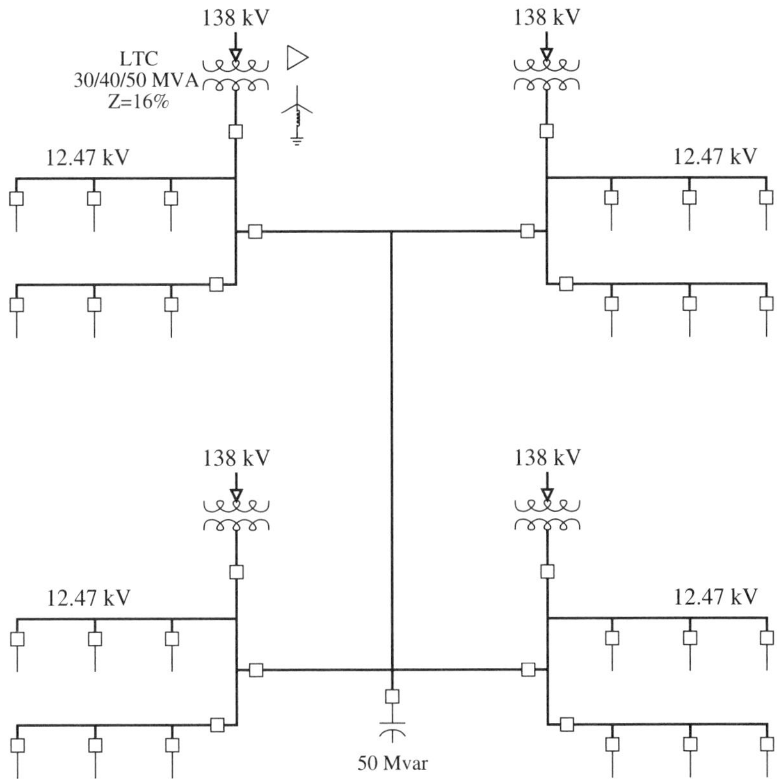

FIGURE 1.14
Example urban distribution substation.

Most utilities normally use a split bus: a bus tie between the two buses is normally left open in distribution substations. The advantages of a split bus are:

- *Lower fault current* — This is the main reason that bus ties are open. For a two-bank station with equal transformers, opening the bus tie cuts fault current in half.
- *Circulating current* — With a split bus, current cannot circulate through both transformers.
- *Bus regulation* — Bus voltage regulation is also simpler with a split bus. With the tie closed, control of paralleled tap changers is more difficult.

Having the bus tie closed has some advantages, and many utilities use closed ties under some circumstances. A closed bus tie is better for

- *Secondary networks* — When feeders from each bus supply either spot or grid secondary networks, closed bus ties help prevent circulating current through the secondary networks.

- *Unequal loading* — A closed bus tie helps balance the loading on the transformers. If the set of feeders on one bus has significantly different loading patterns (either seasonal or daily), then a closed bus tie helps even out the loading (and aging) of the two transformers.

Whether the bus tie is open or closed has little impact on reliability. In the uncommon event that one transformer fails, both designs allow the station to be reconfigured so that one transformer supplies both bus feeders. The closed-tie scenario is somewhat better in that an automated system can reconfigure the ties without total loss of voltage to customers (customers do see a very large voltage sag). In general, both designs perform about the same for voltage sags.

Urban substations are more likely to have more complicated bus arrangements. These could include ring buses or breaker-and-a-half schemes. Figure 1.14 shows an example of a large urban substation with feeders supplying secondary networks. If feeders are supplying secondary networks, it is not critical to maintain continuity to each feeder, but it is important to prevent loss of any one bus section or piece of equipment from shutting down the network (an *N*-1 design).

For more information on distribution substations, see (RUS 1724E-300, 2001; Westinghouse Electric Corporation, 1965).

1.5 Subtransmission Systems

Subtransmission systems are those circuits that supply distribution substations. Several different subtransmission systems can supply distribution substations. Common subtransmission voltages include 34.5, 69, 115, and 138 kV. Higher voltage subtransmission lines can carry more power with less losses over greater distances. Distribution circuits are occasionally supplied by high-voltage transmission lines such as 230 kV; such high voltages make for expensive high-side equipment in a substation. Subtransmission circuits are normally supplied by bulk transmission lines at subtransmission substations. For some utilities, one transmission system serves as both the subtransmission function (feeding distribution substations) and the transmission function (distributing power from bulk generators). There is much crossover in functionality and voltage. One utility may have a 23-kV subtransmission system supplying 4-kV distribution substations. Another utility right next door may have a 34.5-kV distribution system fed by a 138-kV subtransmission system. And within utilities, one can find a variety of different voltage combinations.

Of all of the subtransmission circuit arrangements, a radial configuration is the simplest and least expensive (see Figure 1.15). But radial circuits provide the most unreliable supply; a fault on the subtransmission circuit

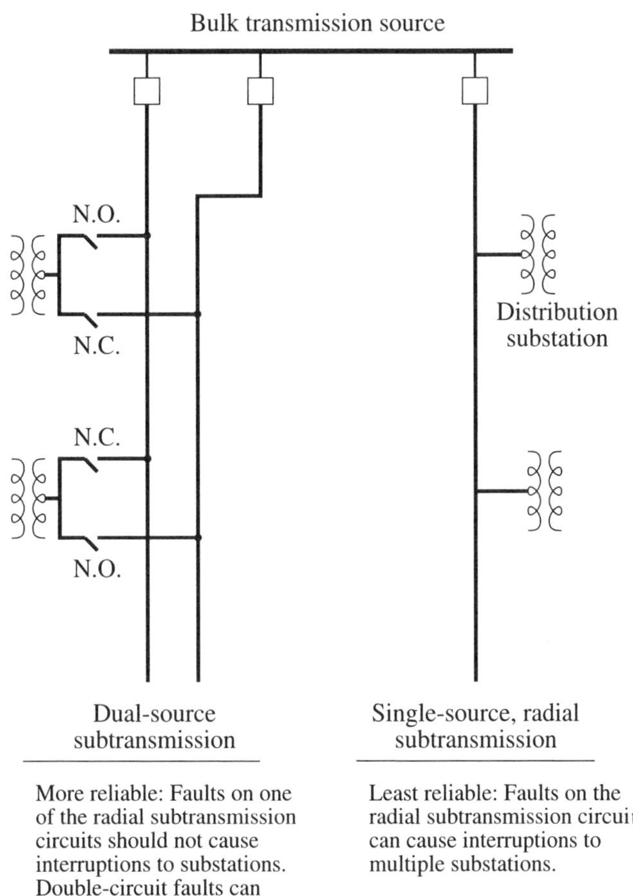
Bulk transmission source
N.O.
N.C.
N.C.
N.O.
Distribution substation
Dual-source subtransmission
Single-source, radial subtransmission
More reliable: Faults on one of the radial subtransmission circuits should not cause interruptions to substations. Double-circuit faults can cause multiple station interruptions.
Least reliable: Faults on the radial subtransmission circuit can cause interruptions to multiple substations.

FIGURE 1.15
Radial subtransmission systems.

can force an interruption of several distribution substations and service to many customers. A variety of redundant subtransmission circuits are available, including dual circuits and looped or meshed circuits (see Figure 1.16). The design (and evolution) of subtransmission configurations depends on how the circuit developed, where the load is needed now and in the future, what the distribution circuit voltages are, where bulk transmission is available, where rights-of-way are available, and, of course, economic factors.

Most subtransmission circuits are overhead. Many are built right along roads and streets just like distribution lines. Some — especially higher voltage subtransmission circuits — use a private right-of-way such as bulk transmission lines use. Some new subtransmission lines are put underground, as development of solid-insulation cables has made costs more reasonable.

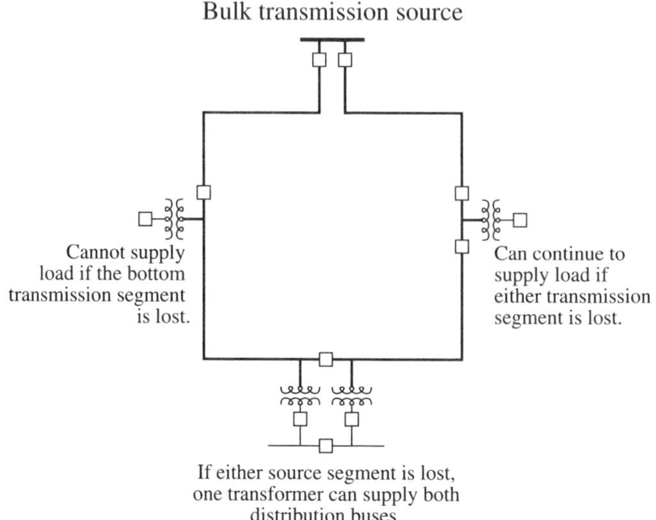
Bulk transmission source
Cannot supply load if the bottom transmission segment is lost.
Can continue to supply load if either transmission segment is lost.
If either source segment is lost, one transformer can supply both distribution buses.

FIGURE 1.16
Looped subtransmission system.

Lower voltage subtransmission lines (69, 34.5, and 23 kV) tend to be designed and operated as are distribution lines, with radial or simple loop arrangements, using wood-pole construction along roads, with reclosers and regulators, often without a shield wire, and with time-overcurrent protection. Higher voltage transmission lines (115, 138, and 230 kV) tend to be designed and operated like bulk transmission lines, with loop or mesh arrangements, tower configurations on a private right-of-way, a shield wire or wires for lightning protection, and directional or pilot-wire relaying from two ends. Generators may or may not interface at the subtransmission level (which can affect protection practices).

1.6 Differences between European and North American Systems

Distribution systems around the world have evolved into different forms. The two main designs are North American and European. This book deals mainly with North American distribution practices; for more information on European systems, see Lakervi and Holmes (1995). For both forms, hardware is much the same: conductors, cables, insulators, arresters, regulators, and transformers are very similar. Both systems are radial, and voltages and power carrying capabilities are similar. The main differences are in layouts, configurations, and applications.

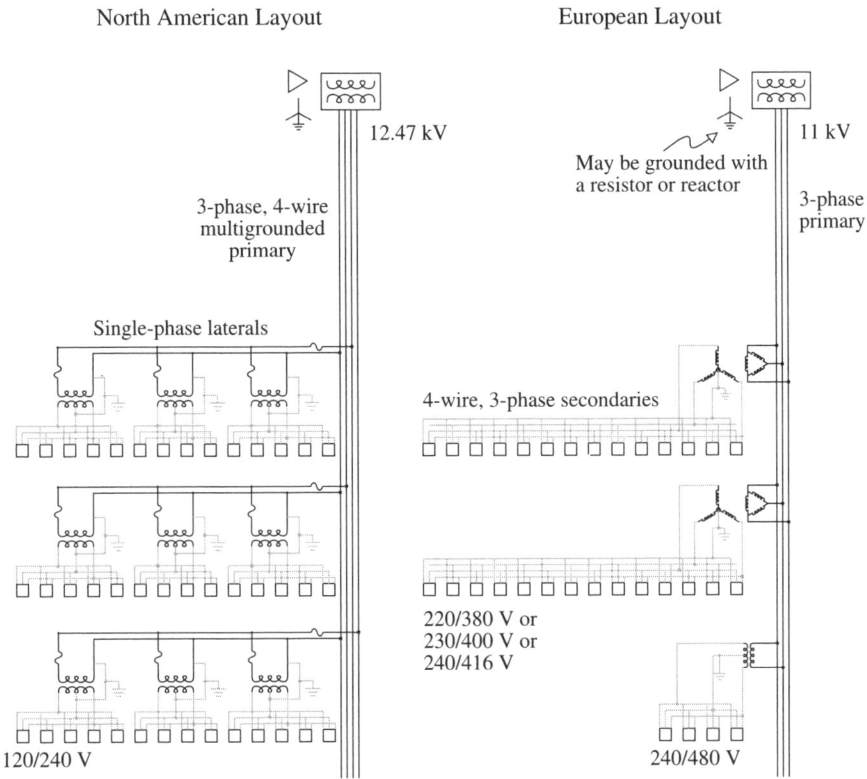

FIGURE 1.17
North American versus European distribution layouts.

Figure 1.17 compares the two systems. Relative to North American designs, European systems have larger transformers and more customers per transformer. Most European transformers are three-phase and on the order of 300 to 1000 kVA, much larger than typical North American 25- or 50-kVA single-phase units.

Secondary voltages have motivated many of the differences in distribution systems. North America has standardized on a 120/240-V secondary system; on these, voltage drop constrains how far utilities can run secondaries, typically no more than 250 ft. In European designs, higher secondary voltages allow secondaries to stretch to almost 1 mi. European secondaries are largely three-phase and most European countries have a standard secondary voltage of 220, 230, or 240 V, twice the North American standard. With twice the voltage, a circuit feeding the same load can reach four times the distance. And because three-phase secondaries can reach over twice the length of a single-phase secondary, overall, a European secondary can reach eight times the length of an American secondary for a given load and voltage drop. Although it is rare, some European utilities supply rural areas with single-phase taps made of two phases with single-phase transformers connected phase to phase.

In the European design, secondaries are used much like primary laterals in the North American design. In European designs, the primary is not tapped frequently, and primary-level fuses are not used as much. European utilities also do not use reclosing as religiously as North American utilities.

Some of the differences in designs center around the differences in loads and infrastructure. In Europe, the roads and buildings were already in place when the electrical system was developed, so the design had to "fit in." Secondary is often attached to buildings. In North America, many of the roads and electrical circuits were developed at the same time. Also, in Europe houses are packed together more and are smaller than houses in America.

Each type of system has its advantages. Some of the major differences between systems are the following (see also Carr and McCall, 1992; Meliopoulos et al., 1998; Nguyen et al., 2000):

- *Cost* — The European system is generally more expensive than the North American system, but there are so many variables that it is hard to compare them on a one-to-one basis. For the types of loads and layouts in Europe, the European system fits quite well. European primary equipment is generally more expensive, especially for areas that can be served by single-phase circuits.
- *Flexibility* — The North American system has a more flexible primary design, and the European system has a more flexible secondary design. For urban systems, the European system can take advantage of the flexible secondary; for example, transformers can be sited more conveniently. For rural systems and areas where load is spread out, the North American primary system is more flexible. The North American primary is slightly better suited for picking up new load and for circuit upgrades and extensions.
- *Safety* — The multigrounded neutral of the North American primary system provides many safety benefits; protection can more reliably clear faults, and the neutral acts as a physical barrier, as well as helping to prevent dangerous touch voltages during faults. The European system has the advantage that high-impedance faults are easier to detect.
- *Reliability* — Generally, North American designs result in fewer customer interruptions. Nguyen et al. (2000) simulated the performance of the two designs for a hypothetical area and found that the average frequency of interruptions was over 35% higher on the European system. Although European systems have less primary, almost all of it is on the main feeder backbone; loss of the main feeder results in an interruption for all customers on the circuit.

European systems need more switches and other gear to maintain the same level of reliability.

- *Power quality* — Generally, European systems have fewer voltage sags and momentary interruptions. On a European system, less primary exposure should translate into fewer momentary interruptions compared to a North American system that uses fuse saving. The three-wire European system helps protect against sags from line-to-ground faults. A squirrel across a bushing (from line to ground) causes a relatively high impedance fault path that does not sag the voltage much compared to a bolted fault on a well-grounded system. Even if a phase conductor faults to a low-impedance return path (such as a well-grounded secondary neutral), the delta – wye customer transformers provide better immunity to voltage sags, especially if the substation transformer is grounded through a resistor or reactor.
- *Aesthetics* — Having less primary, the European system has an aesthetic advantage: the secondary is easier to underground or to blend in. For underground systems, fewer transformer locations and longer secondary reach make siting easier.
- *Theft* — The flexibility of the European secondary system makes power much easier to steal. Developing countries especially have this problem. Secondaries are often strung along or on top of buildings; this easy access does not require great skill to attach into.

Outside of Europe and North America, both systems are used, and usage typically follows colonial patterns with European practices being more widely used. Some regions of the world have mixed distribution systems, using bits of North American and bits of European practices. The worst mixture is 120-V secondaries with European-style primaries; the low-voltage secondary has limited reach along with the more expensive European primary arrangement.

Higher secondary voltages have been explored (but not implemented to my knowledge) for North American systems to gain flexibility. Higher secondary voltages allow extensive use of secondary, which makes undergrounding easier and reduces costs. Westinghouse engineers contended that both 240/480-V three-wire single-phase and 265/460-V four-wire three-phase secondaries provide cost advantages over a similar 120/240-V three-wire secondary (Lawrence and Griscom, 1956; Lokay and Zimmerman, 1956). Higher secondary voltages do not force higher utilization voltages; a small transformer at each house converts 240 or 265 V to 120 V for lighting and standard outlet use (air conditioners and major appliances can be served directly without the extra transformation). More recently, Bergeron et al. (2000) outline a vision of a distribution system where primary-level distribution voltage is stepped down to an extensive 600-V, three-phase secondary system. At each house, an electronic transformer converts 600 V to 120/240 V.

1.7 Loads

Distribution systems obviously exist to supply electricity to end users, so loads and their characteristics are important. Utilities supply a broad range of loads, from rural areas with load densities of 10 kVA/mi^2 to urban areas with 300 MVA/mi^2. A utility may feed houses with a 10- to 20-kVA peak load on the same circuit as an industrial customer peaking at 5 MW. The electrical load on a feeder is the sum of all individual customer loads. And the electrical load of a customer is the sum of the load drawn by the customer's individual appliances. Customer loads have many common characteristics. Load levels vary through the day, peaking in the afternoon or early evening. Several definitions are used to quantify load characteristics at a given location on a circuit:

- *Demand* — The load average over a specified time period, often 15, 20, or 30 min. Demand can be used to characterize real power, reactive power, total power, or current. Peak demand over some period of time is the most common way utilities quantify a circuit's load. In substations, it is common to track the current demand.
- *Load factor* — The ratio of the average load over the peak load. Peak load is normally the maximum demand but may be the instantaneous peak. The load factor is between zero and one. A load factor close to 1.0 indicates that the load runs almost constantly. A low load factor indicates a more widely varying load. From the utility point of view, it is better to have high load-factor loads. Load factor is normally found from the total energy used (kilowatt-hours) as:

  $$LF = \frac{kWh}{d_{kW} \times h}$$

  where

  LF = load factor

  kWh = energy use in kilowatt-hours

  d_{kW} = peak demand in kilowatts

  h = number of hours during the time period
- *Coincident factor* — The ratio of the peak demand of a whole system to the sum of the individual peak demands within that system. The

peak demand of the whole system is referred to as the peak *diversified* demand or as the peak *coincident* demand. The individual peak demands are the *noncoincident* demands. The coincident factor is less than or equal to one. Normally, the coincident factor is much less than one because each of the individual loads do not hit their peak at the same time (they are not coincident).

- *Diversity factor* — The ratio of the sum of the individual peak demands in a system to the peak demand of the whole system. The diversity factor is greater than or equal to one and is the reciprocal of the coincident factor.
- *Responsibility factor* — The ratio of a load's demand at the time of the system peak to its peak demand. A load with a responsibility factor of one peaks at the same time as the overall system. The responsibility factor can be applied to individual customers, customer classes, or circuit sections.

The loads of certain customer classes tend to vary in similar patterns. Commercial loads are highest from 8 a.m. to 6 p.m. Residential loads peak in the evening. Weather significantly changes loading levels. On hot summer days, air conditioning increases the demand and reduces the diversity among loads. At the transformer level, load factors of 0.4 to 0.6 are typical (Gangel and Propst, 1965).

Several groups have evaluated coincidence factors as a function of the number of customers. Nickel and Braunstein (1981) determined that one curve fell roughly in the middle of several curves evaluated. Used by Arkansas Power and Light, this curve fits the following:

$$F_{co} = \frac{1}{2}\left(1 + \frac{5}{2n+3}\right)$$

where n is the number of customers (see Figure 1.18).

At the substation level, coincidence is also apparent. A transformer with four feeders, each peaking at 100 A, will peak at less than 400 A because of diversity between feeders. The coincident factor between four feeders is normally higher than coincident factors at the individual customer level. Expect coincident factors to be above 0.9. Each feeder is already highly diversified, so not much more is gained by grouping more customers together if the sets of customers are similar. If the customer mix on each feeder is different, then multiple feeders can have significant differences. If some feeders are mainly residential and others are commercial, the peak load of the feeders together can be significantly lower than the sum of the peaks. For distribution transformers, the peak responsibility factor ranges from 0.5 to 0.9 with 0.75 being typical (Nickel and Braunstein, 1981).

Different customer classes have different characteristics (see Figure 1.19 for an example). Residential loads peak more in the evening and have a

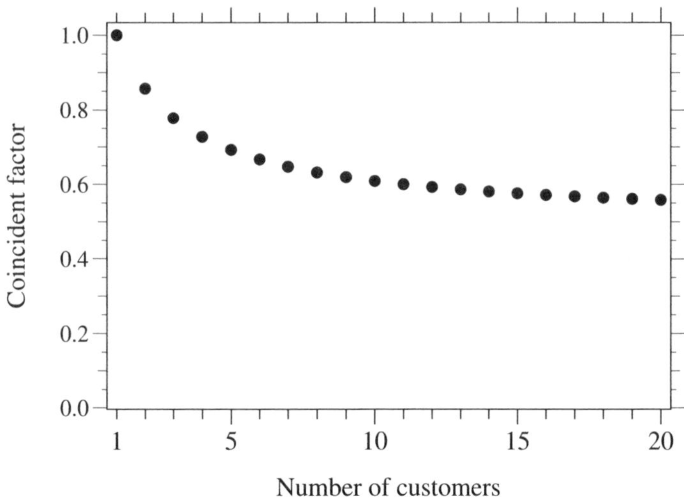

FIGURE 1.18
Coincident factor average curve for utilities.

relatively low load factor. Commercial loads tend to be more 8 a.m. to 6 p.m., and the industrial loads tend to run continuously and, as a class, they have a higher load factor.

1.8 The Past and the Future

Looking at Seelye's *Electrical Distribution Engineering* book (1930), we find more similarities to than differences from present-day distribution systems. The basic layout and operations of distribution infrastructure at the start of the 21st century are much the same as in the middle of the 20th century. Equipment has undergone steady improvements; transformers are more efficient; cables are much less expensive and easier to use; and protection equipment is better (see Figure 1.20 for some development milestones). Utilities operate more distribution circuits at higher voltages and use more underground circuits. But the concepts are much the same: ac, three-phase systems, radial circuits, fused laterals, overcurrent relays, etc. Advances in computer technology have opened up possibilities for more automation and more effective protection.

How will future distribution systems evolve? Given the fact that distribution systems of the year 2000 look much the same as distribution systems in 1950, a good guess is that the distribution system of 2050 (or at least 2025) will look much like today's systems. More and more of the electrical infrastructure will be placed underground. Designs and equipment will continue

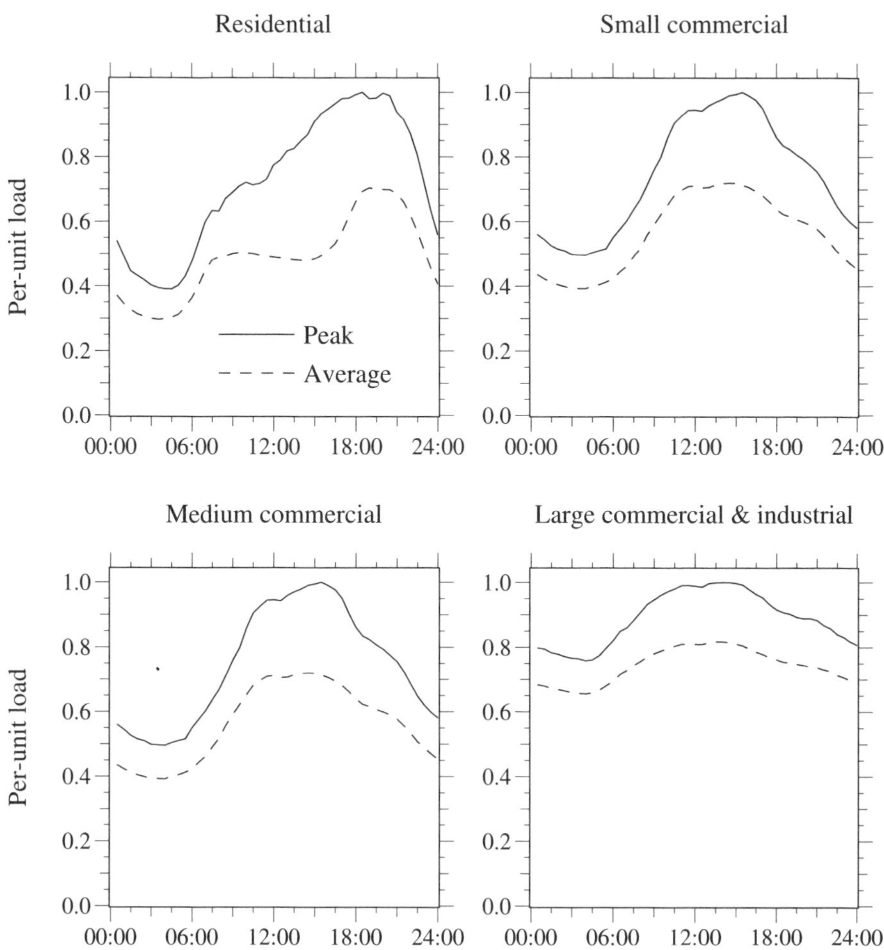

FIGURE 1.19
Daily load profiles for Pacific Gas and Electric (2002 data).

to be standardized. Gradually, the distribution system will evolve to take advantage of computer and communication gains: more automation, more communication between equipment, and smarter switches and controllers. EPRI outlined a vision of a future distribution system that was no longer radial, a distribution system that evolves to support widespread distributed generation and storage along with the ability to charge electric vehicles (EPRI TR-111683, 1998). Such a system needs directional relaying for reclosers, communication between devices, regulators with advanced controls, and information from and possibly control of distributed generators.

Advances in power electronics make more radical changes such as conversion to dc possible. Advances in power electronics allow flexible conversion between different frequencies, phasings, and voltages while still

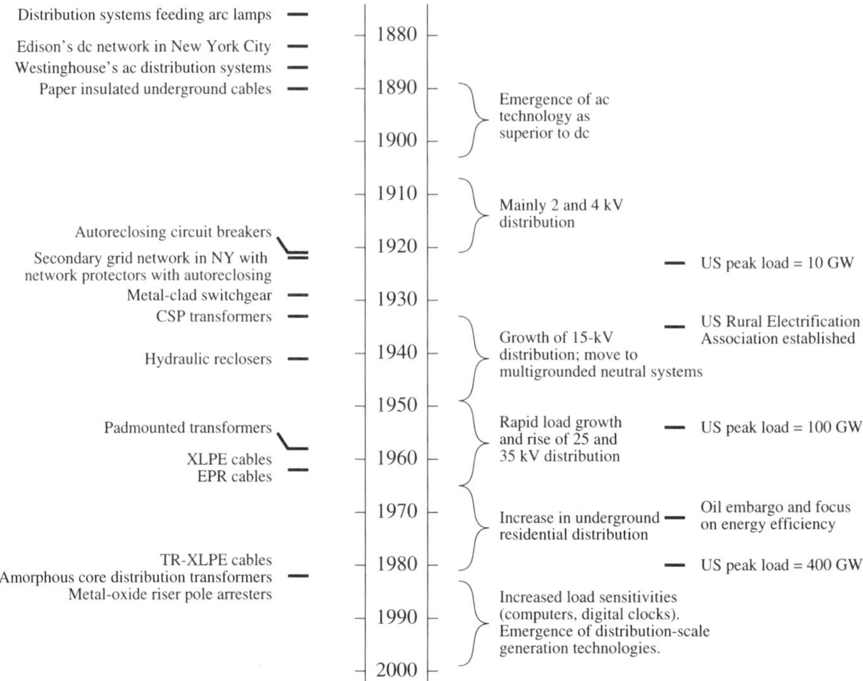

FIGURE 1.20
Electric power distribution development timeline.

producing ac voltage to the end user at the proper voltage. While possible, radical changes are unlikely, given the advantages to evolving an existing system rather than replacing it. Whatever the approach, the future has challenges; utilities will be expected to deliver more reliable power with minimal pollution while keeping the distribution system hidden from view and causing the least disruption possible. And of course, costs are expected to stay the same or go down.

References

Bergeron, R., Slimani, K., Lamarche, L., and Cantin, B., "New Architecture of the Distribution System Using Electronic Transformer," ESMO-2000, Panel on Distribution Transformer, Breakers, Switches and Arresters, 2000.

Carr, J. and McCall, L.V., "Divergent Evolution and Resulting Characteristics Among the World's Distribution Systems," *IEEE Transactions on Power Delivery*, vol. 7, no. 3, pp. 1601–9, July 1992.

CEA, *CEA Distribution Planner's Manual*, Canadian Electrical Association, 1982.

EPRI 1000419, *Engineering Guide for Integration of Distributed Generation and Storage Into Power Distribution Systems*, Electric Power Research Institute, Palo Alto, CA, 2000.

EPRI TR-109178, *Distribution Cost Structure — Methodology and Generic Data*, Electric Power Research Institute, Palo Alto, CA, 1998.

EPRI TR-111683, *Distribution Systems Redesign*, Electric Power Research Institute, Palo Alto, CA, 1998.

Gangel, M.W. and Propst, R.F., "Distribution Transformer Load Characteristics," *IEEE Transactions on Power Apparatus and Systems*, vol. 84, pp. 671–84, August 1965.

IEEE Std. C57.12.44–2000, IEEE Standard Requirements for Secondary Network Protectors.

IEEE Tutorial Course, *Power Distribution Planning*, 1992. Course text 92 EHO 361–6-PWR.

IEEE Working Group on Distribution Protection, "Distribution Line Protection Practices Industry Survey Results," *IEEE Transactions on Power Delivery*, vol. 10, no. 1, pp. 176–86, January 1995.

Jones, A.I., Smith, B.E., and Ward, D.J., "Considerations for Higher Voltage Distribution," *IEEE Transactions on Power Delivery*, vol. 7, no. 2, pp. 782–8, April 1992.

Lakervi, E. and Holmes, E.J., *Electricity Distribution Network Design*, IEE Power Engineering Series 21, Peter Peregrinius, 1995.

Lawrence, R.F. and Griscom, S.B., "Residential Distribution — An Analysis of Systems to Serve Expanding Loads," *AIEE Transactions, Part III*, vol. 75, pp. 533–42, 1956.

Lokay, H.E. and Zimmerman, R.A., "Economic Comparison of Secondary Voltages: Single and Three Phase Distribution for Residential Areas," *AIEE Transactions, Part III*, vol. 75, pp. 542–52, 1956.

Meliopoulos, A.P. S., Kennedy, J., Nucci, C.A., Borghetti, A., and Contaxis, G., "Power Distribution Practices in USA and Europe: Impact on Power Quality," 8th International Conference on Harmonics and Quality of Power, 1998.

Nguyen, H.V., Burke, J.J., and Benchluch, S., "Rural Distribution System Design Comparison," IEEE Power Engineering Society Winter Meeting, 2000.

Nickel, D.L. and Braunstein, H.R., "Distribution. Transformer Loss Evaluation. II. Load Characteristics and System Cost Parameters," *IEEE Transactions on Power Apparatus and Systems*, vol. PAS-100, no. 2, pp. 798–811, February 1981.

Pagel, B., "Energizing International Drive," *Transmission & Distribution World*, April 2000.

RUS 1724E-300, *Design Guide for Rural Substations*, United States Department of Agriculture, Rural Utilities Service, 2001.

Seelye, H.P., *Electrical Distribution Engineering*, McGraw-Hill New York, 1930.

Westinghouse Electric Corporation, *Distribution Systems*, vol. 3, 1965.

Willis, H.L., *Power Distribution Planning Reference Book*, Marcel Dekker, New York, 1997.

No matter how long you've been a Power Lineman, you still notice it when people refer to your poles as "telephone poles."

Powerlineman law #46, By CD Thayer and other Power Linemen, http://www.cdthayer.com/lineman.htm

2

Overhead Lines

Along streets, alleys, through woods, and in backyards, many of the distribution lines that feed customers are overhead structures. Because overhead lines are exposed to trees and animals, to wind and lightning, and to cars and kites, they are a critical component in the reliability of distribution circuits. This chapter discusses many of the key electrical considerations of overhead lines: conductor characteristics, impedances, ampacity, and other issues.

2.1 Typical Constructions

Overhead constructions come in a variety of configurations (see Figure 2.1). Normally one primary circuit is used per pole, but utilities sometimes run more than one circuit per structure. For a three-phase circuit, the most common structure is a horizontal layout with an 8- or 10-ft wood crossarm on a pole (see Figure 2.2). Armless constructions are also widely found where fiberglass insulator standoffs or post insulators are used in a tighter configuration. Utilities normally use 30- to 45-ft poles, set 6 to 8 ft deep. Vertical construction is also occasionally used. Span lengths vary from 100 to 150 ft in suburban areas to as much as 300 or 400 ft in rural areas.

Distribution circuits normally have an underbuilt neutral — the neutral acts as a safety ground for equipment and provides a return path for unbalanced loads and for line-to-ground faults. The neutral is 3 to 5 ft below the phase conductors. Utilities in very high lightning areas may run the neutral wire above the phase conductors to act as a shield wire. Some utilities also run the neutral on the crossarm. Secondary circuits are often run under the primary. The primary and the secondary may share the neutral, or they may each have their own neutral. Many electric utilities share their space with other utilities; telephone or cable television cables may run under the electric secondary.

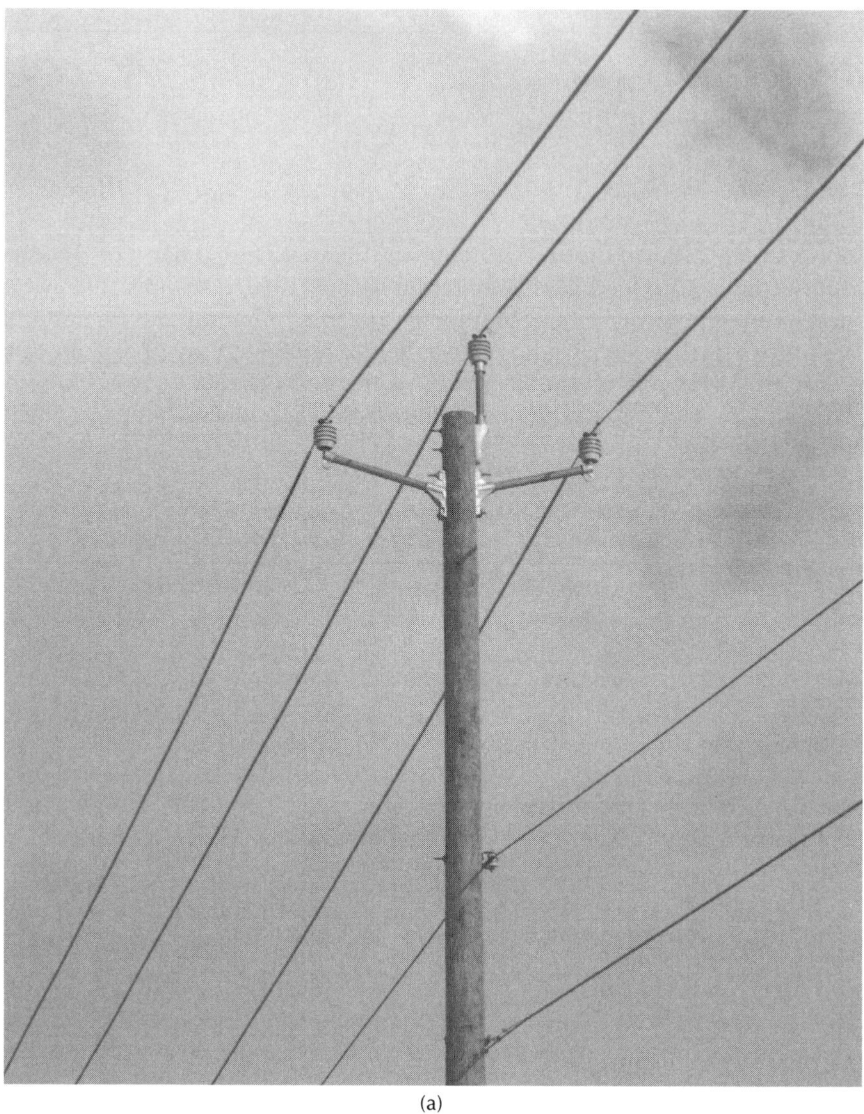

(a)

FIGURE 2.1
Example overhead distribution structures. (a) Three-phase 34.5-kV armless construction with covered wire. (b) Single-phase circuit, 7.2 kV line-to-ground. (c) Single-phase, 4.8-kV circuit. (d) 13.2-kV spacer cable.

(b)

FIGURE 2.1
Continued.

(c)

FIGURE 2.1
Continued.

(d)

FIGURE 2.1
Continued.

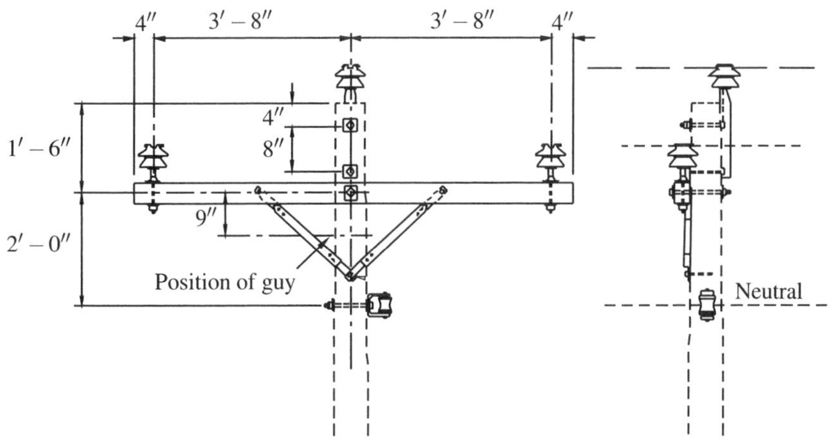

FIGURE 2.2
Example crossarm construction. (From [RUS 1728F-803, 1998].)

Wood is the main pole material, although steel, concrete, and fiberglass are also used. Treated wood lasts a long time, is easy to climb and attach equipment to, and also augments the insulation between the energized conductors and ground. Conductors are primarily aluminum. Insulators are pin type, post type, or suspension, either porcelain or polymer.

The National Electrical Safety Code (IEEE C2-2000) governs many of the safety issues that play important roles in overhead design issues. Poles must have space for crews to climb them and work safely in the air. All equipment must have sufficient strength to stand up to "normal" operations. Conductors must carry their weight, the weight of any accumulated ice, plus withstand the wind pressure exerted on the wire. We are not going to discuss mechanical and structural issues in this book. For more information, see the *Lineman's and Cableman's Handbook* (Kurtz et al., 1997), the *Mechanical Design Manual for Overhead Distribution Lines* (RUS 160-2, 1982), the *NESC* (IEEE C2-2000), and the *NESC Handbook* (Clapp, 1997).

Overhead construction can cost $10,000/mi to $250,000/mi, depending on the circumstances. Some of the major variables are labor costs, how developed the land is, natural objects (including rocks in the ground and trees in the way), whether the circuit is single or three phase, and how big the conductors are. Suburban three-phase mains are typically about $60,000 to $150,000/mi; single-phase laterals are often in the $40,000 to $75,000/mi range. Construction is normally less expensive in rural areas; in urban areas, crews must deal with traffic and set poles in concrete. As Willis (1997) notes, upgrading a circuit normally costs more than building a new line. Typically this work is done live: the old conductor has to be moved to standoff brackets while the new conductor is strung, and the poles may have to be reinforced to handle heavier conductors.

2.2 Conductor Data

A *wire* is metal drawn or rolled to long lengths, normally understood to be a solid wire. Wires may or may not be insulated. A *conductor* is one or more wires suitable for carrying electric current. Often the term *wire* is used to mean conductor. Table 2.1 shows some characteristics of common conductor metals.

Most conductors are either aluminum or copper. Utilities use aluminum for almost all new overhead installations. Aluminum is lighter and less expensive for a given current-carrying capability. Copper was installed more in the past, so significant lengths of copper are still in service on overhead circuits.

Aluminum for power conductors is alloy 1350, which is 99.5% pure and has a minimum conductivity of 61.0% IACS [for more complete characteristics, see the *Aluminum Electrical Conductor Handbook* (Aluminum Association, 1989)]. Pure aluminum melts at 660°C. Aluminum starts to anneal

TABLE 2.1

Nominal or Minimum Properties of Conductor Wire Materials

Property	International Annealed Copper Standard	Commercial Hard-Drawn Copper Wire	Standard 1350-H19 Aluminum Wire	Standard 6201-T81 Aluminum Wire	Galvanized Steel Core Wire	Aluminum Clad Steel
Conductivity,% IACS at 20°C	100.0	97.0	61.2	52.5	8.0	20.3
Resistivity at 20°C, Ω·in.²/1000 ft	0.008145	0.008397	0.013310	0.015515	0.101819	0.04007
Ratio of weight for equal dc resistance and length	1.00	1.03	0.50	0.58	9.1	3.65
Temp. coefficient of resistance, per °C at 20°C	0.00393	0.00381	0.00404	0.00347	0.00327	0.00360
Density at 20°C, lb/in.³	0.3212	0.3212	0.0977	0.0972	0.2811	0.2381
Coefficient of linear expansion, 10^{-6} per °C	16.9	16.9	23.0	23.0	11.5	13.0
Modulus of elasticity, 10^6 psi	17	17	10	10	29	23.5
Specific heat at 20°C, cal/gm-°C	0.0921	0.0921	0.214	0.214	0.107	0.112
Tensile strength, 10^3 psi	62.0	62.0	24.0	46.0	185	175
Minimum elongation,%	1.1	1.1	1.5	3.0	3.5	1.5

Source: Southwire Company, *Overhead Conductor Manual*, 1994.

(soften and lose strength) above 100°C. It has good corrosion resistance; when exposed to the atmosphere, aluminum oxidizes, and this thin, invisible film of aluminum oxide protects against most chemicals, weathering conditions, and even acids. Aluminum can corrode quickly through electrical contact with copper or steel. This galvanic corrosion (dissimilar metals corrosion) accelerates in the presence of salts.

Several variations of aluminum conductors are available:

- *AAC — all-aluminum conductor* — Aluminum grade 1350-H19 AAC has the highest conductivity-to-weight ratio of all overhead conductors. See Table 2.2 for characteristics.
- *ACSR — aluminum conductor, steel reinforced* — Because of its high mechanical strength-to-weight ratio, ACSR has equivalent or higher ampacity for the same size conductor (the kcmil size designation is determined by the cross-sectional area of the aluminum; the steel is neglected). The steel adds extra weight, normally 11 to 18% of the weight of the conductor. Several different strandings are available to provide different strength levels. Common distribution sizes of ACSR have twice the breaking strength of AAC. High strength means the conductor can withstand higher ice and wind loads. Also, trees are less likely to break this conductor. See Table 2.3 for characteristics.
- *AAAC — all-aluminum alloy conductor* — This alloy of aluminum, the 6201-T81 alloy, has high strength and equivalent ampacities of AAC or ACSR. AAAC finds good use in coastal areas where use of ACSR is prohibited because of excessive corrosion.
- *ACAR — aluminum conductor, alloy reinforced* — Strands of aluminum 6201-T81 alloy are used along with standard 1350 aluminum. The alloy strands increase the strength of the conductor. The strands of both are the same diameter, so they can be arranged in a variety of configurations.

For most urban and suburban applications, AAC has sufficient strength and has good thermal characteristics for a given weight. In rural areas, utilities can use smaller conductors and longer pole spans, so ACSR or another of the higher-strength conductors is more appropriate.

Copper has very low resistivity and is widely used as a power conductor, although use as an overhead conductor has become rare because copper is heavier and more expensive than aluminum. It has significantly lower resistance than aluminum by volume — a copper conductor has equivalent ampacity (resistance) of an aluminum conductor that is two AWG sizes larger. Copper has very good resistance to corrosion. It melts at 1083°C, starts to anneal at about 100°C, and anneals most rapidly between 200 and 325°C (this range depends on the presence of impurities and amount of hardening). When copper anneals, it softens and loses tensile strength.

TABLE 2.2

Characteristics of All-Aluminum Conductor (AAC)

					Resistance, Ω/1000 ft					
					dc	60-Hz ac				
AWG	kcmil	Strands	Diameter, in.	GMR, ft	20°C	25°C	50°C	75°C	Breaking. Strength, lb	Weight, lb/1000 ft
6	26.24	7	0.184	0.0056	0.6593	0.6725	0.7392	0.8059	563	24.6
4	41.74	7	0.232	0.0070	0.4144	0.4227	0.4645	0.5064	881	39.1
2	66.36	7	0.292	0.0088	0.2602	0.2655	0.2929	0.3182	1350	62.2
1	83.69	7	0.328	0.0099	0.2066	0.2110	0.2318	0.2527	1640	78.4
1/0	105.6	7	0.368	0.0111	0.1638	0.1671	0.1837	0.2002	1990	98.9
2/0	133.1	7	0.414	0.0125	0.1299	0.1326	0.1456	0.1587	2510	124.8
3/0	167.8	7	0.464	0.0140	0.1031	0.1053	0.1157	0.1259	3040	157.2
4/0	211.6	7	0.522	0.0158	0.0817	0.0835	0.0917	0.1000	3830	198.4
	250	7	0.567	0.0171	0.0691	0.0706	0.0777	0.0847	4520	234.4
	250	19	0.574	0.0181	0.0693	0.0706	0.0777	0.0847	4660	234.3
	266.8	7	0.586	0.0177	0.0647	0.0663	0.0727	0.0794	4830	250.2
	266.8	19	0.593	0.0187	0.0648	0.0663	0.0727	0.0794	4970	250.1
	300	19	0.629	0.0198	0.0575	0.0589	0.0648	0.0705	5480	281.4
	336.4	19	0.666	0.0210	0.0513	0.0527	0.0578	0.0629	6150	315.5
	350	19	0.679	0.0214	0.0494	0.0506	0.0557	0.0606	6390	327.9
	397.5	19	0.724	0.0228	0.0435	0.0445	0.0489	0.0534	7110	372.9
	450	19	0.769	0.0243	0.0384	0.0394	0.0434	0.0472	7890	421.8
	477	19	0.792	0.0250	0.0363	0.0373	0.0409	0.0445	8360	446.8
	477	37	0.795	0.0254	0.0363	0.0373	0.0409	0.0445	8690	446.8
	500	19	0.811	0.0256	0.0346	0.0356	0.0390	0.0426	8760	468.5
	500	37	0.813	0.0260	0.0346	0.0356	0.0390	0.0426	9110	468.3
	556.5	19	0.856	0.0270	0.0311	0.0320	0.0352	0.0383	9750	521.4
	556.5	37	0.858	0.0275	0.0311	0.0320	0.0352	0.0383	9940	521.3
	600	37	0.891	0.0285	0.0288	0.0297	0.0326	0.0356	10700	562.0
	636	37	0.918	0.0294	0.0272	0.0282	0.0309	0.0335	11400	596.0
	650	37	0.928	0.0297	0.0266	0.0275	0.0301	0.0324	11600	609.8
	700	37	0.963	0.0308	0.0247	0.0256	0.0280	0.0305	12500	655.7
	700	61	0.964	0.0310	0.0247	0.0256	0.0280	0.0305	12900	655.8
	715.5	37	0.974	0.0312	0.0242	0.0250	0.0275	0.0299	12800	671.0
	715.5	61	0.975	0.0314	0.0242	0.0252	0.0275	0.0299	13100	671.0
	750	37	0.997	0.0319	0.0230	0.0251	0.0263	0.0286	13100	703.2
	750	61	0.998	0.0321	0.0230	0.0251	0.0263	0.0286	13500	703.2
	795	37	1.026	0.0328	0.0217	0.0227	0.0248	0.0269	13900	745.3
	795	61	1.028	0.0331	0.0217	0.0227	0.0248	0.0269	14300	745.7
	874.5	37	1.077	0.0344	0.0198	0.0206	0.0227	0.0246	15000	820.3
	874.5	61	1.078	0.0347	0.0198	0.0206	0.0227	0.0246	15800	820.6
	900	37	1.092	0.0349	0.0192	0.0201	0.0220	0.0239	15400	844.0
	900	61	1.094	0.0352	0.0192	0.0201	0.0220	0.0239	15900	844.0
	954	37	1.124	0.0360	0.0181	0.0191	0.0208	0.0227	16400	894.5
	954	61	1.126	0.0362	0.0181	0.0191	0.0208	0.0225	16900	894.8
	1000	37	1.151	0.0368	0.0173	0.0182	0.0199	0.0216	17200	937.3
	1000	61	1.152	0.0371	0.0173	0.0182	0.0199	0.0216	17700	936.8

Source: Aluminum Association, *Aluminum Electrical Conductor Handbook*, 1989; Southwire Company, *Overhead Conductor Manual*, 1994.

Different sizes of conductors are specified with gage numbers or area in circular mils. Smaller wires are normally referred to using the American wire gage (AWG) system. The gage is a numbering scheme that progresses geometrically. A number 36 solid wire has a defined diameter of 0.005 in. (0.0127

TABLE 2.3

Characteristics of Aluminum Conductor, Steel Reinforced (ACSR)

AWG	kcmil	Strands	Diameter, in.	GMR, ft	Resistance, Ω/1000 ft dc 20°C	60-Hz ac 25°C	60-Hz ac 50°C	60-Hz ac 75°C	Breaking. Strength, lb	Weight, lb/1000 ft
6	26.24	6/1	0.198	0.0024	0.6419	0.6553	0.7500	0.8159	1190	36.0
4	41.74	6/1	0.250	0.0033	0.4032	0.4119	0.4794	0.5218	1860	57.4
4	41.74	7/1	0.257	0.0045	0.3989	0.4072	0.4633	0.5165	2360	67.0
2	66.36	6/1	0.316	0.0046	0.2534	0.2591	0.3080	0.3360	2850	91.2
2	66.36	7/1	0.325	0.0060	0.2506	0.2563	0.2966	0.3297	3640	102
1	83.69	6/1	0.355	0.0056	0.2011	0.2059	0.2474	0.2703	3550	115
1/0	105.6	6/1	0.398	0.0071	0.1593	0.1633	0.1972	0.2161	4380	145
2/0	133.1	6/1	0.447	0.0077	0.1265	0.1301	0.1616	0.1760	5300	183
3/0	167.8	6/1	0.502	0.0090	0.1003	0.1034	0.1208	0.1445	6620	230
4/0	211.6	6/1	0.563	0.0105	0.0795	0.0822	0.1066	0.1157	8350	291
	266.8	18/1	0.609	0.0197	0.0644	0.0657	0.0723	0.0788	6880	289
	266.8	26/7	0.642	0.0217	0.0637	0.0652	0.0714	0.0778	11300	366
	336.4	18/1	0.684	0.0221	0.0510	0.0523	0.0574	0.0625	8700	365
	336.4	26/7	0.721	0.0244	0.0506	0.0517	0.0568	0.0619	14100	462
	336.4	30/7	0.741	0.0255	0.0502	0.0513	0.0563	0.0614	17300	526
	397.5	18/1	0.743	0.0240	0.0432	0.0443	0.0487	0.0528	9900	431
	397.5	26/7	0.783	0.0265	0.0428	0.0438	0.0481	0.0525	16300	546
	477	18/1	0.814	0.0263	0.0360	0.0369	0.0405	0.0441	11800	517
	477	24/7	0.846	0.0283	0.0358	0.0367	0.0403	0.0439	17200	614
	477	26/7	0.858	0.0290	0.0357	0.0366	0.0402	0.0438	19500	655
	477	30/7	0.883	0.0304	0.0354	0.0362	0.0389	0.0434	23800	746
	556.5	18/1	0.879	0.0284	0.0309	0.0318	0.0348	0.0379	13700	603
	556.5	24/7	0.914	0.0306	0.0307	0.0314	0.0347	0.0377	19800	716
	556.5	26/7	0.927	0.0313	0.0305	0.0314	0.0345	0.0375	22600	765
	636	24/7	0.977	0.0327	0.0268	0.0277	0.0300	0.0330	22600	818
	636	26/7	0.990	0.0335	0.0267	0.0275	0.0301	0.0328	25200	873
	795	45/7	1.063	0.0352	0.0216	0.0225	0.0246	0.0267	22100	895
	795	26/7	1.108	0.0375	0.0214	0.0222	0.0242	0.0263	31500	1093
	954	45/7	1.165	0.0385	0.0180	0.0188	0.0206	0.0223	25900	1075
	954	54/7	1.196	0.0404	0.0179	0.0186	0.0205	0.0222	33800	1228
	1033.5	45/7	1.213	0.0401	0.0167	0.0175	0.0191	0.0208	27700	1163

Sources: Aluminum Association, *Aluminum Electrical Conductor Handbook*, 1989; Southwire Company, *Overhead Conductor Manual*, 1994.

cm), and the largest size, a number 0000 (referred to as 4/0 and pronounced "four-ought") solid wire has a 0.46-in. (1.17-cm) diameter. The larger gage sizes in sequence of increasing conductor size are: 4, 3, 2, 1, 0 (1/0), 00 (2/0), 000 (3/0), 0000 (4/0). Going to the next bigger size (smaller gage number) increases the diameter by 1.1229. Some other useful rules are:

- An increase of three gage sizes doubles the area and weight and halves the dc resistance.
- An increase of six gage sizes doubles the diameter.

Larger conductors are specified in circular mils of cross-sectional area. One circular mil is the area of a circle with a diameter of one mil (one mil is one-thousandth of an inch). Conductor sizes are often given in kcmil, thousands of circular mils. In the past, the abbreviation MCM was used, which also means thousands of circular mils (M is thousands, not mega, in this case). By definition, a solid 1000-kcmil wire has a diameter of 1 in. The diameter of a solid wire in mils is related to the area in circular mils by $d = \sqrt{A}$.

Outside of America, most conductors are specified in mm². Some useful conversion relationships are:

$$1 \text{ kcmil} = 1000 \text{ cmil} = 785.4 \times 10^{-6} \text{ in}^2 = 0.5067 \text{ mm}^2$$

Stranded conductors increase flexibility. A two-layer arrangement has seven wires; a three-layer arrangement has 19 wires, and a four-layer arrangement has 37 wires. The cross-sectional area of a stranded conductor is the cross-sectional area of the metal, so a stranded conductor has a larger diameter than a solid conductor of the same area.

The area of an ACSR conductor is defined by the area of the aluminum in the conductor.

Utilities with heavy tree cover often use covered conductors — conductors with a thin insulation covering. The covering is not rated for full conductor line-to-ground voltage, but it is thick enough to reduce the chance of flashover when a tree branch falls between conductors. Covered conductor is also called *tree wire* or weatherproof wire. Tree wire also helps with animal faults and allows utilities to use armless or candlestick designs or other tight configurations. Tree wire is available with a variety of covering types. The insulation materials polyethylene, XLPE, and EPR are common. Insulation thicknesses typically range from 30 to 150 mils (1 mil = 0.001 in. = 0.00254 cm); see Table 2.4 for typical thicknesses. From a design and operating viewpoint, covered conductors must be treated as bare conductors according to the National Electrical Safety Code (NESC) (IEEE C2-2000), with the only difference that tighter conductor spacings are allowed. It is also used in Australia to reduce the threat of bush fires (Barber, 1999).

While covered wire helps with trees, it has some drawbacks compared with bare conductors. Covered wire is much more susceptible to burndowns caused by fault arcs. Covered-wire systems increase the installed cost somewhat. Covered conductors are heavier and have a larger diameter, so the ice and wind loading is higher than a comparable bare conductor. The covering may be susceptible to degradation due to ultraviolet radiation, tracking, and mechanical effects that cause cracking. Covered conductors are more susceptible to corrosion, primarily from water. If water penetrates the covering, it settles at the low points and causes corrosion (it cannot evaporate). On bare conductors, corrosion is rare; rain washes bare conductors periodically, and evaporation takes care of moisture. The Australian experience has been that complete corrosion can occur with covered wires in 15 to 20 years of operation (Barber, 1999). Water enters the conductor at pinholes caused by lightning strikes, at cover damage caused by abrasion or erosion, and at holes pierced by connectors. Temperature changes then cause water to be

TABLE 2.4

Typical Covering Thicknesses of Covered All-Aluminum Conductor

Size AWG or kcmil	Strands	Cover Thickness, mil	Diameter, in.	
			Bare	Covered
6	7	30	0.184	0.239
4	7	30	0.232	0.285
2	7	45	0.292	0.373
1	7	45	0.328	0.410
1/0	7	60	0.368	0.480
2/0	7	60	0.414	0.524
3/0	7	60	0.464	0.574
4/0	7	60	0.522	0.629
266.8	19	60	0.593	0.695
336.4	19	60	0.666	0.766
397.5	19	80	0.724	0.857
477	37	80	0.795	0.926
556.5	37	80	0.858	0.988
636	61	95	0.918	1.082
795	61	95	1.028	1.187

pumped into the conductor. Because of corrosion concerns, water-blocked conductors are better.

Spacer cables and aerial cables are also alternatives that perform well in treed areas. Spacer cables are a bundled configuration using a messenger wire holding up three phase wires that use covered wire. Because the spacer cable has significantly smaller spacings than normal overhead constructions, its reactive impedance is smaller.

Guy wires, messenger wires, and other wires that require mechanical strength but not current-carrying capability are often made of steel. Steel has high strength (see Table 2.5). Steel corrodes quickly, so most applications use galvanized steel to slow down the corrosion. Because steel is a magnetic material, steel conductors also suffer hysteresis losses. Steel conductors have much higher resistances than copper or aluminum. For some applications requiring strength and conductivity, steel wires coated with copper or aluminum are available. A copperweld conductor has copper-coated steel strands, and an alumoweld conductor aluminum-coated steel strands. Both have better corrosion protection than galvanized steel.

2.3 Line Impedances

Overhead lines have resistance and reactance that impedes the flow of current. These impedance values are necessary for voltage drop, power flow, short circuit, and line-loss calculations.

TABLE 2.5
Characteristics of Steel Conductors

Size	Diameter, in.	Conductor Area, in.²	Weight, lb/1000 ft	Strength, lb	Resistance, Ω/1000 ft dc 25°C	60-Hz ac at the given current level 10A	40A	70A	100A
High-Strength Steel — Class A Galvanizing									
5/8	0.621	0.2356	813	29,600	0.41	0.42	0.43	0.46	0.49
1/2	0.495	0.1497	517	18,800	0.65	0.66	0.68	0.73	0.77
7/16	0.435	0.1156	399	14,500	0.84	0.85	0.88	0.94	1.00
3/8	0.360	0.0792	273	10,800	1.23	1.25	1.28	1.38	1.46
Utilities Grade Steel									
7/16	0.435	0.1156	399	18,000	0.87	0.88	0.90	0.95	1.02
3/8	0.380	0.0882	273	11,500	1.23	1.25	1.28	1.38	1.46
Extra-High-Strength Steel — Class A Galvanizing									
5/8	0.621	0.2356	813	42,400	0.43	0.43	0.44	0.47	0.50
1/2	0.495	0.1497	517	26,900	0.67	0.68	0.69	0.73	0.78
7/16	0.435	0.1156	399	20,800	0.87	0.88	0.90	0.95	1.02
3/8	0.360	0.0792	273	15,400	1.28	1.29	1.31	1.39	1.48
Extra-High-Strength Steel — Class C Galvanizing									
7/16	0.435	0.1156	399	20,800		0.70	0.70	0.71	0.71
3/8	0.360	0.0792	273	15,400		1.03	1.03	1.03	1.04
5/16	0.312	0.0595	205	11,200		1.20	1.30	1.30	1.30

Source: EPRI, *Transmission Line Reference Book: 345 kV and Above,* 2nd ed., Electric Power Research Institute, Palo Alto, CA, 1982.

The dc resistance is inversely proportional to the area of a conductor; doubling the area halves the resistance. Several units are used to describe a conductor's resistance. Conductivity is often given as %IACS, the percent conductivity relative to the International Annealed Copper Standard, which has the following volume resistivities:

$$0.08145\ \Omega\text{-in.}^2/1000\ \text{ft} = 17.241\ \Omega\text{-mm}^2/\text{km} = 10.37\ \Omega\text{-cmil/ft}$$

And with a defined density of 8.89 g/cm^3 at 20°C, the copper standard has the following weight resistivities:

$$875.2\ \Omega\text{-lb/mi}^2 = 0.15328\ \Omega\text{-g/m}^2$$

Hard-drawn copper has 97.3%IACS. Aluminum varies, depending on type; alloy 1350-H19 has 61.2% conductivity.

Temperature and frequency — these change the resistance of a conductor. A hotter conductor provides more resistance to the flow of current. A higher frequency increases the internal magnetic fields. Current has a difficult time flowing in the center of a conductor at high frequency, as it is being opposed by the magnetic field generated by current flowing on all sides of it. Current flows more easily near the edges. This *skin effect* forces the current to flow in a smaller area of the conductor.

Resistance changes with temperature as

$$R_{t2} = R_{t1} \frac{M + t_2}{M + t_1}$$

where

R_{t2} = resistance at temperature t_2 given in °C
R_{t1} = resistance at temperature t_1 given in °C
M = a temperature coefficient for the given material
= 228.1 for aluminum
= 241.5 for annealed hard-drawn copper

For a wide range of temperatures, resistance rises almost linearly with temperature for both aluminum and copper. The effect of temperature is simplified as a linear equation as

$$R_{t2} = R_{t1}[1 + \alpha(t_2 - t_1)]$$

where

α = a temperature coefficient of resistance
= 0.00404 for 61.2% IACS aluminum at 20°C
= 0.00347 for 6201-T81 aluminum alloy at 20°C
= 0.00383 for hard-drawn copper at 20°C
= 0.0036 for aluminum-clad steel at 20°C

So, the resistance of aluminum with a 61.2% conductivity rises 4% for every 10°C rise in temperature.

We can also linearly interpolate using resistances provided at two different temperatures as

$$R(T_c) = R(T_{low}) + \frac{R(T_{high}) - R(T_{low})}{T_{high} - T_{low}}(T_c - T_{low})$$

where

$R(T_c)$ = conductor resistance at temperature T_c
$R(T_{high})$ = resistance at the higher temperature T_{high}
$R(T_{low})$ = resistance at the lower temperature T_{low}

With alternating current, skin effects raise the resistance of a conductor relative to its dc resistance. At 60 Hz, the resistance of a conductor is very close to its dc resistance except for very large conductors. Skin effects are much more important for high-frequency analysis such as switching surges and power-line carrier problems. They play a larger role in larger conductors.

The internal resistance of a solid round conductor including skin effects is [for details, see Stevenson (1962)]:

$$\frac{R_{ac}}{R_{dc}} = \frac{x}{2}\frac{\mathrm{ber}(x)\mathrm{bei}'(x) - \mathrm{bei}(x)\mathrm{ber}'(x)}{(\mathrm{bei}'(x))^2 + (\mathrm{ber}'(x))^2}$$

where

$x = 0.02768\sqrt{\frac{f\mu}{R_{dc}}}$

f = frequency in Hz

μ = relative permeability = 1 for nonmagnetic conductors (including aluminum and copper)

R_{dc} = dc resistance of the conductor in ohms/1000 ft

ber, bei, ber′, and bei′ = real and imaginary modified Bessel functions and their derivatives (also called Kelvin functions)

For x greater than 3 (frequencies in the kilohertz range), the resistance increases linearly with x (Clarke, 1950) approximately as

$$\frac{R_{ac}}{R_{dc}} = \frac{x}{2\sqrt{2}} + \frac{1}{4} = 0.009786\sqrt{\frac{f\mu}{R_{dc}}} + 0.25$$

So, for higher frequencies, the ac resistance increases as the square root of the frequency. For most distribution power-frequency applications, we can ignore skin effects (and they are included in ac resistance tables).

For most cases, we can model a stranded conductor as a solid conductor with the same cross-sectional area. ACSR with a steel core is slightly different. Just as in a transformer, the steel center conductor has losses due to hysteresis and eddy currents. If an ACSR conductor has an even number of layers, the axial magnetic field produced by one layer tends to cancel that produced by the next layer. We can model these as a tubular conductor for calculating skin effect. For odd numbers of layers, especially single-layered conductors like 6/1 or 7/1, the 60-Hz/dc ratio is higher than normal, especially at high current densities. These effects are reflected in the resistances included in tables (such as Table 2.3).

The reactance part of the impedance usually dominates the impedances on overhead circuit for larger conductors; below 4/0, resistance plays more of a role. For all-aluminum conductors on a 10-ft crossarm, the resistance approximately equals the reactance for a 2/0 conductor. Reactance is proportional to inductance; and inductance causes a voltage that opposes the change in the flow of current. Alternating current is always changing, so a reactance always creates a voltage due to current flow.

Distance between conductors determines the external component of reactance. Inductance is based on the area enclosed by a loop of current; a larger area (more separation between conductors) has more inductance. On overhead circuits, reactance of the line is primarily based on the separations between conductors — not the size of the conductor, not the type of metal in the conductor, not the stranding of the conductor.

The reactance between two parallel conductors in ohms per mile is:

$$X_{ab} = 0.2794 \frac{f}{60} \log_{10} \frac{d_{ab}}{GMR}$$

where

f = frequency in hertz
d_{ab} = distance between the conductors
GMR = geometric mean radius of both conductors

d_{ab} and GMR must have the same units, normally feet. More separation — a bigger loop — gives larger impedances.

The geometric mean radius (GMR) quantifies a conductor's internal inductance — by definition, the GMR is the radius of an infinitely thin tube having the same internal inductance as the conductor out to a one-foot radius. The GMR is normally given in feet to ease calculations with distances measured in feet. GMR is less than the actual conductor radius. Many conductor tables provide x_a, the inductive reactance due to flux in the conductor and outside the conductor out to a one-foot radius. The GMR in feet at 60 Hz relates to x_a as:

$$x_a = 0.2794 \log_{10} \frac{1}{GMR}$$

where GMR is in feet, and x_a is in ohms/mile.

For a solid, round, nonmagnetic conductor, the relationship between the actual radius and the GMR is

$$\frac{GMR}{r} = e^{-1/4} = 0.779$$

For stranded conductors, the GMR is

$$GMR = k \cdot r$$

TABLE 2.6

GMR Factor

Strands	GMR Factor, k
1 (solid)	0.7788
3	0.6778
7	0.7256
19	0.7577
37	0.7678
61	0.7722

Source: Aluminum Association, *Ampacities for Aluminum and ACSR Overhead Electrical Conductors*, 1986.

where

k = the *GMR* factor from Table 2.6

r = conductor radius

For ACSR conductors (which are layered), the *GMR* factor is more complicated.

Current flowing in a conductor induces a reactive voltage drop on the conductor it is flowing through. By the same induction, current flow in one conductor creates a voltage gradient along parallel conductors (see Figure 2.3). This voltage is of the same polarity as the voltage on the current-carrying conductor. Closer conductors have larger induced voltages. This induction is significant for several reasons:

- *Opposite flow* — Current flows more easily when a parallel conductor has flow in the opposite direction. The magnetic field from the other conductor creates a voltage drop that encourages flow in the opposite direction. Conductors carrying current in opposite directions have lower impedance when they are closer together.
- *Parallel flow* — A conductor carrying current in the same direction as a parallel conductor faces more impedance because of the current in the other conductor. Conductors carrying current in the same direction have higher impedance when they are closer together.
- *Circulating current* — Current flow in the vicinity of a shorted current loop induces currents to circulate in the loop.

For balanced conditions — balanced voltages, balanced loads, and balanced impedances — we can analyze power systems just with positive-sequence voltages, currents, and impedances. This is regularly done in transmission-planning and industrial load-flow studies. Using just positive-sequence quantities simplifies analysis; it's like a single-phase circuit rather than a three-phase circuit. For distribution circuits, unbalanced loading is quite common, so we normally need more than just positive-sequence parameters — we need the zero-sequence parameters as well. We also need unbalanced analysis approaches for phase-to-ground or phase-to-phase faults.

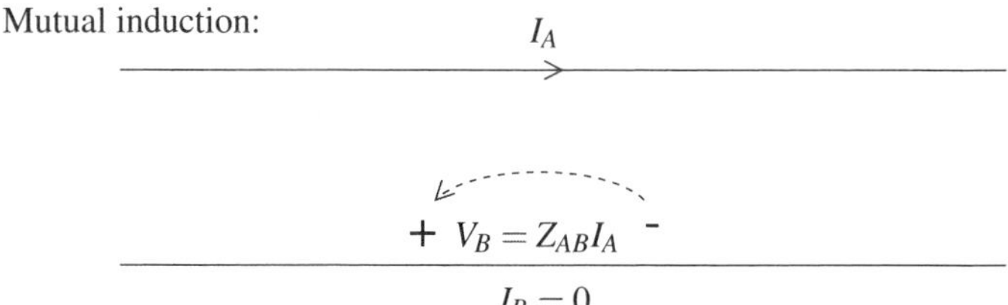

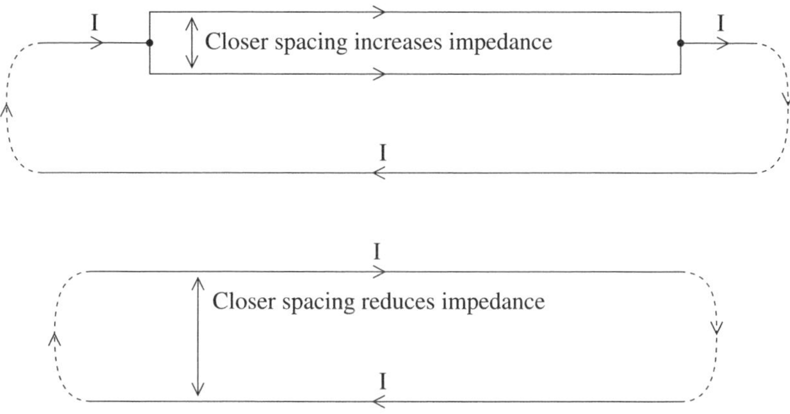

FIGURE 2.3
Mutual induction.

With symmetrical components, the phasors of circuit quantities on each of the three phases resolve to three sets of phasors. For voltage, the symmetrical components relate to the phase voltages as:

$$V_a = V_0 + V_1 + V_2 \qquad V_0 = 1/3\ (V_a + V_b + V_c)$$

$$V_b = V_0 + a^2V_1 + aV_2 \quad \text{and} \quad V_1 = 1/3\ (V_a + aV_b + a^2V_c)$$

$$V_c = V_0 + aV_1 + a^2V_2 \qquad V_2 = 1/3\ (V_a + a^2V_b + aV_c)$$

where $a = 1\angle 120°$ and $a^2 = 1\angle 240°$.

These phase-to-symmetrical conversions apply for phase-to-ground as well as phase-to-phase voltages. The same conversions apply for converting line currents to sequence currents:

$$I_a = I_0 + I_1 + I_2 \qquad I_0 = 1/3\ (I_a + I_b + I_c)$$

$$I_b = I_0 + a^2I_1 + aI_2 \quad \text{and} \quad I_1 = 1/3\ (I_a + aI_b + a^2I_c)$$

$$I_c = I_0 + aI_1 + a^2I_2 \qquad I_2 = 1/3\ (I_a + a^2I_b + aI_c)$$

The voltage drop along each of the phase conductors depends on the currents in each of the phase conductors and the self impedances (such as Z_{aa}) and the mutual impedances (such as Z_{ab}) as

$$V_a = Z_{aa}I_a + Z_{ab}I_b + Z_{ac}I_c$$

$$V_b = Z_{ba}I_a + Z_{bb}I_b + Z_{bc}I_c$$

$$V_c = Z_{ca}I_a + Z_{cb}I_b + Z_{cc}I_c$$

Likewise, when we use sequence components, we have voltage drops of each sequence in terms of the sequence currents and sequence impedances:

$$V_0 = Z_{00}I_0 + Z_{01}I_1 + Z_{02}I_2$$

$$V_1 = Z_{10}I_0 + Z_{11}I_1 + Z_{12}I_2$$

$$V_2 = Z_{20}I_0 + Z_{21}I_1 + Z_{22}I_2$$

This is not much of a simplification until we assume that all of the self-impedance terms are equal ($Z_S = Z_{aa} = Z_{bb} = Z_{cc}$) and all of the mutual impedances are equal ($Z_M = Z_{ab} = Z_{ac} = Z_{bc} = Z_{ba} = Z_{ca} = Z_{cb}$). With this assumption, the sequence impedances decouple; the mutual terms of the zero-sequence matrix (such as Z_{12}) become zero. Zero-sequence current only causes a zero-sequence voltage drop. This is a good enough approximation for many distribution problems and greatly simplifies hand and computer calculations. Now, the sequence voltage drop equations are:

$$V_0 = Z_{00}I_0 = (Z_S + 2Z_M)I_0$$

$$V_1 = Z_{11}I_1 = (Z_S - Z_M)I_1$$

$$V_2 = Z_{22}I_2 = (Z_S - Z_M)I_2$$

Now, we have the sequence terms as

$$Z_0 = Z_S + 2Z_M$$

$$Z_1 = Z_2 = Z_S - Z_M$$

And likewise,

$$Z_S = (Z_0 + 2Z_1)/3$$

$$Z_M = (Z_0 - Z_1)/3$$

Note Z_S, the self-impedance term. Z_S is also the "loop impedance" — the impedance to current through one phase wire and returning through the ground return path. This loop impedance is important because it is the impedance for single-phase lines and the impedance for single line-to-ground faults.

Engineers normally use three methods to find impedances of circuits. In order of least to most accurate, these are:

- Table lookup
- Hand calculations
- Computer calculations

This book provides data necessary for the first two approaches. Table lookups are quite common. Even though table lookup is not the most accurate approach, its accuracy is good enough for analyzing most common distribution problems. Computer calculations are quite accessible and allow easier analysis of more complicated problems.

2.4 Simplified Line Impedance Calculations

The positive-sequence impedance of overhead lines is

$$Z_1 = R_\phi + jk_1 \log_{10} \frac{GMD_\phi}{GMR_\phi}$$

where

R_ϕ = resistance of the phase conductor in Ω/distance
k_1 = $0.2794f/60$ for outputs in Ω/mi
= $0.0529f/60$ for outputs in Ω/1000 ft
f = frequency in hertz
GMR_ϕ =geometric mean radius of the phase conductor in ft
GMD_ϕ = geometric mean distance between the phase conductors in ft
$GMD_\phi = \sqrt[3]{d_{AB}d_{BC}d_{CA}}$ for three-phase lines
GMD_ϕ = 1.26 d_{AB} for a three-phase line with flat configuration, either horizontal or vertical, where $d_{AB} = d_{BC} = 0.5d_{CA}$
$GMD_\phi = d_{AB}$ for two-phase lines*

* The two-phase circuit has two out of the three phases; the single-phase circuit has one phase conductor with a neutral return. While it may seem odd to look at the positive-sequence impedance of a one- or two-phase circuit, the analysis approach is useful. This approach uses fictitious conductors for the missing phases to model the one- or two-phase circuit as an equivalent three-phase circuit (no current flows on these fictitious phases).

$GMD_\phi = d_{AN}$ for single-phase lines*
d_{ij} = distance between the center of conductor i and the center of conductor j, in feet

For 60 Hz and output units of Ω/1000 ft, this is

$$Z_1 = R_\phi + j0.0529 \log_{10} \frac{GMD_\phi}{GMR_\phi}$$

Zero-sequence impedance calculations are more complicated than positive-sequence calculations. Carson's equations are the most common way to account for the ground return path in impedance calculations of overhead circuits. Carson (1926) derived an expression including the earth return path. We'll use a simplification of Carson's equations; it includes the following assumptions (Smith, 1980)

- Since distribution lines are relatively short, the height-dependent terms in Carson's full model are small, so we neglect them.
- The multigrounded neutral is perfectly coupled to the earth (this has some drawbacks for certain calculations as discussed in Chapter 13).
- End effects are neglected.
- The current at the sending end equals that at the receiving end (no leakage currents).
- All phase conductors have the same size conductor.
- The ground is infinite and has uniform resistivity.

Consider the circuit in Figure 2.4; current flows in conductor a and returns through the earth. The voltage on conductor a equals the current times Z_{aa}, which is the self-impedance with an earth return path. The current in conductor a induces a voltage drop along conductor b equaling the phase-a current times Z_{ab}, which is the mutual impedance with an earth return path. These two impedances are found (Smith, 1980) with

$$Z_{aa} = R_\phi + R_e + jk_1 \log_{10} \frac{D_e}{GMR_\phi}$$

$$Z_{ab} = R_e + jk_1 \log_{10} \frac{D_e}{d_{ab}}$$

where
R_e = resistance of the earth return path
= $0.0954f/60$ Ω/mi
= $0.01807f/60$ Ω/1000 ft

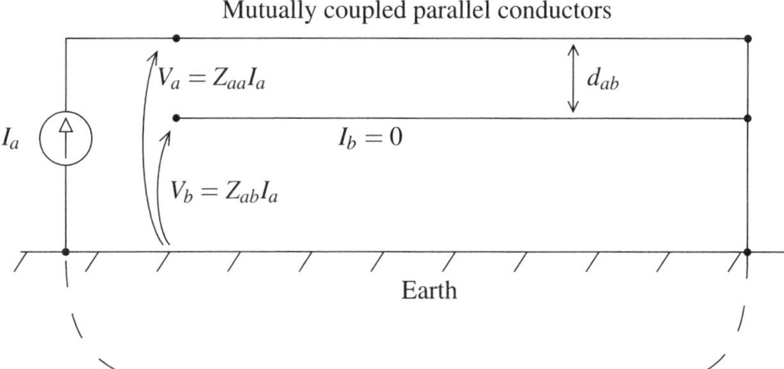

FIGURE 2.4
Depiction of Carson's impedances with earth return.

$D_e = 2160\sqrt{\rho / f}$ = equivalent depth of the earth return current in ft
ρ = earth resistivity in Ω-m
d_{ab} = distance between the centers of conductors a and b

For 60 Hz and output units of Ω/1000 ft,

$$Z_{aa} = R_\phi + 0.01807 + j0.0529 \log_{10} \frac{278.9\sqrt{\rho}}{GMR_\phi}$$

$$Z_{ab} = \quad 0.01807 + j0.0529 \log_{10} \frac{278.9\sqrt{\rho}}{d_{ab}}$$

These equations lead to different formulations for the zero-sequence impedance of circuits depending on the grounding configuration. They are also useful in their own right in many circumstances. Single-phase circuits with a phase and a neutral are often easier to analyze using these equations rather than using sequence components. Consider a single-phase circuit that is perfectly grounded with a current of I_A in the phase conductor. As Figure 2.5 shows, we can find the neutral current as a function of the mutual impedance between the two conductors divided by the self-impedance of the neutral conductor.

Now, let's look at the zero-sequence impedances — these change depending on grounding configuration. Figure 2.6 shows the configurations that we will consider.

A three-wire overhead line has a zero-sequence impedance of (Smith, 1980):

$$Z_0 = R_\phi + 3R_e + j3k_1 \log_{10} \frac{D_e}{\sqrt[3]{GMR_\phi \cdot GMD_\phi^{\,2}}}$$

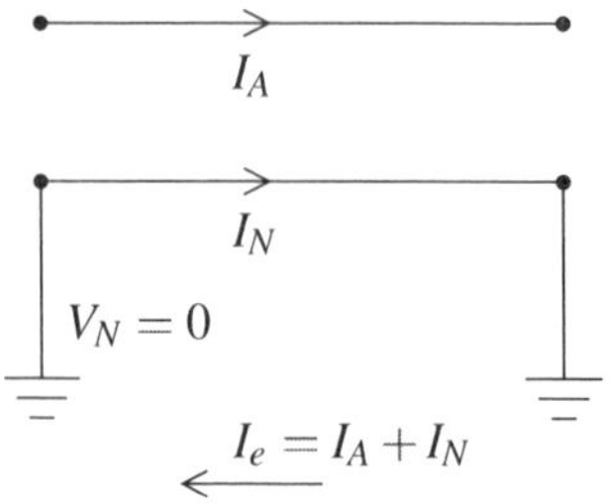

I_A

I_N

$V_N = 0$

$I_e = I_A + I_N$

$V_N = 0 = Z_{AN} I_A + Z_{NN} I_N$

So, $I_N = -\dfrac{Z_{AN}}{Z_{NN}} I_A$

FIGURE 2.5
Current flow in a neutral conductor based on self-impedances and mutual impedances.

For a four-wire multigrounded system, the zero-sequence self-impedance is:

$$Z_0 = R_\phi + 3R_e + j3k_1 \log_{10} \frac{D_e}{\sqrt[3]{GMR_\phi \cdot GMD_\phi^{\;2}}} - 3\frac{Z_{\phi N}^2}{Z_{NN}}$$

where Z_{NN} is the self-impedance of the neutral conductor with an earth return, and $Z_{\phi N}$ is the mutual impedance between the phase conductors as a group and the neutral. For 60 Hz and output units of Ω/1000 ft, the zero-sequence self-impedance is

$$Z_0 = R_\phi + 0.0542 + j0.1587 \log_{10} \frac{278.9\sqrt{\rho}}{\sqrt[3]{GMR_\phi \cdot GMD_\phi^{\;2}}} - 3\frac{Z_{\phi N}^2}{Z_{NN}}$$

$$Z_{NN} = R_N + 0.01807 + j0.0529 \log_{10} \frac{278.9\sqrt{\rho}}{GMR_N}$$

$$Z_{\phi N} = 0.01807 + j0.0529 \log_{10} \frac{278.9\sqrt{\rho}}{GMD_{\phi N}}$$

where

GMR_N = geometric mean radius of the neutral conductor in ft
$GMD_{\phi N}$ = geometric mean distance between the phase conductors as a group and the neutral in ft
$GMD_{\phi N} = \sqrt[3]{d_{AN} d_{BN} d_{CN}}$ for three-phase lines
$GMD_{\phi N} = \sqrt{d_{AN} d_{BN}}$ for two-phase lines
$GMD_{\phi N} = d_{AN}$ for single-phase lines

A special case is for a four-wire unigrounded circuit where the return current stays in the neutral, which has a zero-sequence impedance of (Ender et al., 1960)

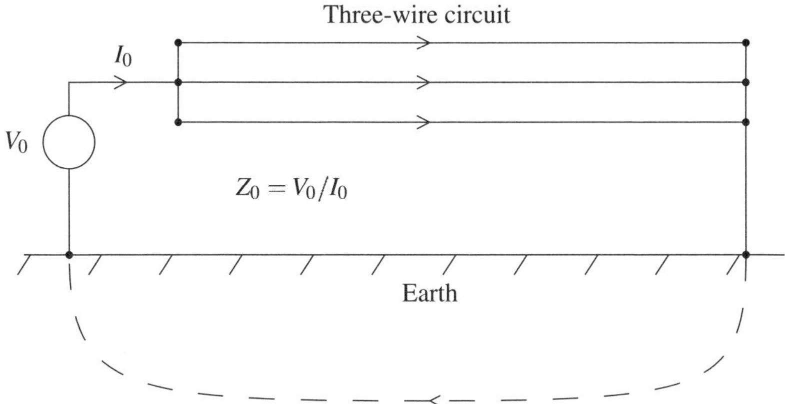

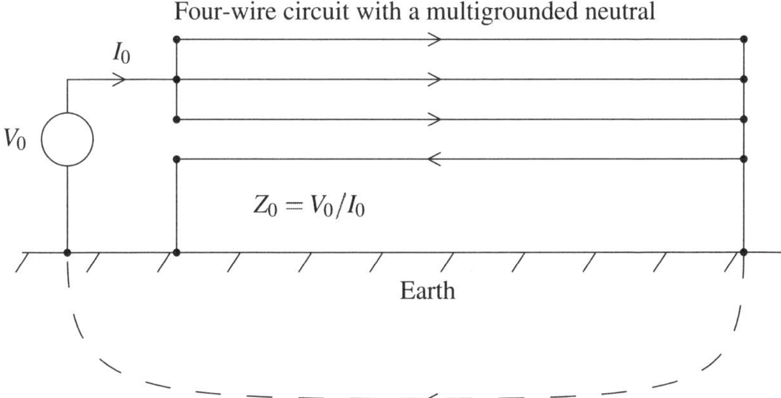

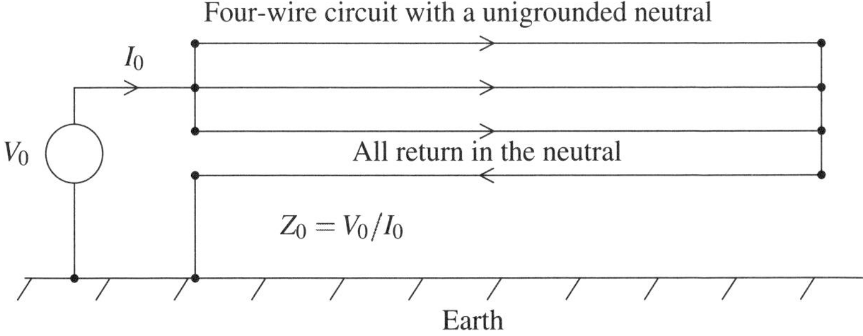

FIGURE 2.6
Different zero-sequence impedances depending on the grounding configuration.

$$Z_0 = R_\phi + 3R_n + jR_n + j3k_1 \log_{10} \frac{GMD_{\phi N}^2}{GMR_N \sqrt[3]{GMR_\phi \cdot GMD_\phi^2}}$$

This is for a four-wire unigrounded circuit where there are no connections between the neutral conductor and earth. We can also use this as an approximation for a multigrounded neutral line that is very poorly grounded. Remember that the equation given above for a multigrounded circuit assumes perfect grounding. For some calculations, that is not accurate. This is the opposite extreme, which is appropriate for some calculations. Lat (1990) used this as one approach to estimating the worst-case overvoltage on unfaulted phases during a line-to-ground fault.

So, what does all of this mean? Some of the major effects are:

- *Conductor size* — Mainly affects resistance — larger conductors have lower positive-sequence resistance. Positive-sequence reactance also lowers with larger conductor size, but since it changes with the logarithm of conductor radius, changes are small.
- *Conductor spacings* — Increasing spacing (higher GMD_ϕ) increases Z_1. Increasing spacing reduces Z_0. Both of these changes with spacing are modest given the logarithmic effect.
- *Neutral* — Adding the neutral always reduces the total zero-sequence impedance, $|Z_0|$. Adding a neutral always reduces the reactive portion of this impedance. But adding a neutral may increase or may decrease the resistive portion of Z_0. Adding a small neutral with high resistance increases the resistance component of Z_0.
- *Neutral spacing* — Moving the neutral closer to the phase conductors reduces the zero-sequence impedance (but may increase the resistive portion, depending on the size of the neutral).
- *Earth resistivity* — The earth resistivity does not change the earth return resistance (R_e only depends on frequency). The current spreads to wider areas of the earth in high-resistivity soil. Earth resistivity does change the reactance. Higher earth resistivities force current deeper into the ground (larger D_e), raising the reactance.
- *Grounding* — The positive-sequence impedance Z_1 stays the same regardless of the grounding, whether four-wire multigrounded, ungrounded, or unigrounded.
- *Negative sequence* — Equals the positive-sequence impedance.

Figure 2.7 and Figure 2.8 show the effects of various parameters on the positive and zero-sequence impedances. Many of the outputs are not particularly sensitive to changes in the inputs. Since many parameters are functions of the logarithm of the variable, major changes induce only small changes in the impedance.

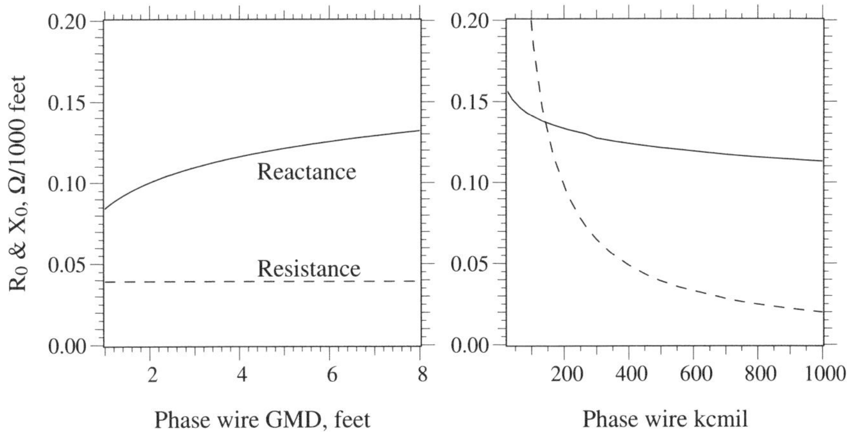

FIGURE 2.7
Effect of spacings and conductor size on the positive-sequence impedance with 500-kcmil AAC phases (GMR = 0.0256 ft) and GMD_{ϕ}=5 ft.

When do we need more accuracy or more sophistication? For power flows, fault calculations, voltage flicker calculations, and voltage sag analysis, we normally don't need more sophistication. For switching surges, lightning, or other higher frequency transient analysis, we normally need more sophisticated line models.

Most unbalanced calculations can be done with this approach, but some cases require more sophistication. Distribution lines and most lower-voltage subtransmission lines are not transposed. On some long circuits, even with balanced loading, the unbalanced impedances between phases creates voltage unbalance.

2.5 Line Impedance Tables

This section has several tables of impedances for all-aluminum, ACSR, and copper constructions. All are based on the equations in the previous section and assume GMD = 4.8 ft, conductor temperature = 50°C, $GMD_{\phi N}$ = 6.3 ft, and earth resistivity = 100 Ω-m. All zero-sequence values are for a four-wire multigrounded neutral circuit.

2.6 Conductor Sizing

We have an amazing variety of sizes and types of conductors. Several electrical, mechanical, and economic characteristics affect conductor selection:

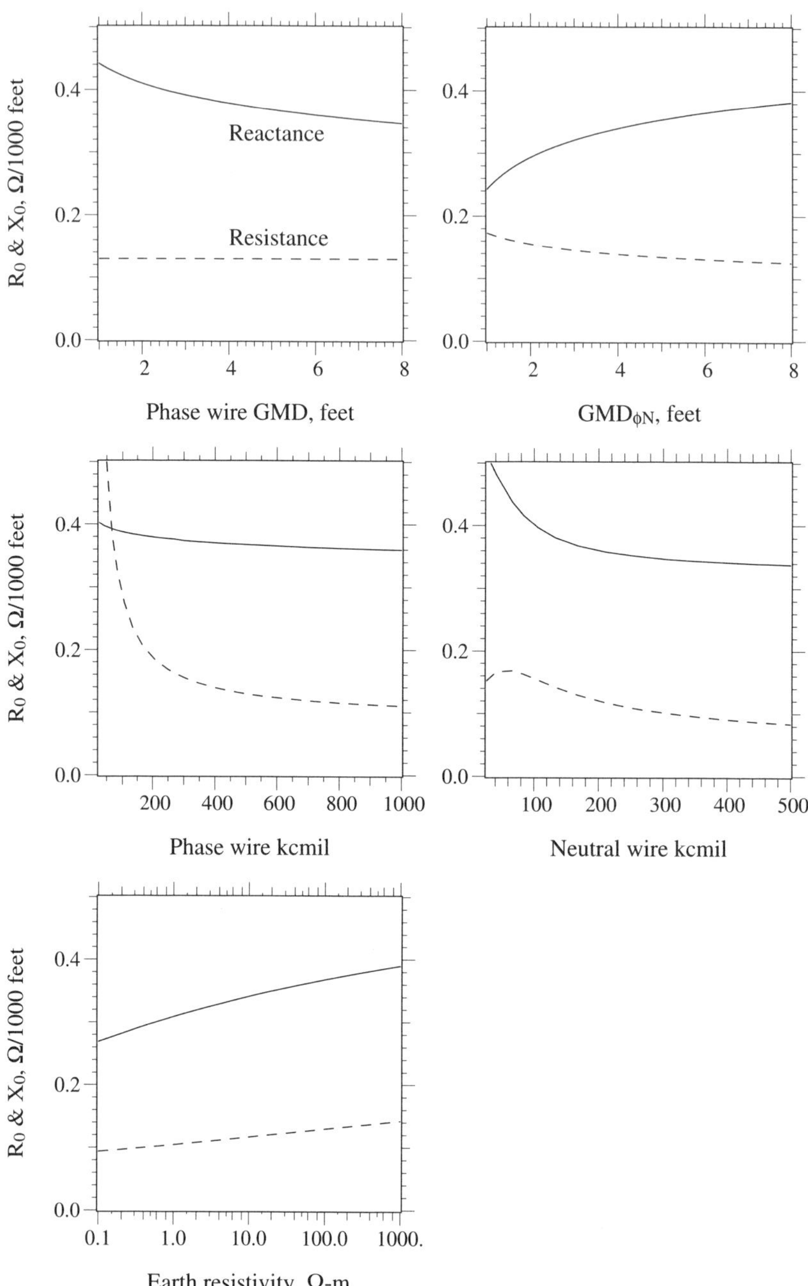

FIGURE 2.8
Effect of various parameters on the zero-sequence impedance with a base case of AAC 500-kcmil phases, 3/0 neutral (168 kcmil), $GMD_{\phi} = 5$ ft, $GMD_{\phi N} = 6.3$ ft, and $\rho = 100$ Ω-m.

TABLE 2.7

Positive-Sequence Impedances of All-Aluminum Conductor

Phase Size	Strands	R_1	X_1	Z_1
6	7	0.7405	0.1553	0.7566
4	7	0.4654	0.1500	0.4890
2	7	0.2923	0.1447	0.3262
1	7	0.2323	0.1420	0.2723
1/0	7	0.1839	0.1394	0.2308
2/0	7	0.1460	0.1367	0.2000
3/0	7	0.1159	0.1341	0.1772
4/0	7	0.0920	0.1314	0.1604
250	7	0.0778	0.1293	0.1509
266.8	7	0.0730	0.1286	0.1478
300	19	0.0649	0.1261	0.1418
336.4	19	0.0580	0.1248	0.1376
350	19	0.0557	0.1242	0.1361
397.5	19	0.0490	0.1229	0.1323
450	19	0.0434	0.1214	0.1289
477	19	0.0411	0.1208	0.1276
500	19	0.0392	0.1202	0.1265
556.5	19	0.0352	0.1189	0.1240
700	37	0.0282	0.1159	0.1192
715.5	37	0.0277	0.1157	0.1190
750	37	0.0265	0.1151	0.1181
795	37	0.0250	0.1146	0.1173
874.5	37	0.0227	0.1134	0.1157
900	37	0.0221	0.1130	0.1152
954	37	0.0211	0.1123	0.1142
1000	37	0.0201	0.1119	0.1137

Note: Impedances, Ω/1000 ft (× 5.28 for Ω/mi or × 3.28 for Ω/km). GMD = 4.8 ft, Conductor temp. = 50°C.

- *Ampacity* — The peak current-carrying capability of a conductor limits the current (and power) carrying capability.
- *Economics* — Often we will use a conductor that normally operates well below its ampacity rating. The cost of the extra aluminum pays for itself with lower I^2R losses; the conductor runs cooler. This also leaves room for expansion.
- *Mechanical strength* — Especially on rural lines with long span lengths, mechanical strength plays an important role in size and type of conductor. Stronger conductors like ACSR are used more often. Ice and wind loadings must be considered.
- *Corrosion* — While not usually a problem, corrosion sometimes limits certain types of conductors in certain applications.

As with many aspects of distribution operations, many utilities standardize on a set of conductors. For example, a utility may use 500-kcmil AAC

TABLE 2.8

AAC Zero-Sequence, Z_0, and Ground-Return Loop Impedances, $Z_S = (2Z_1 + Z_0)/3$

Phase Size	Neutral Size	R_0	X_0	Z_0	R_S	X_S	Z_S
6	6	0.8536	0.5507	1.0158	0.7782	0.2871	0.8294
2	6	0.4055	0.5401	0.6754	0.3301	0.2765	0.4306
2	2	0.4213	0.4646	0.6272	0.3353	0.2513	0.4190
1	6	0.3454	0.5374	0.6389	0.2700	0.2738	0.3845
1	1	0.3558	0.4405	0.5662	0.2734	0.2415	0.3648
1/0	6	0.2971	0.5348	0.6117	0.2216	0.2712	0.3502
1/0	1/0	0.2981	0.4183	0.5136	0.2220	0.2323	0.3213
2/0	6	0.2591	0.5321	0.5919	0.1837	0.2685	0.3253
2/0	2/0	0.2487	0.3994	0.4705	0.1802	0.2243	0.2877
3/0	2	0.2449	0.4540	0.5158	0.1589	0.2407	0.2884
3/0	3/0	0.2063	0.3840	0.4359	0.1461	0.2174	0.2619
4/0	1	0.2154	0.4299	0.4809	0.1331	0.2309	0.2665
4/0	4/0	0.1702	0.3716	0.4088	0.1180	0.2115	0.2422
250	2/0	0.1805	0.3920	0.4316	0.1120	0.2169	0.2441
250	250	0.1479	0.3640	0.3929	0.1012	0.2075	0.2309
266.8	2/0	0.1757	0.3913	0.4289	0.1072	0.2161	0.2413
266.8	266.8	0.1402	0.3614	0.3877	0.0954	0.2062	0.2272
300	2/0	0.1675	0.3888	0.4234	0.0991	0.2137	0.2355
300	300	0.1272	0.3552	0.3773	0.0856	0.2025	0.2198
336.4	2/0	0.1607	0.3875	0.4195	0.0922	0.2123	0.2315
336.4	336.4	0.1157	0.3514	0.3699	0.0772	0.2003	0.2147
350	2/0	0.1584	0.3869	0.4181	0.0899	0.2118	0.2301
350	350	0.1119	0.3499	0.3674	0.0744	0.1994	0.2129
397.5	2/0	0.1517	0.3856	0.4143	0.0832	0.2105	0.2263
397.5	397.5	0.1005	0.3463	0.3606	0.0662	0.1974	0.2082
450	2/0	0.1461	0.3841	0.4109	0.0776	0.2089	0.2229
450	450	0.0908	0.3427	0.3545	0.0592	0.1951	0.2039
477	2/0	0.1438	0.3835	0.4096	0.0753	0.2084	0.2216
477	477	0.0868	0.3414	0.3522	0.0563	0.1943	0.2023
500	4/0	0.1175	0.3605	0.3791	0.0653	0.2003	0.2107
500	500	0.0835	0.3401	0.3502	0.0540	0.1935	0.2009
556.5	4/0	0.1135	0.3591	0.3766	0.0613	0.1990	0.2082
556.5	556.5	0.0766	0.3372	0.3458	0.0490	0.1917	0.1978
700	4/0	0.1064	0.3561	0.3717	0.0542	0.1960	0.2033
700	700	0.0639	0.3310	0.3371	0.0401	0.1876	0.1918
715.5	4/0	0.1060	0.3559	0.3714	0.0538	0.1958	0.2030
715.5	715.5	0.0632	0.3306	0.3366	0.0395	0.1873	0.1915
750	4/0	0.1047	0.3553	0.3705	0.0526	0.1952	0.2022
750	750	0.0609	0.3295	0.3351	0.0380	0.1866	0.1904
795	4/0	0.1033	0.3548	0.3695	0.0511	0.1946	0.2012
795	795	0.0582	0.3283	0.3335	0.0361	0.1858	0.1893
874.5	4/0	0.1010	0.3536	0.3678	0.0488	0.1935	0.1996
874.5	874.5	0.0540	0.3261	0.3305	0.0332	0.1843	0.1873
900	4/0	0.1004	0.3533	0.3672	0.0482	0.1931	0.1990
900	900	0.0529	0.3254	0.3296	0.0324	0.1838	0.1866
954	4/0	0.0993	0.3525	0.3662	0.0471	0.1924	0.1981

TABLE 2.8 (Continued)

AAC Zero-Sequence, Z_0, and Ground-Return Loop Impedances, $Z_S = (2Z_1 + Z_0)/3$

Phase Size	Neutral Size	R_0	X_0	Z_0	R_S	X_S	Z_S
954	954	0.0510	0.3239	0.3279	0.0310	0.1828	0.1854
1000	4/0	0.0983	0.3521	0.3656	0.0462	0.1920	0.1975
1000	1000	0.0491	0.3232	0.3269	0.0298	0.1823	0.1847

Note: Impedances, Ω/1000 ft (× 5.28 for Ω/mi or × 3.28 for Ω/km). GMD = 4.8 ft, $GMD_{\phi N}$ = 6.3 ft, Conductor temp. = 50°C, Earth resistivity = 100 Ω-m.

TABLE 2.9

Positive-Sequence Impedances of ACSR

Phase Size	Strands	R_1	X_1	Z_1
6	6/1	0.7500	0.1746	0.7700
4	6/1	0.4794	0.1673	0.5077
2	6/1	0.3080	0.1596	0.3469
1	6/1	0.2474	0.1551	0.2920
1/0	6/1	0.1972	0.1496	0.2476
2/0	6/1	0.1616	0.1478	0.2190
3/0	6/1	0.1208	0.1442	0.1881
4/0	6/1	0.1066	0.1407	0.1765
266.8	18/1	0.0723	0.1262	0.1454
336.4	18/1	0.0574	0.1236	0.1362
397.5	18/1	0.0487	0.1217	0.1311
477	18/1	0.0405	0.1196	0.1262
556.5	18/1	0.0348	0.1178	0.1228
636	18/1	0.0306	0.1165	0.1204
795	36/1	0.0247	0.1140	0.1167

Notes: Impedances, Ω/1000 ft (× 5.28 for Ω/mi or × 3.28 for Ω/km). GMD = 4.8 ft, Conductor temp. = 50°C.

for all mainline spans and 1/0 AAC for all laterals. While many circuit locations are overdesigned, the utility saves from reduced stocking, fewer tools, and standardized connectors. While many utilities have more than just two conductors, most use just a handful of standard conductors; four to six economically covers the needs of most utilities.

2.7 Ampacities

The ampacity is the maximum designed current of a conductor. This current carrying capacity is normally given in amperes. A given conductor has

TABLE 2.10

ACSR Zero-Sequence, Z_0, and Ground-Return Loop Impedances, $Z_S = (2Z_1 + Z_0)/3$

Phase Size	Neutral Size	R_0	X_0	Z_0	R_S	X_S	Z_S
4	4	0.6030	0.5319	0.8040	0.5206	0.2888	0.5953
2	4	0.4316	0.5242	0.6790	0.3492	0.2812	0.4483
2	2	0.4333	0.4853	0.6505	0.3498	0.2682	0.4407
1	4	0.3710	0.5197	0.6385	0.2886	0.2766	0.3998
1	1	0.3684	0.4610	0.5901	0.2877	0.2571	0.3858
1/0	4	0.3208	0.5143	0.6061	0.2384	0.2712	0.3611
1/0	1/0	0.3108	0.4364	0.5357	0.2351	0.2452	0.3397
2/0	2	0.2869	0.4734	0.5536	0.2034	0.2563	0.3272
2/0	2/0	0.2657	0.4205	0.4974	0.1963	0.2387	0.3090
3/0	2	0.2461	0.4698	0.5304	0.1626	0.2527	0.3005
3/0	3/0	0.2099	0.4008	0.4524	0.1505	0.2297	0.2746
4/0	1	0.2276	0.4465	0.5012	0.1469	0.2426	0.2836
4/0	4/0	0.1899	0.3907	0.4344	0.1344	0.2240	0.2612
266.8	2/0	0.1764	0.3990	0.4362	0.1070	0.2171	0.2421
266.8	266.8	0.1397	0.3573	0.3836	0.0948	0.2032	0.2242
336.4	2/0	0.1615	0.3963	0.4280	0.0921	0.2145	0.2334
336.4	336.4	0.1150	0.3492	0.3676	0.0766	0.1988	0.2130
397.5	2/0	0.1528	0.3944	0.4230	0.0834	0.2126	0.2284
397.5	397.5	0.1002	0.3441	0.3584	0.0659	0.1958	0.2066
477	2/0	0.1446	0.3923	0.4181	0.0752	0.2105	0.2235
477	477	0.0860	0.3391	0.3498	0.0557	0.1927	0.2006
556.5	2/0	0.1389	0.3906	0.4145	0.0695	0.2087	0.2200
556.5	556.5	0.0759	0.3351	0.3436	0.0485	0.1902	0.1963

Note: Impedances, Ω/1000 ft (× 5.28 for Ω/mi or × 3.28 for Ω/km). GMD = 4.8 ft, $GMD_{\phi N}$ = 6.3 ft, Conductor temp. = 50°C, Earth resistivity = 100 Ω-m.

TABLE 2.11

Positive-Sequence Impedances of Hard-Drawn Copper

Phase Size	Strands	R_1	X_1	Z_1
4	3	0.2875	0.1494	0.3240
2	3	0.1809	0.1441	0.2313
1	7	0.1449	0.1420	0.2029
1/0	7	0.1150	0.1393	0.1807
2/0	7	0.0911	0.1366	0.1642
3/0	12	0.0723	0.1316	0.1501
4/0	12	0.0574	0.1289	0.1411
250	12	0.0487	0.1270	0.1360
300	12	0.0407	0.1250	0.1314
350	19	0.0349	0.1243	0.1291
400	19	0.0307	0.1227	0.1265
450	19	0.0273	0.1214	0.1244
500	19	0.0247	0.1202	0.1227

Note: Impedances, Ω/1000 ft (× 5.28 for Ω/mi or × 3.28 for Ω/km). GMD = 4.8 ft, Conductor temp. = 50°C.

TABLE 2.12

Copper Zero-Sequence, Z_0, and Ground-Return Loop Impedances, $Z_S = (2Z_1 + Z_0)/3$

Phase Size	Neutral Size	R_0	X_0	Z_0	R_S	X_S	Z_S
3	3	0.3515	0.4459	0.5678	0.2707	0.2468	0.3663
3	3	0.3515	0.4459	0.5678	0.2707	0.2468	0.3663
6	6	0.5830	0.5157	0.7784	0.4986	0.2751	0.5695
6	6	0.5830	0.5157	0.7784	0.4986	0.2751	0.5695
4	6	0.4141	0.5104	0.6572	0.3297	0.2697	0.4260
4	4	0.4146	0.4681	0.6253	0.3299	0.2556	0.4173
2	6	0.3075	0.5051	0.5913	0.2231	0.2644	0.3460
2	2	0.2924	0.4232	0.5143	0.2181	0.2371	0.3221
1	4	0.2720	0.4606	0.5349	0.1873	0.2482	0.3109
1	1	0.2451	0.4063	0.4745	0.1783	0.2301	0.2911
1/0	4	0.2421	0.4580	0.5180	0.1574	0.2455	0.2916
1/0	1/0	0.2030	0.3915	0.4410	0.1443	0.2234	0.2660
2/0	3	0.2123	0.4352	0.4842	0.1315	0.2362	0.2703
2/0	2/0	0.1672	0.3796	0.4148	0.1165	0.2176	0.2468
3/0	2	0.1838	0.4106	0.4498	0.1095	0.2246	0.2498
3/0	3/0	0.1383	0.3662	0.3915	0.0943	0.2098	0.2300
4/0	2	0.1689	0.4080	0.4415	0.0946	0.2219	0.2412
4/0	4/0	0.1139	0.3584	0.3760	0.0762	0.2054	0.2191
250	1	0.1489	0.3913	0.4187	0.0821	0.2151	0.2303
250	250	0.0993	0.3535	0.3671	0.0656	0.2025	0.2128
300	1	0.1409	0.3893	0.4140	0.0741	0.2131	0.2256
300	300	0.0856	0.3487	0.3590	0.0557	0.1995	0.2071
350	1	0.1351	0.3886	0.4115	0.0683	0.2124	0.2231
350	350	0.0754	0.3468	0.3549	0.0484	0.1985	0.2043
400	1	0.1309	0.3871	0.4086	0.0641	0.2109	0.2204
400	400	0.0680	0.3437	0.3503	0.0431	0.1964	0.2011
450	1/0	0.1153	0.3736	0.3910	0.0566	0.2055	0.2131
450	450	0.0620	0.3410	0.3466	0.0389	0.1946	0.1984
500	1/0	0.1127	0.3724	0.3891	0.0540	0.2043	0.2113
500	500	0.0573	0.3387	0.3435	0.0356	0.1930	0.1963

Note: Impedances, Ω/1000 ft (× 5.28 for Ω/mi or × 3.28 for Ω/km). GMD = 4.8 ft, $GMD_{\phi N}$ = 6.3 ft, Conductor temp. = 50°C, Earth resistivity = 100 Ω-m.

several ampacities, depending on its application and the assumptions used. House and Tuttle (1958) derive the ampacity calculations described below, which are used in IEEE Std. 738-1993 and most other published ampacity tables (Aluminum Association, 1986; Southwire Company, 1994).

Sun, wind, and ambient temperature change a conductor's ampacity. A conductor's temperature depends on the thermal balance of heat inputs and losses. Current driven through a conductor's resistance creates heat (I^2R). The sun is another source of heat into the conductor. Heat escapes from the conductor through radiation and from convection. Considering the balance of inputs and outputs, the ampacity of a conductor is

$$I = \sqrt{\frac{q_c + q_r - q_s}{R_{ac}}}$$

where

q_c = convected heat loss, W/ft
q_r = radiated heat loss, W/ft
q_s = solar heat gain, W/ft
R_{ac} = Nominal ac resistance at operating temperature t, Ω/ft

The convected heat loss with no wind is

$$q_c = 0.283\sqrt{\rho_f}D^{0.75}(t_c - t_a)^{1.25}$$

Wind increases convection losses. The losses vary based on wind speed. The IEEE method uses the maximum q_c from the following two equations:

$$q_{c1} = \left[1.01 + 0.371\left(\frac{D\rho_f V}{\mu_f}\right)^{0.52}\right]K_f(t_c - t_a)$$

$$q_{c2} = 0.1695\left(\frac{D\rho_f V}{\mu_f}\right)^{0.6} K_f(t_c - t_a)$$

where

D = conductor diameter, in.
t_c = conductor operating temperature, °C
t_a = ambient temperature, °C
$t_f = (t_c + t_c)/2$
V = air velocity, ft/h
ρ_f = air density at t_f, lb/ft^3
μ_f = absolute viscosity of air at t_f, lb/h-ft
K_f = thermal conductivity of air at t_f, W/ft^2/°C

The density, viscosity, and thermal conductivity of air all depend on temperature (actually the film temperature T_f near the surface of the conductor, which is taken as the average of the conductor and ambient temperatures). Tables of these are available in the references (IEEE Std. 738-1993; Southwire Company, 1994). We may also use the following polynomial approximations (IEEE Std. 738-1993):

$$\rho_f = \frac{0.080695 - (0.2901 \times 10^{-5})H_c + (0.37 \times 10^{-10})H_c^2}{1 + 0.00367T_f}$$

where H_c is the altitude above sea level in feet.

$$k_f = 0.007388 + 2.27889 \times 10^{-5} T_f - 1.34328 \times 10^{-9} T_f^2$$

$$\mu_f = 0.0415 + 1.2034 \times 10^{-4} T_f - 1.1442 \times 10^{-7} T_f^2 + 1.9416 \times 10^{-10} T_f^3$$

A conductor radiates heat as the absolute temperature to the fourth power as

$$q_r = 0.138 D\varepsilon \left[\left(\frac{T_c + 273}{100} \right)^4 - \left(\frac{T_a + 273}{100} \right)^4 \right]$$

where

D = conductor diameter, in.
ε = emissivity (normally 0.23 to 0.91 for bare wires)
T_c = conductor temperature, °C
T_a = ambient temperature, °C

A conductor absorbs heat from the sun as

$$q_s = \alpha Q_s \frac{D}{12} \sin \theta$$

where

α = solar absorptivity
Q_s = total solar heat in W/ft^2
θ = effective angle of incidence of the sun's rays
D = conductor diameter, in.

The angles and total solar heat depend on the time of day and the latitude. Since the solar input term does not change the output significantly, we can use some default values. For a latitude of 30°N at 11 a.m. in clear atmosphere, Q_s = 95.2 W/ft,^{2} and θ = 78.6°.

Emissivity (ε) is the ability of a conductor to radiate heat into the air. Absorptivity (α) quantifies how much heat a conductor can absorb. Emissivity and absorptivity are interrelated; a nice shiny conductor reflects away much of the sun's heat but does not radiate heat well. Commonly, both are assumed to be 0.5 for bare wire. More conservative assumptions, possibly overconservative, are 0.7 for emissivity and 0.9 for absorptivity.

Some of the main factors impacting ampacity are

- *Allowable conductor temperature* — Ampacity increases significantly with higher allowed temperatures.

TABLE 2.13

Ampacities of All-Aluminum Conductor

		Conductor Temp. = 75°C				Conductor Temp. = 100°C			
		Ambient = 25°C		Ambient = 40°C		Ambient = 25°C		Ambient = 40°C	
Conductor	Stranding	No Wind	Wind	No Wind	Wind	No Wind	Wind	No Wind	Wind
6	7	60	103	46	85	77	124	67	111
4	7	83	138	63	114	107	166	92	148
2	7	114	185	86	152	148	223	128	199
1	7	134	214	101	175	174	258	150	230
1/0	7	157	247	118	203	204	299	176	266
2/0	7	184	286	139	234	240	347	207	309
3/0	7	216	331	162	271	283	402	243	358
4/0	7	254	383	190	313	332	466	286	414
250	7	285	425	213	347	373	518	321	460
250	19	286	427	214	348	375	519	322	462
266.8	7	298	443	223	361	390	539	335	479
266.8	19	299	444	224	362	392	541	337	481
300	19	325	479	243	390	426	584	367	519
336.4	19	351	515	262	419	461	628	397	559
350	19	361	527	269	428	474	644	408	572
397.5	19	394	571	293	464	517	697	445	619
450	19	429	617	319	501	564	755	485	671
477	19	447	640	332	519	588	784	506	697
477	37	447	641	333	520	589	785	507	697
500	19	461	658	342	534	606	805	521	716
556.5	19	496	704	368	571	654	864	562	767
556.5	37	496	705	369	571	654	864	563	768
600	37	522	738	388	598	688	905	592	804
636	37	545	767	404	621	720	943	619	838
650	37	556	782	413	633	737	965	634	857
700	37	581	814	431	658	767	1000	660	888
715.5	37	590	825	437	667	779	1014	670	901
715.5	61	590	825	437	667	780	1014	671	901
750	37	609	848	451	686	804	1044	692	927
795	37	634	881	470	712	840	1086	722	964
795	61	635	882	470	713	840	1087	723	965
800	37	636	884	471	714	841	1087	723	965
874.5	61	676	933	500	754	896	1152	770	1023
874.5	61	676	934	500	754	896	1152	771	1023
900	37	689	950	510	767	913	1172	785	1041
954	37	715	983	529	793	946	1210	813	1074
954	61	719	988	532	797	954	1221	821	1084
1000.0	37	740	1014	547	818	981	1252	844	1111

- *Ambient temperature* — Ampacity increases about 1% for each 1°C decrease in ambient temperature.
- *Wind speed* — Even a small wind helps cool conductors significantly. With no wind, ampacities are significantly lower than with a 2-ft/sec crosswind.

Table 2.13 through Table 2.15 show ampacities of all-aluminum, ACSR, and copper conductors. All assume the following:

TABLE 2.14

Ampacities of ACSR

		Conductor Temp. = 75°C				Conductor Temp. = 100°C			
		Ambient = 25°C		Ambient = 40°C		Ambient = 25°C		Ambient = 40°C	
Conductor	Stranding	No Wind	Wind	No Wind	Wind	No Wind	Wind	No Wind	Wind
6	6/1	61	105	47	86	79	126	68	112
4	6/1	84	139	63	114	109	167	94	149
4	7/1	85	141	64	116	109	168	94	149
2	6/1	114	184	86	151	148	222	128	197
2	7/1	117	187	88	153	150	224	129	199
1	6/1	133	211	100	173	173	255	149	227
1/0	6/1	156	243	117	199	202	294	174	261
2/0	6/1	180	277	135	227	235	337	203	300
3/0	6/1	208	315	156	258	262	370	226	329
4/0	6/1	243	363	182	296	319	443	274	394
266.8	18/1	303	449	227	366	398	547	342	487
266.8	26/7	312	458	233	373	409	559	352	497
336.4	18/1	356	520	266	423	468	635	403	564
336.4	26/7	365	530	272	430	480	647	413	575
336.4	30/7	371	536	276	435	487	655	419	582
397.5	18/1	400	578	298	469	527	708	453	629
397.5	26/7	409	588	305	477	538	719	463	639
477	18/1	453	648	337	525	597	793	513	705
477	24/7	461	656	343	532	607	804	523	714
477	26/7	464	659	345	534	611	808	526	718
477	30/7	471	667	350	540	615	810	529	720
556.5	18/1	504	713	374	578	664	874	571	777
556.5	24/7	513	722	380	585	677	887	582	788
556.5	26/7	517	727	383	588	682	893	587	793
636	24/7	562	785	417	635	739	962	636	854
636	26/7	567	791	420	639	748	972	644	863
795	45/7	645	893	478	721	855	1101	735	977
795	26/7	661	910	489	734	875	1122	753	996
954	45/7	732	1001	541	807	971	1238	835	1099
954	54/7	741	1010	547	814	983	1250	846	1109
1033.5	45/7	769	1048	568	844	1019	1294	877	1148

- Emissivity = 0.5, absorptivity = 0.5
- 30°N at 11 a.m. in clear atmosphere
- Wind speed = 2 ft/sec
- Elevation = sea level

The solar heating input has modest impacts on the results. With no sun, the ampacity increases only a few percent.

Some simplifying equations help for evaluating some of the significant impacts on ampacity. We can estimate changes in ambient and allowable temperature variations (Black and Rehberg, 1985; Southwire Company, 1994) with

$$I_{new} = I_{old}\sqrt{\frac{T_{c,new} - T_{a,new}}{T_{c,old} - T_{a,old}}}$$

TABLE 2.15

Ampacities of Copper Conductors

		Conductor Temp. = 75°C				Conductor Temp. = 100°C			
		Ambient = 25°C		Ambient = 40°C		Ambient = 25°C		Ambient = 40°C	
Conductor	Stranding	No Wind	Wind	No Wind	Wind	No Wind	Wind	No Wind	Wind
6	3	83	140	63	116	107	169	92	150
6	1	76	134	58	110	98	160	85	143
5	3	97	162	73	134	125	195	108	174
5	1	90	155	68	127	115	185	99	165
4	3	114	188	86	154	147	226	127	201
4	1	105	179	80	147	136	214	117	191
3	7	128	211	97	174	166	254	143	226
3	3	133	217	101	178	173	262	149	233
3	1	123	206	93	170	159	248	137	221
2	7	150	244	114	201	195	294	168	262
2	3	157	251	118	206	203	303	175	270
2	1	145	239	110	196	187	287	161	256
1	3	184	291	138	238	239	351	206	313
1	7	177	282	133	232	229	340	197	303
1/0	7	207	326	156	267	269	394	232	351
2/0	7	243	378	183	309	317	457	273	407
3/0	12	292	444	219	362	381	539	328	479
3/0	7	285	437	214	357	373	530	321	472
4/0	19	337	507	252	414	440	617	379	549
4/0	12	342	513	256	418	448	624	386	555
4/0	7	335	506	251	413	438	615	377	547
250	19	377	563	282	459	493	684	424	608
250	12	384	569	287	464	502	692	432	615
300	19	427	630	319	513	559	767	481	682
300	12	435	637	324	519	569	776	490	690
350	19	475	694	355	565	624	847	537	753
350	12	484	702	360	571	635	858	546	763
400	19	520	753	387	612	682	920	587	817
450	19	564	811	420	659	742	993	639	883
500	37	606	865	450	702	798	1061	686	942
500	19	605	865	450	701	797	1059	685	941
600	37	685	968	509	784	905	1190	779	1057
700	37	759	1062	563	860	1003	1308	863	1161
750	37	794	1107	588	895	1051	1364	904	1211
800	37	826	1147	612	927	1092	1412	939	1253
900	37	894	1233	662	995	1189	1527	1023	1356
1000	37	973	1333	719	1075	1313	1676	1129	1488

where I_{new} is the new ampacity based on a new conductor limit $T_{c,new}$ and a new ambient temperature $T_{a,new}$. Likewise, I_{old} is the original ampacity based on a conductor limit $T_{c,old}$ and an ambient temperature $T_{a,old}$.

This approach neglects solar heating and the change in conductor resistance with temperature (both have small impacts). Doing this simplifies the ampacity calculation to a constant (dependent on weather and conductor characteristics) times the difference between the conductor temperature and the ambient temperature: $I^2 = K(T_c - T_a)$. We do not use this simplification for the original ampacity calculation, but it helps us evaluate changes in temperatures or currents.

We use this approach in Figure 2.9 to show the variation in ampacity with ambient conductor assumptions along with two conductor operating limits.

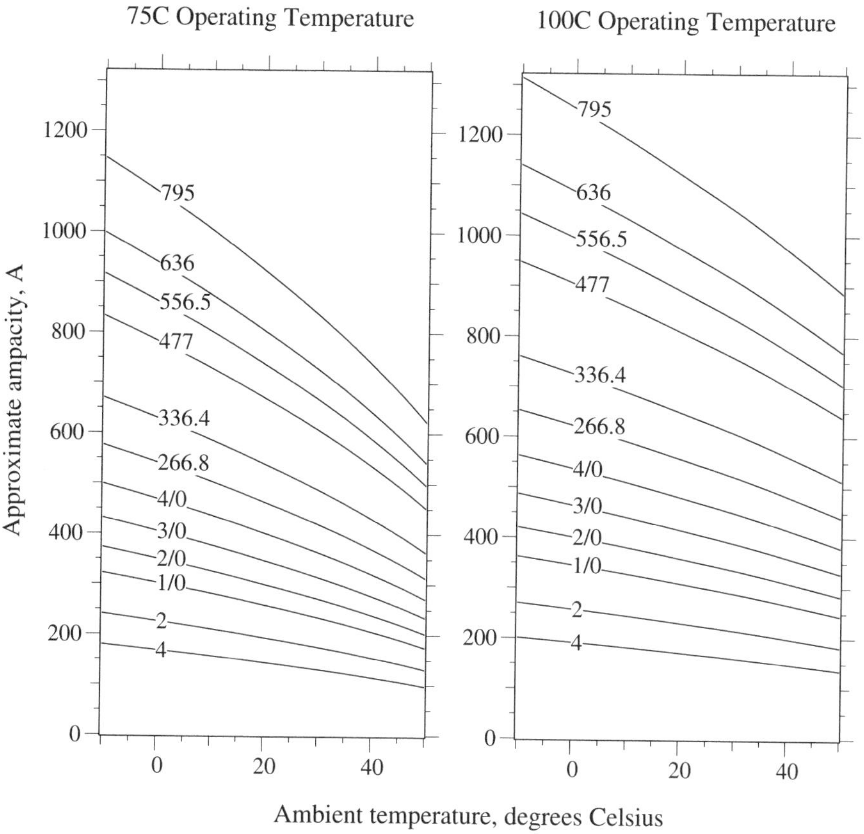

FIGURE 2.9
AAC ampacity with ambient temperature variations, using adjustments from base ampacity data in Table 2.13 (2 ft/sec wind, with sun).

Also, Figure 2.10 shows the conductor temperature vs. loading for several AAC conductors. This graph highlights the major impact of operating temperature on ampacity. If we are overly conservative on a conductor limit, we end up with an overly restrictive ampacity.

We can also use the simplified ampacity equation to estimate the conductor temperature at a current higher or lower than the rated ampacity as (and at a different ambient temperature if we wish):

$$T_{c,new} = T_{a,new} + \left(\frac{I_{new}}{I_{old}}\right)^2 (T_{c,old} - T_{a,old})$$

When examining a line's ampacity, always remember that the overhead wire may not be the weakest link; substation exit cables, terminations, reclosers, or other gear may limit a circuit's current before the conductors do. Also, with currents near a conductor's rating, voltage drop is high.

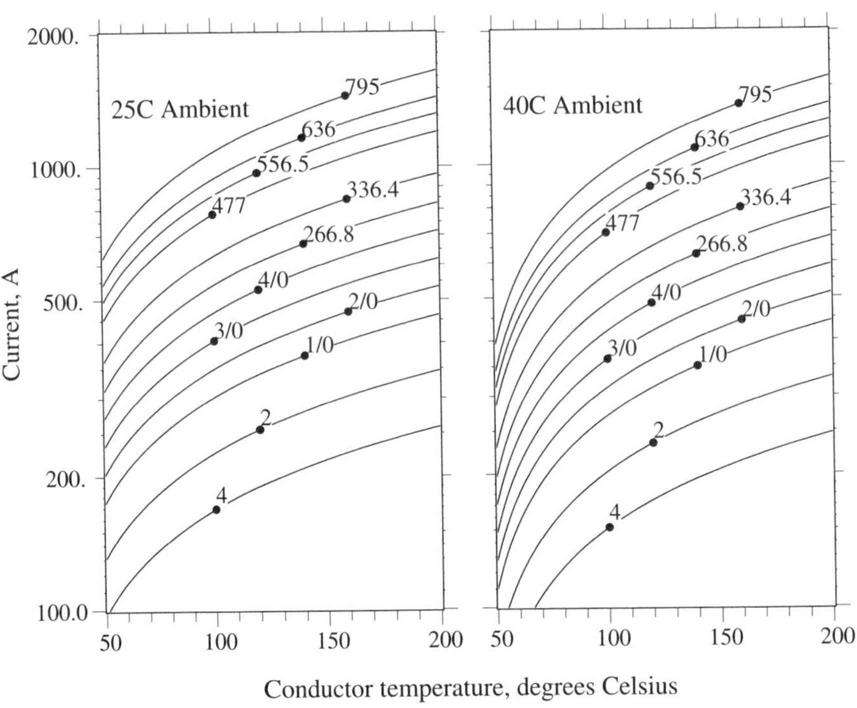

FIGURE 2.10
Conductor temperatures based on the given currents for selected AAC conductors, using adjustments from base ampacity data in Table 2.13 (2 ft/sec wind).

The maximum operating temperature is an important consideration. Higher designed operating temperatures allow higher currents. But at higher temperatures, we have a higher risk of damage to the conductors. Aluminum strands are strain hardened during manufacturing (the H19 in aluminum's 1350-H19 designation means "extra hard"). Heating relaxes the strands — the aluminum elongates and weakens. This damage is called *annealing*. As aluminum anneals, it reverts back to its natural, softer state: fully annealed 1350 aluminum wire elongates by 30% and loses 58% of its strength (10,000 psi vs. 24,000 psi fully hardened). Even fully annealed, failure may not be immediate; the next ice load or heavy winds may break a conductor. Slow annealing begins near 100°C. Aluminum anneals rapidly above 200°C. Annealing damage is permanent and accumulates over time. Remaining strength for AAC conductors varies with conductor temperature and duration of exposure as approximately (Harvey, 1971)

$$R_S = k_1 t^{-\frac{0.1}{d}(0.001T_c - 0.095)}$$

where

R_S = remaining strength, percent of initial strength

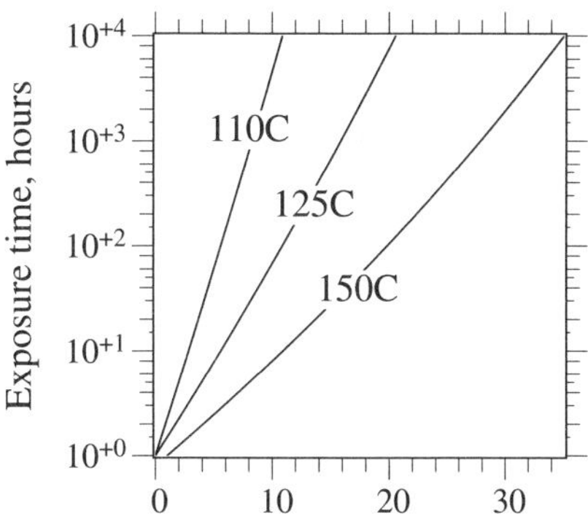

FIGURE 2.11
Loss of strength of all-aluminum conductors due to exposure to high temperatures.

d = strand diameter, in.
t = exposure time, h
T_c = conductor temperature, °C
$k_1 = (-0.24T_c + 135)$, but if $k_1 > 100$, use $k_1 = 100$

Figure 2.11 shows the loss of strength with time for high-temperature operation using this approximation.

ACSR may be loaded higher than the same size AAC conductor. As the aluminum loses strength, the steel carries more of the tension. The steel does not lose strength until reaching higher temperatures.

Covered conductors are darker, so they absorb more heat from the sun but radiate heat better; the Aluminum Association (1986) uses 0.91 for both the emissivity and the absorptivity of covered wire. Table 2.16 shows ampacities of covered wire. Covered conductors have ampacities that are close to bare-conductor ampacities. The most significant difference is that covered conductors have less ability to withstand higher temperatures; the insulation degrades. Polyethylene is especially prone to damage, so it should not be operated above 75°C. EPR and XLPE may be operated up to 90°C.

Some utilities use two ratings, a "normal" ampacity with a 75°C design temperature and an "emergency" ampacity with a 90 or 100°C design. Conductors are selected for the normal rating, but operation is allowed to the emergency rating. Overhead circuits have considerable capability for overload, especially during cooler weather. We do not use relaying to trip "overloaded" circuits. At higher temperatures, conductors age more quickly but do not usually fail immediately.

2.7.1 Neutral Conductor Sizing

Because the neutral conductor carries less current than the phase conductors, utilities can use smaller neutral conductors. On three-phase circuits with

TABLE 2.16

Ampacities of all-Aluminum Conductor Covered with PE, XLPE, or EPR

AWG or kcmil	Stranding	Cover Thickness (mil)	Conductor Temp. = 75°C 25°C Ambient	Conductor Temp. = 75°C 40°C Ambient	Conductor Temp. = 90°C 25°C Ambient	Conductor Temp. = 90°C 40°C Ambient
6	7	30	105	85	120	105
4	7	30	140	110	160	135
2	7	45	185	145	210	180
1	7	45	210	170	245	210
1/0	7	60	240	195	280	240
2/0	7	60	280	225	325	280
3/0	7	60	320	255	375	320
4/0	7	60	370	295	430	370
4/0	19	60	375	295	430	370
266.8	19	60	430	340	500	430
336.4	19	60	500	395	580	495
397.5	19	80	545	430	635	545
477	37	80	615	480	715	610
556.5	37	80	675	530	785	675
636	61	95	725	570	850	725
795	61	95	835	650	980	835
1033.5	61	95	980	760	1150	985

Notes: Emissivity = 0.91, absorptivity = 0.91; 30°N at 12 noon in clear atmosphere; wind speed = 2 ft/sec; elevation = sea level.

Source: Aluminum Association, *Ampacities for Aluminum and ACSR Overhead Electrical Conductors*, 1986.

balanced loading, the neutral carries almost no current. On single-phase circuits with a multigrounded neutral, the neutral normally carries 40 to 60% of the current (the earth carries the remainder).

On single-phase circuits, some utilities use fully rated neutrals, where the neutral and the phase are the same size. Some use reduced neutrals. The resistance of the neutral should be no more than twice the resistance of the phase conductor, and we are safer with a resistance less than 1.5 times the phase conductor, which is a conductivity or cross-sectional area of 2/3 the phase conductor. Common practice is to drop one to three gage sizes for the neutral: a 4/0 phase has a 2/0 neutral, or a 1/0 phase has a number 2 neutral. Dropping three gage sizes doubles the resistance, so we do not want to go any smaller than that.

On three-phase circuits, most utilities use reduced neutrals, dropping the area to about 25 to 70% of the phase conductor (and multiplying the resistance by 1.4 to 4).

Several other factors besides ampacity play a role in how small neutral conductors are:

- *Grounding* — A reduced neutral increases the overvoltages on the unfaulted phases during single line-to-ground faults (see Chapter 13). It also increases stray voltages.

- *Faults* — A reduced neutral reduces the fault current for single line-to-ground faults, which makes it more difficult to detect faults at far distances. Also, the reduced neutral is subjected to the same fault current as the phase, so impacts on burning down the neutral should be considered for smaller neutrals.
- *Secondary* — If the primary and secondary neutral are shared, the neutral must handle the primary and secondary unbalanced current (and have the mechanical strength to hold up the secondary phase conductors in triplex or quadraplex construction).
- *Mechanical* — On longer spans, the sag of the neutral should coordinate with the sag of the phases and the minimum ground clearances to ensure that spacing rules are not violated.

2.8 Secondaries

Utilities most commonly install *triplex* secondaries for overhead service to single-phase customers, where two insulated phase conductors are wrapped around the neutral. The neutral supports the weight of the conductors. Phase conductors are normally all-aluminum, and the neutral is all-aluminum, aluminum-alloy, or ACSR, depending on strength needs. Insulation is normally polyethylene, high-molecular weight polyethylene, or cross-linked polyethylene with thickness ranging from 30 to 80 mils (1.1 to 2 mm) rated for 600 V. Similarly for three-phase customers, quadraplex has three insulated phase conductors wrapped around a bare neutral. Table 2.17 shows characteristics of polyethylene triplex with an AAC neutral.

Triplex secondary ampacities depend on the temperature capability of the insulation. Polyethylene can operate up to 75°C. Cross-linked polyethylene and EPR can operate higher, up to 90°C. Table 2.18 shows ampacities for triplex when operated to each of these maximum temperatures. Quadraplex has ampacities that are 10 to 15% less than triplex of the same size conductor. Ampacities for open-wire secondary are the same as that for bare primary conductors.

Table 2.19 shows impedances of triplex. Two impedances are given: one for the 120-V loop and another for a 240-V loop. The 240-V loop impedance is the impedance to current flowing down one hot conductor and returning on the other. The 120-V loop impedance is the impedance to current down one hot conductor and returning in the neutral (and assuming no current returns through the earth). If the phase conductor and the neutral conductor are the same size, these impedances are the same. With a reduced neutral, the 120-V loop impedance is higher. Table 2.19 shows impedances for the reduced neutral size given; for a fully-rated neutral, use the 240-V impedance for the 120-V impedance.

TABLE 2.17

Typical Characteristics of Polyethylene-Covered AAC Triplex

Phase Conductor		Neutral Options (Bare)					
		ACSR Neutral Messenger		Reduced ACSR Neutral Messenger		AAC Neutral Messenger	
Size (Stranding)	Insulation Thickness, mil	Size (Stranding)	Rated Strength, lb	Size (Stranding)	Rated Strength, lb	Size (Stranding)	Rated Strength, lb
6 (1)	45	6 (6/1)	1190				
6 (7)	45	6 (6/1)	1190				
4 (1)	45	4 (6/1)	1860	6 (6/1)	1190	6 (7)	563
4 (7)	45	4 (6/1)	1860	6 (6/1)	1190	4 (7)	881
2 (7)	45	2 (6/1)	2850	4 (6/1)	1860	2 (7)	1350
1/0 (7)	60	1/0 (6/1)	4380	2 (6/1)	2853	1/0 (7)	1990
1/0 (19)	60	1/0 (6/1)	4380	2 (6/1)	2853	1/0 (7)	1990
2/0 (7)	60	2/0 (6/1)	5310	1 (6/1)	3550	2/0 (7)	2510
2/0 (19)	60	2/0 (6/1)	5310	1 (6/1)	3550		
3/0 (19)	60	3/0 (6/1)	6620	1/0 (6/1)	4380	3/0 (19)	3310
4/0 (19)	60	4/1 (6/1)	8350	2/0 (6/1)	5310	4/0 (19)	4020
336.4 (19)	80	336.4 (18/1)	8680	4/0 (6/1)	8350	336.4 (19)	6146

TABLE 2.18

Ampacities of All-Aluminum Triplex

Phase Conductor AWG	Strands	Conductor temp = 75°C		Conductor temp = 90°C	
		25°C Ambient	40°C Ambient	25°C Ambient	40°C Ambient
6	7	85	70	100	85
4	7	115	90	130	115
2	7	150	120	175	150
1/0	7	200	160	235	200
2/0	7	230	180	270	230
3/0	7	265	210	310	265
4/0	7	310	240	360	310

Note: Emissivity = 0.91, absorptivity = 0.91; 30°N at 12 noon in clear atmosphere; wind speed = 2 ft/sec; elevation = sea level.

Source: Aluminum Association, *Ampacities for Aluminum and ACSR Overhead Electrical Conductors*, 1986.

2.9 Fault Withstand Capability

When a distribution line short circuits, very large currents can flow for a short time until a fuse or breaker or other interrupter breaks the circuit. One important aspect of overcurrent protection is to ensure that the fault arc and fault currents do not cause further, possibly more permanent, damage. The two main considerations are:

TABLE 2.19

Typical Impedances of All-Aluminum Triplex Secondaries, Ω/1000 ft

Phase		Neutral		120-V Loop Impedance*		240-V Loop Impedance	
Size	Strands	Size	Strands	R_{S1}	X_{S1}	R_S	X_S
2	7	4	7	0.691	0.0652	0.534	0.0633
1	19	3	7	0.547	0.0659	0.424	0.0659
1/0	19	2	7	0.435	0.0628	0.335	0.0616
2/0	19	1	19	0.345	0.0629	0.266	0.0596
3/0	19	1/0	19	0.273	0.0604	0.211	0.0589
4/0	19	2/0	19	0.217	0.0588	0.167	0.0576
250	37	3/0	19	0.177	0.0583	0.142	0.0574
350	37	4/0	19	0.134	0.0570	0.102	0.0558
500	37	300	37	0.095	0.0547	0.072	0.0530

* With a full-sized neutral, the 120-V loop impedance is equal to the 240-V loop impedance.

Source: ABB Inc., *Distribution Transformer Guide*, 1995.

- *Conductor annealing* — From the substation to the fault location, all conductors in the fault-current path must withstand the heat generated by the short-circuit current. If the relaying or fuse does not clear the fault in time, the conductor anneals and loses strength.
- *Burndowns* — Right at the fault location, the hot fault arc can burn the conductor. If a circuit interrupter does not clear the fault in time, the arc will melt the conductor until it breaks apart.

For both annealing and arcing damage, we should design protection to clear faults before more damage is done. To do this, make sure that the time-current characteristics of the relay or fuse are faster than the time-current damage characteristics. Characteristics of annealing and arcing damage are included in the next two sections.

2.9.1 Conductor Annealing

During high currents from faults, conductors can withstand significant temperatures for a few seconds without losing strength. For all-aluminum conductors, assuming a maximum temperature of 340°C during faults is common. ACSR conductors can withstand even higher temperatures because short-duration high temperature does not affect the steel core. An upper limit of 645°C, the melting temperature of aluminum, is often assumed. For short-duration events, we ignore convection and radiation heat losses and assume that all heat stays in the conductor. With all heat staying in the conductor, the temperature is a function of the specific heat of the conductor material. Specific heat is the heat per unit mass required to raise the temperature by one degree Celsius (the specific heat of aluminum is 0.214 cal/g-°C). Considering the heat inputs and the conductor characteristics, the

TABLE 2.20

Conductor Thermal Data for Short-Circuit Limits

Conductor Material	λ, °C	K
Copper (97%)	234.0	0.0289
Aluminum (61.2%)	228.1	0.0126
6201 (52.5%)	228.1	0.0107
Steel	180.0	0.00327

Source: Southwire Company, *Overhead Conductor Manual*, 1994.

conductor temperature during a fault is related to the current (Southwire Company, 1994) as

$$\left(\frac{I}{1000A}\right)^2 t = K\log_{10}\left(\frac{T_2+\lambda}{T_1+\lambda}\right)$$

where

I = fault current, A
t = fault duration, sec
A = cross-sectional area of the conductor, kcmil
T_2 = conductor temperature from the fault, °C
T_1 = conductor temperature before the fault, °C
K = constant depending on the conductor, which includes the conductor's resistivity, density, and specific heat (see Table 2.20)
λ = inferred temperature of zero resistance, °C below zero (see Table 2.20)

If we set T_2 to the maximum allowable conductor temperature, we can find the maximum allowable I^2t characteristic for a given conductor. For all-aluminum conductors, with a maximum temperature, T_2 = 340°C, and an ambient of 40°C, the maximum allowable time-current characteristic for a given conductor size (Southwire Company, 1994) is

$$I^2t = (67.1A)^2$$

For ACSR with a maximum temperature of 640°C, the maximum allowable time-current characteristic for a given conductor size (Southwire Company, 1994) is

$$I^2t = (86.2A)^2$$

Covered conductors have more limited short-circuit capability because the insulation is damaged at lower temperatures. Thermoplastic insulations like polyethylene have a maximum short-duration temperature of 150°C. The thermoset insulations EPR and XLPE have a maximum short-duration tem-

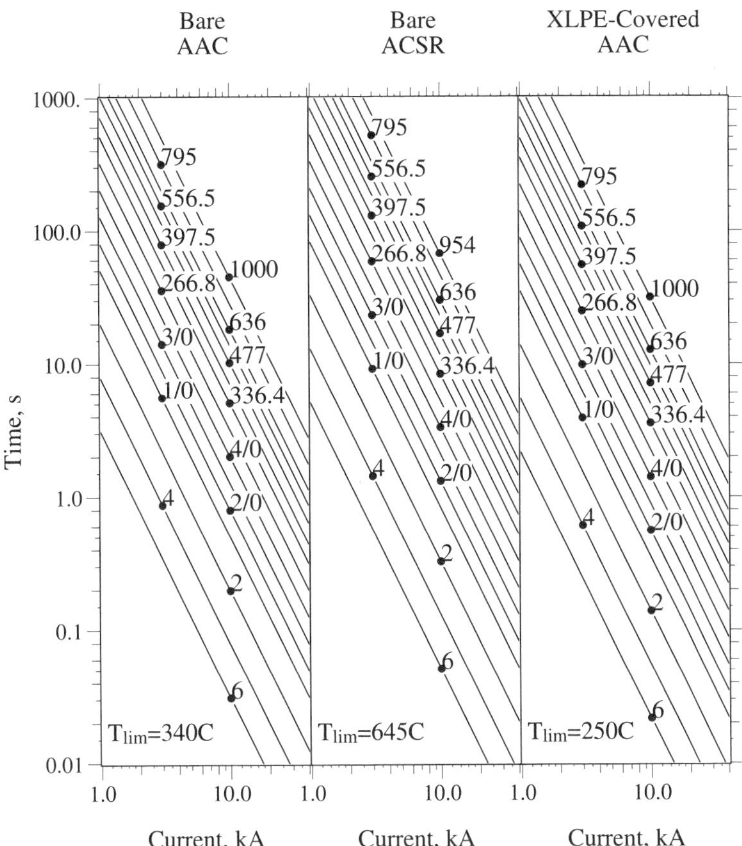

FIGURE 2.12
Annealing curves of bare AAC, ACSR, and covered AAC.

perature of 250°C. With these upper temperature limits (and $T_1 = 40°C$), the allowable time-current characteristics of aluminum conductors are:

$$\text{Polyethylene:} \quad I^2t = (43A)^2$$

$$\text{XLPE or EPR:} \quad I^2t = (56A)^2$$

Figure 2.12 compares short-circuit damage curves for various conductors.

2.9.2 Burndowns

Fault-current arcs can damage overhead conductors. The arc itself generates tremendous heat, and where an arc attaches to a conductor, it can weaken or burn conductor strands. On distribution circuits, two problem areas stand out:

1. *Covered conductor* — Covered conductor (also called tree wire or weatherproof wire) holds an arc stationary. Because the arc cannot move, burndowns happen faster than with bare conductors.
2. *Small bare wire on the mains* — Small bare wire (less than 2/0) is also susceptible to wire burndowns, especially if laterals are not fused.

Covered conductors are widely used to limit tree faults. Several utilities have had burndowns of covered conductor circuits when the instantaneous trip was not used or was improperly applied (Barker and Short, 1996; Short and Ammon, 1997). If a burndown on the main line occurs, all customers on the circuit will have a long interruption. In addition, it is a safety hazard. After the conductor breaks and falls to the ground, the substation breaker may reclose. After the reclosure, the conductor on the ground will probably not draw enough fault current to trip the station breaker again. This is a high-impedance fault that is difficult to detect.

A covered conductor is susceptible to burndowns because when a fault current arc develops, the covering prevents the arc from moving. The heat from the arc is what causes the damage. Although ionized air is a fairly good conductor, it is not as good as the conductor itself, so the arc gets very hot. On bare conductors, the arc is free to move, and the magnetic forces from the fault cause the arc to move (in the direction away from the substation; this is called *motoring*). The covering constricts the arc to one location, so the heating and melting is concentrated on one part of the conductor. If the covering is stripped at the insulators and a fault arcs across an insulator, the arc motors until it reaches the covering, stops, and burns the conductor apart at the junction. A party balloon, lightning, a tree branch, a squirrel — any of these can initiate the arc that burns the conductor down. Burndowns are most associated with lightning-caused faults, but it is the fault current arc, not the lightning, that burns most of the conductor.

Conductor damage is a function of the duration of the fault and the current magnitude. Burndown damage occurs much more quickly than conductor annealing that was analyzed in the previous section.

Although they are not as susceptible as covered conductors, bare conductors can also have burndowns. In tests of smaller bare conductors, Florida Power & Light Co. (FP&L) found that the hot gases from the arc anneal the conductor (Lasseter, 1956). They found surprisingly little burning from the arc; in fact, arcs could seriously degrade conductor strength even when there is no visible damage. Objects like insulators or tie wires absorb heat from the ionized gases and reduce the heat to the conductor.

What we would like to do is plot the arc damage characteristic as a function of time and current along with the time-current characteristics of the protective device (whether it be a fuse or a recloser or a breaker). Doing this, we can check that the protective device will clear the fault before the conductor is damaged. Figure 2.13 shows burndown damage characteristics for small bare ACSR conductors along with a 100 K lateral fuse element and a typical ground relay element. The fuse protects the conductors shown, but

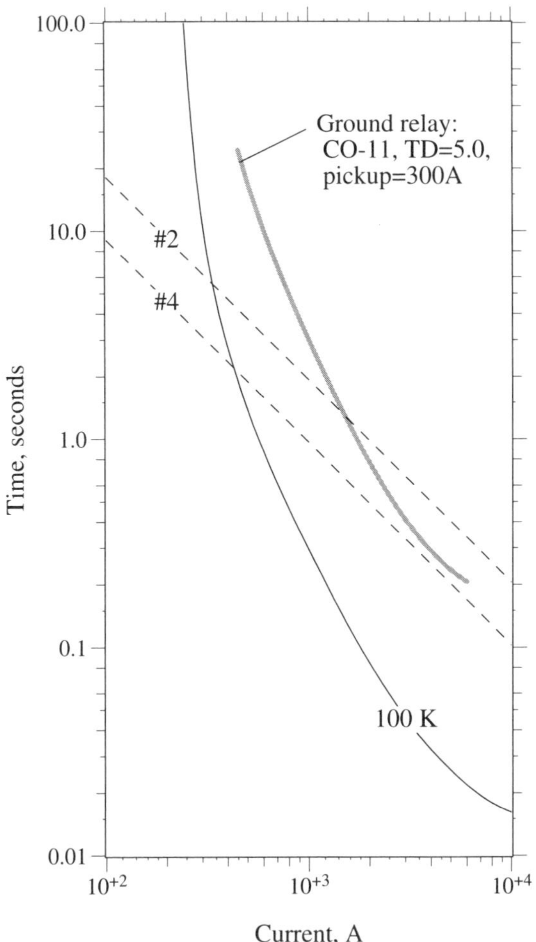

FIGURE 2.13
Bare-conductor ACSR threshold-of-damage curves along with the 100-K lateral fuse total clearing time and a ground relay characteristic. (Damage curves from [Lasseter, 1956].)

the ground relay does not provide adequate protection against damage for these conductors. These damage curves are based on FP&L's tests, where Lasseter reported that the threshold-of-damage was 25 to 50% of the average burndown time (see Table 2.21).

Such arc damage data for different conductor sizes as a function of time and current is limited. Table 2.22 summarizes burndown characteristics of some bare and covered conductors based on tests by Baltimore Gas & Electric (Goode and Gaertner, 1965). Figure 2.14 shows this same data on time-current plots along with a 100 K fuse total clearing characteristic. For conductor sizes not given, take the closest size given in Table 2.22, and scale the burndown time by the ratio of the given conductor area to the area of the desired conductor.

TABLE 2.21

The Burndown Characteristics of Several Small Bare Conductors

Conductor	Threshold of Damage	Average Burndown Time
#4 AAAC	$t = \frac{4375}{I^{1.235}}$	$t = \frac{17500}{I^{1.235}}$
#4 ACSR	$t = \frac{800}{I^{0.973}}$	$t = \frac{3350}{I^{0.973}}$
#2 ACSR	$t = \frac{1600}{I^{0.973}}$	$t = \frac{3550}{I^{0.973}}$
#6 Cu	$t = \frac{410}{I^{0.909}}$	$t = \frac{1440}{I^{0.909}}$
#4 Cu	$t = \frac{500}{I^{0.909}}$	$t = \frac{1960}{I^{0.909}}$

Note: I = rms fault current, A; t = fault duration, sec.

Source: Lasseter, J.A., "Burndown Tests on Bare Conductor," *Electric Light and Power*, pp. 94–100, December 1956.

If covered conductor is used, consider the following options to limit burndowns:

- *Fuse saving* — Using a fuse blowing scheme can increase burndowns because the fault duration is much longer on the time-delay relay elements than on the instantaneous element. With fuse saving, the instantaneous relay element trips the circuit faster and reduces conductor damage.
- *Arc protective devices* (APDs) — These sacrificial masses of metal attach to the ends where the covering is stripped (Lee et al., 1980). The arc end attaches to the mass of metal, which has a large enough volume to withstand much more arcing than the conductor itself.
- *Fuse all taps* — Leaving smaller covered conductors unprotected is a sure way of burning down conductors.
- *Tighter fusing* — Not all fuses protect some of the conductor sizes used on taps. Faster fuses reduce the chance of burndowns.
- *Bigger conductors* — Bigger conductors take longer to burn down. Doubling the conductor cross-sectional area approximately doubles the time it takes to burn the conductor down.

Larger bare conductors are fairly immune to burndown. Smaller conductors used on taps are normally safe if protected by a fuse. The solutions for small bare conductor are

- *Fuse all taps* — This is the best option.

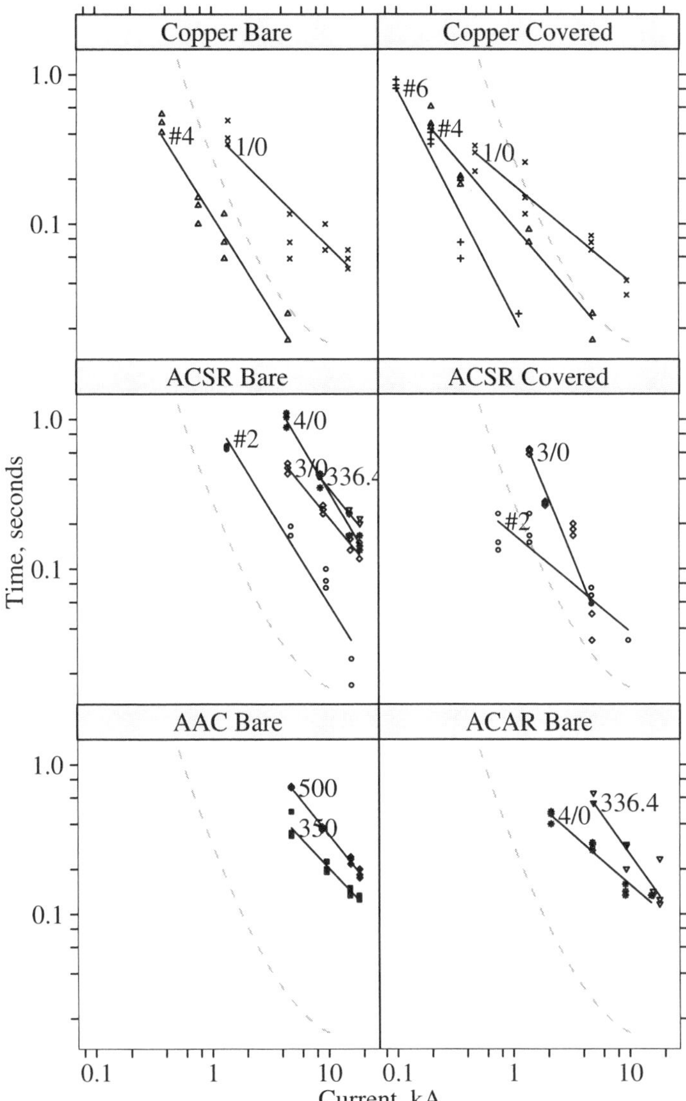

FIGURE 2.14
Burndown characteristics of various conductors. The dashed line is the total clearing time for a 100-K fuse. (Data from [Goode and Gaertner, 1965].)

- *Fuse saving* — The time-delay relay element may not protect smaller tap conductors. Faults cleared by an instantaneous element with fuse saving will not damage bare conductor. If fuse blowing is used, consider an alternative such as a high-set instantaneous or a delayed instantaneous (see Chapter 8 for more information).

TABLE 2.22

Burndown Characteristics of Various Conductors

		Duration, 60-Hz Cycles			
	Current, A	Min	Max	Other	Curvefit
#6 Cu covered	100	48.5	55.5	51	$t = 858/I^{1.51}$
	200	20.5	24.5	22	
	360	3.5	4.5	4.5	
	1140	1.5	1.5	1.5	
#4 Cu covered	200	26.5	36.5	28	$t = 56.4/I^{0.92}$
	360	11	12.5	12	
	1400	4.5	5.5	5.5	
	4900	1	1.5	1.5	
#4 Cu bare	380	24.5	32.5	28.5	$t = 641/I^{1.25}$
	780	6	9	8	
	1300	3.5	7	4.5	
	4600	1	1.5	1	
#2 ACSR covered	750	8	9	14	$t = 15.3/I^{0.65}$
	1400	10	9	14	
	4750	3.5	4.5	4	
	9800	2	2	NA	
#2 ACSR bare	1350	38	39	40	$t = 6718/I^{1.26}$
	4800	10	11.5	10	
	9600	4.5	5	6	
	15750	1	1.5	NA	
1/0 Cu covered	480	13.5	20	18	$t = 16.6/I^{0.65}$
	1300	7	15.5	9	
	4800	4	5	4.5	
	9600	2	2.5	2.5	
1/0 Cu bare	1400	20.5	29.5	22.5	$t = 91/I^{0.78}$
	4800	3.5	7	4.5	
	9600	4	6	6	
	15000	3	4	3.5	
3/0 ACSR covered	1400	35	38	37	$t = 642600/I^{1.92}$
	1900	16	17	16.5	
	3300	10	12	11	
	4800	2	3	3	
3/0 ACSR bare	4550	26	30.5	28.5	$t = 1460/I^{0.95}$
	9100	14	16	15	
	15500	8	9.5	8	
	18600	7	9	7	
4/0 ACAR bare	2100	24	29	28	$t = 80.3/I^{0.68}$
	4800	16	18	17.5	
	9200	8	9.5	8.5	
	15250	8	8	NA	
4/0 ACSR bare	4450	53	66	62	$t = 68810/I^{1.33}$
	8580	21	26	25	
	15250	10	14	NA	
	18700	8	10	8.5	
336.4-kcmil ACAR bare	4900	33	38.5	33	$t = 6610/I^{1.10}$
	9360	12	17.5	17	
	15800	8	8.5	8	
	18000	7	14	7.5	

TABLE 2.22 (Continued)

Burndown Characteristics of Various Conductors

	Current, A	Duration, 60-Hz Cycles Min	Max	Other	Curvefit
336.4-kcmil ACSR bare	8425	25	26	26	$t = 2690/I^{0.97}$
	15200	10	15	14	
	18800	12	13	12	
350-kcmil AAC bare	4800	29	21	20	$t = 448/I^{0.84}$
	9600	11.5	13.5	12	
	15200	8	9	8.5	
	18200	8	7.5	7.5	
500-kcmil AAC bare	4800	42	43	42.5	$t = 2776/I^{0.98}$
	8800	22.5	23	22	
	15400	13	14.5	14	
	18400	11	12	10.5	

Source: Goode, W.B. and Gaertner, G.H., "Burndown Tests and Their Effect on Distribution Design," EEI T&D Meeting, Clearwater, FL, Oct. 14–15, 1965.

2.10 Other Overhead Issues

2.10.1 Connectors and Splices

Connectors and splices are often weak links in the overhead system, either due to hostile environment or bad designs or, most commonly, poor installation. Utilities have had problems with connectors, especially with higher loadings (Jondahl et al., 1991).

Most primary connectors use some sort of compression to join conductors (see Figure 2.15 for common connectors). Compression splices join two conductors together — two conductors are inserted in each end of the sleeve, and a compression tool is used to tighten the sleeve around the conductors. For conductors under tension, automatic splices are also available. Crews just insert the conductors in each end, and serrated clamps within the splice grip the conductor; with higher tension, the wedging action holds tighter.

For tapping a smaller conductor off of a larger conductor, many options are available. Hot-line clamps use a threaded bolt to hold the conductors together. Wedge connectors have a wedge driven between conductors held by a C-shaped body. Compression connectors (commonly called *squeeze-ons*) use dies and compression tools to squeeze together two conductors and the connector.

Good cleaning is essential to making a good contact between connector surfaces. Both copper and aluminum develop a hard oxide layer on the surface when exposed to air. While very beneficial in preventing corrosion, the oxide layer has high electrical resistance. Copper is relatively easy to brush clean. Aluminum is tougher; crews need to work at it harder, and a

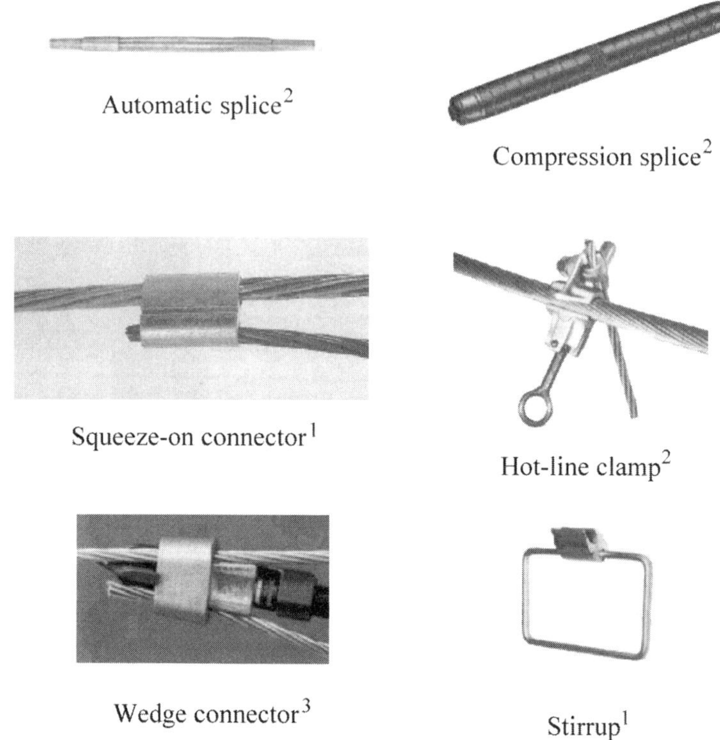

FIGURE 2.15
Common distribution connectors. [1] Reprinted with the permission of Cooper Industries, Inc. [2] Reprinted with the permission of Hubbell Power Systems, Inc. [3] Reprinted with the permission of Tyco Electronics Corporation.

shiny surface is no guarantee of a good contact. Aluminum oxidizes quickly, so crews should clean conductors just before attaching the connector. Without good cleaning, the temperatures developed at the hotspot can anneal the conductor, possibly leading to failure. Joint compounds are important; they inhibit oxidation and help maintain a good contact between joint surfaces.

Corrosion at interfaces can prematurely fail connectors and splices. Galvanic corrosion can occur quickly between dissimilar metals. For this reason, aluminum connectors are used to join aluminum conductors. Waterproof joint compounds protect conductors and joints from corrosion.

Aluminum expands and contracts with temperature, so swings in conductor temperature cause the conductor to creep with respect to the connector. This can loosen connectors and allow oxidation to develop between the connector and conductor. ANSI specifies a standard for connectors to withstand thermal cycling and mechanical stress (ANSI C119.4-1998).

Poor quality work leads to failures. Not using joint compound (or not using enough), inadequate conductor cleaning, misalignments, not fully inserting the conductor prior to compression, or using the wrong dies — any of these mistakes can cause a joint to fail prematurely.

Infrared thermography is the primary way utilities spot bad connectors. A bad connection with a high contact resistance operates at significantly higher temperatures than the surrounding conductor. While infrared inspections are easy for crews to do, they are not foolproof; they can miss bad connectors and falsely target good conductors. Infrared measurements are very sensitive to sunlight, line currents, and background colors. Temperature differences are most useful (but still not perfect indicators). Experience and visual checks of the connector can help identify false readings (such as glare due to sunlight reflection). A bad connector can become hot enough to melt the conductor, but often the conductor can resolidify, temporarily at a lower resistance. Infrared inspections can miss these bad connectors if they are in the resolidified stage. For compression splices, EPRI laboratory tests and field inspections found high success rates using hotstick-mounted resistance measuring devices that measure the resistance across a short section of the conductor (EPRI 1001913, 2001).

Short-circuit current can also damage inline connectors. Mechanical stresses and high currents can damage some splices and connectors. If an inline connector does not make solid contact at its interfaces to the conductor, hotspots can weaken and possibly break the connector or conductor. If the contact is poor enough to cause arcing, the arcing can quickly eat the connection away. Mechanical forces can also break an already weakened or corroded connector.

Hot-line clamps are popular connectors that crews can easily apply with hot-line tools. Threaded bolts provide compression between conductors. Hot-line clamps can become loose, especially if not installed correctly. Utilities have had problems with hot-line clamps attached directly to primary conductors, especially in series with the circuit (rather than tapped for a jumper to equipment) where they are subjected to the heat and mechanical forces of fault currents. Loose or high-resistance hot-line clamps can arc across the interface, quickly burning away the primary conductor.

Stirrups are widely used between the main conductor and a jumper to a transformer or capacitor bank. A stirrup is a bail or loop of wire attached to the main conductor with one or two compression connectors or hot-line connectors. Crews can quickly make a connection to the stirrup with hot-line clamps. The main reason for using the stirrup is to protect the main conductor from burndown. If tied directly to the main conductor, arcing across a poor connection can burn the main conductor down. If a poor hot-line clamp is connected to a stirrup, the stirrup may burn down, but the main line is protected. Also, any arcing when crews attach or detach the connector does not damage the main conductor, so stirrups are especially useful where jumpers may be put on and taken off several times. Using stirrups is reliable; a survey by the National Rural Electric Cooperative Association (NRECA) found that less than 10% of utilities have annual failure rates between 1 and 5%, and almost all of the remainder have failure rates less than 1% (RUS, 1996).

2.10.2 Radio Frequency Interference

Distribution line hardware can generate radio-frequency interference (RFI). Such interference can impact the AM and FM bands as well as VHF television broadcasts. Ham radio frequencies are also affected.

Most power-line noise is from arcs — arcs across gaps on the order of 1 mm, usually at poor contacts. These arcs can occur between many metallic junctions on power-line equipment. Consider two metal objects in close proximity but not quite touching. The capacitive voltage divider between the conducting parts determines the voltage differences between them. The voltage difference between two metallic pieces can cause an arc across a small gap. After arcing across, the gap can clear easily, and after the capacitive voltage builds back up, it can spark over again. These sparkovers radiate radio-frequency noise. Stronger radio-frequency interference is more likely from hardware closer to the primary conductors.

Arcing generates broadband radio-frequency noise from several kilohertz to over 1000 MHz. Above about 50 MHz, the magnitude of arcing RFI drops off. Power-line interference affects lower frequency broadcasts more than higher frequencies. The most common from low to high frequency are: AM radio (0.54 to 1.71 MHz), low-band VHF TV (channels 2 to 6, 54 to 88 MHz), FM radio (88.1 to 107.9 MHz), and high-band VHF TV (channels 7 to 13, 174 to 216 MHz). UHF (ultra-high frequencies, about 500 MHz) are only created right near the sparking source.

On an oscilloscope, arcing interference looks like a series of noise spikes clustered around the peaks of the sinusoidal power-frequency driving voltage (see Figure 2.16). Often power-line noise causes a raspy sound, usually with a 120-Hz characteristic. The "sound" of power-line noise varies depending on the length of the arcing gap, so interference cannot always be identified by a specific characteristic sound (Loftness, 1997).

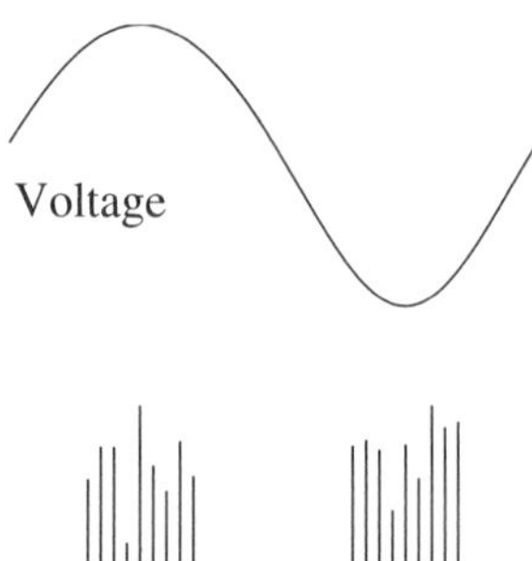

FIGURE 2.16
Arcing source creating radio-frequency interference.

Arcing across small gaps accounts for almost all radio-frequency interference created by utility equipment on distribution circuits. Arcing from corona can also cause interference, but distribution circuit voltages are too low for corona to cause significant interference. Radio interference is more common at higher distribution voltages.

Some common sources and solutions include [for more detail, see (Loftness, 1996; NRECA 90-30, 1992)]:

- *Loose or corroded hot-line clamps* — Replace the connector. After cleaning the conductor and applying fresh inhibitor replace the clamp with a new hot-line clamp or a wedge connector or a squeeze-on connector.
- *Loose nut and washer on a through bolt* — Commonly a problem on double-arming bolts between two crossarms; use lock washers and tighten.
- *Loose or broken insulator tie wire or incorrect tie wire* — Loose tie wires can cause arcing, and conducting ties on covered conductors generate interference; in either case, replace the tie wire.
- *Loose dead-end insulator units* — Replace, preferably with single-unit types. Semiconductive grease provides a short-term solution.
- *Loose metal staples on bonding or ground wires, especially near the top* — Replace with insulated stables (hammering in existing staples may only help for the short term).
- *Loose crossarm lag screw* — Replace with a larger lag screw or with a through bolt and lock washers.
- *Bonding conductors touching or nearly touching other metal hardware* — Separate by at least 1 in. (2.54 cm).
- *Broken or contaminated insulators* — Clean or replace.
- *Defective lightning arresters, especially gapped units* — Replace.

Most of these problems have a common characteristic: gaps between metals, often from loose hardware. Crews can fix most problems by tightening connections, separating metal hardware by at least 1 in., or bonding hardware together. Metal-to-wood interfaces are less likely to cause interference; a tree branch touching a conductor usually does not generate radio-frequency interference.

While interference is often associated with overhead circuits, underground lines can also generate interference. Again, look for loose connections, either primary or secondary such as in load-break elbows.

Interference from an arcing source can propagate in several ways: radiation, induction, and conduction. RFI can radiate from the arcing source just like a radio transmitter. It can conduct along a conductor and also couple inductively from one conductor to parallel conductors. Lower frequencies propagate farther; AM radio is affected over larger distances. Interference is roughly in inverse proportion to frequency at more than a few poles from the source.

Many different interference detectors are available; most are radios with directional antennas. Closer to the source, instruments can detect radio-frequency noise at higher and higher frequencies, so higher frequencies can help pinpoint sources. As you get closer to the source, follow the highest frequency that you can receive. (If you cannot detect interference at higher and higher frequencies as you drive along the line, you are probably going in the wrong direction.) Once a problem pole is identified, an ultrasonic detector with a parabolic dish can zero in on problem areas to identify where the arcing is coming from. Ultrasonic detectors measure ultra-high frequency sound waves (about 20 to 100 kHz) and give accurate direction to the source. Ultrasonic detectors almost require line-of-sight to the arcing source, so they do not help if the arcing is hidden. In such cases, the sparking may be internal to an enclosed device, or the RF could be conducted to the pole by a secondary conductor or riser pole. For even more precise location, crews can use hot-stick mounted detectors to identify exactly what's arcing.

Note that many other nonutility sources of radio-frequency interference exist. Many of these also involve intermittent arcing. Common power-frequency type sources include fans, light dimmers, fluorescent lights, loose wiring within the home or facility, and electrical tools such as drills. Other sources include defective antennas, amateur or CB radios, spark-plug ignitions from vehicles or lawn mowers, home computers, and garage door openers.

References

ABB, *Distribution Transformer Guide*, 1995.

Aluminum Association, *Ampacities for Aluminum and ACSR Overhead Electrical Conductors*, 1986.

Aluminum Association, *Aluminum Electrical Conductor Handbook*, 1989.

ANSI C119.4-1998, *Electric Connectors for Use Between Aluminum-to-Aluminum or Aluminum-to-Copper Bare Overhead Conductors.*

Barber, K., "Improvements in the Performance and Reliability of Covered Conductor Distribution Systems," International Covered Conductor Conference, Cheshire, U.K., January 1999.

Barker, P.P. and Short, T.A., "Findings of Recent Experiments Involving Natural and Triggered Lightning," IEEE/PES Transmission and Distribution Conference, Los Angeles, CA, 1996.

Black, W.Z. and Rehberg, R.L., "Simplified Model for Steady State and Real-Time Ampacity of Overhead Conductors," *IEEE Transactions on Power Apparatus and Systems*, vol. 104, pp. 29–42, October 1985.

Carson, J.R., "Wave Propagation in Overhead Wires with Ground Return," Bell System Technical Journal, vol. 5, pp. 539–54, 1926.

Clapp, A.L., *National Electrical Safety Code Handbook*, The Institute of Electrical and Electronics Engineers, Inc., 1997.

Clarke, E., Circuit Analysis of AC Power Systems, II, General Electric Company, 1950.

Ender, R.C., Auer, G.G., and Wylie, R.A., "Digital Calculation of Sequence Impedances and Fault Currents for Radial Primary Distribution Circuits," *AIEE Transactions on Power Apparatus and Systems*, vol. 79, pp. 1264–77, 1960.

EPRI 1001913, *Electrical, Mechanical, and Thermal Performance of Conductor Connections*, Electric Power Research Institute, Palo Alto, CA, 2001.

EPRI, *Transmission Line Reference Book: 345 kV and Above*, 2nd ed., Electric Power Research Institute, Palo Alto, CA, 1982.

Goode, W.B. and Gaertner, G.H., "Burndown Tests and Their Effect on Distribution Design," EEI T&D Meeting, Clearwater, FL, Oct. 14–15, 1965.

Harvey, J.R., "Effect of Elevated Temperature Operation on the Strength of Aluminum Conductors," IEEE/PES Winter Power Meeting Paper T 72 189–4, 1971. As cited by Southwire (1994).

House, H.E. and Tuttle, P.D., "Current Carrying Capacity of ACSR," *IEEE Transactions on Power Apparatus and Systems*, pp. 1169–78, February 1958.

IEEE C2-2000, *National Electrical Safety Code Handbook*, The Institute of Electrical and Electronics Engineers, Inc.

IEEE Std. 738-1993, *IEEE Standard for Calculating the Current-Temperature Relationship of Bare Overhead Conductors.*

Jondahl, D.W., Rockfield, L.M., and Cupp, G.M., "Connector Performance of New vs. Service Aged Conductor," IEEE/PES Transmission and Distribution Conference, 1991.

Kurtz, E.B., Shoemaker, T.M., and Mack, J.E., *The Lineman's and Cableman's Handbook*, McGraw Hill, New York, 1997.

Lasseter, J.A., "Burndown Tests on Bare Conductor," *Electric Light & Power*, pp. 94–100, December 1956.

Lat, M.V., "Determining Temporary Overvoltage Levels for Application of Metal Oxide Surge Arresters on Multigrounded Distribution Systems," *IEEE Transactions on Power Delivery*, vol. 5, no. 2, pp. 936–46, April 1990.

Lee, R.E., Fritz, D.E., Stiller, P.H., Kilar, L.A., and Shankle, D.F., "Prevention of Covered Conductor Burndown on Distribution Circuits," American Power Conference, 1980.

Loftness, M.O., *AC Power Interference Field Manual*, Percival Publishing, Tumwater, WA, 1996.

Loftness, M.O., "Power Line RF Interference — Sounds, Patterns, and Myths," *IEEE Transactions on Power Delivery*, vol. 12, no. 2, pp. 934–40, April 1997.

NRECA 90-30, *Power Line Interference: A Practical Handbook*, National Rural Electric Cooperative Association, 1992.

RUS 160-2, *Mechanical Design Manual for Overhead Distribution Lines*, United States Department of Agriculture, Rural Utilities Service, 1982.

RUS 1728F-803, *Specifications and Drawings for 24.9/14.4 kV Line Construction*, United States Department of Agriculture, Rural Utilities Service, 1998.

RUS, "Summary of Items of Engineering Interest," United States Department of Agriculture, Rural Utilities Service, 1996.

Short, T.A. and Ammon, R.A., "Instantaneous Trip Relay: Examining Its Role," *Transmission and Distribution World*, vol. 49, no. 2, 1997.

Smith, D.R., "System Considerations — Impedance and Fault Current Calculations," *IEEE Tutorial Course on Application and Coordination of Reclosers, Sectionalizers, and Fuses*, 1980. Publication 80 EHO157-8-PWR.

Southwire Company, *Overhead Conductor Manual*, 1994.

Stevenson, W.D., *Elements of Power System Analysis*, 2nd ed., McGraw Hill, New York, 1962.

Willis, H.L., *Power Distribution Planning Reference Book*, Marcel Dekker, New York, 1997.

Saying "You can't do that" to a Lineman is the same as saying "Hey, how about a contest?"

Powerlineman law #23, By CD Thayer and other Power Linemen, http://www.cdthayer.com/lineman.htm

3

Underground Distribution

Much new distribution is underground. Underground distribution is much more hidden from view than overhead circuits, and is more reliable. Cables, connectors, and installation equipment have advanced considerably in the last quarter of the 20th century, making underground distribution installations faster and less expensive.

3.1 Applications

One of the main applications of underground circuits is for underground residential distribution (URD), underground branches or loops supplying residential neighborhoods. Utilities also use underground construction for substation exits and drops to padmounted transformers serving industrial or commercial customers. Other uses are crossings: river crossings, highway crossings, or transmission line crossings. All-underground construction — widely used for decades in cities — now appears in more places.

Underground construction is expensive, and costs vary widely. Table 3.1 shows extracts from one survey of costs done by the CEA; the two utilities highlighted differ by a factor of ten. The main factors that influence underground costs are:

- *Degree of development* — Roads, driveways, sidewalks, and water pipes — these and other obstacles slow construction and increase costs.
- *Soil condition* — Rocks and frozen ground increase overtime pay for cable crews.
- *Urban, suburban, or rural* — Urban construction is more difficult not only because of concrete, but also because of traffic. Rural construction is generally the least expensive per length, but lengths are long.
- *Conduit* — Concrete-encased ducts cost more than direct-buried conduits, which cost more than preassembled flexible conduit, which cost more than directly buried cable with no conduits.

TABLE 3.1

Comparison of Costs of Different Underground Constructions at Different Utilities

Utility	Construction	$/ft[a]
TAU	Rural or urban, 1 phase, #2 Al, 25 kV, trenched, direct buried	6.7
	Rural, 3 phase, #2 Al, 25 kV, trenched, direct buried	13.4
	Urban commercial, 3 phase, #2 Al, 25 kV, trenched, direct buried	13.4
	Urban express, 3 phase, 500-kcmil Al, 25 kV, trenched, direct buried	23.5
WH	Urban, 1 phase, 1/0 Al, 12.5 kV, trenched, conduit	84.1
	Urban commercial, 3 phase, 1/0 Al, 12.5 kV, trenched, conduit	117.7
	Urban express, 3 phase, 500-kcmil Cu, 12.5 kV, trenched, conduit	277.4

[a] Converted assuming that one 1991 Canadian dollar equals 1.1 U.S. dollars in 2000.

Source: CEA 274 D 723, *Underground Versus Overhead Distribution Systems*, Canadian Electrical Association, 1992.

- *Cable size and materials* — The actual cable cost is a relatively small part of many underground applications. A 1/0 aluminum full-neutral 220-mil TR-XLPE cable costs just under $2 per ft; with a 500-kcmil conductor and a one-third neutral, the cable costs just under $4 per ft.
- *Installation equipment* — Bigger machines and machines more appropriate for the surface and soil conditions ease installations.

3.1.1 Underground Residential Distribution (URD)

A classic underground residential distribution circuit is an underground circuit in a loop arrangement fed at each end from an overhead circuit (see Figure 3.1). The loop arrangement allows utilities to restore customers more quickly; after crews find the faulted section, they can reconfigure the loop and isolate any failed section of cable. This returns power to all customers. Crews can delay replacing or fixing the cable until a more convenient time or when suitable equipment arrives. Not all URD is configured in a loop. Utilities sometimes use purely radial circuits or circuits with radial taps or branches.

Padmounted transformers step voltage down for delivery to customers and provide a sectionalizing point. The elbow connectors on the cables (pistol grips) attach to bushings on the transformer to maintain a dead-front — no exposed, energized conductors. To open a section of cable, crews can simply pull an elbow off of the transformer bushing and place it on a parking stand, which is an elbow bushing meant for holding an energized elbow connector.

Elbows and other terminations are available with continuous-current ratings of 200 or 600 A (IEEE Std. 386-1995). Load-break elbows are designed to break load; these are only available in 200-A ratings. Without load-break capability, crews should of course only disconnect the elbow if the cable is deenergized. Elbows normally have a test point where crews can check if

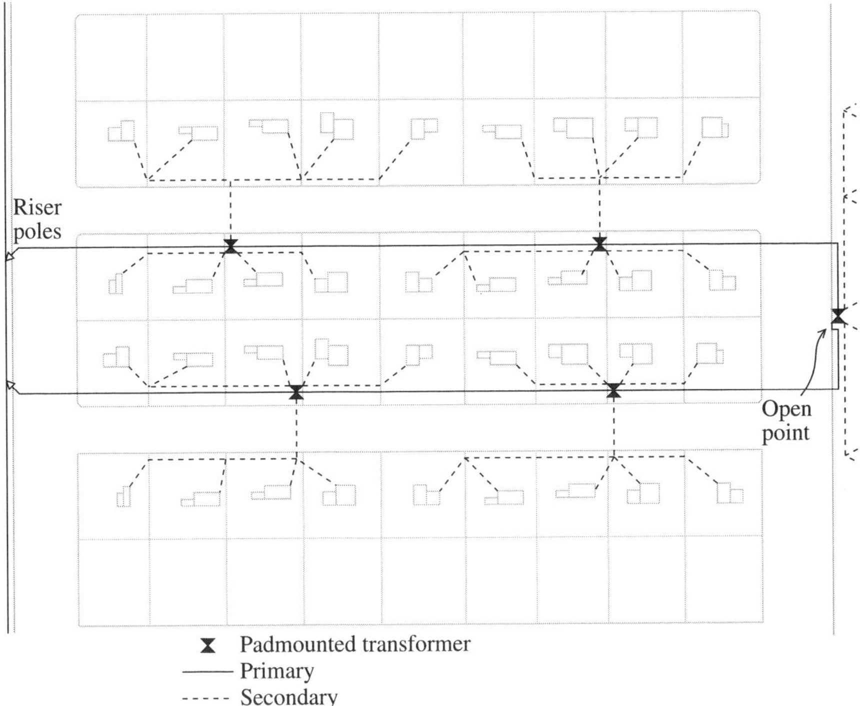

FIGURE 3.1
An example front-lot underground residential distribution (URD) system.

the cable is live. Elbows are also tested to withstand ten cycles of fault current, with 200-A elbows tested at 10 kA and 600-A elbows tested at 25 kA (IEEE Std. 386-1995).

The interface between the overhead circuit and the URD circuit is the riser pole. At the riser pole (or a dip pole or simply a dip), cable terminations provide the interface between the insulated cable and the bare overhead conductors. These pothead terminations grade the insulation to prevent excessive electrical stress on the insulation. Potheads also keep water from entering the cable, which is critical for cable reliability. Also at the riser pole are expulsion fuses, normally in cutouts. Areas with high short-circuit current may also have current-limiting fuses. To keep lightning surges from damaging the cable, the riser pole should have arresters right across the pothead with as little lead length as possible.

Underground designs for residential developments expanded dramatically in the 1970s. Political pressure coupled with technology improvements were the driving forces behind underground distribution. The main developments — direct-buried cables and padmounted transformers having loadbreak elbows — dramatically reduced the cost of underground distribution to close to that of overhead construction. In addition to improving the visual landscape, underground construction improves reliability. Underground residential distribution has had difficulties, especially high cable failure rates. In the late 1960s and early 1970s, given the durability of plastics, the polyethylene cables installed at that time were thought to have a life of at least 50 years. In practice, cables failed at a much higher rate than expected, enough so that many utilities had to replace large amounts of this cable.

According to Boucher (1991), 72% of utilities use front-lot designs for URD. With easier access and fewer trees and brush to clear, crews can more easily install cables along streets in the front of yards. Customers prefer rear-lot service, which hides padmounted transformers from view. Back-lot placement can ease siting issues and may be more economical if lots share rear property lines. But with rear-lot design, utility crews have more difficulty accessing cables and transformers for fault location, sectionalizing, and repair.

Of those utilities surveyed by Boucher (1991), 85% charge for underground residential service, ranging from $200 to $1200 per lot (1991 dollars). Some utilities charge by length, which ranges from $5.80 to $35.00 per ft.

3.1.2 Main Feeders

Whether urban, suburban, or even rural, all parts of a distribution circuit can be underground, including the main feeder. For reliability, utilities often configure an underground main feeder as a looped system with one or more tie points to other sources. Switching cabinets or junction boxes serve as tie points for tapping off lateral taps or branches to customers. These can be in handholes, padmounted enclosures, or pedestals above ground. Three-phase circuits can also be arranged much like URD with sections of cable run between three-phase padmounted transformers. As with URD, the padmounted transformers serve as switching stations.

Although short, many feeders have an important underground section — the substation exit. Underground substation exits make substations easier to design and improve the aesthetics of the substation. Because they are at the substation, the source of a radial circuit, substation exits are critical for reliability. In addition, the loading on the circuit is higher at the substation exit than anywhere else; the substation exit may limit the entire circuit's ampacity. Substation exits are not the place to cut corners. Some strategies to reduce the risks of failures or to speed recovery are: concrete-enclosed ducts to help protect cables, spare cables, overrated cables, and good surge protection.

While not as critical as substation exits, utilities use similar three-phase underground dips to cross large highways or rivers or other obstacles. These are designed in much the same way as substation exits.

3.1.3 Urban Systems

Underground distribution has reliably supplied urban systems since the early 1900s. Cables are normally installed in concrete-encased duct banks beneath streets, sidewalks, or alleys. A duct bank is a group of parallel ducts, usually with four to nine ducts but often many more. Ducts may be precast concrete sections or PVC encased in concrete. Duct banks carry both primary and secondary cables. Manholes every few hundred feet provide access to cables. Transformers are in vaults or in the basements of large buildings.

Paper-insulated lead-covered (PILC) cables dominated urban applications until the late 20th century. Although a few utilities still install PILC, most use extruded cable for underground applications. In urban applications, copper is more widely used than in suburban applications. Whether feeding secondary networks or other distribution configurations, urban circuits may be subjected to heavy loads.

"Vertical" distribution systems are necessary in very tall buildings. Medium-voltage cable strung up many floors feed transformers within a building. Submarine cables are good for this application since their protective armor wire provides support when a cable is suspended for hundreds of feet.

3.1.4 Overhead vs. Underground

Overhead or underground? The debate continues. Both designs have advantages (see Table 3.2). The major advantage of overhead circuits is cost; an underground circuit typically costs anywhere from 1 to 2.5 times the equivalent overhead circuit (see Table 3.3). But the cost differences vary wildly, and it's often difficult to define "equivalent" systems in terms of performance. Under the right conditions, some estimates of cost report that cable installations can be less expensive than overhead lines. If the soil is easy to dig, if the soil has few rocks, if the ground has no other obstacles like water pipes or telephone wires, then crews may be able to plow in cable faster and for less cost than an overhead circuit. In urban areas, underground is almost the only choice; too many circuits are needed, and above-ground space is too expensive or just not available. But urban duct-bank construction is expensive on a per-length basis (fortunately, circuits are short in urban appli-

TABLE 3.2

Overhead vs. Underground: Advantages of Each

Overhead	Underground
Cost — Overhead's number one advantage. Significantly less cost, especially initial cost.	*Aesthetics* — Underground's number one advantage. Much less visual clutter.
Longer life — 30 to 50 years vs. 20 to 40 for new underground works.	*Safety* — Less chance for public contact.
Reliability — Shorter outage durations because of faster fault finding and faster repair.	*Reliability* — Significantly fewer short and long-duration interruptions.
Loading — Overhead circuits can more readily withstand overloads.	*O&M* — Notably lower maintenance costs (no tree trimming).
	Longer reach — Less voltage drop because reactance is lower.

TABLE 3.3

Comparison of Underground Construction Costs with Overhead Costs

Utility		Construction	$/ft[a]	Underground to overhead ratio
Single-Phase Lateral Comparisons				
NP	Overhead	1/0 AA, 12.5 kV, phase and neutral	8.4	
NP	Underground	1/0 AA, 12.5 kV, trenched, in conduit	10.9	1.3
APL	Overhead	Urban, #4 ACSR, 14.4 kV	2.8	
APL	Underground	Urban, #1 AA, 14.4 kV, trenched, direct buried	6.6	2.4
Three-Phase Mainline Comparisons				
NP	Overhead	Rural, 4/0 AA, 12.5 kV	10.3	
NP	Underground	Rural, 1/0 AA, 12.5 kV, trenched, in conduit	17.8	1.7
NP	Overhead	Urban, 4/0 AA, 12.5 kV	10.9	
NP	Underground	Urban, 4/0 AA, 12.5 kV, trenched, in conduit	17.8	1.6
APL	Overhead	Urban, 25 kV, 1/0 ACSR	8.5	
APL	Underground	Urban, 25 kV, #1 AA, trenched, direct buried	18.8	2.2
EP	Overhead	Urban, 336 ACSR, 13.8 kV	8.7	
EP	Underground	Urban residential, 350 AA, 13.8 kV, trenched, direct buried	53.2	6.1
EP	Underground	Urban commercial, 350 AA, 13.8 kV, trenched, direct buried	66.8	7.6

[a] Converted assuming that one 1991 Canadian dollar equals 1.1 U.S. dollars in 2000.

Source: CEA 274 D 723, *Underground Versus Overhead Distribution Systems*, Canadian Electrical Association, 1992.

cations). On many rural applications, the cost of underground circuits is difficult to justify, especially on long, lightly loaded circuits, given the small number of customers that these circuits feed.

Aesthetics is the main driver towards underground circuits. Especially in residential areas, parks, wildlife areas, and scenic areas, visual impact is important. Undergrounding removes a significant amount of visual clutter. Overhead circuits are ugly. It is possible to make overhead circuits less ugly with tidy construction practices, fiberglass poles instead of wood, keeping poles straight, tight conductor configurations, joint use of poles to reduce the number of poles, and so on. Even the best though, are still ugly, and many older circuits look awful (weathered poles tipped at odd angles, crooked crossarms, rusted transformer tanks, etc.).

Underground circuits get rid of all that mess, with no visual impacts in the air. Trees replace wires, and trees don't have to be trimmed. At ground level, instead of poles every 150 ft (many having one or more guy wires) urban construction has no obstacles, and URD-style construction has just padmounted transformers spaced much less frequently. Of course, for maximum benefit, all utilities must be underground. There is little improvement to undergrounding electric circuits if phone and cable television are still strung on poles (i.e., if the telephone wires are overhead, you might as well have the electric lines there, too).

While underground circuits are certainly more appealing when finished, during installation construction is messier than overhead installation. Lawns, gardens, sidewalks, and driveways are dug up; construction lasts longer; and the installation "wounds" take time to heal. These factors don't matter much when installing circuits into land that is being developed, but it can be upsetting to customers in an existing, settled community.

Underground circuits are more reliable. Overhead circuits typically fault about 90 times/100 mi/year; underground circuits fail less than 10 times/100 mi/year. Because overhead circuits have more faults, they cause more voltage sags, more momentary interruptions, and more long-duration interruptions. Even accounting for the fact that most overhead faults are temporary, overhead circuits have more permanent faults that lead to long-duration circuit interruptions. The one disadvantage of underground circuits is that when they do fail, finding the failure is harder, and fixing the damage or replacing the equipment takes longer. This can partially be avoided by using loops capable of serving customers from two directions, by using conduits for faster replacement, and by using better fault location techniques. Underground circuits are much less prone to the elements. A major hurricane may drain an overhead utility's resources, crews are completely tied up, customer outages become very long, and cleanup costs are a major cost to utilities. However, underground circuits are not totally immune from the elements. In "heat storms," underground circuits are prone to rashes of failures. Underground circuits have less overload capability than overhead circuits; failures increase with operating temperature.

In addition to less storm cleanup, underground circuits require less periodic maintenance. Underground circuits don't require tree trimming, easily the largest fraction of most distribution operations and maintenance budgets. The CEA (1992) estimated that underground system maintenance averaged 2% of system plant investment whereas overhead systems averaged 3 to 4%, or as much as twice that of underground systems.

Underground circuits are safer to the public than overhead circuits. Overhead circuits are more exposed to the public. Kites, ladders, downed wires, truck booms — despite the best public awareness campaigns, these still expose the public to electrocution from overhead lines. Don't misunderstand; underground circuits still have dangers, but they're much less than on overhead circuits. For the public, dig-ins are the most likely source of contact. For utility crews, both overhead and underground circuits offer dangers that proper work practices must address to minimize risks.

We cannot assume that underground infrastructure will last as long as overhead circuits. Early URD systems failed at a much higher rate than expected. While most experts believe that modern underground equipment is more reliable, it is still prudent to believe that an overhead circuit will last 40 years, while an underground circuit will only last 30 years.

Overhead vs. underground is not an all or nothing proposition. Many systems are hybrids; some schemes are:

- *Overhead mainline with underground taps* — The larger, high-current conductors are overhead. If the mains are routed along major roads, they have less visual impact. Lateral taps down side roads and into residential areas, parks, and shopping areas are underground. Larger primary equipment like regulators, reclosers, capacitor banks, and automated switches are installed where they are more economical — on the overhead mains. Because the mainline is a major contributor to reliability, this system is still less reliable than an all-underground system.
- *Overhead primary with underground secondary* — Underground secondary eliminates some of the clutter associated with overhead construction. Eliminating much of the street and yard crossings keeps the clutter to the pole-line corridor. Costs are reasonable because the primary-level equipment is still all overhead.

Converting from overhead to underground is costly, yet there are locations and situations where it is appropriate for utilities and their customers. Circuit extensions, circuit enhancements to carry more load, and road-rebuilding projects — all are opportunities for utilities and communities to upgrade to underground service.

3.2 Cables

At the center of a cable is the phase conductor, then comes a semiconducting conductor shield, the insulation, a semiconducting insulation shield, the neutral or shield, and finally a covering jacket. Most distribution cables are single conductor. Two main types of cable are available: concentric-neutral cable and power cable. Concentric-neutral cable normally has an aluminum conductor, an extruded insulation, and a concentric neutral (Figure 3.2 shows a typical construction). A concentric neutral is made from several copper wires wound concentrically around the insulation; the concentric neutral is a true neutral, meaning it can carry return current on a grounded system. Underground residential distribution normally has concentric-neutral cables; concentric-neutral cables are also used for three-phase mainline applications and three-phase power delivery to commercial and industrial customers. Because of their widespread use in URD, concentric-neutral cables are often called URD cables. Power cable has a copper or aluminum phase

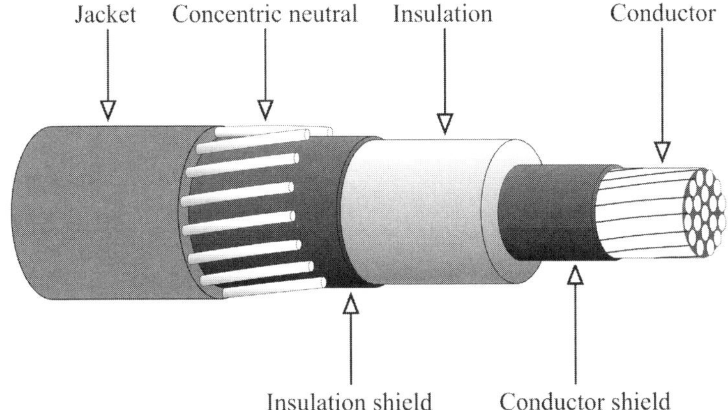

FIGURE 3.2
A concentric neutral cable, typically used for underground residential power delivery.

conductor, an extruded insulation, and normally a thin copper tape shield. On utility distribution circuits, power cables are typically used for mainline feeder applications, network feeders, and other high current, three-phase applications. Many other types of medium-voltage cable are available. These are sometimes appropriate for distribution circuit application: three-conductor power cables, armored cables, aerial cables, fire-resistant cables, extra flexible cables, and submarine cables.

3.2.1 Cable Insulation

A cable's insulation holds back the electrons; the insulation allows cables with a small overall diameter to support a conductor at significant voltage. A 0.175-in. (4.5-mm) thick polymer cable is designed to support just over 8 kV continuously; that's an average stress of just under 50 kV per in. (20 kV/cm). In addition to handling significant voltage stress, insulation must withstand high temperatures during heavy loading and during short circuits and must be flexible enough to work with. For much of the 20th century, paper insulation dominated underground application, particularly PILC cables. The last 30 years of the 20th century saw the rise of polymer-insulated cables, polyethylene-based insulations starting with high-molecular weight polyethylene (HMWPE), then cross-linked polyethylene (XLPE), then tree-retardant XLPE and also ethylene-propylene rubber (EPR) compounds.

Table 3.4 compares properties of TR-XLPE, EPR, and other insulation materials. Some of the key properties of cable insulation are:

- *Dielectric constant* (ε, also called permittivity) — This determines the cable's capacitance: the dielectric constant is the ratio of the capacitance with the insulation material to the capacitance of the same

TABLE 3.4
Properties of Cable Insulations

	Dielectric Constant 20°C	Loss Angle Tan δ at 20°C	Volume Resistivity Ω-m	Annual Dielectric Loss[a] W/1000 ft	Unaged Impulse Strength V/mil	Water Absorption ppm
PILC	3.6	0.003	10^{11}	N/A	1000–2000	25
PE	2.3	0.0002	10^{14}	N/A		100
XLPE	2.3	0.0003	10^{14}	8	3300	350
TR-XLPE	2.4	0.001	10^{14}	10	3000	<300
EPR	2.7–3.3	0.005–0.008	10^{13}–10^{14}	28–599	1200–2000	1150–3200

[a] For a typical 1/0 15-kV cable.

configuration in free space. Cables with higher capacitance draw more charging current.

- *Volume resistivity* — Current leakage through the insulation is a function of the insulation's dc resistivity. Resistivity decreases as temperature increases. Modern insulation has such high resistivity that very little resistive current passes from the conductor through the insulation.
- *Dielectric losses* — Like a capacitor, a cable has dielectric losses. These losses are due to dipole movements within the polymer or by the movement of charge carriers within the insulation. Dielectric losses contribute to a cable's resistive leakage current. Dielectric losses increase with frequency and temperature and with operating voltage.
- *Dissipation factor* (also referred to as the loss angle, loss tangent, tan δ, and approximate power factor) — The dissipation factor is the ratio of the resistive current drawn by the cable to the capacitive current drawn (I_R/I_X). Because the leakage current is normally low, the dissipation factor is approximately the same as the power factor:

$$\text{pf} = I_R / |I| = I_R / \sqrt{I_R^2 + I_X^2} \approx I_R / I_X = \text{dissipation factor}$$

Paper-Insulated Lead-Covered (PILC) Cables. Paper-insulated cables have provided reliable underground power delivery for decades. Paper-insulated lead-sheathed cable has been the dominant cable configuration, used mainly in urban areas. PILC cables have kraft-paper tapes wound around the conductor that are dried and impregnated with insulating oil. A lead sheath is one of the best moisture blocks: it keeps the oil in and keeps water out. Paper cables are normally rated to 85°C with an emergency rating up to 105°C (EPRI TR-105502, 1995). PILC cables have held up astonishingly well; many 50-year-old cables are still in service with almost new insulation capability. While PILC has had very good reliability, some utilities are concerned about its present day failure, not because of bad design or application, but because the in-service stock is so old. Moisture ingress, loss of oil, and thermal stresses — these are the three main causes of PILC failure (EPRI 1000741, 2000). Water decreases the dielectric strength (especially when the cable is hot) and increases the dielectric losses (further heating the cable). Heat degrades the insulating capability of the paper, and if oil is lost, the paper's insulating capability declines. PILC use has declined but still not disappeared. Some utilities continue to use it, especially to supply urban networks. Utilities use less PILC because of its high cost, work difficulties, and environmental concerns. Splicing also requires significant skill, and working with the lead sheath requires environmental and health precautions.

Polyethylene (PE). Most modern cables have polymer insulation extruded around the conductor — either polyethylene derivatives or ethylene-propylene properties. Polyethylene is a tough, inexpensive polymer with good electrical properties. Most distribution cables made since 1970 are based on some variation of polyethylene. Polyethylene is an ethylene polymer, a long string or chain of connected molecules. In polyethylene, some of the polymer chains align in crystalline regions, which give strength and moisture resistance to the material. Other regions have nonaligned polymer chains — these amorphous regions give the material flexibility but are permeable to gas and moisture and are where impurities locate. Polyethylene is a thermoplastic. When heated and softened, the polymer chains break apart (becoming completely amorphous); as it cools, the crystalline regions reform, and the material returns to its original state. Polyethylene naturally has high density and excellent electrical properties with a volume resistivity of greater than 10^{14} Ω-m and an impulse insulation strength of over 2700 V/mil.

High-Molecular Weight Polyethylene (HMWPE). High-molecular weight polyethylene is polyethylene that is stiffer, stronger, and more resistant to chemical attack than standard polyethylene. Insulations with higher molecular weights (longer polymer chains) generally have better electrical properties. As with standard polyethylene, HMWPE insulation is a thermoplastic rated to 75°C. Polyethylene softens considerably as temperature increases. Since plastics are stable and seem to last forever, when utilities first installed HMWPE in the late 1960s and early 1970s, utilities and manufacturers expected long life for polyethylene cables. In practice, failure rates increased dramatically after as little as 5 years of service. The electrical insulating strength (the dielectric strength) of HMWPE was degraded by water treeing, an electrochemical degradation driven by the presence of water and voltage. Polyethylene also degrades quickly under partial discharges; once partial discharges start, they can quickly eat away the insulation. Because of high failure rates, HMWPE insulation is off the market now, but utilities still have many miles of this cable in the ground.

Cross-Linked Polyethylene (XLPE). Cross-linking agents are added that form bonds between polymer chains. The cross-linking bonds interconnect the chains and make XLPE semi-crystalline and add stiffness. XLPE is a thermoset: the material is vulcanized (also called "cured"), irreversibly creating the cross-linking that sets when the insulation cools. XLPE has about the same insulation strengths as polyethylene, is more rigid, and resists water treeing better than polyethylene. Although not as bad as HMWPE, pre-1980s XLPE has proven susceptible to premature failures because of water treeing. XLPE has higher temperature ratings than HWMPE; cables are rated to 90°C under normal conditions and 130°C for emergency conditions.

Tree-Retardant Cross-Linked Polyethylene (TR-XLPE). This has adders to XLPE that slow the growth of water trees. Tree-retardant versions of XLPE have almost totally displaced XLPE in medium-voltage cables. Various compounds when added to XLPE reduce its tendency to grow water trees under voltage. These additives tend to slightly reduce XLPE's electrical properties, slightly increase dielectric losses, and slightly lower initial insulation strength (but much better insulation strength when aged). While there is no standard industry definition of TR-XLPE, different manufacturers offer XLPE compounds with various adders that reduce tree growth. The oldest and most widely used formulation was developed by Union Carbide (now Dow); their HFDA 4202 tree-retardant XLPE maintains its insulation strength better in accelerated aging tests (EPRI TR-108405-V1, 1997) and in field service (Katz and Walker, 1998) than standard XLPE.

Ethylene-Propylene Rubber (EPR). EPR compounds are polymers made from ethylene and propylene. Manufacturers offer different ethylene-propylene formulations, which collectively are referred to as EPR. EPR compounds are thermoset, normally with a high-temperature steam curing process that sets cross-linking agents. EPR compounds have high concentrations of clay fillers that provide its stiffness. EPR is very flexible and rubbery. When new, EPR only has half of the insulation strength as XLPE, but as it ages, its insulation strength does not decrease nearly as much as that of XLPE. EPR is naturally quite resistant to water trees, and EPR has a proven reliable record in the field. EPR has very good high-temperature performance. Although soft, it deforms less at high temperature than XLPE and maintains its insulation strength well at high temperature (Brown, 1983). Most new EPR cables are rated to 105°C under normal conditions and to 140°C for emergency conditions, the MV-105 designation per UL Standard 1072. (Historically, both XLPE and EPR cables were rated to 90°C normal and 130°C emergency.) In addition to its use as cable insulation, most splices and joints are made of EPR compounds. EPR has higher dielectric losses than XLPE; depending on the particular formulation, EPR can have two to three times the losses of XLPE to over ten times the losses of XLPE. These losses increase the cost of operation over its lifetime. While not as common or as widely used as XLPE in the utility market, EPR dominates for medium-voltage industrial applications.

TR-XLPE vs. EPR: which to use? Of the largest investor-owned utilities 56% specify TR-XLPE cables, 24% specify EPR, and the remainder specify a mix (Dudas and Cochran, 1999). Trends are similar at rural cooperatives. In a survey of the co-ops with the largest installed base of underground cable, 42% specify TR-XLPE, 34% specify EPR, and the rest specify both (Dudas and Rodgers, 1999). When utilities specify both EPR and TR-XLPE, commonly EPR is used for 600-A three-phase circuits, and TR-XLPE is used for 200-A applications like URD. Each cable type has advocates. TR-XLPE is less expensive and has lower losses. EPR's main feature is its long history of reliability and water-tree resistance. EPR is also softer (easier to handle) and has a higher temperature rating (higher ampacity). Boggs and Xu (2001) show how EPR and TR-XLPE are becoming more similar: EPR compounds are being designed that have fewer losses; tree-retardant additives to XLPE make the cable more tree resistant at the expense of increasing its water absorption and slightly increasing losses.

Cables have a voltage rating based on the line-to-line voltage. Standard voltage ratings are 5, 8, 15, 25, and 35 kV. A single-phase circuit with a nominal voltage of 7.2 kV from line to ground must use a 15-kV cable, not an 8-kV cable (because the line-to-line voltage is 12.47 kV).

Within each voltage rating, more than one insulation thickness is available. Standards specify three levels of cable insulation based on how the cables are applied. The main factor is grounding and ability to clear line-to-ground faults in order to limit the overvoltage on the unfaulted phases. The standard levels are (AEIC CS5-94, 1994):

- *100 percent level* — Allowed where line-to-ground faults can be cleared quickly (at least within one minute); normally appropriate for grounded circuits
- *133 percent level* — Where line-to-ground faults can be cleared within one hour; normally can be used on ungrounded circuits

Standards also define a 173% level for situations where faults cannot be cleared within one hour, but manufacturers typically offer the 100 and 133% levels as standard cables; higher insulation needs can be met by a custom order or going to a higher voltage rating. Table 3.5 shows standard insulation thicknesses for XLPE and EPR for each voltage level. In addition to protecting against temporary overvoltages, thicker insulations provide higher insulation to lightning and other overvoltages and reduce the chance of failure from water tree growth. For 15-kV class cables, Boucher (1991) reported that 59% of utilities surveyed in North America use 100% insulation (175-mil).

TABLE 3.5

Usual Insulation Thicknesses for XLPE or EPR Cables Based on Voltage and Insulation Level

Voltage Rating, kV	Insulation Thickness, Mil (1 mil = 0.001 in. = 0.00254 cm)	
	100% Level	133% Level
8	115	140
15	175	220
25	260	320
35	345	420

At 25 and 35 kV, the surveyed utilities more universally use 100% insulation (88 and 99%, respectively). Dudas and Cochran (1999) report similar trends in a survey of practices of the 45 largest investor-owned utilities: at 15 kV, 69% of utilities specified 100% insulation; at 25 and 35 kV, over 99% of utilities specified 100% insulation.

3.2.2 Conductors

For underground residential distribution (URD) applications, utilities normally use aluminum conductors; Boucher (1991) reported that 80% of utilities use aluminum (alloy 1350); the remainder, copper (annealed, soft). Copper is more prevalent in urban duct construction and in industrial applications. Copper has lower resistivity and higher ampacity for a given size; aluminum is less expensive and lighter. Cables are often stranded to increase their flexibility (solid conductor cables are available for less than 2/0). ASTM class B stranding is the standard stranding. Class C has more strands for applications requiring more flexibility. Each layer of strands is wound in an opposite direction. Table 3.6 shows diameters of available conductors.

3.2.3 Neutral or Shield

A cable's shield, the metallic barrier that surrounds the cable insulation, holds the outside of the cable at (or near) ground potential. It also provides a path for return current and for fault current. The shield also protects the cable from lightning strikes and from current from other fault sources. The metallic shield is also called the *sheath.*

A concentric neutral — a shield capable of carrying unbalanced current — has copper wires wound helically around the insulation shield. The concentric neutral is expected to carry much of the unbalanced load current, with the earth carrying the rest. For single-phase cables, utilities normally use a "full neutral," meaning that the resistance of the neutral equals that of the phase conductor. Also common is a "one-third neutral," which has a resistance that is three times that of the phase conductor. In a survey of underground distribution practices, Boucher (1991) reported that full neutrals dominated for residential application, and reduced neutrals are used more for commercial and feeder applications (see Figure 3.3).

Power cables commonly have 5-mil thick copper tape shields. These are wrapped helically around the cable with some overlap. In a tape-shield cable, the shield is not normally expected to carry unbalanced load current. As we will see, there is an advantage to having a higher resistance shield: the cable ampacity can be higher because there is less circulating current. Shields are also available that are helically wound wires (like a concentric neutral but with smaller wires).

Whether wires or tapes, cable shields and neutrals are copper. Aluminum corrodes too quickly to perform well in this function. Early unjacketed cables

TABLE 3.6

Conductor Diameters

	Solid	Class B stranding	
Size	**Diameter, in.**	**Strands**	**Diameter, in.**
24	0.0201	7	0.023
22	0.0253	7	0.029
20	0.032	7	0.036
19	0.035	7	0.041
18	0.0403	7	0.046
16	0.0508	7	0.058
14	0.0641	7	0.073
12	0.0808	7	0.092
10	0.1019	7	0.116
9	0.1144	7	0.13
8	0.1285	7	0.146
7	0.1443	7	0.164
6	0.162	7	0.184
5	0.1819	7	0.206
4	0.2043	7	0.232
3	0.2294	7	0.26
2	0.2576	7	0.292
1	0.2893	19	0.332
1/0	0.3249	19	0.373
2/0	0.3648	19	0.419
3/0	0.4096	19	0.47
4/0	0.46	19	0.528
250		37	0.575
300		37	0.63
350		37	0.681
400		37	0.728
500		37	0.813
600		61	0.893
750		61	0.998
1000		61	1.152
1250		91	1.289
1500		91	1.412
1750		127	1.526
2000		127	1.632
2500		127	1.824

normally had a coating of lead-tin alloy to prevent corrosion. Cable neutrals still corroded. Dudas (1994) reports that in 1993, 84% of utilities specified a bare copper neutral rather than a coated neutral.

The longitudinally corrugated (LC) shield improves performance for fault currents and slows down water entry. The folds of a corrugated copper tape are overlapped over the cable core. The overlapping design allows movement and shifting while also slowing down water entry. The design performs better for faults because it is thicker than a tape shield, so it has less resistance, and it tends to distribute current throughout the shield rather than keeping it in a few strands.

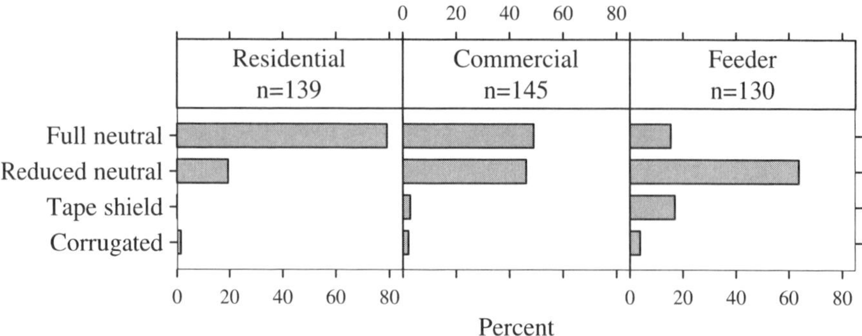

FIGURE 3.3
Surveyed utility use of cable neutral configurations for residential, commercial, and feeder applications. (Data from [Boucher, 1991].)

3.2.4 Semiconducting Shields

In this application, semiconducting means "somewhat conducting": the material has some resistance (limited to a volume resistivity of 500 Ω-m [ANSI/ICEA S-94-649-2000, 2000; ANSI/ICEA S-97-682-2000, 2000]), more than the conductor and less than the insulation. Semiconducting does not refer to nonlinear resistive materials like silicon or metal oxide; the resistance is fixed; it does not vary with voltage. Also called screens or semicons, these semiconducting shields are normally less than 80 mil. The resistive material evens out the electric field at the interface between the phase conductor and the insulation and between the insulation and the neutral or shield. Without the shields, the electric field gradient would concentrate at the closest interfaces between a wire and the insulation; the increased localized stress could break down the insulation. The shields are made by adding carbon to a normally insulating polymer like EPR or polyethylene or cross-linked polyethylene. The conductor shield is normally about 20 to 40 mil thick; the insulation shield is normally about 40 to 80 mil thick. Thicker shields are used on larger diameter cables.

Semiconducting shields are important for smoothing out the electric field, but they also play a critical role in the formation of water trees. The most dangerous water trees are vented trees, those that start at the interface between the insulation and the semiconducting shield. Treeing starts at voids and impurities at this boundary. "Supersmooth" shield formulations have been developed to reduce vented trees (Burns, 1990). These mixtures use finer carbon particles to smooth out the interface. Under accelerated aging tests, cables with supersmooth semiconducting shields outperformed cables with standard semiconducting shields.

Modern manufacturing techniques can extrude the semiconducting conductor shield, the insulation, and the semiconducting insulation shield in one pass. Using this *triple* extrusion provides cleaner, smoother contact between layers than extruding each layer in a separate pass.

A note on terminology: a *shield* is the conductive layer surrounding another part of the cable. The conductor shield surrounds the conductor; the insulation shield surrounds the insulation. Used generically, shield refers to the metallic shield (the sheath). Commonly, the metallic shield is called the neutral, the shield, or the sheath. Sometimes, the sheath is used to mean the outer part of the cable, whether conducting or not conducting.

3.2.5 Jacket

Almost all new cables are jacketed, and the most common jacket is an encapsulating jacket (it is extruded between and over the neutral wires). The jacket provides some (but not complete) protection against water entry. It also provides mechanical protection for the neutral. Common LLDPE jackets are 50 to 80 mil thick.

Bare cable, used frequently in the 1970s, had a relatively high failure rate (Dedman and Bowles, 1990). Neutral corrosion was often cited as the main reason for the higher failure rate. At sections with a corroded neutral, the ground return current can heat spots missing neutral strands. Dielectric failure, not neutral corrosion, is still the dominant failure mode (Gurniak, 1996). Without the jacket, water enters easily and accelerates water treeing, which leads to premature dielectric failure.

Several materials are used for jackets. Polyvinyl chloride (PVC) was one of the earliest jacketing materials and is still common. The most common jacket material is made from linear low-density polyethylene (LLDPE). PVC has good jacketing properties, but LLDPE is even better in most regards: mechanical properties, temperature limits, and water entry. Moisture passes through PVC jacketing more than ten times faster than it passes through LLDPE. LLDPE starts to melt at 100°C; PVC is usually more limited, depending on composition. Low-density polyethylene resists abrasion better and also has a lower coefficient of friction, which makes it easier to pull through conduit.

Semiconducting jackets are also available. Semiconducting jackets provide the grounding advantages of unjacketed cable, while also blocking moisture and physically protecting the cable. When direct buried, an exposed neutral provides an excellent grounding conductor. The neutral in contact with the soil helps improve equipment grounding and improves protection against surges. A semiconducting jacket has a resistivity equivalent to most soils (less than 100 Ω-m), so it transfers current to the ground the same as an unjacketed cable. NRECA (1993) recommends not using a semiconducting jacket for two reasons. First, semiconducting jackets let more water pass through than LLDPE jackets. Second, the semiconducting jacket could contribute to corrosion. The carbon in the jacket (which makes the jacket semiconducting) is galvanic to the neutral and other nearby metals; especially with water in the cable, the carbon accelerates neutral corrosion. Other nearby objects in the ground such as ground rods or pipes can also corrode more rapidly from the carbon in the jacket.

3.3 Installations and Configurations

Just as there are many different soil types and underground applications, utilities have many ways to install underground cable. Some common installation methods include [see NRECA RER Project 90-8 (1993) for more details]:

- *Trenching* — This is the most common way to install cables, either direct-buried or cables in conduit. After a trench is dug, cable is installed, backfill is added and tamped, and the surface is restored. A trenching machine with different cutting chains is available for use on different soils. Backhoes also help with trenching.
- *Plowing* — A cable plow blade breaks up and lifts the earth as it feeds a cable into the furrow. Plowing eliminates backfilling and disturbs the surface less than trenching. NRECA reports that plowing is 30 to 50% less expensive than trenching (NRECA RER Project 90-8, 1993). Plowed cables may have lower ampacity because of air pockets between the cable and the loose soil around the cable. Heat cannot transfer as effectively from the cable to the surrounding earth.
- *Boring* — A number of tunneling technologies are available to drill under roads or even over much longer distances with guided, fluid-assisted drill heads.

Utilities also have a number of installation options, each with tradeoffs:

- *Direct buried* — Cables are buried directly in the earth. This is the fastest and least expensive installation option. Its major disadvantage is that cable replacement or repair is difficult.
- *Conduit* — Using conduit allows for quicker replacement or repair. Rigid PVC conduit is the most common conduit material; steel and HDPE and fiberglass are also used. Cables in conduit have less ampacity than direct-buried cables.
- *Direct buried with a spare conduit* — Burying a cable with a spare conduit provides provisions for repair or upgrades. Crews can pull another cable through the spare conduit to increase capacity or, if the cable fails, run a replacement cable through the spare conduit and abandon the failed cable. Normally, when the cable is plowed in, the conduit is coilable polyethylene.
- *Concrete-encased conduit* — Most often used in urban construction, conduit is encased in concrete. Concrete protects the conduit, resisting collapse due to shifting earth. The concrete also helps prevent dig-ins.
- *Preassembled cable in conduit* — Cable with flexible conduit can be purchased on reels, which crews can plow into the ground together.

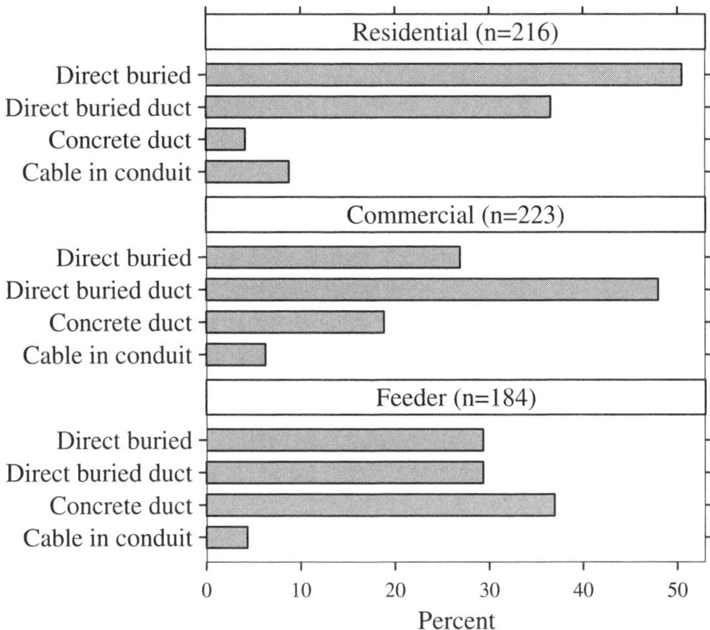

FIGURE 3.4
Surveyed utility cable installation configurations for residential, commercial, and feeder applications. (Data from [Boucher, 1991].)

> The flexible conduit is likely to be more difficult to pull cable through, especially if the conduit is not straight. Flexible conduit is also not as strong as rigid conduit; the conduit can collapse due to rocks or other external forces.

Utilities are split between using direct-buried cable and conduits or ducts for underground residential applications. Conduits are used more for three-phase circuits, for commercial service, and for main feeder applications (see Figure 3.4). Conduit use is rising as shown by a more recent survey in Table 3.7. In a survey of the rural cooperatives with the most underground distribution, Dudas and Rodgers (1999) reported that 80% directly bury cable.

With conduits, customers have less outage time because cables can be replaced or repaired more quickly. In addition, replacement causes much less trouble for customers. Replacement doesn't disturb driveways, streets, or lawns; crews can concentrate their work at padmounted gear, rather than spread out along entire cable runs; and crews are less likely to tie up traffic. Conduit costs more than direct buried cable initially, typically from 25 to 50% more for PVC conduit (but this ranges widely depending on soil conditions and obstacles in or on the ground). Cable in flexible conduit may be slightly less than cable in rigid conduit. While directly buried cable has lower initial costs, lifetime costs can be higher than conduit depending on economic

TABLE 3.7

Surveyed Utility Use of Cable Duct Installations

	Percent of Cable Miles with Each Configuration	
	1998 Installed	**Planned for the Future**
Direct buried	64.6	46.9
Installed in conduit sections	25.5	37.9
Preassembled cable in conduit	7.1	11.7
Direct buried with a spare conduit	1.1	0.5
Continuous lengths of PE tubing	1.7	3.0

Source: Tyner, J. T., "Getting to the Bottom of UG Practices," *Transmission & Distribution World*, vol. 50, no. 7, pp. 44–56, July 1998.

assumptions and assumptions on how long cables will last or if they will need to be upgraded. Some utilities use a combination approach; most cable is direct buried, but ducts are used for road crossings and other obstacles.

The National Electrical Safety Code requires that direct-buried cable have at least 30 in. (0.75 m) of cover (IEEE C2-1997). Typically, trench depths are at least 36 in.

If communication cables are buried with primary power cables, extra rules apply. For direct-buried cable with an insulating jacket, the NESC requires that the neutral must have at least one half of the conductivity of the phase conductor (IEEE C2-1997) (it must be a one-half neutral or a full neutral).

Some urban applications are constrained by small ducts: 3, 3.5, or 4-in. diameters. These ducts were designed to hold three-conductor paper-insulated lead-sheathed cables which have conductors squashed in a sector shape for a more compact arrangement. Insulation cannot be extruded over these shapes, so obtaining an equivalent replacement cable with extruded insulation is difficult. Manufacturers offer thinner cables to meet these applications. For triplex cable, the equivalent outside diameter is 2.155 times the diameter of an individual cable. So, to fit in a 3-in. duct, an individual cable must be less than 1.16 in. in diameter to leave a 1/2-in. space (see Table 3.8 for other duct sizes). Some cable offered as "thin-wall" cable has slightly reduced insulation. For 15-kV cable, the smallest insulation thicknesses range between 150 and 165 mil as compared to the standard 175 mil (EPRI 1001734, 2002) (the ICEA allows 100% 15-kV cable insulation to range from 165 to 205

TABLE 3.8

Maximum Cable Diameters for Small Conduits Using PILC or Triplexed Cables that Leave 1/2-in. Pulling Room

Duct Size, in.	Largest Three-Conductor 15-kV PILC	Maximum Cable Diameter for Triplex Construction, in.	Largest Standard Construction Triplexed 15-kV Copper Cable
3.0	350 kcmil	1.16	3/0
3.5	750 kcmil	1.39	350 kcmil
4.0	1000 kcmil	1.62	500 kcmil

mil (ANSI/ICEA S-97-682-2000, 2000)). One manufacturer has proposed reduced insulation thicknesses based on the fact that larger conductors have lower peak voltage stress on the insulation than smaller conductors (Cinquemani et al., 1997), for example, 110-mil insulation at 15 kV for 4/0 through 750 kcmil. The maximum electric field (EPRI 1001734, 2002) is given by

$$E_{max} = \frac{2V}{d \ln(D / d)}$$

where

E_{max} = maximum electric field, V/mil (or other distance unit)
V = operating or rated voltage to neutral, V
d = inside diameter of the insulation, mil (or other distance unit)
D = outside diameter of the insulation in the same units as d

So, a 750-kcmil cable with 140-mil insulation has about the same maximum voltage stress as a 1/0 cable with 175-mil insulation at the same voltage. Nevertheless, most manufacturers are reluctant to trim the primary insulation too much, fearing premature failure due to water treeing. In addition to slightly reduced insulation, thin-wall cables are normally compressed copper and have thinner jackets and thinner semiconducting shields around the conductor and insulation. EPRI has also investigated other polymers for use in thin-wall cables (EPRI TR-111888, 2000). Their investigations found promising results with novel polymer blends that could achieve insulation strengths that are 30 to 40% higher than XLPE. These tests suggest promise, but more work must be done to improve the extrusion of these materials.

3.4 Impedances

3.4.1 Resistance

Cable conductor resistance is an important part of impedance that is used for fault studies and load flow studies. Resistance also greatly impacts a cable's ampacity. The major variable that affects resistance is the conductor's temperature; resistance rises with temperature. Magnetic fields from alternating currents also reduce a conductor's resistance relative to its dc resistance. At power frequencies, skin effect is only apparent for large conductors and proximity effect only occurs for conductors in very tight configurations. The starting point for resistance calculations is the dc resistance. From there, we can adjust for temperature and for frequency effects. Table 3.9 shows the dc resistances of several common conductors used for cables.

Resistance increases with temperature as

TABLE 3.9

dc Resistance at 25°C in **Ω**/1000 ft

	Aluminum		Uncoated Copper		Coated Copper	
Size	Solid	Class-B Stranded	Solid	Class-B Stranded	Solid	Class-B Stranded
24			26.2		27.3	
22			16.5		17.2	
20			10.3	10.5	10.7	11.2
19			8.21		8.53	
18			6.51	6.64	6.77	7.05
16			4.1	4.18	4.26	4.44
14	4.22		2.57	2.62	2.68	2.73
12	2.66	2.7	1.62	1.65	1.68	1.72
10	1.67	1.7	1.02	1.04	1.06	1.08
9	1.32	1.35	0.808	0.824	0.831	0.857
8	1.05	1.07	0.641	0.654	0.659	0.679
7	0.833	0.85	0.508	0.518	0.523	0.539
6	0.661	0.674	0.403	0.41	0.415	0.427
5	0.524	0.535	0.319	0.326	0.329	0.339
4	0.415	0.424	0.253	0.259	0.261	0.269
3	0.33	0.336	0.201	0.205	0.207	0.213
2	0.261	0.267	0.159	0.162	0.164	0.169
1	0.207	0.211	0.126	0.129	0.13	0.134
1/0	0.164	0.168	0.1	0.102	0.103	0.106
2/0	0.13	0.133	0.0795	0.0811	0.0814	0.0843
3/0	0.103	0.105	0.063	0.0642	0.0645	0.0668
4/0	0.082	0.0836	0.05	0.0509	0.0512	0.0525
250		0.0708		0.0431		0.0449
300		0.059		0.036		0.0374
350		0.0505		0.0308		0.032
400		0.0442		0.027		0.0278
500		0.0354		0.0216		0.0222
600		0.0295		0.018		0.0187
750		0.0236		0.0144		0.0148
1000		0.0177		0.0108		0.0111
1250		0.0142		0.00863		0.00888
1500		0.0118		0.00719		0.0074
1750		0.0101		0.00616		0.00634
2000		0.00885		0.00539		0.00555
2500		0.00715		0.00436		0.00448

Note: × 5.28 for Ω/mi or × 3.28 for Ω/km.

$$R_{t2} = R_{t1}\frac{M + t_2}{M + t_1}$$

where

R_{t2} = resistance at temperature t_2 given, °C
R_{t1} = resistance at temperature t_1 given, °C
M = a temperature coefficient for the given material
= 228.1 for aluminum
= 234.5 for soft-drawn copper

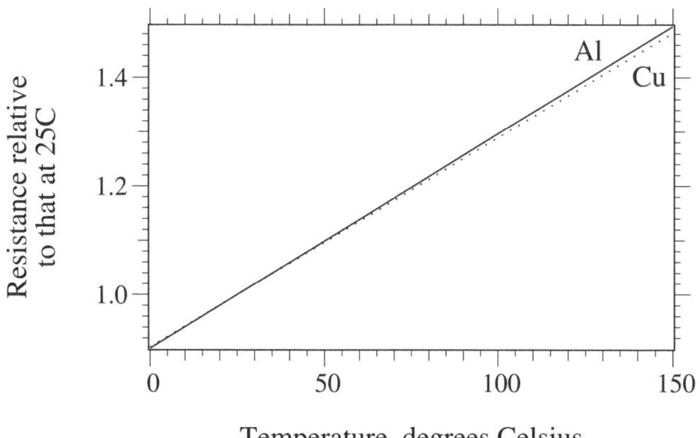

FIGURE 3.5
Resistance change with temperature.

Both copper and aluminum change resistivity at about the same rate as shown in Figure 3.5.

The ac resistance of a conductor is the dc resistance increased by a skin effect factor and a proximity effect factor

$$R = R_{dc}(1 + Y_{cs} + Y_{cp})$$

where

R_{dc} = dc resistance at the desired operating temperature, Ω/1000 ft
Y_{cs} = skin-effect factor
Y_{cp} = proximity effect factor

The skin-effect factor is a complex function involving Bessel function solutions. The following polynomial approximates the skin-effect factor (Anders, 1998):

$$Y_{cs} = \frac{x_s^4}{192 + 0.8x_s^4} \qquad \text{for } x_s \le 2.8$$

$$Y_{cs} = -0.136 - 0.0177x_s + 0.0563x_s^2 \qquad \text{for } 2.8 < x_s \le 3.8$$

$$Y_{cs} = \frac{x_s}{2\sqrt{2}} - \frac{11}{15} \qquad \text{for } 3.8 < x_s$$

where

$$x_s = 0.02768\sqrt{\frac{f \cdot k_s}{R_{dc}}}$$

f = frequency, Hz

k_s = skin effect constant = 1 for typical conductors in extruded cables, may be less than one for paper cables that are dried and impregnated and especially those with round segmental conductors [see Neher and McGrath (1957) or IEC (1982)].

R_{dc} = dc resistance at the desired operating temperature, Ω/1000 ft

For virtually all applications at power frequency, x_s is < 2.8.

With a conductor in close proximity to another current-carrying conductor, the magnetic fields from the adjacent conductor force current to flow in the portions of the conductor most distant from the adjacent conductor (with both conductors carrying current in the same direction). This magnetic field effect increases the effective ac resistance. The proximity effect factor is approximately (Anders, 1998; IEC 287, 1982):

$$Y_{cp} = ay^2\left(0.312y^2 + \frac{1.18}{a+0.27}\right)$$

where

$$a = \frac{x_p^4}{192+0.8x_p^4}, \quad y = \frac{d_c}{s}$$

$$x_p = 0.02768\sqrt{\frac{f \cdot k_p}{R_{dc}}}$$

d_c = conductor diameter

s = distance between conductor centers

k_p = proximity effect constant = 1 for typical conductors in extruded cables; may be < 1 for paper cables that are dried and impregnated and especially those with round segmental conductors [see Neher and McGrath (1957) or IEC (1982)].

At power frequencies, we can ignore proximity effect if the spacing exceeds ten times the conductor diameter (the effect is less than 1%).

Table 3.10 and Table 3.11 show characteristics of common cable conductors.

3.4.2 Impedance Formulas

Smith and Barger (1972) showed that we can treat a multi-wire concentric neutral as a uniform sheath; further work by Lewis and Allen (1978) and by Lewis, Allen, and Wang (1978) simplified the calculation of the representation of the concentric neutral. Following the procedure and nomenclature of Smith (1980) and Lewis and Allen (1978), we can find a cable's sequence impedances from the self and mutual impedances of the cable phase and neutral conductors as

TABLE 3.10

Characteristics of Aluminum Cable Conductors

Conductor	Stranding	GMR, in.	ac/dc Resistance Ratio	Resistances, Ω/1000 ft		
				dc at 25°C	ac at 25°C	ac at 90°C
2	7	0.105	1	0.2660	0.2660	0.3328
1	19	0.124	1	0.2110	0.2110	0.2640
1/0	19	0.139	1	0.1680	0.1680	0.2102
2/0	19	0.156	1	0.1330	0.1330	0.1664
3/0	19	0.175	1	0.1050	0.1050	0.1314
4/0	19	0.197	1	0.0836	0.0836	0.1046
250	37	0.216	1.01	0.0707	0.0714	0.0893
350	37	0.256	1.01	0.0505	0.0510	0.0638
500	37	0.305	1.02	0.0354	0.0361	0.0452
750	61	0.377	1.05	0.0236	0.0248	0.0310
1000	61	0.435	1.09	0.0177	0.0193	0.0241

TABLE 3.11

Characteristics of Copper Cable Conductors

Conductor	Stranding	GMR, in.	ac/dc Resistance Ratio	Resistances, Ω/1000 ft		
				dc at 25°C	ac at 25°C	ac at 90°C
2	7	0.105	1	0.1620	0.1620	0.2027
1	19	0.124	1	0.1290	0.1290	0.1614
1/0	19	0.139	1	0.1020	0.1020	0.1276
2/0	19	0.156	1.01	0.0810	0.0818	0.1023
3/0	19	0.175	1.01	0.0642	0.0648	0.0811
4/0	19	0.197	1.01	0.0510	0.0515	0.0644
250	37	0.216	1.01	0.0431	0.0435	0.0545
350	37	0.256	1.03	0.0308	0.0317	0.0397
500	37	0.305	1.06	0.0216	0.0229	0.0286
750	61	0.377	1.13	0.0144	0.0163	0.0204
1000	61	0.435	1.22	0.0108	0.0132	0.0165

$$Z_{11} = Z_{aa} - Z_{ab} - \frac{(Z_{ax} - Z_{ab})^2}{Z_{xx} - Z_{ab}}$$

$$Z_{00} = Z_{aa} + 2Z_{ab} - \frac{(Z_{ax} + 2Z_{ab})^2}{Z_{xx} + 2Z_{ab}}$$

The self and mutual impedances in the sequence equations are found with

$$Z_{aa} = R_{\phi} + R_e + jk_1 \log_{10} \frac{D_e}{GMR_{\phi}}$$

$$Z_{ab} = R_e + jk_1 \log_{10} \frac{D_e}{GMD_\phi}$$

$$Z_{xx} = R_N + R_e + jk_1 \log_{10} \frac{D_e}{GMR_N}$$

$$Z_{ax} = R_e + jk_1 \log_{10} \frac{D_e}{DN2}$$

where the self and mutual impedances with earth return are:

Z_{aa} = self impedance of each phase conductor
Z_{ab} = the mutual impedance between two conductors (between two phases, between two neutral, or between a phase and a neutral)
Z_{ax} = the mutual impedance between a phase conductor and its concentric neutral (or sheath)
Z_{xx} = self impedance of each concentric neutral (or shield)

and

R_ϕ = resistance of the phase conductor, Ω/distance
R_N = resistance of the neutral (or shield), Ω/distance
k_1 = $0.2794f/60$ for outputs in Ω/mi
= $0.0529f/60$ for outputs in Ω/1000 ft
f = frequency, Hz
GMR_ϕ = geometric mean radius of the phase conductor, in. (see Table 3.12)
GMD_ϕ = geometric mean distance between the phase conductors, in.
= $\sqrt[3]{d_{AB}d_{BC}d_{CA}}$
= 1.26 d_{AB} for a three-phase line with flat configuration, either horizontal or vertical, when $d_{AB} = d_{BC} = 0.5d_{CA}$
= the cable's outside diameter for triplex cables
= 1.15 times the cable's outside diameter for cables cradled in a duct
d_{ij} = distance between the center of conductor i and the center of conductor j, in. (see Figure 3.6)
R_e = resistance of the earth return path
= $0.0954(f/60)$Ω/mi
= $0.01807(f/60)$Ω/1000 ft
$D_e = 25920\sqrt{\rho/f}$ = equivalent depth of the earth return current, in.
ρ = earth resistivity, Ω-m
GMR_N = geometric mean radius of the sheath or neutral. For single-conductor cables with tape or lead sheaths, set GMR_N equal to the average radius of the sheath. For cables with a multi-wire concentric neutral, use $GMR_N = \sqrt[n]{0.7788nDN2^{(n-1)}r_n}$ where n is the number of neutrals and r_n is the radius of each neutral, in.

TABLE 3.12

Geometric Mean Radius (GMR) of Class B Stranded Copper and Aluminum Conductors

Size	Stranding	GMR, in. Round	Compressed	Compact
8	7	0.053		
6	7	0.067		
4	7	0.084		
2	7	0.106	0.105	
1	19	0.126	0.124	0.117
1/0	19	0.141	0.139	0.131
2/0	19	0.159	0.156	0.146
3/0	19	0.178	0.175	0.165
4/0	19	0.200	0.197	0.185
250	37	0.221	0.216	0.203
350	37	0.261	0.256	0.240
500	37	0.312	0.305	0.287
750	61	0.383	0.377	0.353
1000	61	0.442	0.435	0.413

Source: Southwire Company, *Power Cable Manual*, 2nd ed., 1997.

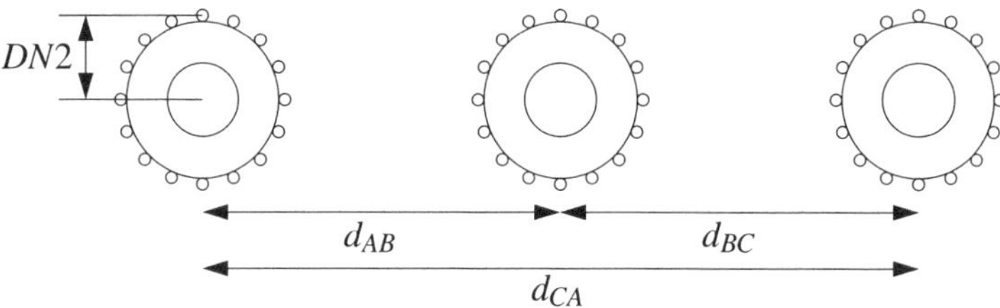

FIGURE 3.6
Cable dimensions for calculating impedances.

*DN*2 = effective radius of the neutral = the distance from the center of the phase conductor to the center of a neutral strand, in.

Smith (1980) reported that assuming equal GMR_N and *DN*2 for cables from 1/0 to 1000 kcmil with one-third neutrals is accurate to 1%.

For single-phase circuits, the zero and positive-sequence impedances are the same:

$$Z_{11} = Z_{00} = Z_{aa} - \frac{Z_{ax}^2}{Z_{xx}}$$

This is the loop impedance, the impedance to current flow through the phase conductor that returns in the neutral and earth. The impedances of two-phase circuits are more difficult to calculate (see Smith, 1980).

The sheath resistances depend on whether it is a concentric neutral, a tape shield, or some other configuration. For a concentric neutral, the resistance is approximately (ignoring the lay of the neutral):

$$R_{neutral} = \frac{R_{strand}}{n}$$

where

R_{strand} = resistance of one strand, in Ω/unit distance

n = number of strands

A tape shield's resistance (Southwire Company, 1997) is

$$R_{shield} = \frac{\rho_c}{A_s}$$

where

ρ_c = resistivity of the tape shield, Ω-cmil/ft = 10.575 for uncoated copper at 25°C

A_s = effective area of the shield in circular mil

$$A_s = 4b \cdot d_m \cdot \sqrt{\frac{50}{100-L}}$$

b = thickness of the tape, mil

d_m = mean diameter outside of the metallic shield, mil

L = lap of the tape shield in percent (normally 10 to 25%)

Normally, we can use dc resistance as the ac resistance for tape shields or concentric neutrals. The skin effect is very small because the shield conductors are thin (skin effect just impacts larger conductors). We should adjust the sheath resistance for temperature; for copper conductors, the adjustment is:

$$R_{t2} = R_{t1}\frac{234.5+t_2}{234.5+t_1}$$

where

R_{t2} = resistance at temperature t_2 given in °C

R_{t1} = resistance at temperature t_1 given in °C

These calculations are simplifications. More advanced models, normally requiring a computer, can accurately find each element in the full impedance matrix. For most load-flow calculations, this accuracy is not needed, though access to user-friendly computer models allows quicker results than calculating the equations shown here. For evaluating switching transients and some ampacity problems or configurations with several cables, we sometimes need more sophisticated models [see Amateni (1980) or Dommel (1986) for analytical details].

In a cable, the neutral tightly couples with the phase. Phase current induces neutral voltages that force circulating current in the neutrals. With balanced, positive-sequence current in the three phases and with symmetrical conductors, the neutral current (Lewis and Allen, 1978; Smith and Barger, 1972) is

$$I_{X1} = -\frac{Z_{ax} - Z_{ab}}{Z_{xx} - Z_{ab}} I_a$$

which is

$$I_{X1} = -\frac{j0.0529 \log_{10} \dfrac{d_{ab}}{DN2}}{R_N + j0.0529 \log_{10} \dfrac{d_{ab}}{GMR_N}} I_a$$

Since $DN2$ and GMR_N are almost equal, if R_N is near zero, the neutral (or shield) current (I_{X1}) almost equals the phase current (I_a). Higher neutral resistances actually reduce positive-sequence resistances.

Significant effects on positive and zero-sequence impedances include:

- *Cable separation* — Larger separations increase Z_1; spacing does not affect Z_0. Triplex cables have the lowest positive-sequence impedance.
- *Conductor size* — Larger conductors have much less resistance; reactance drops somewhat with increasing size.
- *Neutral/shield resistance* — Increasing the neutral resistance increases the reactive portion of the positive and zero-sequence impedances. Beyond a certain point, increasing neutral resistances decreases the resistive portion of Z_1 and Z_0.
- *Other cables or ground wires* — Adding another grounded wire nearby has similar impacts to lowering sheath resistances. Zero-sequence resistance and reactance usually drop. Positive-sequence reactance is likely to decrease, but positive-sequence resistance may increase.

Figure 3.7 and Figure 3.8 show the impact of the most significant variables on impedances for three-phase and single-phase circuits. None of the following significantly impacts either the positive or zero-sequence impedances: insulation thickness, insulation type, depth of burial, and earth resistivity.

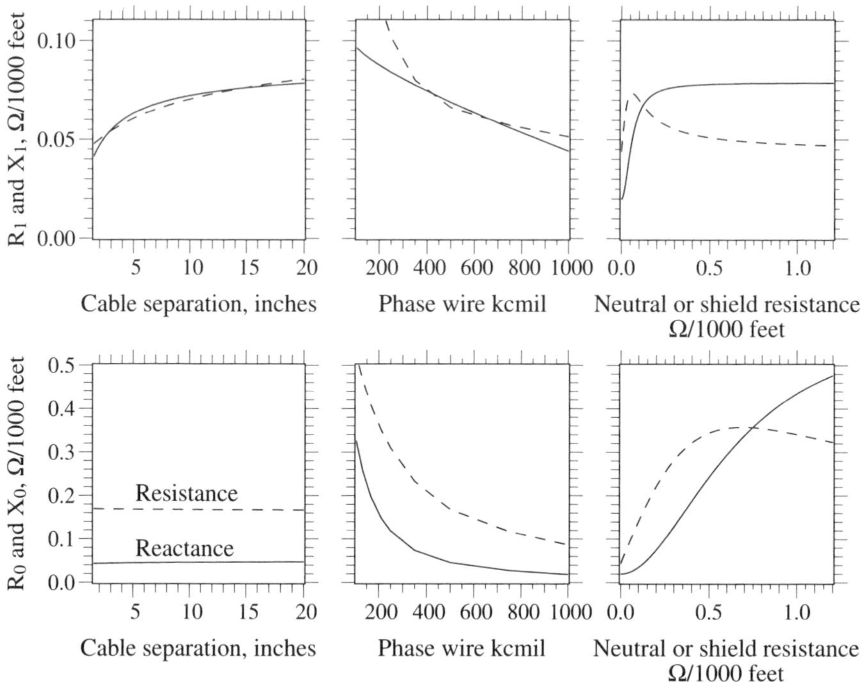

FIGURE 3.7
Effect of various parameters on the positive-sequence (top row) and zero-sequence impedances (bottom row) with a base case having 500-kcmil aluminum cables with 1/3 neutrals, 220-mil insulation, a horizontal configuration with 7.5 in. between cables, and ρ = 100 Ω-m.

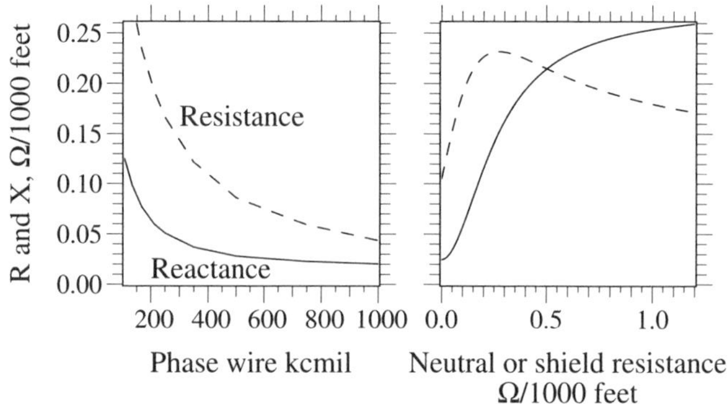

FIGURE 3.8
Resistance and reactance of a single-phase cable ($R = R_0 = R_1$ and $X = X_0 = X_1$) as the size of the cable and neutral varies with a base case having a 4/0 aluminum cable with a full neutral, 220-mil insulation, and ρ = 100 Ω-m.

TABLE 3.13
Loop Impedances of Single-Phase Concentric-Neutral Aluminum Cables

Conductor	Full Neutral			1/3 Neutral		
Size	Neutral	R	X	Neutral	R	X
2	10#14	0.4608	0.1857			
1	13#14	0.3932	0.1517			
1/0	16#14	0.3342	0.1259	6#14	0.3154	0.2295
2/0	13#12	0.2793	0.0974	7#14	0.2784	0.2148
3/0	16#12	0.2342	0.0779	9#14	0.2537	0.1884
4/0	13#10	0.1931	0.0613	11#14	0.2305	0.1645
250	16#10	0.1638	0.0493	13#14	0.2143	0.1444
350	20#10	0.1245	0.0387	18#14	0.1818	0.1092
500				16#12	0.1447	0.0726
750				15#10	0.1067	0.0462
1000				20#10	0.0831	0.0343

Notes: Impedances, Ω/1000 ft (× 5.28 for Ω/mi or × 3.28 for Ω/km). Conductor temperature = 90°C, neutral temperature = 80°C, 15-kV class, 220-mil insulation, ρ = 100 Ω-m. For the neutral, 10#14 means 10 strands of 14-gage wire.

3.4.3 Impedance Tables

This section contains tables of several common cable configurations found on distribution circuits. All values are for a multigrounded circuit. Many other cable configurations are possible, with widely varying impedances. For PILC cables, refer to impedances in the Westinghouse (1950) T&D book. For additional three-phase power cable configurations, refer to the IEEE Red Book (IEEE Std. 141-1993), St. Pierre (2001), or Southwire Company (1997).

3.4.4 Capacitance

Cables have significant capacitance, much more than overhead lines. A single-conductor cable has a capacitance given by:

$$C = \frac{0.00736\varepsilon}{\log_{10}\left(\frac{D}{d}\right)}$$

where

- C = capacitance, μF/1000 ft
- ε = dielectric constant (2.3 for XLPE, 3 for EPR, see Table 3.4 for others)
- d = inside diameter of the insulation, mil (or other distance unit)
- D = outside diameter of the insulation in the same units as d

TABLE 3.14

Impedances of Three-Phase Circuits Made of Three Single-Conductor Concentric-Neutral Aluminum Cables

Conductor Size	Neutral Size	R_1	X_1	R_0	X_0	R_S	X_S
Full Neutral							
2	10#14	0.3478	0.1005	0.5899	0.1642	0.4285	0.1217
1	13#14	0.2820	0.0950	0.4814	0.1166	0.3484	0.1022
1/0	16#14	0.2297	0.0906	0.3956	0.0895	0.2850	0.0902
2/0	13#12	0.1891	0.0848	0.3158	0.0660	0.2314	0.0785
3/0	16#12	0.1578	0.0789	0.2573	0.0523	0.1910	0.0701
4/0	13#10	0.1331	0.0720	0.2066	0.0423	0.1576	0.0621
250	16#10	0.1186	0.0651	0.1716	0.0356	0.1363	0.0553
350	20#10	0.0930	0.0560	0.1287	0.0294	0.1049	0.0471
1/3 Neutral							
1/0	6#14	0.2180	0.0959	0.5193	0.2854	0.3185	0.1591
2/0	7#14	0.1751	0.0930	0.4638	0.2415	0.2713	0.1425
3/0	9#14	0.1432	0.0896	0.4012	0.1787	0.2292	0.1193
4/0	11#14	0.1180	0.0861	0.3457	0.1375	0.1939	0.1032
250	13#14	0.1034	0.0833	0.3045	0.1103	0.1704	0.0923
350	18#14	0.0805	0.0774	0.2353	0.0740	0.1321	0.0762
500	16#12	0.0656	0.0693	0.1689	0.0468	0.1000	0.0618
750	15#10	0.0547	0.0584	0.1160	0.0312	0.0752	0.0494
1000	20#10	0.0478	0.0502	0.0876	0.0248	0.0611	0.0417

Notes: Impedances, Ω/1000 ft (× 5.28 for Ω/mi or × 3.28 for Ω/km). Resistances for a conductor temperature = 90°C and a neutral temperature = 80°C, 220-mil insulation (15 kV), ρ = 100 Ω-m. Flat spacing with a 7.5-in. separation between cables. For the neutral, 10#14 means 10 strands of 14-gage wire.

The vars provided by cable are

$$Q_{var} = 2\pi \cdot f \cdot C \cdot V_{LG,kV}^2$$

where

Q_{var} = var/1000 ft/phase
f = frequency, Hz
C = capacitance, μF/1000 ft
$V_{LG,kV}$ = line-to-ground voltage, kV

Table 3.17 shows capacitance values and reactive power produced by cables for typical cables. The table results are for XLPE cable with a dielectric constant (ε) of 2.3. For other insulation, both the capacitance and the reactive power scale linearly. For example, for EPR with ε = 3, multiply the values in Table 3.17 by 1.3 (3/2.3 = 1.3).

TABLE 3.15

Impedances of Single-Conductor Aluminum Power Cables with Copper Tape Shields

Conductor Size	R_1	X_1	R_0	X_0	R_S	X_S
Flat spacing with a 7.5-in. separation						
2	0.3399	0.1029	0.6484	0.4088	0.4427	0.2049
1	0.2710	0.0990	0.5808	0.3931	0.3743	0.1971
1/0	0.2161	0.0964	0.5268	0.3790	0.3196	0.1906
2/0	0.1721	0.0937	0.4833	0.3653	0.2759	0.1842
3/0	0.1382	0.0911	0.4494	0.3493	0.2419	0.1771
4/0	0.1113	0.0883	0.4217	0.3314	0.2148	0.1693
250	0.0955	0.0861	0.4037	0.3103	0.1982	0.1609
350	0.0696	0.0822	0.3734	0.2827	0.1709	0.1490
500	0.0508	0.0781	0.3483	0.2557	0.1499	0.1373
750	0.0369	0.0732	0.3220	0.2185	0.1319	0.1216
1000	0.0290	0.0698	0.3018	0.1915	0.1200	0.1104
Triplex						
2	0.3345	0.0531	0.7027	0.4244	0.4573	0.1769
1	0.2655	0.0501	0.6330	0.4060	0.3880	0.1687
1/0	0.2105	0.0483	0.5767	0.3893	0.3326	0.1620
2/0	0.1666	0.0465	0.5310	0.3734	0.2880	0.1554
3/0	0.1326	0.0448	0.4944	0.3550	0.2532	0.1482
4/0	0.1056	0.0432	0.4636	0.3346	0.2249	0.1403
250	0.0896	0.0424	0.4418	0.3109	0.2070	0.1319
350	0.0637	0.0403	0.4067	0.2807	0.1780	0.1204
500	0.0447	0.0381	0.3769	0.2518	0.1554	0.1093
750	0.0308	0.0359	0.3443	0.2129	0.1353	0.0949
1000	0.0228	0.0348	0.3197	0.1853	0.1218	0.0850

Note: Impedances, Ω/1000 ft (× 5.28 for Ω/mi or × 3.28 for Ω/km). Resistances for a conductor temperature = 90°C and a shield temperature = 50°C, 220-mil insulation (15 kV), ρ = 100 Ω-m, 5-mil copper tape shield with a lap of 20%.

3.5 Ampacity

A cable's ampacity is the maximum continuous current rating of the cable. We should realize that while we may derive one number, say 480 A, for ampacity during normal operations for a given conductor, there is nothing magic about 480 A. The cable will not burst into flames at 481 A; the 480 A is simply a design number. We don't want to exceed that current during normal operations.

The insulation temperature is normally the limiting factor. By operating below the ampacity of a given cable, we keep the cable insulation below its

TABLE 3.16

Impedances of Single-Conductor Copper Power Cables

Conductor Size	R_1	X_1	R_0	X_0	R_S	X_S
Flat spacing with a 7.5-in. separation						
2	0.2083	0.1029	0.5108	0.4401	0.3092	0.2153
1	0.1671	0.0991	0.4718	0.4267	0.2687	0.2083
1/0	0.1334	0.0965	0.4405	0.4115	0.2358	0.2015
2/0	0.1082	0.0938	0.4171	0.3967	0.2112	0.1948
3/0	0.0871	0.0911	0.3975	0.3794	0.1906	0.1872
4/0	0.0705	0.0884	0.3816	0.3626	0.1742	0.1798
250	0.0607	0.0862	0.3719	0.3471	0.1644	0.1732
350	0.0461	0.0823	0.3558	0.3181	0.1493	0.1609
500	0.0352	0.0782	0.3411	0.2891	0.1372	0.1485
750	0.0272	0.0732	0.3241	0.2490	0.1261	0.1318
1000	0.0234	0.0699	0.3104	0.2196	0.1191	0.1198
Triplex						
2	0.2032	0.0508	0.5707	0.4642	0.3257	0.1886
1	0.1619	0.0477	0.5301	0.4480	0.2846	0.1811
1/0	0.1281	0.0460	0.4966	0.4295	0.2509	0.1738
2/0	0.1028	0.0442	0.4709	0.4116	0.2255	0.1667
3/0	0.0816	0.0426	0.4485	0.3910	0.2039	0.1587
4/0	0.0649	0.0409	0.4299	0.3713	0.1866	0.1510
250	0.0551	0.0398	0.4175	0.3532	0.1759	0.1442
350	0.0403	0.0377	0.3962	0.3202	0.1589	0.1319
500	0.0292	0.0355	0.3765	0.2882	0.1450	0.1197
750	0.0211	0.0333	0.3524	0.2450	0.1315	0.1039
1000	0.0173	0.0322	0.3336	0.2142	0.1227	0.0929

Note: Impedances, Ω/1000 ft (× 5.28 for Ω/mi or × 3.28 for Ω/km). Resistances for a conductor temperature = 90°C and a shield temperature = 50°C, 220-mil insulation (15 kV), ρ = 100 Ω-m, 5-mil copper tape shield with a lap of 20%.

TABLE 3.17

Cable Capacitance for Common Cable Sizes and Voltages

	Capacitance, μF/1000 ft				Reactive power, kvar/1000 ft			
					12.5 kV	12.5 kV	25 kV	34.5 kV
Size	175 mil	220 mil	260 mil	345 mil	175 mil	220 mil	260 mil	345 mil
2	0.0516	0.0441	0.0396	0.0333	1.01	0.862	3.09	4.98
1	0.0562	0.0479	0.0428	0.0358	1.1	0.936	3.35	5.35
1/0	0.0609	0.0516	0.046	0.0383	1.19	1.01	3.6	5.72
2/0	0.0655	0.0553	0.0492	0.0407	1.28	1.08	3.84	6.09
3/0	0.0712	0.0599	0.0531	0.0437	1.39	1.17	4.15	6.54
4/0	0.078	0.0654	0.0578	0.0473	1.52	1.28	4.52	7.08
250	0.0871	0.0727	0.064	0.0521	1.7	1.42	5.00	7.79
350	0.0995	0.0826	0.0725	0.0586	1.94	1.61	5.67	8.76
500	0.113	0.0934	0.0817	0.0656	2.21	1.83	6.38	9.81
750	0.135	0.111	0.0969	0.0772	2.65	2.18	7.57	11.5
1000	0.156	0.127	0.111	0.0875	3.04	2.49	8.64	13.1

Note: For XLPE cable with ε = 2.3.

recommended maximum temperature. Cross-linked polyethylene cables are rated for a maximum operating temperature of 90°C during normal operations. Operating cables above their ampacity increases the likelihood of premature failures: water trees may grow faster, thermal runaway-failures are more likely, and insulation strength may decrease. In addition to absolute temperature, thermal cycling also ages cable more quickly.

Ampacity most often limits the loading on a cable; rarely, voltage drop or flicker limits loadings. Relative to overhead lines, cables of a given size have lower impedance and lower ampacities. So cable circuits are much less likely than overhead circuits to be voltage-drop limited. Only very long cable runs on circuits with low primary voltages are voltage-drop limited. Ampacity is not the only consideration for cable selection; losses and stocking considerations should also factor into cable selection. Choosing the smallest cable that meets ampacity requirements has the lowest initial cost, but since the cable is running hotter, the cost over its life may not be optimal because of the losses. Also allow for load growth when selecting cables.

Ampacity calculations follow simple principles: the temperature at the conductor is a function of the heat generated in a cable (I^2R) and the amount of heat conducted away from the cable. We can model the thermal performance with a thermal circuit analogous to an electric circuit: heat is analogous to current; temperature to voltage; and thermal resistance to electrical resistance. Heat flow through a thermal resistance raises the temperature between the two sides of the thermal material. Higher resistance soils or insulations trap the heat and cause higher temperatures. Using the thermal equivalent of Ohm's law, the temperature difference is:

$$\Delta T = T_C - T_A = R_{TH} H = R_{TH}(I^2 R)$$

where

T_C = conductor temperature, °C

T_A = ambient earth temperature, °C

R_{TH} = total thermal resistance between the cable conductor and the air, thermal Ω-ft

H = heat generated in the cable, W (= I^2R)

I = electric current in the conductor, A

R = electric resistance of the conductor, Ω/ft

Most ampacity tables and computer calculation routines are based on the classic paper by Neher and McGrath (1957). The original paper is an excellent reference. Ander's book (1998) provides a detailed discussion of cable ampacity calculations, including the Neher–McGrath method along with IEC's method that is very similar (IEC 287, 1982). Hand calculations or spreadsheet calculations of the Neher–McGrath equations are possible, but tiresome; while straightforward in principle, the calculations are very detailed. A review of the Neher–McGrath procedure — the inputs, the tech-

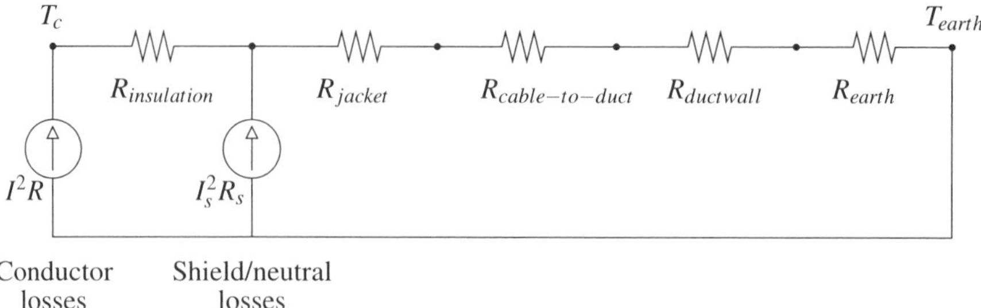

FIGURE 3.9
Thermal circuit model of a cable for ampacity calculations.

niques, the assumptions — provides a better understanding of ampacity calculations to better use computer ampacity calculations.

The Neher–McGrath procedure solves for the current in the equation above. Figure 3.9 shows a simplified model of the thermal circuit. The two main sources of heat within the cable are the I^2R losses in the phase conductor and the I^2R losses in the neutral or shield. The cable also has dielectric losses, but for distribution-class voltages, these are small enough that we can neglect them. The major thermal resistances are the insulation, the jacket, and the earth. If the cable system is in a duct, the air space within the duct and the duct walls adds thermal resistance. These thermal resistances are calculated from the thermal resistivities of the materials involved. For example, the thermal resistance of the insulation, jacket, and duct wall are all calculated with an equation of the following form:

$$R = 0.012\rho \log_{10}(D / d)$$

where

R = thermal resistance of the component, thermal Ω-ft
ρ = thermal resistivity of the component material, °C-cm/W
D = outside diameter of the component
d = inside diameter of the component

Thermal resistivity quantifies the insulating characteristics of a material. A material with ρ = 1°C-cm/W has a temperature rise of 1°C across two sides of a 1-cm^3 cube for a flow of one watt of heat through the cube. As with electrical resistivity, the inverse of thermal resistivity is thermal conductivity. Table 3.18 shows resistivities commonly used for cable system components. The thermal resistance of a material quantifies the radial temperature rise from the center outward. One thermal Ω-ft has a radial temperature rise of 1°C for a heat flow of 1 W per ft of length (length along the conductor). Mixing of metric (SI) units with English units comes about for historical reasons.

TABLE 3.18

Thermal Resistivities of Common Components

Component	Thermal Resistivity, °C-cm/W
XLPE insulation	350
EPR insulation	500
Paper insulation	700
PE jackets	350
PVC jackets	500
Plastic ducts	480
Concrete	85
Thermal fill	60
Soil	90
Water	160
Air	4000

Sources: IEC 287, *Calculation of the Continuous Current Rating of Cables (100% Load Factor)*, 2nd ed., International Electrical Commission (IEC), 1982; Neher, J. H. and McGrath, M. H., "The Calculation of the Temperature Rise and Load Capability of Cable Systems," *AIEE Transactions*, vol. 76, pp. 752–64, October 1957.

TABLE 3.19

Ampacities of Single-Phase Circuits of Full-Neutral Aluminum Conductor Cables

	Direct Buried Load Factor		In Conduit Load Factor	
Size	100%	75%	100%	75%
2	187	201	146	153
1	209	225	162	170
1/0	233	252	180	188
2/0	260	282	200	210
3/0	290	316	223	234
4/0	325	356	249	262
250	359	395	276	291
350	424	469	326	345

Note: 90°C conductor temperature, 25°C ambient earth temperature, ρ = 90°C-cm/W.

The Neher–McGrath calculations also account for multiple cables, cables with cyclic daily load cycles, external heat sources, duct arrangements, and shield resistance and grounding variations.

Often, the easiest way to find ampacities for a given application is with ampacity tables. Table 3.19 and Table 3.20 show ampacities for common distribution configurations. Of the many sources of ampacity tables, the IEEE publishes the most exhaustive set of tables (IEEE Std. 835-1994). The National Electrical Code (NFPA 70, 1999) and manufacturer's publications (Okonite, 1990; Southwire Company, 1997) are also useful. Ampacity tables provide a

TABLE 3.20

Ampacities of Three-Phase Circuits Made of Single-Conductor, One-Third Neutral Aluminum Cables

	Direct Buried Load factor		In Conduit Load factor	
Size	**100%**	**75%**	**100%**	**75%**
Flat spacing (7.5-in. separation)				
1/0	216	244	183	199
2/0	244	277	207	226
3/0	274	312	233	255
4/0	308	352	262	287
250	336	386	285	315
350	392	455	334	370
500	448	525	382	426
750	508	601	435	489
1000	556	664	478	541
Triplex				
1/0	193	224	158	173
2/0	220	255	180	197
3/0	249	290	204	225
4/0	283	330	232	256
250	312	365	257	284
350	375	442	310	345
500	452	535	375	419
750	547	653	457	514
1000	630	756	529	598

Note: 90°C conductor temperature, 25°C ambient earth temperature, ρ = 90°C-cm/W.

good starting point for determining the ampacity of a specific cable application. When using tables, be careful that the assumptions match your particular situation; if not, ampacity results can be much different than expected.

Conductor temperature limits, sheath resistance, thermal resistivity of the soil — these are some of the variables that most impact ampacity (see Figure 3.10). These and other effects are discussed in the next few paragraphs [see also (CEA, 1982; NRECA RER Project 90-8, 1993) for more discussions].

Sheath resistance — On a three-phase circuit, the resistance of the sheath (or shield or neutral) plays an important role in ampacity calculations. Because a cable's phase conductor and sheath couple so tightly, current through the phase induces a large voltage along the sheath. With the cable sheath grounded periodically, circulating current flows to counter the induced voltage. The circulating current is a function of the resistance of the sheath. This circulating current leads to something counterintuitive: sheaths with higher resistance have more ampacity. Higher resistance sheaths reduce the circulating current and reduce the I^2R losses in the sheath. This effect is

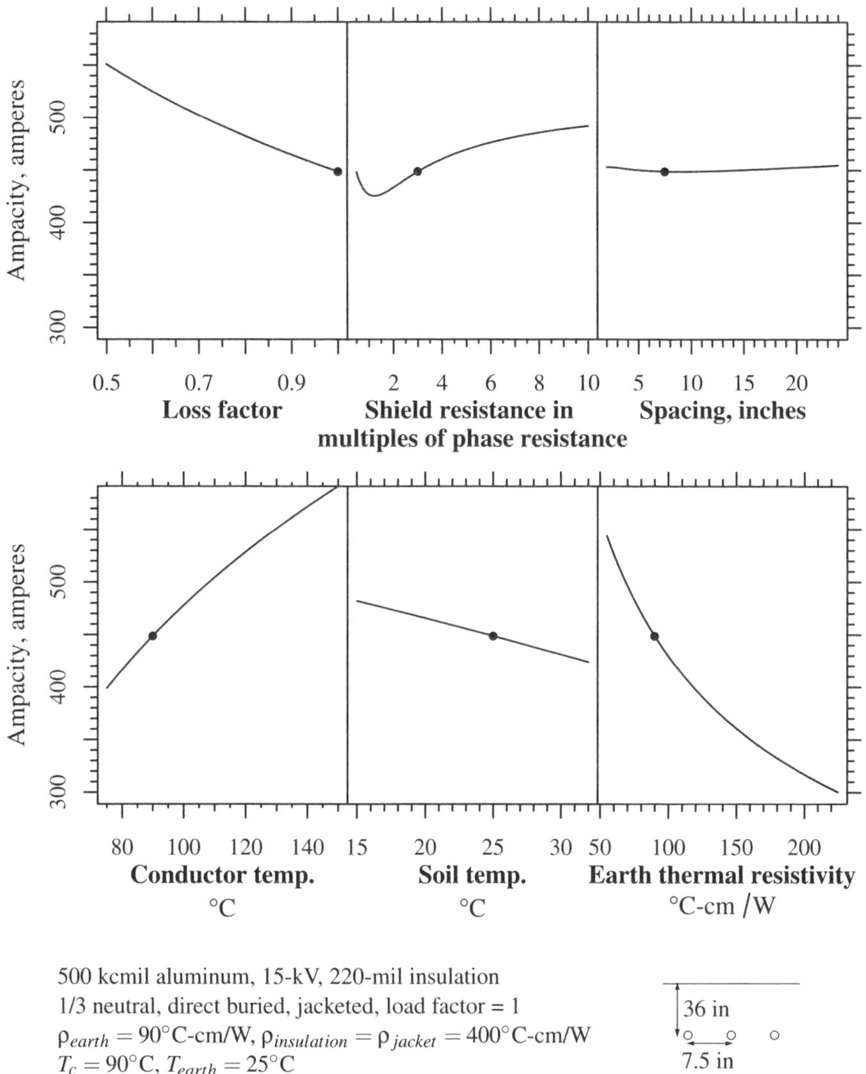

FIGURE 3.10
Effect of variables on ampacity for an example cable.

most pronounced in larger conductors. Many ampacity tables assume that cable sheaths are open circuited, this eliminates the sheath losses and increases the ampacity. The open-circuit sheath values can be approximately corrected to account for circulating currents (Okonite, 1990) by

$$k = \sqrt{\frac{I^2 R}{I_S^2 R_S + I^2 R}}$$

where

k = ampacity multiplier to account for sheath losses, i.e., $I_{\text{grounded sheath}} = k \cdot I_{\text{open sheath}}$

I = phase conductor current, A

I_S = sheath current, A

I^2R = phase conductor losses, W/unit of length

$I_S^2R_S$ = sheath losses, W/unit of length

The sheath losses are a function of the resistance of the sheath and the mutual inductance between the sheath and other conductors. For a triangular configuration like triplex, the shield losses are

$$I_S^2 R_S = I^2 R_S \frac{X_M^2}{R_S^2 + X_M^2}$$

where

$X_M = 2\pi f(0.1404)\log_{10}(2S / d_S)$

and

X_M = mutual inductance of the sheath and another conductor, mΩ/1000 ft

R_S = resistance of the sheath, mΩ/1000 ft

f = frequency, Hz

S = spacing between the phase conductors, in.

d_S = mean diameter of the sheath, in.

For configurations other than triplex, see Southwire Company (1997) or Okonite (1990). Figure 3.11 shows how sheath losses vary with conductor size and with spacing. Spacing has a pronounced effect. Steel ducts can significantly increase heating from circulating currents. In fact, even nearby steel pipes can significantly reduce ampacity.

Spacings — Separating cables separates the heat sources. But at larger spacings, circulating currents are higher. Optimal spacings involve balancing these effects. For smaller cables, separating cables provides the best ampacity. For larger cables (with larger circulating currents), triplex or other tight spacing improves ampacity. For one-third neutral, aluminum cables, NRECA (1993) shows that a flat spacing with 7.5 in between cables has better ampacity than triplex for conductors 500 kcmil and smaller. For copper cables, the threshold is lower: conductors larger than 4/0 have better ampacity with a triplex configuration.

Conductor temperature — If we allow a higher conductor temperature, we can operate a cable at higher current. If we know the ampacity for a given conductor temperature, at a different conductor temperature we can find the ampacity with the following approximation:

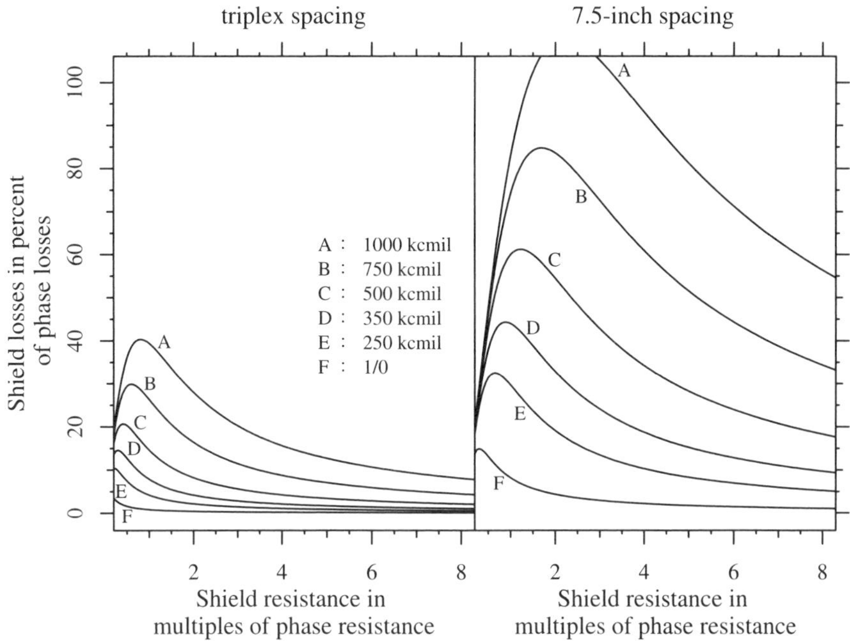

FIGURE 3.11
Shield losses as a function of shield resistance for aluminum cables (triplex configuration).

$$I' = I\sqrt{\frac{T_C' - T_A'}{T_C - T_A}\,\frac{228.1 + T_C}{228.1 + T_C'}} \qquad \text{(Aluminum conductor)}$$

$$I' = I\sqrt{\frac{T_C' - T_A'}{T_C - T_A}\,\frac{234.5 + T_C}{234.5 + T_C'}} \qquad \text{(Copper conductor)}$$

where

I' = ampacity at a conductor temperature of T_C' and an ambient earth temperature T_A'

I = ampacity at a conductor temperature of T_C and an ambient earth temperature T_A (all temperatures are in °C)

We can use these equations to find emergency ampacity ratings of cables. In an emergency, XLPE can be operated to 130°C. Some EPR cables can be operated to 140°C (MV-105 cables). ICEA standards allow emergency overload for 100 hours per year with five such periods over the life of the cable. Polyethylene cables, including HMWPE, have little overload capability. Their maximum recommended emergency temperature is 95°C. Table 3.21 shows common ampacity multipliers; these are valid for both copper and aluminum conductors within the accuracy shown. We can also use the appro-

TABLE 3.21

Common Ampacity Rating Conversions (with T_A = 25°C)

Original Temperature, °C	New Temperature, °C	Ampacity Multiplier
75	95	1.15
90	75	0.90
90	105	1.08
90	130	1.20
105	140	1.14

priate temperature-adjustment equation to adjust for different ambient earth temperatures.

Loss factor — The earth has a high thermal storage capability; it takes considerable time to heat (or cool) the soil surrounding the cable. Close to the cable, the peak heat generated in the cable determines the temperature drop; farther out, the average heat generated in the cable determines the temperature drop. As discussed in Chapter 5, we normally account for losses using the loss factor, which is the average losses divided by the peak losses. Since this number is not normally available, we find the loss factor from the load factor (the load factor is the average load divided by the peak load). Assuming a 100% load factor (continuous current) is most conservative but can lead to a cable that is larger than necessary. We should try to err on the high side when estimating the load factor. A 75% load factor is commonly used.

Conduits — The air space in conduits or ducts significantly reduces ampacity. The air insulation barrier traps more heat in the cable. Direct-buried cables may have 10 to 25% higher ampacities. Although, the less air the better, there is little practical difference in the thermal performance between the sizes of ducts commonly used. Concrete duct banks have roughly the same thermal performance as direct-buried conduits (concrete is more consistent and less prone to moisture fluctuations).

Soil thermal resistivity and temperature — Soils with lower thermal resistivity more readily conduct heat away from cables. Moisture is an important component, moist soil has lower thermal resistivity (see Figure 3.12). Dense soil normally has better conductivity. More so than any other single factor, soil resistivity impacts the conductor's temperature and the cable's ampacity. A resistivity of 90°C-cm/W is often assumed for ampacity calculations. This number is conservative enough for many areas, but if soil resistivities are higher, cable temperatures can be much higher than expected. For common soils, Table 3.22 shows typical ranges of thermal resistivities. At typical installation depths, resistivity varies significantly with season as moisture content changes. Unfortunately in many locations, just when we need ampacity the most — during peak load in the summer — the soil is close to its hottest and driest. Seasonal changes can be significant, but daily

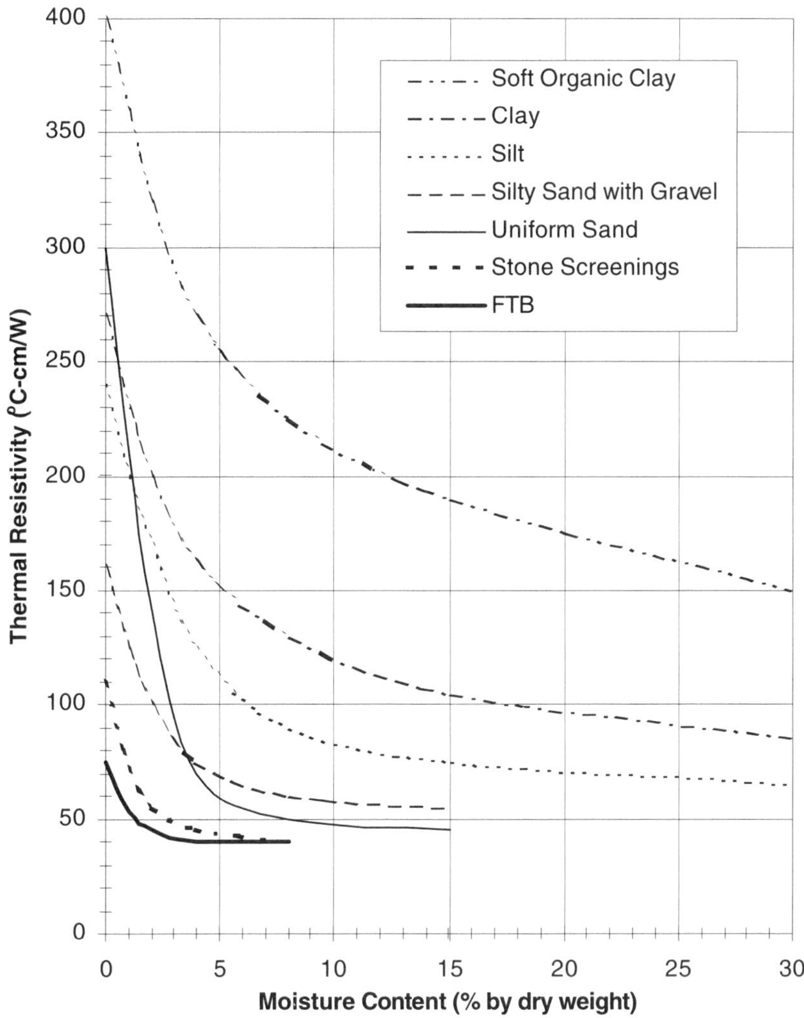

FIGURE 3.12
Effect of moisture on the thermal resistivity of various soils. (Copyright © 1997. Electric Power Research Institute. TR-108919. *Soil Thermal Properties Manual for Underground Power Transmission.* Reprinted with permission.)

changes are not; soil temperature changes lag air temperature changes by 2 to 4 weeks.

The depth of burial can affect ampacity. With a constant resistivity and soil temperature, deeper burial decreases ampacity. But deeper, the soil tends to have lower temperature, more moisture, and soil is more stable seasonally. To go deep enough to take advantage of this is not cost effective though.

For areas with poor soil (high clay content in a dry area, for example), one of several thermal backfills can give good performance, with stable

TABLE 3.22

Typical Thermal Resistivities of Common Soils

USCS	Soil	Dry Density (g/cm³)	Range of Moisture Contents (%) Above Water Table	Saturated Moisture Content (%)	Thermal Resistivity (°C-cm/W) Wet–Dry
GW	well graded gravel	2.1	3–8	10	40–120
GP	poor graded gravel	1.9	2–6	15	45–190
GM	silty gravel	2.0	4–9	12	50–140
GC	clayey gravel	1.9	5–12	15	55–150
SW	well graded sand	1.8	4–12	18	40–130
SP	uniform sand	1.6	2–8	25	45–300
SM	silty sand	1.7	6–16	20	55–170
SC	clayey sand	1.6	8–18	25	60–180
ML	Silt	1.5	8–24	30	65–240
CL	silty clay	1.6	10–22	25	70–210
OL	organic silt	1.2	15–35	45	90–350
MH	micaceous silt	1.3	12–30	40	75–300
CH	clay	1.3	20–35	40	85–270
OH	soft organic clay	0.9	30–70	75	110–400
Pt	silty peat	0.4			150–600+

resistivities below 60°C-cm/W even when moisture content drops below one percent.

Earth interface temperature — Because soil conductivity depends on moisture, the temperature at the interface between the cable or duct and the soil is important. Unfortunately, heat tends to push moisture away. High interface temperatures can dry out the surrounding soil, which further increases the soil's thermal resistivity. Soil drying can lead to a runaway situation; hotter cable temperatures dry the soil more, raising the cable temperature more and so on. Some soils, especially clay, shrink significantly as it dries; the soil can pull away from the cable, leaving an insulating air layer. Thermal runaway can lead to immediate failure. Direct-buried cables are the most susceptible; ducts provide enough of a barrier that temperature is reduced by the time it reaches the soil.

Depending on the soil drying characteristics in an area, we may decide to limit earth interface temperatures. Limiting earth interface temperatures to 50 to 60°C reduces the risk of thermal runaway. But doing this also significantly decreases the ampacity of direct-buried cable to about that of cables in conduit. In fact, using the conduit ampacity values is a good approximation for the limits needed to keep interface temperatures in the 50 to 60°C range.

Current unbalance — Almost every ampacity table (including those in this section) assumes balanced, three-phase currents. On multigrounded distribution systems, this assumption is rarely true. An ampacity of 100 A means a limit of 100 A on each conductor. Unbalance restricts the power a three-

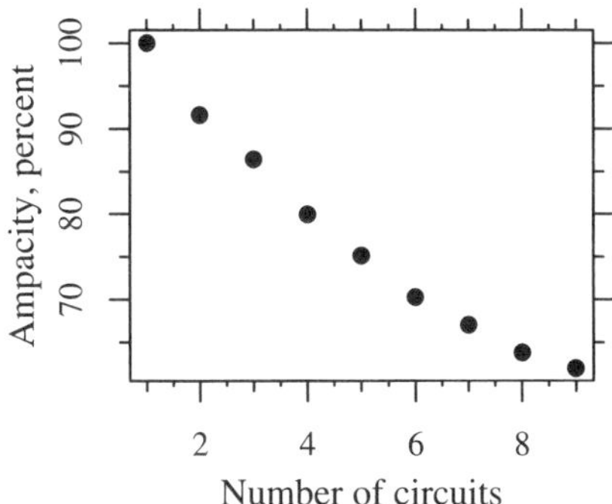

FIGURE 3.13
Ampacity reduction with multiple cable circuits in a duct bank (15 kV, aluminum, 500 kcmil, tape shield power cables, triplex configuration).

phase cable circuit can carry ($I_A = I_B = I_C = 100$ A carries more power than $I_A = 100$ A, $I_B = I_C = 70$ A). In addition, the unbalanced return current may increase the heating in the cable carrying the highest current. It may or it may not; it depends on phase relationships and the phase angle of the unbalanced current. If the unbalances are just right, the unbalanced return current can significantly increase the neutral current on the most heavily loaded phases. Unbalance also depends on the placement of the cables. In a flat configuration, the middle cable is the most limiting because the outer two cables heat the middle cable.

Just as higher sheath resistances reduce circulating currents, higher sheath resistances reduce unbalance currents in the sheath. Higher sheath resistances force more of the unbalanced current to return in the earth. The heat generated in the sheath from unbalance current also decreases with increasing sheath resistance (except for very low sheath resistances, where the sheath has less resistance than the phase conductor).

System voltage and insulation thickness — Neither significantly impacts the ampacity of distribution cables. Ampacity stays constant with voltage; 5-kV cables have roughly the same ampacity as 35-kV cables. At higher voltages, insulation is thicker, but this rise in the thermal resistance of the insulation reduces the ampacity just slightly. Higher operating voltages also cause higher dielectric losses, but again, the effect is small (it is more noticeable with EPR cable).

Number of cables — Cables in parallel heat each other, which restricts ampacity. Figure 3.13 shows an example for triplex power cables in duct banks.

Cable crossings and other hotspots — Tests have found that cable crossings can produce significant hotspots (Koch, 2001). Other hotspots can occur in locations where cables are paralleled for a short distance like taps to padmounted transformers or other gear. Differences in surface covering (such as asphalt roads) can also produce hot spots. Anders and Brakelmann (1999a, 1999b) provide an extension to the Neher–McGrath model that includes the effects of cable crossings at different angles. They conclude: "the derating of

3 to 5% used by some utilities may be insufficient, especially for cables with smaller conductors."

Riser poles — Cables on a riser pole require special attention. The protective vertical conduit traps air, and the sun adds external heating. Hartlein and Black (1983) tested a specific riser configuration and developed an analytical model. They concluded that the size of the riser and the amount of venting were important. Large diameter risers vented at both ends are the best. With three cables in one riser, they found that the riser portion of the circuit limits the ampacity. This is especially important in substation exit cables and their riser poles. In a riser pole application, ampacity does not increase for lower load factors; a cable heats up much faster in the air than when buried in the ground (the air has little thermal storage). NRECA (1993) concluded that properly vented risers do not need to be derated, given that venting can increase ampacity between 10 and 25%. If risers are not vented, then the riser becomes the limiting factor. Additional work in this area has been done by Cress (1991) (tests and modeling for submarine cables in riser poles) and Anders (1996) (an updated analytical model).

3.6 Fault Withstand Capability

Short-circuit currents through a conductor's resistance generates tremendous heat. All cable between the source and the fault is subjected to the same phase current. For cables, the weakest link is the insulation; both XLPE and EPR have a short-duration upper temperature limit of 250°C. The short-circuit current injects energy as a function of the fault duration multiplied by the square of the current.

For aluminum conductors and XLPE or EPR insulation, the maximum allowable time-current characteristic is given by

$$I^2t = (48.4A)^2$$

where

I = fault current, A

t = fault duration, sec

A = cross-sectional area of the conductor, kcmil

This assumes an upper temperature limit of 250°C and a 90°C starting temperature. For copper, the upper limit is defined by

$$I^2t = (72.2A)^2$$

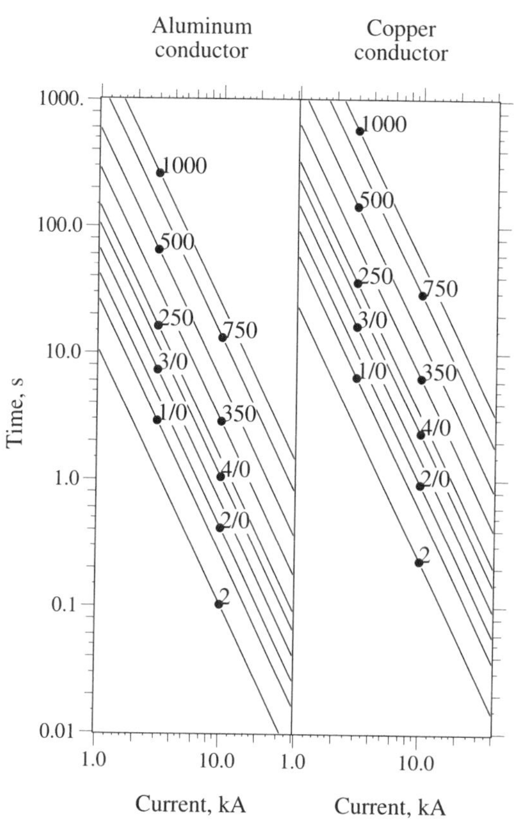

FIGURE 3.14
Short-circuit limit of cables with EPR or XLPE insulation.

We can plot these curves along with the time-current characteristics of the protecting relay, fuse, or recloser to ensure that the protective devices protect our cables.

Damage to the shield or the neutral is more likely than damage to the phase conductor. During a ground fault, the sheath may conduct almost as much current as the phase conductor, and the sheath is normally smaller. With a one-third neutral, the cable neutral's I^2t withstand is approximately 2.5 times less than the values for the phase conductor indicated in Figure 3.14 (this assumes a 65°C starting temperature). Having more resistance, a tape shield is even more vulnerable. A tape shield has a limiting time-current characteristic of

$$I^2t = (z \cdot A)^2$$

where z is 79.1 for sheaths of copper, 58.2 for bronze, 39.2 for zinc, 23.7 for copper-nickel, and 15 for lead [with a 65°C starting temperature and an

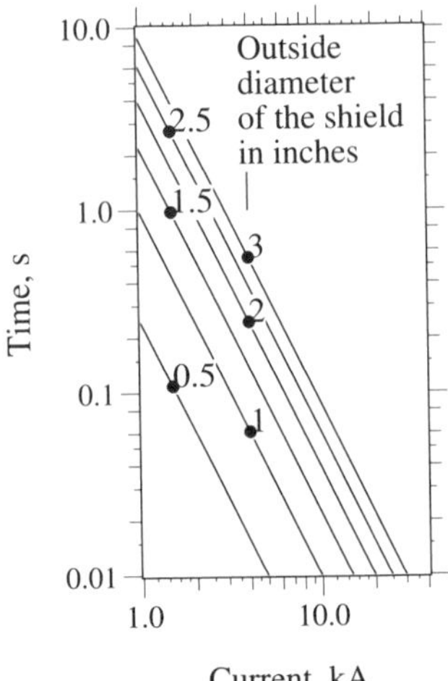

FIGURE 3.15
Short-circuit insulation limit of copper tape shields based on outside diameter (starting temperature is 65°C, final temperature is 250°C, 20% lap on the shield).

upper limit of 250°C; using data from (Kerite Company)]. Figure 3.15 shows withstand characteristics for a 5-mil copper tape shield. The characteristic changes with cable size because larger diameter cables have a shield with a larger circumference and more cross-sectional area. If a given fault current lasts longer than five times the insulation withstand characteristic (at 250°C), the shield reaches its melting point.

In the vicinity of the fault, the fault current can cause considerably more damage to the shield or neutral. With a concentric neutral, the fault current may only flow on a few strands of the conductor until the cable has a grounding point where the strands are tied together. Excessive temperatures can damage the insulation shield, the insulation, and the jacket. In addition, the temperature may reach levels that melt the neutral strands. A tape shield can suffer similar effects: where tape layers overlap, oxidation can build up between tape layers, which insulate the layers from each other. This can restrict the fault current to a smaller portion of the shield. Additionally, where the fault arc attaches, the arc injects considerable heat into the shield or neutral, causing further damage at the failure point. Some additional damage at the fault location must be tolerated, but the arc can burn one or more neutral strands several feet back toward the source.

Martin et al. (1974) reported that longitudinally corrugated sheaths perform better than wire or tape shields for high fault currents. They also reported that a semiconducting jacket helped spread the fault current to the sheaths of other cables (the semiconducting material breaks down).

Pay special attention to substation exit cables in areas with high fault currents (especially since exit cables are critical for circuit reliability). During a close fault, where currents are high, a reduced neutral or tape shield is most prone to damage.

3.7 Cable Reliability

3.7.1 Water Trees

The most common failure cause of solid-dielectric cables has been *water treeing*. Water trees develop over a period of many years and accelerate the failure of solid dielectric cables. Excessive treeing has led to the premature failure of many polyethylene cables. Cable insulation can tree two ways:

- *Electrical trees* — These hollow tubes develop from high electrical stress; this stress creates partial discharges that eat away at the insulation. Once initiated, electrical trees can grow fast, failing cable within hours or days.
- *Water trees* — Water trees are small discrete voids separated by insulation. Water trees develop slowly, growing over a period of months or years. Much less electrical stress is needed to cause water trees. Water trees actually look more like fans, blooms, or bushes whereas electrical trees look more like jagged branched trees. As its name indicates, water trees need moisture to grow; water that enters the dielectric accumulates in specific areas (noncrystalline regions) and causes localized degradation. Voids, contaminants, temperature, and voltage stress — all influence the rate of growth.

The formation of water trees does not necessarily mean the cable will fail. A water tree can even bridge the entire dielectric without immediate failure. Failure occurs when a water tree converts to an electrical tree. One explanation of the initiation of electrical trees is from charges trapped in the cable insulation. In Thue's words (1999), "they can literally bore a tunnel from one void or contaminant to the next." Impulses and dc voltage (in a hi-pot test) can trigger electrical treeing in a cable that is heavily water treed.

The growth rate of water trees tends to reduce with time; as trees fan out, the electrical stress on the tree reduces. Trees that grow from contaminants near the boundary of the conductor shield are most likely to keep growing. These are "vented" trees. Bow-tie trees (those that originate inside the cable) tend to grow to a critical length and then stop growing.

The electrical breakdown strength of aged cable has variation, a variation that has a skewed probability distribution. Weibull or lognormal distributions are often used to characterize this probability and predict future failure probabilities.

Polyethylene insulation systems have been plagued by early failures caused by water trees. Early XLPE and especially HMWPE had increasing failure rates that have led utilities to replace large quantities of cable. By most accounts, polyethylene-based insulation systems have become much more resistant to water treeing and more reliable for many reasons (Dudas, 1994; EPRI 1001894, 2001; Thue, 1999):

- *Extruded semiconducting shields* — Rather than taped conductor and insulation shields, manufacturers extrude both semiconducting shields as they are extruding the insulation. This one-pass extrusion provides a continuous, smooth interface. The most dangerous water trees are those that initiate from imperfections at the interface between the insulation and the semiconducting shield. Reducing these imperfections reduces treeing.
- *Cleaner insulation* — AEIC specifications for the allowable number and size of contaminants and protrusions have steadily improved. Both XLPE compound manufacturers and cable manufacturers have reduced contaminants by improving their production and handling processes.
- *Fewer voids* — Dry curing reduces the number and size of voids in the cable. Steam-cured cables pass through a long vulcanizing tube filled with 205°C steam pressurized at 20 atm. Cables cured with steam have sizeable voids in the insulation. Instead of steam, dry curing uses nitrogen gas pressurized to 10 atm; an electrically heated tube radiates infrared energy that heats the cable. Dry curing has voids, but these voids have volumes 10 to 100 times less than with steam curing.
- *Tree-retardant formulations* — Tree-retardant formulations of XLPE perform much better in accelerated aging tests, tests of field-aged cables, and also in field experience.

EPR insulation has proven to be naturally water tree resistant; EPR cables have performed well in service since the 1970s. EPR insulation can and does have water trees, but they tend to be smaller. EPR cable systems have also improved by having cleaner insulation compounds, jackets, and extruded semiconductor shields.

Several accelerated aging tests have been devised to predict the performance of insulation systems. The tests use one of two main methods to quantify performance: (1) loss of insulation strength or (2) time to failure. In accelerated aging, testers normally submerge cables in water, operate the cables at a continuous overvoltage, and possibly subject the cables to thermal cycling. The accelerated water treeing test (AWTT) is a protocol that measures the loss of insulation strength of a set of samples during one year of testing (ANSI/ICEA S-94-649-2000, 2000). The wet aging as part of this test includes application of three times rated voltage and current sufficient to heat the water to 60°C. In another common test protocol, the accelerated cable life test (ACLT), cables are submerged in water, water is injected into the conductor strands, cables are operated to (commonly) four times nominal voltage, and cables are brought to 90°C for eight hours each day. The cables are operated to failure. Brown (1991) reported that under such a test, XLPE and TR-XLPE cables had geometric mean failure times of 53 and 161 days, respectively. Two EPR constructions did not fail after 597 days of testing. Because EPR and XLPE age differently depending on the type of stress, EPR can come out better or worse than TR-XLPE, depending on the test conditions. There is no consensus on the best accelerated-aging test. Normally such tests are used to compare two types of cable constructions. Bernstein concludes, "… there is still no acceptable means of relating service and laboratory aging to 'remaining life' " (EPRI 1000273, 2000).

Even without voltage, XLPE cable left outdoors can age. EPRI found that XLPE cables left in the Texas sun for 10 years lost over 25% of their ac insulation strength (EPRI 1001389, 2002). These researchers speculate that heating from the sun led to a loss of peroxide decomposition by-products, which is known to result in loss of insulation strength.

Since water promotes water treeing, a few utilities use different forms of water blocking (Powers, 1993). Water trees grow faster when water enters the insulation from both sides: into the conductor strands and through the cable sheath. The most common water-protection method is a filled strand conductor; moisture movement or migration is minimized by the filling, which can be a semiconducting or an insulating filler. Another variation uses water absorbing powders; as the powder absorbs water it turns to a gel that blocks further water movement. An industry standard water blocking test is provided (ICEA Publication T-31-610, 1994; ICEA Publication T-34-664, 1996). In addition to reducing the growth and initiation of water trees, a strand-blocked conductor reduces corrosion of aluminum phase conductors. We can also use solid conductors to achieve the same effect (on smaller cables).

Another approach to dealing with water entry and treeing in existing cable is to use a silicone injection treatment (Nannery et al., 1989). After injection into the stranded conductor, the silicone diffuses out through the conductor shield and into the insulation. The silicone fills water-tree voids and reacts with water such that it dries the cable. This increases the dielectric strength and helps prevent further treeing and loss of life.

Another way to increase the reliability is to increase the insulation thickness. As an example, the maximum electrical stress in a cable with an insulation thickness of 220 mil (1 mil = 0.001 in. = 0.00254 cm) is 14% lower than a 175-mil cable (Mackevich, 1988).

Utilities and manufacturers have taken steps to reduce the likelihood of cable degradation. Table 3.23 shows trends in cable specifications for underground residential cable. Tree-retardant insulation and smooth semiconductor shields, jackets and filled conductors, and dry curing and triple extrusion are features specified by utilities to improve reliability.

TABLE 3.23

Trends in URD Cable Specifications

Characteristic	1983	1988	1993	1998
XLPE insulation	84	52	20	0
TR-XLPE insulation		36	52	68
EPR insulation	12	12	28	32
Protective jacket	64	80	92	93[a]
Filled strand conductor	4	32	60	68
Dry cure for XLPE and TR-XLPE		24	56	52
Triple extrusion		44	64	67[a]
Supersmooth semicon shields		0	44	56
Bare copper neutrals		72	84	

Note: Percentage of the 25 largest investor-owned utilities in the US that specify the given characteristic.

[a] Somewhat different data set: percentages from the top 45 largest investor-owned utilities.

Sources: Dudas, J. H., "Technical Trends in Medium Voltage URD Cable Specifications," *IEEE Electrical Insulation Magazine*, vol. 10, no. 2, pp. 7–16, March/April 1994; Dudas, J. H. and Cochran, W. H., "Technical Advances in the Underground Medium-Voltage Cable Specification of the Largest Investor-Owned Utilities in the U.S.," *IEEE Electrical Insulation Magazine*, vol. 15, no. 6, pp. 29–36, November/December 1999.

Good lightning protection also reduces cable faults. This requires surge arresters at the riser pole and possibly arresters at the cable open point (depending on the voltage). Keep arrester lead lengths as short as possible. Surges are a known cause of dielectric failures. Surges that do not fail the insulation may cause aging. Accelerated aging tests have found that 15-kV XLPE cables tested with periodic surges applied with magnitudes of 40, 70, and 120 kV failed more often and earlier than samples that were not surged (EPRI EL-6902, 1990; EPRI TR-108405-V1, 1997; Hartlein et al., 1989; Hartlein et al., 1994). Very few of the failures occurred during the application of a surge; this follows industry observations that cables often fail after a thunderstorm, not during the storm.

Rather than continue patching, many utilities regularly replace cable. Program policies are done based on the number of failures (the most common approach), cable inspection, customer complaints, or cable testing. High-molecular weight polyethylene and older XLPE are the most likely candidates for replacement. Most commonly, utilities replace cable after two or three electrical failures within a given time period (see Table 3.24).

3.7.2 Other Failure Modes

Cable faults can be caused by several events including:

TABLE 3.24

Typical Cable Replacement Criteria

Replacement Criteria	Responses (n = 51)
One failure	2%
Two failures	31%
Three failures	41%
Four failures	4%
Five failures	6%
Based on evaluation procedures	16%

Source: Tyner, J. T., "Getting to the Bottom of UG Practices," *Transmission & Distribution World*, vol. 50, no. 7, pp. 44–56, July 1998.

- Dig-ins
- Cable failures
- Cable equipment failures — splices, elbows, terminations

Better public communications reduces dig-ins into cables. The most common way is with one phone number that can be used to coordinate marking of underground facilities before digging is done. Physical methods of reducing dig-ins include marker tape, surface markings, or concrete covers. Marker tape identifies cable. A few utilities use surface marking to permanently identify the location of underground facilities. Concrete covers above underground facilities physically block dig-ins.

Temporary faults are unusual in underground facilities. Faults are normally bolted, permanent short circuits. Reclosing will just do additional damage to the cable. Occasionally, animals or water will temporarily fault a piece of live-front equipment. Recurring temporary faults like these can be very difficult to find.

Another type of temporary, self-clearing fault can occur on a cable splice (Stringer and Kojovic, 2001). Figure 3.16 shows a typical waveform of an impending splice failure. This type of fault has some distinguishing characteristics: it self-clears in 1/4 cycle, the frequency of occurrence increases with time, and faults occur near the peak of the voltage. The author has observed this type of fault during monitoring (but never identified the culprit). This type of fault can occur in a cable splice following penetration of water into the splice. The water breaks down the insulation, then the arc energy melts the water and creates vapor at high pressure. Finally, the high-pressure vapor extinguishes the arc. The process can repeat when enough water accumulates again until the failure is permanent. This type of self-clearing fault can go unnoticed until it finally fails. The downside is that it causes a short-duration voltage sag that may affect sensitive equipment. Another problem, the fault may have enough current to blow a fuse; but since the fault self-clears, it can be much harder to find. Crews may just replace the fuse (successfully) and leave without replacing the damaged equipment.

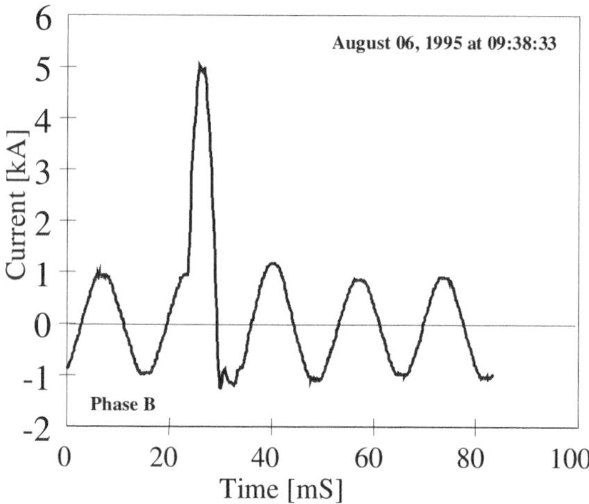

FIGURE 3.16
Self-clearing fault signature on an incipient cable-splice failure. (From Stringer, N. T. and Kojovic, L. A., "Prevention of Underground Cable Splice Failures," *IEEE Trans. Industry App.*, 37(1), 230-9, Jan./Feb. 2001. With permission. ©2001 IEEE.)

3.7.3 Failure Statistics

The annual failures of cables is on the order of 6 to 7 failures per 100 mi per year (3.7 to 4.3 failures per 100 km per year) according to survey data from the Association of Edison Illuminating Companies from 1965 through 1991 (Thue, 1999). Figure 3.17 shows cable failure data from a variety of sources; experience varies widely. Application, age, and type of cable markedly change the results. Utilities have experienced high failures of HMWPE, especially those that installed in the early 1970s. An EPRI database of 15 utilities showed a marked increase in failure rates for HMWPE cables with time (Stember et al., 1985). XLPE also shows a rise in failure rates with time, but not as dramatic (see Figure 3.18). The EPRI data showed failure rates increased faster with a higher voltage gradient on the dielectric for both HMWPE and XLPE.

Much of the failure data in Figure 3.17 is dominated by earlier polyethylene-based cable insulation technologies. Not as much data is available on the most commonly used insulation materials: TR-XLPE and EPR. The AEIC survey reported results in 1991 — both had fewer than 0.5 failures per 100 cable mi during that year. TR-XLPE results were better (0.2 vs. 0.4 failures/100 mi/year for EPR), but the installed base of TR-XLPE would have been newer than EPR at that time. Jacketed cable has had fewer failures than unjacketed cable as shown in Table 3.25.

Another consideration for underground circuits is the performance of connectors and other cable accessories. 200-A elbows have failed at high rates (and they tend to fail when switching under load) (Champion, 1986).

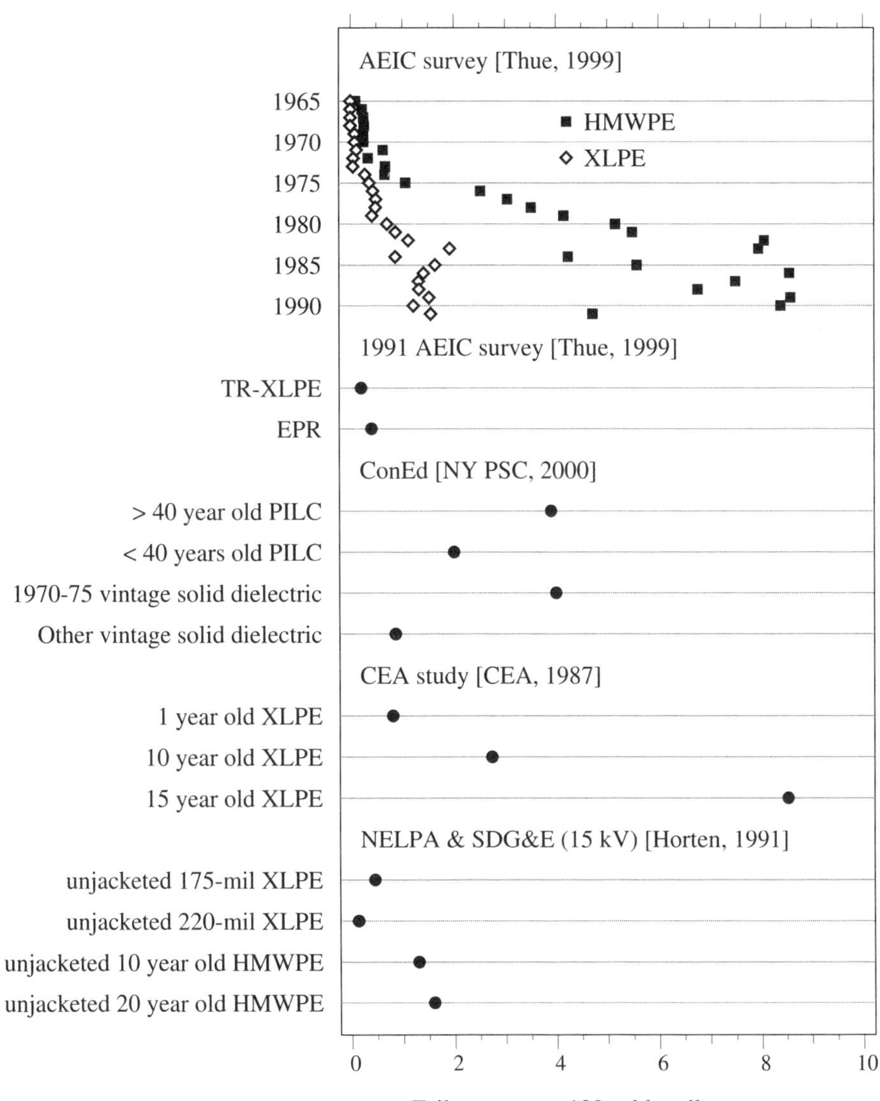

FIGURE 3.17
Cable failure rates found in different studies and surveys (in cable miles, not circuit miles). (Data from [CEA 117 D 295, 1987; Horton and Golberg, 1991; State of New York Department of Public Service, 2000; Thue, 1999].)

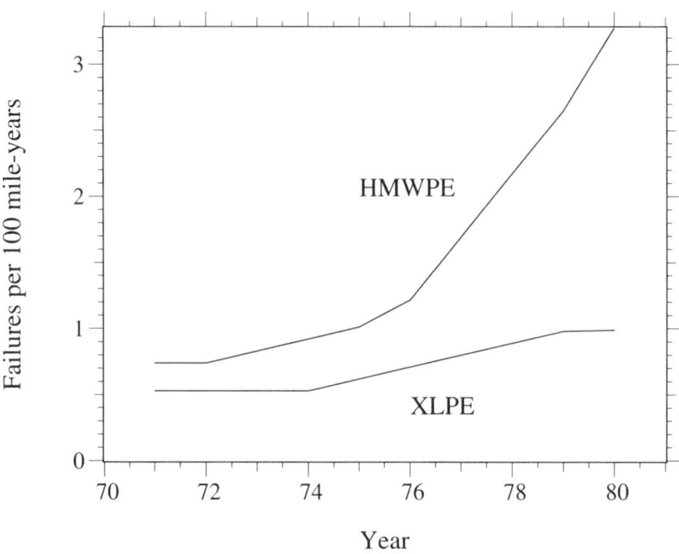

FIGURE 3.18
Cumulative service-time failure rates for HMWPE and XLPE cable. (From Stember, L. H., Epstein, M. M., Gaines, G. V., Derringer, G. C., and Thomas, R. E., "Analysis of Field Failure Data on HMWPE- and XLPE-Insulated High-Voltage Distribution Cable," *IEEE Trans. Power Apparatus Sy.*, PAS-104(8), 1979-85, August 1985. With permission. ©1985 IEEE.)

TABLE 3.25
Comparison of the Median of the Average Yearly Failure Rates of XLPE Found by AEIC from 1983 to 1991

Configuration	Failures per 100 Cable miles/year
No jacket	3.1
Jacketed	0.2
Direct buried	2.6
Duct	0.2

Source: Thue, W. A., *Electrical Power Cable Engineering*, Marcel Dekker, New York, 1999.

One important factor is that the type of splice should be correctly matched with the type of cable (Mackevich, 1988). Table 3.26 shows annual failure rates for some common underground components that were developed based on data from the Northwest Underground Distribution Committee of the Northwest Electric Light and Power Association (Horton and Golberg, 1990; Horton and Golberg, 1991). Table 3.27 shows failure rates of splices for New York City.

An EPRI review of separable connector reliability found mixed results (EPRI 1001732, 2002). Most utilities do not track these failures. One utility that did keep records found that failure rates of separable connectors ranged

TABLE 3.26

Annual Underground-Component Failure Rates

Component	Annual Failure Rate, %
Load-break elbows	0.009*t*
15-kV molded rubber splices	0.31
25-Kv molded rubber splices	0.18
35-Kv molded rubber splices	0.25
Single-phase padmounted transformers	0.3

Note: *t* is the age of the elbow in years.

Sources: Horton, W. F. and Golberg, S., "The Failure Rates of Underground Distribution System Components," Proceedings of the Twenty-Second Annual North American Power Symposium, 1990; Horton, W. F. and Golberg, S., "Determination of Failure Rates of Underground Distribution System Components from Historical Data," IEEE/PES Transmission and Distribution Conference, 1991.

TABLE 3.27

Underground Network Component Failure Rates in New York City (Con Edison)

Component	Annual Failure Rate, %
Splices connecting paper to solid cables (stop joints)	1.20
Splices connecting similar cables (straight joints)	0.51
Network transformers	0.58

Source: State of New York Department of Public Service, "A Report on Consolidated Edison's July 1999 System Outages," March 2000.

from 0.1 to 0.4% annually. Of these failures, an estimated 3 to 20% are from overheating. They also suggested that thermal monitoring is a good practice, but effectiveness is limited because the monitoring is often done when the loadings and temperatures are well below their peak.

3.8 Cable Testing

A common approach to test cable and determine insulation integrity is to use a hi-pot test. In a hi-pot test, a dc voltage is applied for 5 to 15 min. IEEE-400 specifies that the hi-pot voltage for a 15-kV class cable is 56 kV for an acceptance test and 46 kV for a maintenance test (ANSI/IEEE Std. 400-1980). Other industry standard tests are given in (AEIC CS5-94, 1994; AEIC CS6-96, 1996; ICEA S-66-524, 1988). High-pot testing is a brute-force test; imminent failures are detected, but the amount of deterioration due to aging is not quantified (it is a go/no–go test).

The dc test is controversial — some evidence has shown that hi-pot testing may damage XLPE cable (Mercier and Ticker, 1998). EPRI work has shown that dc testing accelerates treeing (EPRI TR-101245, 1993; EPRI TR-101245-V2, 1995). For hi-pot testing of 15-kV, 100% insulation (175-mil, 4.445-mm) XLPE cable, EPRI recommended:

- Do not do testing at 40 kV (228 V/mil) on cables that are aged (especially those that failed once in service and then are spliced). Above 300 V/mil, deterioration was predominant.
- New cable can be tested at the factory at 70 kV. No effect on cable life was observed for testing of new cable.
- New cable can be tested at 55 kV in the field prior to energization if aged cable has not been spliced in.
- Testing at lower dc voltages (such as 200 V/mil) will not pick out bad sections of cable.

Another option for testing cable integrity: ac testing does not degrade solid dielectric insulation (or at least degrades it more slowly). The use of very low frequency ac testing (at about 0.1 Hz) may cause less damage to aged cable than dc testing (Eager et al., 1997) (but utilities have reported that it is not totally benign, and ac testing has not gained widespread usage). The low frequency has the advantage that the equipment is much smaller than 60-Hz ac testing equipment.

3.9 Fault Location

Utilities use a variety of tools and techniques to locate underground faults. Several are described in the next few paragraphs [see also EPRI TR-105502 (1995)].

Divide and conquer — On a radial tap where the fuse has blown, crews narrow down the faulted section by opening the cable at locations. Crews start by opening the cable near the center, then they replace the fuse. If the fuse blows, the fault is upstream; if it doesn't blow, the fault is downstream. Crews then open the cable near the center of the remaining portion and continue bisecting the circuit at appropriate sectionalizing points (usually padmounted transformers). Of course, each time the cable faults, more damage is done at the fault location, and the rest of the system has the stress of carrying the fault currents. Using current-limiting fuses reduces the fault-current stress but increases the cost.

Fault indicators — Faulted circuit indicators (FCIs) are small devices clamped around a cable that measure current and signal the passage of fault current. Normally, these are applied at padmounted transformers. Faulted circuit indicators do not pinpoint the fault; they identify the fault to a cable section. After identifying the failed section, crews must use another method

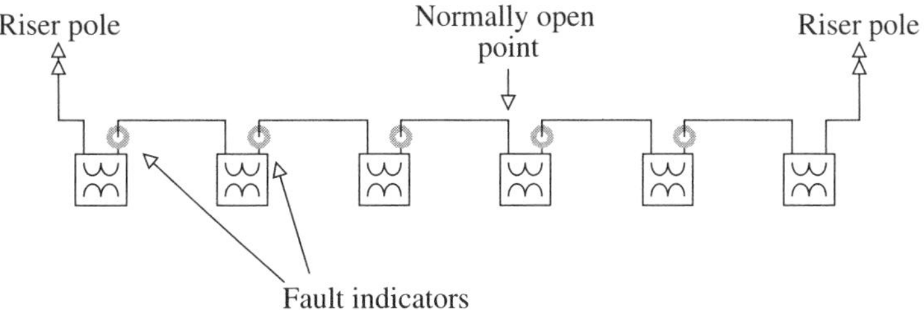

FIGURE 3.19
Typical URD fault indicator application.

such as the thumper to precisely identify the fault. If the entire section is in conduit, crews don't need to pinpoint the location; they can just pull the cable and replace it (or repair it if the faulted portion is visible from the outside). Cables in conduit require less precise fault location; a crew only needs to identify the fault to a given conduit section.

Utilities' main justification for faulted circuit indicators is reducing the length of customer interruptions. Faulted circuit indicators can significantly decrease the fault-finding stage relative to the divide-and-conquer method. Models that make an audible noise or have an external indicator decrease the time needed to open cabinets.

Utilities use most fault indicators on URD loops. With one fault indicator per transformer (see Figure 3.19), a crew can identify the failed section and immediately reconfigure the loop to restore power to all customers. The crew can then proceed to pinpoint the fault and repair it (or even delay the repair for a more convenient time). For larger residential subdivisions or for circuits through commercial areas, location is more complicated. In addition to transformers, fault indicators should be placed at each sectionalizing or junction box. On three-phase circuits, either a three-phase fault indicator or three single-phase indicators are available; single-phase indicators identify the faulted phase (a significant advantage). Other useful locations for fault indicators are on either end of cable sections of overhead circuits, which are common at river crossings or under major highways. These sections are not fused, but fault indicators will show patrolling crews whether the cable section has failed.

Fault indicators may be reset in a variety of ways. On manual reset units, crews must reset the devices once they trip. These units are less likely to reliably indicate faults. Self-resetting devices are more likely to be accurate as they automatically reset based on current, voltage, or time. Current-reset is most common; after tripping, if the unit senses current above a threshold, it resets [standard values are 3, 1.5, and 0.1 A (NRECA RER Project 90-8, 1993)]. With current reset, the minimum circuit load at that point must be above the threshold, or the unit will never reset. On URD loops, when applying current-reset indicators, consider that the open point might change. This changes the current that the fault indicator sees. Again, make sure the circuit load is enough to reset the fault indicator. Voltage reset models provide a voltage sensor; when the voltage exceeds some value (the voltage sensor senses at secondary voltage or at an elbow's capacitive test point). Time-reset units simply reset after a given length of time.

Fault indicators should only operate for faults — not for load, not for inrush, not for lightning, and not for backfeed currents. False readings can send crews on wild chases looking for faults. Reclose operations also cause loads and transformers to draw inrush, which can falsely trip a fault indicator. An inrush restraint feature disables tripping for up to one second following energization. On single-phase taps, inrush restraint is really only needed for manually-reset fault indicators (the faulted phase with the blown fuse will not have inrush that affects downstream fault indicators). Faults in adjacent cables can also falsely trip indicators; the magnetic fields couple into the pickup coil. Shielding can help prevent this. Several scenarios cause backfeed that can trip fault indicators. Downstream of a fault, the stored charge in the cable will rush into the fault, possibly tripping fault indicators. McNulty (1994) reported that 2000 ft of 15-kV cable created an oscillatory current transient that peaked at 100 A and decayed in 0.15 msec. Nearby capacitor banks on the overhead system can make outrush worse. Motors and other rotating equipment can also backfeed faults. To avoid false trips, use a high set point. Equipment with filtering that reduces the indicator's sensitivity to transient currents also helps, but too much filtering may leave the faulted-circuit indicator unable to detect faults cleared rapidly by current-limiting fuses.

Self-resetting fault indicators can also falsely reset. Backfeed currents and voltages can reset fault indicators. On a three-phase circuit with one phase tripped, the faulted phase can backfeed through three-phase transformer connections (see Chapter 4), providing enough current or enough voltage to reset faulted-circuit indicators. On single-phase circuits, these are not a problem. In general, single-phase application is much easier; we do not have backfeed problems or problems with indicators tripping from faults on nearby cables. For single-phase application guidelines, see (IEEE Std 1216-2000).

Fault indicators may have a threshold-type trip characteristic like an instantaneous relay (any current above the set point trips the flag), or they may have a time-overcurrent characteristic which trips faster for higher currents. Those units with time-overcurrent characteristics should be coordinated with minimum clearing curves of current-limiting fuses to ensure that they operate. Another type of fault indicator uses an adaptive setting that trips based on a sudden increase in current followed by a loss of current.

Set the trip level on fault indicators to less than 50% of the available fault current or 500 A, whichever is less (IEEE P1610/D03, 2002). This trip threshold should be at least two to three times the load on the circuit to minimize false indications. These two conditions will almost never conflict, only at the end of a very long feeder (low fault currents) on a cable that is heavily loaded.

Normally, fault indicators are fixed equipment, but they can be used for targeted fault location. When crews arrive at a faulted and isolated section, they first apply fault indicators between sections (normally at padmounted transformers). Crews reenergize the failed portion and then check the fault indicators to identify the faulted section. Only one extra fault is applied to the circuit, not multiple faults as with the divide and conquer method.

Section testing — Crews isolate a section of cable and apply a dc hi-pot voltage. If the cable holds the hi-pot voltage, crews proceed to the next section and repeat until finding a cable that cannot hold the hi-pot voltage. Because the voltage is dc, the cable must be isolated from the transformer. In a faster variation of this, high-voltage sticks are available that use the ac line voltage to apply a dc voltage to the isolated cable section.

Thumper — The thumper applies a pulsed dc voltage to the cable. As its name implies, at the fault the thumper discharges sound like a thumping noise as the gap at the failure point repeatedly sparks over. The thumper charges a capacitor and uses a triggered gap to discharge the capacitor's charge into the cable. Crews can find the fault by listening for the thumping noise. Acoustic enhancement devices can help crews locate weak thumping noises; antennas that pick up the radio-frequency interference from the arc discharge also help pinpoint the fault. Thumpers are good for finding the exact fault location so that crews can start digging. On a 15-kV class system, utilities typically thump with voltages from 10 to 15 kV, but utilities sometimes use voltages to 25 kV.

While pulsed discharges are thought to be less damaging to cable than a steady dc voltage, utilities have concern that thumping can damage the unfailed sections of cable. When a thumper pulse breaks down the cable, the incoming surge shoots past the fault. When it reaches the open point, the voltage doubles, then the voltage pulse bounces back and forth between the open point and the fault, switching from +2 to $-2E$ (where E is the thumper pulse voltage). In tests, EPRI research found that thumping can reduce the life of aged cable (EPRI EL-6902, 1990; EPRI TR-108405-V1, 1997; Hartlein et al., 1989; Hartlein et al., 1994). The thumping discharges at the failure point can also increase the damage at the fault point. Most utilities try to limit the voltage or discharge energy, and a few don't use a thumper for fear of additional damage to cables and components (Tyner, 1998). A few utilities also disconnect transformers from the system during thumping to protect the transformer and prevent surges from propagating through the transformer (these surges should be small). If the fault has no gap, and if the fault is a solid short circuit, then no arc forms, and the thumper will not create its characteristic thump (fortunately, solid short circuits are rare in cable faults). Some crews keep thumping in an effort to burn the short circuit apart enough to start arcing. With cable in conduit, the thumping may be louder near the conduit ends than at the fault location. Generally, crews should start with the voltage low and increase as needed. A dc hi-pot voltage can help determine how much voltage the thumper needs.

Radar — Also called time-domain reflectometry (TDR), a radar set injects a very short-duration current pulse into the cable. At discontinuities, a portion of the pulse will reflect back to the set; knowing the velocity of wave propagation along cable gives us an estimate of the distance to the fault. Depending on the test set and settings, radar pulses can be from 5 ns to 5 μs wide. Narrower pulses give higher resolution, so users can better differentiate between faults and reflections from splices and other discontinuities (Banker et al., 1994).

Radar does not give pinpoint accuracy; its main use is to narrow the fault to a certain section. Then, crews can use a thumper or other pinpoint technique to find the failure. Taking a radar pulse from either end of a cable and averaging the results can lead to an improved estimate of the location. Radar location on circuits with taps can be complicated, especially those with multiple taps; the pulse will reflect off the taps, and the reflection from the actual fault will be less than it otherwise would be. Technology has been developed to use above-ground antennas to sense and pinpoint faults based on the radar signals.

Radar and thumper — After a fuse or other circuit interrupter clears a fault in a cable, the area around the fault point recovers some insulation strength. Checking the cable with an ohm meter would show an open circuit. Likewise, the radar pulse passes right by the fault, so the radar set alone cannot detect the fault. Using radar with a thumper solves this problem. A thumper pulse breaks down the gap, and the radar superimposes a pulse that reflects off the fault arc. The risetime of the thumper waveshape is on the order of a few microseconds; the radar pulse total width may be less than 0.05 μsec. Another less attractive approach is to use a thumper to continually burn the cable until the fault resistance becomes low enough to get a reading on a radar set (this is less attractive because it subjects the cable to many more thumps, especially if crews use high voltages).

Boucher (1991) reported that fault indicators were the most popular fault locating approach, but most utilities use a variety of techniques (see Figure 3.20). Depending on the type of circuit, the circuit layout, and the equipment available, different approaches are sometimes better.

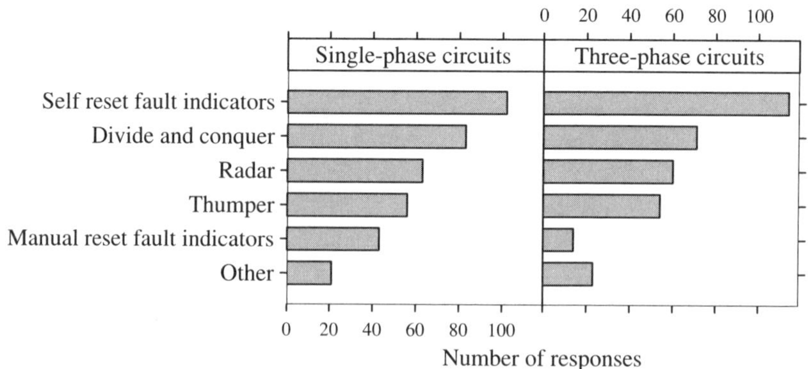

FIGURE 3.20
Utility use of fault-locating techniques (204 utilities surveyed, multiple responses allowed). (Data from [Boucher, 1991].)

When applying test voltages to cables, crews must be mindful that cables can hold significant charge. Cables have significant capacitance, and cables can maintain charge for days.

References

AEIC CS5-94, *Specification for Cross-Linked Polyethylene Insulated, Shielded Power Cables Rated 5 through 46 kV,* Association of Edison Illuminating Companies, 1994.

AEIC CS6-96, *Specification for Ethylene Propylene Rubber Shielded Power Cables Rated 5–69 kV,* Association of Edison Illuminating Companies, 1996.

Ametani, A., "A General Formulation of Impedance and Admittance of Cables," *IEEE Transactions on Power Apparatus and Systems,* vol. PAS-99, no. 3, pp. 902–10, May/June 1980.

Anders, G. and Brakelmann, H., "Cable Crossings-Derating Considerations. I. Derivation of Derating Equations," *IEEE Transactions on Power Delivery,* vol. 14, no. 3, pp. 709–14, July 1999a.

Anders, G. and Brakelmann, H., "Cable Crossings-Derating Considerations. II. Example of Derivation of Derating Curves," *IEEE Transactions on Power Delivery,* vol. 14, no. 3, pp. 715–20, July 1999b.

Anders, G. J., "Rating of Cables on Riser Poles, in Trays, in Tunnels and Shafts — A Review," *IEEE Transactions on Power Delivery,* vol. 11, no. 1, pp. 3–11, January 1996.

Anders, G. J., Rating of Electric Power Cables: Ampacity Computations for Transmission, Distribution, and Industrial Applications, IEEE Press, McGraw-Hill, New York, 1998.

ANSI/ICEA S-94-649-2000, *Standard for Concentric Neutral Cables Rated 5 through 46 kV,* Insulated Cable Engineers Association, 2000.

ANSI/ICEA S-97-682-2000, *Standard for Utility Shielded Power Cables Rated 5 through 46 kV,* Insulated Cable Engineers Association, 2000.

ANSI/IEEE Std. 400-1980, *IEEE Guide for Making High-Direct-Voltage Tests on Power Cable Systems in the Field.*

Banker, W. A., Nannery, P.R., Tarpey, J. W., Meyer, D.F., and Piesinger, G.H., "Application of High Resolution Radar to Provide Non-destructive Test Techniques for Locating URD Cable Faults and Splices," *IEEE Transactions on Power Delivery,* vol. 9, no. 3, pp. 1187–94, July 1994.

Boggs, S. and Xu, J. J., "Water Treeing — Filled vs. Unfilled Cable Insulation," *IEEE Electrical Insulation Magazine,* vol. 17, no. 1, pp. 23–9, January/February 2001.

Boucher, R., "A Summary Of The Regional Underground Distribution Practices For 1991," Regional Underground Distribution Practices (IEEE paper 91 TH0398-8-PWR), 1991.

Brown, M., "EPR Insulation Cuts Treeing and Cable Failures," *Electrical World,* vol. 197, no. 1, pp. 105–6, January 1983.

Brown, M., "Accelerated Life Testing of EPR-Insulated Underground Cable," *IEEE Electrical Insulation Magazine,* vol. 7, no. 4, pp. 21–6, July/August 1991.

Burns, N. M., Jr., "Performance of Supersmooth Extra Clean Semiconductive Shields in XLPE Insulated Power Cables," IEEE International Symposium on Electrical Insulation, 1990.

CEA 117 D 295, *Survey of Experience with Polymer Insulated Power Cable in Underground Service*, Canadian Electrical Association, 1987.

CEA 274 D 723, *Underground Versus Overhead Distribution Systems*, Canadian Electrical Association, 1992.

CEA, *CEA Distribution Planner's Manual*, Canadian Electrical Association, 1982.

Champion, T., "Elbow Failures Cast Doubt on Reliability," *Electrical World*, vol. 200, no. 6, pp. 71–3, June 1986.

Cinquemani, P. L., Yingli, W., Kuchta, F. L., and Doench, C., "Performance of Reduced Wall EPR Insulated Medium Voltage Power Cables. I. Electrical Characteristics," *IEEE Transactions on Power Delivery*, vol. 12, no. 2, pp. 571–8, April 1997.

Cress, S. L. and Motlis, H., "Temperature Rise of Submarine Cable on Riser Poles," *IEEE Transactions on Power Delivery*, vol. 6, no. 1, pp. 25–33, January 1991.

Dedman, J. C. and Bowles, H. L., "A Survey of URD Cable Installed on Rural Electric Systems and Failures of That Cable," IEEE Rural Electric Power Conference, 1990.

Dommel, H. W., "Electromagnetic Transients Program Reference Manual (EMTP Theory Book)," prepared for Bonneville Power Administration, 1986.

Dudas, J. H., "Technical Trends in Medium Voltage URD Cable Specifications," *IEEE Electrical Insulation Magazine*, vol. 10, no. 2, pp. 7–16, March/April 1994.

Dudas, J. H. and Cochran, W. H., "Technical Advances in the Underground Medium-Voltage Cable Specification of the Largest Investor-Owned Utilities in the U.S.," *IEEE Electrical Insulation Magazine*, vol. 15, no. 6, pp. 29–36, November/December 1999.

Dudas, J. H. and Rodgers, J. R., "Underground Cable Technical Trends for the Largest Rural Electric Co-ops," *IEEE Transactions on Industry Applications*, vol. 35, no. 2, pp. 324–31, March/April 1999.

Eager, G. S., Katz, C., Fryszczyn, B., Densley, J., and Bernstein, B. S., "High Voltage VLF Testing of Power Cables," *IEEE Transactions on Power Delivery*, vol. 12, no. 2, pp. 565–70, April 1997.

EPRI 1000273, *Estimation of Remaining Life of XLPE-Insulated Cables*, Electric Power Research Institute, Palo Alto, CA, 2000.

EPRI 1000741, *Condition Assessment of Distribution PILC Cable Assets*, Electric Power Research Institute, Palo Alto, CA, 2000.

EPRI 1001389, *Aging of Extruded Dielectric Distribution Cable: Phase 2 Service Aging*, Electric Power Research Institute, Palo Alto, CA, 2002.

EPRI 1001732, *Thermal Issues and Ratings of Separable Insulated Connectors*, Electric Power Research Institute, Palo Alto, CA, 2002.

EPRI 1001734, *State of the Art of Thin Wall Cables Including Industry Survey*, Electric Power Research Institute, Palo Alto, CA, 2002.

EPRI 1001894, *EPRI Power Cable Materials Selection Guide*, EPRI, Palo Alto, CA, 2001.

EPRI EL-6902, *Effects of Voltage Surges on Solid-Dielectric Cable Life*, Electric Power Research Institute, Palo Alto, CA, 1990.

EPRI TR-101245, *Effect of DC Testing on Extruded Cross-Linked Polyethylene Insulated Cables*, Electric Power Research Institute, Palo Alto, CA, 1993.

EPRI TR-101245-V2, Effect of DC Testing on Extruded Cross-Linked Polyethylene Insulated Cables — Phase II, Electric Power Research Institute, Palo Alto, CA, 1995.

EPRI TR-105502, *Underground Cable Fault Location Reference Manual*, Electric Power Research Institute, Palo Alto, CA, 1995.

EPRI TR-108405-V1, *Aging Study of Distribution Cables at Ambient Temperatures with Surges*, Electric Power Research Institute, Palo Alto, CA, 1997.

EPRI TR-108919, *Soil Thermal Properties Manual for Underground Power Transmission*, Electric Power Research Institute, Palo Alto, CA, 1997.

EPRI TR-111888, *High-Ampacity, Thin-Wall, Novel Polymer Cable*, Electric Power Research Institute, Palo Alto, CA, 2000.

Gurniak, B., "Neutral Corrosion Problem Overstated," *Transmission & Distribution World*, pp. 152–8, August 1996.

Hartlein, R. A. and Black, W. Z., "Ampacity of Electric Power Cables in Vertical Protective Risers," *IEEE Transactions on Power Apparatus and Systems*, vol. PAS-102, no. 6, pp. 1678–86, June 1983.

Hartlein, R. A., Harper, V. S., and Ng, H. W., "Effects of Voltage Impulses on Extruded Dielectric Cable Life," *IEEE Transactions on Power Delivery*, vol. 4, no. 2, pp. 829–41, April 1989.

Hartlein, R. A., Harper, V. S., and Ng, H. W., "Effects of Voltage Surges on Extruded Dielectric Cable Life Project Update," *IEEE Transactions on Power Delivery*, vol. 9, no. 2, pp. 611–9, April 1994.

Horton, W. F. and Golberg, S., "The Failure Rates of Underground Distribution System Components," Proceedings of the Twenty-Second Annual North American Power Symposium, 1990.

Horton, W. F. and Golberg, S., "Determination of Failure Rates of Underground Distribution System Components from Historical Data," IEEE/PES Transmission and Distribution Conference, 1991.

ICEA Publication T-31-610, *Guide for Conducting a Longitudinal Water Penetration Resistance Test for Sealed Conductor*, Insulated Cable Engineers Association, 1994.

ICEA Publication T-34-664, *Conducting Longitudinal Water Penetration Resistance Tests on Cable*, Insulated Cable Engineers Association, 1996.

ICEA S-66-524, *Cross-Linked Thermosetting Polyethylene Insulated Wire and Cable for Transmission and Distribution of Electrical Energy*, Insulated Cable Engineers Association, 1988.

IEC 287, *Calculation of the Continuous Current Rating of Cables (100% Load Factor)*, 2nd ed., International Electrical Commission (IEC), 1982.

IEEE C2-1997, National Electrical Safety Code.

IEEE P1610/D03, *Draft Guide for the Application of Faulted Circuit Indicators for 200/600 A, Three-Phase Underground Distribution*, 2002.

IEEE Std 1216-2000, *IEEE Guide for the Application of Faulted Circuit Indicators for 200 A, Single-Phase Underground Residential Distribution (URD).*

IEEE Std. 141-1993, *IEEE Recommended Practice for Electric Power Distribution for Industrial Plants.*

IEEE Std. 386-1995, *IEEE Standard for Separable Insulated Connector Systems for Power Distribution Systems Above 600 V.*

IEEE Std. 835-1994, *IEEE Standard Power Cable Ampacities.*

Katz, C. and Walker, M., "Evaluation of Service Aged 35 kV TR-XLPE URD Cables," *IEEE Transactions on Power Delivery*, vol. 13, no. 1, pp. 1–6, January 1998.

Kerite Company, "Technical Application Support Data." Downloaded from www.kerite.com, June 2002.

Koch, B., "Underground Lines: Different Problems, Practical Solutions," *Electrical World T&D*, March/April 2001.

Lewis, W. A. and Allen, G. D., "Symmetrical Component Circuit Constants and Neutral Circulating Currents for Concentric Neutral Underground Distribution Cables," *IEEE Transactions on Power Apparatus and Systems*, vol. PAS-97, no. 1, pp. 191–9, January/February 1978.

Lewis, W. A., Allen, G. D., and Wang, J. C., "Circuit Constants for Concentric Neutral Underground Distribution Cables on a Phase Basis," *IEEE Transactions on Power Apparatus and Systems*, vol. PAS-97, no. 1, pp. 200–7, January/February 1978.

Mackevich, J. P., "Trends in Underground Residential Cable Systems," IEEE Rural Electric Power Conference, 1988.

Martin, M. A., Silver, D. A., Lukac, R. G., and Suarez, R., "Normal and Short Circuit Operating Characteristics of Metallic Shielding Solid Dielectric Power Cable," *IEEE Transactions on Power Apparatus and Systems*, vol. PAS-93, no. 2, pp. 601–13, March/April 1974.

McNulty, W. J., "False Tripping of Faulted Circuit Indicators," IEEE/PES Transmission and Distribution Conference, 1994.

Mercier, C. D. and Ticker, S., "DC Field Test for Medium-Voltage Cables: Why Can No One Agree?," *IEEE Transactions on Industry Applications*, vol. 34, no. 6, pp. 1366–70, November/December 1998.

Nannery, P. R., Tarpey, J. W., Lacenere, J. S., Meyer, D. F., and Bertini, G., "Extending the Service Life of 15 kV Polyethylene URD Cable Using Silicone Liquid," *IEEE Transactions on Power Delivery*, vol. 4, no. 4, pp. 1991–6, October 1989.

Neher, J. H. and McGrath, M. H., "The Calculation of the Temperature Rise and Load Capability of Cable Systems," *AIEE Transactions*, vol. 76, pp. 752–64, October 1957.

NFPA 70, *National Electrical Code*, National Fire Protection Association, 1999.

NRECA RER Project 90-8, *Underground Distribution System Design and Installation Guide*, National Rural Electric Cooperative Association, 1993.

Okonite, *Engineering Data for Copper and Aluminum Conductor Electrical Tables*, Okonite Company, publication EHB-90, 1990.

Powers, W. F., "An Overview of Water-Resistant Cable Designs," *IEEE Transactions on Industry Applications*, vol. 29, no. 5, 1993.

Smith, D. R., "System Considerations — Impedance and Fault Current Calculations," IEEE Tutorial Course on Application and Coordination of Reclosers, Sectionalizers, and Fuses, 1980. Publication 80 EHO157-8-PWR.

Smith, D. R. and Barger, J. V., "Impedance and Circulating Current Calculations for UD Multi-Wire Concentric Neutral Circuits," *IEEE Transactions on Power Apparatus and Systems*, vol. PAS-91, no. 3, pp. 992–1006, May/June 1972.

Southwire Company, *Power Cable Manual*, 2nd ed, 1997.

St. Pierre, C., *A Practical Guide to Short-Circuit Calculations*, Electric Power Consultants, Schenectady, NY, 2001.

State of New York Department of Public Service, "A Report on Consolidated Edison's July 1999 System Outages," March 2000.

Stember, L. H., Epstein, M. M., Gaines, G. V., Derringer, G. C., and Thomas, R. E., "Analysis of Field Failure Data on HMWPE- and XLPE-Insulated High-Voltage Distribution Cable," *IEEE Transactions on Power Apparatus and Systems*, vol. PAS-104, no. 8, pp. 1979–85, August 1985.

Stringer, N. T. and Kojovic, L. A., "Prevention of Underground Cable Splice Failures," *IEEE Transactions on Industry Applications*, vol. 37, no. 1, pp. 230–9, January/February 2001.

Thue, W. A., *Electrical Power Cable Engineering*, Marcel Dekker, New York, 1999.

Tyner, J. T., "Getting to the Bottom of UG Practices," *Transmission & Distribution World*, vol. 50, no. 7, pp. 44–56, July 1998.

Westinghouse Electric Corporation, *Electrical Transmission and Distribution Reference Book*, 1950.

I was down in the hole and pulled on one of the splices thinking that I might find the faulted one by pulling it apart, well you can only guess what happens next! KA-BOOOM, I was really pissed off at that point and still am. BUT YOU KNOW SOMETHING, ITS MY FAULT FOR TAKING SOME OTHER HALF ASS LINEMAN'S WORD FOR IT BEING DEAD AND NOT CHECKING IT OUT FOR MY SELF!

Anonymous poster, about beginning work after another lineman told him that the cables were disconnected at the source end.

www.powerlineman.com

4

Transformers

ac transformers are one of the keys to allowing widespread distribution of electric power as we see it today. Transformers efficiently convert electricity to higher voltage for long distance transmission and back down to low voltages suitable for customer usage. The distribution transformer normally serves as the final transition to the customer and often provides a local grounding reference. Most distribution circuits have hundreds of distribution transformers. Distribution feeders may also have other transformers: voltage regulators, feeder step banks to interface circuits of different voltages, and grounding banks.

4.1 Basics

A transformer efficiently converts electric power from one voltage level to another. A transformer is two sets of coils coupled together through a magnetic field. The magnetic field transfers all of the energy (except in an autotransformer). In an ideal transformer, the voltages on the input and the output are related by the turns ratio of the transformer:

$$V_1 = \frac{N_1}{N_2} V_2$$

where N_1 and N_2 are the number of turns and V_1 and V_2 are the voltage on windings 1 and 2.

In a real transformer, not all of the flux couples between windings. This *leakage* flux creates a voltage drop between windings, so the voltage is more accurately described by

$$V_1 = \frac{N_1}{N_2} V_2 - X_L I_1$$

where X_L is the leakage reactance in ohms as seen from winding 1, and I_1 is the current out of winding 1.

The current also transforms by the turns ratio, opposite of the voltage as

$$I_1 = \frac{N_2}{N_1} I_2 \quad \text{or} \quad N_1 I_1 = N_2 I_2$$

The "ampere-turns" stay constant at $N_1 I_1 = N_2 I_2$; this fundamental relationship holds well for power and distribution transformers.

A transformer has a magnetic core that can carry large magnetic fields. The cold-rolled, grain-oriented steels used in cores have permeabilities of over 1000 times that of air. The steel provides a very low-reluctance path for magnetic fields created by current through the windings.

Consider voltage applied to the *primary* side (source side, high-voltage side) with no load on the *secondary* side (load side, low-voltage side). The winding draws *exciting* current from the system that sets up a sinusoidal magnetic field in the core. The flux in turn creates a back emf in the coil that limits the current drawn into the transformer. A transformer with no load on the secondary draws very little current, just the exciting current, which is normally less than 0.5% of the transformer's full-load current. On the unloaded secondary, the sinusoidal flux creates an open-circuit voltage equal to the primary-side voltage times the turns ratio.

When we add load to the secondary of the transformer, the load pulls current through the secondary winding. The magnetic coupling of the secondary current pulls current through the primary winding, keeping constant ampere-turns. Normally in an inductive circuit, higher current creates more flux, but not in a transformer (except for the leakage flux). The increasing force from current in one winding is countered by the decreasing force from current through the other winding (see Figure 4.1). The flux in the core on a loaded transformer is the same as that on an unloaded transformer, even though the current is much higher.

The voltage on the primary winding determines the flux in the transformer (the flux is proportional to the time integral of voltage). The flux in the core determines the voltage on the output-side of the transformer (the voltage is proportional to the time derivative of the flux).

Figure 4.2 shows models with the significant impedances in a transformer. The detailed model shows the series impedances, the resistances and the reactances. The series resistance is mainly the resistance of the wires in each winding. The series reactance is the leakage impedance. The shunt branch is the magnetizing branch, current that flows to magnetize the core. Most of the magnetizing current is reactive power, but it includes a real power component. Power is lost in the core through:

- *Hysteresis* — As the magnetic dipoles change direction, the core heats up from the friction of the molecules.

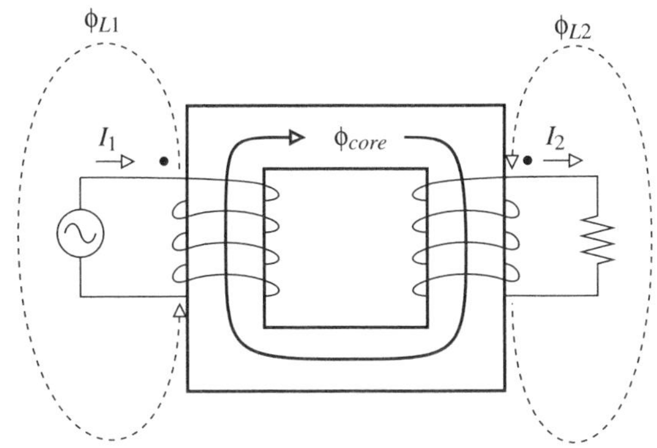

Magnetic equivalent circuit

$\mathcal{R} \approx 0$

$N_1 I_1$ ϕ_{core} $N_2 I_2$

Since $\mathcal{R} \approx 0$, $N_1 I_1 = N_2 I_2$

Electric circuit

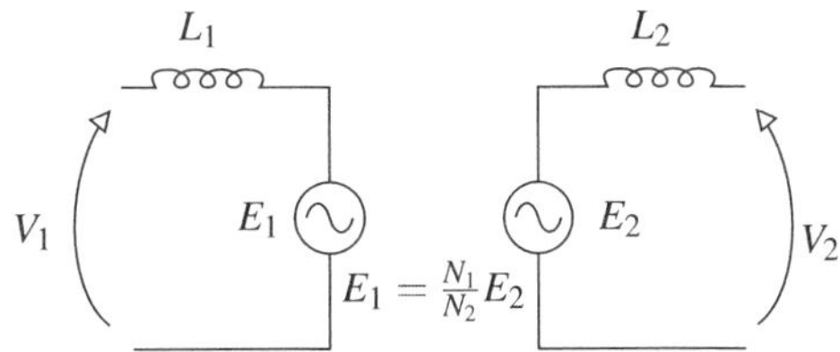

L_1 and L_2 are from the leakage fluxes, ϕ_{L1} and ϕ_{L2}

FIGURE 4.1
Transformer basic function.

Detailed transformer model

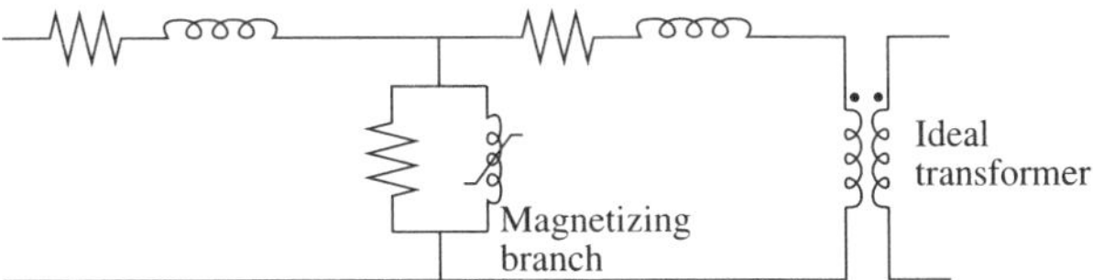

Simplified model

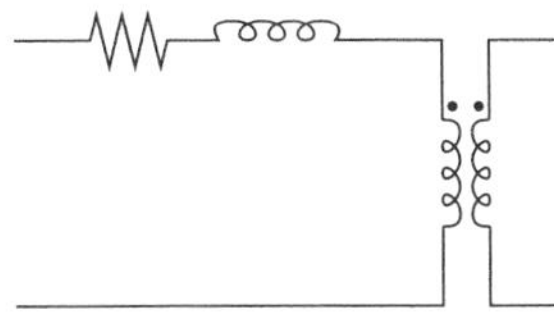

FIGURE 4.2
Transformer models.

TABLE 4.1

Common Scaling Ratios in Transformers

Quantity	Relative to kVA	Relative to a Reference Dimension, l
Rating	kVA	l^4
Weight	K $kVA^{3/4}$	K l^3
Cost	K $KVA^{3/4}$	K (% Total Loss)$^{-3}$
Length	K $kVA^{1/4}$	K l
Width	K $kVA^{1/4}$	K l
Height	K $kVA^{1/4}$	K l
Total losses	K $kVA^{3/4}$	K l^3
No-load losses	K $kVA^{3/4}$	K l^3
Exciting current	K $kVA^{3/4}$	K l^3
% Total loss	K $kVA^{-1/4}$	K l^{-1}
% No-load loss	K $kVA^{-1/4}$	K l^{-1}
% Exciting current	K $kVA^{-1/4}$	K l^{-1}
% R	K $kVA^{-1/4}$	K l^{-1}
% X	K $kVA^{1/4}$	K l
Volts/turn	K $kVA^{1/2}$	K l^2

Source: Arthur D. Little, "Distribution Transformer Rulemaking Engineering Analysis Update," Report to U.S. Department of Energy Office of Building Technology, State, and Community Programs. Draft. December 17, 2001.

- *Eddy currents* — Eddy currents in the core material cause resistive losses. The core flux induces the eddy currents tending to oppose the change in flux density.

The magnetizing branch impedance is normally above 5,000% on a transformer's base, so we can neglect it in many cases. The core losses are often referred to as iron losses or no-load losses. The load losses are frequently called the wire losses or copper losses. The various parameters of transformers scale with size differently as summarized in Table 4.1.

The simplified transformer model in Figure 4.2 with series resistance and reactance is sufficient for most calculations including load flows, short-circuit calculations, motor starting, or unbalance. Small distribution transformers have low leakage reactances, some less than 1% on the transformer rating, and X/R ratios of 0.5 to 5. Larger power transformers used in distribution substations have higher impedances, usually on the order of 7 to 10% with X/R ratios between 10 and 40.

The leakage reactance causes voltage drop on a loaded transformer. The voltage is from flux that doesn't couple from the primary to the secondary winding. Blume et al. (1951) describes leakage reactance well. In a real transformer, the windings are wound around a core; the high- and low-voltage windings are adjacent to each other. Figure 4.3 shows a configuration; each winding contains a number of turns of wire. The sum of the current in each wire of the high-voltage winding equals the sum of the currents in the

Side View of Windings

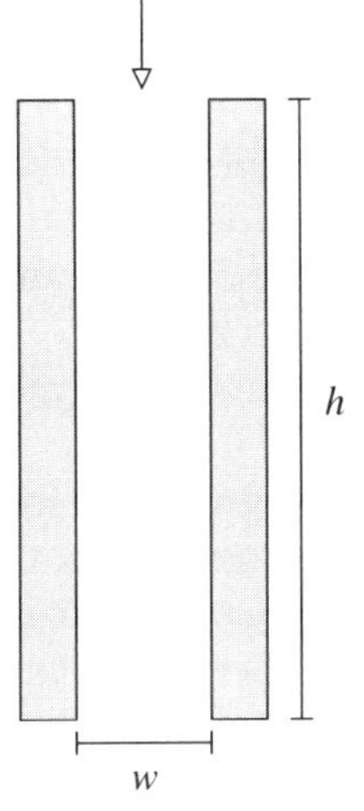

Top View of Windings

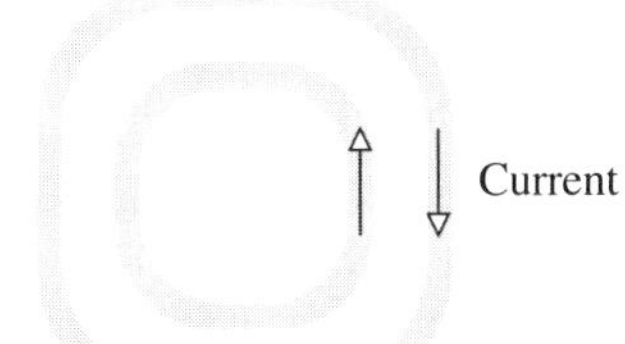

Equivalent Circuit

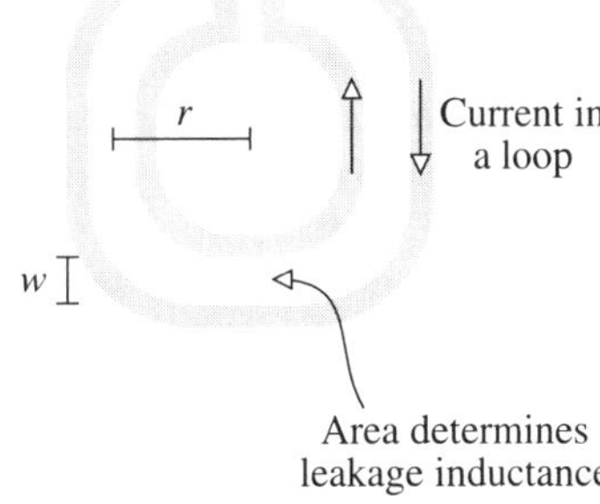

FIGURE 4.3
Leakage reactance.

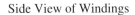

low-voltage winding ($N_1I_1 = N_2I_2$), so each winding is equivalent to a busbar. Each busbar carries equal current, but in opposite directions. The opposing currents create flux in the gap between the windings (this is called *leakage flux*). Now, looking at the two windings from the top, we see that the windings are equivalent to current flowing in a loop encompassing a given area. This area determines the leakage inductance.

The leakage reactance in percent is based on the coil parameters and separations (Blume et al., 1951) as follows:

$$X_{\%} = \frac{126 f (NI)^2 rw}{10^{11} h S_{kVA}}$$

where

f = system frequency, Hz
N = number of turns on one winding
I = full load current on the winding, A
r = radius to the windings, in.
w = width between windings, in.
h = height of the windings, in.
S_{kVA} = transformer rating, kVA

In general, leakage impedance increases with:

- Higher primary voltage (thicker insulation between windings)
- kVA rating
- Larger core (larger diameter leads to more area enclosed)

Leakage impedances are under control of the designer, and companies will make transformers for utilities with customized impedances. Large distribution substation transformers often need high leakage impedance to control fault currents, some as high as 30% on the base rating.

Mineral oil fills most distribution and substation transformers. The oil provides two critical functions: conducting heat and insulation. Because the oil is a good heat conductor, an oil-filled transformer has more load-carrying capability than a dry-type transformer. Since it provides good electrical insulation, clearances in an oil-filled transformer are smaller than a dry-type transformer. The oil conducts heat away from the coils into the larger thermal mass of the surrounding oil and to the transformer tank to be dissipated into the surrounding environment. Oil can operate continuously at high temperatures, with a normal operating temperature of 105°C. It is flammable; the flash point is 150°C, and the fire point is 180°C. Oil has high dielectric strength, 220 kV/in. (86.6 kV/cm), and evens out voltage stresses since the dielectric constant of oil is about 2.2, which is close to that of the insulation. The oil also coats and protects the coils and cores and other metal surfaces from corrosion.

4.2 Distribution Transformers

From a few kVA to a few MVA, distribution transformers convert primary-voltage to low voltage that customers can use. In North America, 40 million distribution transformers are in service, and another one million are installed each year (Alexander Publications, 2001). The transformer connection determines the customer's voltages and grounding configuration.

Distribution transformers are available in several standardized sizes as shown in Table 4.2. Most installations are single phase. The most common

TABLE 4.2

Standard Distribution Transformer Sizes

	Distribution Transformer Standard Ratings, kVA
Single phase	5, 10, 15, 25, 37.5, 50, 75, 100, 167, 250, 333, 500
Three phase	30, 45, 75, 112.5, 150, 225, 300, 500

TABLE 4.3
Insulation Levels for Distribution Transformers

Low-Frequency Test Level, kV rms	Basic Lightning Impulse Insulation Level, kV Crest	Chopped-Wave Impulse Levels	
		Minimum Voltage, kV Crest	Minimum Time to Flashover, μs
10	30	36	1.0
15	45	54	1.5
19	60	69	1.5
26	75	88	1.6
34	95	110	1.8
40	125	145	2.25
50	150	175	3.0
70	200	230	3.0
95	250	290	3.0
140	350	400	3.0

Source: IEEE Std. C57.12.00-2000.

overhead transformer is the 25-kVA unit; padmounted transformers tend to be slightly larger where the 50-kVA unit is the most common.

Distribution transformer impedances are rather low. Units under 50 kVA have impedances less than 2%. Three-phase underground transformers in the range of 750 to 2500 kVA normally have a 5.75% impedance as specified in (ANSI/IEEE C57.12.24-1988). Lower impedance transformers provide better voltage regulation and less voltage flicker for motor starting or other fluctuating loads. But lower impedance transformers increase fault currents on the secondary, and secondary faults impact the primary side more (deeper voltage sags and more fault current on the primary).

Standards specify the insulation capabilities of distribution transformer windings (see Table 4.3). The low-frequency test is a power-frequency (60 Hz) test applied for one minute. The basic lightning impulse insulation level (BIL) is a fast impulse transient. The front-of-wave impulse levels are even shorter-duration impulses.

The through-fault capability of distribution transformers is also given in IEEE C57.12.00-2000 (see Table 4.4). The duration in seconds of the short-circuit capability is:

$$t = \frac{1250}{I^2}$$

where I is the symmetrical current in multiples of the normal base current from Table 4.4.

Overhead and padmounted transformer tanks are normally made of mild carbon steel. Corrosion is one of the main concerns, especially for anything on the ground or in the ground. Padmounted transformers tend to corrode

TABLE 4.4

Through-Fault Capability of Distribution Transformers

Single-Phase Rating, kVA	Three-Phase Rating, kVA	Withstand Capability in per Unit of Base Current (Symmetrical)
5–25	15–75	40
37.5–110	112.5–300	35
167–500	500	25

Source: IEEE Std. C57.12.00-2000, *IEEE Standard General Requirements for Liquid-Immersed Distribution, Power, and Regulating Transformers.*

near the base (where moisture and dirt and other debris may collect). Submersible units, being highly susceptible to corrosion, are often stainless steel.

Distribution transformers are "self cooled"; they do not have extra cooling capability like power transformers. They only have one kVA rating. Because they are small and because customer peak loadings are relatively short duration, overhead and padmounted distribution transformers have significant overload capability. Utilities regularly size them to have peak loads exceeding 150% of the nameplate rating.

Transformers in underground vaults are often used in cities, especially for network transformers (feeding secondary grid networks). In this application, heat can be effectively dissipated (but not as well as with an overhead or padmounted transformer).

Subsurface transformers are installed in an enclosure just big enough to house the transformer with a grate covering the top. A "submersible" transformer is normally used, one which can be submerged in water for an extended period (ANSI/IEEE C57.12.80-1978). Heat is dissipated through the grate at the top. Dirt and debris in the enclosure can accelerate corrosion. Debris blocking the grates or vents can overheat the transformer.

Direct-buried transformers have been attempted over the years. The main problems have been overheating and corrosion. In soils with high electrical and thermal resistivity, overheating is the main concern. In soils with low electrical and thermal resistivity, overheating is not as much of a concern, but corrosion becomes a problem. Thermal conductivity in a direct-buried transformer depends on the thermal conductivity of the soil. The buried transformer generates enough heat to dry out the surrounding soil; the dried soil shrinks and creates air gaps. These air gaps act as insulating layers that further trap heat in the transformer.

4.3 Single-Phase Transformers

Single-phase transformers supply single-phase service; we can use two or three single-phase units in a variety of configurations to supply three-phase

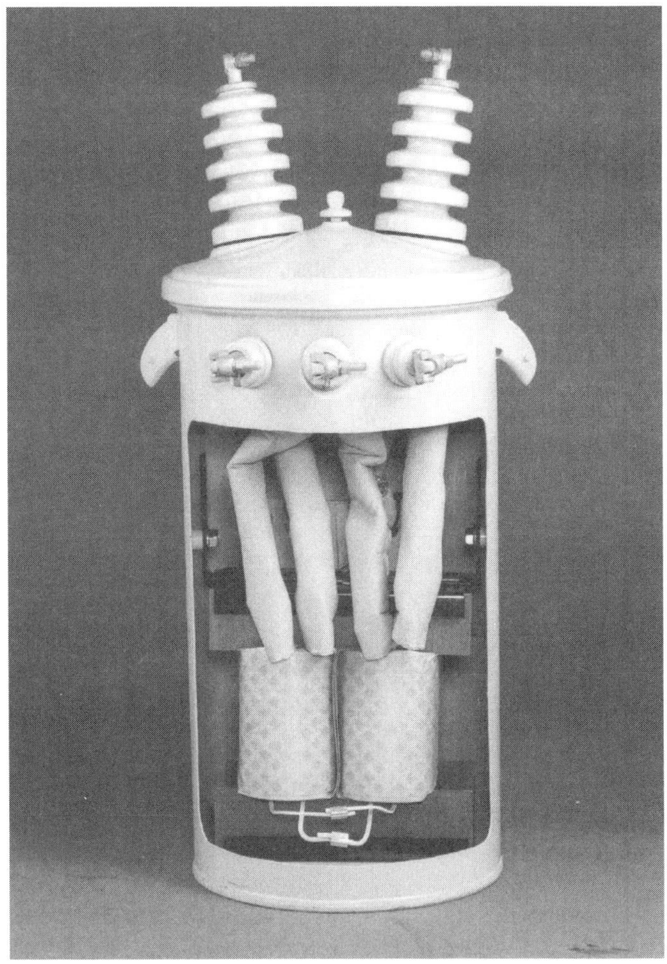

FIGURE 4.4
Single-phase distribution transformer. (Photo courtesy of ABB, Inc. With permission.)

service. A transformer's nameplate gives the kVA ratings, the voltage ratings, percent impedance, polarity, weight, connection diagram, and cooling class. Figure 4.4 shows a cutaway view of a single-phase transformer.

For a single-phase transformer supplying single-phase service, the load-full current in amperes is

$$I = \frac{S_{kVA}}{V_{kV}}$$

where

S_{kVA} = Transformer kVA rating

V_{kV} = Line-to-ground voltage rating in kV

TABLE 4.5

Winding Designations for Single-Phase Primary and Secondary Transformer Windings with One Winding

Nomenclature	Examples	Description
E	13800	E shall indicate a winding of E volts that is suitable for Δ connection on an E volt system.
E/E_1Y	2400/4160Y	E/E_1Y shall indicate a winding of E volts that is suitable for Δ connection on an E volt system or for Y connection on an E_1 volt system.
E/E_1GrdY	7200/12470GrdY	E/E_1GrdY shall indicate a winding of E volts having reduced insulation that is suitable for Δ connection on an E volt system or Y connection on an E_1 volt system, transformer, neutral effectively grounded.
E_1GrdY/E	12470GrdY/7200 480GrdY/277	E_1GrdY/E shall indicate a winding of E volts with reduced insulation at the neutral end. The neutral end may be connected directly to the tank for Y or for single-phase operation on an E_1 volt system, provided the neutral end of the winding is effectively grounded.
$E_1 = \sqrt{3}\ E$		

Note: E is line-to-neutral voltage of a Y winding, or line-to-line voltage of a Δ winding.

Source: IEEE Std. C57.12.00-2000.

So, a single-phase 50-kVA transformer with a high-voltage winding of 12470GrdY/7200 V has a full-load current of 6.94 A on the primary. On a 240/120-V secondary, the full-load current across the 240-V winding is 208.3 A.

Table 4.5 and Table 4.6 show the standard single-phase winding connections for primary and secondary windings. High-voltage bushings are labeled H*, starting with H1 and then H2 and so forth. Similarly, the low-voltage bushings are labeled X1, X2, X3, and so on.

The standard North American single-phase transformer connection is shown in Figure 4.5. The standard secondary load service is a 120/240-V three-wire service. This configuration has two secondary windings in series with the midpoint grounded. The secondary terminals are labeled X1, X2, and X3 where the voltage X1-X2 and X2-X3 are each 120 V. X1-X3 is 240 V.

Power and distribution transformers are assigned polarity dots according to the terminal markings. Current entering H1 results in current leaving X1. The voltage from H1 to H2 is in phase with the voltage from X1 to X3.

On overhead distribution transformers, the high-voltage terminal H1 is always on the left (when looking into the low-voltage terminals; the terminals are not marked). On the low-voltage side, the terminal locations are different, depending on size. If X1 is on the right, it is referred to as *additive polarity* (if X3 is on the right, it is *subtractive polarity*). Polarity is additive if the voltages add when the two windings are connected in series around the transformer (see Figure 4.6). Industry standards specify the polarity of a

TABLE 4.6

Two-Winding Transformer Designations for Single-Phase Primaries and Secondaries

Nomenclature	Examples	Description
E/2E	120/240 240/280 X4 X3 X2 X1	E/2E shall indicate a winding, the sections of which can be connected in parallel for operation at E volts, or which can be connected in series for operation at 2E volts, or connected in series with a center terminal for three-wire operation at 2E volts between the extreme terminals and E volts between the center terminal and each of the extreme terminals.
2E/E	240/120 X3 X2 X1	2E/E shall indicate a winding for 2E volts, two-wire full kilovoltamperes between extreme terminals, or for 2E/E volts three-wire service with 1/2 kVA available only, from midpoint to each extreme terminal.
E × 2E	240 × 480 X4 X3 X2 X1	E × 2E shall indicate a winding for parallel or series operation only but not suitable for three-wire service.

Source: IEEE Std. C57.12.00-2000. Copyright 2000 IEEE. All rights reserved.

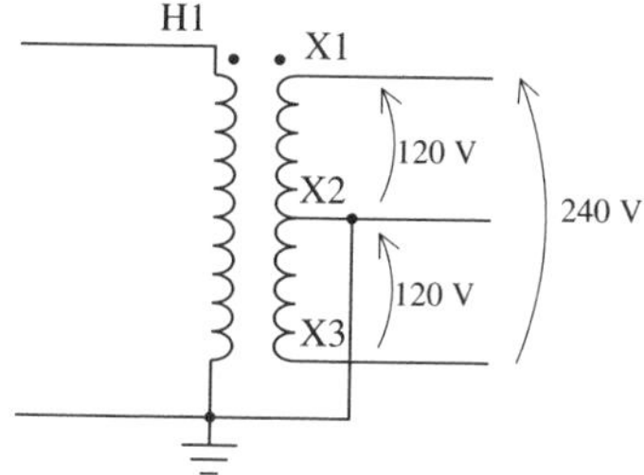

FIGURE 4.5
Single-phase distribution transformer diagram.

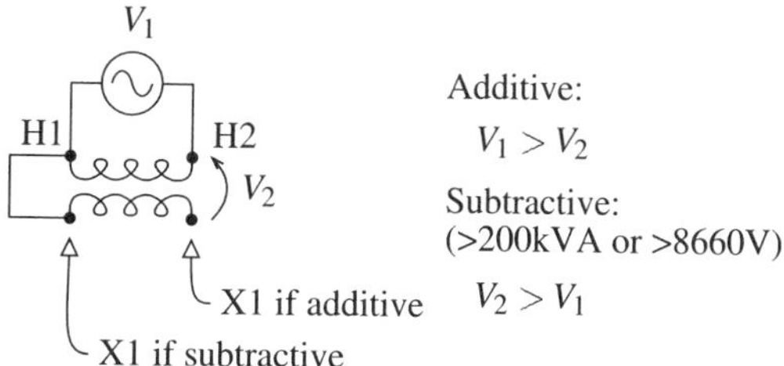

FIGURE 4.6
Additive and subtractive polarity.

transformer, which depends on the size and the high-voltage winding. Single-phase transformers have additive polarity if (IEEE C57.12.00-2000):

$$\text{kVA} \leq 200 \text{ and } V \leq 8660$$

All other distribution transformers have subtractive polarity. The reason for the division is that originally all distribution transformers had additive polarity and all power transformers had subtractive polarity. Increasing sizes of distribution transformers caused overlap between "distribution" and "power" transformers, so larger distribution transformers were made with subtractive polarity for consistency. Polarity is important when connecting single-phase units in three-phase banks and for paralleling units.

Manufacturers make single-phase transformers as either shell form or core form (see Figure 4.7). Core-form designs prevailed prior to the 1960s; now, both shell- and core-form designs are available. Single-phase core-form transformers must have *interlaced* secondary windings (the low-high-low design). Every secondary leg has two coils, one wrapped around each leg of the core. The balanced configuration of the interlaced design allows unbalanced loadings on each secondary leg. Without interlacing, unbalanced secondary loads excessively heat the tank. An unbalanced secondary load creates an unbalanced flux in the iron core. The core-form construction does not have a return path for the unbalanced flux, so the flux returns outside of the iron core (in contrast, the shell-form construction has a return path for such flux). Some of the stray flux loops through the transformer tank and heats the tank.

The shell-form design does not need to have interlaced windings, so the *noninterlaced* configuration is normally used on shell-form transformers since it is simpler. The noninterlaced secondary has two to four times the reactance: the secondary windings are separated by the high-voltage winding and the insulation between them. Interlacing reduces the reactance since the low-voltage windings are right next to each other.

Using a transformer's impedance magnitude and load losses, we can find the real and reactive impedance in percent as

$$R = \frac{W_{CU}}{10 S_{kVA}}$$

$$X = \sqrt{Z^2 - R^2}$$

where

S_{kVA} = transformer rating, kVA

$W_{CU} = W_{TOT} - W_{NL}$ = load loss at rated load, W

W_{TOT} = total losses at rated load, W

W_{NL} = no-load losses, W

Z = nameplate impedance magnitude, %

Core form, interlaced

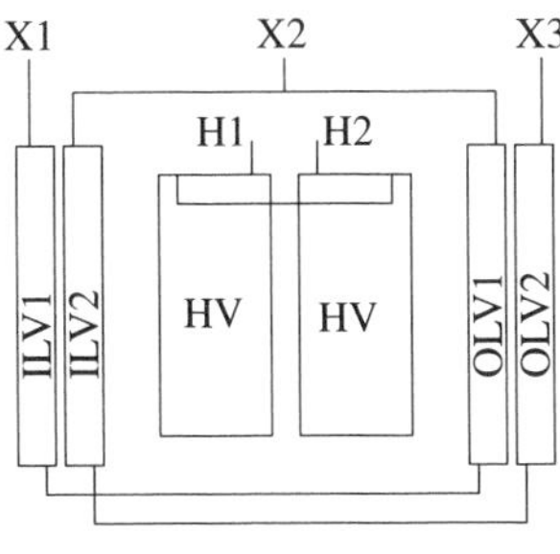

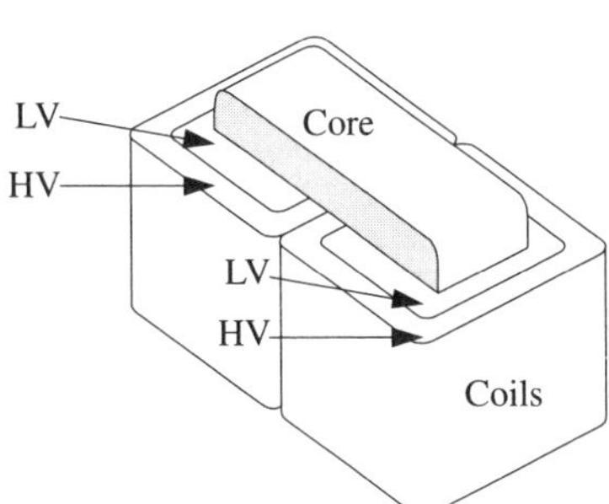

Shell form, non-interlaced

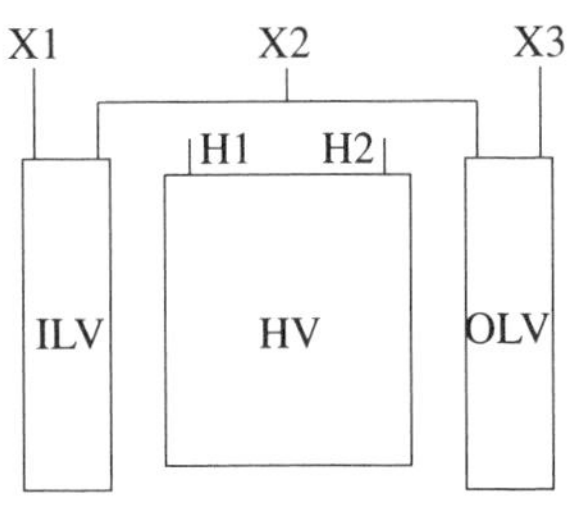

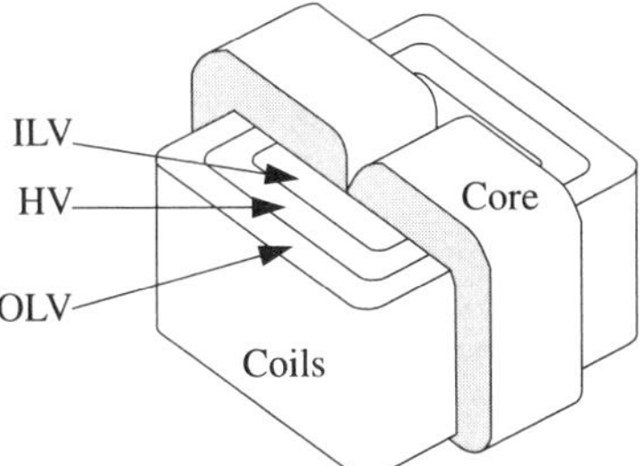

FIGURE 4.7
Core-form and shell-form single-phase distribution transformers. (From IEEE Task Force Report, "Secondary (Low-Side) Surges in Distribution Transformers," *IEEE Trans. Power Delivery,* 7(2), 746–756, April 1992. With permission. ©1992 IEEE.)

The nameplate impedance of a single-phase transformer is the *full-winding* impedance, the impedance seen from the primary when the full secondary winding is shorted from X1 to X3. Other impedances are also important; we need the two *half-winding* impedances for secondary short-circuit calculations and for unbalance calculations on the secondary. One impedance is the impedance seen from the primary for a short circuit from X1 to X2. Another is from X2 to X3. The half-winding impedances are not provided on the nameplate; we can measure them or use the following approximations. Figure 4.8 shows a model of a secondary winding for use in calculations.

The half-winding impedance of a transformer depends on the construction. In the model in Figure 4.8, one of the half-winding impedances in percent equals $Z_A + Z_1$; the other equals $Z_A + Z_2$. A core- or shell-form transformer with an interlaced secondary winding has an impedance in percent of approximately:

$$Z_{\mathrm{HX1\text{-}2}} = Z_{\mathrm{HX2\text{-}3}} = 1.5\,R + \mathrm{j}\,1.2\,X$$

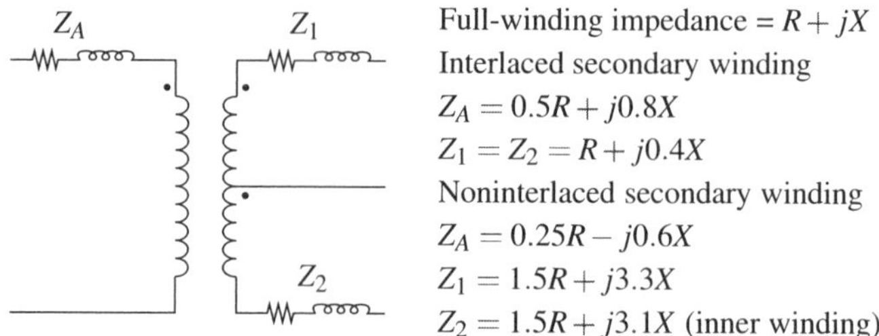

FIGURE 4.8
Model of a 120/240-V secondary winding with all impedances in percent. (Impedance data from [Hopkinson, 1976].)

where R and X are the real and reactive components of the full-winding impedance (H1 to H2 and X1 to X3) in percent. A noninterlaced shell-form transformer has an impedance in percent of approximately:

$$Z_{HX1-2} = Z_{HX2-3} = 1.75\ R + \mathrm{j}\ 2.5\ X$$

In a noninterlaced transformer, the two half-winding impedances are not identical; the impedance to the inner low-voltage winding is less than the impedance to the outer winding (the radius to the gap between the outer secondary winding and the primary winding is larger, so the gap between windings has more area).

A secondary fault across one 120-V winding at the terminals of a noninterlaced transformer has current about equal to the current for a fault across the 240-V winding. On an interlaced transformer, the lower relative impedance causes higher currents for the 120-V fault.

Consider a 50-kVA transformer with Z = 2%, 655 W of total losses, no-load losses of 106 W, and a noninterlaced 120/240-V secondary winding. This translates into a full-winding percent impedance of 1.1 + j1.67. For a fault across the 240-V winding, the current is found as

$$Z_{\Omega,240\mathrm{V}} = (R + jX)\frac{10(0.24\mathrm{kV})^2}{S_{kVA}} = (1.1 + j1.67)\frac{10(0.24\mathrm{kV})^2}{50\mathrm{kVA}} = 0.013 + j0.019\Omega$$

$$I_{240\mathrm{V}} = \left|\frac{0.24\mathrm{kV}}{Z_{\Omega,240\mathrm{V}}}\right| = 10.4\mathrm{kA}$$

For a fault across the 120-V winding on this noninterlaced transformer, the current is found as

$$Z_{\Omega,120\text{V}} = (1.75R + j2.5X)\frac{10(0.12\text{kV})^2}{S_{kVA}} = (1.93 + j4.18)\frac{10(0.12\text{kV})^2}{50\text{kVA}}$$

$$= 0.0055 + j0.0120\Omega$$

$$I_{120\text{V}} = \left|\frac{0.12\text{kV}}{Z_{\Omega,120\text{V}}}\right| = 9.06\text{kA}$$

Consider the same transformer characteristics on a transformer with an interlaced secondary and Z = 1.4%. The 240-V and 120-V short-circuit currents are found as

$$Z_{\Omega,240\text{V}} = (R + jX)\frac{10(0.24\text{kV})^2}{S_{kVA}} = (1.1 + j0.87)\frac{10(0.24\text{kV})^2}{50\text{kVA}} = 0.013 + j0.01\Omega$$

$$I_{240\text{V}} = \left|\frac{0.24\text{kV}}{Z_{\Omega,240\text{V}}}\right| = 14.9\text{kA}$$

$$Z_{\Omega,120\text{V}} = (1.5R + j1.2X)\frac{10(0.12\text{kV})^2}{S_{kVA}} = (1.65 + j1.04)\frac{10(0.12\text{kV})^2}{50\text{kVA}}$$

$$= 0.0048 + j0.003\Omega$$

$$I_{120\text{V}} = \left|\frac{0.12\text{kV}}{Z_{\Omega,120\text{V}}}\right| = 21.4\text{kA}$$

The fault current for a 120-V fault is significantly higher than the 240-V current.

Completely self-protected transformers (CSPs) are a widely used single-phase distribution transformer with several built-in features (see Figure 4.9):

- Tank-mounted arrester
- Internal "weak-link" fuse
- Secondary breaker

CSPs do not need a primary-side cutout with a fuse. The internal primary fuse protects against an internal failure in the transformer. The weak link has less fault-clearing capability than a fuse in a cutout, so they need external current-limiting fuses where fault currents are high.

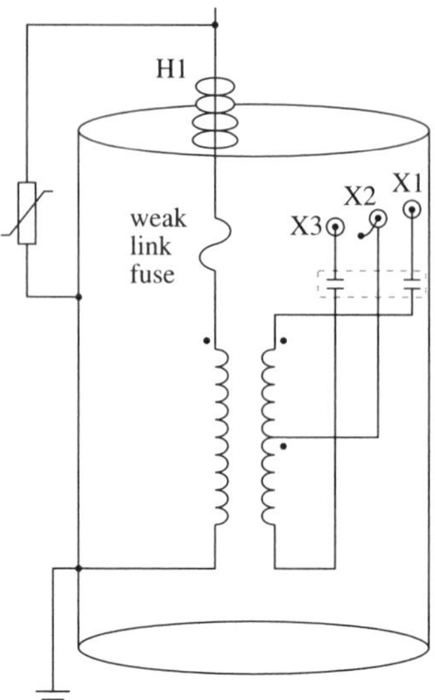

FIGURE 4.9
Completely self-protected transformer.

Secondary breakers provide protection against overloads and secondary faults. The breaker responds to current and oil temperature. Tripping is controlled by deflection of bimetallic elements in series. The oil temperature and current through the bimetallic strips heat the bimetal. Past a critical temperature, the bimetallic strips deflect enough to operate the breaker. Figure 4.10 shows trip characteristics for secondary breakers inside two size transformers. The secondary breaker has an emergency position to allow extra overload without tripping (to allow crews time to replace the unit). Crews can also use the breaker to drop the secondary load.

Some CSPs have overload-indicating lights that signal an overload. The indicator light doesn't go off until line crews reset the breaker. The indicator lights are not ordered as often (and crews often disable them in the field) because they generate a fair number of nuisance phone calls from curious/helpful customers.

4.4 Three-Phase Transformers

Three-phase overhead transformer services are normally constructed from three single-phase units. Three-phase transformers for underground service (either padmounted, direct buried, or in a vault or building or manhole) are normally single units, usually on a three- or five-legged core. Three-phase distribution transformers are usually core construction (see Figure 4.11), with

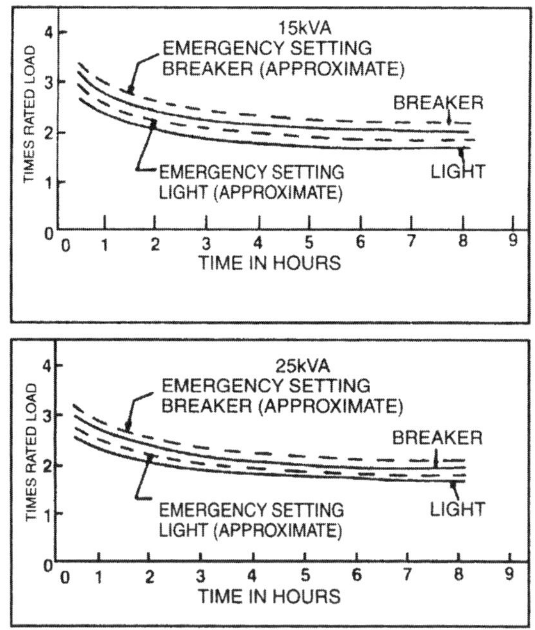

FIGURE 4.10
Clearing characteristics of a secondary breaker. (From ERMCO, Inc. With permission.)

either a three-, four-, or five-legged core construction (shell-type construction is rarely used). The five-legged wound core transformer is very common. Another option is *triplex* construction, where the three transformer legs are made from single individual core/coil assemblies (just like having three separate transformers).

The kVA rating for a three-phase bank is the total of all three phases. The full-load current in amps in each phase of a three-phase unit or bank is

$$I = \frac{S_{kVA}}{3V_{LG,kV}} = \frac{S_{kVA}}{\sqrt{3}V_{LL,kV}}$$

where

S_{kVA} = Transformer three-phase kVA rating
$V_{LG,kV}$ = Line-to-ground voltage rating, kV
$V_{LL,kV}$ = Line-to-line voltage rating, kV

A three-phase, 150-kVA transformer with a high-voltage winding of 12470GrdY/7200 V has a full-load current of 6.94 A on the primary (the same current as one 50-kVA single-phase transformer).

There are many types of three-phase connections used to serve three-phase load on distribution systems (ANSI/IEEE C57.105-1978; Long, 1984; Rusch

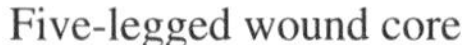

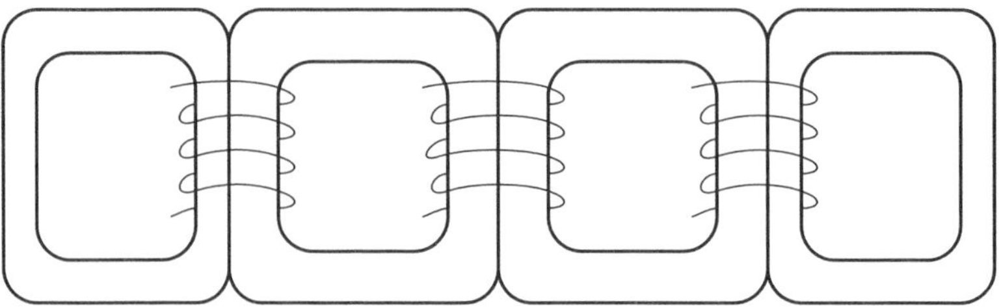

Four-legged stacked core

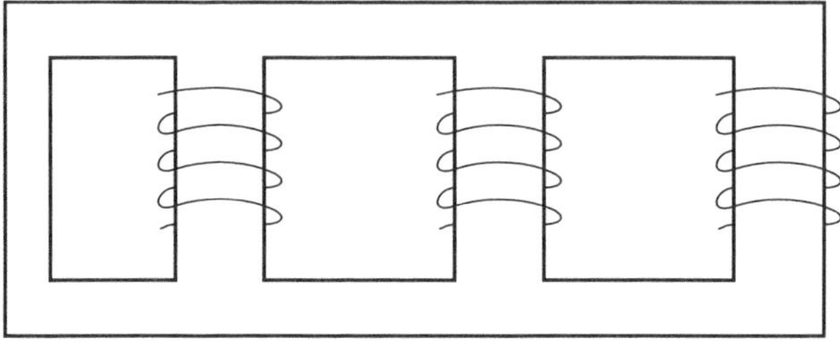

Three-legged stacked core

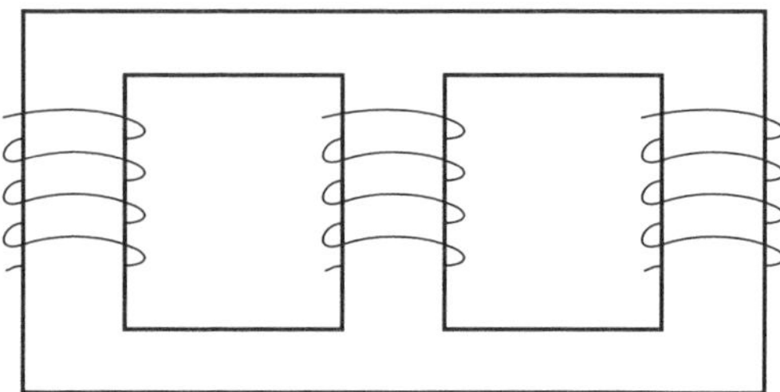

FIGURE 4.11
Three-phase core constructions.

and Good, 1989). Both the primary and secondary windings may be connected in different ways: delta, floating wye, or grounded wye. This notation describes the connection of the transformer windings, not the configuration of the supply system. A "wye" primary winding may be applied on a "delta" distribution system. On the primary side of three-phase distribution transformers, utilities have a choice between grounded and ungrounded winding connections. The tradeoffs are:

- *Ungrounded primary* — The delta and floating-wye primary connections are suitable for ungrounded and grounded distribution systems. Ferroresonance is more likely with ungrounded primary connections. Ungrounded primary connections do not supply ground fault current to the primary distribution system.

- *Grounded primary* — The grounded-wye primary connection is only suitable on four-wire grounded systems (either multigrounded or unigrounded). It is not for use on ungrounded systems. Grounded-wye primaries may provide an unwanted source for ground fault current.

Customer needs play a role in the selection of the secondary configuration. The delta configuration and the grounded-wye configuration are the two most common secondary configurations. Each has advantages and disadvantages:

- *Grounded-wye secondary* — Figure 4.12 shows the most commonly used transformers with a grounded-wye secondary winding: grounded wye – grounded wye and the delta – grounded wye. The

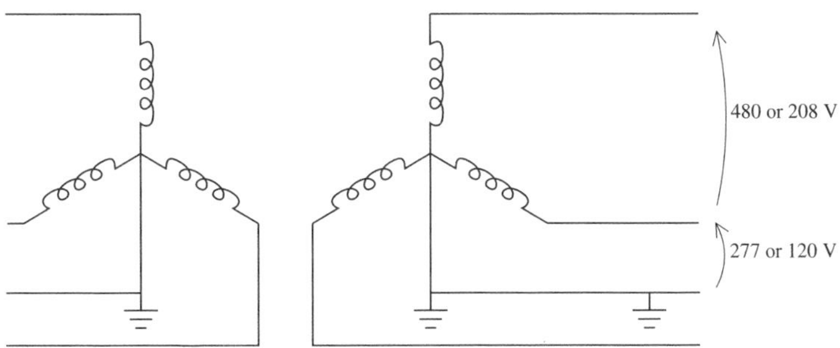

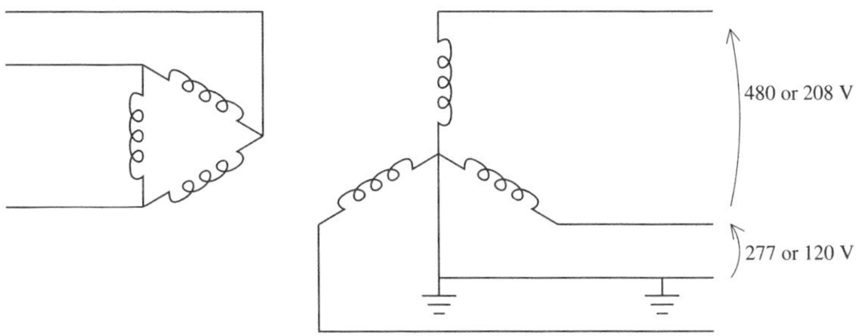

FIGURE 4.12
Three-phase distribution transformer connections with a grounded-wye secondary.

standard secondary voltages are 480Y/277 V and 208Y/120 V. The 480Y/277-V connection is suitable for driving larger motors; lighting and other 120-V loads are normally supplied by dry-type transformers. A grounded-wye secondary adeptly handles single-phase loads on any of the three phases with less concerns about unbalances.

- *Delta secondary* — An ungrounded secondary system like the delta can supply three-wire ungrounded service. Some industrial facilities prefer an ungrounded system, so they can continue to operate with line-to-ground faults. With one leg of the delta grounded at the midpoint of the winding, the utility can supply 240/120-V service. End-users can use more standard 230-V motors (without worrying about reduced performance when run at 208 V) and still run lighting and other single-phase loads. This tapped leg is often called the *lighting* leg (the other two legs are the *power* legs). Figure 4.13 shows the most commonly used connections with a delta secondary windings. This is commonly supplied with overhead transformers.

Many utilities offer a variety of three-phase service options and, of course, most have a variety of existing transformer connections. Some utilities restrict choices in an effort to increase consistency and reduce inventory. A restrictive utility may only offer three choices: 480Y/277-V and 208Y/120-V four-wire, three-phase services, and 120/240-V three-wire single-phase service.

For supplying customers requiring an ungrounded secondary voltage, either a three-wire service or a four-wire service with 120 and 240 V, the following provides the best connection:

- Floating wye – delta

For customers with a four-wire service, either of the following are normally used:

- Grounded wye – grounded-wye
- Delta – grounded wye

Choice of preferred connection is often based on past practices and equipment availability.

A wye – delta transformer connection shifts the phase-to-phase voltages by 30° with the direction dependent on how the connection is wired. The phase angle difference between the high-side and low-side voltage on delta – wye and wye – delta transformers is 30°; by industry definition, the low voltage lags the high voltage (IEEE C57.12.00-2000). Figure 4.14 shows wiring diagrams to ensure proper phase connections of popular three-phase connections.

Table 4.7 shows the standard winding designations shown on the nameplate of three-phase units.

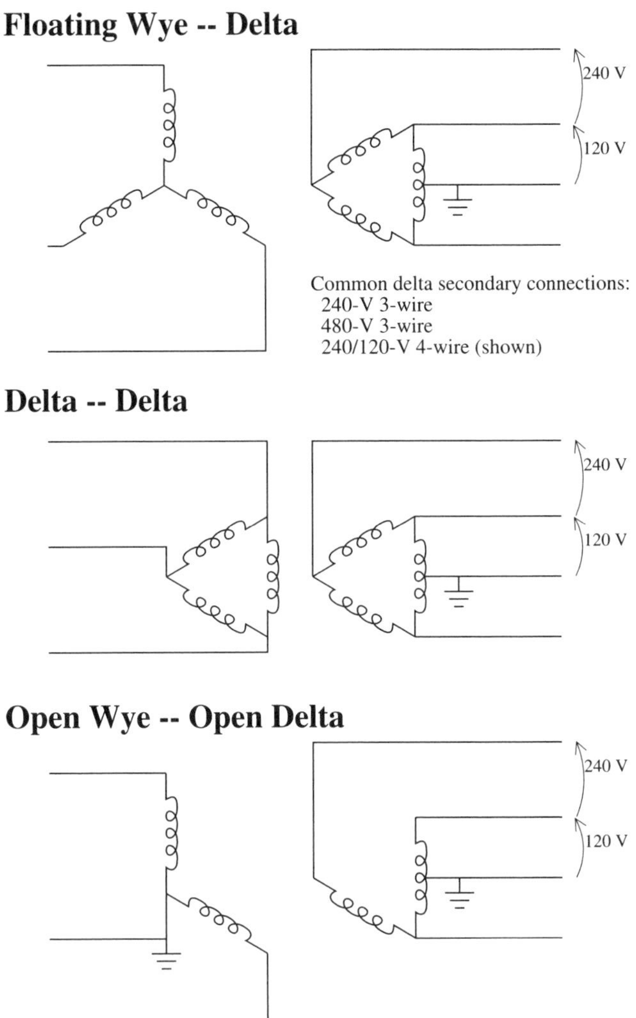

FIGURE 4.13
Common three-phase distribution transformer connections with a delta-connected secondary.

4.4.1 Grounded Wye – Grounded Wye

The most common three-phase transformer supply connection is the grounded wye – grounded wye connection. Its main characteristics are:

- *Supply* — Must be a grounded 4-wire system
- *Service*
  - Supplies grounded-wye service, normally either 480Y/277 V or 208Y/120 V.

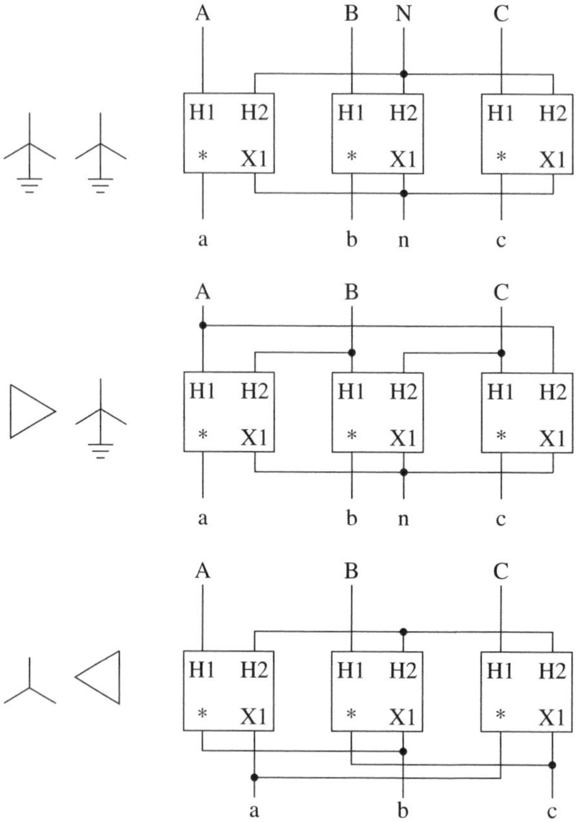

FIGURE 4.14
Wiring diagrams for common transformer connections with additive units. Subtractive units have the same secondary connections, but the physical positions of X1 and * are reversed on the transformer.

  - Cannot supply 120 and 240 V.
  - Does not supply ungrounded service. (But a grounded wye – floating wye connection can.)

- *Tank heating* — Probable with three-legged core construction; less likely, but possible under severe unbalance with five-legged core construction. Impossible if made from three single-phase units.
- *Zero sequence* — All zero-sequence currents — harmonics, unbalance, and ground faults — transfer to the primary. It also acts as a high-impedance ground source to the primary.
- *Ferroresonance* — No chance of ferroresonance with a bank of single-phase units or triplex construction; some chance with a four- or five-legged core construction.

TABLE 4.7

Three-Phase Transformer Designations

Nomenclature	Examples	Description
E	2400 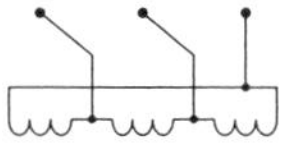	E shall indicate a winding that is permanently Δ connected for operation on an E volt system.
E_1Y	4160Y 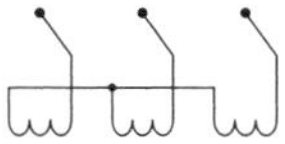	E_1Y shall indicate a winding that is permanently Y connected without a neutral brought out (isolated) for operation on an E_1 volt system.
E_1Y/E	4160Y/2400 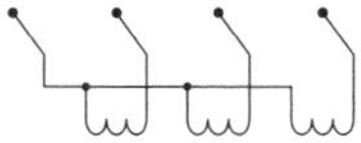	E_1Y/E shall indicate a winding that is permanently Y connected with a fully insulated neutral brought out for operation on an E_1 volt system, with E volts available from line to neutral.
E/E_1Y	2400/4160Y 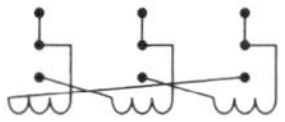	E/E_1Y shall indicate a winding that may be Δ connected for operation on an E volt system, or may be Y connected without a neutral brought out (isolated) for operation on an E_1 volt system.
$E/E_1Y/E$	2400/4160Y/2400 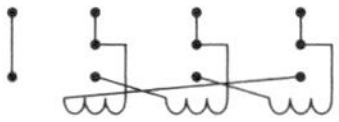	$E/E_1Y/E$ shall indicate a winding that may be Δ connected for operation on an E volt system or may be Y connected with a fully insulated neutral brought out for operation on an E_1 volt system with E volts available from line to neutral.
E_1GrdY/E	12470GrdY/7200 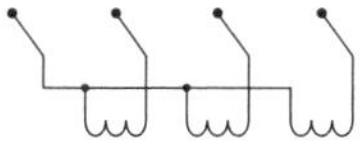	E_1GrdY/E shall indicate a winding with reduced insulation and permanently Y connected, with a neutral brought out and effectively grounded for operation on an E_1 volt system with E volts available from line to neutral.
$E/E_1GrdY/E$	7200/12470GrdY/7200 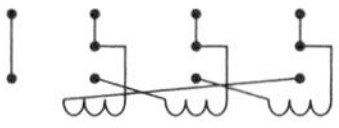	$E/E_1GrdY/E$ shall indicate a winding, having reduced insulation, which may be Δ connected for operation on an E volt system or may be connected Y with a neutral brought out and effectively grounded for operation on an E_1 volt system with E volts available from line to neutral.
$V \times V_1$	7200 × 14400 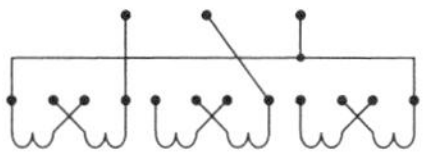	$V \times V_1$ shall indicate a winding, the sections of which may be connected in parallel to obtain one of the voltage ratings (as defined in a–g) of V, or may be connected in series to obtain one of the voltage ratings (as defined in a–g) of V_1. Winding are permanently Δ or Y connected.

Source: IEEE Std. C57.12.00-2000.

- *Coordination* — Because ground faults pass through to the primary, larger transformer services and local protective devices should be coordinated with utility ground relays.

The grounded wye – grounded wye connection has become the most common three-phase transformer connection. Reduced ferroresonance is the main reason for the shift from the delta – grounded wye to the grounded wye – grounded wye.

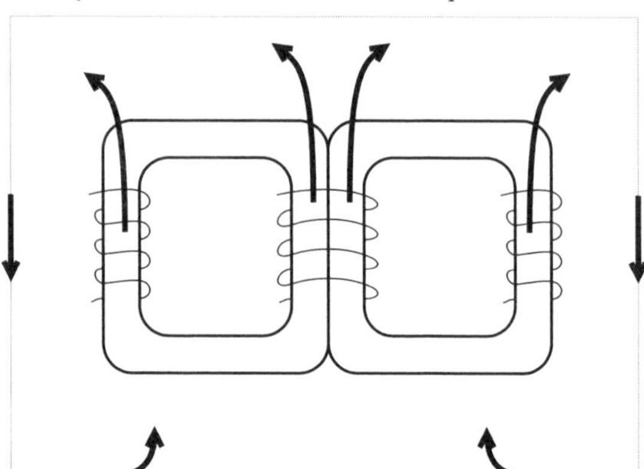

Stray flux in the tank due to zero sequence current

FIGURE 4.15
Zero-sequence flux caused by unbalanced voltages or unbalanced loads.

A grounded wye – grounded wye transformer with three-legged core construction is not suitable for supplying four-wire service. Unbalanced secondary loading and voltage unbalance on the primary system, these unbalances heat the transformer tank. In a three-legged core design, zero-sequence flux has no iron-core return path, so it must return via a high-reluctance path through the air gap and partially through the transformer tank (see Figure 4.15). The zero-sequence flux induces eddy currents in the tank that heat the tank.

A four- or five-legged core transformer greatly reduces the problem of tank heating with a grounded wye – grounded wye connection. The extra leg(s) provide an iron path for zero-sequence flux, so none travels into the tank. Although much less of a problem, tank heating can occur on four and five-legged core transformers under certain conditions; very large voltage unbalances may heat the tank. The outer leg cores normally do not have full capacity for zero-sequence flux (they are smaller than the inner leg cores), so under very high voltage unbalance, the outer legs may saturate. Once the legs saturate, some of the zero-sequence flux flows in the tank causing heating. The outer legs may saturate for a zero-sequence voltage of about 50 to 60% of the rated voltage. If a fuse or single-phase recloser or single-pole switch opens upstream of the transformer, the unbalance may be high enough to heat the tank, depending on the loading on the transformer and whether faults still exist. The worst conditions are when a single-phase interrupter clears a line-to-line or line-to-line-to-line fault (but not to ground) and the transformer is energized through one or two phases.

To completely eliminate the chance of tank heating, do not use a core-form transformer. Use a bank made of three single-phase transformers, or use triplex construction.

A wye – wye transformer with the primary and secondary neutrals tied together internally causes high line-to-ground secondary voltages if the neutral is not grounded. This connection cannot supply three-wire ungrounded service. Three-phase padmounted transformers with an H0X0 bushing have the neutrals bonded internally. If the H0X0 bushing is floated, high voltages can occur from phase to ground on the secondary.

To supply ungrounded secondary service with a grounded-wye primary, use a grounded wye – floating wye connection: the secondary should be floating-wye with no connection between the primary and secondary neutral points.

4.4.2 Delta – Grounded Wye

The delta – grounded wye connection has several interesting features, many related to its delta winding, which establishes a local grounding reference and blocks zero-sequence currents from entering the primary.

- *Supply* — 3-wire or 4-wire system.
- *Service*
  - Supplies grounded-wye service, normally either 480Y/277 V or 208Y/120 V.
  - Cannot supply both 120 and 240 V.
  - Does not supply ungrounded service.
- *Ground faults* — This connection blocks zero sequence, so upstream ground relays are isolated from line-to-ground faults on the secondary of the customer transformer.
- *Harmonics* — The delta winding isolates the primary from zero-sequence harmonics created on the secondary. Third harmonics and other zero-sequence harmonics cannot get through to the primary (they circulate in the delta winding).
- *No primary ground source* — For line-to-ground faults on the primary, the delta – grounded wye connection cannot act as a grounding source.
- *Secondary ground source* — Provides a grounding source for the secondary, independent of the primary-side grounding configuration.
- *No tank heating* — The delta connection ensures that zero-sequence flux will not flow in the transformer's core. We can safely use a three-legged core transformer.
- *Ferroresonance* — Highly susceptible.

4.4.3 Floating Wye – Delta

The floating-wye–delta connection is popular for supplying ungrounded service and 120/240-V service. This type of connection may be used from either a grounded or ungrounded distribution primary. The main characteristics of this supply are:

- *Supply* — 3-wire or 4-wire system.
- *Service*
  - Can supply ungrounded service.
  - Can supply four-wire service with 240/120-V on one leg with a midtapped ground.
  - Cannot supply grounded-wye four-wire service.
- *Unit failure* — Can continue to operate if one unit fails if it is rewired as an open wye – open delta.
- *Voltage unbalance* — Secondary-side unbalances are more likely than with a wye secondary connection.
- *Ferroresonance* — Highly susceptible.

Do not use single-phase transformers with secondary breakers (CSPs) in this connection. If one secondary breaker opens, it breaks the delta on the secondary. Now, the primary neutral can shift wildly. The transformer may be severely overloaded by load unbalance or single phasing on the primary.

Facilities should ensure that single-phase loads only connect to the lighting leg; any miswired loads have overvoltages. The phase-to-neutral connection from the neutral to the opposite phase (where both power legs come together) is 208 V on a 240/120-V system.

The floating wye – delta is best used when supplying mainly three-phase load with a smaller amount of single-phase load. If the single-phase load is large, the three transformers making up the connection are not used as efficiently, and voltage unbalances can be high on the secondary.

In a conservative loading guideline, size the lighting transformer to supply all of the single-phase load plus 1/3 of the three-phase load (ANSI/IEEE C57.105-1978). Size each power leg to carry 1/3 of the three-phase load plus 1/3 of the single-phase load. ABB (1995) describes more accurate loading equations:

Lighting leg loading in kVA:

$$kVA_{bc} = \frac{1}{3}\sqrt{k_3^2 + 4k_1^2 + 4k_3k_1\cos\alpha}$$

Lagging power leg loading in kVA:

$$kVA_{ca} = \frac{1}{3}\sqrt{k_3^2 + k_1^2 - 2k_3k_1\cos(120^\circ + \alpha)}$$

Leading power leg loading in kVA:

$$kVA_{ab} = \frac{1}{3}\sqrt{k_3^2 + k_1^2 - 2k_3k_1\cos(120° - \alpha)}$$

where

k_1 = single-phase load, kVA
k_3 = balanced three-phase load, kVA
$\alpha = \theta_3 - \theta_1$
θ_3 = phase angle in degrees for the three-phase load
θ_1 = phase angle in degrees for the single-phase load

For wye – delta connections, the wye on the primary is normally intentionally ungrounded. If it is grounded, it creates a grounding bank. This is normally undesirable because it may disrupt the feeder protection schemes and cause excessive circulating current in the delta winding. Utilities sometimes use this connection as a grounding source or for other unusual reasons.

Delta secondary windings are more prone to voltage unbalance problems than a wye secondary winding (Smith et al., 1988). A balanced three-phase load can cause voltage unbalance if the impedances of each leg are different. With the normal practice of using a larger lighting leg, the lighting leg has a lower impedance. Voltage unbalance is worse with longer secondaries and higher impedance transformers. High levels of single-phase load also aggravate unbalances.

4.4.4 Other Common Connections

4.4.4.1 Delta – Delta

The main features and drawbacks of the delta – delta supply are:

- *Supply* — 3-wire or 4-wire system.
- *Service*
  - Can supply ungrounded service.
  - Can supply four-wire service with 240/120-V on one leg with a midtapped ground.
  - Cannot supply grounded-wye four-wire service.
- *Ferroresonance* — Highly susceptible.
- *Unit failure* — Can continue to operate if one unit fails (as an open delta – open delta).
- *Circulating current* — Has high circulating current if the turns ratios of each unit are not equal.

A delta – delta transformer may have high circulating current if any of the three legs has unbalance in the voltage ratio. A delta winding forms a series loop. Two windings are enough to fix the three phase-to-phase voltage vectors. If the third winding does not have the same voltage as that created by the series sum of the other two windings, large circulating currents flow to offset the voltage imbalance. ANSI/IEEE C57.105-1978 provides an example where the three phase-to-phase voltages summed to 1.5% of nominal as measured at the open corner of the delta winding (this voltage should be zero for no circulating current). With a 5% transformer impedance, a current equal to 10% of the transformer rating circulates in the delta when the open corner is connected. The voltage sees an impedance equal to the three winding impedances in series, resulting in a circulating current of 100% × 1.5% / (3×5%) = 10%. This circulating current directly adds to one of the three windings, possibly overloading the transformer.

Single-phase units with secondary breakers (CSPs) should not be used for the lighting leg. If the secondary breaker on the lighting leg opens, the load loses its neutral reference, but the phase-to-phase voltages are maintained by the other two legs (like an open delta – open delta connection). As with the loss of the neutral connection to a single-phase 120/240-V customer, unbalanced single-phase loads shift the neutral and create low voltages on one leg and high voltages on the lightly loaded leg.

4.4.4.2 Open Wye – Open Delta

The main advantage of the open wye – open delta transformer configuration is that it can supply three-phase load from a two-phase supply (but the supply must have a neutral).

The main features and drawbacks of the open wye – delta supply are:

- *Supply* — 2 phases and the neutral of a 4-wire grounded system.
- *Service*
  - Can supply ungrounded service.
  - Can supply four-wire service with 240/120-V on one leg with a midtapped ground.
  - Cannot supply grounded-wye four-wire service.
- *Ferroresonance* — Immune.
- *Voltage unbalance* — May have high secondary voltage unbalance.
- *Primary ground current* — Creates high primary-side current unbalance. Even with balanced loading, high currents are forced into the primary neutral.

Open wye – open delta connections are most efficiently applied when the load is predominantly single phase with some three-phase load, using one large lighting-leg transformer and another smaller unit. This connection is easily upgraded if the customer's three-phase load grows by adding a second power-leg transformer.

For sizing each bank, size the power leg for $1/\sqrt{3} = 0.577$ times the balanced three-phase load, and size the lighting leg for all of the single-phase load plus 0.577 times the three-phase load (ANSI/IEEE C57.105-1978). The following equations more accurately describe the split in loading on the two transformers (ABB, 1995). The load on the lighting leg in kVA is

$$kVA_L = \frac{k_3^2}{3} + {k_1}^2 + \frac{2k_3k_1}{\sqrt{3}}\cos(\alpha + 30°) \text{ for a leading lighting leg}$$

$$kVA_L = \frac{k_3^2}{3} + k_1^2 + \frac{2k_3k_1}{\sqrt{3}}\cos(\alpha - 30°) \text{ for a lagging lighting leg}$$

The power leg loading in kVA is:

$$kVA_L = \frac{k_3}{\sqrt{3}}$$

where
k_1 = single-phase load, kVA
k_3 = balanced three-phase load, kVA
$\alpha = \theta_3 - \theta_1$
θ_3 = phase angle in degrees for the three-phase load
θ_1 = phase angle in degrees for the single-phase load

The lighting leg may be on the leading or lagging leg. In the open wye – open delta connection shown in Figure 4.13, the single-phase load is on the leading leg. For a lagging connection, switch the lighting and the power leg. Having the lighting connection on the leading leg reduces the loading on the lighting leg. Normally, the power factor of the three-phase load is less than that of the single-phase load, so α is positive, which reduces the loading on the lighting leg.

On the primary side, it is important that the two high-voltage primary connections are not made to the same primary phase. If this is accidentally done, the phase-to-phase voltage across the open secondary is two times the normal phase-to-phase voltage.

The open wye – open delta connection injects significant current into the neutral on the four-wire primary. Even with a balanced three-phase load, significant current is forced into the ground as shown in Figure 4.16. The extra unbalanced current can cause primary-side voltage unbalance and may trigger ground relays.

Open-delta secondary windings are very prone to voltage unbalance, which can cause excessive heating in end-use motors (Smith et al., 1988). Even balanced three-phase loads significantly unbalance the voltages. Voltage unbalance is less with lower-impedance transformers. Voltage unbalance

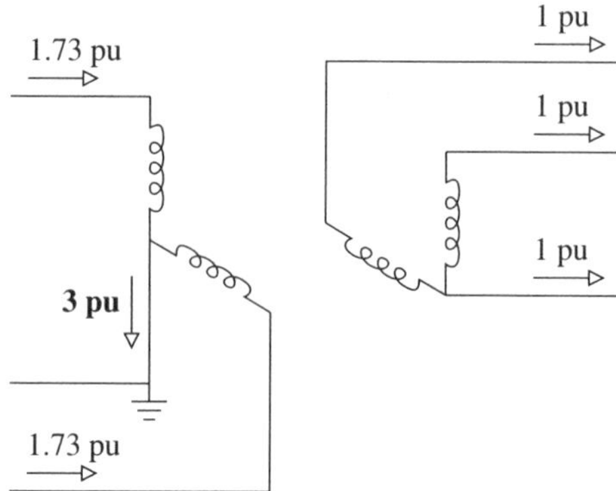

FIGURE 4.16
Current flow in an open wye – open delta transformer with balanced three-phase load.

reduces significantly if the connection is upgraded to a floating wye – closed delta connection. In addition, the component of the negative-sequence voltage on the primary (which is what really causes motor heating) can add to that caused by the transformer configuration to sometimes cause a negative-sequence voltage above 5% (which is a level that significantly increases heating in a three-phase induction motor).

While an unusual connection, it is possible to supply a balanced, grounded four-wire service from an open-wye primary. This connection (open wye – partial zig-zag) can be used to supply 208Y/120-V service from a two-phase line. One of the 120/240-V transformers must have four bushings; X2 and X3 are not tied together but connected as shown in Figure 4.17. Each of the transformers must be sized to supply 2/3 of the balanced three-phase load. If four-bushing transformers are not available, this connection can be made with three single-phase transformers. Instead of the four-bushing transformer, two single-phase transformers are placed in parallel on the primary, and the secondary terminals of each are configured to give the arrangement in Figure 4.17.

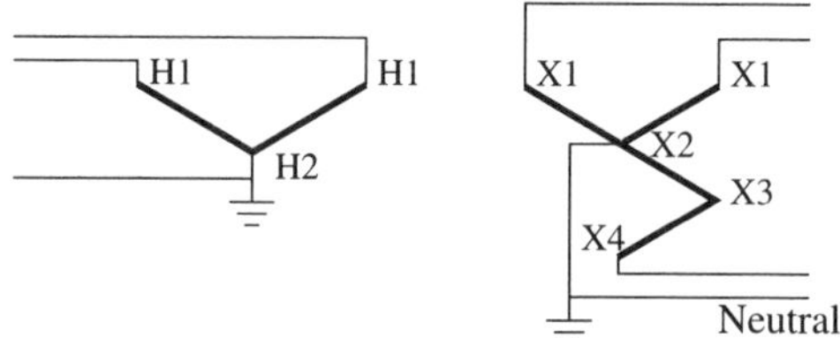

FIGURE 4.17
Quasiphasor diagram of an open-wye primary connection supplying a wye four-wire neutral service such as 208Y/120 V. (From ANSI/IEEE Std. C57.105-1978. Copyright 1978 IEEE. All rights reserved.)

4.4.4.3 Other Suitable Connections

While not as common, several other three-phase connections are used at times:

- *Open delta – Open delta* — Can supply a three-wire ungrounded service or a four-wire 120/240-V service with a midtapped ground on one leg of the transformer. The ungrounded high-side connection is susceptible to ferroresonance. Only two transformers are needed, but it requires all three primary phases. This connection is less efficient for supplying balanced three-phase loads; the two units must total 115% of the connected load. This connection is most efficiently applied when the load is predominantly single phase with some three-phase load, using one large lighting-leg transformer and another smaller unit.
- *Delta – Floating wye* — Suitable for supplying a three-wire ungrounded service. The ungrounded high-side connection is susceptible to ferroresonance.
- *Grounded wye – Floating wye* — Suitable for supplying a three-wire ungrounded service from a multigrounded primary system. The grounded primary-side connection reduces the possibility of ferroresonance.

4.4.5 Neutral Stability with a Floating Wye

Some connections with a floating-wye winding have an unstable neutral, which we should avoid. Unbalanced single-phase loads on the secondary, unequal magnetizing currents, and suppression of third harmonics — all can shift the neutral.

Consider a *floating wye – grounded wye* connection. In a wye – wye transformer, the primary and secondary voltages have the same vector relationships. The problem is that the neutral point does not have a grounding source; it is free to float. Unbalanced loads or magnetizing currents can shift the neutral and create high neutral-to-earth voltages and overvoltages on the phases with less loading. The reverse connection with a grounded wye – floating wye works because the primary-side neutral is connected to the system neutral, which has a grounding source. The grounding source fixes the neutral voltage.

In a floating wye, current in one branch is dependent on the currents in the other two branches. What flows in one branch must flow out the other two branches. This creates conditions that shift the neutral (Blume et al., 1951):

- *Unbalanced loads* — Unequal single-phase loads shift the neutral point. Zero-sequence current has no path to flow (again, the ground source is missing). Loading one phase drops the voltage on that phase and raises the voltage on the other two phases. Even a small unbalance significantly shifts the neutral.

- *Unequal magnetizing currents* — Just like unequal loads, differences in the amount of magnetizing current each leg needs can shift the floating neutral. In a four- or five-legged core, the asymmetry of the core causes unequal magnetizing requirements on each phase.
- *Suppression of third harmonics* — Magnetizing currents contain significant third harmonics that are zero sequence. But, the floating wye connection has no ground source to absorb the zero-sequence currents, so they are suppressed. The suppression of the zero-sequence currents generates a significant third-harmonic voltage in each winding, about 50% of the phase voltage on each leg according to Blume et al. (1951). With the neutral grounded in the floating wye – grounded wye, a significant third-harmonic voltage adds to each phase-to-ground load. If the neutral is floating (on the wye–wye transformer with the neutrals tied together), the third-harmonic voltage appears between the neutral and ground.

In addition to the floating wye – grounded wye, avoid these problem connections that have an unstable neutral:

- *Grounded wye – grounded wye on a three-wire system* — The grounded-wye on the primary does not have an effective grounding source, so it acts the same as a floating-wye–grounded-wye.
- *A wye – wye transformer with the primary and secondary neutrals tied together internally (the H0X0 bushing) but with the neutral left floating* — Again, the neutral point can float. Unbalanced loading is not a problem, but magnetizing currents and suppression of third harmonics are. These can generate large voltages between the neutral point and ground (and between the phase wires and ground). If the secondary neutral is isolated from the primary neutral, each neutral settles to a different value. But when the secondary neutral is locked into the primary neutral, the secondary neutral follows the neutral shift of the primary and shifts the secondary phases relative to ground.

Another poor connection is the floating wye – floating wye. Although not as bad as the floating-wye–grounded-wye connection, the neutral can shift if the connection is made of three units of different magnetizing characteristics. The neutral shift can lead to an overvoltage across one of the windings. Also, high harmonic voltage appears on the primary-side neutral (which is okay if the neutral is properly insulated from the tank).

Three-legged core transformers avoid some of the problems with a floating wye. The phantom tertiary acts as a mini ground source, stabilizes the neutral, and even allows some unbalance of single-phase loads. But as it stabilizes the neutral, the unbalances heat the tank. Given that, it is best to avoid these transformer connections. They provide no features or advantages over other transformer connections.

4.4.6 Sequence Connections of Three-Phase Transformers

The connection determines the effect on zero sequence, which impacts unbalances and response to line-to-ground faults and zero-sequence harmonics. Figure 4.18 shows how to derive sequence connections along with common examples. In general, three-phase transformers may affect the zero-sequence circuit as follows:

- *Isolation* — A floating wye – delta connection isolates the primary from the secondary in the zero sequence.
- *Pass through* — The grounding of the grounded wye – grounded wye connection is determined by the grounding upstream of the transformer.
- *Ground source* — A delta – grounded wye connection provides a ground source on the secondary. (And, the delta – grounded wye connection also isolates the primary from the secondary.)

4.5 Loadings

Distribution transformers are *output rated;* they can deliver their rated kVA without exceeding temperature rise limits when the following conditions apply:

- The secondary voltage and the voltage/frequency do not exceed 105% of rating. So, a transformer is a constant kVA device for a voltage from 100 to 105% (the standards are unclear below that, so treat them as constant current devices).
- The load power factor ≥ 80%.
- Frequency ≥ 95% of rating.

The transformer loading and sizing guidelines of many utilities are based on ANSI/IEEE C57.91-1981.

Modern distribution transformers are 65°C rise units, meaning they have normal life expectancy when operated with an average winding temperature rise above ambient of not more than 65°C and a hottest spot winding temperature rise of not more than 80°C. Some older units are 55°C rise units, which have less overloading capability.

At an ambient temperature of 30°C, the 80°C hottest-spot rise for 65°C rise units gives a hottest-spot winding temperature of 110°C. The hot-spot tem-

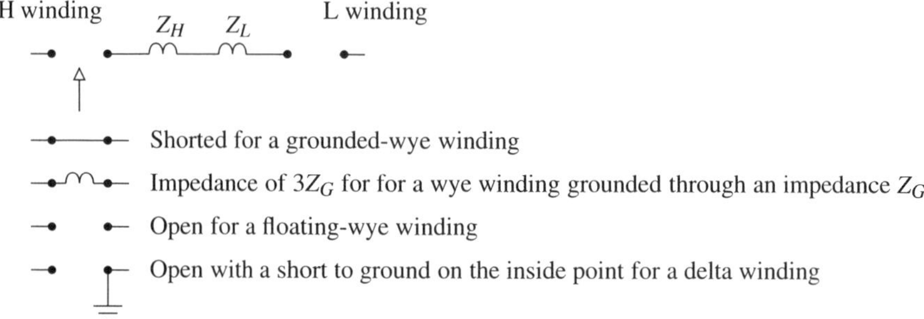

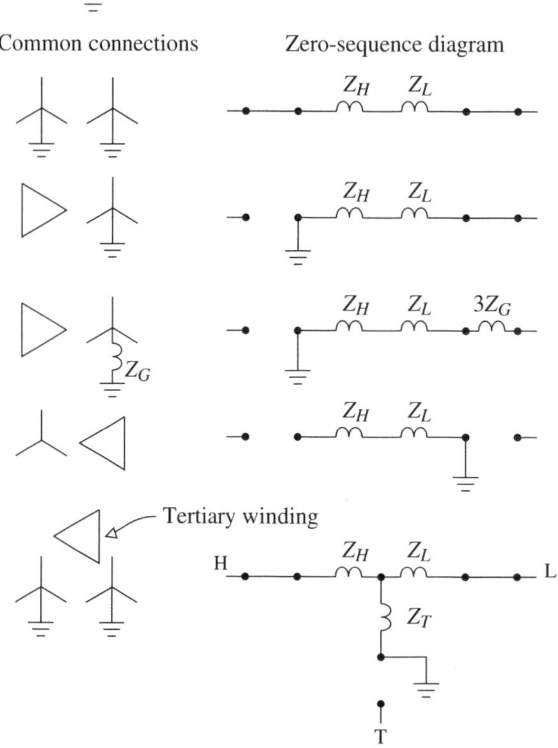

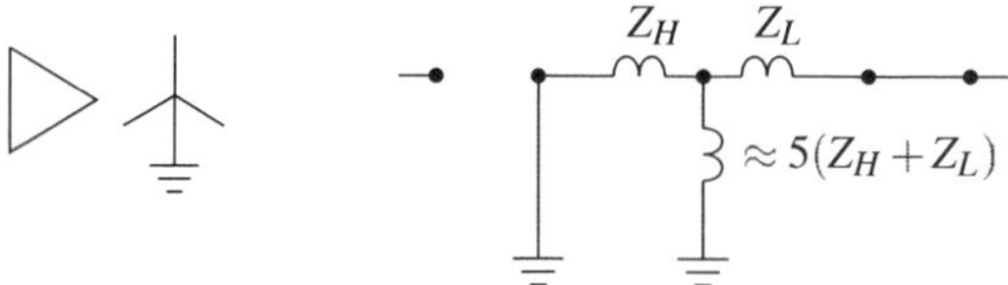

FIGURE 4.18
Zero-sequence connections of various three-phase transformer connections.

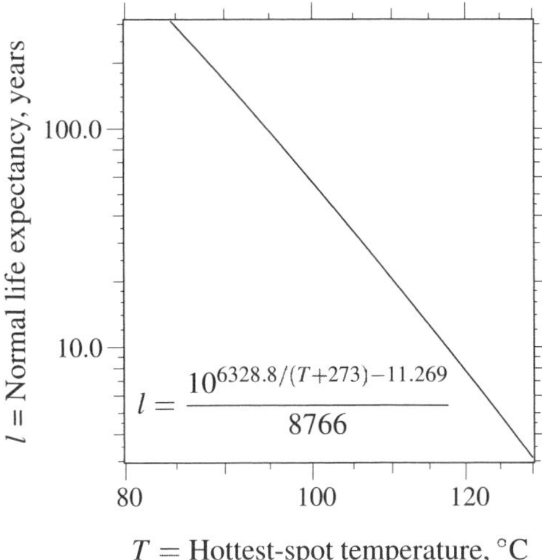

FIGURE 4.19
Transformer life as a function of the hottest-spot winding temperature.

perature on the winding is critical; that's where insulation degrades. The insulation's life exponentially relates to hot-spot winding temperature as shown in Figure 4.19. At 110°C, the normal life expectancy is 20 years. Because of daily and seasonal load cycles, most of the time temperatures are nowhere near these values. Most of the time, temperatures are low enough not to do any significant insulation degradation. We can even run at temperatures above 110°C for short periods. For the most economic operation of distribution transformers, they are normally sized to operate at significant overloads for short periods of the year.

We can load distribution transformers much more heavily when it is cold. Locations with winter-peaking loads can have smaller transformers for a given loading level. The transformer's kVA rating is based on an ambient temperature of 30°C. For other temperatures, ANSI/IEEE C57.91-1981 suggests the following adjustments to loading capability:

- > 30°C: decrease loading capability by 1.5% of rated kVA for each °C above 30°C.
- < 30°C: increase loading capability by 1% of rated kVA for each °C above 30°C.

Ambient temperature estimates for a given region can be found using historical weather data. For loads with normal life expectancy, ANSI/IEEE C57.91-1981 recommends the following estimate of ambient temperature:

- *Average daily temperature for the month involved* — As an approximation, the average can be approximated as the average of the daily highs and the daily lows.

For short-time loads where we are designing for a moderate sacrifice of life, use:

- *Maximum daily temperature*

In either case, the values should be averaged over several years for the month involved. C57.91-1981 also suggests adding 5°C to be conservative. These values are for outdoor overhead or padmounted units. Transformers installed in vaults or other cases with limited air flow may require some adjustments.

Transformers should also be derated for altitudes above 3300 ft (1000 m). At higher altitudes, the decreased air density reduces the heat conducted away from the transformer. ANSI/IEEE C57.91-1981 recommends derating by 0.4% for each 330 ft (100 m) that the altitude is above 3300 ft (1000 m).

Load cycles play an important role in determining loading. ANSI/IEEE C57.91-1981 derives an equivalent load cycle with two levels: the peak load and the initial load. The equivalent two-step load cycle may be derived from a more detailed load cycle. The guide finds a continuous load and a short-duration peak load. Both are found using the equivalent load value from a more complicated load cycle:

$$L = \sqrt{\frac{L_1^2 t_1 + L_2^2 t_2 + L_3^2 t_3 + \cdots + L_n^2 t_n}{t_1 + t_2 + t_3 + \cdots + t_n}}$$

where

L = equivalent load in percent, per unit, or actual kVA

$L_1, L_2, \ldots,$ = The load steps in percent, per unit, or actual kVA

$t_1, t_2, \ldots,$ = The corresponding load durations

The continuous load is the equivalent load found using the equation above for 12 h preceding and 12 h following the peak and choosing the higher of these two values. The guide suggests using 1-h time blocks. The peak is the equivalent load from the equation above where the irregular peak exists.

The C57.91 guide has loading guidelines based on the peak duration and continuous load prior to the peak. Table 4.8 shows that significant overloads are allowed depending on the preload and the duration of the peak.

Because a region's temperature and loading patterns vary significantly, there is no universal transformer application guideline. Coming up with standardized tables for initial loading is based on a prediction of peak load, which for residential service normally factors in the number of houses, average size (square footage), central air conditioner size, and whether electric heat is used. Once the peak load is estimated, it is common to pick a transformer with a kVA rating equal to or greater than the peak load kVA estimate. With this arrangement, some transformers may operate significantly above their ratings for short periods of the year. Load growth can push the peak

TABLE 4.8

Transformer Loading Guidelines

Peak Load Duration, Hours	Extra Loss of Life[a], %	Equivalent Peak Loading in Per Unit of Rated kVA with the Percent Preload and Ambient Temperatures Given Below										
		50% Preload				75% Preload				90% Preload		
		Ambient Temp., °C				Ambient Temp., °C				Ambient Temp., °C		
		20	30	40	50	20	30	40	50	20	30	40
1	Normal	2.26	2.12	1.96	1.79	2.12	1.96	1.77	1.49	2.02	1.82	1.43
	0.05	2.51	2.38	2.25	2.11	2.40	2.27	2.12	1.95	2.31	2.16	1.97
	0.10	2.61	2.49	2.36	2.23	2.50	2.37	2.22	2.07	2.41	2.27	2.11
	0.50	2.88	2.76	2.64	2.51	2.77	2.65	2.52	2.39	2.70	2.57	2.43
2	Normal	1.91	1.79	1.65	1.50	1.82	1.68	1.52	1.26	1.74	1.57	1.26
	0.05	2.13	2.02	1.89	1.77	2.05	1.93	1.80	1.65	1.98	1.85	1.70
	0.10	2.22	2.10	1.99	1.87	2.14	2.02	1.90	1.75	2.07	1.95	1.81
	0.50	2.44	2.34	2.23	—	2.37	2.26	2.15	—	2.31	2.20	2.08
4	Normal	1.61	2.50	1.38	1.25	1.56	1.44	1.30	1.09	1.50	1.36	1.13
	0.05	1.80	1.70	1.60	1.48	1.76	1.65	1.54	1.40	1.71	1.60	1.47
	0.10	1.87	1.77	1.67	—	1.83	1.72	1.62	1.50	1.79	1.68	1.56
	0.50	2.06	1.97	—	—	2.02	1.93	—	—	1.99	1.89	—
8	Normal	1.39	1.28	1.18	1.05	1.36	1.25	1.13	0.96	1.33	1.21	1.02
	0.05	1.55	1.46	1.36	1.25	1.53	1.43	1.33	1.21	1.51	1.41	1.29
	0.10	1.61	1.53	1.43	1.33	1.59	1.50	1.41	1.30	1.57	1.47	1.38
	0.50	1.78	1.69	1.61	—	1.76	1.67	1.58	—	1.74	1.65	1.56
24	Normal	1.18	1.08	0.97	0.86	1.17	1.07	0.97	0.84	1.16	1.07	0.95
	0.05	1.33	1.24	1.15	1.04	1.33	1.24	1.13	1.04	1.32	1.23	1.13
	0.10	1.39	1.30	1.21	1.11	1.38	1.29	1.20	1.10	1.38	1.29	1.20
	0.50	1.54	1.45	1.37	1.28	1.53	1.45	1.37	1.28	1.53	1.45	1.36

[a] Extra loss of life in addition to 0.0137% per day loss of life for normal life expectancy.

Source: ANSI/IEEE C57.91-1981, *IEEE Guide for Loading Mineral-Oil-Immersed Overhead and Pad-Mounted Distribution Transformers Rated 500 kVA and Less with 65 Degrees C or 55 Degrees C Average Winding Rise.*

load above the peak kVA estimate, and inaccuracy of the load prediction will mean that some units are going to be loaded more than expected. The load factor (the ratio of average demand to peak demand) for most distribution transformers is 40 to 60%. Most distribution transformers are relatively lightly loaded most of the time, but some have peak loads well above their rating. In analysis of data from three utilities, the Oak Ridge National Laboratory found that distribution transformers have an average load of 15 to 40% of their rating, with 30% being most typical (ORNL-6925, 1997).

The heat input into the transformer is from no-load losses and from load losses. The economics of transformer application and purchasing involve consideration of the thermal limitations as well as the operating costs of the losses. Transformer stocking considerations also play a role. For residential customers, a utility may limit inventory to 15, 25, and 50-kVA units (5, 10, 15, 25, 37.5, 50-kVA units are standard sizes).

Some utilities use transformer load management programs to more precisely load transformers to get the most economic use of each transformer's

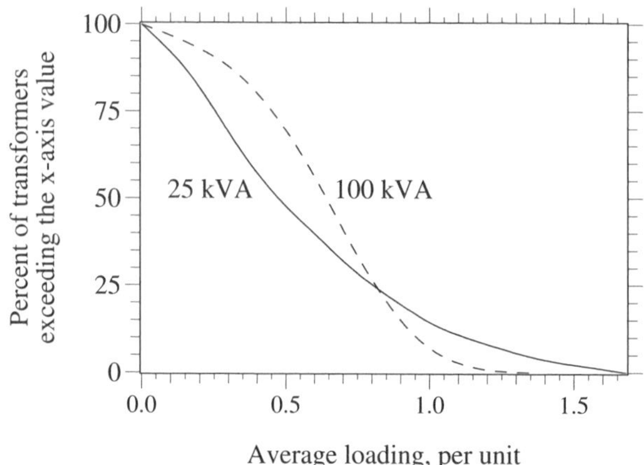

FIGURE 4.20
Distributions of average loadings of two transformer sizes at one utility. (From [ORNL-6927, 1998])

life. These programs take billing data for the loads from each transformer to estimate that transformer's loading. These programs allow the utility to more aggressively load transformers because those needing changeout can be targeted more precisely. Load management programs require data setup and maintenance. Most important, each meter must be tied to a given transformer (many utilities have this information infrastructure, but some do not).

Transformer loadings vary considerably. Figure 4.20 shows the distribution of average loadings on two sizes of transformers at one typical utility. Most transformers are not heavily loaded: in this case, 85% of units have average loadings less than the nameplate. Many units are very lightly loaded, and 10% are quite heavily loaded. Smaller units have more spread in their loading.

Seevers (1995) demonstrates a simple approach to determining transformer loading. Their customers (in the southern US) had 1 kW of demand for every 400 kWh's, regardless of whether the loads peaked in the winter or summer. Seevers derived the ratio by comparing substation demand with kWh totals for all customers fed from the substation (after removing primary-metered customers and other large loads). To estimate the load on a given transformer, sum the kWh for the month of highest usage for all customers connected to the transformer and convert to peak demand, in this case by dividing by 400 kWh per kW-demand. While simple, this method identifies grossly undersized or oversized transformers. Table 4.9 shows guidelines for replacement of underloaded transformers.

Transformers with an internal secondary breaker (CSPs) are a poor-man's form of transformer load management. If the breaker trips from overload, replace the transformer (unless there are extraordinary weather and loading conditions that are unlikely to be repeated).

TABLE 4.9

One Approach to a Transformer Replacement Program

Existing Transformer kVA	Loading Estimate in kVA	Recommended Size in kVA
25	10 or less	10
37.5	15 or less	10
50	20 or less	15
75	37.5 or less	37.5
100	50 or less	50
167	100 or less	100
	75 or less	75 or 50

Source: Seevers, O. C., *Management of Transmission and Distribution Systems*, Penn Well Publishing Company, Tulsa, OK, 1995.

Especially in high-lightning areas, consider the implications of reduction of insulation capability. At hottest-spot temperatures above 140°C, the solid insulation and the oil may release gasses. While not permanently reducing insulation, the short-term loss of insulation strength can make the transformer susceptible to damage from lightning-caused voltage surges. The thermal time constant of the winding is very short, 5 to 15 min. On this time scale, loads on distribution transformers are quite erratic with large, short-duration overloads (well above the 20- or 30-min demand loadings). These loads can push the winding hottest-spot temperature above 140°C.

Padmounted transformers have a special concern related to loading: case temperatures. Under heavy loading on a hot day, case temperatures can become hot. ABB measured absolute case temperatures of 185 to 200°F (85 to 95°C) and case temperature rises above ambient of 50 to 60°C on 25 and 37.5-kVA transformers at 180% loadings and on a 50-kVA transformer at 150% continuous load (NRECA RER Project 90-8, 1993). The hottest temperatures were on the sides of the case where the oil was in contact with the case (the top of the case was significantly cooler). While these temperatures sound quite high, a person's pain-withdrawal reflexes will normally protect against burns for normal loadings that would be encountered. Reflexes will protect against blistering and burning for case temperatures below 300°F (149°C). Skin contacts must be quite long before blistering occurs. For a case temperature of 239°F (115°C), NRECA reported that the skin-contact time to blister is 6.5 sec (which is more than enough time to pull away). At 190°F (88°C), the contact time to blister is 19 sec.

4.6 Losses

Transformer losses are an important purchase criteria and make up an appreciable portion of a utility's overall losses. The Oak Ridge National Laboratory estimates that distribution transformers account for 26% of transmission and distribution losses and 41% of distribution and subtransmission losses (ORNL-6804/R1, 1995). At one utility, Grainger and Kendrew (1989) estimated that distribution transformers were 55% of distribution losses and 2.14% of electricity sales; of the two main contributors to losses, 86% were no-load losses, and 14% were load losses.

Load losses are also called copper or wire or winding losses. Load losses are from current through the transformer's windings generating heat through the winding resistance as I^2R.

No-load losses are the continuous losses of a transformer, regardless of load. No-load losses for modern silicon-steel-core transformers average about 0.2% of the transformer rating (a typical 50-kVA transformer has no-load losses of 100 W), but designs vary from 0.15 to 0.4% depending on the needs of the utility. No-load losses are also called iron or core losses because they are mainly a function of the core materials. The two main components of no-load losses are eddy currents and hysteresis. Hysteresis describes the memory of a magnetic material. More force is necessary to demagnetize magnetic material than it takes to magnetize it; the magnetic domains in the material resist realignment. Eddy current losses are small circulating currents in the core material. The steel core is a conductor that carries an alternating magnetic field, which induces circulating currents in the core. These currents through the resistive conductor generate heat and losses. Cores are typically made from cold-rolled, grain-oriented silicon steel laminations. Manufacturers limit eddy currents by laminating the steel core in 9- to 14-mil thick layers, each insulated from the other. Core losses increase with steady-state voltage.

Hysteresis losses are a function of the volume of the core, the frequency, and the maximum flux density (Sankaran, 2000):

$$P_h \propto V_e f B^{1.6}$$

where

V_e = volume of the core

f = frequency

B = maximum flux density

The eddy-current losses are a function of core volume, frequency, flux density, lamination thicknesses, and resistivity of the core material (Sankaran, 2000):

$$P_e \propto V_e B^2 f^2 t^2 / r$$

where

t = thickness of the laminations

r = resistivity of the core material

TABLE 4.10

Loss Reduction Alternatives

	No-Load Losses	Load Losses	Cost
To Decrease No-Load Losses			
Use lower-loss core materials	Lower	No change[a]	Higher
Decrease flux density by:			
(1) increasing core CSA[b]	Lower	Higher	Higher
(2) decreasing volts/turn	Lower	Higher	Higher
Decrease flux path length by decreasing conductor CSA	Lower	Higher	Lower
To Decrease Load Losses			
Use lower-loss conductor materials	No change	Lower	Higher
Decrease current density by increasing conductor CSA	Higher	Lower	Higher
Decrease current path length by:			
(1) decreasing core CSA	Higher	Lower	Lower
(2) increasing volts/turn	Higher	Lower	Higher

[a] Amorphous core materials would result in higher load losses.
[b] CSA=cross-sectional area

Source: ORNL-6847, *Determination Analysis of Emergy Conservation Standards for Distribution Transformers,* Oak Ridge National Laboratory, U.S. Department of Emergy, 1996.

Amorphous core metals significantly reduce core losses — as low as one quarter of the losses of silicon-steel cores — on the order of 0.005 to 0.01% of the transformer rating. Amorphous cores do not have a crystalline structure like silicon-steel cores; their iron atoms are randomly distributed. Amorphous materials are made by rapidly cooling a molten alloy, so crystals do not have a chance to form. Such core materials have low hysteresis loss. Eddy current-losses are very low because of the high resistivity of the material and very thin laminations (1-mil thick). Amorphous-core transformers are larger for the same kVA rating and have higher initial costs.

Load losses, no-load losses, and purchase price are all interrelated. Approaches to reduce load losses tend to increase no-load losses and vice versa. For example, a larger core cross-sectional area decreases no-load losses (the flux density core is less), but this requires longer winding conductors and more I^2R load losses. Table 4.10 shows some of the main tradeoffs.

Information from transformer load management programs can help with transformer loss analysis. Table 4.11 shows typical transformer loading data from one utility. The average load on most transformers is relatively low (25 to 30% of transformer rating), which highlights the importance of no-load losses. The total equivalent losses on a transformer are

$$L_{total} = P^2 F_{ls} L_{load} + L_{no-load}$$

where

L_{total} = average losses, kW (multiply this by 8760 to find the annual kilowatt-hours)

P = peak transformer load, per unit

F_{ls} = loss factor, per unit

$L_{no\text{-}load}$ = rated no-load losses, kW

L_{load} = rated load losses, kW

Many utilities evaluate the total life-cycle cost of distribution transformers, accounting for the initial purchase price and the cost of losses over the life of the transformer (the total owning cost or TOC). The classic work done by Gangel and Propst (1965) on transformer loads and loss evaluation provides the foundation for much of the later work. Many utilities follow the Edison Electric Association's economic evaluation guidelines (EEI, 1981). To evaluate the total owning cost, the utility's cost of losses are evaluated using transformer loading assumptions, including load factor, coincident factor, and responsibility factor. Utilities typically assign an equivalent present value for the costs of no-load losses and another for the cost of load losses. Loss values typically range from \$2 to \$4/W of no-load losses and \$0.50 to \$1.50/W of load losses (ORNL-6847, 1996). Utilities that evaluate the life costs of transformers purchase lower-loss transformers. For example, a 50-

TABLE 4.11

Summary of the Loading of One Utility's Single-Phase Pole-Mounted Distribution Transformers

Size (kVA)	No. of Installed Transformers	MWh/ Transformer	Annual PU Avg. Load	Annual PU Load Factor	Calculated Loss Factor
10	59,793	21	0.267	0.405	0.200
15	106,476	34	0.292	0.430	0.221
25	118,584	60	0.309	0.444	0.234
37	77,076	96	0.329	0.445	0.235
50	50,580	121	0.308	0.430	0.222
75	24,682	166	0.281	0.434	0.225
100	8,457	220	0.280	0.463	0.252
167	3,820	372	0.283	0.516	0.304
250	592	631	0.320	0.568	0.360
333	284	869	0.331	0.609	0.407
500	231	1,200	0.304	0.598	0.394
667	9	1,666	0.317	0.476	0.264
833	51	2,187	0.333	0.629	0.431

Note: PU = per unit

Source: ORNL-6925, *Supplement to the "Determination Analysis" (ORNL-6847) and Analysis of the NEMA Efficiency Standard for Distribution Transformers*, Oak Ridge National Laboratory, U.S. Department of Emergy, 1997.

kVA single-phase, non-loss-evaluated transformer would have approximately 150 W of no-load losses and 675 W of load losses; the same loss-evaluated transformer would have approximately 100 W of no-load losses and 540 W of load losses (ORNL-6925, 1997). Nickel (1981) describes an economic approach in detail and compares it to the EEI method. The IEEE has developed a more recent guide (C57.12.33).

4.7 Network Transformers

Network transformers, the distribution transformers that serve grid and spot networks, are large three-phase units. Network units are normally vault-types or subway types, which are defined as (ANSI C57.12.40-1982):

- *Vault-type transformers* — Suitable for occasional submerged operation
- *Subway-type transformers* — Suitable for frequent or continuous submerged operation

Network transformers are often housed in vaults. Vaults are underground rooms accessed through manholes that house transformers and other equipment. Vaults may have sump pumps to remove water, air venting systems, and even forced-air circulation systems. Network transformers are also used in buildings, usually in the basement. In these, vault-type transformers may be used (as long as the room is properly built and secured for such use). Utilities may also use dry-type units and units with less flammable insulating oils.

A network transformer has a three-phase, primary-side switch that can open, close, or short the primary-side connection to ground. The standard secondary voltages are 216Y/125 V and 480Y/277 V. Table 4.12 shows standard sizes. Transformers up to 1000 kVA have a 5% impedance; above 1000 kVA, 7% is standard. *X/R* ratios are generally between 3 and 12. Lower impedance transformers (say 4%) have lower voltage drop and higher secondary fault currents. (Higher secondary fault currents help on a network to burn clear faults.) Lower impedance has a price though — higher circulating currents and less load balance between transformers. Network trans-

TABLE 4.12

Standard Network Transformer Sizes

	Standard Ratings, kVA
216Y/125 V	300, 500, 750, 1000
480Y/277 V	500, 750, 1000, 1500, 2000, 2500

formers may also be made out of standard single-phase distribution transformers, but caution is warranted if the units have very low leakage impedances (which could cause very high circulating currents and secondary fault levels higher than network protector ratings).

Most network transformers are connected delta – grounded wye. By blocking zero sequence, this connection keeps ground currents low on the primary cables. Then, we can use a very sensitive ground-fault relay on the substation breaker. Blocking zero sequence also reduces the current on cable neutrals and cable sheaths, including zero-sequence harmonics, mainly the third harmonic. One disadvantage of this connection is with combination feeders — those that feed network loads as well as radial loads. For a primary line-to-ground fault, the feeder breaker opens, but the network transformers will continue to backfeed the fault until all of the network protectors operate (and some may stick). Now, the network transformers backfeed the primary feeder as an ungrounded circuit. An ungrounded circuit with a single line-to-ground fault on one phase causes a neutral shift that raises the line-to-neutral voltage on the unfaulted phases to line-to-line voltage. The non-network load connected phase-to-neutral is subjected to this overvoltage.

Some networks use grounded wye – grounded wye connections. This connection fits better for combination feeders. For a primary line-to-ground fault, the feeder breaker opens. Backfeeds to the primary through the network still have a grounding reference with the wye – wye connection, so chances of overvoltages are limited. The grounded wye – grounded wye connection also reduces the change of ferroresonance in cases where a transformer has single-pole switching.

Most network transformers are core type, either a three- or five-legged core. The three-legged core, either with a stacked or wound core, is suitable for a delta – grounded wye connection (but not a grounded wye – grounded wye connection because of tank heating). A five-legged core transformer is suitable for either connection type.

4.8 Substation Transformers

In a distribution substation, power-class transformers provide the conversion from subtransmission circuits to the distribution primary. Most are connected delta – grounded wye to provide a ground source for the distribution neutral and to isolate the distribution ground system from the subtransmission system.

Station transformers can range from 5 MVA in smaller rural substations to over 80 MVA at urban stations (base ratings). Stations with two banks, each about 20 MVA, are common. Such a station can serve about six to eight feeders.

Power transformers have multiple ratings, depending on cooling methods. The base rating is the self-cooled rating, just due to the natural flow to the surrounding air through radiators. The transformer can supply more load with extra cooling turned on. Normally, fans blow air across the radiators and/or oil circulating pumps. Station transformers are commonly supplied with OA/FA/FOA ratings. The OA is open air, FA is forced air cooling, and FOA is forced air cooling plus oil circulating pumps.

The ANSI ratings were revised in the year 2000 to make them more consistent with IEC designations. This system has a four-letter code that indicates the cooling (IEEE C57.12.00-2000):

- *First letter* — Internal cooling medium in contact with the windings:
  - **O** mineral oil or synthetic insulating liquid with fire point = 300°C
  - **K** insulating liquid with fire point > 300°C
  - **L** insulating liquid with no measurable fire point
- *Second letter* — Circulation mechanism for internal cooling medium:
  - **N** natural convection flow through cooling equipment and in windings
  - **F** forced circulation through cooling equipment (i.e., coolant pumps); natural convection flow in windings (also called nondirected flow)
  - **D** forced circulation through cooling equipment, directed from the cooling equipment into at least the main windings
- *Third letter* — External cooling medium:
  - **A** air
  - **W** water
- *Fourth letter* — Circulation mechanism for external cooling medium:
  - **N** natural convection
  - **F** forced circulation: fans (air cooling), pumps (water cooling)

So, OA/FA/FOA is equivalent to ONAN/ONAF/OFAF. Each cooling level typically provides an extra one-third capability: 21/28/35 MVA. Table 4.13 shows equivalent cooling classes in the old and new naming schemes.

Utilities do not overload substation transformers as much as distribution transformers, but they do run them hot at times. As with distribution transformers, the tradeoff is loss of life versus the immediate replacement cost of the transformer. Ambient conditions also affect loading. Summer peaks are much worse than winter peaks. IEEE Std. C57.91-1995 provides detailed loading guidelines and also suggests an approximate adjustment of 1% of the maximum nameplate rating for every degree C above or below 30°C. The hottest spot conductor temperature is the critical point where insulation degrades. Above a hot-spot conductor temperature of 110°C, life expectancy decreases exponentially. The life halves for every 8°C increase in operating temperature. Most of the time, the hottest temperatures are nowhere near

TABLE 4.13

Equivalent Cooling Classes

Year 2000 Designations	Designations Prior to Year 2000
ONAN	OA
ONAF	FA
ONAN/ONAF/ONAF	OA/FA/FA
ONAN/ONAF/OFAF	OA/FA/FOA
OFAF	FOA
OFWF	FOW

Source: IEEE Std. C57.12.00-2000. Copyright 2000 IEEE. All rights reserved.

this. Tillman (2001) provides the loading guide for station transformers shown in Table 4.14.

The impedance of station transformers is normally about 7 to 10%. This is the impedance on the base rating, the self-cooled rating (OA or ONAN). The impedance is normally higher for voltages on the high-side of the transformer that are higher (like 230 kV). Transformer impedance can be specified when ordering. Large stations with 50 plus MVA transformers are normally provided with extra impedance to control fault currents, some as high as 30% on the transformer's base rating.

The positive and zero-sequence impedances are the same for a shell-type transformer, so the bolted fault currents on the secondary of the transformer are the same for a three-phase fault and for a line-to-ground fault (provided that both are fed from an infinite bus). In a three-legged core type transformer, the zero-sequence impedance is lower than the positive-sequence impedance (typically $Z_0 = 0.85Z_1$), so ground faults can cause higher currents. With a three-legged core transformer design, there is no path for zero-sequence flux. Therefore, zero-sequence current will meet a lower-impedance branch. This makes the core-type transformer act as if it had a delta-connected tertiary winding. This is the magnetizing branch (from line to ground), and this effectively reduces the zero-sequence impedance. In a shell-type transformer, there is a path through the iron for flux to flow, so the excitation impedance to zero sequence is high.

Because most distribution circuits are radial, the substation transformer is a critical component. Power transformers normally have a failure rate between 1 to 2% annually (CEA 485 T 1049, 1996; CIGRE working group 12.05, 1983; IEEE Std. 493-1997). Many distribution stations are originally designed with two transformers, where each is able to serve all of the substation's feeders if one of the transformers fails. Load growth in some areas has severely reduced the ability of one transformer to supply the whole station. To ensure transformer reliability, use good lightning protection and thermal management. Do not use reduced-BIL designs (BIL is the basic lightning impulse insulation level). Also, reclosing and relaying practices should ensure that excessive through faults do not damage transformers.

TABLE 4.14

Example Substation Transformer Loading Guide

Type of Load	FA (ONAF) Max Top Oil Temp (°C)	NDFOA (OFAF) Max Top Oil Temp (°C)	NDFOA (OFAF) Max Winding Temp (°C)	Max % Load
Normal summer load	105	95	135	130
Normal winter load	80	70	115	140
Emergency summer load	115	105	150	140
Emergency winter load	90	80	130	150
Non-cyclical load	95	85	115	110

Alarm Settings	FA 65°C Rise	NDFOA 65°C Rise
Top Oil	105°C	95°C
Hot Spot	135°C	135°C
Load Amps	130%	130%

Note: (1) The normal summer loading accounts for periods when temperatures are abnormally high. These might occur every 3 to 5 years. For every degree C that the normal ambient temperature during the hottest month of the year exceeds 30°C, de-rate the transformer 1% (i.e., 129% loading for 31°C average ambient). (2) The % load is given on the basis of the current rating. For MVA loading, multiply by the per unit output voltage. If the output voltage is 0.92 per unit, the recommended normal summer MVA loading is 120%. (3) Exercise caution if the load power factor is less than 0.95 lagging. If the power factor is less than 0.92 lagging, then lower the recommended loading by 10% (i.e., 130 to 120%). (4) Verify that cooling fans and pumps are in good working order and oil levels are correct. (5) Verify that the soil condition is good: moisture is less than 1.5% (1.0% preferred) by dry weight, oxygen is less than 2%, acidity is less than 0.5, and CO gas increases after heavy load seasons are not excessive. (6) Verify that the gauges are reading correctly when transformer loads are heavy. If correct field measurements differ from manufacturer's test report data, then investigate further before loading past nameplate criteria. (7) Verify with infrared camera or RTD during heavy load periods that the LTC top oil temperature relative to the main tank top oil temperature is correct. For normal LTC operation, the LTC top oil is cooler than the main tank top oil. A significant deviation from this indicates LTC abnormalities. (8) If the load current exceeds the bushing rating, do not exceed 110°C top oil temperature (IEEE, 1995). If bushing size is not known, perform an infrared scan of the bushing terminal during heavy load periods. Investigate further if the temperature of the top terminal cap is excessive. (9) Use winding power factor tests as a measure to confirm the integrity of a transformer's insulation system. This gives an indication of moisture and other contaminants in the system. High BIL transformers require low winding power factors (<0.5%), while low BIL transformers can tolerate higher winding power factors (<1.5%). (10) If the transformer is extremely dry (less than 0.5% by dry weight) and the load power factor is extremely good (0.99 lag to 0.99 lead), then add 10% to the above recommendations.

Source: Tillman, R. F., Jr, "Loading Power Transformers," in *The Electric Power Engineering Handbook*, L. L. Grigsby, Ed.: CRC Press, Boca Raton, FL, 2001.

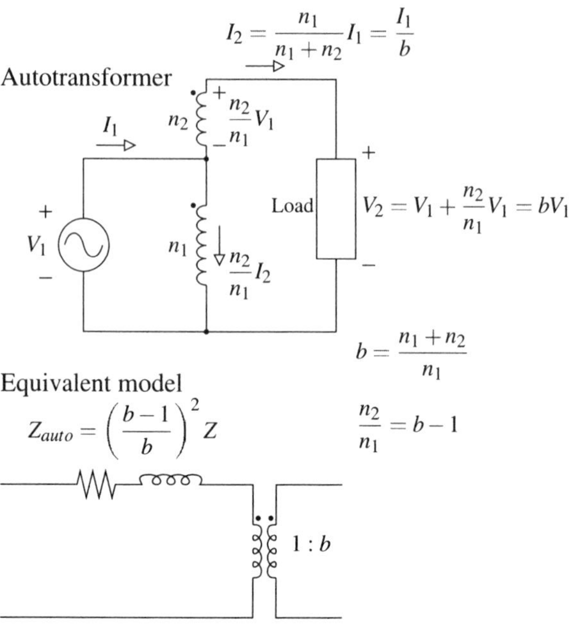

FIGURE 4.21
Autotransformer with an equivalent circuit.

4.9 Special Transformers

4.9.1 Autotransformers

An autotransformer is a winding on a core with a tap off the winding that provides voltage boost or buck. This is equivalent to a transformer with one winding in series with another (see Figure 4.21).

For small voltage changes, autotransformers are smaller and less costly than standard transformers. An autotransformer transfers much of the power directly through a wire connection. Most of the current passes through the lower-voltage series winding at the top, and considerably less current flows through the shunt winding.

Autotransformers have two main applications on distribution systems:

- *Voltage regulators* — A regulator is an autotransformer with adjustable taps that is normally capable of adjusting the voltage by ±10%.
- *Step banks* — Autotransformers are often used instead of traditional transformers on step banks and even substation transformers where

the relative voltage change is moderate. This is normally voltage changes of less than a factor of three such as a 24.94Y/14.4 kV–12.47Y/7.2 kV bank.

The required rating of an autotransformer depends on the voltage change between the primary and secondary. The rating of each winding as a percentage of the load is

$$S = \frac{b-1}{b}$$

where
b = voltage change ratio, per unit

To obtain a 10% voltage change (b = 1.1), an autotransformer only has to be rated at 9% of the load kVA. For a 2:1 voltage change (b = 2), an autotransformer has to be rated at 50% of the load kVA. By comparison, a standard transformer must have a kVA rating equal to the load kVA.

The series impedance of autotransformers is less than an equivalent standard transformer. The equivalent series impedance of the autotransformer is

$$Z_{auto} = \left(\frac{b-1}{b}\right)^2 Z$$

where Z is the impedance across the entire winding. A 5%, 100-kVA conventional transformer has an impedance of 25.9 Ω at 7.2 kV line to ground. A 2:1 autotransformer (b = 2) with a load-carrying capability of 100 kVA and a winding rating of 50 kVA and also a 5% winding impedance has an impedance of 6.5 Ω, one-fourth that of a conventional transformer.

For three-phase applications on grounded systems, autotransformers are often connected in a grounded wye. Other possibilities are delta (each winding is phase to phase), open delta (same as a delta, but without one leg), and open wye. Because of the direct connection, it is not possible to provide ground isolation between the high- and low-voltage windings.

4.9.2 Grounding Transformers

Grounding transformers are sometimes used on distribution systems. A grounding transformer provides a source for zero-sequence current. Grounding transformers are sometimes used to convert a three-wire, ungrounded circuit into a four-wire, grounded circuit. Figure 4.22 shows the two most common grounding transformers. The zig-zag connection is the most widely used grounding transformer. Figure 4.23 shows how a grounding bank supplies current to a ground fault. Grounding transformers

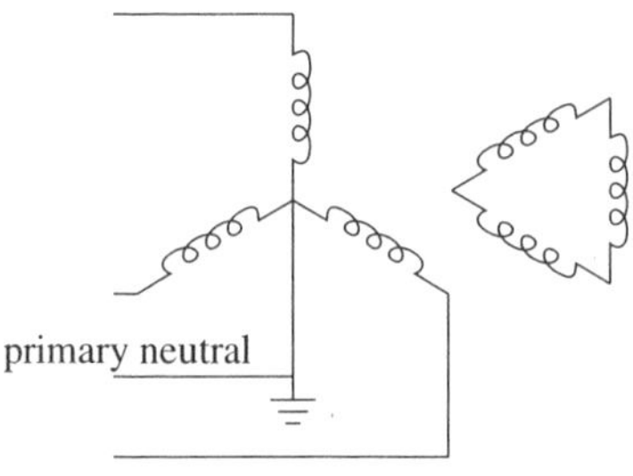

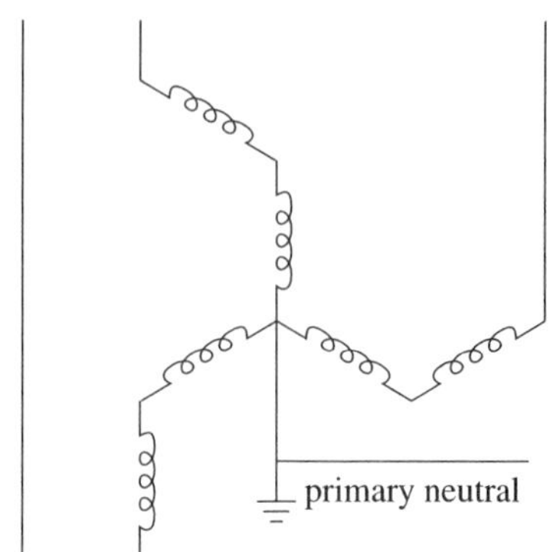

FIGURE 4.22
Grounding transformer connections.

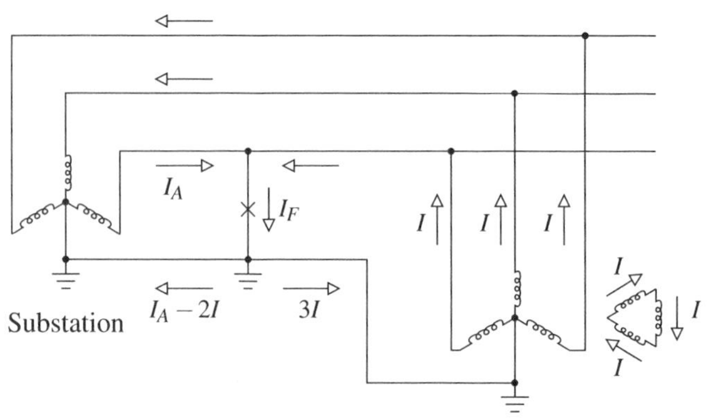

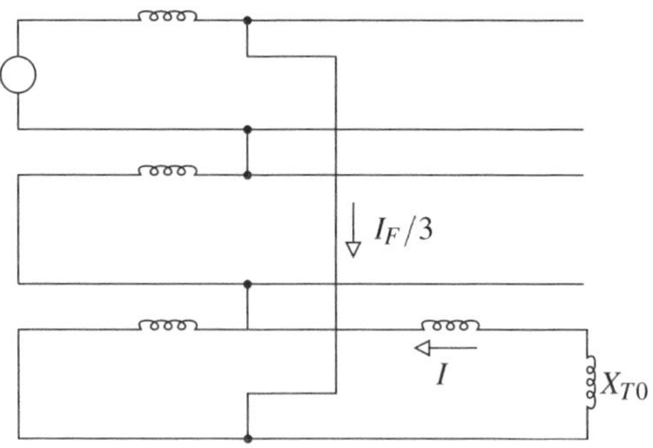

FIGURE 4.23
A grounding transformer feeding a ground fault.

used as the only ground source to a distribution circuit should be in service whenever the three-phase power source is in service. If the grounding transformer is lost, a line-to-ground causes high phase-to-neutral voltages on the unfaulted phases, and load unbalances can also cause neutral shifts and overvoltages.

A grounding transformer must handle the unbalanced load on the circuit as well as the duty during line-to-ground faults. If the circuit has minimal unbalance, then we can drastically reduce the rating of the transformer. It only has to be rated to carry short-duration (but high-magnitude) faults, normally a 10-sec or 1-min rating is used. We can also select the impedance of the grounding transformer to limit ground-fault currents.

Each leg of a grounding transformer carries one-third of the neutral current and has line-to-neutral voltage. So in a grounded wye – delta transformer, the total power rating including all three phases is the neutral current times the line-to-ground voltage:

$$S = V_{LG} I_N$$

A zig-zag transformer is more efficient than a grounded wye – delta transformer. In a zig-zag, each winding has less than the line-to-ground voltage, by a factor of $\sqrt{3}$, so the bank may be rated lower:

$$S = V_{LG} I_N / \sqrt{3}$$

ANSI/IEEE Std. 32-1972 requires a continuous rating of 3% for a 10-sec rated unit (which means the short-time rating is 33 times the continuous rating). A 1-min rated bank has a continuous current rating of 7%. On a 12.47-kV system supplying a ground-fault current of 6000 A, a zig-zag would need a 24.9-MVA rating. We will size the bank to handle the 24.9 MVA for 10 sec, which is equivalent to a 0.75-MVA continuous rating, so this bank could handle 180 A of neutral current continuously.

For both the zig-zag and the grounded wye – delta, the zero-sequence impedance equals the impedance between one transformer primary and its secondary.

Another application of grounding transformers is in cases of telephone interference due to current flow in the neutral/ground. By placing a grounding bank closer to the source of the neutral current, the grounding bank shifts some of the current from the neutral to the phase conductors to lower the neutral current that interferes with the telecommunication wires.

Grounding transformers are also used where utilities need a ground source during abnormal conditions. One such application is for a combination feeder that feeds secondary network loads and other non-network line-to-ground connected loads. If the network transformers are delta – grounded wye connected, the network will backfeed the circuit during a line-to-ground fault. If that happens while the main feeder breaker is open, the single-phase load on the unfaulted phases will see an overvoltage because the circuit is being back fed through the network loads as an ungrounded system. A grounding bank installed on the feeder prevents the overvoltage during backfeed conditions. Another similar application is found when applying distributed generators. A grounded wye – delta transformer is often specified as the interconnection transformer to prevent overvoltages if the generator drives an island that is separated from the utility source.

Even if a grounding bank is not the only ground source, it must be sized to carry the voltage unbalance. The zero-sequence current drawn by a bank is the zero-sequence voltage divided by the zero-sequence impedance:

$$I_0 = V_0/Z_0$$

Severe voltage unbalance can result when one phase voltage is opened upstream (usually from a blown fuse or a tripped single-phase recloser). In this case, the zero-sequence voltage equals the line-to-neutral voltage. The grounding bank will try to hold up the voltage on the opened phase and supply all of the load on that phase, which could severely overload the transformer.

4.10 Special Problems

4.10.1 Paralleling

Occasionally, crews must install distribution transformers, either at a changeover or for extra capacity. If a larger bank is being installed to replace an existing unit, paralleling the banks during the changeover eliminates the customer interruption. In order to parallel transformer banks, several criteria should be met:

- *Phasing* — The high and low-voltage connections must have the same phasing relationship. On three-phase units, banks of different connection types can be paralleled as long as they have compatible outputs: a delta – grounded wye may be paralleled with a grounded wye – grounded wye.
- *Polarity* — If the units have different polarity, they should be wired accordingly. (Flip one of the secondary connections.)
- *Voltage* — The phase-to-phase and phase-to-ground voltages on the outputs should be equal. Differences in turns ratios between the transformers will cause circulating current to flow through the transformers (continuously, even with zero load).

Before connecting the second transformer, crews should ensure that the secondary voltages are all zero or very close to zero (phase A to phase A, B to B, C to C, and the neutral to neutral).

If the percent impedances of the transformers are unequal, the load will not split in the same proportion between the two units. Note that this is the percent impedance, not the impedance in ohms. The unit with the lower percent impedance takes more of the current relative to its rating. For unequal impedances, the total bank must be derated (ABB, 1995) as

$$d = \frac{\frac{Z_2}{Z_1} K_1 + K_2}{K_1 + K_2}$$

where

K_1 = Capacity of the unit or bank with the *larger* percent impedance
K_2 = Capacity of the unit or bank with the *smaller* percent impedance
Z_1 = Percent impedance of unit or bank 1
Z_2 = Percent impedance of unit or bank 2

4.10.2 Ferroresonance

Ferroresonance is a special form of series resonance between the magnetizing reactance of a transformer and the system capacitance. A common form of ferroresonance occurs during single phasing of three-phase distribution transformers (Hopkinson, 1967). This most commonly happens on cable-fed transformers because of the high capacitance of the cables. The transformer connection is also critical for ferroresonance. An ungrounded primary connection (see Figure 4.24) leads to the highest magnitude ferroresonance. During single phasing (usually when line crews energize or deenergize the transformer with single-phase cutouts at the cable riser pole) a ferroresonant circuit between the cable capacitance and the transformer's magnetizing

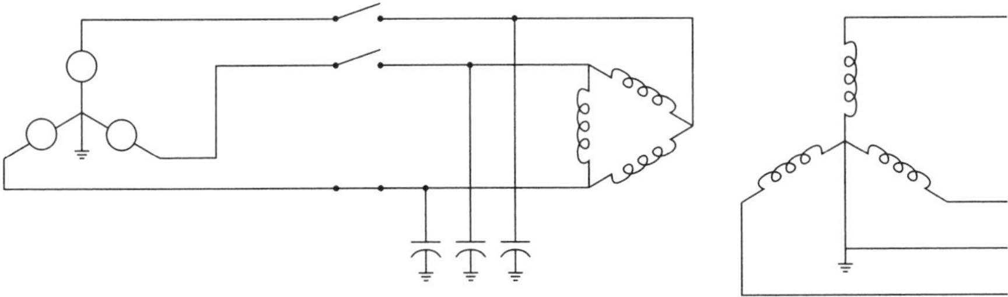

FIGURE 4.24
Ferroresonant circuit with a cable-fed transformer with an ungrounded high-side connection.

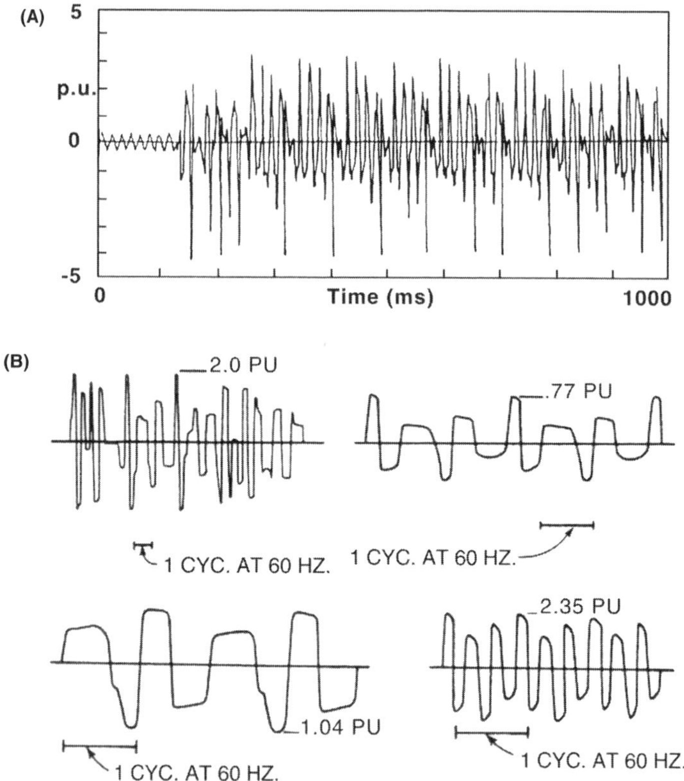

FIGURE 4.25
Examples of ferroresonance. (A) From Walling, R. A., Hartana, R. K., and Ros, W. J., "Self-Generated Overvoltages Due to Open-Phasing of Ungrounded-Wye Delta Transformer Banks," *IEEE Trans. Power Delivery*, 10(1), 526-533, January 1995. With permission. ©1995 IEEE. (B) Smith, D. R., Swanson, S. R., and Borst, J. D., "Overvoltages with Remotely-Switched Cable-Fed Grounded Wye-Wye Transformers," *IEEE Trans. Power Apparatus Sys.*, PAS-94(5), 1843-1853, 1975. With permission. ©1975 IEEE.

reactance drives voltages to as high as five per unit on the open legs of the transformer. The voltage waveform is normally distorted and often chaotic (see Figure 4.25).

Ferroresonance drove utilities to use three-phase transformer connections with a grounded-wye primary, especially on underground systems.

The chance of ferroresonance is determined by the capacitance (cable length) and by the core losses and other resistive load on the transformer (Walling et al., 1993). The core losses are an important part of the ferroresonant circuit.

Walling (1994) breaks down ferroresonance in a way that highlights several important aspects of this complicated phenomenon. Consider the simplified ferroresonant circuit in Figure 4.26. The transformer magnetizing branch has the core-loss resistance in parallel with a switched inductor. When the trans-

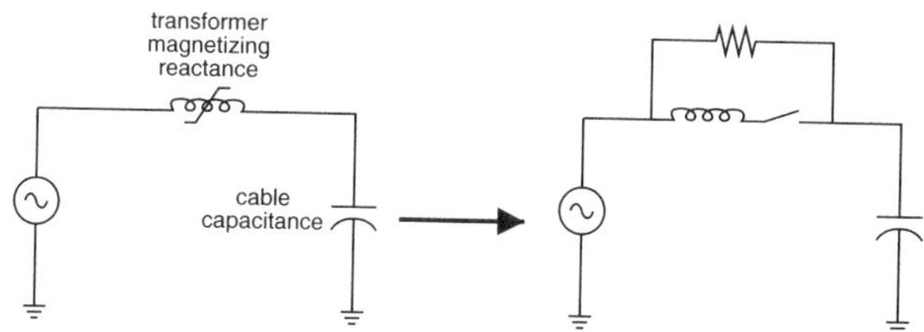

FIGURE 4.26
Simplified equivalent circuit of ferroresonance on a transformer with an ungrounded high-side connection.

former is unsaturated, the switched inductance is open, and the only connection between the capacitance and the system is through the core-loss resistance. When the core saturates, the capacitive charge dumps into the system (the switch in Figure 4.26 closes). The voltage overshoots and, as the core comes out of saturation, charge is again trapped on the capacitor (but of opposite polarity). This happens every half cycle (see Figure 4.27 for waveforms). If the core loss is large enough (or the resistive load on the transformer is large enough), the charge on the capacitor drains off before the next half cycle, and ferroresonance does not occur. The transformer core does not stay saturated long during each half cycle, just long enough to

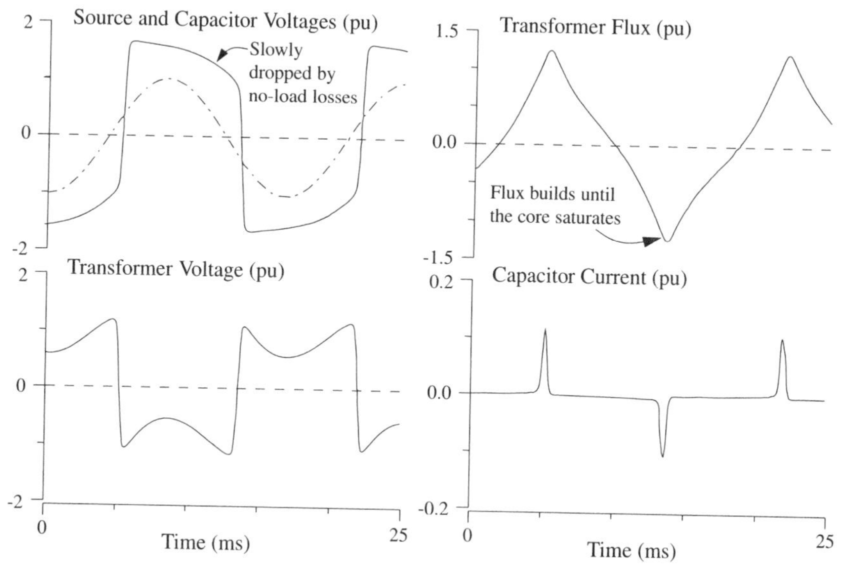

FIGURE 4.27
Voltages, currents and transformer flux during ferroresonance. (Adapted from Walling, R. A., "Ferroresonant Overvoltages in Today's Loss-Evaluated Distribution Transformers," IEEE/PES Transmission and Distribution Conference, 1994. With permission of the General Electric Company.)

TABLE 4.15

Transformer Primary Connections Susceptible to Ferroresonance

Susceptible Connections	Not Susceptible
Floating-wye Delta Grounded-wye with 3, 4, or 5-legged core construction Line-to-line connected single-phase units	Grounded wye made of three individual units or units of triplex construction Open wye – open delta Line-to-ground connected single-phase units

release the trapped charge on the capacitor. If the cable susceptance or even just the transformer susceptance is greater than the transformer core loss conductance, then ferroresonant overvoltages may occur.

In modern silicon-steel distribution transformers, the flux density at rated voltage is typically between 1.3 and 1.6 T. These operating flux densities slightly saturate the core (magnetic steel fully saturates at about 2 T). Because the core is operated near saturation, a small transient (such as switching) is enough to saturate the core. Once started, the ferroresonance self-sustains. The resonance repeatedly saturates the transformer every half cycle.

Table 4.15 shows what types of transformer connections are susceptible to ferroresonance. To avoid ferroresonance on floating wye – delta transformers, some utilities temporarily ground the wye on the primary side of floating wye – delta connections during switching operations.

Ferroresonance can occur on transformers with a grounded primary connection if the windings are on a common core such as the five-legged core transformer [the magnetic coupling between phases completes the ferroresonant circuit (Smith et al., 1975)]. The five-legged core transformer connected as a grounded wye – grounded wye is the most common underground transformer configuration. Ferroresonant overvoltages involving five-legged core transformers normally do not exceed two per unit.

Ferroresonance is a function of the cable capacitance and the transformer no-load losses. The lower the losses relative to the capacitance, the higher the ferroresonant overvoltage can be. For transformer configurations that are susceptible to ferroresonance, ferroresonance can occur approximately when

$$B_C \geq P_{NL}$$

where

B_C = capacitive reactive power per phase, vars

P_{NL} = core loss per phase, W

The capacitive reactive power on one phase in vars depends on the voltage and the capacitance as

$$B_C = \frac{V_{kV}^2}{3} 2\pi f C$$

where

V_{kV} = rated line-to-line voltage, kV

f = frequency, Hz

C = capacitance from one phase to ground, μF

Normally, ferroresonance occurs without equipment failure if the crew finishes the switching operation in a timely manner. The loud banging, rumbling, and rattling of the transformer during ferroresonance may alarm line crews. Occasionally, ferroresonance is severe enough to fail a transformer. The overvoltage stresses the transformer insulation, and the repeated saturation may cause tank heating as flux leaves the core (although many modes of ferroresonance barely saturate the transformer and do not cause significant tank heating). Surge arresters are the most likely equipment casualty. In attempting to limit the ferroresonant overvoltage, an arrester may absorb more current than it can handle and thermally run away. Gapped silicon-carbide arresters were particularly prone to failure, as the gap could not reseal the repeated sparkovers from a long-duration overvoltage. Gapless metal-oxide arresters are much more resistant to failure from ferroresonance and help hold down the overvoltages. Ferroresonant overvoltages may also fail customer's equipment from high secondary voltages. Small end-use arresters are particularly susceptible to damage.

Ferroresonance is more likely with

- *Unloaded transformers* — Ferroresonance disappears with load as little as a few percent of the transformer rating.
- *Higher primary voltages* — Shorter cable lengths are required for ferroresonance. Resonance is more likely even without cables, just due to the internal capacitance of the transformer. With higher voltages, the capacitances do not change significantly (cable capacitance increases just slightly because of thicker insulation), but vars are much higher for the same capacitance.
- *Smaller transformers* — Smaller no-load losses.
- *Low-loss transformers* — Smaller no-load losses.

Severe ferroresonance with voltages reaching peaks of 4 or 5 per unit occurs on three-phase transformers with an ungrounded high-voltage winding during single-pole switching. If the transformer is fed by underground cables and crews switch the transformer remotely, ferroresonance is likely.

On overhead circuits, ferroresonance is common with ungrounded primary connections on 25- and 35-kV distribution systems. At these voltages, the internal capacitance of most transformers is enough to ferroresonate. The use of low-loss transformers has caused ferroresonance to appear on overhead 15-kV distribution systems as well. Amorphous core and low-loss silicon-steel core transformers have much lower core losses than previous designs. With less core losses, ferroresonance happens with lower amounts of capacitance. Tests by the Southern California Edison Company on three-phase transformers with ungrounded primary connections found that ferroresonance occurred when the capacitive power per phase exceeded the transformer's no-load losses per phase by the following relationship (Jufer, 1994):

$$B_C \geq 1.27 P_{NL}$$

The phase-to-ground capacitance of overhead transformers is primarily due to the capacitance between the primary and secondary windings (the secondary windings are almost at zero potential). A typical 25-kVA transformer has a phase-to-ground capacitance of about 2 nF (Walling et al., 1995). For a 7.2-kV line-to-ground voltage, 0.002 μF is 39 vars. So, if the no-load losses are less than 39 vars/1.27 = 30.7 W per phase, the transformer may ferroresonate under single-pole switching.

Normally, ferroresonance occurs on three-phase transformers, but ferroresonance can occur on single-phase transformers if they are connected phase to phase, and one of the phases is opened either remotely or at the transformer. Jufer (1994) found that small single-phase padmounted transformers connected phase to phase ferroresonate when remotely switched with relatively short cable lengths. Their tests of silicon-steel core transformers found that a 25-kVA transformer resonated with 50 ft (15 m) of 1/0 XLPE cable at 12 kV. A 50-kVA transformer resonated with 100 ft of cable, and a 75-kVA unit resonated with 150 ft of the cable. Peak primary voltages reached 3 to 4 per unit. Secondary-side peaks were all under 2 per unit. Longer cables produced slightly higher voltages during ferroresonance. Jufer found that ferroresonance didn't occur if the resistive load in watts per phase (including the transformer's no-load losses and the resistive load on the secondary) exceeded 1.15 times the capacitive vars per phase ($P_{NL} + P_L > 1.15B_C$). Bohmann et al. (1991) describes a feeder where single-phase loads were switched to a phase-to-phase configuration, and the reconfiguration caused a higher-than-normal arrester failure rate that was attributed to ferroresonant conditions on the circuit.

It is widely believed that a grounded-wye primary connection eliminates ferroresonance. This is not true if the three-phase transformer has windings on a common core. The most common underground three-phase distribution transformer has a five-legged wound core. The common core couples the phases. With the center phase energized and the outer phases open, the coupling induces 50% voltage in the outer phases. Any load on the outer two phases is effectively in series with the voltage induced on the center phase. Because the coupling is indirect and the open phase capacitance is in parallel with a transformer winding to ground, this type of ferroresonance is not as severe as ferroresonance on configurations with an ungrounded primary winding. Overvoltages rarely exceed 2.5 per unit.

Five-legged core ferroresonance also depends on the core losses of the transformer and the phase-to-ground capacitance. If the capacitive vars exceed the resistive load in watts, ferroresonance may occur. Higher capac-

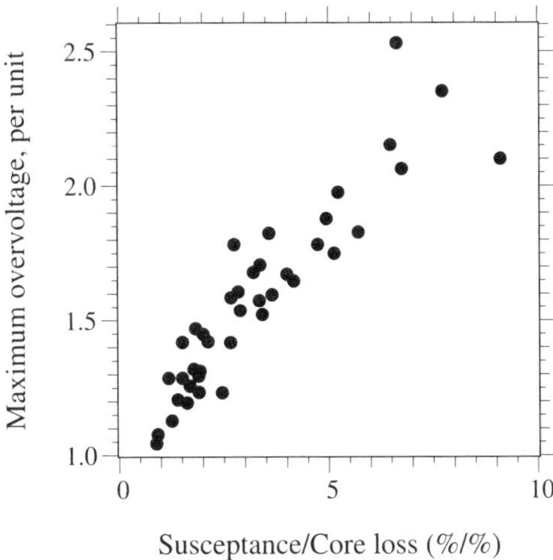

FIGURE 4.28
Five-legged core ferroresonance as a function of no-load losses and line-to-ground capacitance. (Adapted from Walling, R. A., Barker, K. D., Compton, T. M., and Zimmerman, L. E., "Ferroresonant Overvoltages in Grounded Wye-Wye Padmount Transformers with Low-Loss Silicon Steel Cores," *IEEE Trans. Power Delivery*, 8(3), 1647-60, July 1993. With permission. ©1993 IEEE.)

itances — longer cable lengths — generally cause higher voltages (see Figure 4.28). To limit peak voltages to below 1.25 per unit, the capacitive power must be limited such that [equivalent to that proposed by Walling (1992)]:

$$B_C \leq 1.86 P_{NL}$$

with B_C in vars and P_{NL} in watts; both are per phase.

Ferroresonance can occur with five-legged core transformers, even when switching at the transformer terminals, due to the transformer's internal line-to-ground capacitance. On 34.5-kV systems, transformers smaller than 500 kVA may ferroresonate if single-pole switched right at the transformer terminals. Even on 15-kV class systems where crews can safely switch all but the smallest 5-legged core transformers at the terminals, we should include the transformer's capacitance in any cable length calculation; the transformer's capacitance is equivalent to several feet (meters) of cable. The capacitance from line-to-ground is mainly due to the capacitance between the small paper-filled layers of the high-voltage winding. This capacitance is very difficult to measure since it is in parallel with the coil. Walling (1992) derived an empirical equation to estimate the line-to-ground transformer capacitance per phase in μF:

$$C = \frac{0.000469 S_{kVA}^{0.4}}{V_{kV}^{0.25}}$$

where

S_{kVA} = transformer three-phase kVA rating
V_{kV} = rated line-to-line voltage in kV

In vars, this is

$$B_C = 0.000982 f V_{kV}^{1.75} S_{kVA}^{0.4}$$

where f is the system frequency, Hz.

To determine whether the transformer no-load losses exceed the capacitive power, the transformer's datasheet data is most accurate. For coming up with generalized guidelines, using such data is not realistic since so many different transformer makes and models are ordered. Walling (1992) offered the following approximation between the three-phase transformer rating and the no-load losses in watts per phase:

$$P_{NL} = S_{kVA}\left(4.54 - 1.13\log_{10}\left(S_{kVA}\right)\right)/3$$

Walling (1992) used his approximations of transformer no-load losses and transformer capacitance to find cable length criteria for remote single-pole switching. Consider a 75-kVA 3-phase 5-legged core transformer at 12.47 kV. Using these approximations, the no-load losses are 60.5 W per phase, and the transformer's capacitance is 27.4 vars per phase. To keep the voltage under 1.25 per unit, the total vars allowed per phase is 1.85(60.5W) = 111.9 vars. So, the cable can add another 84.5 vars before we exceed the limit. At 12.47 kV, a 4/0 175-mil XLPE cable has a capacitance of 0.412 μF/mi, which is 1.52 vars per foot. For this cable, 56 ft is the maximum length that we should switch remotely. Beyond that, we may have ferroresonance above 1.25 per unit. Table 4.16 shows similar criteria for several three-phase transformers and voltages. The table shows critical lengths for 4/0 cables; smaller cables have less capacitance, so somewhat longer lengths are permissible. At 34.5 kV, crews should only remotely switch larger banks.

Another situation that can cause ferroresonance is when a secondary has ungrounded power factor correction capacitors. Resonance can even occur on a grounded wye – grounded wye connection with three separate transformers. With one phase open on the utility side, the ungrounded capacitor bank forms a series resonance with the magnetizing reactance of the open leg of the grounded-wye transformer.

Ferroresonance most commonly happens when switching an unloaded transformer. It also usually happens with manual switching; ferroresonance can occur because a fault clears a single-phase protective device, but this is much less common. The main reason that ferroresonance is unlikely for most situations using a single-phase protective device is that either the fault or the existing load on the transformer prevents ferroresonance.

TABLE 4.16

Cable Length Limits in Feet for Remote Single-Pole Switching to Limit Ferroresonant Overvoltages to Less than 1.25 per Unit

	Critical Cable Lengths, ft		
Transformer Rating kVA	12.47 kV 4/0 XLPE 175 mil 0.412 μF/mi 1.52 vars/ft	24.94 kV 4/0 XLPE 260 mil 0.261 μF/mi 4.52 vars/ft	34.5 kV 4/0 XLPE 345 mil 0.261 μF/mi 7.08 vars/ft
75	56	5	0
112.5	81	10	0
150	103	16	0
225	144	26	1
300	181	36	6
500	265	59	16
750	349	82	27
1000	417	100	36
1500	520	128	49
2000	592	146	56

If the fuse is a tap fuse and several customers are on a section, the transformers will have somewhat different characteristics, which lowers the probability of ferroresonance (and ferroresonance is less likely with larger transformers).

Solutions to ferroresonance include

- Using a higher-loss transformer
- Using a three-phase switching device instead of a single-phase device
- Switching right at the transformer rather than at the riser pole
- Using a transformer connection not susceptible to ferroresonance
- Limiting remote switching of transformers to cases where the capacitive vars of the cable are less than the transformer's no load losses

Arrester application on transformer connections susceptible to ferroresonance brings up several interesting points. Ferroresonance can slowly heat arresters until failure. Ferroresonance is a weak source; even though the per-unit magnitudes are high, the voltage collapses when the arrester starts to conduct (we cannot use the arresters time-overvoltage curve [TOV] to predict failure). Normally, extended ferroresonance of several minutes can occur before arresters are heated enough to enter thermal runaway. The most vulnerable arresters are those that are tightly applied relative to the voltage rating. Tests by the DSTAR group for ferroresonance on 5-legged core transformers in a grounded wye – grounded wye connection (Lunsford, 1994; Walling et al., 1994) found

- Arrester currents were always less than 2 A.
- Under-oil arresters, which have superior thermal characteristics, reached thermal stability and did not fail.
- Porcelain-housed arresters showed slow heating — sometimes enough to fail, sometimes not, depending on the transformer type, cable lengths, and arrester type. Elbow arresters showed slow heating — slower than the riser-pole arresters. Failure times for either type were typically longer than 30 min.

With normal switching times of less than one minute, arresters do not have enough time to heat and fail. Crews should be able to safely switch transformers under most circumstances. Load — even 5% of the transformer rating — prevents ferroresonance in most cases. The most danger is with unloaded transformers. If an arrester fails, the failure may not operate the disconnect, which can lead to a dangerous scenario. When a line worker recloses the switch, the stiff power-frequency source will fail the arrester. The disconnect should operate and draw an arc. On occasion, the arrester may violently shatter.

One option to limit the exposure of the arresters is to put the arresters upstream of the switch. At a cable riser pole this is very difficult to do without seriously compromising the lead length of the arrester.

4.10.3 Switching Floating Wye – Delta Banks

Floating wye – delta banks present special concerns. As well as being prone to ferroresonance, single-pole switching can cause overvoltages due to a neutral shift. On a floating wye – delta, the secondary delta connection fixes the transformer's primary neutral close to ground potential. After one phase of the primary wye is opened, the neutral can float far from ground. This causes overvoltages, both on the secondary side and the primary side. The severity depends on the balance of the load.

When crews open one of the power-leg phases, if there is no three-phase load and only the single-phase load on the lighting leg of the transformer, the open primary voltage V_{open} reaches 2.65 times normal as shown in Figure 4.29. The voltage across the open switch also sees high voltage. The voltage from B to B′ in Figure 4.29 can reach over 2.75 per unit. Secondary line-to-line voltages on the power legs can reach 1.73 per unit. The secondary delta forces the sum of the three primary line-to-neutral voltages to be equal. With single-phase load on phase C and no other load, the neutral shifts to the C-phase voltage. The delta winding forces $V_{B'N}$ to be equal to $-V_{AN}$, which significantly shifting the potential of point B′.

The line-to-ground voltage on the primary-side of the transformer on the open phase is a function of the load unbalance on the secondary. Given the ratio of the single-phase load to the three-phase load, this voltage is [assum-

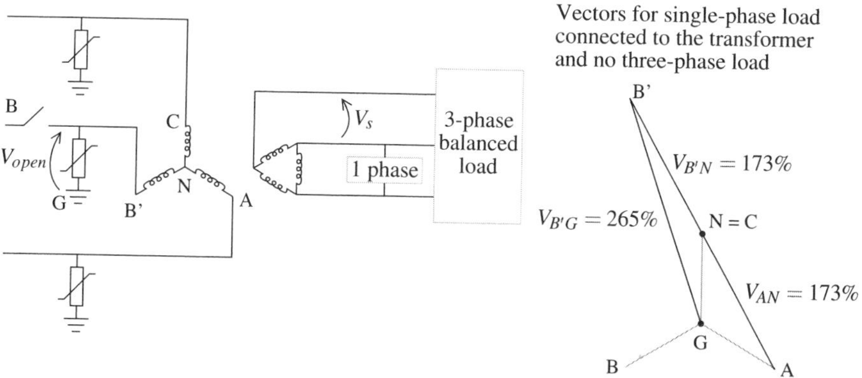

FIGURE 4.29
Neutral-shift overvoltages on a floating wye – delta transformer during single-pole switching.

ing passive loads and that the power factor of the three-phase load equals that of the single-phase load (Walling et al., 1995)]

$$V_{open} = \frac{\sqrt{7K^2 + K + 1}}{K + 2}$$

where

$$K = \frac{\text{Single-phase load}}{\text{Balanced three-phase load}}$$

On the secondary side, the worst of the two line-to-line voltages across the power legs have the following overvoltages depending on loading balance (PTI, 1999):

$$V_s = \sqrt{3}\,\frac{K + 1}{K + 2}$$

Figure 4.30 shows these voltages as a function of the ratio K.

Contrary to a widespread belief, transformer saturation does not significantly reduce the overvoltage. Walling et al.'s (1995) EMTP simulations showed that saturation did not significantly reduce the peak voltage magnitude. Saturation does distort the waveforms significantly and reduces the energy into a primary arrester.

Some ways to avoid these problems are

- *Use another connection* — The best way to avoid problems with this connection is to use some other connection. Some utilities do not

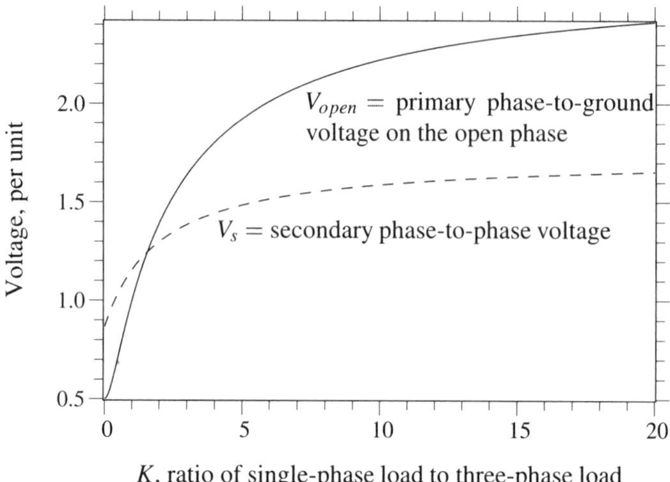

FIGURE 4.30
Neutral-shift overvoltages as a function of the load unbalance.

offer an open wye – delta connection and instead move customers to grounded-wye connections.

- *Neutral grounding* — Ground the primary-wye neutral during switching operations, either with a temporary grounding jumper or install a cutout. This prevents the neutral-shift and ferroresonant overvoltage. The ground-source effects during the short-time switching are not a problem. The line crew must remove the neutral jumper after switching. Extended operation as a grounding bank can overheat the transformer and interfere with a circuit's ground-fault protection schemes.
- *Switching order* — Neutral shifts (but not ferroresonance) are eliminated by always switching in the lighting leg last and taking it out first.

Arrester placement is a sticky situation. If the arrester is upstream of the switch, it does not see the neutral-shift/ferroresonant overvoltage. But the transformer is not protected against the overvoltages. Arresters downstream of the switch protect the transformer but may fail. One would rather have an arrester failure than a transformer failure, unless the failure is near a line crew (since an arrester is smaller, it is more likely than a transformer to explode violently — especially porcelain-housed arresters). Another concern was reported by Walling (2000): during switching operations, 10-per-unit overvoltage bursts for 1/4 cycle ringing at about 2 kHz when closing in the second phase. These were found in measurements during full-scale tests and also in simulations. This transient repeats every cycle with a declining peak magnitude for more than one second. If arresters are downstream from the switches, they can easily control the overvoltage. But if they are upstream of the switches, this high voltage stresses the transformer insulation.

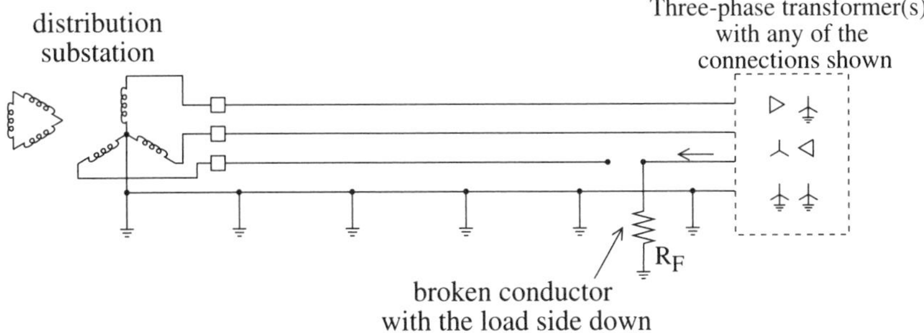

FIGURE 4.31
Backfeed to a downed conductor.

Overall, grounding the transformer's primary neutral is the safest approach.

4.10.4 Backfeeds

During a line-to-ground fault where a single-phase device opens, current may backfeed through a three-phase load (see Figure 4.31). It is a common misconception that this type of backfeed can only happen with an ungrounded transformer connection. Backfeed can also occur with a grounded three-phase connection. This creates hazards to the public in downed wire situations. Even though it is a weak source, the backfed voltage is just as dangerous. Lineworkers also have to be careful. A few have been killed after touching wires downstream of open cutouts that they thought were deenergized.

The general equations for the backfeed voltage and current based on the sequence impedances of the load (Smith, 1994) are

$$I_F = \frac{(A - 3Z_0Z_2)V}{3Z_0Z_1Z_2 + R_FA}$$

$$V_F = R_FI_F$$

where

$A = Z_0Z_1 + Z_1Z_2 + Z_0Z_2$
Z_1 = positive-sequence impedance of the load, Ω
Z_2 = negative-sequence impedance of the load, Ω
Z_0 = zero-sequence impedance of the load, Ω
R_F = fault resistance, Ω
V = line-to-neutral voltage, V

The line and source impedances are left out of the equations because they are small relative to the load impedances. Under an open circuit with no fault ($R_F = \infty$), the backfeed voltage is

$$V_F = \frac{\left(A - 3Z_0 Z_2\right)V}{A}$$

For an ungrounded transformer connection ($Z_0 = \infty$), the backfeed current is

$$I_F = \frac{\left(Z_1 - 2Z_2\right)V}{3Z_1 Z_2 + R_F\left(Z_1 + Z_2\right)}$$

The backfeed differs depending on the transformer connection and the load:

- Grounded wye – grounded wye transformer connection
  - Will not backfeed the fault when the transformer is unloaded or has balanced line-to-ground loads (no motors). It will backfeed the fault with line-to-line connected load (especially motors).
- Ungrounded primary transformer
  - Will backfeed the fault under no load. It may not be able to provide much current with no load, but there can be significant voltage on the conductor. Motor load will increase the backfeed current available.

Whether it is a grounded or ungrounded transformer, the available backfeed current depends primarily on the connected motor load. Motors dominate since they have much lower negative-sequence impedance; typically it is equal to the locked-rotor impedance or about 15 to 20%. With no fault impedance ($R_F = 0$), the backfeed current is approximately:

$$I_F = \frac{M_{kVA}}{9V_{LG,kV} \cdot Z_{2,pu}}$$

where M_{kVA} is the three-phase motor power rating in kVA (and we can make the common assumption that 1 hp = 1 kVA), $V_{LG,kV}$ is the line-to-ground voltage in kV, and $Z_{2,pu}$ is the per-unit negative-sequence (or locked-rotor) impedance of the motor(s). Figure 4.32 shows the variation in backfeed current versus motor kVA on the transformer for a 12.47-kV system (assuming $Z_{2,pu} = 0.15$).

The voltage on the open phases depends on the type of transformer connection and the portion of the load that is motors. Figure 4.33 shows the backfeed voltage for an open circuit and for a typical high-impedance fault ($R_F = 200\ \Omega$).

As discussed in Chapter 7, the maximum sustainable arc length in inches is roughly $l = \sqrt{I \cdot V}$ where I is the rms current in amperes, and V is the

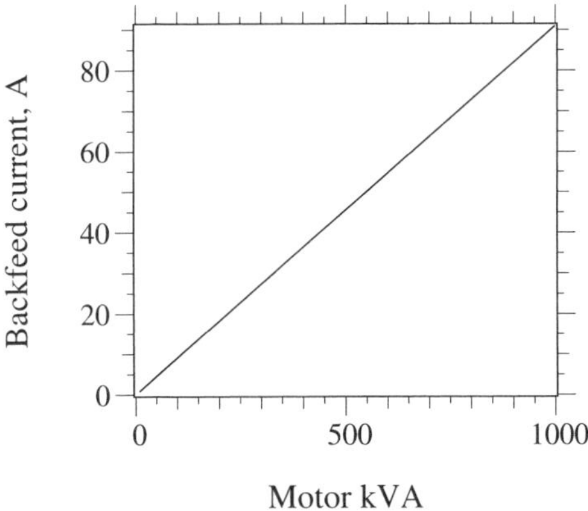

FIGURE 4.32
Available backfeed current on a 12.47-kV circuit (grounded wye – grounded wye or an ungrounded connection, $R_F = 0$).

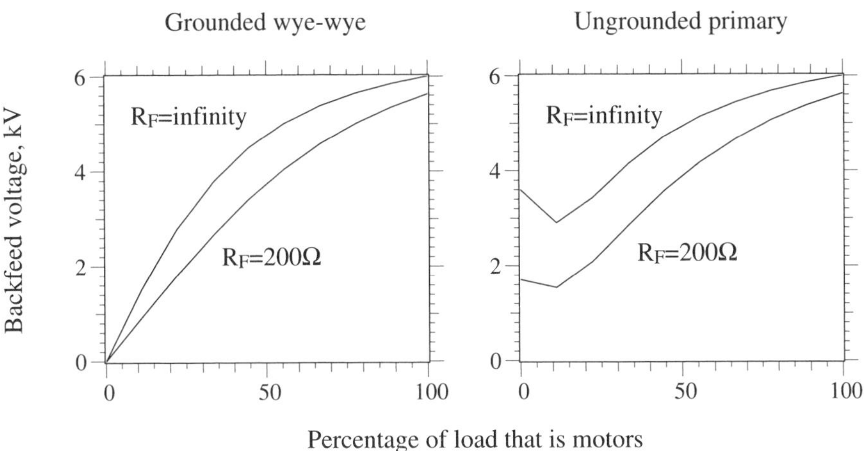

FIGURE 4.33
Available backfeed voltage on a 12.47-kV circuit.

voltage in kV. For a line-to-ground fault on a 12.47-kV circuit, if the backfeed voltage is 4 kV with 50 A available (typical values from Figure 4.32 and Figure 4.33), the maximum arc length is 28 in. (0.7 m). Even though the backfeed source is weak relative to a traditional fault source, it is still strong enough to maintain a significant arc during backfeeds.

In summary, the backfeed voltage is enough to be a safety hazard to workers or the public (e.g., in a wire down situation). The available backfeed is a stiff enough source to maintain an arc of significant length. The arc can continue to cause damage at the fault location during a backfeed condition. It may also spark and sputter at a low level. Options to reduce the chances of backfeed problems include:

- Make sure crews follow safety procedures (if it is not grounded, it is not dead).
- Follow standard practices regarding downed conductors including proper line designs and maintenance, public education, and worker training.

Another option is to avoid single-pole protective devices (switches, fuses, or single-phase reclosers) upstream of three-phase transformer banks. Most utilities have found that backfeeding problems are not severe enough to warrant not using single-pole protective devices.

To analyze more complicated arrangements, use a steady-state circuit analysis program (EMTP has this capability). Most distribution fault analysis programs cannot handle this type of complex arrangement.

4.10.5 Inrush

When a transformer is first energized or reenergized after a short interruption, the transformer may draw *inrush* current from the system due to the core magnetization being out of sync with the voltage. The inrush current may approach short-circuit levels, as much as 40 times the transformer's full-load current. Inrush may cause fuses, reclosers, or relays to falsely operate. It may also falsely operate faulted-circuit indicators or cause sectionalizers to misoperate.

When the transformer is switched in, if the system voltage and the transformer core magnetization are not in sync, a magnetic transient occurs. The transient drives the core into saturation and draws a large amount of current into the transformer.

The worst inrush occurs with residual flux left on the transformer core. Consider Figure 4.34 and Figure 4.35, which shows the worst-case scenario. A transformer is deenergized near the peak core flux density (B_{max}), when the voltage is near zero. The flux decays to about 70% of the maximum and holds there (the residual flux, B_r). Some time later, the transformer is reenergized at a point in time when the flux would have been at its negative peak; the system voltage is crossing through zero and rising positively. The positive voltage creates positive flux that adds to the residual flux already on the transformer core (remember, flux is the time integral of the voltage). This quickly saturates the core; the effective magnetizing branch drops to the air-core impedance of the transformer.

The air core impedance is roughly the same magnitude as the transformer's leakage impedance. Flux controls the effective impedance, so when the core saturates, the small impedance pulls high-magnitude current from the system. The core saturates in one direction, so the transformer draws pulses of inrush every other half cycle with a heavy dc component. The dc offset introduced by the switching decays away relatively quickly. Figure

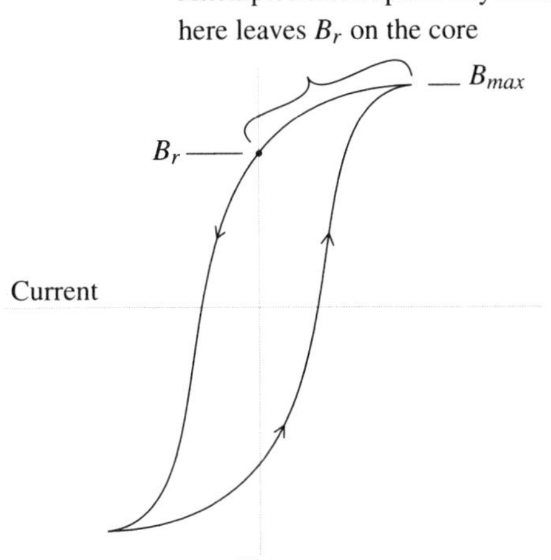

FIGURE 4.34
Hysteresis curve showing the residual flux during a circuit interruption.

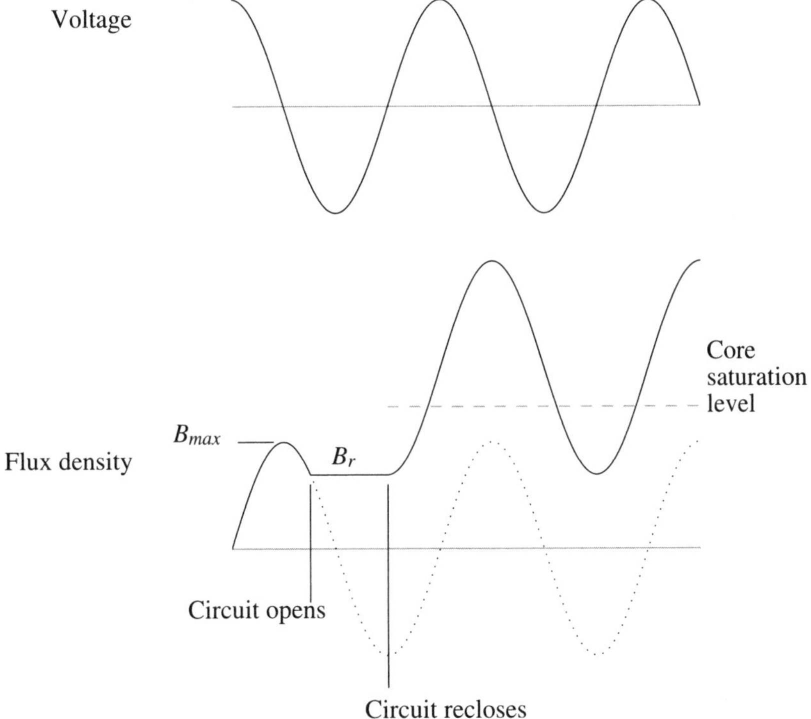

FIGURE 4.35
Voltage and flux during worst-case inrush.

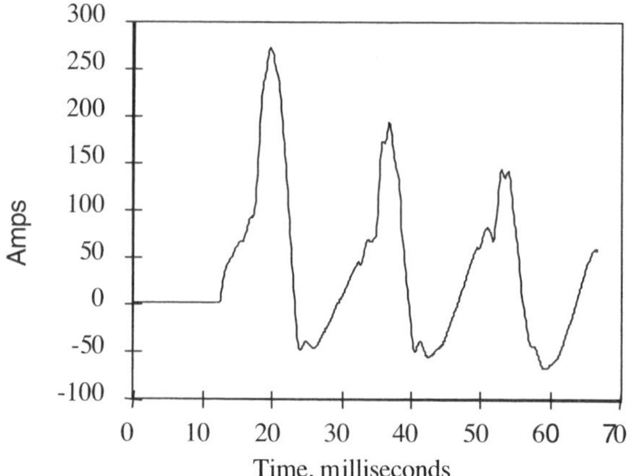

FIGURE 4.36
Example inrush current measured at a substation (many distribution transformers together). (Copyright © 1996. Electric Power Research Institute. TR-106294-V3. *An Assessment of Distribution System Power Quality: Volume 3: Library of Distribution System Power Quality Monitoring Case Studies.* Reprinted with permission.)

4.36 shows an example of inrush following a reclose operation measured at the distribution substation breaker.

Several factors significantly impact inrush:

- *Closing point* — The point where the circuit closes back in determines how close the core flux can get to its theoretical maximum. The worst case is when the flux is near its peak. Fortunately, this is also when the voltage is near zero, and switches tend to engage closer to a voltage peak (an arc tends to jump the gap).
- *Design flux* — A transformer that is designed to operate lower on the saturation curve draws less inrush. Because there is more margin between the saturation point and the normal operating region, the extra flux during switching is less likely to push the core into saturation.
- *Transformer size* — Larger transformers draw more inrush. Their saturated impedances are smaller. But, on a per-unit basis relative to their full-load capability, smaller transformers draw more inrush. The inrush into smaller transformers dies out more quickly.
- *Source impedance* — Higher source impedance relative to the transformer size limits the current that the transformer can pull from the system. The peak inrush with significant source impedance (Westinghouse Electric Corporation, 1950) is

$$i_{peak} = \frac{i_0}{1 + i_0 X}$$

where

i_0 = peak inrush without source impedance in per unit of the transformer rated current

X = source impedance in per unit on the transformer kVA base

Other factors have less significance. The load on the transformer does not significantly change the inrush. For most typical loading conditions, the current into the transformer will interrupt at points that still leave about 70% of the peak flux on the core.

While interruptions generally cause the most severe inrush, other voltage disturbances may cause inrush into a transformer. Voltage transients and especially voltage with a dc component can saturate the transformer and cause inrush. Some examples are:

- *Voltage sags* — Upon recovery from a voltage sag from a nearby fault, the sudden rise in voltage can drive a transformer into saturation.
- *Sympathetic inrush* — Energizing a transformer can cause a nearby transformer to also draw inrush. The inrush into the switched transformer has a significant dc component that causes a dc voltage drop. The dc voltage can push the other transformer into saturation and draw inrush.
- *Lightning* — A flash to the line near the transformer can push the transformer into saturation.

References

ABB, *Distribution Transformer Guide*, 1995.

Alexander Publications, *Distribution Transformer Handbook*, 2001.

ANSI C57.12.40-1982, *American National Standard Requirements for Secondary Network Transformers, Subway and Vault Types (Liquid Immersed).*

ANSI/IEEE C57.12.24-1988, *American National Standard Underground-type Three-Phase Distribution Transformers, 2500 kVA and Smaller; High Voltage 34 500 GrdY/19 200 V and Below; Low Voltage 480 V and Below — Requirements.*

ANSI/IEEE C57.12.80-1978, *IEEE Standard Terminology for Power and Distribution Transformers.*

ANSI/IEEE C57.91-1981, *IEEE Guide for Loading Mineral-Oil-Immersed Overhead and Pad-Mounted Distribution Transformers Rated 500 kVA and Less with 65 Degrees C Or 55 Degrees C Average Winding Rise.*

ANSI/IEEE C57.105-1978, *IEEE Guide for Application of Transformer Connections in Three-Phase Distribution Systems.*

ANSI/IEEE Std. 32-1972, *IEEE Standard Requirements, Terminology, and Test Procedure for Neutral Grounding Devices.*

Blume, L. F., Boyajian, A., Camilli, G., Lennox, T. C., Minneci, S., and Montsinger, V. M., *Transformer Engineering*, Wiley, New York, 1951.

Bohmann, L. J., McDaniel, J., and Stanek, E. K., "Lightning Arrester Failures and Ferroresonance on a Distribution System," IEEE Rural Electric Power Conference, 1991.

CEA 485 T 1049, *On-line Condition Monitoring of Substation Power Equipment Utility Needs*, Canadian Electrical Association, 1996.

CIGRE working group 12.05, "An International Survey on Failure in Large Power Transformer Service," *Electra*, no. 88, pp. 21–48, 1983.

EEI, "A Method for Economic Evaluation of Distribution Transformers," March, 28–31, 1981.

EPRI TR-106294-V3, *An Assessment of Distribution System Power Quality: Volume 3: Library of Distribution System Power Quality Monitoring Case Studies*, Electric Power Research Institute, Palo Alto, CA, 1996.

Gangel, M. W. and Propst, R. F., "Distribution Transformer Load Characteristics," *IEEE Transactions on Power Apparatus and Systems*, vol. 84, pp. 671–84, August 1965.

Grainger, J. J. and Kendrew, T. J., "Evaluation of Technical Losses on Electric Distribution Systems," CIRED, 1989.

Hopkinson, F. H., "Approximate Distribution Transformer Impedances," General Electric Internal Memorandum, 1976. As cited by Kersting, W. H. and Phillips, W. H., "Modeling and Analysis of Unsymmetrical Transformer Banks Serving Unbalanced Loads," Rural Electric Power Conference, 1995.

Hopkinson, R. H., "Ferroresonant Overvoltage Control Based on TNA Tests on Three-Phase Delta-Wye Transformer Banks," *IEEE Transactions on Power Apparatus and Systems*, vol. 86, pp. 1258–65, October 1967.

IEEE C57.12.00-2000, *IEEE Standard General Requirements for Liquid-Immersed Distribution, Power, and Regulating Transformers.*

IEEE Std. 493-1997, *IEEE Recommended Practice for the Design of Reliable Industrial and Commercial Power Systems (Gold Book).*

IEEE Std. C57.91-1995, *IEEE Guide for Loading Mineral-Oil-Immersed Transformers.*

IEEE Task Force Report, "Secondary (Low-Side) Surges in Distribution Transformers," *IEEE Transactions on Power Delivery*, vol. 7, no. 2, pp. 746–56, April 1992.

Jufer, N. W., "Southern California Edison Co. Ferroresonance Testing of Distribution Transformers," IEEE/PES Transmission and Distribution Conference, 1994.

Long, L. W., "Transformer Connections in Three-Phase Distribution Systems," in Power Transformer Considerations of Current Interest to the Utility Engineer, 1984. IEEE Tutorial Course, 84 EHO 209-7-PWR.

Lunsford, J., "MOV Arrester Performance During the Presence of Ferroresonant Voltages," IEEE/PES Transmission and Distribution Conference, 1994.

Nickel, D. L., "Distribution Transformer Loss Evaluation. I. Proposed Techniques," *IEEE Transactions on Power Apparatus and Systems*, vol. PAS-100, no. 2, pp. 788–97, February 1981.

NRECA RER Project 90-8, *Underground Distribution System Design and Installation Guide*, National Rural Electric Cooperative Association, 1993.

ORNL-6804/R1, *The Feasibility of Replacing or Upgrading Utility Distribution Transformers During Routine Maintenance*, Oak Ridge National Laboratory, U.S. Department of Energy, 1995.

ORNL-6847, *Determination Analysis of Energy Conservation Standards for Distribution Transformers*, Oak Ridge National Laboratory, U.S. Department of Energy, 1996.

ORNL-6925, *Supplement to the "Determination Analysis" (ORNL-6847) and Analysis of the NEMA Efficiency Standard for Distribution Transformers*, Oak Ridge National Laboratory, U.S. Department of Energy, 1997.

ORNL-6927, *Economic Analysis of Efficient Distribution Transformer Trends*, Oak Ridge National Laboratory, U.S. Department of Energy, 1998.

PTI, "Distribution Transformer Application Course Notes," Power Technologies, Inc., Schenectady, NY, 1999.

Rusch, R. J. and Good, M. L., "Wyes and Wye Nots of Three-Phase Distribution Transformer Connections," IEEE Rural Electric Power Conference, 1989.

Sankaran, C., "Transformers," in *The Electrical Engineering Handbook*, R. C. Dorf, Ed.: CRC Press, Boca Raton, FL, 2000.

Seevers, O. C., *Management of Transmission & Distribution Systems*, PennWell Publishing Company, Tulsa, OK, 1995.

Smith, D. R., "Impact of Distribution Transformer Connections on Feeder Protection Issues," Texas A&M Annual Conference for Protective Relay Engineers, March 1994.

Smith, D. R., Braunstein, H. R., and Borst, J. D., "Voltage Unbalance in 3- and 4-Wire Delta Secondary Systems," *IEEE Transactions on Power Delivery*, vol. 3, no. 2, pp. 733–41, April 1988.

Smith, D. R., Swanson, S. R., and Borst, J. D., "Overvoltages with Remotely-Switched Cable-Fed Grounded Wye-Wye Transformers," *IEEE Transactions on Power Apparatus and Systems*, vol. PAS-94, no. 5, pp. 1843–53, 1975.

Tillman, R. F., Jr, "Loading Power Transformers," in *The Electric Power Engineering Handbook*, L. L. Grigsby, Ed.: CRC Press, Boca Raton, FL, 2001.

Walling, R. A., "Ferroresonance Guidelines for Modern Transformer Applications," in Final Report to the Distribution Systems Testing, Application, and Research (DSTAR) Consortium: General Electric, Industrial and Power Systems, Power Systems Engineering Department, 1992. As cited in NRECA RER Project 90-8, 1993.

Walling, R. A., "Ferroresonant Overvoltages in Today's Loss-Evaluated Distribution Transformers," IEEE/PES Transmission and Distribution Conference, 1994.

Walling, R. A., 2000. Verbal report at the fall IEEE Surge Protective Devices Committee Meeting.

Walling, R. A., Barker, K. D., Compton, T. M., and Zimmerman, L. E., "Ferroresonant Overvoltages in Grounded Wye-Wye Padmount Transformers with Low-Loss Silicon Steel Cores," *IEEE Transactions on Power Delivery*, vol. 8, no. 3, pp. 1647–60, July 1993.

Walling, R. A., Hartana, R. K., Reckard, R. M., Sampat, M. P., and Balgie, T. R., "Performance of Metal-Oxide Arresters Exposed to Ferroresonance in Padmount Transformers," *IEEE Transactions on Power Delivery*, vol. 9, no. 2, pp. 788–95, April 1994.

Walling, R. A., Hartana, R. K., and Ros, W. J., "Self-Generated Overvoltages Due to Open-Phasing of Ungrounded-Wye Delta Transformer Banks," *IEEE Transactions on Power Delivery*, vol. 10, no. 1, pp. 526–33, January 1995.

Westinghouse Electric Corporation, *Electrical Transmission and Distribution Reference Book*, 1950.

All hell broke loose, we had a ball of fire that went phase to phase shooting fire out the xfmer vents like a flame thrower showering slag on the linemen and sent the monster galloping down the line doing the Jacobs ladder effect for 2 spans before it broke …

The next time you're closing in on that new shiny xfmer out of the shop, think about the night we got a lemon

anonymous poster
www.powerlineman.com

5

Voltage Regulation

One of a utility's core responsibilities is to deliver voltage to customers within a suitable range, so utilities must regulate the voltage. On distribution circuits, voltage drops due to current flowing through the line impedances. Primary and secondary voltage drop can be allocated as necessary along the circuit to provide end users with suitable voltage. Voltage regulators — in the substation or on feeders — can adjust primary voltage. This chapter discusses voltage regulators and regulation standards and techniques.

5.1 Voltage Standards

Most regulatory bodies and most utilities in America follow the ANSI voltage standards (ANSI C84.1-1995). This standard specifies acceptable operational ranges at two locations on electric power systems:

- *Service voltage* — The service voltage is the point where the electrical systems of the supplier and the user are interconnected. This is normally at the meter. Maintaining acceptable voltage at the service entrance is the *utility's* responsibility.
- *Utilization voltage* — The voltage at the line terminals of utilization equipment. This voltage is the *facility's* responsibility. Equipment manufacturers should design equipment which operates satisfactorily within the given limits.

The standard allows for some voltage drop within a facility, so service voltage requirements are tighter than utilization requirements.

The standard also defines two ranges of voltage:

- *Range A* — Most service voltages are within these limits, and utilities should design electric systems to provide service voltages within

TABLE 5.1

ANSI C84.1 Voltage Ranges for 120 V

	Service Voltage		Utilization Voltage	
	Minimum	**Maximum**	**Minimum**	**Maximum**
Range A	114 (–5%)	126 (+5%)	110 (–8.3%)	125 (+4.2%)
Range B	110 (–8.3%)	127 (+5.8%)	106 (–11.7%)	127 (+5.8%)

these limits. As the standard says, voltage excursions "should be infrequent."

- *Range B* — These requirements are more relaxed than Range A limits. According to the standard: "Although such conditions are a part of practical operations, they shall be limited in extent, frequency, and duration. When they occur, corrective measures shall be undertaken within a reasonable time to improve voltages to meet Range A requirements." Utilization equipment should give acceptable performance when operating within the Range B utilization limits, "insofar as practical" according to the standard.

These limits only apply to sustained voltage levels and not to momentary excursions, sags, switching surges, or short-duration interruptions.

Table 5.1 shows the most important limits, the limits on low-voltage systems. The table is given on a 120-V base; it applies at 120 V but also to any low-voltage system up to and including 600 V. The main target for utilities is the Range A service voltage, 114 to 126 V.

ANSI C84.1 defines three voltage classes: low voltage (1 kV or less), medium voltage (greater than 1 kV and less than 100 kV), and high voltage (greater than or equal to 100 kV). Within these classes, ANSI provides standard nominal system voltages along with the voltage ranges. A more detailed summary of the ANSI voltages is shown in Table 5.2 and Table 5.3.

For low-voltage classes, two nominal voltages are given — one for the electric system and a second, somewhat lower, nominal for the utilization equipment (for low-voltage motors and controls; other utilization equipment may have different nominal voltages). In addition, the standard gives common nameplate voltage ratings of equipment as well as information on what nominal system voltages the equipment is applicable to. As the standard points out, there are many inconsistencies between equipment voltage ratings and system nominal voltages.

For medium-voltage systems, ANSI C84.1 gives tighter limits for Ranges A and B. Range A is –2.5 to +5%, and Range B is –5 to +5.8%. However, most utilities do not follow these as limits for their primary distribution systems (utilities use the ANSI service voltage guidelines and set their primary voltage limits to meet the service voltage guidelines based on their practices). The three-wire voltages of 4,160, 6,900, and 13,800 V are mainly suited for industrial customers with large motors. Industrial facilities use motors on these systems with ratings of 4,000, 6,600, and 13,200 V, respectively.

TABLE 5.2

ANSI Standard Nominal System Voltages and Voltage Ranges for Low-Voltage Systems

		Range A			Range B		
		Maximum	Minimum		Maximum	Minimum	
Nominal System Voltage	Nominal Utilization Voltage	Utilization and Service Voltage[a]	Service Voltage	Utilization Voltage	Utilization and Service Voltage	Service Voltage	Utilization Voltage
Two Wire, Single Phase							
120	115	126	114	110	127	110	106
Three Wire, Single Phase							
120/240	**115/230**	**126/252**	**114/228**	**110/220**	**127/254**	**110/220**	**106/212**
Four Wire, Three Phase							
208Y/120	**200**	**218/126**	**197/114**	**191/110**	**220/127**	**191/110**	**184/106**
240/120	**230/115**	**252/126**	**228/114**	**220/110**	**254/127**	**220/110**	**212/106**
480Y/277	**460**	**504/291**	**456/263**	**440/254**	**508/293**	**440/254**	**424/245**
Three Wire, Three Phase							
240	230	252	228	220	254	220	212
480	**460**	**504**	**456**	**440**	**508**	**440**	**424**
600	575	630	570	550	635	550	530

Note: Bold entries show preferred system voltages.

[a] The maximum utilization voltage for Range A is 125 V or the equivalent (+4.2%) for other nominal voltages through 600 V.

Improper voltage regulation can cause many problems for end users. Sustained overvoltages or undervoltages can cause the following end-use impacts:

- *Improper or less-efficient equipment operation* — For example, lights may give incorrect illumination or a machine may run fast or slow.
- *Tripping of sensitive loads* — For example, an uninterruptible power supply (UPS) may revert to battery storage during high or low voltage. This may drain the UPS batteries and cause an outage to critical equipment.

In addition, undervoltages can cause

- *Overheating of induction motors* — For lower voltage, an induction motor draws higher current. Operating at 90% of nominal, the full-load current is 10 to 50% higher, and the temperature rises by 10 to 15%. With less voltage, the motor has reduced motor starting torque.

TABLE 5.3

ANSI Standard Nominal System Voltages and Voltage Ranges for Medium-Voltage Systems

	Range A			Range B		
	Maximum	Minimum		Maximum	Minimum	
Nominal System Voltage	Utilization and Service Voltage	Service Voltage	Utilization Voltage	Utilization and Service Voltage	Service Voltage	Utilization Voltage
Four Wire, Three Phase						
4160Y/2400	4370/2520	4050/2340	3740/2160	4400/2540	3950/2280	3600/2080
8320Y/4800	8730/5040	8110/4680		8800/5080	7900/4560	
12000Y/6930	12600/7270	11700/6760		12700/7330	11400/6580	
12470Y/7200	**13090/7560**	**12160/7020**		**13200/7620**	**11850/6840**	
13200Y/7620	**13860/8000**	**12870/7430**		**13970/8070**	**12504/7240**	
13800Y/7970	14490/8370	13460/7770		14520/8380	13110/7570	
20780Y/1200	21820/12600	20260/11700		22000/12700	19740/11400	
22860Y/13200	24000/13860	22290/12870		24200/13970	21720/12540	
24940Y/14400	**26190/15120**	**24320/14040**		**26400/15240**	**23690/13680**	
34500Y/19920	**36230/20920**	**33640/19420**		**36510/21080**	**32780/18930**	
Three Wire, Three Phase						
2400	2520	2340	2160	2540	2280	2080
4160	**4370**	**4050**	**3740**	**4400**	**3950**	**3600**
4800	5040	4680	4320	5080	4560	4160
6900	7240	6730	6210	7260	6560	5940
13800	**14490**	**13460**	**12420**	**14520**	**13110**	**11880**
23000	24150	22430		24340	21850	
34500	36230	33640		36510	32780	

Notes: Bold entries show preferred system voltages. Some utilization voltages are blank because utilization equipment normally does not operate directly at these voltages.

Also, overvoltages can cause

- *Equipment damage or failure* — Equipment can suffer insulation damage. Incandescent light bulbs wear out much faster at higher voltages.
- *Higher no-load losses in transformers* — Magnetizing currents are higher at higher voltages.

5.2 Voltage Drop

We can approximate the voltage drop along a circuit as

$$V_{drop} = |V_s| - |V_r| \approx I_R \cdot R + I_X \cdot X$$

where

V_{drop} = voltage drop along the feeder, V

R = line resistance, Ω

X = line reactance, Ω

I_R = line current due to real power flow (in phase with the voltage), A

I_X = line current due to reactive power flow (90° out of phase with the voltage), A

In terms of the load power factor, *pf*, the real and reactive line currents are

$$I_R = I \cdot pf = I \cos\theta$$

$$I_X = I \cdot qf = I \sin\theta = I \sin(\cos^{-1}(pf))$$

where

I = magnitude of the line current, A

pf = load power factor

qf = load reactive power factor = $\sin(\cos^{-1}(pf))$

θ = angle between the voltage and the current

While just an approximation, Brice (1982) showed that $I_R \cdot R + I_X \cdot X$ is quite accurate for most distribution situations. The largest error occurs under heavy current and leading power factor. The approximation has an error less than 1% for an angle between the sending and receiving end voltages up to 8° (which is unlikely on a distribution circuit). Most distribution programs use the full complex phasor calculations, so the error is mainly a consideration for hand calculations.

This approximation highlights two important aspects about voltage drop:

- *Resistive load* — At high power factors, the voltage drop strongly depends on the resistance of the conductors. At a power factor of 0.95, the reactive power factor (*qf*) is 0.31; so even though the resistance is normally smaller than the reactance, the resistance plays a major role.
- *Reactive load* — At moderate to low power factors, the voltage drop depends mainly on the reactance of the conductors. At a power factor of 0.8, the reactive power factor is 0.6, and because the reactance is usually larger than the resistance, the reactive load causes most of the voltage drop. Poor power factor significantly increases voltage drop.

Voltage drop is higher with lower voltage distribution systems, poor power factor, single-phase circuits, and unbalanced circuits. The main ways to reduce voltage drop are to:

- Increase power factor (add capacitors)
- Reconductor with a larger size

- Balance circuits
- Convert single-phase sections to three-phase sections
- Reduce load
- Reduce length

In many cases, we can live with significant voltage drop as long as we have enough voltage regulation equipment to adjust for the voltage drop on the circuit.

5.3 Regulation Techniques

Distribution utilities have several ways to control steady-state voltage. The most popular regulation methods include:

- Substation load tap-changing transformers (LTCs)
- Substation feeder or bus voltage regulators
- Line voltage regulators
- Fixed and switched capacitors

Most utilities use LTCs to regulate the substation bus and supplementary feeder regulators and/or switched capacitor banks where needed.

Taps on distribution transformers are another tool to provide proper voltage to customers. Distribution transformers are available with and without no-load taps (meaning the taps are to be changed without load) with standard taps of ±2.5 and ±5%. Utilities can use this feature to provide a fixed boost for customers on a circuit with low primary voltage. This also allows the primary voltage to go lower than most utilities would normally allow. Remember, the service entrance voltage is most important. Most distribution transformers are sold without taps, so this practice is not widespread. It also requires consistency; an area of low primary voltage may have several transformers to adjust — if one is left out, the customers fed by that transformer could receive low voltage.

5.3.1 Voltage Drop Allocation and Primary Voltage Limits

Most utilities use the ANSI C84.1 ranges for the service entrance, 114 to 126 V. How they control voltage and allocate voltage drop varies. Consider the voltage profile along the circuit in Figure 5.1. The substation LTC or bus regulator controls the voltage at the source. Voltage drops along the primary line, the distribution transformer, and the secondary. We must consider the customers at the start and end of the circuit:

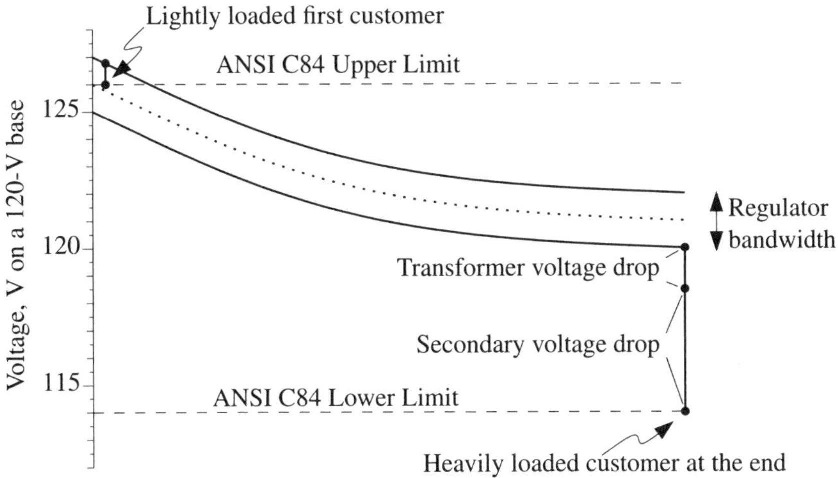

FIGURE 5.1
Voltage drop along a radial circuit with no capacitors or line regulators.

- *End — Heavily loaded* — Low voltages are a concern, so we consider a heavily loaded transformer and secondary. The allocation across the secondary depends on the utility's design practices as far as allowable secondary lengths and conductor sizes are concerned.
- *Source — Lightly loaded* — Near the source, we can operate the primary above 126 V, but we must ensure that the first customer does not have overvoltages when that customer is lightly loaded. Commonly, utilities assume that the secondary and transformer drop to this lightly loaded customer is 1 V. With that, the upper primary voltage limit is 127 V.

In the voltage drop along the primary, we must consider the regulator bandwidth (and bandwidths for capacitors if they are switched based on voltage). Voltage regulators allow the voltage to deviate by half the bandwidth in either direction. So, if we have a 2-V bandwidth and a desired range of 7 V of primary drop, subtracting the 2-V bandwidth only leaves 5 V of actual drop (see Figure 5.1). Likewise, if we choose 127 V as our upper limit on the primary, our maximum set voltage is 126 V with a 2-V regulator bandwidth.

Normally, utilities use standardized practices to allocate voltage drop. Deviations from the standard are possible but often not worth the effort.

If we have an express feeder at the start of a circuit, we can regulate the voltage much higher than 126 V as long as the voltage drops enough by the time the circuit reaches the first customer.

Primary voltage allocation affects secondary allocation and vice versa. A rural utility may have to allow a wide primary voltage range to run long

TABLE 5.4

Primary Voltage Ranges at Several Utilities

Service Area Type	Minimum	Maximum	Percent Range
Dense urban area	120	127	5.4
Dense urban area	117	126	7.5
Urban/suburban	114	126	10.0
Urban/suburban	115	125	8.3
Urban/suburban			
No conservation reduction	119	126	5.8
With conservation reduction	119	123	3.3
Multi-state area	117	126	7.5
Multi-state area			
Urban standard	123	127	3.3
Rural standard	119	127	6.6
Suburban and rural	113	125	10.0
Suburban and rural			
Urban standard	116	125	7.5
Rural standard	112	125	10.8
Urban and rural	115	127	10.0
Rural, mountainous	116	126	8.3
Rural, mountainous	113	127	11.7

Source: Willis, H. L., *Power Distribution Planning Reference Book,* Marcel Dekker, New York, 1997b, with additional utilities added.

circuits, which leaves little voltage drop left for the transformer and secondary. Since rural loads are typically each fed by their own transformer, rural utilities can run the primary almost right to the service entrance. Using low-impedance distribution transformers and larger-than-usual transformers also helps reduce the voltage drop beyond the primary. For the secondary conductors, triplex instead of open wire and larger size conductors help reduce secondary drop. Utilities that allow less primary voltage drop can run longer secondaries.

Utility practices on voltage limits on the primary range widely, as shown in Table 5.4. The upper range is more consistent — most are from 125 to 127 V — unless the utility uses voltage reduction (for energy conservation or peak shaving). The lower range is more variable, anywhere from 112 to 123 V. Obviously, the utility that uses a 112-V lower limit is not required to abide by the ANSI C84.1 limits.

5.3.2 Load Flow Models

Load flows provide voltage profiles that help when planning new distribution circuits, adding customers, and tracking down and fixing voltage problems. Most distribution load-flow programs offer a function to plot the voltage as a function of distance from the source.

We can model a distribution circuit at many levels of detail. Many utilities are modeling more of their systems in more detail. For most load flows, utilities normally just model the primary. Modeling the secondary is occasionally useful for modeling specific problems at a customer. We can still have very good models with simplifications. Modeling long laterals or branches is normally a good idea, but we can lump most laterals together as a load where they tie into the main line. Modeling each transformer as a load is rarely worth the effort; we can combine loads together and maintain accuracy with some common sense. Most mainline circuits can be accurately modeled if broken into 10 to 20 sections with load lumped with each section. Of course, accurate models of capacitors and line regulators are a good idea.

Correctly modeling load phasing provides a better voltage profile on each phase. Unbalanced loads cause more voltage drop because of:

- *Higher loop impedance* — The impedance seen by unbalanced loads, the loop impedance including the zero-sequence impedance, is higher than the positive-sequence impedance seen by balanced loads.
- *Higher current on the loaded phases* — If the current splits unevenly by phases, the more heavily loaded phases see more voltage drop.

Utilities often do not keep accurate phasing information, but it helps improve load-flow results. We do not need the phasing on every transformer, but we will have better accuracy if we know the phasing of large single-phase taps.

Of the data entered into the load flow model, the load allocation is the trickiest. Most commonly, loads are entered in proportion to the transformer kVA. If a circuit has a peak load equal to the sum of the kVA of all of the connected transformers divided by 2.5, then each load is modeled as the given transformer size in kVA divided by 2.5. Incorporating metering data is another more sophisticated way to allocate load. If a utility has a transformer load management system or other system that ties metered kilowatt-hour usage to a transformer to estimate loadings, feeding this data to the load flow can yield a more precise result. In most cases, all of the loads are given the same power factor, usually what is measured at the substation. Additional measurements could be used to fine-tune the allocation of power factor. Some utilities also assign power factor by customer class.

Most distribution load flow programs offer several load types, normally constant power, constant current, and constant impedance:

- *Constant power load* — The real and reactive power stays constant as the voltage changes. As voltage decreases, this load draws more current, which increases the voltage drop. A constant power model is good for induction motors.
- *Constant current load* — The current stays constant as the voltage changes, and the power increases with voltage. As voltage decreases, the current draw stays the same, so the voltage drop does not change.

TABLE 5.5

Load Modeling Approximations Recommended by Willis (1997a)

Feeder Type	Percent Constant Power	Percent Constant Impedance
Residential and commercial, summer peaking	67	33
Residential and commercial, winter peaking	40	60
Urban	50	50
Industrial	100	0
Developing countries	25	75

Source: Willis, H. L., "Characteristics of Distribution Loads," in *Electrical Transmission and Distribution Reference Book*. Raleigh, NC, ABB Power T&D Company, 1997.

- *Constant impedance load* — The impedance is constant as the voltage changes, and the power increases as the square of the voltage. As voltage decreases, the current draw drops off linearly; so the voltage drop decreases. The constant impedance model is good for incandescent lights and other resistive loads.

Normally, we can model most circuits as something like 40 to 60% constant power and 40 to 60% constant impedance (see Table 5.5 for one set of recommendations). Modeling all loads as constant current is a good approximation for many circuits. Modeling all loads as constant power is conservative for voltage drop.

5.3.3 Voltage Problems

Voltage complaints (normally undervoltages) are regular trouble calls for utilities. Some are easy to fix; others are not. First, check the secondary. Before tackling the primary, confirm that the voltage problem is not isolated to the customers on the secondary. If secondary voltage drop is occurring, check loadings, make sure the transformer is not overloaded, and check for a loose secondary neutral.

If the problem is on the primary, some things to look for include:

- *Excessive unbalance* — Balancing currents helps reduce voltage drop.
- *Capacitors* — Look for blown fuses, incorrect time clock settings, other incorrect control settings, or switch malfunctions.
- *Regulators* — Check settings. See if more aggressive settings can improve the voltage profile enough: a higher set voltage, more line drop compensation, and/or a tighter bandwidth.

These problems are relatively easy to fix. If it is not these, and if there is too much load for the given amount of impedance, we will have to add equipment to fix the problem. Measure the primary voltage (and if possible the loadings) at several points along the circuit. An easy way to measure the primary voltage is to find a lightly loaded distribution transformer and measure the secondary voltage. Measure the power factor at the substation. A poor power factor greatly increases the voltage drop.

Load flows are a good tool to try out different options to improve voltage on a circuit. If possible, match voltage profiles with measurements on the circuit. Measurements provide a good sanity check. Try to measure during peak load conditions. Regulator and capacitor controllers can provide extra information if they have data logging capability. Normally, we allocate the load for the model equally by transformer kVA. This may not always be right, and measurements can help "tweak" the model. A load flow can help determine the best course of action. Where do we need a supplementary line regulator? How many? Can fixed capacitors do the job? Do we need switched capacitors? Circuits with poor power factor are the best candidates for capacitors as they will help reduce line losses as well as improve voltage.

In addition to extra regulating equipment, consider other options. Sometimes, we can move one or more circuit sections to a different feeder to reduce the loading on the circuit. If transformers have taps, investigate changing the transformer taps. Though it is expensive, we can also build new circuits, upgrade to a higher voltage, or reconductor.

5.3.4 Voltage Reduction

Utilities can use voltage adjustments as a way to manage system load. Voltage reduction can reduce energy consumption and/or reduce peak demand. Several studies have shown roughly a linear response relationship between voltage and energy use — a 1% reduction in voltage reduces energy usage by 1% (or just under 1%, depending on the study). Kirshner and Giorsetto (1984) analyzed trials of conservation voltage reduction (CVR) at several utilities. While results varied significantly, most test circuits had energy savings of between 0.5 and 1% for each 1% voltage reduction. Their regression analysis of the feeders found that residential energy savings were 0.76% for each 1% reduction in voltage, while commercial and industrial loads had reductions of 0.99% and 0.41% (but the correlations between load class and energy reduction were fairly small).

Voltage reduction works best with resistive loads because the power drawn by a resistive load decreases with the voltage squared. Lighting and resistive heating loads are the dominant resistive loads; these are not ideal resistive loads. For example, the power on incandescent lights varies as the voltage to the power of about 1.6, which is not quite to the power of 2 but close. Residential and commercial loads have higher percentages of resistive load. For water heaters and other devices that regulate to a temperature, reducing voltage does not reduce overall energy usage; the devices just run more often.

Voltage reduction to reduce demand has even more impact than that on energy reduction. The most reduction occurs right when the voltage is reduced, and then some of the reduction is lost as some loads keep running longer than normal to compensate for lower voltage. For example, Priess and Warnock (1978) found that during a 4-h, 5% voltage reduction, the demand on one typical residential circuit dropped by 4% initially and diminished to a 3% drop by the end of the 4-h period.

Voltage reduction works best on short feeders — those that do not have much voltage drop. On these, we can control reduction just through adjustments of the station LTC regulator settings. It is straightforward to set up a system where operators can change the station set voltage through SCADA. On longer circuits, we need extra measures. Some strategies include:

- *Extra regulators* — Extra regulators can help flatten the voltage profile along the circuit. Each regulator is set with a set voltage and compensation settings appropriate for a tighter voltage range. This approach is most appropriate for energy conservation. Controlling the regulators to provide peak shaving is difficult; the communications and controls add significantly to the cost.
- *Feeder capacitors* — The vars injected by capacitors help flatten the voltage profile and allow a lower set voltage on the station LTC. On many circuits, just fixed capacitors can flatten the profile enough to reduce the station set voltage. McCarthy (2000) reported how Georgia Power used this strategy to reduce peak loads by 500 kW on circuits averaging approximately 18 MW.
- *Tighter bandwidth* — With a smaller regulator bandwidth, the voltage spread on the circuit is smaller. A smaller bandwidth requires more frequent regulator or LTC maintenance (the regulator changes taps more often) but not drastic differences. Kirshner (1990) reported that reducing the bandwidth from 3 to 1.5 V doubled the number of regulator tap changes.
- *Aggressive line drop compensation* — An aggressive line-drop compensation scheme can try to keep the voltage at the low end (say, at 114 V) for the last customer at all times. The set voltage in the station may be 115 to 117 V, depending on the circuit voltage profile. Aggressive compensation boosts the voltage during heavy loads, while trying to keep voltages low at the ends of circuits. During light loads, the station voltage may drop to well under 120 V. This strategy helps the least at heavy load periods, so it is more useful for energy conservation than for peak shaving. Aggressive compensation makes low voltages more likely at the end of circuits. If any of the planning assumptions are wrong, especially power factor and load placement, customers at the end of circuits can have low voltages.
- *Others* — Other voltage profile improvement options help when implementing a voltage reduction program, although some of these options, such as reconductoring, undergrounding, load balancing, and increasing primary voltage levels, are quite expensive.

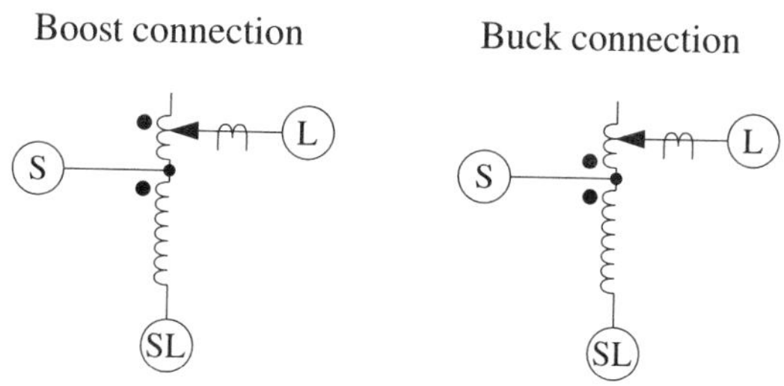

FIGURE 5.2
ANSI type A single-phase regulator, meaning taps on the load bushing.

5.4 Regulators

Voltage regulators are autotransformers with automatically adjusting taps. Commonly, regulators provide a range from −10 to +10% with 32 steps. Each step is 5/8%, which is 0.75 V on a 120-V scale.

A single-phase regulator has three bushings: the source (S), the load (L), and the source-load (SL). The series winding is between S and L. Figure 5.2 shows a straight regulator (ANSI type A) with the taps on the load side. An ANSI type B, the inverted design, has the taps on the source bushing. The regulator controller measures current with a CT at the L bushing and measures the voltage with a PT between L and SL. Regulators have a reversing switch that can flip the series winding around to change back and forth between the boost and the buck connection.

Regulators are rated on current (IEEE Std. C57.15-1999). Regulators also have a kVA rating which is the two-winding transformer rating and not the load-carrying capability. A regulator at 7.62 kV line to ground with a ±10% range and a load current rating of 100 A has a kVA rating of 0.1(7.62 kV)(100A) = 76 kVA. The load-carrying capability is ten times the regulator's kVA rating.

By reducing the range of regulation, we can extend the rating of the regulator. Reducing the range from ±10 to ±5% increases the rating by 60% (see Figure 5.3).

The impedance is the two-winding impedance times a base value about ten times as large. Because the impedance is so small, we can normally neglect it.

Three-phase regulators, often used in stations, are used on wye or delta systems. A three-phase regulator controls all three phases simultaneously. These are normally larger units. The normal connection internally is a wye connection with the neutral point floating.

Commonly, utilities use single-phase units, even for regulating three-phase circuits. We can connect single-phase regulators in several ways [see Figure 5.4 and (Bishop et al., 1996)]:

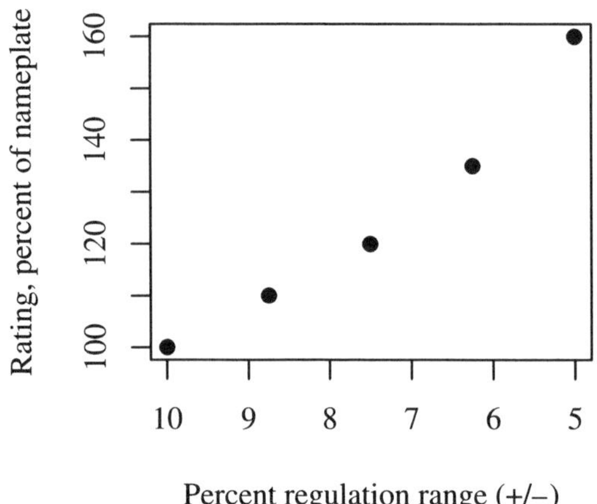

FIGURE 5.3
Increased regulator ratings with reduced regulation range.

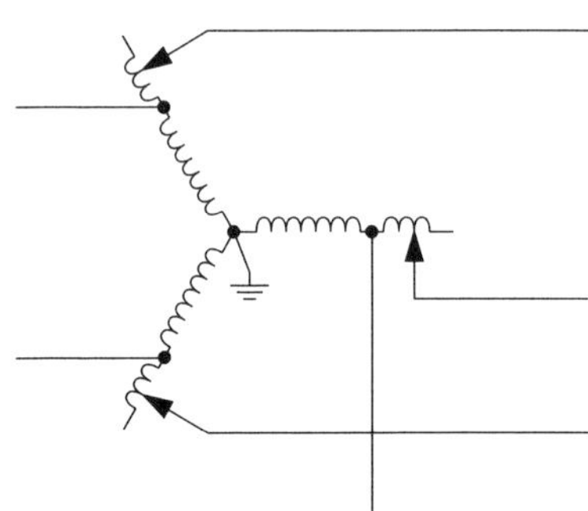

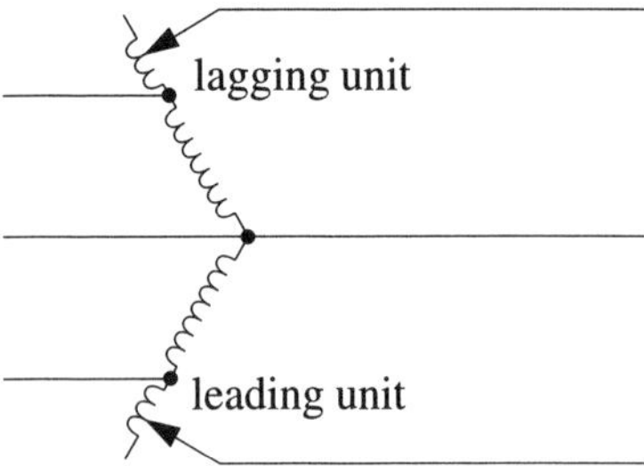

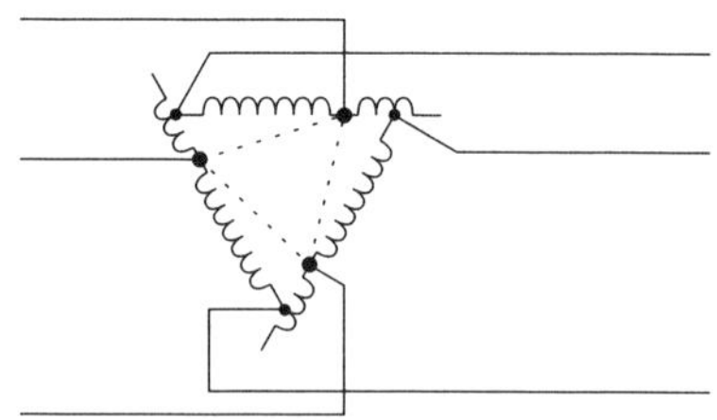

FIGURE 5.4
Three-phase regulator connections.

- *Line to neutral* — On four-wire systems, three-phase circuits normally have three single-phase regulators connected line to neutral. Line-to-neutral connections are also appropriate for single-phase and two-phase circuits. Each regulator independently controls voltage, which helps control voltage unbalance as well as steady-state voltage.
- *Open delta* — Only two single-phase regulators are needed, each connected phase to phase.
- *Closed delta* — Three regulators are connected phase to phase. Using the closed delta extends the regulation range by 50%, from ±10 to ±15%.

In both of the delta connections, the regulators see a current phase-shifted relative to the voltage. In the leading connection with unity power factor loads, the line current through the regulator leads the line-to-line voltage by 30°. The lagging connection has the current reversed: for a unit power factor load, the line current lags the line-to-line voltage by 30°. In the open-delta configuration, one of the units is leading and the other is lagging. In the closed-delta arrangement, all three units are either leading or all three are lagging. Although uncommon, both of the delta connections can be applied on four-wire systems.

Regulators have a voltage regulating relay that controls tap changes. This relay has three basic settings that control tap changes (see Figure 5.5):

- *Set voltage* — Also called the set point or bandcenter, the set voltage is the desired output of the regulator.
- *Bandwidth* — Voltage regulator controls monitor the difference between the measured voltage and the set voltage. Only when the difference exceeds one half of the bandwidth will a tap change start. Use a bandwidth of at least two times the step size, 1.5 V for ±10%, 32-step regulators. Settings of 2 and 2.5 V are common.
- *Time delay* — This is the waiting time between the time when the voltage goes out of band and when the controller initiates a tap

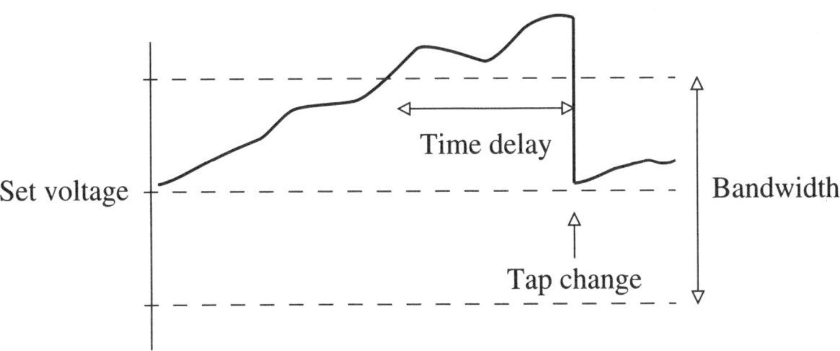

FIGURE 5.5
Regulator tap controls based on the set voltage, bandwidth, and time delay.

change. Longer time delays reduce the number of tap changes. Typical time delays are 30 to 60 sec.

If the voltage is still out of bounds after a tap change, the controller makes additional tap changes until the voltage is brought within bounds. The exact details vary by controller, and some provide programmable modes. In some modes, controllers make one tap change at a time. In other modes, the controller may initiate the number of tap changes it estimates are needed to bring the voltage back within bounds. The time delay relay resets if the voltage is within bounds for a certain amount of time.

A larger bandwidth reduces the number of tap changes, but at a cost. With larger bandwidth, the circuit is not as tightly regulated. We should include the bandwidth in voltage profile calculations to ensure that customers are not given over or under voltages. Voltage that was used for bandwidth can be used for voltage drop along the circuit. With a higher bandwidth we may need more regulators on a given line. So, use at least two times the step size, but do not use excessively high bandwidths such as 3 or 3.5 V.

In addition to these basics, regulator controllers also have line-drop compensation to boost voltages more during heavy load. Controllers also may have high and low voltage limits to prevent regulation outside of a desired range of voltages. In addition to the regulator and control application information provided here, see Beckwith (1998), Cooper Power Systems (1978), General Electric (1979), and Westinghouse (1965).

Many regulators are bi-directional units; they can regulate in either direction, depending on the direction of power flow. A bi-directional regulator measures voltage on the source side using an extra PT or derives an estimate from the current. If the regulator senses reverse power flow, it switches to regulating the side that is normally the source side. We need reverse mode for a regulator on circuits that could be fed by an alternate source in the reverse direction. Without a reverse mode, the regulator can cause voltage problems during backfeeds. If a unidirectional regulator is fed "backwards," the regulator PT is now on the side of the source. Now, if the voltage drops, the regulator initiates a tap raise. However, the voltage the PT sees does not change because it is on the source side (very stiff). What happened was the voltage on the load side went down (but the regulator controller does not know that because it is not measuring that side). The controller still sees low voltage, so it initiates another tap raise which again lowers the voltage on the other side of the regulator. The controller keeps trying to raise the voltage until it reaches the end of its regulation range. So, we have an already low voltage that got dropped by an extra 10% by the unidirectional regulator. If the controller initially sees a voltage above its set voltage, it ratchets all the way to the high end causing a 10% overvoltage. Also, if the incoming voltage varies above and below the bandwidth, the regulator can run back and forth between extremes. A bi-directional regulator prevents these runaways. Depending on its mode, under reverse power, a bi-directional regulator can regulate in the reverse direction, halt tap changes, or move to the neutral point (these last two do not require PTs on both sides but just power direction sensing).

Regulators also have an operations counter. The counter helps identify when a regulator is due for refurbishment. Regulators are designed to perform many tap changes, often over one million tap changes over the life of a regulator. A regulator might change taps 70 times per day, which is 25,000 times per year (Sen and Larson, 1994). A regulator counter also provides a good warning indicator; excessive operations suggest that something is wrong, such as wrong line drop compensation settings, a bandwidth or time delay that is too small, or widely fluctuating primary voltages.

Regulators have "drag hands" — markers on the tap position indicator that show the maximum and minimum tap positions since the drag hands were last reset. The drag hands are good indicators of voltage problems. If maintenance reviews continually show the drag upper hand pegging out at +10%, the upstream voltage is probably too low. More work is needed to correct the circuit's voltage profile. Advanced controllers record much more information, including tap change records and demand metering to profile voltages, currents, and power factors.

5.4.1 Line-Drop Compensation

LTC transformer and regulator controls can be augmented with line-drop compensation. During heavy load, the controller boosts voltage the most, and during light load, voltage is boosted the least. The line-drop compensator uses an internal model of the impedance of the distribution line to match the line impedance. The user can set the *R* and *X* values in the compensator to adjust the compensation. The controller adjusts taps based on the voltage at the voltage regulating relay, which is the PT voltage plus the voltage across the line-drop compensator circuit (see Figure 5.6). With no compensation, the voltage regulating relay adjusts the taps based on the PT voltage.

Since load on a typical distribution line is distributed, *R* and *X* compensator settings are chosen so that the maximum desired boost is obtained

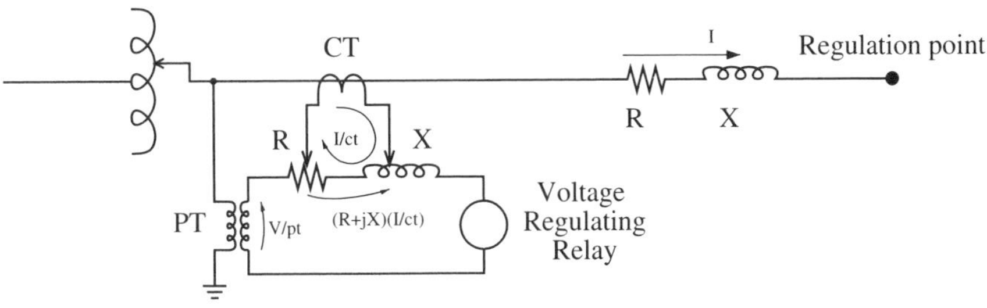

FIGURE 5.6
Line drop compensator circuit.

under heavy load while a given voltage is obtained under light load. There are two main approaches for selecting settings:

- *Load center* — The settings are chosen to regulate the voltage at a given point downstream of the regulator.
- *Voltage spread* — The R and X settings are chosen to keep the voltage within a chosen band when operating from light load to full load. The R and X settings may or may not be proportional to the line's R and X.

The main complication of all of the methods is that the load and power factors change (especially with downstream capacitor banks). Many regulators are set up without line drop compensation. It is obviously easier and less prone to mistakes, but we are losing out on some significant capability. If we set the regulator set voltage at 120 V, and we do not get enough boost along the line, we will need more regulators. With a higher set voltage such as 126 V, we do not need as many regulators, but we have high voltages at light load and possibly overvoltages if the circuit has capacitors. With line drop compensation, we have boost when we need it during heavy load, but not during light load (see Figure 5.7). Line-drop compensation also normally leads to a smaller range of fluctuations in voltage through the day for customers along the circuit.

5.4.1.1 Load-Center Compensation

The classic way to set compensator settings is to use the *load-center* method. Consider a line with impedances R_L and X_L with a load at the end. Now, if we pick the R_{set} and X_{set} of the compensator to match those of the line, as the load changes the regulator responds and adjusts the regulator taps to keep the voltage constant, not at the regulator but at the load. To achieve this, we can set the R_{set} and X_{set} of the regulator as

$$R_{set} = \frac{I_{CT}}{N_{PT}} R_L$$

$$X_{set} = \frac{I_{CT}}{N_{PT}} X_L$$

where

R_{set} = regulator setting for resistive compensation, V
X_{set} = regulator setting for reactive compensation, V
I_{CT} = primary rating of the current transformer, A
N_{PT} = potential transformer ratio (primary voltage/secondary voltage)
R_L = primary line resistance from the regulator to the regulation point, Ω
X_L = primary line reactance from the regulator to the regulation point, Ω

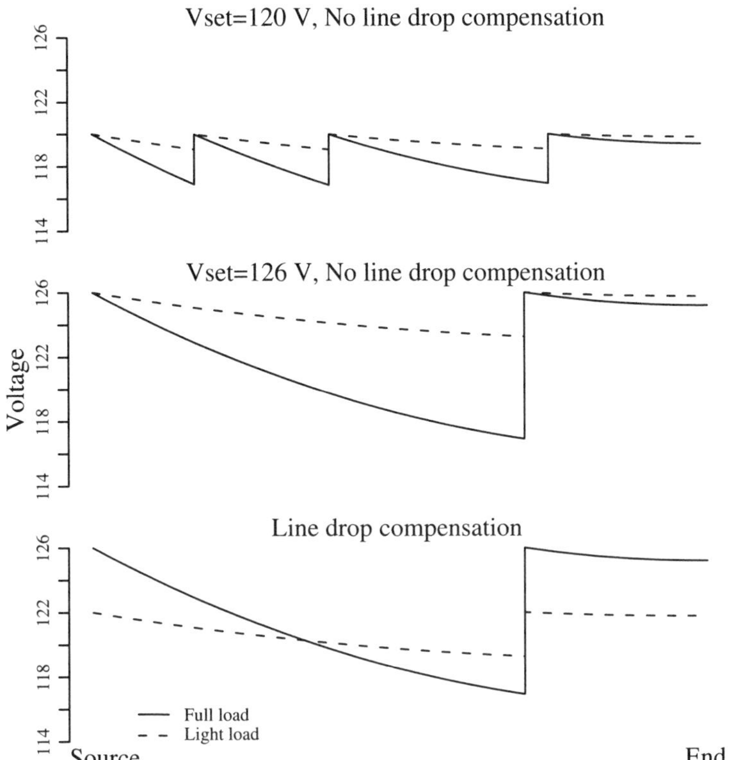

FIGURE 5.7
Voltage profiles on a circuit with various forms of regulation.

A regulator's R and X compensator settings are in units of volts. By using volts as units, we can directly see the impact of the regulator on a 120-V scale. Consider an example where the set voltage is 120 V. With a current at unity power factor and R_{set} = 6 V (X_{set} does not matter at unity power factor), the controller regulates the voltage to 120 + 6 = 126 V when the current is at the peak CT rating. If the current is at half of the CT rating, the controller regulates to the set voltage plus 3 or 123 V. Available compensator settings are normally from –24 to +24 V.

Note that the primary CT rating is an important part of the conversion to compensator settings. The CT rating may be the same as the regulator rating or it may be higher. The CT rating is given on the nameplate. Table 5.6 shows the regulator ratings and primary CT current rating for one manufacturer. Regulators may be applied where the nameplate voltage does not match the system voltage if they are close enough to still allow the desired regulation range at the given location. Also, some regulators have taps that allow them to be used at several voltages. Make sure to use the appropriate PT ratio for the tap setting selected.

TABLE 5.6

Regulator and Primary CT Ratings in Amperes

Regulator Current Ratings	CT Primary Current
25	25
50	50
75	75
100	100
150	150
167, 200	200
219, 231, 250	250
289, 300	300
328, 334, 347, 400	400
418, 438, 463, 500	500
548, 578, 656, 668	600
833, 875, 1000, 1093	1000
1332, 1665	1600

When specifying impedances for the line-drop compensator, use the correct line impedances. For a three-phase circuit, use the positive-sequence impedance. For a single-phase line, use the loop impedance Z_S which is about twice the positive-sequence impedance.

On a delta regulator, either an open delta or a closed delta, divide the PT ratio by $\sqrt{3}$. On a delta regulator the PT connects from phase to phase, but the internal circuit model of the line-drop compensator is phase to ground, so we need the $\sqrt{3}$ factor to correct the voltage.

Line-drop compensation works perfectly for one load at the end of a line, but how do we set it for loads distributed along a line? If loads are uniformly distributed along a circuit that has uniform impedance, we can hold the voltage constant at the midpoint of the section by using:

- 3/8 *rule* — For a uniformly distributed load, a regulator can hold the voltage constant at the midpoint of the circuit if we use line-drop compensation settings based on 3/8 of the total line impedance. A circuit with a uniformly distributed load has a voltage drop to the end of the circuit of one half of the drop had all of the loads been lumped into one load at the end of the circuit. Three-fourths of this drop is on the first half of the circuit, so (1/2)(3/4) = 3/8 is the equivalent voltage drop on a uniformly distributed load.

Make sure not to allow excessive voltages. We can only safely compensate a certain amount, and we will have overvoltages just downstream of the regulator if we compensate too much. Check the voltage to the voltage regulating relay to ensure that it is not over limits. The maximum voltage is

$$V_{max} = V_{set} + (pf \cdot R_{set} + qf \cdot X_{set})\, I_{max}$$

where

V_{set} = regulator set voltage
R_{set} = resistive setting for compensation, V
X_{set} = reactive setting for compensation, V
pf = load power factor
qf = load reactive power factor = $\sin(\cos^{-1}(pf))$
I_{max} = maximum load current in per unit relative to the regulator CT rating

If V is more than what you desired, reduce R_{set} and X_{set} appropriately to meet your desired limit.

5.4.1.2 Voltage-Spread Compensation

In another method, the *voltage-spread* method, we find compensator settings by specifying the band over which the load-side voltage should operate. For example, we might want the regulator to regulate to 122 V at light load and 126 V at full load. If we know or can estimate the light-load and full-load current, we can find R and X compensator settings to keep the regulated voltage within the proper range. If we want the regulator to operate over a given compensation range C, we can choose settings to satisfy the following:

$$C = V - V_{set} = pf \cdot R_{set} + qf \cdot X_{set}$$

where

R_{set} = resistive setting for compensation, V
X_{set} = reactive setting for compensation, V
pf = load power factor
qf = load reactive power factor = $\sin(\cos^{-1}(pf))$
C = total desired compensation voltage, V
V_{set} = regulator set voltage, V
V = voltage that the controller will try to adjust the regulator to, V

With line current operating to the regulator CT rating limit (which is often the regulator size) and the current at the given power factor, these settings will boost the regulator by C volts on a 120-V scale. Any number of settings for R_{set} and X_{set} are possible to satisfy this equation. If we take $X_{set} = \frac{X}{R} R_{set}$ where the X/R ratio is selectable, the settings are

$$R_{set} = \frac{C}{pf + \frac{X}{R} qf}$$

$$X_{set} = \frac{\frac{X}{R} C}{pf + \frac{X}{R} qf} = \frac{X}{R} R_{set}$$

where

$\frac{X}{R}$ = X/R ratio of the compensator settings

Note that C must be given as seen on the regulator PT secondaries, on a 120-V base. As an example, if the feeder voltage should be not more than 126 V at the limit of the regulator, and the desired voltage at no load is 122 V, set the regulator set voltage at 122 V and find R_{set} and X_{set} to give $C = 4$ V. For a power factor of 0.85 and $\frac{X}{R} = 3$, the equations above give $R_{set} = 1.64$ V and $X_{set} = 4.94$ V.

To control the voltage range for a light load other than zero and for a peak load other than the regulator CT rating, we can use the following to find the voltage swing from light load to full load as

$$V_{max} - V_{min} = (pf \cdot R_{set} + qf \cdot X_{set})I_{max} - (pf \cdot R_{set} + qf \cdot X_{set})I_{min}$$

where

V_{max} = desired voltage at the maximum load current on a 120-V base, V

V_{min} = desired voltage at the minimum load current on a 120-V base, V

I_{max} = maximum load current in per-unit relative to the regulator CT rating

I_{min} = minimum load current in per-unit relative to the regulator CT rating

Now, the R and X settings are

$$R_{set} = \frac{V_{max} - V_{min}}{(pf + \frac{X}{R} qf)(I_{max} - I_{min})}$$

$$X_{set} = \frac{X}{R} R_{set}$$

And, the regulator set voltage is

$$V_{set} = V_{min} - (pf \cdot R_{set} + qf \cdot X_{set})I_{min} = V_{min} - \frac{V_{max} - V_{min}}{I_{max} - I_{min}} I_{min}$$

With a compensator X/R ratio equal to the line X/R ratio, these equations move the effective load center based on the choice of voltage and current minimums and maximums.

Just like we can choose to have the compensator X/R ratio equal the line X/R ratio, we can choose other values as well. There are good reasons why we might want to use other ratios; this is done mainly to reduce the sensitivity to power factor changes. The *zero reactance* method of selecting compensator makes $X_{set} = 0$ (and the compensator $X/R = 0$) but otherwise uses the same equations as the voltage spread method (General Electric, 1979).

By making X_{set} zero, the compensator is not sensitive to variations in power factor caused by switched capacitors or load variation; only real power changes cause regulator movement. This method also simplifies application of regulators. The equations become

$$R_{set} = \frac{V_{max} - V_{min}}{pf(I_{max} - I_{min})}$$

$$X_{set} = 0$$

And, the regulator set voltage is

$$V_{set} = V_{min} - (pf \cdot R_{set})I_{min}$$

The equations simplify more if we assume that $I_{min} = 0$ (our error with this is that voltages run on the high side during light load). A further simplification is to assume that the power factor is one. If the power factor is less than that at full load, the regulator will not boost the voltage quite as much. Often, we do not know the power factor at the regulator location anyway.

This method is useful with switched capacitor banks close to the regulator. It does not perform well for low power factors if we have assumed a power factor near unity. With this control, the regulator will not provide enough boost with poor power-factor load.

Another option is to take $X/R = 0.6$, which weights the real power flow more than the reactive power flow, but not as extremely as the zero reactance compensation method. So, although the controller is somewhat desensitized to changes in power factor, the regulator provides some action based on reactive power. Figure 5.8 shows several X/R compensator settings chosen to provide an operating band from 121 V at light load to 127 V at full load. The settings were chosen based on a power factor of 0.9, and the curves show the voltage as the power factor varies. The middle graph with $X/R = 0.6$ performs well over a wide range of power factors. The graph on the left, where $X/R = 3$ which is the line X/R ratio, has the most variation with changes in power factor. If power factor is lower than we expected, the compensator will cause high voltages.

With $X/R = 0.6$ and $pf = 0.9$, the voltage spread equations are

$$R_{set} = 0.86\frac{V_{max} - V_{min}}{(I_{max} - I_{min})}$$

$$X_{set} = 0.6R_{set}$$

And, the regulator set voltage is

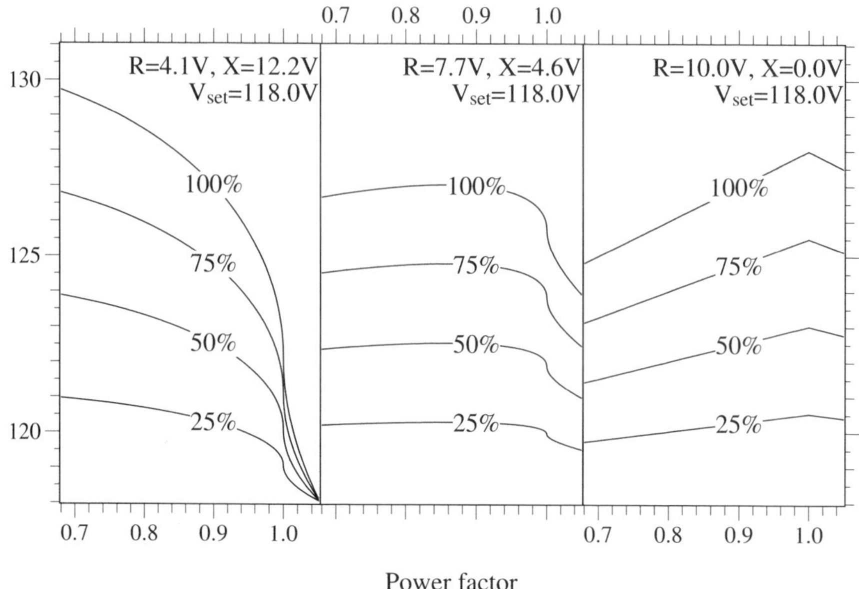

FIGURE 5.8
Regulated voltage based on different compensator settings and power factors with the percentage loadings given on the graph. All settings are chosen to operate from 121 V at light load (33%) to 127 V at full load (100% of the primary CT ratio) at a power factor of 0.9.

$$V_{set} = V_{min} - \frac{V_{max} - V_{min}}{I_{max} - I_{min}} I_{min}$$

The *universal compensator* method fixes compensation at R_{set} = 5 V and X_{set} = 3 V to give a 6-V compensation range with current ranging up to the regulator CT rating (General Electric, 1979). For other voltage ranges and maximum currents, we can use:

$$R_{set} = \frac{5}{I_{max}} \frac{(V_{max} - V_{min})}{6}$$

$$X_{set} = \frac{3}{I_{max}} \frac{(V_{max} - V_{min})}{6}$$

And we assume that I_{min} = 0, so the regulator set voltage is

$$V_{set} = V_{min}$$

To make this even more "cookbook," we can standardize on values of V_{max} and V_{min}, for example, values of 126 V and 120 V. If the full-load is the CT rating (which we might want in order to be conservative), the default settings become $R_{set} = 5$ V and $X_{set} = 3$ V. The universal compensation method is easy yet relatively robust.

With any of the voltage-spread methods of setting the R and X line-drop compensation, the peak current is an important parameter. If we underestimate the load current, the regulator can overcompensate and cause high voltages (if we do not have a voltage override limiter or if it is disabled). Check regulator loadings regularly to ensure that the compensation is appropriate.

5.4.1.3 Effects of Regulator Connections

On an open-delta regulator, one regulator is connected leading, and the other lagging. We need to adjust the compensator settings to account for the 30° phase shift. On the leading regulator, the current leads the voltage by 30°; so we need to subtract 30° from the compensator settings, which is the same as multiplying by $1\angle 30°$ or $(\cos 30° - j \sin 30°)$. Modify the settings for the leading regulator (Cooper Power Systems, 1978; Westinghouse Electric Corporation, 1965) with

$$R'_{set} = 0.866\ R_{set} + 0.5X_{set}$$

$$X'_{set} = 0.866\ X_{set} - 0.5R_{set}$$

And for the lagging regulator we need to add 30°, which gives

$$R'_{set} = 0.866\ R_{set} - 0.5X_{set}$$

$$X'_{set} = 0.866\ X_{set} + 0.5R_{set}$$

For an X/R ratio above 1.67, R'_{set} is negative on the lagging regulator; and for a ratio below 0.58, X'_{set} is negative on the leading regulator. Most controllers allow negative compensation.

In the field, how do we tell between the leading and the lagging regulator? Newer regulator controllers can tell us which is which from phase angle measurements. For older controllers, we can modify the compensator settings to find out (Lokay and Custard, 1954). Set the resistance value on both regulators to zero, and set the reactance setting on both to the same nonzero value. The unit that moves up the most number of tap positions is the lagging unit (with balanced voltages, this is the unit that goes to the highest raise position). If the initial reactance setting is not enough, raise the reactance settings until the leading and lagging units respond differently.

With a closed-delta regulator, all three regulators are connected either leading or lagging. All three regulators have the same set of compensator settings; adjust them all with either the leading or the lagging equations described for the open-delta regulator.

On a three-phase regulator, even on a delta system, the compensator settings do not need adjustment. The controller accounts for any phase shift that might occur inside the regulator.

5.4.2 Voltage Override

Use the *voltage override* feature on the regulator controller. No matter how we select the line-drop compensation settings, an important feature is an upper voltage limit on the regulation action. The regulator keeps the regulated voltage below this limit regardless of the line-drop compensation settings. Always use this feature to protect against overvoltages caused by incorrect line-drop compensation settings or unusually high loadings. This upper voltage limiter is also called "first house protection," as it is the first few customers downstream that could have overvoltages due to regulator action. With a voltage limit, we can set line-drop compensator settings more aggressively and not worry about causing overvoltages to customers. On a regulator without an upper limit (normally older units), increase estimated peak loadings when calculating line-drop compensation settings in order to reduce the risk of creating overvoltages. Voltage override functions usually have a deadband type setting on the voltage limit to prevent repeated tap changes. For example, we might set a 126-V upper limit with a deadband of an extra 2 V. Above 128 V the controller immediately taps the regulator down to 126 V, and between 126 and 128 V the controller prohibits tap raises (different controllers implement this function somewhat differently; some include time delays). Even without line-drop compensation, the voltage override function helps protect against sudden changes in upstream voltages (the out-of-limit response is normally faster than normal time-delay settings programmed into regulators).

5.4.3 Regulator Placement

With no feeder regulators, the entire voltage drop on a circuit must be within the allowed primary voltage range. One feeder regulator can cover primary voltage drops up to twice the allowed voltage variation. Similarly, two supplementary regulators can cover primary voltage drops up to three times the allowed variation. For a uniformly distributed load, optimum locations for two regulators are at distances from the station of approximately 20% of the feeder length for one and 50% for the other. For one feeder regulator, the optimum location for a uniformly distributed load is at 3/8 of the line length from the station.

When placing regulators and choosing compensator settings, allow for some load growth on the circuit. If a regulator is applied where the load is right near its rating, it may not be able to withstand the load growth. However, it is more than just concern about the regulator's capability. If we want to keep the primary voltage above 118 V, and we add a regulator to a circuit right at the point where the primary voltage falls to 118 V, that will correct the voltage profile along the circuit with present loadings. If loadings increase in the future, the voltage upstream of the regulator will drop below 118 V. As previously discussed, when setting line-drop compensator settings, the maximum load on the regulator should allow room for load growth to reduce the chance that the regulator boosts the voltage too much.

Several regulators can be strung together on a circuit. Though this can meet the steady-state voltage requirements of customers, it will create a very weak source for them. Flicker problems from motors and other fluctuating loads are more likely.

Also consider the effect of dropped load on regulators. A common case is a recloser downstream of a line regulator. If the regulator is tapped up because of heavy load and the recloser suddenly drops a significant portion of the load, the voltage downstream of the regulator will pop up until the regulator controller shifts the taps back down.

5.4.4 Other Regulator Issues

Normally, voltage regulators help with voltage unbalance as each regulator independently controls its phase. If we aggressively compensate, the line-drop compensation can cause voltage unbalance. Consider a regulator set to operate between 120 V at no load and 126 V at full load. If one phase is at 50% load and the other two are at 0% load, the line-drop compensator will tap to 123 V on the loaded phase and to 120 V on the unloaded phases. Depending on customer placements, this may be fine if the voltages correct themselves along the line. But if the unbalance is due to a large tapped lateral just downstream of the regulator, the regulator needlessly unbalances the voltages.

Capacitor banks pose special coordination issues with regulators. A fixed capacitor bank creates a constant voltage rise on the circuit and a constant reactive contribution to the current. Either fixed or switched, capacitors upstream of a regulator do not interfere with the regulator's control action. Downstream capacitors pose the problem. A capacitor just downstream of a regulator affects the current that the regulator sees, but it does not measurably change the shape of the voltage profile beyond the regulator. In this case, we would like the line-drop compensation to ignore the capacitor. The voltage-spread compensation with a low compensator X/R or the zero-reactance compensator settings work well because they ignore or almost ignore the reactive current, so it works with fixed or switched banks downstream of the regulator. The load-center approach is more difficult to get to work with capacitors.

We do not want to ignore the capacitor at the end of a circuit section we are regulating because the capacitor significantly alters the profile along the circuit. In this case, we do not want zero-reactance compensation; we want some X to compensate for the capacitive current.

Switched capacitors can interact with the tap-changing controls on regulators upstream of the capacitors. This sort of interaction is rare but can happen if the capacitor is controlled by voltage (not radio, not time of day, not vars). A regulator may respond to an upstream or downstream capacitor switching, but that does not add up to many extra tap changes since the capacitor switches infrequently. Normally, the capacitor cannot cycle back and forth against the regulator. The only case might be if the regulator has negative settings for the reactive line-drop compensation.

With several regulators in series, adjustments to the time delay settings are the proper way to coordinate operations between units. Set the downstream regulator with the longest time delay so it does not change taps excessively. For multiple regulators, increase the time delay with increasing distance from the source. Tap changes by a downstream regulator do not change the voltage upstream, but tap changes by an upstream regulator affect all downstream regulators. If a downstream regulator acts before the upstream regulator, the downstream regulator may have to tap again to meet its set voltage. Making the downstream regulator wait longer prevents it from tapping unnecessarily. Separate the time delays by at least 10 to 15 sec to allow the upstream unit to complete tap change operations.

5.5 Station Regulation

Utilities most commonly use load tap changing transformers (LTCs) to control distribution feeder voltages at the substation. In many cases (short, urban, thermally limited feeders) an LTC is all the voltage support a circuit needs.

An LTC or a stand-alone voltage regulator must compensate for the voltage change on the subtransmission circuit as well as the voltage drop through the transformer. Of these, the voltage drop through the transformer is normally the largest. Normally, the standard ±10% regulator can accomplish this. A regulator can hit the end of its range if the load has especially poor power factor. The voltage drop across a transformer follows:

$$V_{drop} = I_R \cdot R + I_X \cdot X$$

Since a transformer's *X/R* ratio is so high, the reactive portion of the load creates the most voltage drop across the transformer. Consider a 10% impedance transformer at full load with a load power factor of 0.8, which means the reactive power factor is 0.6. In this case, the voltage drop across the transformer is 6%. If the subtransmission voltage is 120 V (on a 120-V scale), the maximum that the regulator can boost the voltage to is 124 V. If this example had a transformer loaded to more than its base open-air rating (OA or ONAN), the regulator would be more limited in range. In most cases, we do not run into this problem as power factors are normally much better than these.

In most cases, bus regulation suffices. For cases where circuits have significant voltage drop, individual feeder regulation can be better. Individual feeder regulation also performs better on circuits with different load cycles. If commercial feeders are on the same bus as residential feeders, it is less likely that a bus regulator can keep voltages in line on all circuits. Normally, we handle this by using bus regulation and supplementary line regulators. In some cases, individual feeder regulation in the station is more appropriate.

The voltage on feeders serving secondary networks is controlled at the primary substation with LTC transformers. These circuits are short enough that feeder regulators are unnecessary. Network feeders are often supplied by parallel station transformers; paralleling LTC units raises several issues that are discussed in the next section.

5.5.1 Parallel Operation

With care, we can parallel regulators. The most common situation is in a substation where a utility wants to parallel two LTC transformers. If two paralleled transformers do not have the same turns ratio, current will circulate to balance the voltages. The circulating current is purely reactive, but it adds extra loading on the transformer.

Some of the methods to operate LTC transformers in parallel (Jauch, 2001; Westinghouse Electric Corporation, 1965) include

- *Negative-reactance control* — The reactance setting in the line-drop compensator is set to a negative value, so higher reactive current forces the control to lower taps. The transformer with the higher tap has more reactive current, and the transformer with the lower tap is absorbing this reactive current (it looks capacitive to this transformer). So, a negative-reactance setting forces the transformer with the highest tap (and most reactive current) to lower its taps and bring it into alignment with the other unit. This method limits the use of line-drop compensation and can lead to lower bus voltages.
- *Master-follower* — One controller, the master, regulates the voltage and signals the other tap changers (the followers) to match the tap setting. The master control normally gets feedback from the followers to confirm their operation.
- *Var balancing* — The controller adjusts taps as required to equalize the vars flowing in parallel transformers. Auxiliary circuitry is required. This method has the advantage that it works with transformers fed from separate transmission sources.
- *Circulating current method* — This is the most common control. Auxiliary circuitry is added to separate the load current through each transformer from the circulating current. Each transformer LTC control is fed the load current. The controller adjusts taps to minimize

the difference in current between parallel units. Removing a unit does not require changing controller settings.

The complications associated with paralleling regulators are another reason utilities normally avoid closed bus ties in distribution substations.

5.5.2 Bus Regulation Settings

Although too often left unused, bus regulators (whether stand-alone regulators or load tap changing transformers) can use line-drop compensation. The concept of a load center rarely has good meaning for a bus supporting several circuits, but the voltage spread methods allow the regulator to boost voltage under heavy load.

The voltage-spread equations assume that the power factor at full load is the same as the power factor at light load. If the power factor is different at light and peak loads, we can use this information to provide more precise settings. We could solve the following to find new R and X settings with different power factors

$$V_{max} - V_{min} = (pf_{max} \cdot R_{set} + qf_{max} \cdot X_{set})I_{max} - (pf_{min} \cdot R_{set} + qf_{min} \cdot X_{set})I_{min}$$

However, it is easier to use the equations in Section 5.4.1.2 and use the average of the power factor at peak load and the power factor at light load. With line-drop compensation for bus regulation, the voltage-override feature helps to ensure that the LTC or regulator does not cause excessive voltages.

Individual substation feeder regulators are set the same as line feeder regulators. We can tune controller settings more precisely based on the individual characteristics of a given feeder. If the first part of the feeder is an express feeder with no load on it, we could boost the voltage higher than normal, especially if the circuit is voltage limited. Our main constraint is making sure that the first customer does not have high voltage.

5.6 Line Loss and Voltage Drop Relationships

Line losses are from the line current flowing through the resistance of the conductors. After distribution transformer losses, primary line losses are the largest cause of losses on the distribution system. Like any resistive losses, line losses are a function of the current squared multiplied by the resistance (I^2R). Ways to reduce line losses include

- Use a higher system voltage
- Balance circuits

- Convert single-phase circuits to three-phase circuits
- Reduce loads
- Increase power factor (capacitors)
- Reconductor with a larger size

Because losses are a function of the current squared, most losses occur on the primary near the substation. Losses occur regardless of the power factor of the circuit. Reducing the reactive portion of current reduces the total current, which can significantly impact losses.

Approximations using uniform load distributions are useful. A uniformly distributed load along a circuit of length l has the same losses as a single lumped load placed at a length of $l/3$ from the source end. For voltage drop, the equivalent circuits are different: a uniformly distributed load along a circuit of length l has the same voltage drop as a single lumped load placed at a length of $l/2$ from the source end. This 1/2 rule for voltage drop and the 1/3 rule for losses are helpful approximations when doing hand calculations or when making simplifications to enter in a load-flow program.

For a uniformly increasing load, the equivalent lumped load is at $0.53l$ of the length from the source. Figure 5.9 shows equivalent circuits for a uniform load and a uniformly increasing load.

Line losses decrease as operating voltage increases because the current decreases. Schultz (1978) derived several expressions for primary feeder I^2R losses on circuits with uniform load densities. His analysis showed that most 15 to 35 kV circuits are not voltage-drop limited — most are thermally limited. As the system voltage varies, the losses change the most for voltage-limited circuits (Schultz, 1978):

$$L_2 = \left(\frac{V_1}{V_2}\right)^2 L_1 \quad \text{for a voltage-limited circuit}$$

$$L_2 = \left(\frac{V_1}{V_2}\right)^{2/3} L_1 \quad \text{for a thermally-limited circuit}$$

where

V_1, V_2 = voltage on circuits 1 and 2

L_1, L_2 = feeder I^2R losses on circuits 1 and 2

On a system-wide basis, losses are expected to change with voltage with an exponent somewhere between 2/3 and 2.

Losses, voltage drop, and capacity are all interrelated. Three-phase circuits have the highest power transfer capacity, the lowest voltage drop, and the lowest losses. Table 5.7 compares capacity, voltage drop, and losses of a balanced three-phase system with several other phasing configurations.

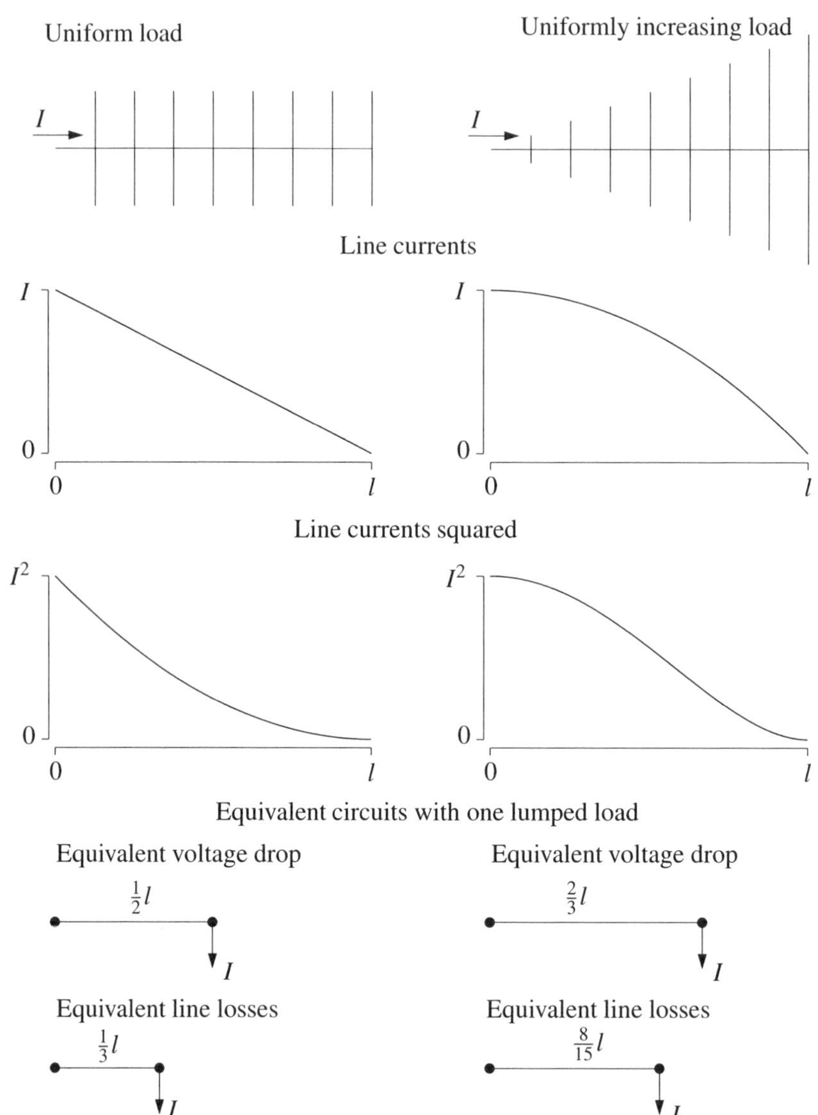

FIGURE 5.9
Equivalent circuits of uniform loads.

TABLE 5.7

Characteristics of Various Systems

System	Capacity in per Unit	Voltage Drop in per Unit for Equal kVA	Line Losses in per Unit For Equal kVA
Balanced three phase	1.0	1.0	1.0
Two phases	0.5	2.0	2.0
Two phases and a multigrounded neutral	0.67	2.0–3.3	1.2–3.0
Two phases and a unigrounded neutral	0.67	2.5–4.5	2.25
One phase and a multigrounded neutral	0.33	3.7–4.5	3.5–4.0
One phase and a unigrounded neutral	0.33	6.0	6.0

Notes: The two-phase circuits assume all load is connected line to ground. Neutrals are the same size as the phases. Reduced neutrals increase voltage drop and (usually) line losses. The voltage drop and line loss ratios for circuits with multigrounded neutrals vary with conductor size.

Utilities consider both peak losses and energy losses. Peak losses are important because they compose a portion of the peak demand; energy losses are the total kilowatt-hours wasted as heat in the conductors. The peak losses are more easily estimated from measurements and models. The average losses can be found from the peak losses using the loss factor F_{ls}:

$$F_{ls} = \frac{\text{Average losses}}{\text{Peak losses}}$$

Normally, we do not have enough information to directly measure the loss factor. We do have the load factor (the average demand over the peak demand). The loss factor is some function of the load factor squared. The most common approximation (Gangel and Propst, 1965) is

$$F_{ls} = 0.15F_{ld} + 0.85F_{ld}^2$$

This is often used for evaluating line losses and transformer load losses (which are also a function of I^2R). Load factors closer to one result in loss factors closer to one. Another common expression is $F_{ls} = 0.3F_{ld} + 0.7F_{ld}^2$. Figure 5.10 shows both relationships.

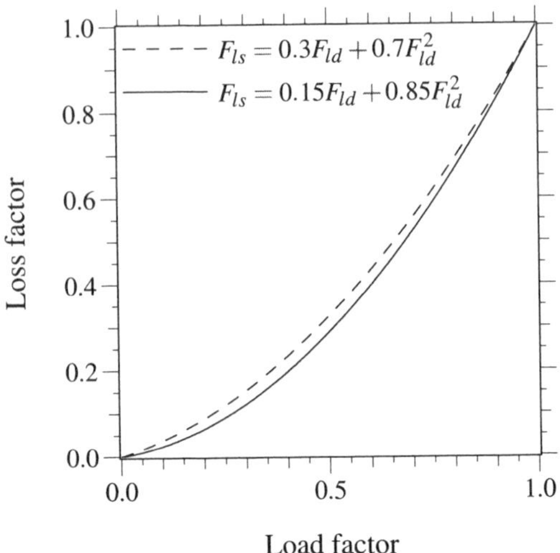

FIGURE 5.10
Relationship between load factor and loss factor.

References

ANSI C84.1-1995, *American National Standards for Electric Power Systems and Equipment — Voltage Ratings (60 Hz).*

Beckwith, *Basic Considerations for the Application of LTC Transformers and Associated Controls*, Beckwith Electric Company, Application Note #17, 1998.

Bishop, M. T., Foster, J. D., and Down, D. A., "The Application of Single-Phase Voltage Regulators on Three-Phase Distribution Systems," *IEEE Industry Applications Magazine*, pp. 38–44, July/August 1996.

Brice, C. W., "Comparison of Approximate and Exact Voltage Drop Calculations for Distribution Lines," *IEEE Transactions on Power Apparatus and Systems*, vol. PAS-101, no. 11, pp. 4428–31, November, 1982.

Cooper Power Systems, "Determination of Regulator Compensator Settings," 1978. Publication R225-10-1.

Gangel, M. W. and Propst, R. F., "Distribution Transformer Load Characteristics," *IEEE Transactions on Power Apparatus and Systems*, vol. 84, pp. 671–84, August 1965.

General Electric, *Omnitext*, 1979. GET-3537B.

IEEE Std. C57.15-1999, *IEEE Standard Requirements, Terminology, and Test Code for Step-Voltage Regulators.*

Jauch, E. T., "Advanced Transformer Paralleling," IEEE/PES Transmission and Distribution Conference and Exposition, 2001.

Kirshner, D., "Implementation of Conservation Voltage Reduction at Commonwealth Edison," *IEEE Transactions on Power Systems*, vol. 5, no. 4, pp. 1178–82, November 1990.

Kirshner, D. and Giorsetto, P., "Statistical Tests of Energy Savings Due to Voltage Reduction," *IEEE Transactions on Power Apparatus and Systems*, vol. PAS-103, no. 6, pp. 1205–10, June 1984.

Lokay, H. E. and Custard, R. L., "A Field Method for Determining the Leading and Lagging Regulator in An Open-Delta Connection," *AIEE Transactions*, vol. 73, Part III, pp. 1684–6, 1954.

McCarthy, C., "CAPS — Choosing the Feeders, Part I," in Systems Engineering Technical Update: Cooper Power Systems, 2000.

Priess, R. F. and Warnock, V. J., "Impact of Voltage Reduction on Energy and Demand," *IEEE Transactions on Power Apparatus and Systems*, vol. PAS-97, no. 5, pp. 1665–71, Sept/Oct 1978.

Schultz, N. R., "Distribution Primary Feeder I^2R Losses," *IEEE Transactions on Power Apparatus and Systems*, vol. PAS-97, no. 2, pp. 603–9, March–April 1978.

Sen, P. K. and Larson, S. L., "Fundamental Concepts of Regulating Distribution System Voltages," IEEE Rural Electric Power Conference, Department of Electrical Engineering, Colorado University, Denver, CO, 1994.

Westinghouse Electric Corporation, *Distribution Systems*, vol. 3, 1965.

Willis, H. L., "Characteristics of Distribution Loads," in *Electrical Transmission and Distribution Reference Book*. Raleigh, NC: ABB Power T&D Company, 1997a.

Willis, H. L., *Power Distribution Planning Reference Book*, Marcel Dekker, New York, 1997b.

Regs? Treat them with respect. they are a transformer. Anyone who has dropped load with a tx knows that you can build a fire if you dont take the load into consideration. The difference with regs is that the load is the feeder. Get it?

anonymous post
www.powerlineman.com

6

Capacitor Application

Capacitors provide tremendous benefits to distribution system performance. Most noticeably, capacitors reduce losses, free up capacity, and reduce voltage drop:

- *Losses; Capacity* — By canceling the reactive power to motors and other loads with low power factor, capacitors decrease the line current. Reduced current frees up capacity; the same circuit can serve more load. Reduced current also significantly lowers the I^2R line losses.
- *Voltage drop* — Capacitors provide a voltage boost, which cancels part of the drop caused by system loads. Switched capacitors can regulate voltage on a circuit.

If applied properly and controlled, capacitors can significantly improve the performance of distribution circuits. But if not properly applied or controlled, the reactive power from capacitor banks can create losses and high voltages. The greatest danger of overvoltages occurs under light load. Good planning helps ensure that capacitors are sited properly. More sophisticated controllers (like two-way radios with monitoring) reduce the risk of improperly controlling capacitors, compared to simple controllers (like a time clock).

Capacitors work their magic by storing energy. Capacitors are simple devices: two metal plates sandwiched around an insulating dielectric. When charged to a given voltage, opposing charges fill the plates on either side of the dielectric. The strong attraction of the charges across the very short distance separating them makes a tank of energy. Capacitors oppose changes in voltage; it takes time to fill up the plates with charge, and once charged, it takes time to discharge the voltage.

On ac power systems, capacitors do not store their energy very long — just one-half cycle. Each half cycle, a capacitor charges up and then discharges its stored energy back into the system. The net real power transfer is zero. Capacitors provide power just when reactive loads need it. Just when a motor with low power factor needs power from the system, the capacitor is there to provide it. Then in the next half cycle, the motor releases its excess energy, and the capacitor is there to absorb it. Capacitors and reactive loads

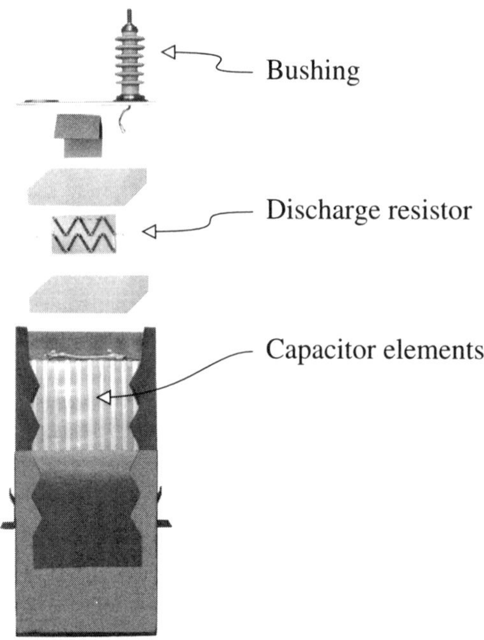

FIGURE 6.1
Capacitor components. (From General Electric Company. With permission.)

exchange this reactive power back and forth. This benefits the system because that reactive power (and extra current) does not have to be transmitted from the generators all the way through many transformers and many miles of lines; the capacitors can provide the reactive power locally. This frees up the lines to carry real power, power that actually does work.

Capacitor units are made of series and parallel combinations of capacitor packs or elements put together as shown in Figure 6.1. Capacitor elements have sheets of polypropylene film, less than one mil thick, sandwiched between aluminum foil sheets. Capacitor dielectrics must withstand on the order of 2000 V/mil (78 kV/mm). No other medium-voltage equipment has such high voltage stress. An underground cable for a 12.47-kV system has insulation that is at least 0.175 in. (4.4 mm) thick. A capacitor on the same system has an insulation separation of only 0.004 in. (0.1 mm).

Utilities often install substation capacitors and capacitors at points on distribution feeders. Most feeder capacitor banks are pole mounted, the least expensive way to install distribution capacitors. Pole-mounted capacitors normally provide 300 to 3600 kvar at each installation. Many capacitors are switched, either based on a local controller or from a centralized controller through a communication medium. A line capacitor installation has the capacitor units as well as other components, possibly including arresters, fuses, a control power transformer, switches, and a controller (see Figure 6.2 for an example).

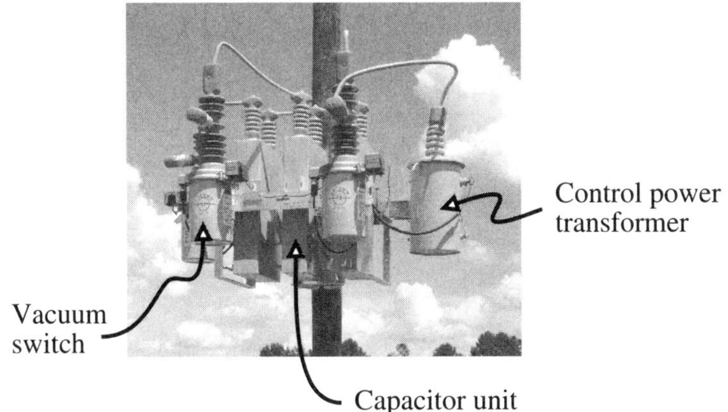

FIGURE 6.2
Overhead line capacitor installation. (From Cooper Power Systems, Inc. With permission.)

While most capacitors are pole mounted, some manufacturers provide padmounted capacitors. As more circuits are put underground, the need for padmounted capacitors will grow. Padmounted capacitors contain capacitor cans, switches, and fusing in a deadfront package following standard padmounted-enclosure integrity requirements (ANSI C57.12.28-1998). These units are much larger than padmounted transformers, so they must be sited more carefully to avoid complaints due to aesthetics. The biggest obstacles are cost and aesthetics. The main complaint is that padmounted capacitors are large. Customers complain about the intrusion and the aesthetics of such a large structure (see Figure 6.3).

FIGURE 6.3
Example padmounted capacitor. (From Northeast Power Systems, Inc. With permission.)

TABLE 6.1

Substation vs. Feeder Capacitors

Advantages	Disadvantages
Feeder Capacitors	
Reduces line losses	More difficult to control reliably
Reduces voltage drop along the feeder	Size and placement important
Frees up feeder capacity	
Lower cost	
Substation Capacitors	
Better control	No reduction in line losses
Best placement if leading vars are needed for system voltage support	No reduction in feeder voltage drop
	Higher cost

Substation capacitors are normally offered as open-air racks. Normally elevated to reduce the hazard, individual capacitor units are stacked in rows to provide large quantities of reactive power. All equipment is exposed. Stack racks require a large substation footprint and are normally engineered for the given substation. Manufacturers also offer metal-enclosed capacitors, where capacitors, switches, and fuses (normally current-limiting) are all enclosed in a metal housing.

Substation capacitors and feeder capacitors both have their uses. Feeder capacitors are closer to the loads — capacitors closer to loads more effectively release capacity, improve voltage profiles, and reduce line losses. This is especially true on long feeders that have considerable line losses and voltage drop. Table 6.1 highlights some of the differences between feeder and station capacitors. Substation capacitors are better when more precise control is needed. System operators can easily control substation capacitors wired into a SCADA system to dispatch vars as needed. Modern communication and control technologies applied to feeder capacitors have reduced this advantage. Operators can control feeder banks with communications just like station banks, although some utilities have found the reliability of switched feeder banks to be less than desired, and the best times for switching in vars needed by the system may not correspond to the best time to switch the capacitor in for the circuit it is located on.

Substation capacitors may also be desirable if a leading power factor is needed for voltage support. If the power factor is leading, moving this capacitor out on the feeder increases losses. Substation capacitors cost more than feeder capacitors. This may seem surprising, but we must individually engineer station capacitors, and the space they take up in a station is often valuable real estate. Pole-mounted capacitor installations are more standardized.

Utilities normally apply capacitors on three-phase sections. Applications on single-phase lines are done but less common. Application of three-phase banks downstream of single-phase protectors is normally not done because of ferroresonance concerns. Most three-phase banks are connected grounded-wye on four-wire multigrounded circuits. Some are connected in floating wye. On three-wire circuits, banks are normally connected as a floating wye.

Most utilities also include arresters and fuses on capacitor installations. Arresters protect capacitor banks from lightning-overvoltages. Fuses isolate failed capacitor units from the system and clear the fault before the capacitor fails violently. In high fault-current areas, utilities may use current-limiting fuses. Switched capacitor units normally have oil or vacuum switches in addition to a controller. Depending on the type of control, the installation may include a control power transformer for power and voltage sensing and possibly a current sensor. Because a capacitor bank has a number of components, capacitors normally are not applied on poles with other equipment.

Properly applied capacitors return their investment very quickly. Capacitors save significant amounts of money in reduced losses. In some cases, reduced loadings and extra capacity can also delay building more distribution infrastructure.

6.1 Capacitor Ratings

Capacitor units rated from 50 to over 500 kvar are available; Table 6.2 shows common capacitor unit ratings. A capacitor's rated kvar is the kvar at rated voltage. Three-phase capacitor banks are normally referred to by the total kvar on all three phases. Distribution feeder banks normally have one or two or (more rarely) three units per phase. Many common size banks only have one capacitor unit per phase.

IEEE Std. 18 defines standards for capacitors and provides application guidelines. Capacitors should not be applied when any of the following limits are exceeded (IEEE Std. 18-2002):

- 135% of nameplate kvar
- 110% of rated rms voltage, and crest voltage not exceeding $1.2\sqrt{2}$ of rated rms voltage, including harmonics but excluding transients
- 135% of nominal rms current based on rated kvar and rated voltage

Capacitor dielectrics must withstand high voltage stresses during normal operation — on the order of 2000 V/mil. Capacitors are designed to withstand overvoltages for short periods of time. IEEE Std. 18-1992 allows up to 300 power-frequency overvoltages within the time durations in Table 6.3 (without transients or harmonic content). New capacitors are tested with at least a 10-sec overvoltage, either a dc-test voltage of 4.3 times rated rms or an ac voltage of twice the rated rms voltage (IEEE Std. 18-2002).

TABLE 6.2

Common Capacitor Unit Ratings

Volts, rms (Terminal-to-Terminal)	kvar	Number of Phases	BIL, kV
216	5, 7 1/2, 13 1/3, 20, and 25	1 and 3	30
240	2.5, 5, 7 1/2, 10, 15, 20, 25, and 50	1 and 3	30
480, 600	5, 10, 15, 20, 25, 35, 50, 60, and 100	1 and 3	30
2400	50, 100, 150, 200, 300, and 400	1 and 3	75, 95, 125, 150, and 200
2770	50, 100, 150, 200, 300, 400, and 500	1 and 3	75, 95, 125, 150, and 200
4160, 4800	50, 100, 150, 200, 300, 400, 500, 600, 700, and 800	1 and 3	75, 95, 125, 150, and 200
6640, 7200, 7620, 7960, 8320, 9540, 9960, 11,400, 12,470, 13,280, 13,800, 14,400	50, 100, 150, 200, 300, 400, 500, 600, 700, and 800	1	95, 125, 150, and 200
15,125	50, 100, 150, 200, 300, 400, 500, 600, 700, and 800	1	125, 150, and 200
19,920	100, 150, 200, 300, 400, 500, 600, 700, and 800	1	125, 150, and 200
20,800, 21,600, 22,800, 23,800, 24,940	100, 150, 200, 300, 400, 500, 600, 700, and 800	1	150 and 200

Source: IEEE Std. 18-2002.

TABLE 6.3

Maximum Permissible Power-Frequency Voltages

Duration	Maximum Permissible Voltage (multiplying factor to be applied to rated voltage rms)
6 cycles	2.20
15 cycles	2.00
1 sec	1.70
15 sec	1.40
1 min	1.30
30 min	1.25
Continuous	1.10

Note: This is not in IEEE Std. 18-2002. Cupdated capacitor application guide.

Source: ANSI/IEEE Std. 18-1992.

Capacitors should withstand various peak voltage and current transients; the allowable peak depends on the number of transients expected per year (see Table 6.4).

The capacitance of a unit in microfarads is

TABLE 6.4

Expected Transient Overcurrent and Overvoltage Capability

Probable Number of Transients per year	Permissible Peak Transient Current (multiplying factor to be applied to rated rms current)	Permissible Peak Transient Voltage (multiplying factor to be applied to rated rms voltage)
4	1500	5.0
40	1150	4.0
400	800	3.4
4000	400	2.9

Note: This is not in IEEE Std. 18-2002, but it will be addressed in IEEE's updated capacitor application guide.

Source: ANSI/IEEE Std. 18-1992.

$$C_{uF} = \frac{2.65Q_{kvar}}{V_{kV}^2}$$

where

V_{kV} = capacitor voltage rating, kV

Q_{kvar} = unit reactive power rating, kvar

Capacitors are made within a given tolerance. The IEEE standard allows reactive power to range between 100 and 110% when applied at rated sinusoidal voltage and frequency (at 25°C case and internal temperature) (IEEE Std. 18-2002). Older units were allowed to range up to 115% (ANSI/IEEE Std. 18-1992). Therefore, the capacitance also must be between 100 and 110% of the value calculated at rated kvar and voltage. In practice, most units are from +0.5 to +4.0%, and a given batch is normally very uniform.

Capacitor losses are typically on the order of 0.07 to 0.15 W/kvar at nominal frequency. Losses include resistive losses in the foil, dielectric losses, and losses in the internal discharge resistor.

Capacitors must have an internal resistor that discharges a capacitor to 50 V or less within 5 min when the capacitor is charged to the peak of its rated voltage $(\sqrt{2}V_{rms})$. This resistor is the major component of losses within a capacitor. The resistor must be low enough such that the *RC* time constant causes it to decay in 300 sec as

$$\frac{50}{\sqrt{2}V} \le e^{-300/RC}$$

where

V = capacitor voltage rating, V

R = discharge resistance, Ω

C = capacitance, F

TABLE 6.5

Maximum Ambient Temperatures for Capacitor Application

Mounting Arrangement	Ambient Air Temperature (°C) 4-h Average[a]
Isolated capacitor	46
Single row of capacitors	46
Multiple rows and tiers of capacitors	40
Metal-enclosed or -housed equipments	40

[a] The arithmetic average of the four consecutive highest hourly readings during the hottest day expected at that location.

Source: IEEE Std. 18-2002.

So, the discharge resistor must continually dissipate at least the following power in watts:

$$P_{watts} = -\frac{Q_{kvar}}{113.2}\ln\left(\frac{35.36}{V}\right)$$

where Q_{kvar} is the capacitor rating (single or three phase). For 7.2-kV capacitors, the lower bound on losses is 0.047 W/kvar.

Some utilities use a shorting bar across the terminals of capacitors during shipping and in storage. The standard recommends waiting for 5 min to allow the capacitor to discharge through the internal resistor.

Capacitors have very low losses, so they run very cool. But capacitors are very sensitive to temperature and are rated for lower temperatures than other power system equipment such as cables or transformers. Capacitors do not have load cycles like transformers; they are always at full load. Also, capacitors are designed to operate at high dielectric stresses, so they have less margin for degraded insulation. Standards specify an upper limit for application of 40 or 46°C depending on arrangement (see Table 6.5). These limits assume unrestricted ventilation and direct sunlight. At the lower end, IEEE standard 18 specifies that capacitors shall be able to operate continuously in a –40°C ambient.

6.2 Released Capacity

In addition to reducing losses and improving voltage, capacitors release capacity. Improving the power factor increases the amount of real-power load the circuit can supply. Using capacitors to supply reactive power reduces the amount of current in the line, so a line of a given ampacity can

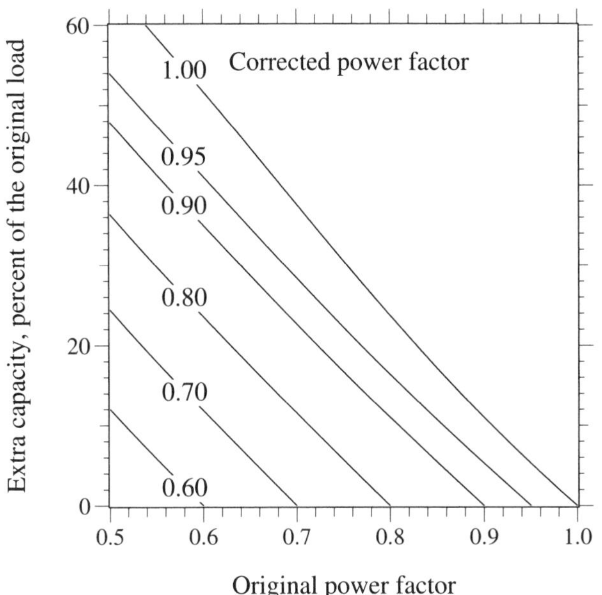

FIGURE 6.4
Released capacity with improved power factor.

carry more load. Figure 6.4 shows that capacitors release significant capacity, especially if the original power factor is low. Figure 6.5 shows another way to view the extra capacity, as a function of the size of capacitor added.

6.3 Voltage Support

Capacitors are constant-impedance devices. At higher voltages, capacitors draw more current and produce more reactive power as

$$I = I_{rated} V_{pu} \quad \text{and} \quad Q_{kvar} = Q_{rated} {V_{pu}}^2$$

where V_{pu} is the voltage in per unit of the capacitor's voltage rating. Capacitors applied at voltages other than their rating provide vars in proportion to the per-unit voltage squared.

Capacitors provide almost a fixed voltage rise. The reactive current through the system impedance causes a voltage rise in percent of

$$V_{rise} = \frac{Q_{kvar} X_L}{10\, V_{kV,l\text{-}l}^2}$$

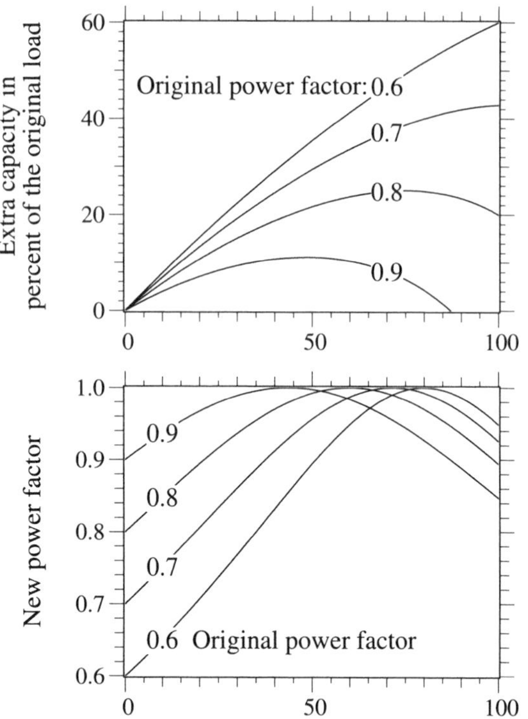

FIGURE 6.5
Extra capacity as a function of capacitor size.

where

X_L = positive-sequence system impedance from the source to the capacitor, Ω

$V_{kV,\,l\text{-}l}$ = line-to-line system voltage, kV

Q_{kvar} = three-phase bank rating, kvar

While this equation is very good for most applications, it is not exactly right because the capacitive current changes in proportion to voltage. At a higher operating voltage, a capacitor creates more voltage rise than the equation predicts.

Since the amount of voltage rise is dependent on the impedance upstream of the bank, to get the voltage boost along the entire circuit, put the capacitor at the end of the circuit. The best location for voltage support depends on where the voltage support is needed. Figure 6.6 shows how a capacitor changes the voltage profile along a circuit. Unlike a regulator, a capacitor changes the voltage profile upstream of the bank.

Table 6.6 shows the percentage voltage rise from capacitors for common conductors at different voltages. This table excludes the station transformer

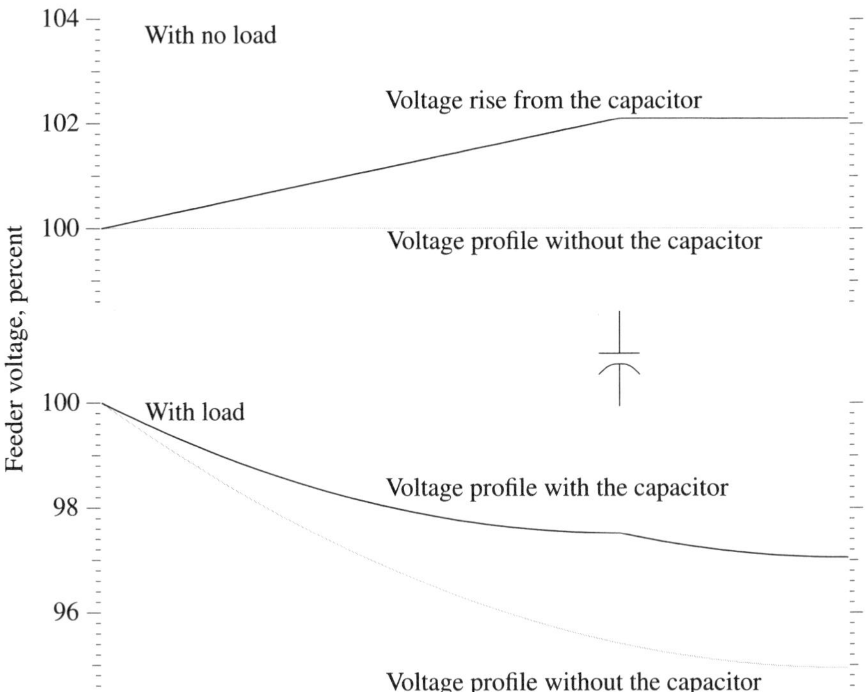

FIGURE 6.6
Voltage profiles after addition of a capacitor bank. (Copyright © 2002. Electric Power Research Institute. 1001691. *Improved Reliability of Switched Capacitor Banks and Capacitor Technology*. Reprinted with permission.)

TABLE 6.6
Percent Voltage Rise for Various Conductors and Voltage Levels

		Percent Voltage Rise per Mile with 100 kvar per Phase			
	X_L	Line-to-Line System Voltage, kV			
Conductor Size	**Ω/mi**	**4.8**	**12.47**	**24.9**	**34.5**
4	0.792	1.031	0.153	0.038	0.020
2	0.764	0.995	0.147	0.037	0.019
1/0	0.736	0.958	0.142	0.036	0.019
4/0	0.694	0.903	0.134	0.034	0.017
350	0.656	0.854	0.127	0.032	0.017
500	0.635	0.826	0.122	0.031	0.016
750	0.608	0.791	0.117	0.029	0.015

Note: Impedance are for all-aluminum conductors with GMD=4.8 ft.

impedance but still provides a useful approximation. Inductance does not change much with conductor size; the voltage change stays the same over a wide range of conductor sizes. For 15-kV class systems, capacitors increase the voltage by about 0.12% per mi per 100 kvar per phase.

On switched capacitor banks, the voltage change constrains the size of banks at some locations. Normally, utilities limit the voltage change to 3 to 4%. On a 12.47-kV circuit, a three-phase 1200-kvar bank boosts the voltage 4% at about 8 mi from the substation. To keep within a 4% limit, 1200-kvar banks must only be used within the first 8 mi of the station.

6.4 Reducing Line Losses

One of the main benefits of applying capacitors is that they can reduce distribution line losses. Losses come from current through the resistance of conductors. Some of that current transmits real power, but some flows to supply reactive power. Reactive power provides magnetizing for motors and other inductive loads. Reactive power does not spin kWh meters and performs no useful work, but it must be supplied. Using capacitors to supply reactive power reduces the amount of current in the line. Since line losses are a function of the current squared, I^2R, reducing reactive power flow on lines significantly reduces losses.

Engineers widely use the "2/3 rule" for sizing and placing capacitors to optimally reduce losses. Neagle and Samson (1956) developed a capacitor placement approach for uniformly distributed lines and showed that the optimal capacitor location is the point on the circuit where the reactive power flow equals half of the capacitor var rating. From this, they developed the 2/3 rule for selecting and placing capacitors. For a uniformly distributed load, the optimal size capacitor is 2/3 of the var requirements of the circuit. The optimal placement of this capacitor is 2/3 of the distance from the substation to the end of the line. For this optimal placement for a uniformly distributed load, the substation source provides vars for the first 1/3 of the circuit, and the capacitor provides vars for the last 2/3 of the circuit (see Figure 6.7).

A generalization of the 2/3 rule for applying n capacitors to a circuit is to size each one to $2/(2n+1)$ of the circuit var requirements. Apply them equally spaced, starting at a distance of $2/(2n+1)$ of the total line length from the substation and adding the rest of the units at intervals of $2/(2n+1)$ of the total line length. The total vars supplied by the capacitors is $2n/(2n+1)$ of the circuit's var requirements. So to apply three capacitors, size each to 2/7 of the total vars needed, and locate them at per unit distances of 2/7, 4/7, and 6/7 of the line length from the substation.

Grainger and Lee (1981) provide an optimal yet simple method for placing fixed capacitors on a circuit with any load profile, not just a uniformly

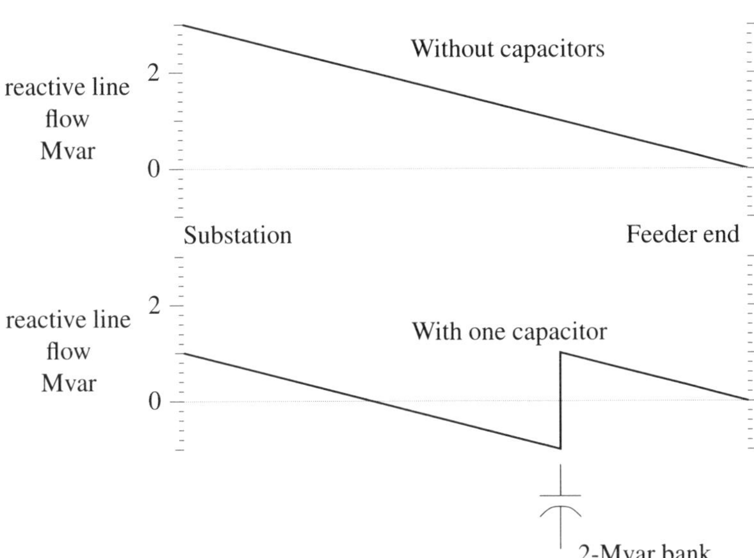

FIGURE 6.7
Optimal capacitor loss reduction using the two-thirds rule. (Copyright © 2002. Electric Power Research Institute. 1001691. *Improved Reliability of Switched Capacitor Banks and Capacitor Technology.* Reprinted with permission.)

distributed load. With the Grainger/Lee method, we use the reactive load profile of a circuit to place capacitors. The basic idea is again to locate banks at points on the circuit where the reactive power equals one half of the capacitor var rating. With this 1/2-*kvar rule*, the capacitor supplies half of its vars downstream, and half are sent upstream. The basic steps of this approach are:

1. *Pick a size* — Choose a standard size capacitor. Common sizes range from 300 to 1200 kvar, with some sized up to 2400 kvar. If the bank size is 2/3 of the feeder requirement, we only need one bank. If the size is 1/6 of the feeder requirement, we need five capacitor banks.
2. *Locate the first bank* — Start from the end of the circuit. Locate the first bank at the point on the circuit where var flows on the line are equal to half of the capacitor var rating.
3. *Locate subsequent banks* — After a bank is placed, reevaluate the var profile. Move upstream until the next point where the var flow equals half of the capacitor rating. Continue placing banks in this manner until no more locations meet the criteria.

There is no reason we have to stick with the same size of banks. We could place a 300-kvar bank where the var flow equals 150 kvar, then apply a 600-

kvar bank where the var flow equals 300 kvar, and finally apply a 450-kvar bank where the var flow equals 225 kvar. Normally, it is more efficient to use standardized bank sizes, but different size banks at different portions of the feeder might help with voltage profiles.

The 1/2-kvar method works for any section of line. If a line has major branches, we can apply capacitors along the branches using the same method. Start at the end, move upstream, and apply capacitors at points where the line's kvar flow equals half of the kvar rating of the capacitor. It also works for lines that already have capacitors (it does not optimize the placement of all of the banks, but it optimizes placement of new banks). For large industrial loads, the best location is often going to be right at the load.

Figure 6.8 shows the optimal placement of 1200-kvar banks on an example circuit. Since the end of the circuit has reactive load above the 600-kvar threshold for sizing 1200-kvar banks, we apply the first capacitor at the end

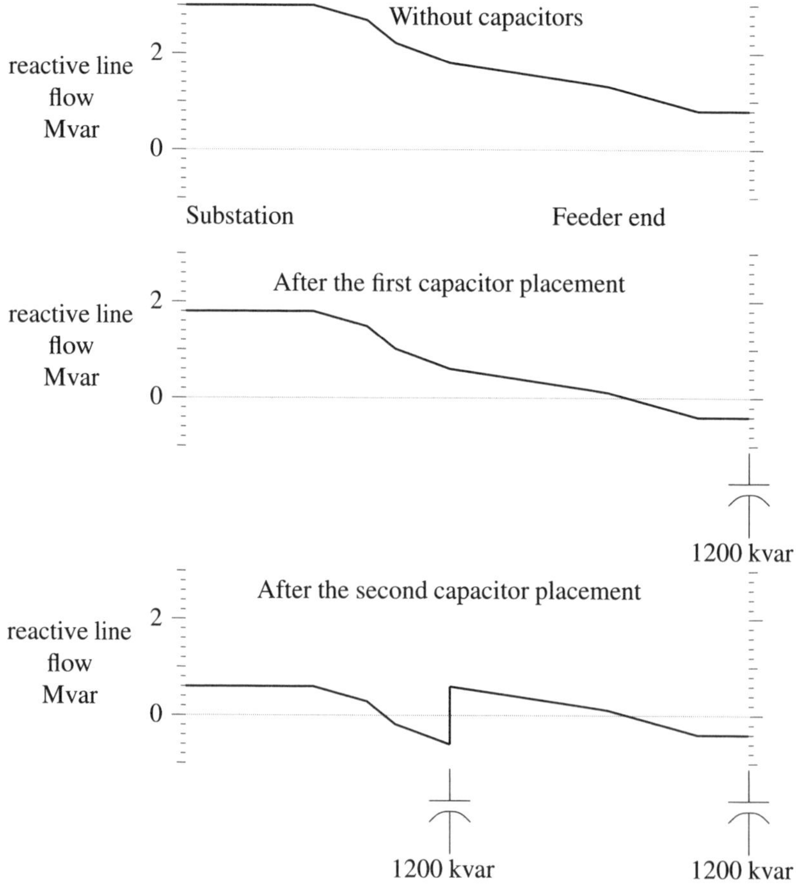

FIGURE 6.8
Placement of 1200-kvar banks using the 1/2-kvar method.

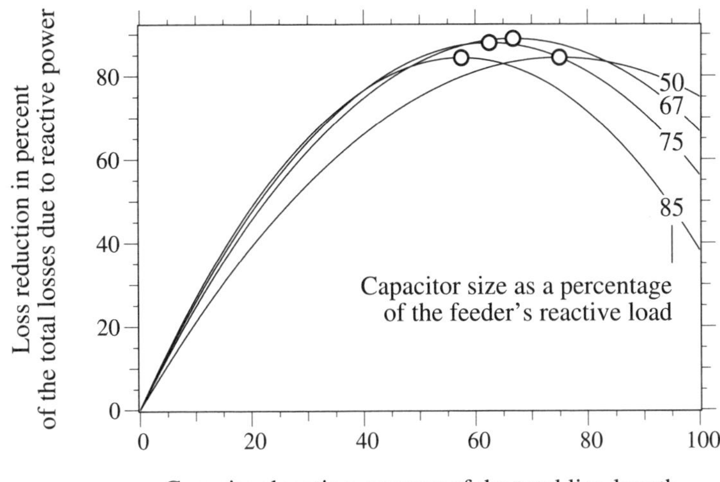

FIGURE 6.9
Sensitivity to losses of sizing and placing one capacitor on a circuit with a uniform load. (The circles mark the optimum location for each of the sizes shown.)

of the circuit. (The circuit at the end of the line could be one large customer or branches off the main line.) The second bank goes near the middle. The circuit has an express feeder near the start. Another 1200-kvar bank could go in just after the express feeder, but that does not buy us anything. The two capacitors total 2400 kvar, and the feeder load is 3000 kvar. We really need another 600-kvar capacitor to zero out the var flow before it gets to the express feeder.

Fortunately, capacitor placement and sizing does not have to be exact. Quite good loss reduction occurs even if sizing and placement are not exactly optimum. Figure 6.9 shows the loss reduction for one fixed capacitor on a circuit with a uniform load. The 2/3 rule specifies that the optimum distance is 2/3 of the distance from the substation and 2/3 of the circuit's var requirement. As long as the size and location are somewhat close (within 10%), the not-quite-optimal capacitor placement provides almost as much loss reduction as the optimal placement.

Consider the voltage impacts of capacitors. Under light load, check that the capacitors have not raised the voltages above allowable standards. If voltage limits are exceeded, reduce the size of the capacitor banks or the number of capacitor banks until voltage limits are not exceeded. If additional loss reduction is desired, consider switched banks as discussed below.

6.4.1 Energy Losses

Use the average reactive loading profile to optimally size and place capacitors for energy losses. If we use the peak-load case, the 1/2-kvar method optimizes losses during the peak load. If we have a load-flow case with the average reactive load, the 1/2-kvar method or the 2/3 rule optimizes energy losses. This leads to more separation between banks and less kvars applied than if we optimize for peak losses.

If an average system case is not available, then we can estimate it by scaling the peak load case by the reactive load factor, *RLF*:

$$RLF = \frac{\text{Average kvar Demand}}{\text{Peak kvar Demand}}$$

The reactive load factor is similar to the traditional load factor except that it only considers the reactive portion of the load. If we have no information on the reactive load factor, use the total load factor. Normally, the reactive load factor is higher than the total load factor. Figure 6.10 shows an example of power profiles; the real power (kW) fluctuates significantly more than the reactive power (kvar).

6.5 Switched Banks

Switched banks provide benefits under the following situations:

- *More loss reduction* — As the reactive loading on the circuit changes, we reduce losses by switching banks on and off to track these changes.
- *Voltage limits* — If optimally applied banks under the average loading scenario cause excessive voltage under light load, then use switched banks.

In addition, automated capacitors — those with communications — have the flexibility to also use distribution vars for transmission support.

Fixed banks are relatively easy to site and size optimally. Switched banks are more difficult. Optimally sizing capacitors, placing them, and deciding when to switch them are difficult tasks. Several software packages are available that can optimize this solution. This is an intensely studied area, and technical literature documents several approaches (among these Carlisle and El-Keib, 2000; Grainger and Civanlar, 1985; Shyh, 2000).

To place switched capacitors using the 1/2-kvar method, again place the banks at the location where the line kvar equals half the bank rating. But instead of using the average reactive load profile (the rule for fixed banks), use the average reactive flow during the time the capacitor is on. With time-switched banks and information on load profiles (or typical load profiles),

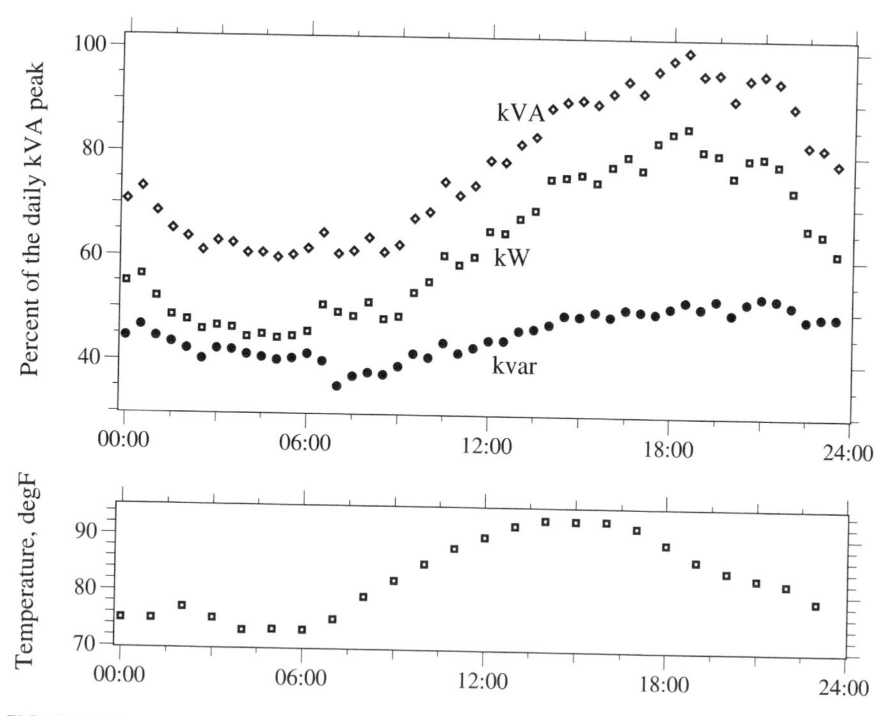

FIGURE 6.10
Example of real and reactive power profiles on a residential feeder on a peak summer day with 95% air conditioning. (Data from East Central Oklahoma Electric Cooperative, Inc. [RUS 1724D-112, 2001].)

we can pick the on time and the off time and determine the proper sizing based on the average reactive flow between the on and off times. Or, we can place a bank and pick the on and off times such that the average reactive line flow while the bank is switched on equals half of the bank rating. In these cases, we have specified the size and either the placement or switching time. To more generally optimize — including sizing, placement, number of banks, and switching time — we must use a computer, which iterates to find a solution [see Lee and Grainger (1981) for one example].

Combinations of fixed and switched banks are more difficult. The following approach is not optimal but gives reasonable results. Apply fixed banks to the circuit with the 1/2-kvar rule based on the light-load case. Check voltages. If there are undervoltages, increase the size of capacitors, use more capacitor banks, or add regulators. Now, look for locations suitable for switched banks. Again, use the average reactive line flows for the time when the capacitor is on (with the already-placed fixed capacitors in the circuit model). When applying switched capacitors, check the light-load case for possible overvoltages, and check the peak-load case for undervoltages.

6.6 Local Controls

Several options for controls are available for capacitor banks:

- *Time clock* — The simplest scheme: the controller switches capacitors on and off based on the time of day. The on time and the off time are programmable. Modern controllers allow settings for weekends and holidays. This control is the cheapest but also the most susceptible to energizing the capacitor at the wrong time (due to loads being different from those expected, to holidays or other unexpected light periods, and especially to mistakenly set or inaccurate clocks). Time clock control is predictable; capacitors switch on and off at known times and the controller limits the number of switching operations (one energization and one deenergization per day).
- *Temperature* — Another simple control; the controller switches the capacitor bank on or off depending on temperature. Normally these might be set to turn the capacitors on in the range of 85 and 90°F and turn them off at temperatures somewhere between 75 and 80°F.
- *Voltage* — The capacitor switches on and off, based on voltage. The user provides the threshold minimum and maximum voltages as well as time delays and bandwidths to prevent excessive operations. Voltage control is most appropriate when the primary role of a capacitor is voltage support and regulation.
- *Vars* — The capacitor uses var measurements to determine switching. This is the most accurate method of ensuring that the capacitor is on at the appropriate times for maximum reduction of losses.
- *Power factor* — Similar to var control, the controller switches capacitors on and off based on the measured power factor. This is rarely used by utilities.
- *Current* — The capacitor switches on and off based on the line current (as measured downstream of the capacitor). While not as effective as var control, current control does engage the capacitor during heavy loads, which usually corresponds to the highest needs for vars.

Many controllers offer many or all of these possibilities. Many are usable in combination; turn capacitors on for low voltage or for high temperature.

Var, power factor, voltage, or current controllers require voltage or current sensing or both. To minimize cost and complexity, controllers often switch all three phases using sensors on just one phase. A control power transformer is often also used to sense voltage. While unusual, Alabama Power switches each phase independently depending on the var requirements of each phase (Clark, 2001); this optimizes loss reduction and helps reduce unbalance. Because capacitor structures are rather busy, some utilities like to use voltage and/or current-sensing insulators. Meter-grade accuracy is not needed for controlling capacitors.

To coordinate more than one capacitor with switched var controls, set the most-distant unit to have the shortest time delay. Increase the time delay on successive units progressing back to the substation. This leaves the unit closest to the substation with the longest time delay. The most distant unit switches first. Upstream units see the change and do not need to respond. This strategy is the opposite of that used for coordinating multiple line voltage regulators.

For var-controlled banks, locate the current sensor on the source (substation) side of the bank. Then, the controller can detect the reactive power change when the capacitor switches. To properly calculate vars, the wiring for the CT and PT must provide correct polarities to the controller.

One manufacturer provides the following rules of thumb for setting var control trip and close settings (Fisher Pierce, 2000):

- Close setpoint: 2/3 × capacitor bank size (in kvar), lagging.
- Trip setpoint: Close set point – 1.25 × bank size, will be leading. (This assumes that the CT is on the source side of the bank.)

For a 600-kvar bank application, this yields

Close setpoint: 2/3 × 600 = +400 kvar (lagging)

Trip setpoint: 400 – 1.25 × 600 = –350 kvar (leading)

For this example, the unit trips when the load kvar drops below +250 kvar (lagging). This effectively gives a bandwidth wide enough (+400 to +250 kvar) to prevent excessive switching operations in most cases.

Voltage-controlled capacitor banks have bandwidths. Normally, we want the bandwidth to be at least 1.5 times the expected voltage change due to the capacitor bank. Ensure that the bandwidth is at least 3 or 4 V (on a 120-V scale). Set the trip setting below the normal light-load voltage (or the bank will never switch off).

If a switched capacitor is located on a circuit that can be operated from either direction, make sure the controller mode can handle operation with power flow in either direction. Time-of-day, temperature, and voltage control are not affected by reverse power flow; var, current, and power factor control are affected. Some controllers can sense reverse power and shift control modes. One model provides several options if it detects reverse power: switch to voltage mode, calculate var control while accounting for the effect of the capacitor bank, inhibit switching, trip and lock out the bank, or close and hold the bank in. If a circuit has distributed generation, we do not want to shift modes based on reverse power flow; the controller should shift modes only for a change in direction to the system source.

Capacitor controllers normally have counters to record the number of operations. The counters help identify when to perform maintenance and can identify control-setting problems. For installations that are excessively switching, modify control settings, time delays, or bandwidths to reduce switching. Some controllers can limit the number of switch operations within a given time period to reduce wear on capacitor switches.

Voltage control provides extra safety to prevent capacitors from causing overvoltages. Some controllers offer types of voltage override control; the primary control may be current, vars, temperature, or time of day, but the controller trips the bank if it detects excessive voltage. A controller may also restrain from switching in if the extra voltage rise from the bank would push the voltage above a given limit.

6.7 Automated Controls

Riding the tide of lower-cost wireless communication technologies, many utilities have automated capacitor banks. Many of the cost reductions and feature improvements in communication systems have resulted from the proliferation of cellular phones, pagers, and other wireless technologies used by consumers and by industry. Controlling capacitors requires little bandwidth, so high-speed connections are unnecessary. This technology changes quickly. The most common communications systems for distribution line capacitors are 900-MHz radio systems, pager systems, cellular phone systems, cellular telemetric systems, and VHF radio. Some of the common features of each are

- *900-MHz radio* — Very common. Several spread-spectrum data radios are available that cover 902–928 MHz applications. A private network requires an infrastructure of transmission towers.
- *Pager systems* — Mostly one-way, but some two-way, communications. Pagers offer inexpensive communications options, especially for infrequent usage. One-way communication coverage is widespread; two-way coverage is more limited (clustered around major cities). Many of the commercial paging networks are suitable for capacitor switching applications.
- *Cellular phone systems* — These use one of the cellular networks to provide two-way communications. Many vendors offer cellular modems for use with several cellular networks. Coverage is typically very good.
- *Cellular telemetric systems* — These use the unused data component of cellular signals that are licensed on existing cellular networks. They allow only very small messages to be sent — enough, though,

to perform basic capacitor automation needs. Coverage is typically very good, the same as regular cellular coverage.

- *VHF radio* — Inexpensive one-way communications are possible with VHF radio communication. VHF radio bands are available for telemetry uses such as this. Another option is a simulcast FM signal that uses extra bandwidth available in the commercial FM band.

Standard communication protocols help ease the building of automated infrastructures. Equipment and databases are more easily interfaced with standard protocols. Common communication protocols used today for SCADA applications and utility control systems include DNP3, IEC 870, and Modbus.

DNP 3.0 (Distributed Network Protocol) is the most widely used standard protocol for capacitor controllers (DNP Users Group, 2000). It originated in the electric industry in America with Harris Distributed Automation Products and was based on drafts of the IEC870-5 SCADA protocol standards (now known as IEC 60870-5). DNP supports master–slave and peer-to-peer communication architectures. The protocol allows extensions while still providing interoperability. Data objects can be added to the protocol without affecting the way that devices interoperate. DNP3 was designed for transmitting data acquisition information and control commands from one computer to another. (It is not a general purpose protocol for hypertext, multimedia, or huge files.)

One-way or two-way — we can remotely control capacitors either way. Two-way communication has several advantages:

- *Feedback* — A local controller can confirm that a capacitor switched on or off successfully. Utilities can use the feedback from two-way communications to dispatch crews to fix capacitor banks with blown fuses, stuck switches, misoperating controllers, or other problems.
- *Voltage/var information* — Local information on line var flows and line voltages allows the control to more optimally switch capacitor banks to reduce losses and keep voltages within limits.
- *Load flows* — Voltage, current, and power flow information from pole-mounted capacitor banks can be used to update and verify load-flow models of a system. The information can also help when tracking down customer voltage, stray voltage, or other power quality problems. Loading data helps utilities monitor load growth and plan for future upgrades. One utility even uses capacitor controllers to capture fault location information helping crews to locate faults.

When a controller only has one-way communications, a local voltage override control feature is often used. The controller blocks energizing a capacitor bank if doing so would push the voltage over limits set by the user.

Several schemes and combinations of schemes are used to control capacitors remotely:

- *Operator dispatch* — Most schemes allow operators to dispatch distribution capacitors. This feature is one of the key reasons utilities automate capacitor banks. Operators can dispatch distribution capacitors just like large station banks. If vars are needed for transmission support, large numbers of distribution banks can be switched on. This control scheme is usually used in conjunction with other controls.
- *Time scheduling* — Capacitors can be remotely switched, based on the time of day and possibly the season or temperature. While this may seem like an expensive time control, it still allows operators to override the schedule and dispatch vars as needed.
- *Substation var measurement* — A common way to control feeder capacitors is to dispatch based on var/power factor measurements in the substation. If a feeder has three capacitor banks, they are switched on or off in some specified order based on the power factor on the feeder measured in the substation.
- *Others* — More advanced (and complicated) algorithms can dispatch capacitors based on a combination of local var measurements and voltage measurements along with substation var measurements.

6.8 Reliability

Several problems contribute to the overall reliability or unreliability of capacitor banks. In a detailed analysis of Kansas City Power & Light's automated capacitor banks, Goeckeler (1999) reported that blown fuses are KCP&L's biggest problem, but several other problems exist (Table 6.7). Their automa-

TABLE 6.7

Maintenance Needs Identified by Kansas City Power & Light's Capacitor Automation System Based on Two Years of Data

Problem	Annual Percent Failures
Primary fuse to capacitor blown (nuisance fuse operation)	9.1
Failed oil switches	8.1
Hardware accidentally set to "Local" or "Manual"	4.2
Defective capacitor unit	3.5
Miscellaneous	2.4
Control power transformer	1.5
TOTAL	28.8

Source: Goeckeler, C., "Progressive Capacitor Automation Yields Economic and Practical Benefits at KCPL," *Utility Automation*, October 1999.

tion with two-way communications allowed them to readily identify bank failures. The failure rates in Table 6.7 are high, much higher than most distribution equipment. Capacitor banks are complicated; they have a lot of equipment to fail; yet, failure rates should be significantly better than this.

An EPRI survey on capacitor reliability found wide differences in utilities' experience with capacitors (EPRI 1001691, 2002). Roughly one-third of survey responses found feeder capacitors "very good," another one-third found them "typical of line equipment," and the final third found them "problematic." The survey along with follow-up contacts highlighted several issues:

- *Misoperation of capacitor fuses* — Many utilities have operations of fuses where the capacitor bank is unharmed. This can unbalance circuit voltages and reduce the number of capacitors available for var support. Review fusing practices to reduce this problem.
- *Controllers* — Controllers were found "problematic" by a significant number of utilities. Some utilities had problems with switches and with the controllers themselves.
- *Lightning and faults* — In high-lightning areas, controllers can fail from lightning. Controllers are quite exposed to lightning and power-supply overvoltages during faults. Review surge protection practices and powering and grounding of controllers.
- *Human element* — Many controllers are set up incorrectly. Some controllers are hard to program. And, field crews often do not have the skills or proper attitudes toward capacitors and their controls. At some utilities, crews often manually switch off nearby capacitors (and often forget to turn them back on after finishing their work). To reduce these problems, properly train crews and drive home the need to have capacitors available when needed.

6.9 Failure Modes and Case Ruptures

Capacitors can fail in two modes:

- *Low current, progressive failure* — The dielectric fails in one of the elements within the capacitor (see Figure 6.11). With one element shorted, the remaining elements in the series string have increased voltage and higher current (because the total capacitive impedance is lower). With more stress, another element may short out. Failures can cascade until the whole string shorts out. In this scenario, the current builds up slowly as elements successively fail.

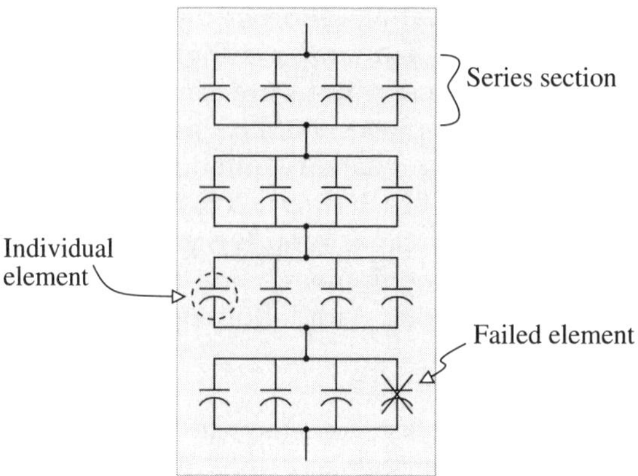

FIGURE 6.11
Capacitor unit with a failed element.

- *High current* — A low-impedance failure develops across the capacitor terminals or from a phase terminal to ground. A broken connector could cause such a fault.

Most failures are progressive. Sudden jumps to high current are rare. To detect progressive failures quickly, fusing must be very sensitive. Film-foil capacitors have few case ruptures — much less than older paper units. An EPRI survey of utilities (EPRI 1001691, 2002) found that film-foil capacitor ruptures were rare to nonexistent. This contrasts sharply with paper capacitors, where Newcomb (1980) reported that film/paper capacitors ruptured in 25% of failures.

Paper and paper-film capacitors have an insulating layer of paper between sheets of foil. When a breakdown in a pack occurs, the arc burns the paper and generates gas. In progressive failures, even though the current is only somewhat higher than normal load current, the sustained arcing can create enough gas to rupture the enclosure. Before 1975, capacitors predominantly used polychlorinated biphenyls (PCB) as the insulating liquid. Environmental regulations on PCB greatly increased the costs of cleanup if these units ruptured (US Environmental Protection Agency 40 CFR Part 761 Polychlorinated Biphenyls (PCBs) Manufacturing, Processing, Distribution in Commerce, and Use Prohibitions). The environmental issues and safety concerns led utilities to tighten up capacitor fusing.

In modern film-foil capacitors, sheets of polypropylene film dielectric separate layers of aluminum foil. When the dielectric breaks down, the heat from the arc melts the film; the film draws back; and the aluminum sheets weld together. With a solid weld, a single element can fail and not create any gas (the current is still relatively low). In film-foil capacitors, the progressive failure mode is much less likely to rupture the case. When all of the packs in series fail, high current flows through the capacitor. This can generate enough heat and gas to rupture the capacitor if it is not cleared quickly.

Figure 6.12 shows capacitor-rupture curves from several sources. Most case-rupture curves are based on tests of prefailed capacitors. The capacitors are failed by applying excessive voltage until the whole capacitor is broken down. The failed capacitor is then subjected to a high-current short-circuit

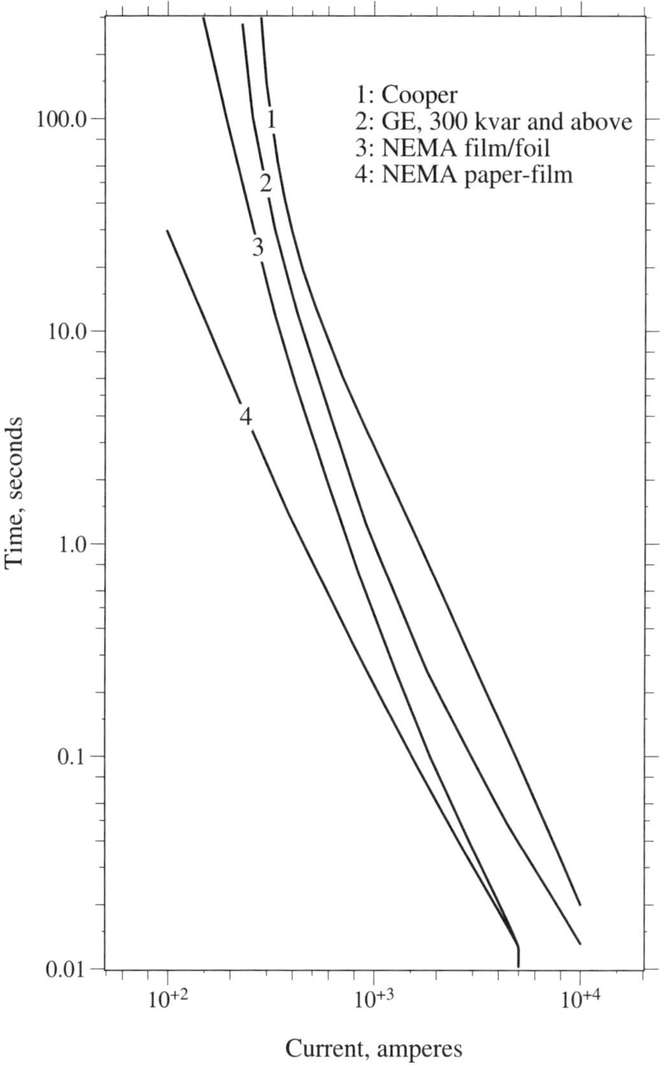

FIGURE 6.12
Capacitor rupture curves. (Data from [ANSI/IEEE Std. 18-1992; Cooper Power Systems, 1990; General Electric, 2001].)

source of known amperage for a given time. Several such samples are tested to develop a case-rupture curve.

The case-rupture curves do not represent all failure modes. Such curves do not show the performance during the most common failures: low-current and progressive element failures (before all elements are punctured). Although, thankfully, rare, high-current faults more severe than those tested for the rupture curves are possible. An arc through the insulating dielectric fluid can generate considerable pressure. Pratt et al. (1977) performed tests on film/foil capacitor units with arc lengths up to 3 in. (7.6 cm) in length. They chose 3 in. as the maximum realistic arc length in a capacitor as the gap spacing between internal series section terminals. Under these conditions, they damaged or ruptured several units for currents and times well below the capacitor rupture curves in Figure 6.12.

Also consider other equipment at a capacitor bank installation. Capacitor switches, especially oil switches, are vulnerable to violent failure. This type of failure has not received nearly the attention that capacitor ruptures or distribution transformer failures have. Potential transformers, current transformers, controller power-supply transformers, and arresters: these too can fail violently. Any failure in which an arc develops inside a small enclosure can rupture or explode. In areas with high fault current, consider applying current-limiting fuses. These will help protect against violent failures of capacitor units, switches, and other accessories in areas with high fault current.

When one element fails and shorts out, the other series sections have higher voltage, and they draw more current. Capacitor packs are designed with a polypropylene film layer less than one mil thick (0.001 in. or 0.025 mm), which is designed to hold a voltage of 2000 V. Table 6.8 shows the number of series sections for several capacitors as reported by Thomas (1990). More recent designs could have even fewer groups. One manufacturer uses three series sections for 7.2 to 7.96 kV units and six series sections for 12.47 to 14.4 kV units. As series sections fail, the remaining elements must hold increasing voltage, and the capacitor draws more current in the same proportion. Figure

TABLE 6.8

Number of Series Sections in Different Voltage Ratings

Unit Voltage, V	Manufacturer A	Manufacturer B	Manufacturer C
2,400	2	2	2
7,200	4	4	4
7,620	5	5	4
13,280	8	8	7
13,800	8	8	—
14,400	8	8	8

Source: Thomas, E. S., "Determination of Neutral Trip Settings for Distribution Capacitor Banks," IEEE Rural Electric Power Conference, 1990. With permission. ©1990 IEEE.

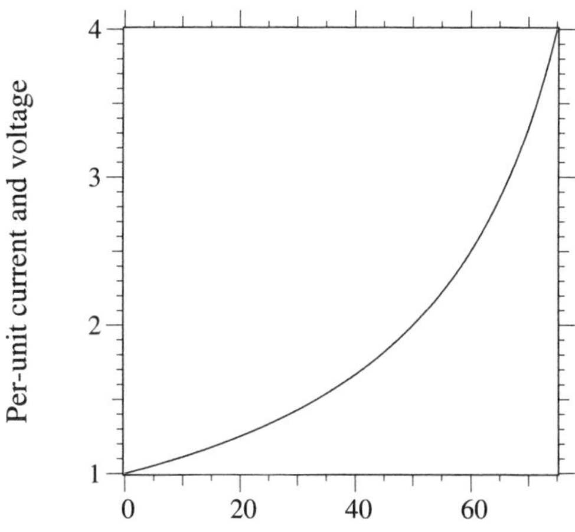

FIGURE 6.13
Per-unit current drawn by a failing bank depending on the portion of the bank that is failed (assuming an infinite bus). This is also the per-unit voltage applied on the series sections still remaining.

6.13 shows the effect on the per-unit current drawn by a failing unit and the per-unit voltage on the remaining series sections.

If a capacitor bank has multiple units on one phase and all units are protected by one fuse (group fusing), the total bank current should be considered. Consider a bank with two capacitor units. If one unit loses half of its series sections, that unit will draw twice its nominal current. The group — the two units together — will draw 1.5 times the nominal bank load. (This is the current that the fuse sees.)

6.10 Fusing and Protection

The main purpose of the fuse on a capacitor bank is to clear a fault if a capacitor unit or any of the accessories fail. The fuse must clear the fault quickly to prevent any of the equipment from failing violently. Ruptures of capacitors have historically been problematic, so fusing is normally tight. Fuses must be sized to withstand normal currents, including harmonics.

A significant number of utilities have problems with nuisance fuse operations on capacitor banks. A fuse is blown, but the capacitors themselves are still functional. These blown fuses may stay on the system for quite some time before they are noticed (see Figure 6.14). Capacitors with blown fuses increase voltage unbalance, can increase stray voltages, and increase losses. Even if the capacitor controller identifies blown fuses, replacement adds extra maintenance that crews must do.

FIGURE 6.14
Capacitor bank with a blown fuse. (Copyright © 2002. Electric Power Research Institute. 1001691. *Improved Reliability of Switched Capacitor Banks and Capacitor Technology.* Reprinted with permission.)

IEEE guides suggest selecting a fuse capable of handling 1.25 to 1.35 times the nominal capacitor current (IEEE Std. C37.48-1997); a 1.35 factor is most common. Three factors can contribute to higher than expected current:

- *Overvoltage* — Capacitive current increases linearly with voltage, and the reactive vars increase as the square of the voltage. When estimating maximum currents, an upper voltage limit of 110% is normally assumed.
- *Harmonics* — Capacitors can act as a sink for harmonics. This can increase the peak and the rms of the current through the capacitor. Additionally, grounded three-phase banks absorb zero-sequence harmonics from the system.
- *Capacitor tolerance* — Capacitors were allowed to have a tolerance to +15% above their rating (which would increase the current by 15%).

Most fusing practices are based on fusing as tightly as possible to prevent case rupture. So, the overload capability of fuse links is included in fuse sizing. This effectively allows a tighter fusing ratio. K and T tin links can be overloaded to 150%, so for these links with a 1.35 safety factor, the smallest size fuse that can be used is

$$I_{min} = \frac{1.35I_1}{1.5} = 0.9I_1$$

where

I_{min} = minimum fuse rating, A

I_1 = capacitor bank current, A

Table 6.9 shows one manufacturer's recommendations based on this tight-fusing approach.

With this tight-fusing strategy, fuses must be used consistently. If silver links are used instead of tin links, the silver fuses can blow from expected levels of current because silver links have no overload capability.

Prior to the 1970s, a fusing factor of 1.65 was more common. Due to concerns about case ruptures and PCBs,* the industry went to tighter fusing factors, 1.35 being the most common. Because of the good performance of all-film capacitors and problems with nuisance fuse operations, consider a

TABLE 6.9

Fusing Recommendations for ANSI Tin Links from One Manufacturer

3-Phase Bank	**System Line-to-Line Voltage, kV**							
kvar	**4.2**	**4.8**	**12.5**	**13.2**	**13.8**	**22.9**	**24.9**	**34.5**
Recommended Fuse Link								
150	20T	20T	8T	6T	6T			
300	40K	40K	15T	12T	12T	8T	8T	5T
450	65K	50K	20T	20T	20T	10T	10T	8T
600	80K	65K	25T	25T	25T	15T	15T	10T
900		100K	40K	40K	40K	20T	20T	15T
1200			50K	50K	50K	30T	25T	20T
1800			80K	80K	80K	40K	40K	30K
2400			100K	100K	100K	65K	50K	40K
Fusing Ratio for the Recommended Link (Link Rating/Nominal Current)								
150	0.96	1.11	1.15	0.91	0.96			
300	0.96	1.11	1.08	0.91	0.96	1.06	1.15	1.00
450	1.04	0.92	0.96	1.02	1.06	0.88	0.96	1.06
600	0.96	0.90	0.90	0.95	1.00	0.99	1.08	1.00
900		0.92	0.96	1.02	1.06	0.88	0.96	1.00
1200			0.90	0.95	1.00	0.99	0.90	1.00
1800			0.96	1.02	1.06	0.88	0.96	1.00
2400			0.90	0.95	1.00	1.07	0.90	1.00

Note: This is not the manufacturer's most up-to-date fusing recommendation. It is provided mainly as an example of a commonly applied fusing criteria for capacitors.

Source: Cooper Power Systems, *Electrical Distribution — System Protection*, 3rd ed., 1990.

looser fusing factor, possibly returning to the 1.65 factor. Slower fuses should also have fewer nuisance fuse operations.

Capacitors are rated to withstand 180% of rated rms current, including fundamental and harmonic currents. Fusing is normally not based on this limit, and is normally much tighter than this, usually from 125 to 165% of rated rms current. Occasionally, fuses in excess of 180% are used. In severe harmonic environments (usually in commercial or industrial applications), normally fuses blow before capacitors fail, but sometimes capacitors fail before the fuse operates. This depends on the fusing strategy.

If a capacitor bank has a blown fuse, crews should test the capacitors before re-fusing. A handheld digital capacitance meter is the most common approach and is accurate. Good multimeters also can measure a capacitance high enough to measure the capacitance on medium-voltage units. There is a chance that capacitance-testers may miss some internal failures requiring high voltage to break down the insulation at the failure. Measuring the capacitance on all three phases helps identify units that may have partial failures. Partial failures show up as a change in capacitance. In a partial failure, one of several series capacitor packs short out; the remaining packs appear as a lower impedance (higher capacitance). As with any equipment about to be energized, crews should visually check the condition of the capacitor unit and make sure there are no bulges, burn marks, or other signs that the unit may have suffered damage.

Some utilities have problems with nuisance fuse operations on distribution transformers. Some of the causes of capacitor fuse operations could be the same as transformer fuse operations, but some differences are apparent:

- Capacitor fuses see almost continuous full load (when the capacitor is switched in).
- Capacitor fuses tend to be bigger. The most common transformer sizes are 25 and 50 kVA, usually with less than a 15 A fuse. Typical capacitor sizes are 300 to 1200 kvar with 15 to 65 A fuses.
- Both have inrush; a capacitor's is quicker.
- Transformers have secondary faults and core saturation that can contribute to nuisance fuse operations; capacitors have neither.

Some possible causes of nuisance fuse operations are

- *Lightning* — Capacitors are a low impedance to the high-frequency lightning surge, so they naturally attract lightning current, which can blow the fuse. Smaller, faster fuses are most prone to lightning. Given that the standard rule of thumb that a fuse at least as big as a 20K or a 15T should prevent nuisance operations, it is hard to see how lightning itself could cause a significant number of fuse operations (as most capacitor bank fuses are larger than this).

- *Outrush to nearby faults* — If a capacitor dumps its stored charge into a nearby fault, the fuse can blow. Capacitor banks also have inrush every time they are switched in, but this is well below the melt point of the fuse.
- *Severe harmonics* — Harmonics increase the current through the fuse.
- *Animal or other bushing faults* — A fault across a bushing due to an animal, contamination on the bushing, or tree contact can blow a fuse. By the time anyone notices the blown fuse, the squirrel or branch has disappeared. Use animal guards and covered jumpers to reduce these.
- *Mechanical damage and deterioration* — Corrosion and vibration can weaken fuse links. On fuse links collected from the field on transformers, Ontario Hydro found that 3% had broken strain wires (CEA 288 D 747, 1998). Another 15% had braids that were brittle and had broken strands. Larger fuses used in capacitors should not have as much of a problem.
- *Installation errors* — Fuses are more likely to blow if crews put in the wrong size fuse or wrong type fuse or do not properly tighten the braid on the fuse.

Outrush is highlighted as a possible failure mode that has been neglected by the industry. Outrush is sometimes considered for station banks to calculate the probability of a fuse operation from a failure of an adjacent parallel unit. But for distribution fuses, nearby faults have not been considered in regard to the effects on fuse operations.

The energy input into the fuse during outrush depends on the line resistance between the capacitor and the fault (see Figure 6.15). The capacitor has stored energy; when the fault occurs, the capacitor discharges its energy into the resistance between the capacitor and the fault. Closer faults cause more energy to go into the fuse. The I^2t that the fuse suffers during outrush to a line-to-ground fault is

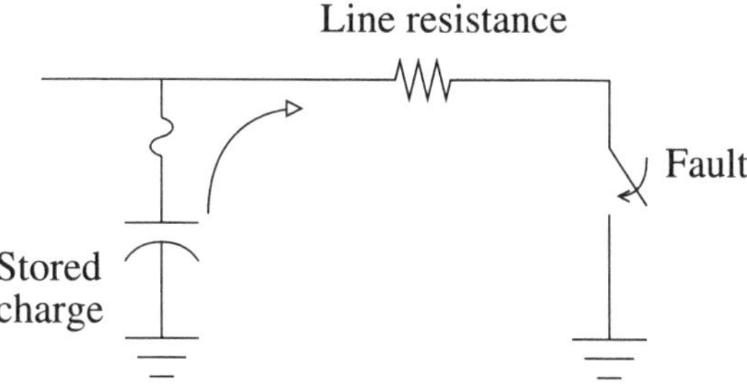

FIGURE 6.15
Outrush from a capacitor to a nearby fault.

$$I^2t = \frac{\frac{1}{2}CV_{pk}^2}{R} = \frac{2.65Q_{kvar}}{R}V_{pu}^2$$

where

C = capacitance of one unit, μF
V_{pk} = peak voltage on the capacitor at the instant of the fault, kV
R = resistance between the capacitor and the fault, Ω
Q_{kvar} = single-phase reactive power, kvar
V_{pu} = voltage at the instant of the fault in per unit of the capacitor's rated voltage

Table 6.10 shows several sources of fuse operations and the I^2t that they generate for a 900-kvar bank at 12.47 kV. The nominal load current is 41.7 A. Utilities commonly use 40 or 50-A fuses for this bank. The table shows the minimum melt I^2t of common fuses. Outrush to nearby faults produces high enough energy to blow common fuses, especially the K links. Of the other possible causes of fuse operation, none are particularly high except for a lightning first stroke. The lightning data is misleading because much of the first stroke will go elsewhere — usually, the line flashes over, and much of the lightning current diverts to the fault.

Use Figure 6.16 to find outrush I^2t for other cases. Two factors make outrush worse:

- *Higher system voltages* — The outrush I^2t stays the same with increases in voltage for the same size capacitor bank. The line impedance stays the same for different voltages. But higher-voltage capac-

TABLE 6.10

Comparison of I^2t of Events that Might Blow a Fuse to the Capability of Common Fuses for a Three-Phase, 900-kvar Bank at 12.47 kV (I_{load} = 41.7 A)

Source	I^2t, A²-sec
Lightning, median 1st stroke	57,000
Lightning, median subsequent stroke	5,500
Inrush at nominal voltage (I_{SC}=5 kA, X/R=8)	4,455
Inrush at 105% voltage	4,911
Outrush to a fault 500-ft away (500-kcmil AAC)	20,280
Outrush to a fault 250-ft away (500-kcmil AAC)	40,560
Outrush to a fault 250-ft away with an arc restrike[a]	162,240
40K fuse, minimum melt I^2t	36,200
50K fuse, minimum melt I^2t	58,700
40T fuse, minimum melt I^2t	107,000

[a] Assumes that the arc transient leaves a voltage of 2 per unit on the capacitor before the arc restrikes.

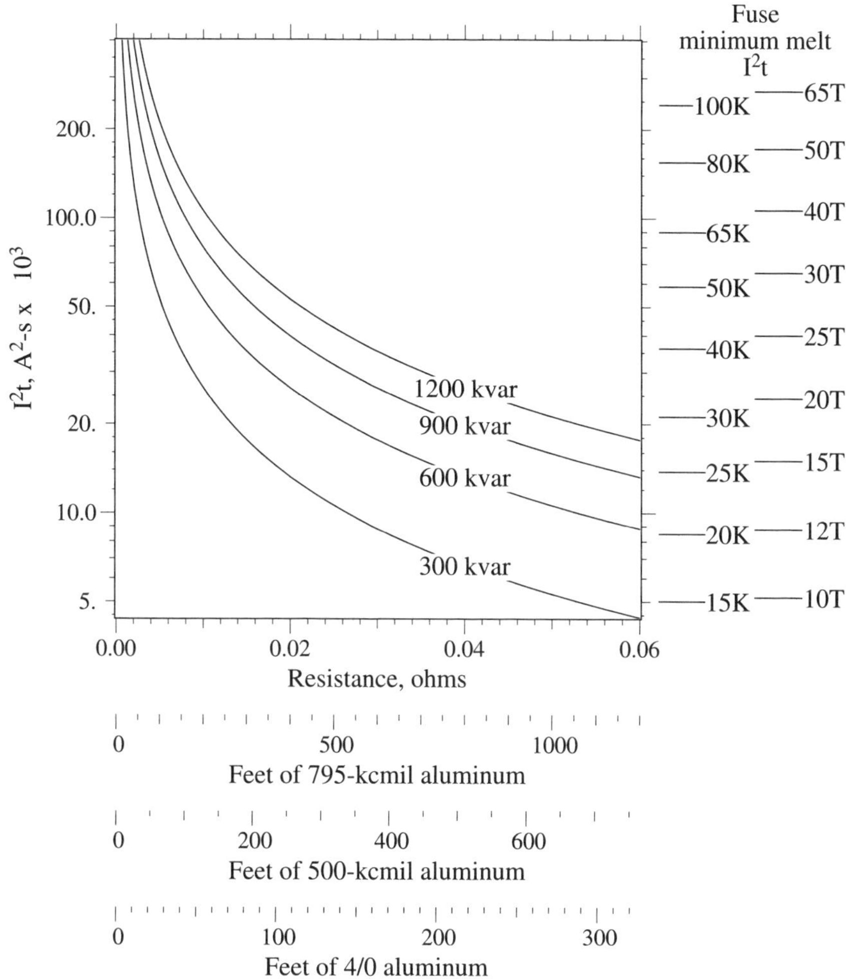

FIGURE 6.16
Outrush as a function of the resistance to the fault for various size capacitor banks (the sizes given are three-phase kvar; the resistance is the resistance around the loop, out and back; the distances are to the fault).

itor banks use smaller fuses, with less I^2t capability. So, a 25-kV capacitor installation is more likely to have nuisance fuse operations than a 12.5-kV system.

- *Larger conductors* — Lower resistance.

Consider a 1200-kvar bank with 500-kcmil conductors. At 12.47 kV (I_{load} = 55.6 A) with a 65K fuse, the fuse exceeds its minimum melt I^2t for faults up to 150 ft away. At 24.94 kV (I_{load} = 27.8 A) with a 30K fuse, the fuse may melt for faults up to 650 ft away. At 34.5 kV (I_{load} = 20.1 A) with a 25 K fuse, the location is off of the chart (it is about 950 ft). Note that the distance scales in Figure 6.16 do not include two important resistances: the capacitor's internal resistance and the fuse's resistance. Both will help reduce the I^2t. Also, the minimum melt I^2t values of the fuses in Figure 6.16 are the 60-Hz values. For high-frequency currents like an outrush discharge, the minimum melt I^2t of expulsion fuses is 30 to 70% of the 60-Hz I^2t (Burrage, 1981).

As an estimate of how much outrush contributes to nuisance fuse operations, consider a 900-kvar bank at 12.47 kV with 40K fuses. We will estimate that the fuse may blow or be severely damaged for faults within 250 ft (76 m). Using a typical fault rate on distribution lines of 90 faults/100 mi/year (56 faults/100 km/year), faults within 250 ft (75 m) of a capacitor occur at the rate of 0.085 per year. This translates into 8.5% fuse operations per capacitor bank per year, a substantial number.

The stored energy on the fault depends on the timing of the fault relative to the point on the voltage wave. Unfortunately, most faults occur at or near the peak of the sinusoid.

Several system scenarios could make individual instances worse; most are situations that leave more than normal voltage on the capacitor before it discharges into the fault:

- *Regulation overvoltages* — Voltages above nominal increase the outrush energy by the voltage squared.
- *Voltage swells* — If a line-to-ground fault on one phase causes a voltage swell on another and the fault jumps to the "swelled" phase, higher-than-normal outrush flows through the fuse.
- *Arc restrikes* — If a nearby arc is not solid but sputters, arc restrikes, much like restrikes of switches, can impress more voltage on the capacitor and subject the fuse to more energy, possibly much larger voltage depending on the severity. (I know of no evidence that this occurs regularly; most arcs are solid, and the system stays faulted once the arc bridges the gap.)
- *Lightning* — A nearby lightning strike to the line can charge up the capacitor (and start the fuse heating). In most cases, the lightning will cause a nearby flashover, and the capacitor's charge will dump right back through the fuse.
- *Multiple-phase faults* — Line-to-line and three-phase faults are more severe for two reasons: the voltage is higher, and the resistance is lower. For example, on a line-to-line fault, the voltage is the line-to-line voltage, and the resistance is the resistance of the phase wires (rather than the resistance of a phase wire and the neutral in series).

These estimates are conservative in that they do not consider skin effects, which have considerable effect at high frequencies. Skin effects increase the conductor's resistance. The transients oscillate in the single-digit kilohertz range. At these frequencies, conductor resistance increases by a factor of two to three. On the negative side, the fuse element is impacted by skin effects, too — higher frequency transients cause the fuse to melt more quickly.

Capacitors also have inrush every time they are energized. Inrush into grounded banks has a peak current (IEEE Std. 1036-1992) of

$$I_{pk} = 1.41\sqrt{I_{SC} I_1}$$

where

I_{pk} = peak value of inrush current, A
I_{SC} = available three-phase fault current, A
I_1 = capacitor bank current, A

The energy into a fuse from inrush is normally very small. It subjects the capacitor fuse to an I^2t (in A^2-s) (Brown, 1979) of

$$I^2t = 2.65\sqrt{1+k^2}\, I_{SC} I_1 \,/\, 1000$$

where

$k = X/R$ ratio at the bank location

Inrush is much worse if a capacitor is switching into a system with a nearby capacitor. The outrush from the already-energized bank dumps into the capacitor coming on line. Fuses at both banks see this transient. In substation applications, this *back-to-back* switching is a major design consideration, often requiring insertion of reactors between banks. For distribution feeder capacitors, the design constraints are not as large. A few hundred feet of separation is enough to prevent inrush/outrush problems. For back-to-back switching, the I^2t is almost the same as that for outrush:

$$I^2t = \frac{\frac{1}{2} C V_{pk}^2}{R} = \frac{2.65 Q_{kvar}}{R} V_{pu}^2$$

The only difference is that the capacitance is the series combination of the two capacitances: $C=C_1C_2/(C_1+C_2)$, and $Q_{kvar}=Q_1Q_2/(Q_1+Q_2)$. For the same size banks, $C=C_1/2$, and $Q_{kvar}=Q_1/2$. Figure 6.16 applies if we double the kvar values on the curves. In most situations, maintaining a separation of 500 ft between capacitor banks prevents fuse operations from this inrush/outrush. Separate capacitor banks by 500 ft (150 m) on 15-kV class circuits to avoid inrush problems. Large capacitor banks on higher voltage distribution systems may require modestly larger separations.

Preventing case ruptures is a primary goal of fusing. The fuse should clear before capacitor cases fail. Figure 6.17 shows capacitor rupture curves compared against fuse clearing curves. The graph shows that there is considerable margin between fuse curves and rupture curves. Consider a 12.47-kV, 900-kvar bank of three 300-kvar units, which has a nominal current of 41.7 A. Utilities commonly use a 40 or 50 K fuse for this bank. Larger fuses for this bank are possible, while still maintaining levels below case rupture curves. An EPRI survey found that case ruptures on modern film-foil capacitors are rare (EPRI 1001691, 2002). This gives us confidence that we can loosen fusing practices without having rupture problems.

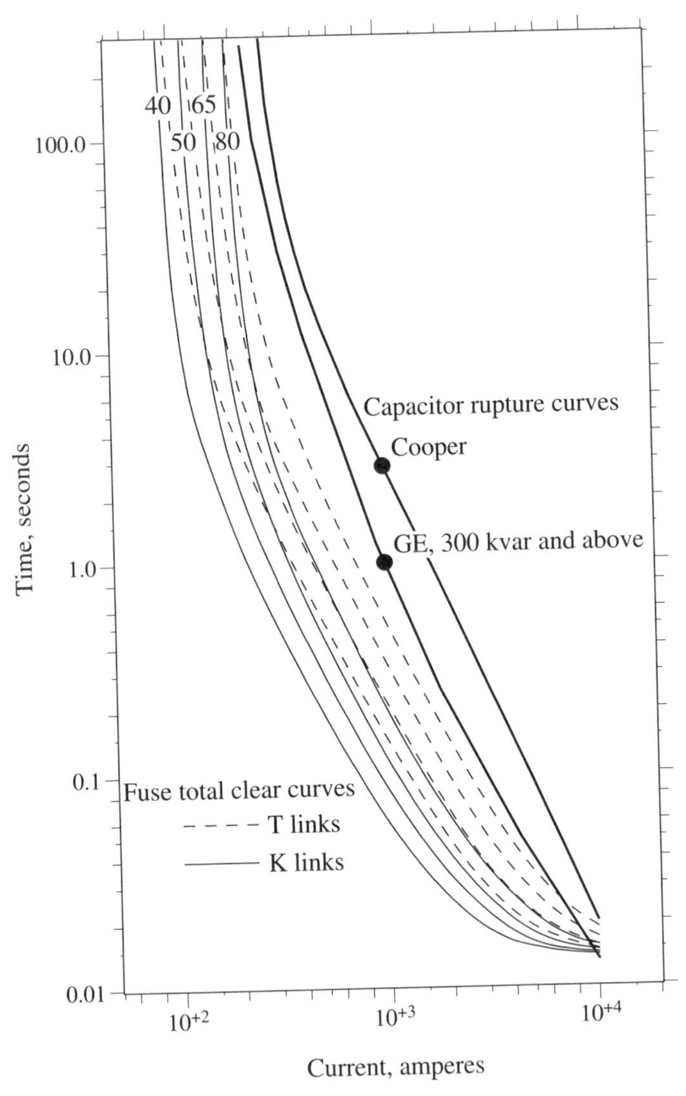

FIGURE 6.17
Fuse curves with capacitor rupture curves.

In areas of high fault current, current-limiting fuses provide extra safety. Either a backup current-limiting fuse in series with an expulsion link or a full-range current-limiting fuse is an appropriate protection scheme in high fault-current areas. While it may seem that expulsion fuses provide adequate protection even to 8 kA (depending on which rupture curve we use), current-limiting fuses provide protection for those less frequent faults with longer internal arcs. They also provide protection against failures in the capacitor switches and other capacitor-bank accessories. Utilities that apply current-limiting fuses on capacitors normally do so for areas with fault currents above 3 to 5 kA.

With backup current-limiting fuses, it is important that crews check the backup fuse whenever the expulsion link operates. On transformers, crews can get away with replacing the expulsion link. If the transformer still does not have voltage, they will quickly know that they have to replace the backup link. But, on capacitors, there is no quick indication that the backup-fuse has operated. Crews must check the voltage on the cutout to see if the backup fuse is operational; or crews should check the capacitor neutral current after replacing the expulsion link to make sure it is close to zero (if all three phases are operational, the balanced currents cancel in the neutral). In addition to not fixing the problem, failing to replace a blown backup fuse could cause future problems. The backup fuse is not designed to hold system voltage continuously — they are not an insulator. Eventually, they will track and arc over.

Because of utility problems with nuisance fuse operations, some loosening of fusing practices is in order. For most of the possible causes of nuisance-fuse operations, increasing the fuse size will decrease the number of false operations. Going to a slower fuse, especially, helps with outrush and other fast transients. If you have nuisance fuse operation problems, consider using T links and/or increase the fuse size one or two sizes. Treat these recommendations as tentative; as of this writing, these fusing issues are the subject of ongoing EPRI research, which should provide more definitive recommendations.

Neutral monitoring (Figure 6.18) is another protection feature that some capacitor controllers offer. Neutral monitoring can detect several problems:

- *Blown fuse* — When one capacitor fuse blows, the neutral current jumps to a value equal to the phase current.
- *Failing capacitor unit* — As a capacitor fails, internal groups of series packs short out. Prior to complete failure, the unit will draw more current than normal. Figure 6.19 shows how the neutral current changes when a certain portion of the capacitor shorts out. Capacitors rated from 7.2 to 7.96 kV normally have three or four series sections, so failure of one element causes neutral currents of 25% (for four in series) or 34% (for three in series) of the phase current.

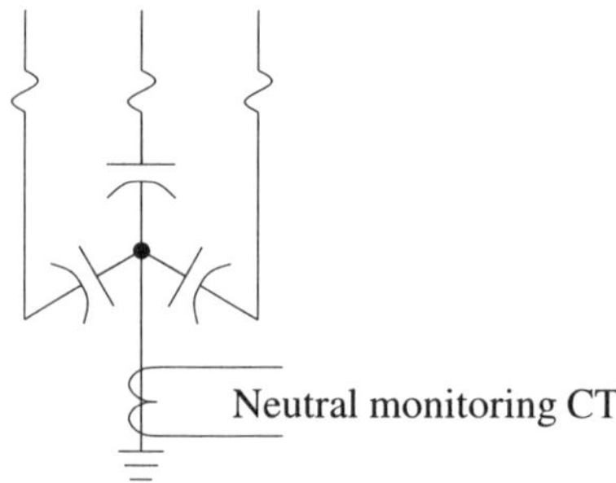

FIGURE 6.18
Neutral monitoring of a capacitor bank.

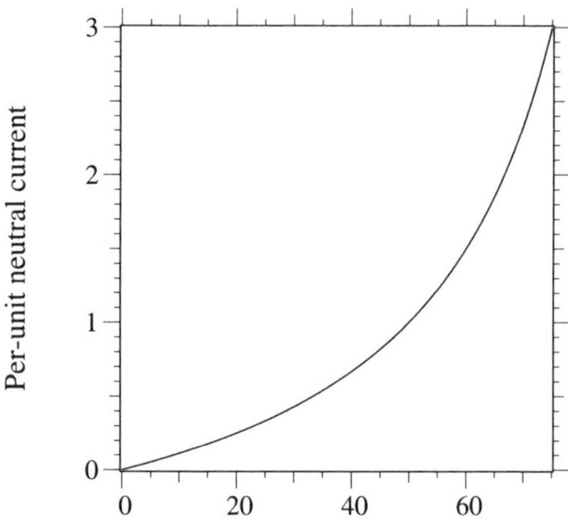

FIGURE 6.19
Neutral current drawn by a failing grounded-wye bank depending on the portion of the bank that is failed (the neutral current is in per-unit of the nominal capacitor current).

Failure of more than half of the series sections causes more than the capacitor's rated current in the neutral.

- *High harmonic current* — Excessive neutral current may also indicate high harmonic currents.

Neutral monitoring is common in substation banks, and many controllers for switched pole-mounted banks have neutral-monitoring capability. Neutral-current monitors for fixed banks are also available, either with a local warning light or a wireless link to a centralized location.

Neutral monitoring can help reduce operations and maintenance by eliminating regular capacitor patrols and field checks. Quicker replacement of blown fuses also reduces the time that excessive unbalance is present (and the extra losses and possibility of stray voltage). This can lead to more reliable var regulation, and even reduce the number of capacitor banks needed.

6.11 Grounding

Three-phase capacitors can be grounded in a wye configuration or ungrounded, either in a floating-wye or a delta. For multigrounded distribution systems, a grounded-wye capacitor bank offers advantages and disadvantages:

- *Unit failure and fault current* — If a unit fails, the faulted phase draws full fault current. This allows the fuse to blow quickly, but requires fuses to be rated for the full fault current.
- *Harmonics* — The grounded-wye bank can attract zero-sequence harmonics (balanced 3rd, 9th, 15th, ...). This problem is often found in telephone interference cases.

Advantages and disadvantages of the floating-wye, ungrounded banks include

- *Unit failure* — The collapse of voltage across a failed unit pulls the floating neutral to phase voltage. Now, the neutral shift stresses the remaining capacitors with line-to-line voltage, 173% of the capacitor's rating.
- *Fault current* — When one unit fails, the circuit does not draw full fault current — it is a high-impedance fault. This is an advantage in some capacitor applications.
- *Harmonics* — Less chance of harmonic problems because the ungrounded, zero-sequence harmonics (balanced 3rd, 9th, 15th, ...) cannot flow to ground through the capacitor.

The response of the floating-wye configuration deserves more analysis. During a progressive failure, when one series section shorts out, the shift of the neutral relieves the voltage stress on the remaining series sections. In the example in Figure 6.20, for a floating-wye bank with half of the series sections shorted, the line-to-neutral voltage becomes 0.75 per unit. The remaining elements normally see 50% of the line-to-neutral voltage, but now they see 75% (1.5 per unit, so the current is also 1.5 times normal). The reduction in voltage stress due to the neutral shift prolongs the failure — not what we want. The excess heating at the failure point increases the risk of gas generation and case rupture. When one element fails, we really want the fuse (or other protection) to trip quickly. The neutral shift also increases the voltage stress on the units on the other phases.

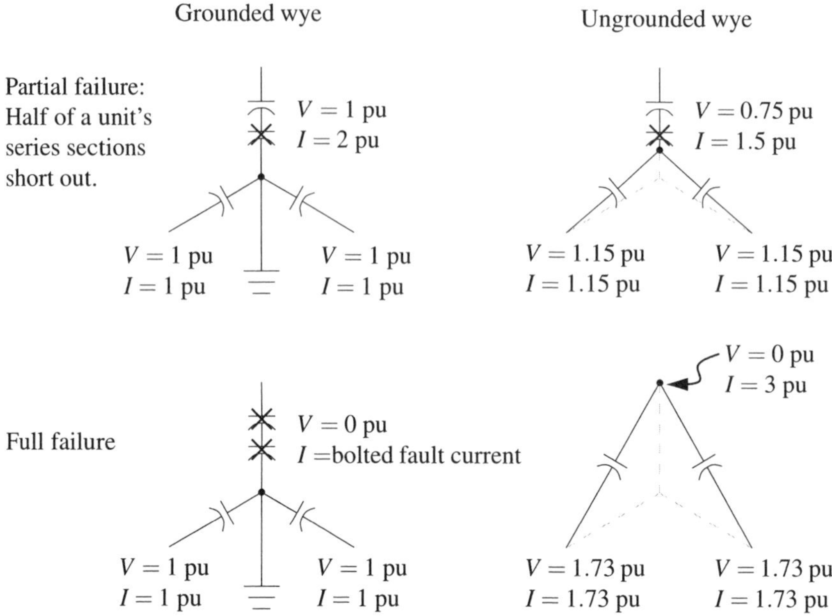

FIGURE 6.20
Comparison of grounded wye and ungrounded-wye banks during a partial and full failure of one unit. (Copyright © 2002. Electric Power Research Institute. 1001691. *Improved Reliability of Switched Capacitor Banks and Capacitor Technology*. Reprinted with permission.)

Floating-wye configurations are best applied with neutral detection — a potential transformer measuring voltage between the floating neutral and ground can detect a failure of one unit. When one unit fails, a relay monitoring the neutral PT should trip the capacitor's oil or vacuum switch (obviously, this only works on switched banks).

Standard utility practice is to ground banks on multigrounded systems. Over 80% of the respondents to an EPRI survey used grounded-wye capacitor connections (EPRI 1001691, 2002). On three-wire systems, utilities use both ungrounded-wye and delta configurations.

Most utilities use two-bushing capacitors, even though most also use a grounded neutral. Having two bushings allows crews to convert capacitor banks to a floating neutral configuration if telephone interference is a problem.

Utilities universally ground capacitor cases on pole-mounted capacitors (even though it is not strictly required by the National Electrical Safety Code [IEEE C2-1997]). In rare cases, banks with single-bushing capacitors are floated when it becomes necessary to convert a bank to a floating-wye. Avoid this if possible.

References

ANSI C57.12.28-1998, *Pad-Mounted Equipment Enclosure Integrity.*

ANSI/IEEE Std. 18-1992, IEEE *Standard for Shunt Power Capacitors.*

Brown, R. A., "Capacitor Fusing," IEEE/PES Transmission and Distribution Conference and Expo, April 1–6, 1979.

Burrage, L. M., "High Frequency Characteristics of Capacitors and Fuses — Applied in High-Voltage Shunt Banks," IEEE/PES Transmission and Distribution Conference, 1981.

Carlisle, J. C. and El-Keib, A. A., "A Graph Search Algorithm for Optimal Placement of Fixed and Switched Capacitors on Radial Distribution Systems," *IEEE Transactions on Power Delivery,* vol. 15, no. 1, pp. 423–8, January 2000.

CEA 288 D 747, *Application Guide for Distribution Fusing,* Canadian Electrical Association, 1998.

Clark, G. L., "Development of the Switched Capacitor Bank Controller for Independent Phase Switching on the Electric Distribution System," Distributech 2001, San Diego, CA, 2001.

Cooper Power Systems, *Electrical Distribution — System Protection,* 3rd ed., 1990.

DNP Users Group, "A DNP3 Protocol Primer," 2000. www.dnp.org.

EPRI 1001691, *Improved Reliability of Switched Capacitor Banks and Capacitor Technology,* Electric Power Research Institute, Palo Alto, CA, 2002.

Fisher Pierce, "POWERFLEX 4400 and 4500 Series Instruction Manual," 2000.

General Electric, "Case Rupture Curves," 2001. Downloaded from http://www.geindustrial.com.

Goeckeler, C., "Progressive Capacitor Automation Yields Economic and Practical Benefits at KCPL," *Utility Automation,* October 1999.

Grainger, J. J. and Civanlar, S., "Volt/VAr control on distribution systems with lateral branches using shunt capacitors and voltage regulators. I. The overall problem," *IEEE Transactions on Power Apparatus and Systems,* vol. PAS-104, no. 11, pp. 3278–83, November 1985.

Grainger, J. J. and Lee, S. H., "Optimum Size and Location of Shunt Capacitors for Reduction of Losses on Distribution Feeders," *IEEE Transactions on Power Apparatus and Systems,* vol. PAS-100, no. 3, pp. 1005–18, March 1981.

IEEE C2-1997, *National Electrical Safety Code.*

IEEE Std. 18-2002, *IEEE Standard for Shunt Power Capacitors.*

IEEE Std. 1036-1992, *IEEE Guide for Application of Shunt Power Capacitors.*

IEEE Std. C37.48-1997, *IEEE Guide for the Application, Operation, and Maintenance of High-Voltage Fuses, Distribution Enclosed Single-pole Air Switches, Fuse Disconnecting Switches, and Accessories.*

Lee, S. H. and Grainger, J. J., "Optimum Placement of Fixed and Switched Capacitors on Primary Distribution Feeders," *IEEE Transactions on Power Apparatus and Systems,* vol. PAS-100, no. 1, pp. 345–52, January 1981.

Neagle, N. M. and Samson, D. R., "Loss Reduction from Capacitors Installed on Primary Feeders," *AIEE Transactions,* vol. 75, pp. 950–9, Part III, October 1956.

Newcomb, G. R., "Film/Foil Power Capacitor," *IEEE International Symposium on Electrical Insulation,* Boston, MA, 1980.

Pratt, R. A., Olive, W. W. J., Whitman, B. D., and Brown, R. W., "Capacitor Case Rupture Withstand Capability and Fuse Protection Considerations," EEI T&D Conference, Chicago, IL, May 5-6, 1977.

RUS 1724D-112, *The Application of Capacitors on Rural Electric Systems*, United States Department of Agriculture, Rural Utilities Service, 2001.

Shyh, J. H., "An Immune-Based Optimization Method to Capacitor Placement in a Radial Distribution System," *IEEE Transactions on Power Delivery*, vol. 15, no. 2, pp. 744–9, April 2000.

Thomas, E. S., "Determination of Neutral Trip Settings for Distribution Capacitor Banks," IEEE Rural Electric Power Conference, 1990.

They quit respecting us when we got soft and started using bucket trucks and now anybody can become a lineman, it is sick.

In response to: Do linemen feel they are respected by management and coworkers for the jobs they are doing, do management and coworkers understand what you do?

www.powerlineman.com

7

Faults

Faults kill. Faults start fires. Faults force interruptions. Faults create voltage sags. Tree trimming, surge arresters, animal guards, cable replacements: these tools reduce faults. We cannot eliminate all faults, but appropriate standards and maintenance practices help in the battle. When faults occur, we have ways to reduce their impacts. This chapter focuses on the general characteristics of faults and specific analysis of common fault types with suggestions on how to reduce them. One of the definitions of a fault is (ANSI/IEEE Std. 100-1992):

Fault: A physical condition that causes a device, a component, or an element to fail to perform in a required manner; for example, a short circuit or a broken wire.

A fault almost always involves a short circuit between energized phase conductors or between a phase and ground. A fault may be a bolted connection or may have some impedance in the fault connection. The term "fault" is often used synonymously with the term "short circuit" defined as (ANSI/IEEE Std. 100-1992):

Short circuit: An abnormal connection (including an arc) of relatively low impedance, whether made accidentally or intentionally, between two points of different potential. (*Note:* The term *fault* or *short-circuit fault* is used to describe a short circuit.)

When a short-circuit fault occurs, the fault path explodes in an intense arc. Local customers endure an interruption, and customers farther away, a voltage sag; faults cause most reliability and power quality problems. Faults kill and injure line operators. Crew operating practices, equipment, and training must account for where fault arcs are likely to occur and must minimize crew exposure.

7.1 General Fault Characteristics

There are many causes of faults on distribution circuits. A large EPRI study was done to characterize distribution faults in the 1980s at 13 utilities mon-

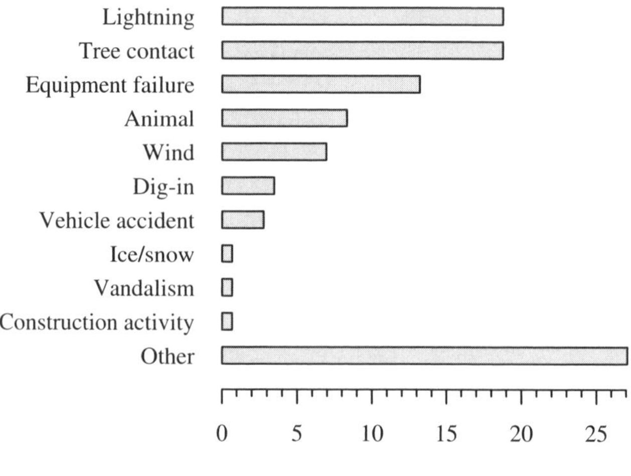

FIGURE 7.1
Fault causes measured in the EPRI fault study. (Data from [Burke and Lawrence, 1984; EPRI 1209-1, 1983].)

itoring 50 feeders (Burke and Lawrence, 1984; EPRI 1209-1, 1983). The distribution of permanent fault causes found in the EPRI study is shown in Figure 7.1. Many of the fault causes are discussed in more detail in this chapter. Approximately 40% of faults in this study occurred during periods of adverse weather which included rain, snow and ice.

Distribution faults occur on one phase, on two phases, or on all three phases. Single-phase faults are the most common. Almost 80% of the faults measured involved only one phase either in contact with the neutral or with ground (see Table 7.1). As another data point, measurements on 34.5-kV

TABLE 7.1

Number of Phases Involved in Each Fault Measured in the EPRI Fault Study

Fault	Percentage
One phase to neutral	63%
Phase to phase	11%
Two phases to neutral	2%
Three phase	2%
One phase on the ground	15%
Two phases on the ground	2%
Three phases on the ground	1%
Other	4%

Source: Burke, J. J. and Lawrence, D. J., "Characteristics of Fault Currents on Distribution Systems," *IEEE Transactions on Power Apparatus and Systems*, vol. PAS-103, no. 1, pp. 1–6, January 1984. EPRI 1209-1 (1983).

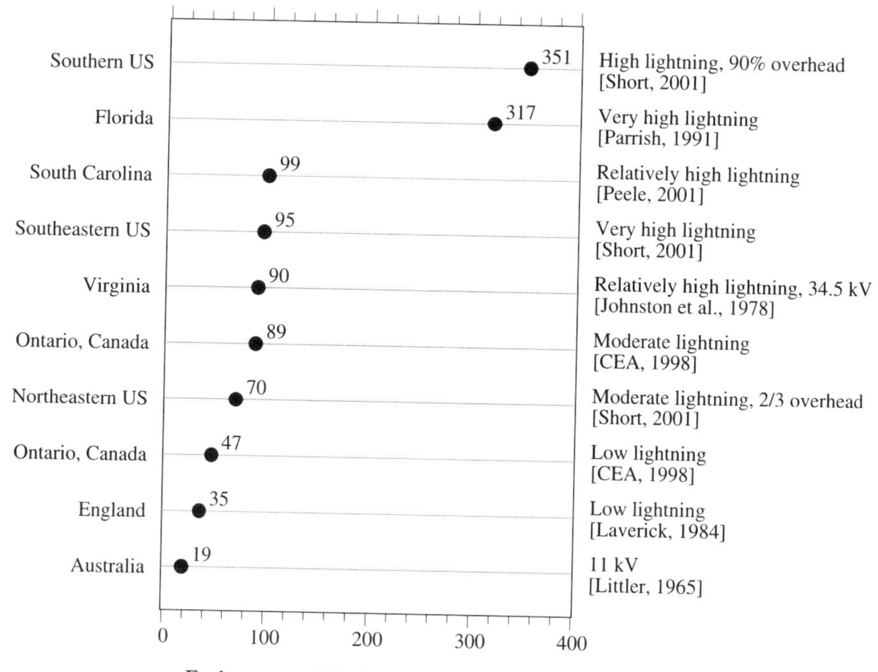

FIGURE 7.2
Fault rates found in different studies. (Data from [CEA 160 D 597, 1998; Johnston et al., 1978; Laverick, 1984; Littler, 1965; Parrish, 1991; Peele, 2001; Short, 2001].)

feeders found that 75% of faults involved ground (also 54% were phase to ground, and 15% were phase to phase) (Johnston et al., 1978). Most faults are single phase because most of the overall length of distribution lines is single phase, so any fault on single-phase sections would only involve one phase. Also, on three-phase sections, many types of faults tend to occur from phase to ground. Equipment faults and animal faults tend to cause line-to-ground faults. Trees can also cause line-to-ground faults on three-phase structures, but line-to-line faults are more common. Lightning faults tend to be two or three phases to ground on three-phase structures.

Figure 7.2 shows fault rates found in various studies for predominantly overhead circuits. Ninety faults per 100 mi per year (55 faults/100 km/year) is common for utilities with moderate lightning. Fault rates increase significantly in higher lightning areas. This type of data is difficult to obtain. Utilities more commonly track faults that cause sustained interruptions, interruptions that contribute to reliability indices such as SAIDI (some data on these faults is shown in Figure 7.3). The actual fault rates are higher than this because many temporary faults are cleared by reclosing circuit breakers or reclosers.

Faults are either *temporary* or *permanent*. A permanent fault is one where permanent damage is done to the system. This includes insulator failures,

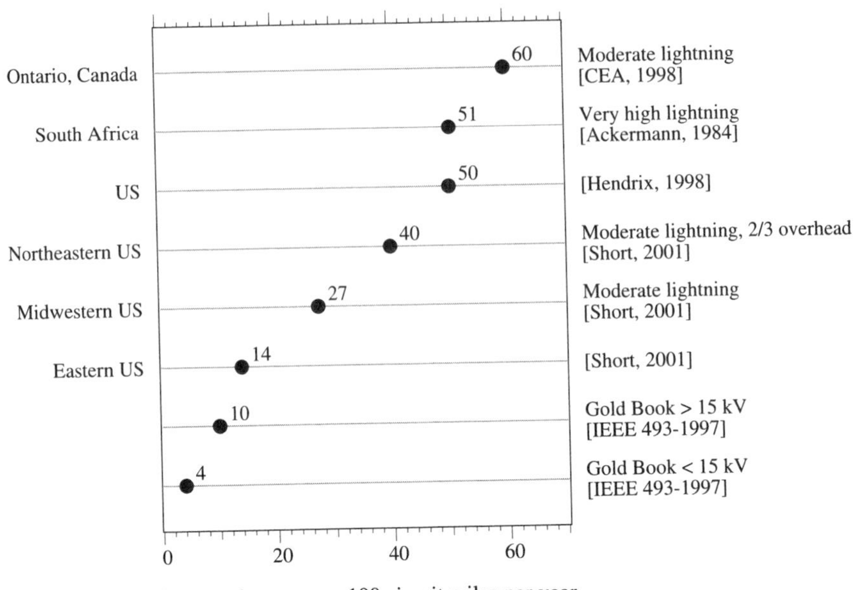

FIGURE 7.3
Fault rates for faults that cause sustained interruptions. (Data from [Ackermann and Baker, 1984; CEA 160 D 597, 1998; Hendrix, 1998; IEEE Std. 493-1997; Johnston et al., 1978; Parrish, 1991; Short, 2001].)

broken wires, or failed equipment such as transformers or capacitors. Virtually all faults on underground equipment are permanent. Most equipment fails to a short circuit. Permanent faults on distribution circuits usually cause sustained interruptions for some customers. To clear the fault, a fuse, recloser, or circuit breaker must operate to interrupt the circuit. The most critical location is the three-phase mains, since a fault on the main feeder will cause an interruption to all customers on the circuit. A permanent fault also causes a voltage sag to customers on the feeder and on adjacent feeders. Permanent faults may cause momentary interruptions for a customer. A common example is a fault on a fused lateral (tap). With *fuse saving* (where an upstream circuit breaker or recloser attempts to open before the tap fuse blows), a permanent fault causes a momentary interruption for customers downstream of the circuit breaker or recloser. After the first attempt to save the fuse, if the fault is still there, the circuit breaker allows the fuse to clear the fault. If a fault is permanent, all customers on the circuit experience a momentary — and the customers on the fused lateral experience a sustained interruption.

A temporary fault does not permanently damage any system equipment. If the circuit is interrupted and then reclosed after a delay, the system operates normally. Temporary (non-damage) faults make up 50 to 90% of faults on overhead distribution systems. The causes of temporary faults include lightning, conductors slapping together in the wind, tree branches that fall across conductors and then fall or burn off, animals that cause faults and

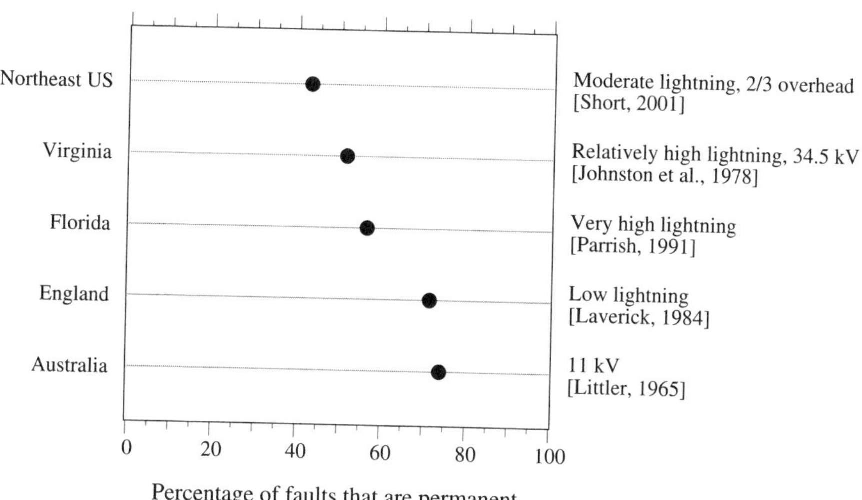

FIGURE 7.4
Percentage of faults that are permanent (on predominantly overhead circuits). (Data from [Johnston et al., 1978; Laverick, 1984; Littler, 1965; Parrish, 1991; Short, 2001].)

fall off, and insulator flashovers caused by pollution. Temporary faults are the main reason that reclosing is used almost universally on distribution circuit breakers and reclosers (on overhead circuits). Temporary faults will cause voltage sags for customers on the circuit with the fault and possibly for customers on adjacent feeders. Temporary faults cause sustained interruptions if the fault is downstream of a fuse, and fuse saving is not used or is not successful. For temporary faults on the feeder backbone, all customers on the circuit are momentarily interrupted. Faults that are normally temporary can turn into permanent faults. If the fault is allowed to remain too long, the fault arc can do permanent damage to conductors, insulators, or other hardware. In addition, the fault current flowing through equipment can do damage. The most common damage of this type is to connectors or circuit interrupters such as fuses.

The majority of faults on overhead distribution circuits are temporary. The data in Figure 7.4 confirms the widely held belief that 50 to 80% of faults are temporary. This very limited data set shows high lightning areas with lower percentages of temporary faults. This contradicts the notion that temporary faults are higher in areas with more lightning. Storms with lightning and wind should cause more temporary faults.

Determining the percentage of faults that are temporary versus permanent is complicated. For faults that operate fuses, it is easy. If it can be successfully re-fused without any repair, the fault is temporary:

$$\text{Percent permanent} = \frac{\text{Fuses replaced after repair}}{\text{Total number of fuse operations}} \times 100\%$$

Circuit breaker or recloser operations are difficult. If *fuse blowing* is used, where tap fuses always operate before the circuit breaker or recloser, the percentage of temporary faults cleared by circuit breakers and reclosers is:

$$\text{Percent permanent} = \frac{\text{Number of lockouts}}{\text{Number of lockouts} + \text{Number of successful reclose sequences}} \times 100\%$$

A SCADA system produces these numbers, but if this information is not available, the percentage can be approximated using circuit breaker count numbers:

$$\text{Percent permanent} = \frac{l}{n\text{-}r \cdot l} \times 100\%$$

where

n = total number of circuit breaker (or recloser) operations
r = number of reclose attempts before lockout (there are r+1 circuit breaker operations during a lockout cycle)
l = number of lockouts

If *fuse saving* is used, where the circuit breaker operates before lateral fuses, then it is more difficult to estimate the number of temporary faults. For the whole circuit (it is not possible to separate the faults on the mains from the faults on the taps), we can estimate the percentage as follows:

$$\text{Percent permanent} = \frac{l + f}{l + s + f_2} \times 100\%$$

where

s = number of successful reclose sequences
f = number of fuses replaced following repair (not including nuisance fuse operations)
f_2 = number of fuse operations that are not coincident with circuit breaker trips

f_2 should be close to zero, since the circuit breaker should operate for all faults. Assuming f_2 is zero (which may have to be done, since this is a difficult number to obtain) implies no nuisance fuse operations without a circuit breaker operation. It is difficult for an outage data management system to properly determine the number of temporary faults.

Faults frequently occur near the peak of the voltage waveform as shown in Figure 7.5. About 60% of the faults in the EPRI fault study occurred when the voltage was within 5% of the peak prefault voltage (where the angle was

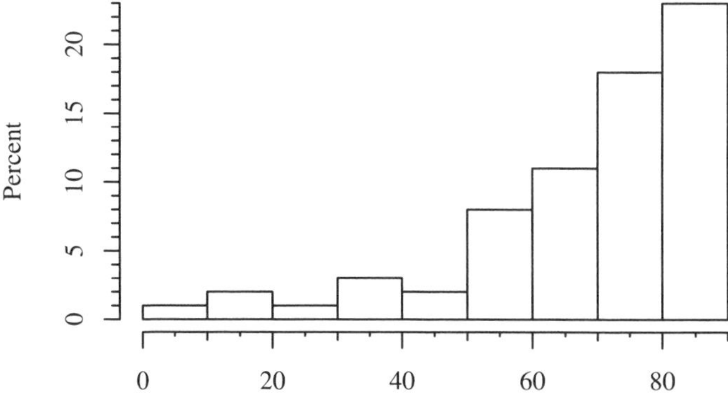

FIGURE 7.5
Point of fault on the voltage waveform. (Data from [Burke and Lawrence 1984; EPRI 1209-1, 1983].)

70 to 90°). This is reasonable. Any insulation failure, whether it be a squirrel breaching a bushing or a failure in a cable, more likely strikes with the voltage at or near its peak. Some faults defy this pattern. Lightning faults happen at any point on the voltage waveform because the fault occurs when the lightning strikes (although lightning can cause a flashover but not a fault if the voltage is very close to a zero-crossing of the power-frequency voltage). Two-phase and three-phase faults create more instances in which the voltage is not near its peak.

7.2 Fault Calculations

The magnitude of fault current is limited only by the system impedance and any fault impedance. The system impedance includes the impedances of wires, cables, and transformers back to the source. For faults involving ground, the impedance includes paths through the earth and through the neutral wire. The impedance of the fault depends on the type of fault.

Most distribution primary circuits are radial, with only one source and one path for fault currents. Figure 7.6 shows equations for calculating fault currents for common distribution faults.

The equations in Figure 7.6 assume that the positive-sequence impedance is equal to the negative-sequence impedance. As an example, the impedance term due to the sequence components for a line-to-line fault is $(Z_1 + Z_2)$, which simplifies to $2Z_1$ when the impedances are assumed to be equal. This is accurate for virtually all distribution circuits. With a large generator nearby, the equivalent circuit may have different positive- and negative-

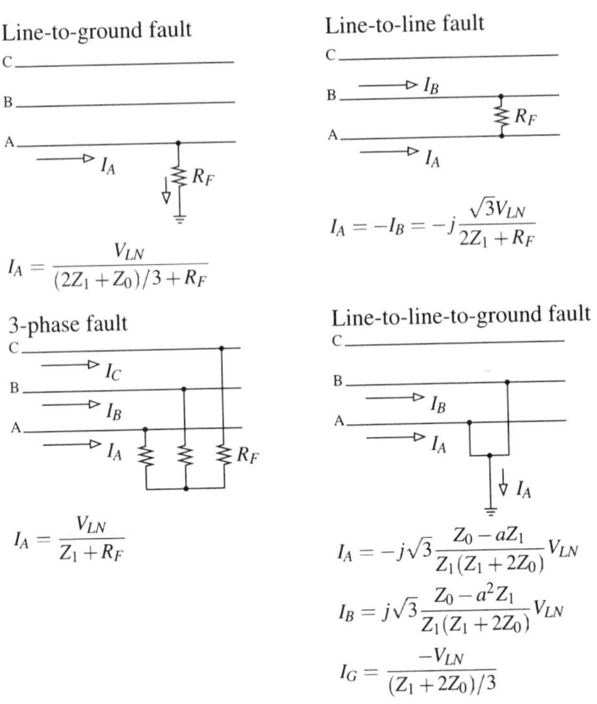

FIGURE 7.6
Fault-current calculations.

sequence impedances (but that case is usually done on the computer and not with hand calculations). The maximum currents occur with a bolted fault where R_F is zero. The maximum current for a line-to-line fault is 86.6% of the maximum three-phase fault current. In all cases, the load current is ignored. In most cases, load will not significantly change results.

The three-phase fault current is almost always the highest magnitude. On most circuits, the zero-sequence impedance is significantly higher than the positive-sequence impedance. One important location where the line-to-ground fault current may be higher is at the substation. There are two reasons for this:

1. A delta – wye transformer is a zero-sequence source. The positive-sequence impedance includes the impedance of the subtransmission and transmission system. The zero-sequence impedance does not. Figure 7.7 shows the sequence diagrams for the positive and zero sequences. The delta-wye connection forms a zero-sequence source while the positive-sequence impedance includes the subtransmission equivalent impedance.
2. If the substation transformer has three-legged core-form construction, the zero-sequence impedance is lower than its positive-

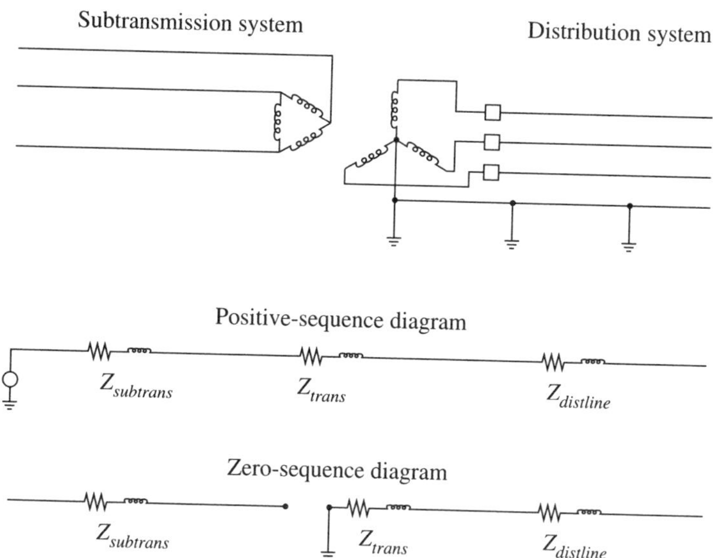

FIGURE 7.7
Positive and zero-sequence diagrams for a delta – wye substation transformer.

sequence impedance. Typically, the zero-sequence impedance is 85% of the positive-sequence impedance, which increases ground-fault currents by 5.2%.

In cases where the zero-sequence impedance is less than the positive-sequence impedance, the line-to-ground fault gives the highest phase current. The double line-to-ground fault produces the highest-magnitude ground current.

In order to reduce fault currents for line-to-ground faults, a neutral reactor on the station transformer is sometimes used. Figure 7.8 shows the equations for faults involving ground for circuits with a neutral reactor (the line-to-line fault and the three-phase fault are not affected). A common value for a neutral reactor is 1 Ω for 15-kV class distribution circuits.

The impedance seen by line-to-ground faults is a function of both the positive and the zero sequence impedances. This important loop impedance is $Z_S = (2Z_1 + Z_0)/3$. The sequence impedances, Z_1 and Z_0, used in the fault calculations include the sum of the impedances with both resistance and reactance along the fault current path. Some of the common branch impedances are given below including some rule-of-thumb values that are useful for hand calculations:

- Overhead lines:
  - $|Z_1| = 0.5\ \Omega/\text{mi}\ (0.3\ \Omega/\text{km})$
  - $|Z_S| = 1\ \Omega/\text{mi}\ (0.6\ \Omega/\text{km})$

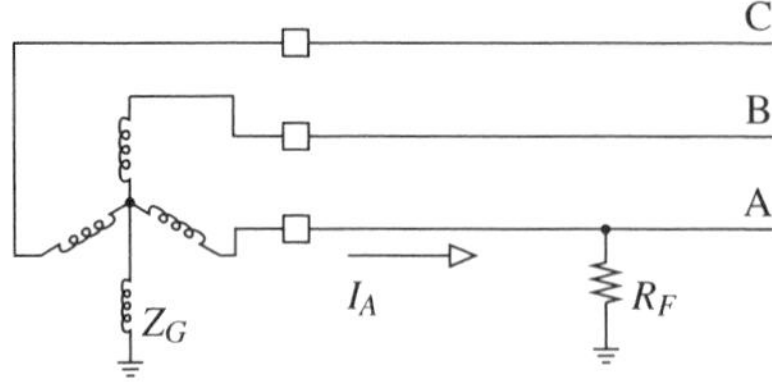

$$I_A = \frac{V_{LN}}{(2Z_1 + Z_0)/3 + R_F + Z_G}$$

Line-to-line-to-ground fault

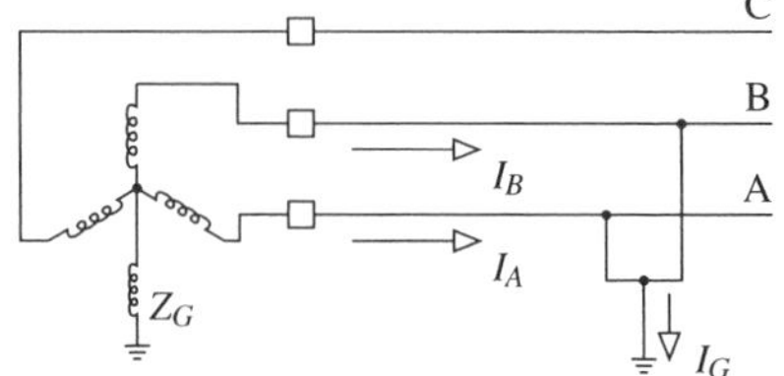

$$I_A = -j\sqrt{3}\frac{Z_0 + 3Z_G - aZ_1}{Z_1(Z_1 + 2Z_0 + 6Z_G)}V_{LN}$$

$$I_B = j\sqrt{3}\frac{Z_0 + 3Z_G - a^2 Z_1}{Z_1(Z_1 + 2Z_0 + 6Z_G)}V_{LN}$$

$$I_G = \frac{-V_{LN}}{(Z_1 + 2Z_0 + 6Z_G)/3}$$

FIGURE 7.8
Fault-current calculations with a neutral reactor on the substation transformer.

- Underground cables:
  - $|Z_1|$ = 0.6 Ω/mi (0.35 Ω/km)
  - $|Z_S|$ = 0.5 Ω/mi (0.3 Ω/km)
- Substation transformer:
  - $|Z_1| = |Z_S|$ = 1 Ω
  - A typical 15-kV substation transformer impedance is 1 Ω, which corresponds to a bus fault current of 7.2 kA for a 12.47-kV circuit.
- Subtransmission equivalent: often can be ignored

See Smith (1980) for an excellent paper on fault calculations for additional information. Include impedances for step-down transformer banks, series reactors, and voltage regulators. Use the rule-of-thumb numbers above for back-of-the-envelope calculations and as checks for computer modeling.

The simplified equation for a transformer impedance is

$$Z_1 = Z_0 = j\frac{kV^2}{MVA}Z_{\%}$$

where

kV = line-to-line voltage
MVA = transformer base rating — open air (OA) rating
$Z_{\%}$ = transformer impedance, per unit

We ignore the resistive component since the *X/R* ratio of station transformers is generally greater than 10 and often in the range of 20 to 30. The transmission/subtransmission equivalent is usually small, and we often ignore it (especially for calculating maximum fault currents). Include the transmission system impedance for weak subtransmission systems such as 34.5-, 46- or 69-kV circuits, or for very large substations. Find the transmission equivalent from the per unit impedances (r_1, x_1, r_0, and x_0) on a given MVA base referred to the distribution voltage (Smith, 1980) as

$$Z_1 = (r_1 + jx_1)\frac{kV_s^2}{MVA_b}\left(\frac{kV_{pb}}{kV_p}\right)^2$$

$$Z_0 = (r_0 + jx_0)\frac{kV_s^2}{MVA_b}\left(\frac{kV_{pb}}{kV_p}\right)^2$$

where

MVA_b = base MVA at which the r and x impedances are given
kV_s = line-to-line voltage in kV on the secondary side of the station transformer
kV_p = line-to-line voltage in kV on the primary
kV_{pb} = base line-to-line voltage on the primary used to calculate MVA_b (often equal to kV_p)

If the transmission impedances are available as a fault MVA with a power factor, find the transmission equivalent (Smith, 1980) with

$$Z_1 = \frac{kV_s^2}{MVA}(pf + j\sqrt{1-pf^2})\left(\frac{kV_{pb}}{kV_p}\right)^2$$

$$Z_0 = \frac{\sqrt{3}kV_s^2}{kI_g \cdot kV_{pb}}(pf_g + j\sqrt{1-pf_g^2})\left(\frac{kV_{pb}}{kV_p}\right)^2 - 2Z_1$$

where

MVA = 3-phase short-circuit MVA at the primary terminals of the station transformer (see Table 7.2 for typical maximum values)
kI_g = available ground fault current in kA at the primary terminals of the station transformer
pf = power factor in per unit for the available three-phase fault current
pf_g = power factor in per unit for the available single-phase fault current

While almost all distribution circuits are radial, there may be other fault current sources. We ignore these other sources most of the time, but occasionally, we consider motors and generators in fault calculations. Synchronous motors and generators contribute large currents relative to their size. On a typical 15-kV class distribution circuit, one or two megawatts worth

TABLE 7.2

Typical Maximum Transmission/Subtransmission Fault Levels

Transmission Voltage, kV	Maximum Symmetrical Fault, MVA
69	3,000
115	5,000
138	6,000
230	10,000

of connected synchronous units are needed to significantly affect fault currents. On weaker circuits, smaller units can impact fault currents. Induction motors and generators also feed faults. Inverter-based distributed generation can contribute fault current, but generally much less than synchronous or induction units. Of course, on feeders that have network load, current through network transformers backfeeds faults until the network protectors operate.

7.2.1 Transformer Connections

The fault current on each side of a three-phase transformer connection can differ in magnitude and phasing. In the common case of a delta – grounded-wye connection, the current on the source side of the transformer differs from the currents on the fault side for line-to-ground or line-to-line faults (see Figure 7.9). For a line-to-ground fault on the primary side of the transformer, the current appears on two phases on the primary with a per unit current of 0.577 (which is $1/\sqrt{3}$).

These differences are often needed when coordinating a primary-side protective device and a secondary-side device. In distribution substations, this is commonly a fuse on the primary side and a relay controlling a circuit breaker on the secondary side. The line-to-line fault must be considered — this gives more per-unit current on one phase in the primary, 1.15 per unit

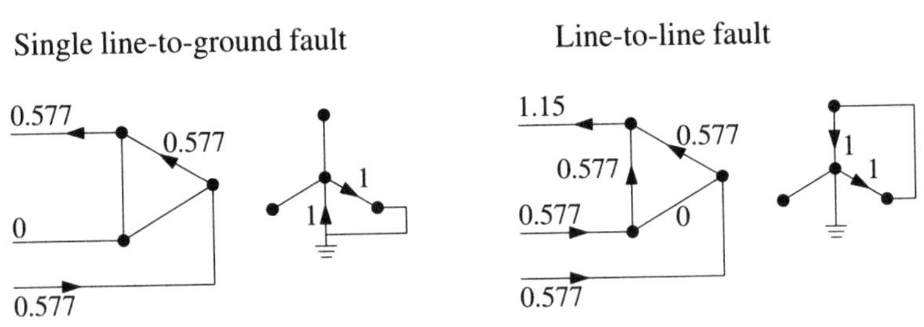

FIGURE 7.9
Per-unit fault-currents on both sides of a delta – grounded-wye transformer.

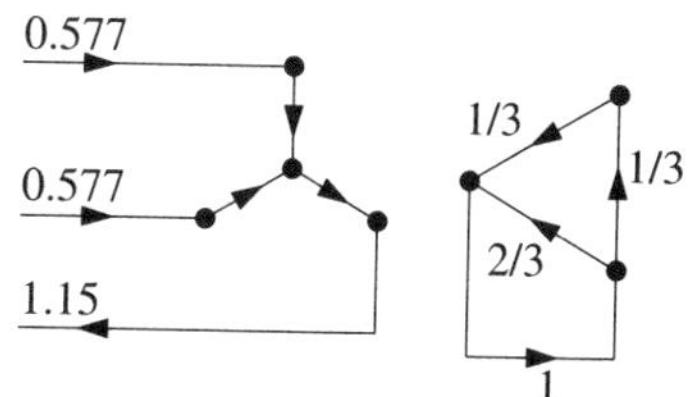

FIGURE 7.10
Per-unit fault-currents on both sides of a wye – delta transformer.

($2/\sqrt{3}$) in one of the phases (see Figure 7.9). To make sure a primary fuse coordinates with a secondary device, shift the minimum-melting time-current curve of the primary-side fuse to the left by a factor of $0.866 = \sqrt{3}/2$ (after also adjusting for the transformer turns ratio). The current differences also mean that the transformer is not protected as well for single-phase faults; a primary-side fuse takes longer to clear the single-phase fault since it sees less current than for a three-phase or line-to-line fault.

Fault currents are only different for unbalanced secondary currents. For a three-phase secondary fault, the per-unit currents on the primary equal those on the secondary (with the actual currents related by the turns ratio of the transformer). A wye – wye transformer does not disturb the current relationships; the per-unit currents on both sides of the transformer are equal.

In a floating wye – delta, similar current relationships exist; a line-to-line secondary fault shows up on the primary side on all three phases, one of which is 1.15 per unit (see Figure 7.10). For a floating wye – delta transformer with a larger center-tapped lighting leg and two power legs, fault current calculations are difficult. Faults can occur from phase to phase and from phase to the secondary neutral, and the lighting transformer will have a different impedance than the power leg transformers. For an approach to modeling this, see Kersting and Phillips (1996).

7.2.2 Fault Profiles

Fault profiles show fault current with distance along a circuit. Determining where thermal or mechanical short-circuit limits on equipment may be exceeded, helping select or check interrupting capabilities of protective equipment, and coordinating protective devices are important uses of fault profiles. Figure 7.11 and Figure 7.12 show typical fault current profiles of distribution circuits.

Some general trends that the fault profiles show are:

- *Distance* — The fault current drops off as the inverse of distance ($1/d$).
- *Ground faults* — On overhead circuits, the ground fault current falls off faster (and the ground fault current is generally lower)

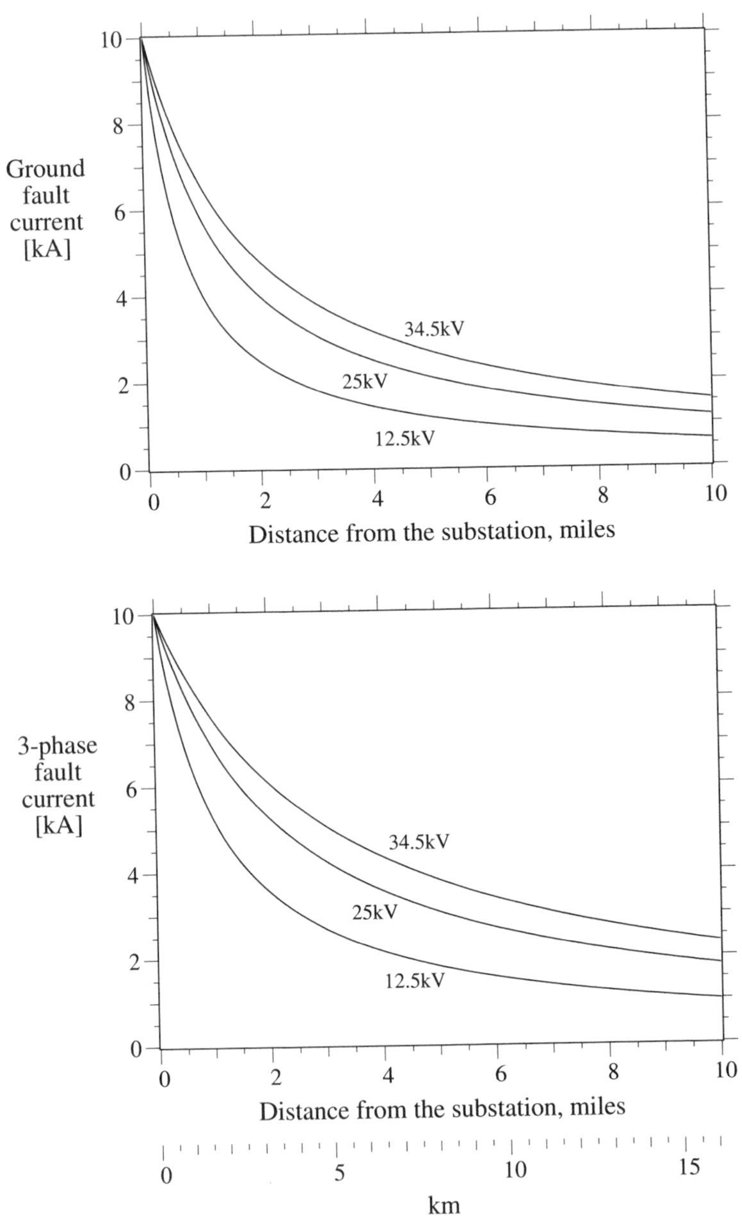

FIGURE 7.11
Fault-current profiles for line-to-ground faults and for three-phase faults for an overhead circuit. Phase characteristics: 500 kcmil, all-aluminum, GMD = 4.69 ft (1.43 m). Neutral characteristics: 3/0 all-aluminum, 4-ft (1.22-m) line-neutral spacing. $Z_1 = 0.207 + j0.628\ \Omega$/mile ($0.1286 + j0.3901\ \Omega$/km), $Z_0 = 0.720 + j1.849\ \Omega$/mile ($0.4475 + j1.1489\ \Omega$/km), $Z_S = 0.378 + j1.035\ \Omega$/mile ($0.2350 + j0.6430\ \Omega$/km).

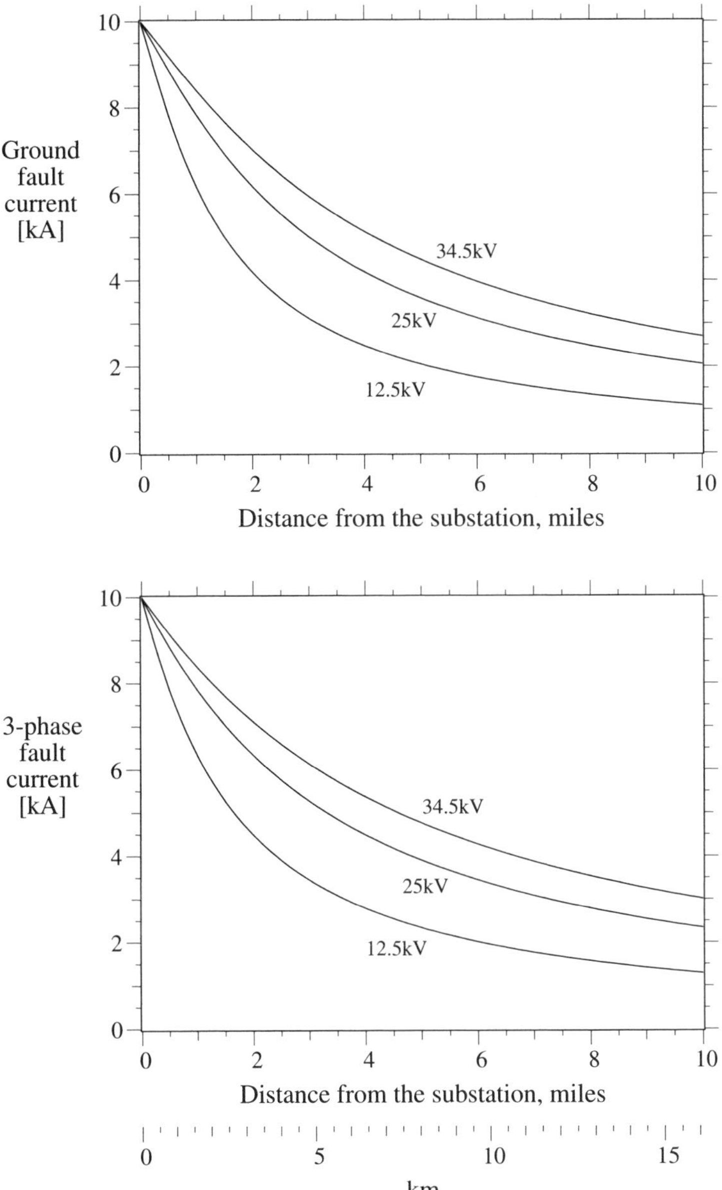

FIGURE 7.12
Fault-current profiles for line-to-ground faults and for three-phase faults for an underground cable circuit. 500-kcmil aluminum conductor, 220-mil XLPE insulation, 1/3 neutrals, flat spacing, 7.5 in. between cables. $Z_1 = 0.3543 + j0.3596\ \Omega/\text{mile}$ $(0.2201 + j0.2234\ \Omega/\text{km})$, $Z_0 = 0.8728 + j0.2344\ \Omega/\text{mile}$ $(0.5423 + j0.1456\ \Omega/\text{km})$, $Z_S = 0.5271 + j0.3178\ \Omega/\text{mile}$ $(0.3275 + j0.1975\ \Omega/\text{km})$.

than the three-phase fault current. The zero-sequence reactance is generally over three times the positive-sequence reactance, and the zero-sequence resistance is also higher than the positive-sequence resistance.

- *System voltage* — On higher-voltage distribution systems, the fault current drops off more slowly. The actual line impedance does not change with voltage ($Z_S \approx 1\ \Omega/\text{mi}$), and since $I = V_{LN}/Z$, it takes more impedance (more circuit length) to reduce the fault current.
- *Cables* — Underground cables have much lower reactance than overhead circuits, so the fault current does not fall off as fast on underground circuits. Also, note that *X/R* ratios are lower on cables.
- *Profiles* — The three-phase and ground-fault profiles of underground cables are similar. The zero-sequence reactance can actually be smaller than the positive-sequence reactance (but the zero-sequence resistance is larger than the positive-sequence resistance).

7.2.3 Effect of *X/R* Ratio

In a reactive circuit (high *X/R* ratio), it is naturally more difficult for a protective device such as a circuit breaker to clear a fault. Protective devices clear a fault at a current zero. Within the interruptor, dielectric strength builds up to prevent the arc from reigniting after the current zero. In a resistive circuit (low *X/R* ratio), the voltage and current are in phase, so after a current zero, a quarter cycle passes before the voltage across the protective device (called the *recovery voltage*) reaches its peak. In a reactive circuit, the fault current naturally lags the voltage by 90°; the voltage peaks at a current zero. Therefore, the recovery voltage across the protective device rises to its peak in much less than a quarter cycle (possibly in 1/20th of a cycle or less), and the fault arc is much more likely to reignite.

Another factor that makes it more difficult for protective devices to clear faults is asymmetry. Circuits with inductance resist a change in current. A short circuit creates a significant change in current, possibly creating an offset. If the fault occurs when the current would naturally be at its negative peak, the current starts at that point on the waveshape but is offset by 1.0 per unit. The dc offset decays, depending on the *X/R* ratio. The offset is described by the following equation:

$$i(t) = \underbrace{\sqrt{2}I_{rms}\sin(2\pi ft + \beta - \theta)}_{\text{ac component}} - \underbrace{\sqrt{2}I_{rms}\sin(\beta - \theta)e^{-\frac{2\pi ft}{X/R}}}_{\text{decaying dc component}}$$

where

$i(t)$ = instantaneous value of current at time t

I_{rms} = root-mean square (rms) value of the ac component of current,

$$I_{rms} = V / \sqrt{R^2 + X^2}$$

β = the closing angle which defines the point on the waveform at which the fault is initiated

θ = system impedance angle = $\tan^{-1} \frac{X}{R}$

f = system frequency, Hz

t = time, sec

Asymmetry is higher with higher *X/R* ratios. The worst case offset with $X/R = \infty$ is 2 per unit. Figure 7.13 shows an example of an offset fault current.

If a phase faults at the natural zero crossing ($\beta = \theta$), no offset occurs. The highest magnitude of the dc component occurs when the fault happens 90° from the natural zero crossing of the circuit (when $\beta = \theta \pm \pi/2$). The highest

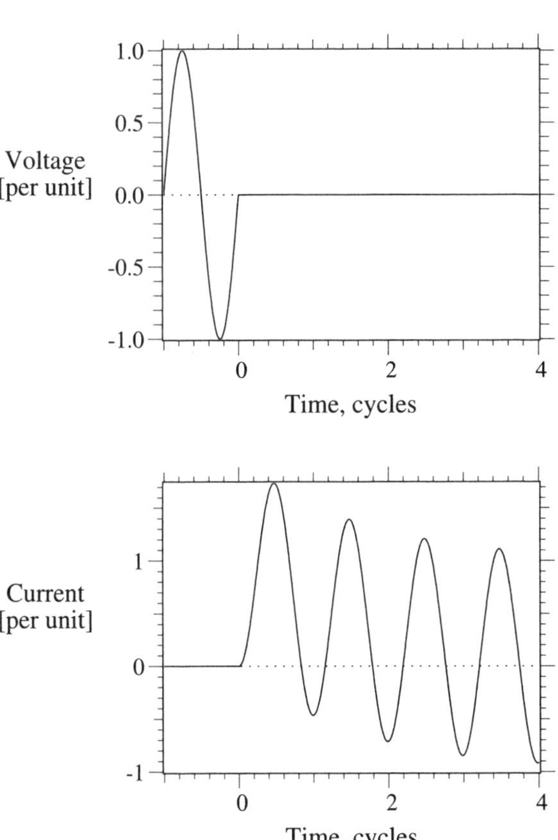

FIGURE 7.13
Example of an asymmetric fault with *X/R* = 10 which initiated when the closing angle β = 0, which is when the voltage crosses zero.

dc offset does not align with the highest peak asymmetric current (which is the sum of the ac and decaying dc component). The peak current occurs when the closing angle $\beta= 0$ for all X/R ratios ($\beta = 0$ when the fault occurs at a voltage zero crossing). The ratio of the peak current I_p to the rms current I_{rms} can be approximated by

$$\frac{I_p}{I_{rms}} = \sqrt{2}\left[1+e^{-(\theta+\pi/2)\frac{R}{X}}\sin\theta\right] = \sqrt{2}\left[1+e^{-\pi\frac{R}{X}}\right] \quad \text{for } \theta = \pi/2$$

This is the most industry-accepted approximation that is used, but it gives an approximation that is slightly low. A more accurate approximation can be found (St. Pierre, 2001) with

$$\frac{I_p}{I_{rms}} = \sqrt{2}\left[1+e^{-2\pi\frac{R}{X}\tau}\right]$$

where τ is a fictitious time found with

$$\tau = 0.49 - 0.1e^{-\frac{1}{3}\frac{X}{R}}$$

In addition to causing a higher peak magnitude, asymmetry also causes a longer first half cycle (important for fuse operating time) and much higher first half cycle $\int I^2\,dt$. The occurrence of asymmetry is reduced by the fact that most faults occur when the voltage is near its peak (Figure 7.5). In a circuit with a high X/R ratio, when the voltage is at its peak, the fault current is naturally near zero. Therefore, for most faults, the asymmetry is small, especially for line-to-ground faults. For two- or three-phase faults where each phase is faulted simultaneously (as can happen with lightning), asymmetry is much more likely.

Asymmetry is important to consider for application of cutouts, circuit breakers, and other equipment with fault current ratings. Equipment is generally tested at a given X/R ratio. If the equipment is applied at a location where the X/R ratio is higher, then the equipment may have less capability than the rating indicates. Equipment often has a momentary duty rating which is the short-time (first-cycle) withstand capability. This is strongly influenced by asymmetry.

Other impacts of asymmetry include:

- Asymmetry can saturate current transformers (CTs). On distribution circuits, overcurrent relays should still operate although they could be more susceptible to miscoordination.
- Fuses respond to $\int I^2 dt$, so asymmetrical current melts the link significantly faster.

- Asymmetry can foul up fault-location algorithms in digital relays and fault recorders.

7.2.4 Secondary Faults

Secondary faults vary depending on the transformer connection and the type of fault on the secondary. For a standard single-phase 120/240-V secondary for residential service, two faults are of interest: a fault from a phase to the neutral and a fault from one of the hot legs to the other across the full 240 V. The impedance to the fault includes the transformer plus the secondary impedance. The secondary current for a bolted fault across the 240-V legs (between the two hot legs) is

$$I_{240} = \frac{240}{\sqrt{\left(R_T + \frac{R_S L}{1000}\right)^2 + \left(X_T + \frac{X_S L}{1000}\right)^2}}$$

where

I_{240} = Secondary current, symmetrical A rms for a 240-V fault (phase-to-phase)
R_T = Transformer full-winding resistance, Ω at 240 V (from terminals X1 to X3)
X_T = Transformer full-winding reactance, Ω at 240 V (from terminals X1 to X3)
R_S = Secondary conductor resistance to a 240-V fault, Ω/1000 ft
X_S = Secondary conductor resistance to a 240-V fault, Ω/1000 ft
L = Distance to the fault, ft

and

$$R_T = 0.0576 \frac{W_{CU}}{S_{kVA}^2}$$

$$Z_T = 0.576 \frac{Z_{\%}}{S_{kVA}}$$

$$X_T = \sqrt{Z_T^2 - R_T^2}$$

where

S_{kVA} = transformer rating, kVA
$W_{CU} = W_{TOT} - W_{NL}$ = load loss at rated load, W
W_{TOT} = total losses at rated load, W
W_{NL} = no-load losses, W
$Z_{\%}$ = nameplate impedance magnitude, %

For a short circuit from one of the hot legs to the neutral, both the transformer and the secondary have different impedances. For the transformer, the half-winding impedance must be used; for the secondary, the loop impedance through the phase and the neutral should be used.

$$I_{120} = \frac{120}{\sqrt{\left(R_{T1} + \frac{R_{S1}L}{1000}\right)^2 + \left(X_{T1} + \frac{X_{S1}L}{1000}\right)^2}}$$

where

I_{120} = Secondary current in symmetrical A rms for a 120-V fault (phase-to-neutral)
R_{T1} = Transformer half-winding resistance, Ω at 120 V (from terminals X1 to X3)
X_{T1} = Transformer half -winding reactance, Ω at 120 V (from terminals X1 to X3)
R_{S1} = Secondary conductor resistance to a 120-V fault, Ω/1000 ft
X_{S1} = Secondary conductor resistance to a 120-V fault, Ω/1000 ft
L = Distance to the fault, ft

In absence of better information, use the following impedances for transformers with an interlaced secondary winding:

$$R_{T1} = 0.375R_T \quad \text{and} \quad X_{T1} = 0.3X_T$$

And use the following impedances for transformers with noninterlaced secondary windings:

$$R_{T1} = 0.4375R_T \quad \text{and} \quad X_{T1} = 0.625X_T$$

Figure 7.14 shows fault profiles for secondary faults on various size transformers. The secondary is triplex with 3/0 aluminum conductors and a reduced neutral. It has impedances of

$$R_S = 0.211\ \Omega/1000\ \text{ft} \qquad X_S = 0.0589\ \Omega/1000\ \text{ft}$$

$$R_{S1} = 0.273\ \Omega/1000\ \text{ft} \qquad X_{S1} = 0.0604\ \Omega/1000\ \text{ft}$$

The secondary has significant impedances; fault currents drop quickly from the transformers. Close to the transformer, line-to-neutral faults are higher magnitude. At large distances from the transformer, the secondary impedances dominate the fault currents. Faults across 240 V are normally higher magnitude than line-to-neutral faults.

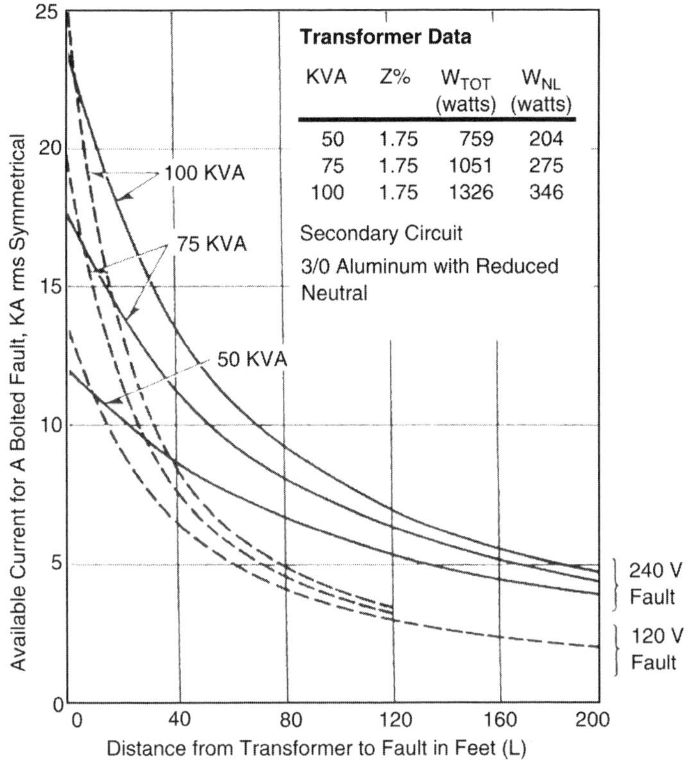

FIGURE 7.14
Fault profiles for 120/240-V secondary faults ($R_{T1} = 0.375R_T$, and $X_{T1} = 0.5X_T$). (From ABB Inc., *Distribution Transformer Guide*, 1995. With permission.)

Normally in secondary calculations, we can ignore the impedance offered by the distribution primary. The primary-system impedance is usually small relative to the transformer impedance, and neglecting it is conservative for most uses. On weak distribution systems or with large, low-impedance distribution transformers, the distribution system impedance plays a greater role.

7.2.5 Primary-to-Secondary Faults

Faults from the distribution primary to the secondary can subject end-use equipment to significant overvoltages. Figure 7.15 shows a circuit diagram of a fault from the primary to a 120/240-V secondary. This type of fault can occur several ways: a high-to-low fault within the transformer, a broken primary wire falling into the secondary, or a broken primary jumper. As we will discuss, the transformer helps limit the overvoltage. Having the primary fall on the secondary does not automatically mean primary-scale voltages in customers' homes and facilities.

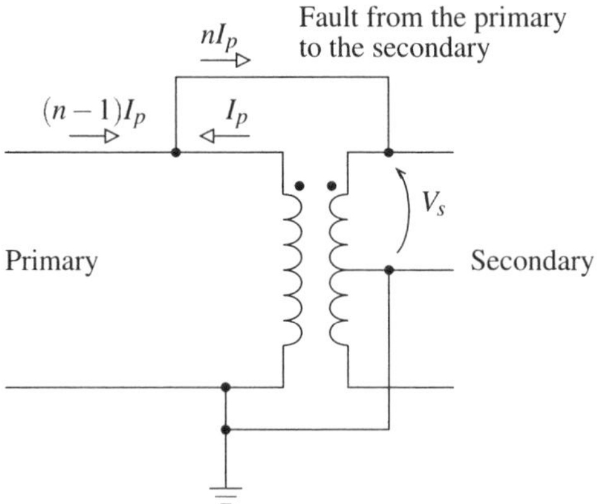

FIGURE 7.15
Fault from the primary to a 120/240-V secondary circuit.

The per-unit secondary voltage for a fault from the primary to the secondary (PTI, 1999) is

$$V_s = \frac{n}{1+(n-1)^2 \dfrac{S_{kVA}}{10\ V_{kV}\ I_{kA}\ Z_{\%}}}$$

where

V_s = secondary voltage, per unit at 120 V
n = transformer turns ratio from the primary voltage to the half-voltage secondary rating (normally 120 V)
I_{kA} = available primary fault current for a single line-to-ground fault, kA
S_{kVA} = transformer rating, kVA
$Z_{\%}$ = half winding impedance of the transformer, %
V_{kV} = primary line-to-ground winding rated voltage, kV

Figure 7.16 shows the per-unit overvoltage for various transformer sizes. Surprisingly, the primary voltage does not impact the overvoltage significantly. The overvoltage equation in per unit reduces (PTI, 1999) to approximately

$$V_s \approx \frac{1.2 Z_{\%} I_{kA}}{S_{kVA}}$$

The overvoltage increases with higher available fault current, higher impedance transformers, and smaller transformers. For all but the smallest transformers with the highest impedance, the overvoltage is not too hazardous. But, if a fuse operates to separate the transformer from the circuit but

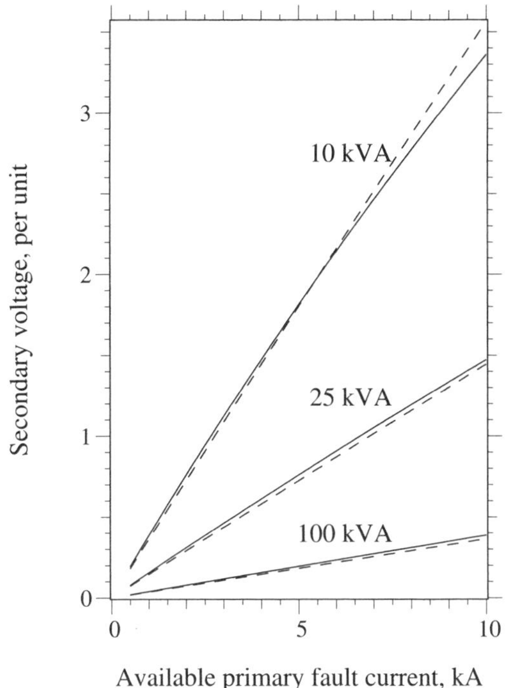

FIGURE 7.16
Secondary voltage during a fault from the primary to a 120/240-V secondary circuit. The solid lines are for a 4.8-kV circuit, and the dashed lines are for a 34.5-kV circuit. The results assume that $Z_{\%} = 3\%$.

leaves the primary-to-secondary fault, the fault imposes full primary voltage on the secondary (at least until the first failure on the secondary system). Such a condition can occur when the fault starts on the primary side above the transformer fuse (see Figure 7.17). If the transformer fuse blows before the upstream line fuse, the secondary voltage rises to the primary voltage. If the fault is below the transformer fuse, it does not matter which fuse blows first; either clears the fault.

The example in Figure 7.15 shows a fault to the secondary leg that is in phase with the primary (off of the *X*1 bushing of the transformer). A fault to the other secondary leg (off of *X*3) has very similar effects; the voltages and currents are almost the same, so the equations and graphs in this section also apply.

Although the transformer helps hold down the overvoltage, the primary-to-secondary fault may initiate a sizeable switching transient that could impact end-use equipment.

With most line fuses and transformer fuses used, the line fuse will clear before the transformer fuse and before the transformer suffers damage (good news on both counts). Even though the upstream fuse is larger, it sees $(n - 1)$

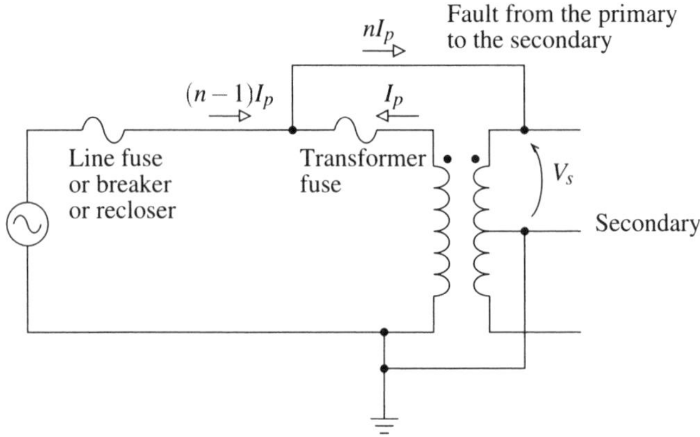

FIGURE 7.17
Fault from the primary to a 120/240-V secondary circuit.

times the fault current. With the primary fault above the transformer fuse, the transformer fuse is more likely to operate before the line fuse with

- *Small transformer fuses* — Another reason not to fuse transformers too tightly; smaller, fast transformer fuses are more likely to clear before an upstream device.
- *Upstream breaker or recloser* — If the upstream device is a circuit breaker or recloser instead of a fuse, the tripping time is much longer, especially on a time-delayed trip (but even a fast trip is relatively long). If a circuit breaker is upstream of the transformer, the transformer fuse is likely to blow before the circuit breaker for locations with high fault currents and with small transformer fuses.

A more detailed analysis of the coordination of the two devices requires using the time-current characteristics of each of the protective devices along with the currents. The current into the primary winding, I_p in kA is

$$I_p = \frac{I_{kA}}{(n-1) + \dfrac{10 I_{kA}}{(n-1)} \dfrac{V_{kV}}{S_{kVA}} Z_{\%}} \approx \frac{I_{kA}}{n}$$

Again, the upstream device sees $(n-1)I_p$, which is almost the full available current for a single line-to-ground fault, I_{kA}.

A transformer with a secondary circuit breaker (as in a completely self-protected transformer, a CSP) has another possible mode where the transformer separates. If the secondary circuit breaker opens before the upstream primary device, the high-to-low fault raises the secondary voltage to the primary voltage. The secondary circuit breaker may not be able to clear the

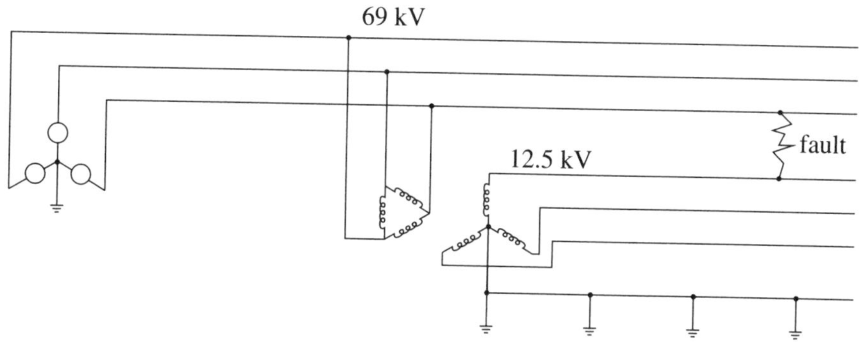

FIGURE 7.18
Example of a fault from a transmission conductor to a distribution conductor.

fault because the arc recovery voltage is much higher than the rating for the secondary circuit breaker; this is good news in that it helps protect end-use equipment from extreme overvoltages, but the secondary circuit breaker may fail trying to clear the fault. If the upstream device is a fuse, the fuse will probably clear before the secondary circuit breaker opens, but if the upstream device is a circuit breaker, the secondary circuit breaker will probably try to open first.

7.2.6 Underbuilt Fault to a Transmission Circuit

Faults from transmission circuits to distribution circuits are another hazard that can subject distribution equipment and customer equipment to extremely high voltages. Consider the example in Figure 7.18 of a fault from a subtransmission circuit to a distribution circuit.

As is the case for primary-to-secondary faults discussed in the previous section, overvoltages are not extremely high as long as the distribution circuit stays connected. But if a distribution interrupter opens the circuit, the voltage on the faulted distribution conductor jumps to the full transmission-line voltage. With voltage at several times normal, something will fail quickly. Such a severe overvoltage is also likely to damage end-use equipment. The distribution interrupter, either a circuit breaker or recloser, may not be able to clear the fault (the recovery voltage is many times normal); it may fail trying.

Faults further from the distribution substation cause higher voltages, with the highest voltage at the fault location. Current flowing back towards the circuit causes a voltage rise along the circuit.

While one can use a computer model for an exact analysis (but it is not possible with most standard distribution short-circuit programs), a simplified single-phase analysis (assuming a wye – wye transformer) helps frame the problem. The fault current is approximately

$$I = \frac{V_S}{\frac{(n-1)}{n} Z_A + \frac{n}{(n-1)} Z_B} \approx \frac{V_S}{Z_A + Z_B}$$

where

n = ratio of the transmission to distribution voltage (n = 69/12.5 = 5.5 in the example)

V_S = rms line-to-ground transmission source voltage (40 kV in the example)

Z_A = loop impedance from the transmission source to the high side of the distribution station

Z_B = loop impedance from the high-side at the distribution station out to the fault and back to the distribution low-side of the distribution substation

And, the 69-kV impedance often dominates, so the fault current is really determined by Z_A. For the distribution and transmission line impedances, Z_A and Z_B, you can use 1 Ω/mi for quick approximations. The worst case is with a small Z_A, a stiff subtransmission system.

The voltage at the fault is

$$V = I \frac{Z_B}{2} + V_d$$

where

V_d = line-to-ground voltage on the distribution circuit at the substation (as a worst case, assume that it is the nominal voltage; it will usually be less because of the sag that pulls down the voltages).

Figure 7.19 shows results from a series of computer simulations on a 12.5-kV circuit for various fault locations and subtransmission source stiffnesses. Results only modestly differ for other configurations: a 69-kV source in the opposite direction, a looped transmission source, a different substation transformer configuration, or different phases faulted. The worst cases are for stiff transmission systems.

If a distribution interrupter opens to leave transmission voltage on the distribution circuit, distribution transformers would saturate and metal-oxide arresters would move into heavy conduction. Transformer saturation distorts the voltage but may not appreciably reduce the peak voltage. Arresters can reduce the peak voltage, but they could still allow quite high voltages to customers. Arresters with an 8.4-kV maximum continuous operating voltage start conducting for power-frequency voltages at about 11 to 12 kV (1.5 to 1.6 times the nominal system line-to-ground voltage). At higher voltages, the arresters will draw more current. Depending on the number of arresters, a stiff transmission source can still push the voltage to between 3 and 4 per

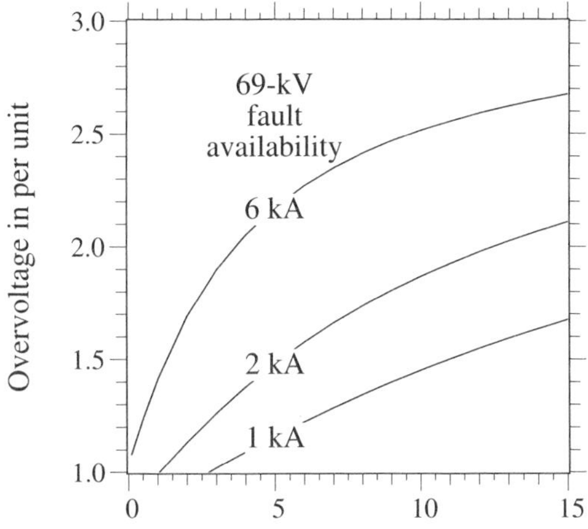

FIGURE 7.19
Results of simulations of a fault from a 69-kV circuit to a 12.5-kV circuit (before the distribution substation breaker trips).

unit, which is 20 to 30 kV (until an arrester or something else fails). In fact, the best protection happens when the arrester fails as fast as possible; the arrester becomes a sacrificial protector. Normal-duty arresters fail faster than heavy-duty arresters, which limits the duration of overvoltages. Goedde et al. (2002) propose using gapped arresters; their lab tests found that gapped arresters clip the overvoltage to a lower magnitude and fail faster during overvoltages.

To reduce the hazard of underbuilt distribution lines, consider the following experimental options:

- *Arresters* — Use normal-duty arresters and possibly gapped arresters.
- *Fuses* — Try to avoid using fuses or reclosers where it leaves significant downstream exposure underbuilt. The fast operation of fuses and reclosers are more likely to clear the distribution circuit before the overbuilt circuit.
- *Directional relays* — In faults to a transmission circuit, the power flows from the fault into the distribution station transformer (the opposite direction of power flow for normal faults). Tripping the distribution circuit breaker only for faults with forward power flow leaves the circuit breaker in for subtransmission faults.
- *Disable the instantaneous* — Without the instantaneous trip on the distribution feeder, the circuit breaker will wait longer before tripping. The transmission circuit is more likely to trip first.
- *Coordinate devices* — Coordinate the transmission-line protection to clear before the distribution circuit operates over the range of fault

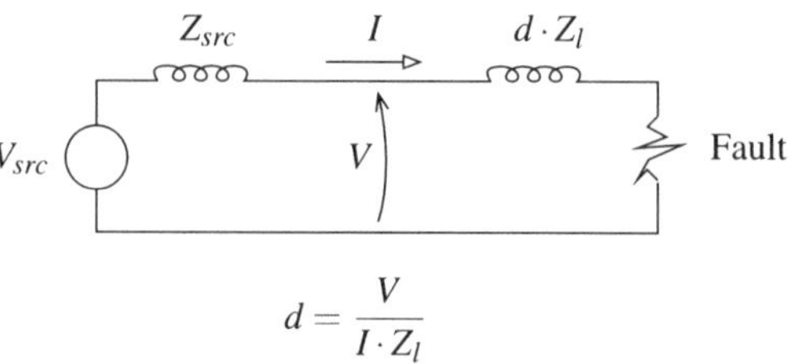

FIGURE 7.20
Fault location calculations.

currents that can occur. Include the effects of multiple reclose operations on the transmission circuit. Evaluate the substation recloser or circuit breaker; and also, consider feeder protective devices (normally reclosers). Try to speed up clearing times on the transmission circuit as much as possible (both ends for looped subtransmission circuits).

- *Ground switch (very experimental)* — Whenever the distribution circuit breaker opens, engage a grounding switch on the load side of the distribution circuit breaker. This grounds the fault, preventing overvoltages and sustaining the fault on the transmission circuit.
- *Structures* — As much as possible, design the common structure to minimize the chance of faults between circuits. Use wide spacings between the two circuits, and build the subtransmission circuit to high mechanical standards to reduce the chance of broken conductors or crossarms or braces. More extreme protection could be provided by stringing a grounded conductor and placing it between the subtransmission and the distribution circuits. If the ground is involved in the fault, it will prevent overvoltages. These options are obviously difficult to retrofit, but these issues should be considered when designing new circuits.

Thoroughly review such options before implementation. Normally, utilities treat underbuilt circuits the same as any other circuit. These are experimental approaches; I do not know of any implementation of these options for circuits with overbuilt transmission. Also, most of these options do not help as much for distribution lines fed from a different transmission source.

7.2.7 Fault Location Calculations

If we know the voltages and currents during a fault, we can use these to estimate the distance to the fault. The equation is very simple, just Ohm's Law (see Figure 7.20):

$$d = \frac{V}{I \cdot Z_l}$$

where

V = voltage during the fault, V

I = current during the fault, A

Z_l = line impedance, Ω/length unit

d = distance to the fault, length unit such as mi

With complex values entered for the voltages and impedances and currents, the distance estimate should come out as a complex number. The real component should be a realistic estimate of the distance to the fault; the imaginary component should be close to zero. If not, then something is wrong.

While the idea is simple, a useful implementation is more difficult. Different fault types are possible (phase-to-phase, phase-to-ground, etc.), and each type of fault sees a different impedance. Fault currents may have offsets. The fault may add impedance. There are uncertainties in the impedances, especially the ground return path. Conductor size changes also make location more difficult.

Many relays or power quality recorders or other instruments record fault waveforms. Some relays have fault-locating algorithms built in.

The Ohm's Law equation is actually overdetermined. We have more information than we really need. The distance is a real quantity, but the voltages, currents, and impedances are complex, so the real part of the result is the distance, and the reactive part is zero. Most fault-locating algorithms use this extra information, allow the fault resistance to vary, and find the distance that provides the optimal fit (Girgis et al., 1993; Santoso et al., 2000). The problem with this approach is that the fault resistance soaks up the error in other parts of the data. It does not necessarily mean a better distance estimation. Most fault arcs have a resistance that is very close to zero. In most cases, we're better off assuming zero fault resistance.

The most critical input to a fault location algorithm is the impedance data. Be sure to use the impedances and voltages and currents appropriate for the type of fault. For line-to-ground faults, use line-to-ground quantities; and for others, use phase-to-phase quantities:

- Line-to-ground fault

$$V = V_a,\ I = I_a,\ Z = Z_S = (2Z_1 + Z_0)/3$$

- Line-to-line, line-to-line-to-ground, or three-phase faults

$$V = V_{ab},\ I = I_a - I_b,\ Z = Z_1$$

Remember that these are all complex quantities. It helps to have software that automatically calculates complex phasors from a waveform. Several methods are available to calculate the rms values from a waveform; a Fourier transform is most common. Some currents have significant offset that can add error to the result. Try to find the magnitudes and angles after the offset has decayed (this is not possible on some faults cleared quickly by fuses).

If potential transformers are connected phase to phase, we can still estimate locations for ground faults if we know the zero-sequence source impedance. Schweitzer (1990) shows that the phase-to-ground voltage is

$$V_a = 1/3(V_{ab} - V_{ca}) - Z_{0,src}I_0$$

where

$Z_{0,src}$ = zero-sequence impedance of the source, Ω

I_0 = zero-sequence current measured during the fault = $I_a/3$ for a single line-to-ground fault on phase A

Although the voltages and currents are complex, we can also estimate the distance just using the absolute values. Although we lose some information on how accurate our solution is because we lose the phase angle information, in many cases it is as good as using the complex quantities. So, the simple fault location solution with absolute values is

$$d = \frac{V}{I \cdot Z_l}$$

where

V = absolute value of the rms voltage during the fault, V

I = absolute value of the rms current during the fault, A

Z_l = absolute value of the line impedance, Ω/length unit

d = distance to the fault, length unit such as mi

With this simple equation, we can estimate answers with voltage and current magnitudes. For a ground fault, Z_l=Z_S is about 1 Ω/mi. If the line-to-ground voltage, V=5000 V, and the fault current, I=1500 A, the fault is at about 3.3 mi (5000/1500). Remember to use the phase-to-phase voltage and $|I_a - I_b|$ (and not $|I_a| - |I_b|$) for faults involving more than one phase.

We can calculate the distance to the fault using only the magnitude of the current (no phase angles needed and only prefault voltage needed) and the line and source impedances involved. If we know the absolute value of the fault current and the prefault voltage and the source impedance, the distance to the fault is a solution to the quadratic equation

$$d = \frac{-b + \sqrt{b^2 - 4ac}}{2a}$$

where

$$a = Z_l^2$$

$$b = 2R_l R_{src} + 2X_l X_{src}$$

$$c = Z_{src}^2 - \left(\frac{V_{prefault}}{I_{fault}}\right)^2$$

and

R_{src} = source resistance, Ω
X_{src} = source reactance, Ω
Z_{src} = absolute value of the source impedance, Ω
R_l = line resistance, Ω/unit distance
X_l = line reactance, Ω/unit distance
Z_l = absolute value of the line impedance, Ω/unit distance
I_{fault} = absolute value of the rms current during the fault, A
$V_{prefault}$ = absolute value of the rms voltage just prior to the fault, V

In this case, we are doing the same thing as taking a fault current profile (such as Figure 7.11) and interpolating the distance. In fact, it is often much easier to use a fault current profile developed from a computer output rather than this messy set of equations. If the prefault voltage is missing, assume that it is equal to the nominal voltage. If we have the prefault voltage, divide the current by the per-unit prefault voltage before interpolating on the fault current profile. Using a fault current profile also allows changes in line impedances along the length of the line. Carolina Power & Light used this approach, and Lampley (2002) reported that their locations were accurate to within 0.5 mi 75% of the time; and in most of the remaining cases, the fault was usually no more than 1 to 2 mi from the estimate.

We can also just use voltages. If we know the source impedance, we do not need current. The distance calculation is another quadratic formula solution, this time with

$$d = \frac{-b - \sqrt{b^2 - 4ac}}{2a}$$ (the negative root because a is negative)

where

$$a = Z_l^2 - Z_l^2\left(\frac{V_{prefault}}{V_{fault}}\right)^2$$

$$b = 2R_l R_{src} + 2X_l X_{src}$$

$$c = Z_{src}^2$$

and

V_{fault} = absolute value of the rms voltage during the fault, V

As with the fault current approach, rather than using this equation, we can interpolate a voltage profile graph to find the distance to the fault (such as those in Figure 10.6). Again, we are assuming that the arc impedance is zero.

Fault locations of line-to-line and three-phase faults are most accurate because the ground path is not included. The ground return path has the most uncertainties. The impedance of the ground return path depends on the number of ground rods, the earth resistivity, and the presence of other objects in the return path (cable TV, buried water pipes, etc.). The ground return path is also nonuniform with length.

This type of fault location is useful for approximate locations. For permanent faults, a location estimate helps shorten the lengths of circuit patrolled. Distance estimates can also help find those irksome recurring temporary faults that cause repeated momentary interruptions. Fault locations are most accurate when the fault is within 5 mi of the measurement; beyond that, the voltage profile and fault current profiles flatten out considerably, which increases error. Fault location is difficult if a circuit has many branches. If a fault is 2 mi from the source on phase B, but there are 12 separate circuits that meet that criteria, the location information is not as useful. Fault location is also difficult on circuits with many wire-size changes; it works best on circuits with uniform mainline impedances with relatively short taps. Impedance-based location methods produce close but not pinpoint results. For underground faults, we would like to know exactly where to dig, but these methods do not have that accuracy.

7.3 Limiting Fault Currents

Limiting fault current has many benefits, which improve the safety and reliability of distribution systems:

- *Failures* — Overhead line burndowns are less likely, cable thermal failures are less likely, violent equipment failures are less likely.
- *Equipment ratings* — We can use reclosers and circuit breakers with less interrupting capability and switches and elbows with less momentary and fault close ratings. Lower fault currents reduce the need for current-limiting fuses and for power fuses and allow the use of cutouts and under-oil fuses.
- *Shocks* — Step and touch potentials are less severe during faults.
- *Conductor movement* — Conductors move less during faults (this provides more safety for workers in the vicinity of the line and makes conductor slapping faults less likely).
- *Coordination* — Fuse coordination is easier. Fuse saving is more likely to work.

At most distribution substations, three-phase fault currents are limited to less than 10 kA, with many sites achieving limits of 7 to 8 kA. The two main ways that utilities manage fault currents are:

- *Transformer impedance* — Specifying a higher-impedance substation transformer limits the fault current. Normal transformer impedances are around 8%, but utilities can specify impedances as high as 20% to reduce fault currents.
- *Split substation bus* — Most distribution substations have an open tie between substation buses, mainly to reduce fault currents (by a factor of two).

Line reactors and a neutral reactor on the substation transformer are two more options used to limit fault currents, especially in large urban stations where fault currents may exceed 40 kA.

There are drawbacks of increasing impedance to reduce fault currents. Higher impedance reduces the stiffness of the system: voltage sags are worse, voltage flicker is worse, harmonics are worse, voltage regulation is more difficult.

A reactor in the substation transformer neutral limits ground fault currents. Even though the neutral reactor provides no help for phase-to-phase or three-phase faults, it provides many of the benefits of other methods of fault reduction. Neutral reactors cost much less than line reactors. Ground faults are the most common fault; and for many types of single-phase equipment, the phase-to-ground fault is the only possible failure mode. A neutral reactor does not cause losses or degrade voltage regulation to the degree of phase reactors. On the downside, a neutral reactor has a cost and uses substation space, and a neutral reactor reduces the effectiveness of the grounding system (see Chapter 13).

Several advanced fault-current limiting devices have been designed (EPRI EL-6903, 1990). Most use some sort of nonlinear elements — arresters, saturating reactors, superconducting elements, or power electronics such as a gate-turn-off thyristor — to limit the fault current either through the physics of the device or through computer control. Since most distribution systems have managed fault currents sufficiently well, these devices have not found a market. Given that, the EPRI study surveyed utilities and found evidence that a market for fault-current limiters exists if a device had low enough cost and was robust enough.

7.4 Arc Characteristics

Many distribution faults involve arcs through the air, either directly through the air or across the surface of hardware. Although a relatively good conductor, the arc is a very hot, explosive fireball that can cause further damage at the fault location (including fires, wire burndowns, and equipment damage). This section discusses some of the physical properties of arcs, along with the ways in which arcs can cause damage.

Normally, the air is a relatively good insulator, but when heavily ionized, the air becomes a low-resistance conductor. An arc stream in the air consists of highly ionized gas particles. The arc ionization is due to *thermal* ionization caused by collisions from the random velocities of particles (between electrons, photons, atoms, or molecules). Thermal ionization increases with increasing temperature and with increasing pressure. The heat produced by the current flow (I^2R) maintains the ionization. The arc stream has very low resistance because there is an abundance of free, charged particles, so current flow can be maintained with little electric field. Another type of ionization caused by acceleration of electrons from the electric field may initially start the ionization during the electric-field breakdown, but once the arc is created, electric-field ionization plays a less significant role than thermal ionization.

One of the characteristics that is useful for estimating arc-related phenomenon is the arc voltage. The voltage across an arc remains constant over a wide range of currents and arc lengths, so the arc resistance decreases as the current increases. The voltage across an arc ranges between 25 and 40 V/in (10 to 16 V/cm) over the current range of 100 A to 80 kA (Goda et al., 2000; Strom, 1946). The arc voltage is somewhat chaotic and varies as the arc length changes. More variation exists at lower currents. As an illustration of the energy in an arc, consider a 3-in. (7.6-cm) arc that has a voltage of about 100 V. If the fault current is 10 kA, the power in the arc is $P = V \cdot I = 100\text{ V} \cdot 10\text{ kA} = 1\text{ MW}$. Yes, 1 MW! Arcs are explosive and as hot as the surface of the sun.

An upper bound of roughly 10,000 to 20,000 K on the temperature of the arc maintains the relatively constant arc voltage per unit length. For larger currents, the arc responds by increasing the volume of gas ionized (the arc expands rather than increasing the arc-stream temperature). Higher currents increase the cross-sectional area of the arc, which reduces the resistance of the arc column; the current density is the same, but the area is larger. So, the voltage drop along the arc stream remains roughly constant. The arc voltage depends on the type of gas and the pressure. One of the reasons an arc voltage under oil has a higher voltage gradient than an arc in air is because the ionizing gas is mainly hydrogen, which has a high heat conductivity. A high heat conductivity causes the arc to restrict and creates a higher-density current flow (and more resistance). The arc voltage gradient is also a function of pressure. For arcs in nitrogen (the main ionizing gas of arcs in air), the arc voltage increases with pressure as $V \propto P^k$, where k is approximately 0.3 (Cobine, 1941).

Another parameter of interest is the arc resistance. A 3-ft (1-m) arc has a voltage of about 1400 V. If the fault current at that point in the line was 1000 A, then the arc resistance is about 1.4 Ω. A 1-ft (0.3-m) arc with the same fault current has a resistance of 0.47 Ω. Most fault arcs have resistances of 0

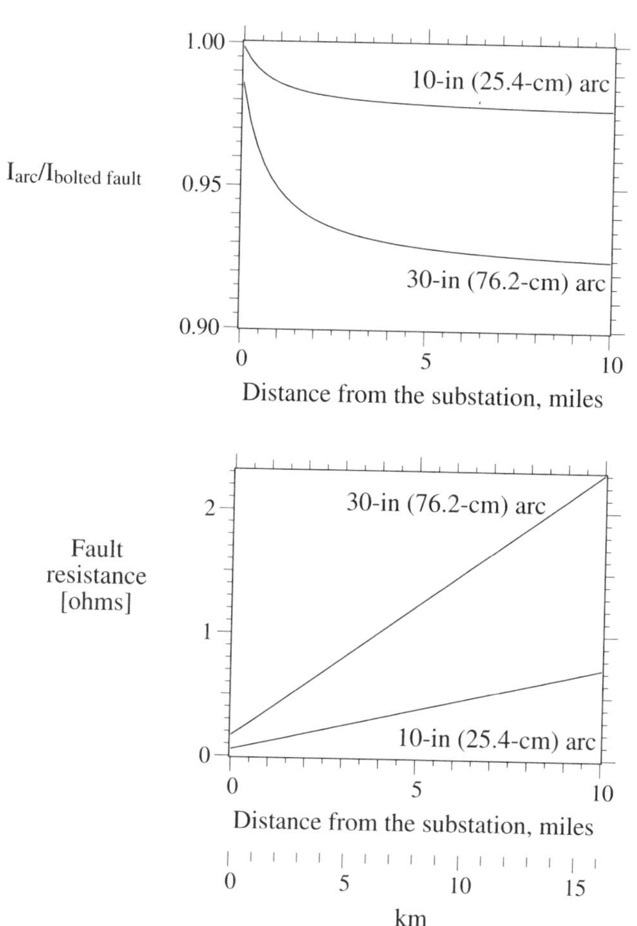

FIGURE 7.21
Ratio of fault current with and without an arc of the given length on a 12.47-kV circuit. This assumes the same system parameters as the fault profile in Figure 7.11 with the following additional assumptions: the arc voltage gradient equals 40 V/in. (16 V/cm), the arc voltage is all resistive, and the nonlinearity of the arc voltage is ignored.

to 2 Ω. Figure 7.21 shows that the impact of an arc on the fault currents along the line is fairly minor.

An arc voltage waveform has distinguishing characteristics. Figure 7.22 shows an arcing fault voltage that was initiated by tree contact on a 13-kV circuit measured during the EPRI DPQ project. The voltage on the arc is in phase with the fault current (it is primarily resistive). When the arc current goes to zero, the arc will extinguish. The recovery voltage builds up quickly because of the stored energy in the system inductance. Voltage builds to a point where it causes arc reignition. The reason for the blip at the start of the waveform (it is not a straight square wave) is that the arc cools off at the current zero. Cooling lowers the ionization rate and increases the arc resis-

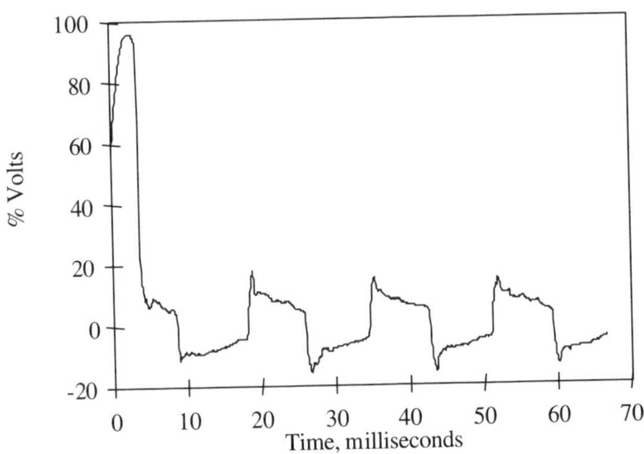

FIGURE 7.22
Arcing fault measured during the EPRI DPQ study. (Copyright © 1996. Electric Power Research Institute. TR-106294-V3. *An Assessment of Distribution System Power Quality: Volume 3: Library of Distribution System Power Quality Monitoring Case Studies.* Reprinted with permission.)

tance. Once it heats up again, the voltage characteristic flattens out. The waveform is high in the odd harmonics and for many purposes can be approximated as a square wave.

The movement and growth of an arc is primarily in the vertical direction. Tests at IREQ in Quebec showed that the primary reason that the arc elongates and moves vertically is the rising hot gases of the arc (Drouet and Nadeau, 1979). The magnetic forces ($J \times B$) did not dominate the direction or elongation. As a first approximation over a range of currents between 1 and 20 kA, arc voltages up to 18 kV, and durations up to 0.5 sec, the arc length can be expressed as a function of the duration only as

$$l = 30t$$

where

l = arc length, m

t = fault duration, sec

The arc movement is a consideration for underbuilt distribution and for vertical construction. The equation above can be used as an approximation to determine if an underbuilt distribution circuit could evolve and fault a distribution or transmission circuit above. It also gives some idea of how faults can evolve to more than one phase. Figure 7.23 shows an example of a fault evolving from one to three phases over the course of about 1.5 sec (the construction type is unknown, but it is probably a horizontal configuration). Given the vertical movement of a fault current arc, vertical designs are more prone to having faults evolve to more than one phase. We might

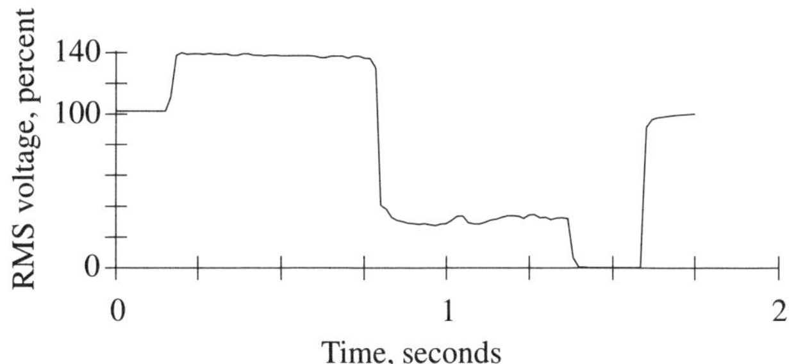

FIGURE 7.23
Voltage waveform from a fault that started as a single-phase fault (indicated by a swell on the phase shown), evolved to a double-line-to-ground fault (the voltage sags to about 35%), and finally evolved to a three-phase-to-ground fault (where the voltage sags to zero). (Copyright © 1996. Electric Power Research Institute. TR-106294-V3. *An Assessment of Distribution System Power Quality: Volume 3: Library of Distribution System Power Quality Monitoring Case Studies.* Reprinted with permission.)

think that a fault evolving to include other phases is not a concern since the three-phase circuit has to be opened anyway, whether it is a single- or three-phase fault. But, the voltage sag during the fault is more severe if more than one phase is involved, which is a good reason to use designs that do not tend to propagate to more than one phase (and to use relaying or fusing that operates quickly enough to prevent it from happening).

The temperature in the arc can be on the order of 10,000 K. This heat creates hazards from burning and from the pressure wave developed during the fault. The longer the arc, the more energy is created. NFPA provides guidelines on safe distances for workers based on arc blasts (NFPA 70E-2000). Several groups have worked to determine the appropriate characteristics of protective clothing (ASTM F1506-94, 1994). Because of the pressure wave, consider hearing protection and fall protection for workers who could be exposed to fault arcs. Arcs can cause fires: pole fires or fires in oil-filled equipment such transformers.

The pressure wave from an arc in an enclosed substation was the probable cause of a collapse of a substation building (important since many distribution stations are required to be indoors because of environmental considerations). Researchers found during tests that the pressure from a fault arc can be approximated (Drouet and Nadeau, 1979) by

$$A = 1.5\frac{I \cdot t}{l}$$

where

A = pressure, kN/m^2 (1 kN/m^2 = 20.9 lb/ft^2)
I = fault current magnitude, kA
t = duration of the fault, sec
l = distance from the source, m (for $l > 1$ m)

Although many electrical damage characteristics are a function of $\int I^2\, dt$, the pressure wave is primarily a function of $\int I\, dt$ (because the voltage along the arc length is constant and relatively independent of the arc current). Where arcs attach to wires, melting weakens wires and can lead to wire burndown. Most tests have shown that the damage is proportional to $\int I^k\, dt$, where k is near one but varies depending on the conductor type. For burndowns or other situations where the arc burns the conductor, the total length of the arc is unimportant, the small portion of the arc near the attachment point is important. The voltage drop near the attachment point is also very constant and does not vary significantly with current. The damage to conductors is very much like that of an electrical arc cutting torch.

Burndowns are much more likely on covered wire (also called tree wire). The covering restricts the movement of the attachment point of the arc to the conductor. On bare wire, the arc will move because of the heating forces on the arc and the magnetic forces (also called motoring).

On bare wire, burndowns are a consideration only on small conductors. Tests (Lasseter, 1956) have shown that the main cause of failure on small aluminum conductors is that the hot gases from the arc anneal the aluminum, which reduces tensile strength. The testers found little evidence of arc burns on the conductors. Failures can occur midspan or at a pole. Motoring is not fast enough to protect the small wire.

Arcs can damage insulators following flashover along the surface of the insulators. This was the primary reason for the development of arcing horns for transmission-line insulators. Arcing horns encourage a flashover away from the insulator rather than along the surface. Arcs can fail distribution insulators. During fault tests across insulators by Florida Power & Light (Lasseter, 1965), the top of the arc moved along the conductor. The point of failure was at the bottom of the insulator where the arc moved up the pin to the bottom edge of the porcelain. The bottom of the insulator gets very hot and can fail from thermal shock. The threshold of chipping was about 360 C (C = coulombs = A-sec = $\int I\, dt$), and the threshold of shattering was about 1125 C (see Figure 7.24). Adding an aluminum or copper washer (but not a steel washer) on top of the crossarm under the flange of the grounded steel pin reduced insulator shattering. The arc attaches to the washer rather than moving up along the pin, increasing the threshold of chipping by a factor of five. Composite insulators perform better for surface arcs than porcelain insulators (Mazurek et al., 2000). Some composite insulators have an external arc withstand test where $I \cdot t$ shall be 150 kA-cycles (2500 C for 60 Hz) (IEEE Std. 1024-1988).

Distribution voltages can sustain very long arcs, but self-clearing faults can occur such as when a conductor breaks and falls to the ground (stretching an arc as it falls). The maximum arc length is important because the longer the arc, the more energy is in the arc. For circuits with fault currents on the order of 1000 A and where the transient rise to the open-circuit voltage is about 10 μs, about 50 V may be interrupted per centimeter of arc length (Slepian, 1930). For a line-to-ground voltage of 7200 V, a line-to-ground arc

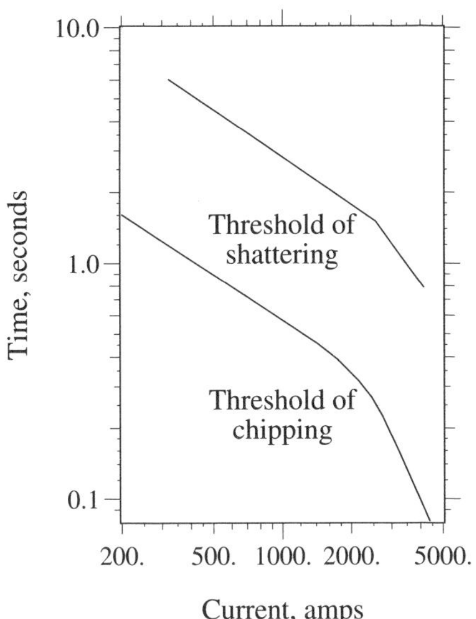

FIGURE 7.24
Insulator damage characteristics. (Data from [Lasseter, 1965].)

can reach a length of about 12 ft (3.7 m) before it clears. As another approximation, the length that an arc can maintain in resistive circuits is [from (Rizk and Nguyen, 1984) with some reformulation]:

$$l = V\sqrt{I}$$

where

l = arc length, in. (1 in. = 2.54 cm)

I = rms current in the previous half cycle, A

V = system rms voltage, kV (line to ground or line to line depending on fault type)

7.5 High-Impedance Faults

When a conductor comes in physical contact with the ground but does not draw enough current to operate typical protective devices, you have a *high-impedance* fault. In the most common scenario, an overhead wire breaks and falls to the ground (a *downed wire*). If the phase wire misses the grounded neutral or another ground as it falls, the circuit path is completed by the high-impedance path provided by the contact surface and the earth.

The return path for a conductor lying on the ground can be a high impedance. The resistance varies depending on the surface of the ground. Table

TABLE 7.3

Typical High-Impedance Fault Current Magnitudes

Surface	Current, A
Dry asphalt	0
Concrete (no rebar)	0
Dry sand	0
Wet sand	15
Dry sod	20
Dry grass	25
Wet sod	40
Wet grass	50
Concrete (with rebar)	75

Source: IEEE Tutorial Course 90EH0310-3-PWR, "Detection of Downed Conductors on Utility Distribution Systems," 1990. With permission. ©1990 IEEE.

7.3 shows typical current values measured for conductors on different surfaces (for 15-kV class circuits).

The frequency of high-impedance faults is uncertain. Most utilities responding to an IEEE survey reported that high-impedance faults made up less than 2% of faults while a sizeable number (15% of those surveyed) suggested that between 2 and 5% of distribution faults were not detectable (IEEE Working Group on Distribution Protection, 1995). Even with small numbers, high-impedance faults pose an important safety hazard.

On distribution circuits, high-impedance faults are still an unsolved problem. It is not for lack of effort; considerable research has been done to find ways to detect high-impedance faults, and progress has been made [see (IEEE Tutorial Course 90EH0310-3-PWR, 1990) for a more in-depth summary]. Research has identified many characteristics of high-impedance faults and have tested them for detection purposes. Efforts have been concentrated on detection at the substation based on phase and ground currents. High-impedance faults usually involve arcing, and arcing generally creates the lower odd harmonics. Arcing faults may also contain significant 2- to 10-kHz components. Arcing also bursts in characteristic patterns. High-impedance faults often cause characteristic changes in the load (for example, a broken conductor will drop the load on that phase). None of these detection methods is perfect, so some detection schemes use more than one method to try to detect high-impedance faults.

We can also detect broken conductors at the ends of radial circuits. Loss of voltage is the simplest method. Communication to an upstream protective device or to a control center is required. A difficulty is that it takes many devices to adequately cover a radial circuit (depending on how many branches occur on the circuit). Also, the "ends" of circuits could change during circuit reconfigurations or sectionalizing due to circuit interruptions. Also, if the loss-of-voltage detector is downstream of a fuse, another detector is needed at the fuse, so we can determine if the fuse operated or the conductor broke.

Practices that help reduce high-impedance faults include

- *Tight construction framings* — If a phase wire breaks, it is more likely to contact a neutral as it falls. (A drawback is that utilities have reported poorer reliability with tighter constructions like the armless design.) A vertical construction is better than a horizontal construction. Single-phase structures are better than three-phase structures.
- *Stronger conductors* — Larger conductors or ACSR instead of all-aluminum conductor are stronger and less likely to break for a given mechanical or arcing condition.
- *Smaller/faster fuses* — Faster fuses are more likely to operate for high-impedance faults. In addition, small fuses are likely to clear before arcing damages wires, which could burn down the wires.
- *Tree trimming* — Clearing trees and trimming reduces the number of trees or branches breaking conductors.
- *Fewer reclose attempts* — Each reclose attempt causes more damage at the fault location.
- *Higher primary voltages* — High-impedance faults are much less likely at 34.5 kV and somewhat less likely at 24.94 kV than 15-kV class voltages.
- *Public education* — Public advertisements warning the public to stay away from downed wires help reduce accidents when high-impedance faults do occur.

Practices to *avoid* include:

- *Covered wire* — Burndowns are more likely with covered wire. If a covered wire does contact the ground, it is less likely to show visible signs that it is energized such as arcing or jumping which would help keep bystanders away.
- *Unfused taps* — Burndowns are more likely with the smaller wire used on lateral taps.
- *Midspan connectors* — Flying taps can cause localized heating and mechanical stress.
- *Rear-lot construction* — Rear-lot construction is not as well maintained as road-side construction, so trees are more likely to break wires. If wires do come down, it is more hazardous since they are coming down in someone's backyard.
- *Neutral on the crossarm* — If a phase wire breaks, it is much less likely to contact the neutral as it falls if the neutral is on the crossarm. Other constructions that may have this same problem are overhead shield wires and spacer cables that do not have an additional neutral below.

Three-wire distribution systems have some advantages and some disadvantages related to high-impedance faults. The main advantage of three-wire systems is that there is no unbalanced load. A sensitive ground relay can be used, which would detect many high-impedance faults. The sensitivity of the ground relay is limited by the line capacitance. The main disadvantage of three-wire systems is that there is no multigrounded neutral. If a phase conductor breaks, there is a high probability that there will be a high-impedance fault. If there is underbuilt secondary or phone or cable TV under the three-wire system, then a high-impedance fault is less likely because a grounded conductor is below the phases.

Spacer cable has some mechanical strength advantages that could help keep phase wires in the air, and it has fewer faults due to trees. A downside is the covering which makes burndowns more likely. Also, it has a messenger wire that may act as the neutral; if it does not also have an underbuilt neutral, a phase conductor is more likely to fall unimpeded.

Backfeeds from three-phase transformer and capacitor installations can cause dangerous situations. If a wire breaks near a pole, at least half of the time, the load side (downstream side) will lie on the ground. Backfeed to the downed wire can occur through three-phase transformers downstream of the fault. The backfeed can provide enough voltage and current to the downed wire to be dangerous (but there will not be enough current to trip protective devices). Note that a grounded-wye – grounded-wye connection does not eliminate backfeeds. Another backfeed scenario is shown in Figure 7.25 where a three-phase load is fed from a fused tap. A bolted, low-impedance fault on one phase will blow the fuse, but backfeed current may flow through the three-phase connection. Even with a low-impedance fault, the backfeed current can be low enough that neither the remaining tap fuses nor the transformer fuses will operate. The fault can continue to arc until the wire burns down. An ungrounded capacitor bank can also provide a backfeed path.

Commercial relays that have high-impedance fault detection capabilities are available. One of the main problems with detection of high-impedance faults is false fault detections. If a detection system in the station detects a fault and a whole feeder is tripped for an event that is not really a high-impedance fault, reliability is severely hurt. Before reenergizing, crews patrol the circuit and make sure there was not a downed conductor. The sensitivity of a device must be traded off against its dependability. If it is too sensitive, many false operations will result. An alternative to tripping is alarming. Operators in control centers may always trip a circuit if it signals a high-impedance fault for fear of discipline if it really was a high-impedance fault and an accident occurred. Each time a high-impedance fault is detected, crews would have to patrol the circuit. If operators have too many false alarms, they may ignore the alarm.

A tree branch in contact with a phase conductor also forms a high-impedance fault. This is not as dangerous as a downed wire. Most of the time the circuit operates normally, without danger to the public (but see Chapter 13

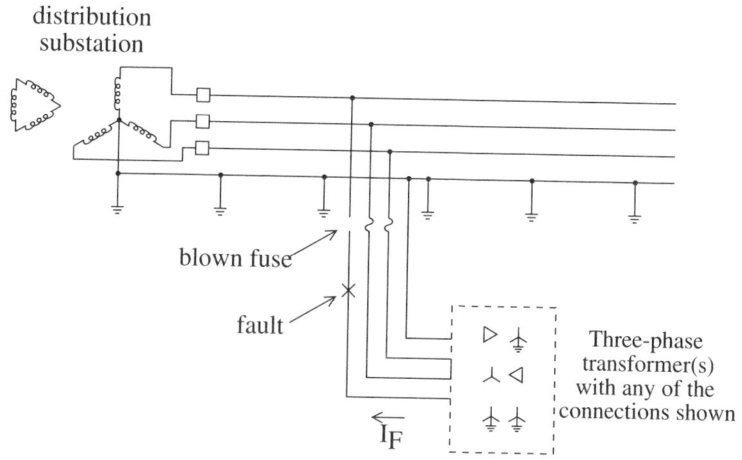

FIGURE 7.25
Backfeed to a fault downstream of a blown fuse.

for more analysis of this situation). A tree branch in contact with a phase conductor can draw enough current to trip a high-impedance fault detector. In a heavily treed area, crews could require many hours to find the location where the tree contact is taking place.

One of the main problems with substation detection is that each feeder usually covers many miles of line. With most faults, lateral fuses provide an effective way to isolate and identify the location of faults within a relatively small area. While it would be nice to have high-impedance fault detection capability on lateral taps, the costs have been prohibitive. Contrast a station detection scheme (one device) with detectors at taps — tens or hundreds of devices.

Another solution to falling conductors is using guards installed on poles below the phase wires to "catch" phase conductors. The guards are connected to the grounded neutral, so when a phase conductor breaks, a low-impedance fault is created. This would be a significant expense to install system-wide, but it may be suitable for isolated locations where it is critical not to have energized downed conductors (a stretch that runs across a school playground or a span that crosses a major road).

7.6 External Fault Causes

7.6.1 Trees

For many utilities, trees cause more faults and more interruptions than any other factor. Tree trimming is expensive — the largest maintenance item for most distribution companies. Tree trimming is also a contentious issue with the public. Residents hate to have their 100-year-old maple trees touched (or even their 30-year-old cottonwoods).

Faults caused by trees generally occur from three conditions:

1. Falling trees knock down poles or break pole hardware.
2. Tree branches blown by the wind push conductors together.
3. A branch falls across the wires and forms a bridge from conductor to conductor (or natural tree growth causes a bridge).

Tree-caused faults can be temporary or permanent. Falling trees or branches can cause permanent faults. Either falling across wires or pushing them together, tree branches can cause temporary faults. Broken tree branches account for the majority of interruptions. In one utility in the northeast U.S., 63% were caused by broken branches compared to 11% from falling trees and only 2% from tree growth (Simpson, 1997). Niagara Mohawk Corporation (NIMO) found that 86% of permanent tree-faults were outside of the right-of-way, and most were from major breakage (see Figure 7.26). Falling trees do the most damage; they often break conductors and even poles in some cases. Tree faults usually occur during storms, primarily from wind. Snow and ice additionally contribute to tree failures.

Several companies have done tests to evaluate the electrical properties of tree branches and how they cause faults (Goodfellow, 2000; Williams, 1999). For a tree branch to cause a fault, the branch must bridge the gap between two conductors, which usually must be sustained for more than 1 min. A tree touching just one conductor will *not* fault. The tree branch must cause a connection between two bare conductors (it can be phase to phase or phase to neutral). A tree branch into one phase conductor normally draws less than one amp of current under most conditions; this may burn some leaves, but it will not fault. On small wires in contact with a tree, the arcing to the tree may be enough to burn the wire down under the right conditions. While the tree in contact with one wire will not fault the circuit, there are some safety issues with trees in contact with overhead conductors (see Chapter 13).

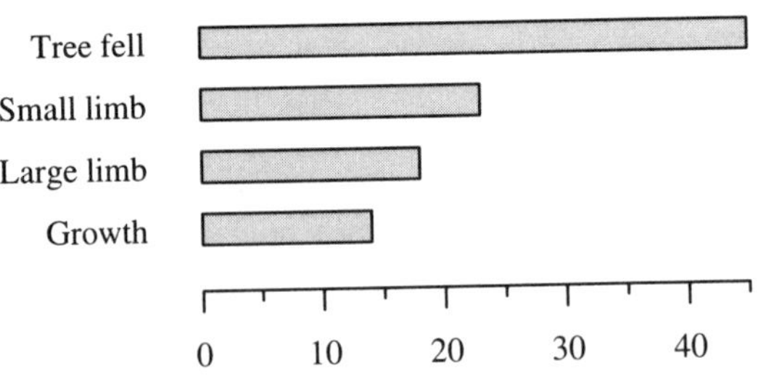

FIGURE 7.26
Tree failure causes for the Niagara Mohawk Power Corporation. (Data from [Finch, 2001].)

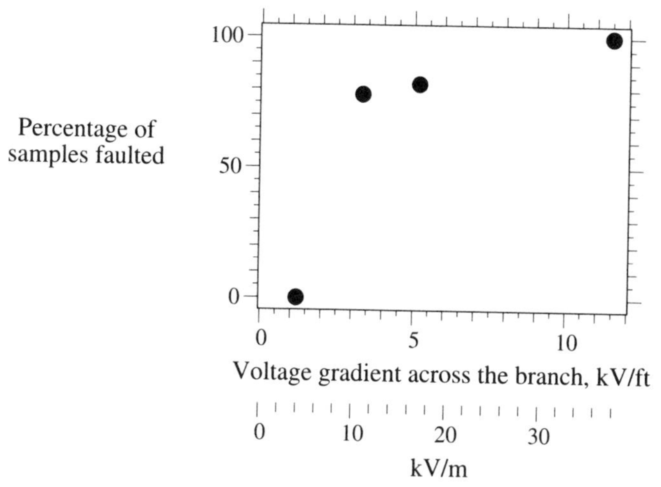

FIGURE 7.27
Percentage of samples faulted based on the voltage gradient across the tree branch. (Data from [Goodfellow, 2000].)

A fault across a tree branch between two conductors takes some time to develop. If a branch falls across two conductors, arcing occurs at each end where the wire is in contact with the branch. At this point in the process, the current is small (the tree branch has a relatively high impedance). The arcing burns the branch and creates carbon by oxidizing organic compounds. The carbon provides a good conducting path. Arcing then occurs from the carbon to the unburned portion of the branch. A carbon track develops at each end and moves inward.

Once the carbon path is established completely across the branch, the fault is a low-impedance path. Now the current is high; it is effectively a bolted fault. It is also a permanent fault. If a circuit breaker or recloser is opened and then reclosed, the low-impedance carbon path will still be there unless the branch burns enough to fall off of the wires.

Some other notable electrical effects include the following:

- It makes little difference if the branch is wet or dry. Live branches are more likely to fault for a given voltage gradient, but dead branches are more likely to break.
- Little branches can burn through and fall off before the full carbon track develops, so minor leaf and branch burning does not cause faults.
- The likelihood of a fault depends on the voltage gradient along the branch (see Figure 7.27).
- The time it takes for a fault to occur also depends on the voltage gradient (see Figure 7.28).
- Lower voltage circuits are much more immune to flashovers from branches across conductors. A 4.8-kV circuit on a 10-ft crossarm has

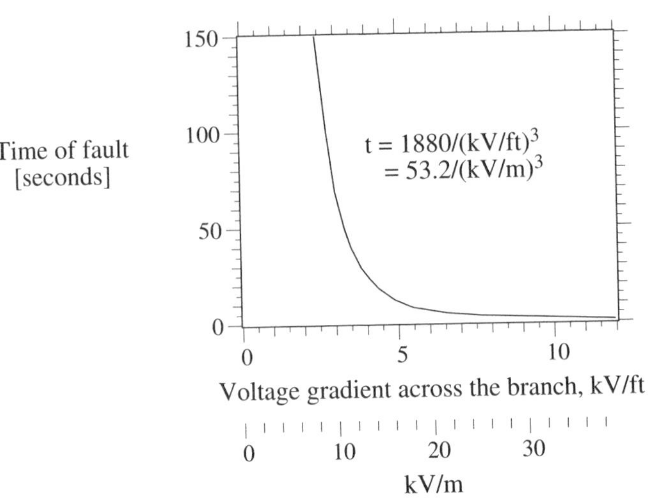

FIGURE 7.28
Time to fault based on the voltage gradient across the tree branch. (Data from [Goodfellow, 2000] with the curvefit added)

about a phase-to-phase voltage gradient of 1 kV/ft, very unlikely to fault from tree contact. A 12.47-kV circuit has a 2.7 kV/ft gradient, which is more likely to fault.

- Candlestick or armless designs are more likely to flashover because of tighter conductor-to-conductor spacings.

These effects reveal some key issues:

- Trimming around the conductors in areas with a heavy canopy does not prevent tree faults. Traditionally, crews trim a "hole" around the conductors with about a 10-ft (3-m) radius. If there is a heavy canopy of trees above the conductors, this trimming strategy performs poorly since most faults are caused by branches falling from above.
- Vertical construction may help since the likelihood of a phase-to-phase contact by falling branches is reduced.
- Three-phase construction is more at risk than single-phase construction.

Tree trimming is expensive. An EPRI survey found that utilities spend an average of about $10 per customer each year on tree trimming (EPRI TR-109178, 1998). Trimming can also irritate communities. It is always a dilemma that people do not want their trees trimmed, but they also do not want interruptions. Consider the following general tree-trimming guidelines:

- *Removal* — This is the most effective fault-prevention strategy, and many homeowners are willing to have trees removed.

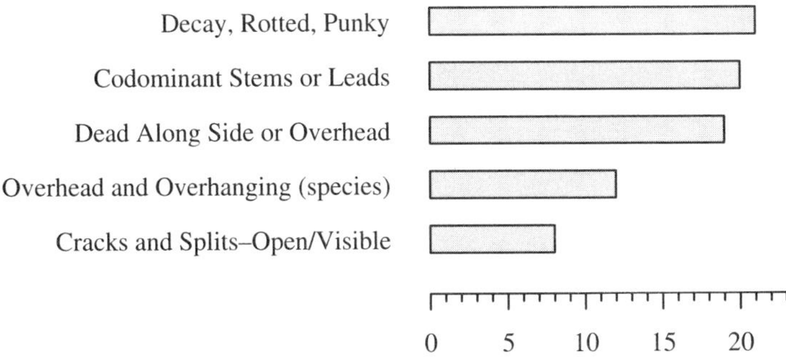

FIGURE 7.29
Defects causing tree failure for the Niagara Mohawk Power Corporation. (Data from [Finch, 2001].)

- *Danger trees* — Trimming/removal is most effective if trees and branches that are likely to fail are removed or trimmed to safe distances. This does take some expertise by tree trimming crews.
- *Target* — As with any fault-reduction program, efforts are best spent on the poorest performing circuits that affect the most customers. Along the same thought, spend more on three-phase mains than on single-phase taps.

Targeting danger trees is especially beneficial but requires expertise. In a careful examination of several cases where broken branches or trees damaged the system, 64% of the trees were living (Finch, 2001). Finch also advises examining trees from the backside, inside the tree line (defects on that side are more likely to fail the tree into the line). Finch describes several defects that help signal danger trees (see Figure 7.29). Dead trees or large splits are easy to spot. Cankers (a fungal disease) or codominant stems (two stems, neither of which dominates, each stem at a branching point is approximately the same size) require more training and experience to detect. It also helps to know the types of trees that are prone to interruptions (this varies by area and types of trees). For Niagara Mohawk, black locusts and aspens are particularly troublesome; large, old roadside maples also caused more than their share of damage (see Table 7.4).

Acceptable tree trimming (that is also still effective) is a public relations battle. Some strategies that help along these lines include:

- Talking to residents prior to/during tree trimming.
- Trimming trees during the winter (or tree-trimming done "under the radar") — The community will not notice tree trimming as much when the leaves are not on the trees.

TABLE 7.4

Comparison of Trees Causing Permanent Faults with the Tree Population

Species	Percent of Outages	Percent of New York State Population
Ash	8	7.9
Aspen	9	0.6
Black Locust	11	0.3
Black Walnut	5	N/A
Red Maple	14	14.7
Silver Maple	5	0.2
Sugar Maple	20	12.0
White Pine	6	3.3

Source: Finch, K., "Understanding Tree Outages," EEI Vegetation Managers Meeting, Palm Springs, CA, May 1, 2001.

- Trimming trees during storm cleanups. Right after outages, residents are more willing to accept their beloved trees being hacked up (this is a form of the often practiced "storm-induced maintenance"; fix it when it falls down).
- Cleaning up after trees are trimmed/removed.
- Offering free firewood.

Choosing a tree-trimming cycle is tricky. Many utilities use a three- to five-year cycle. Longer tree-trimming cycles lead to higher fault rates (see Figure 7.30). The optimal trimming cycle depends on

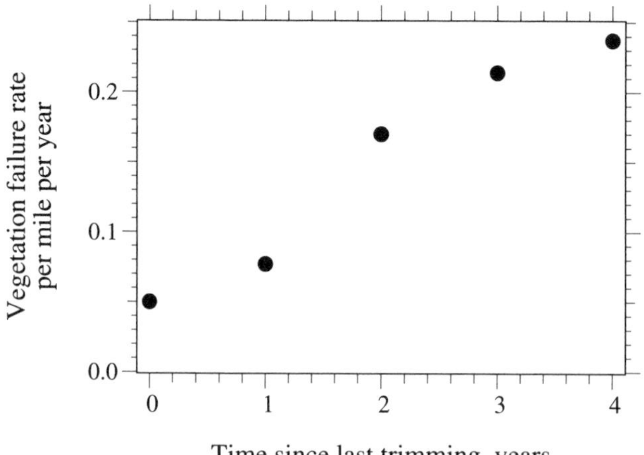

FIGURE 7.30
Tree failure rates vs. time since last trimming for one Midwestern utility. (Data from [Kuntz, 1999].)

TABLE 7.5

Interruption Rates in Outages per 100 Miles per Year Comparing Bare Wire, Tree Wire, and Spacer Cable at One Utility in the U.S. from 1995 to 1997

Fault Type	Bare	Tree Wire	Spacer Cable
Tree related	17.6	6.6	1.8
Animals	12.1	5.9	2.9
Lightning	3.4	1.9	1.0
Unknown	5.9	2.6	1.0
All other	11.3	4.6	5.9
Totals	50.3	21.7	12.5

Source: Hendrix, "Reliability of Overhead Distribution Circuits," Hendrix Wire & Cable, Inc., August 1998.

- Type of trees, growth rates, and growing conditions
- Community tolerance for trimming
- Economic assumptions, especially the chosen time value of money

In heavily treed areas, covered conductors help reduce tree faults. This "tree wire" provides extra insulation that reduces the chance of flashover for a branch between conductors. Good fault data is hard to find comparing fault rates of bare wire with covered wire. One utility whose results are provided in Table 7.5 has shown reductions in interruption rates of greater than 50% for covered wire and even more for spacer cable (the only caveat here is the data is published by a manufacturer, so it may not be unbiased). Tree and animal faults were also reduced by over 50%. European experience with covered conductors suggests that covered-wire fault rates are about 75% less than bare-wire fault rates. In Finland, fault rates on bare lines are about 3 per 100 km/year on bare and 1 per 100 km/year on covered wire (Hart, 1994). Covered wire helps with animal faults as well as tree faults.

Spacer cable and aerial cables are also alternatives that perform well in treed areas. Spacer cables are a bundled configuration using a messenger wire holding up three-phase wires that use covered wire. Aerial cables have fully-rated insulation just like underground cables. In South America, both covered wire and a form of aerial cable have been successfully used in treed areas (Bernis and de Minas Gerais, 2001). The Brazilian company CEMIG found that spacer cable faults were lower than bare-wire circuits by a 10 to 1 ratio (although the article did not specify if this included both temporary and permanent faults). The aerial cable faults were lower than bare wire by a 20 to 1 ratio. The effect on interruption durations is shown in Table 7.6. Several spacer cables or aerial cables can be constructed on a pole. Spacer cables and aerial cables have some of the same burndown considerations as covered wire. Spacer cable construction does have a reputation for being hard to work with. Both spacer cable and aerial cable costs more than bare

TABLE 7.6

Comparison of the Reliability Index SAIDI (Average Hours of Interruption per Customer per Year) of Bare Wire, Spacer Cable, and Aerial Cable in Brazil

Construction	SAIDI, h
Bare wire	9.9
Spacer cable	4.7
Aerial cable	3.0

Source: Bernis, R. A. O. and de Minas Gerais, C. E., "CEMIG Addresses Urban Dilemma," *Transmission and Distribution World*, vol. 53, no. 3, pp. 56–61, March 2001.

wire. CEMIG estimated that the initial investment was returned by the reduction in tree trimming. They did minimal trimming around aerial cable (an estimated factor of 12 reduction in maintenance costs) and only minor trimming around spacer cable (an estimated factor of 6 reduction in maintenance costs).

7.6.2 Weather and Lightning

Many faults on overhead circuits are weather related: icing, wind, and lightning. The fault rate during severe storms increases dramatically. Much of the physical and electrical stresses from these events are well beyond the design capability of distribution circuits.

Overhead circuits are designed to NESC (IEEE C2-1997) mechanical standards and clearances, which prescribe the performance of the line itself to the normal severe weather that the poles and wires and other structures must withstand. Most storm failures are from external causes, usually wind blowing tree limbs or whole trees into wires. These cause faults and can bring down whole structures.

Lightning causes many faults on distribution circuits. While most are temporary and do not do any damage, 5 to 10% of lightning faults permanently damage equipment: transformers, arresters, cables, insulators. Distribution circuits do not have any direct protection against lightning-caused faults since distribution insulation cannot withstand lightning voltages. If lightning hits a line, it causes a fault nearly 100% of the time. Since most lightning-caused faults do not do any permanent damage, reclosing is used to minimize the impact on customers. After the circuit flashes over (and there's a fault), a recloser or reclosing circuit breaker will open and, after a short delay, reclose the circuit.

It is important to properly protect equipment from lightning. Transformers and cables are almost always protected with surge arresters. This prevents most permanent faults caused by lightning. Equipment protection, arresters, and lightning protection are discussed in more detail in Chapter 12.

7.6.3 Animals

Faults caused by animals are often the number two cause of outages for utilities (after trees). Squirrels cause the most faults. Squirrels thrive in suburbs and love trees; utilities have noted increases in squirrel faults following development of wooded areas. Squirrels are usually active in the morning and sleep at night. Squirrel faults usually occur in fair weather. The patterns of animal-caused faults have been used to classify "unknown" faults (Mo and Taylor, 1995).

The two main ways to protect equipment against animals (particularly squirrels) are:

- Bushing protectors
- Covered lead wires

Both of these were rated "very good" at reducing animal-caused interruptions in an EPRI survey (EPRI TE-114915, 1999). Several survey respondents noted that the bushing protectors were susceptible to deterioration and tracking (they rated only "good" for durability). Some of the other comments regarding bushing protectors include:

- Insects nest in the bushing coverings, and birds probing for insects cause bird electrocutions and faults.
- Bushing covers hide loose connections and insulator damage and interfere with infrared inspections.

Bushing protectors and covered lead wires are inexpensive if installed with equipment (but expensive to retrofit). For transformer bushing protectors, have crews leave some room between the bottom of the bushing protector and the tank, so water does not build up and leak down through the bushing. Some additional items that also help include

- Trimming trees — Squirrels get to utility equipment via trees (pole climbing is less common). If trees are kept away from lines, utility equipment is less attractive.
- Good outage tracking — Many outages are repeated, so a good outage tracking system can pinpoint hotspots to identify where to target maintenance.
- Identifying animal — If outages are tracked by animal, it is easier to identify proper solutions.
- Maintaining proper clearances
- Avoiding metal crossarms

Animal faults vary by construction habits within a region. Some common problem areas that can lead to frequent animal faults include

- *Transformer bushings* — A very common animal fault is across a transformer bushing. Insulating paints are available for transformers, but it degrades quickly. Bushing guards and/or insulated lead wires offer the best protection.
- *Arresters* (especially polymer) — Another common animal fault is across an arrester, especially a tank-mounted arrester. Polymer-housed arresters have more problems than porcelain-housed arresters because they are much shorter. Use animal guards on tank-mounted arresters (especially on polymer-housed arresters).
- *Cutouts* — Cutouts are sometimes installed such that there is a low clearance between a phase conductor and a grounded object.

Fusing can also change the impact of animal-caused faults for faults across a distribution transformer bushing or a tank-mounted arrester. If the transformer is externally fused, only the customers on the transformer have an outage. If the transformer is a CSP with an internal fuse, then the tap fuse or upstream circuit breaker operates.

Birds rank second (behind squirrels) as far as the number of outages caused by animals (EPRI TE-114915, 1999; Frazier and Bonham, 1996). Many of the practices listed above can help with birds as well. Additionally, some bird-specific practices include:

- Get rid of nests.
- Track as a separate category.
- Remove nearby roosting areas.

The types of animals causing faults varies considerably by region, and there is also significant variation within a region. Animal faults also ebb and flow with animal populations. Animal population data can be used as one way to determine if "unknown" faults are really being caused by certain animals.

7.6.4 Other External Causes

Automobiles and poles do not always coexist nicely. It is difficult for utilities to prevent car accidents. Sometimes utilities can work with city engineers and police to try to lower speed limits and manage traffic better to avoid bad spots. Reflectors on poles may help. Siting poles further from roads also helps.

Balloons and other debris also cause many interruptions. Covered wire helps. Ladders, cranes, and other tall equipment into primary lines cause dangerous ground-level voltages as well as causing faults. Keep getting the word out. Public awareness campaigns help.

TABLE 7.7

Permanent-Fault Causes

Source	Rural	Urban
Equipment failures	14.1%	18.4%
Loss of supply	7.8%	9.6%
External factors	78.1%	72.0%

Source: Horton, W. F., Golberg, S., and Volkmann, C. A., "The Failure Rates of Overhead Distribution System Components," IEEE Power Engineering Society Transmission and Distribution Conference, 1991. With permission. ©1991 IEEE.

7.7 Equipment Faults

Equipment failures — transformers, capacitors, splices, terminations, insulators, connectors — cause faults. When equipment fails, it is almost always as a short circuit and rarely as an open circuit.

Equipment failures on overhead circuits are usually a small percentage of faults (see Figure 7.1). This is confirmed by another study at Pacific Gas & Electric Co. shown in Table 7.7. These are shown as a percentage of permanent faults. Since equipment faults are almost always permanent faults, the overall percentage of equipment failures is a low percentage of all faults (since most faults are temporary on overhead circuits). On underground circuits, most faults are due to equipment failures.

Distribution transformers are the most common major device, so their failure rate is important. Transformers generally fail at rates of about 0.5% per year. The most common failure mode starts as a breakdown of the turn-to-turn insulation.

Table 7.8 shows equipment failure rates recorded over a 5-year period at PG&E. This data is generic service-time failure rates, which is an estimate to the actual failure rate. Note that this is for California which has very little lightning and mild weather; other areas may have higher equipment failures. The rate of all permanent faults was 0.11 faults/mi/year for rural circuits (0.071 faults/km/year) and 0.16 faults/mi/year for urban circuits (0.102 faults/km/year). The only component where there was a statistical difference between urban and rural at the 90% confidence level was the difference in failure rates of transformers (the sample size for the rest of the numbers was too small to statistically determine a difference).

Another source for equipment failure rate data is the IEEE Gold Book (IEEE Std. 493-1997) (Table 7.9). Note that the Gold Book is for industrial facilities. Application and loading practices may be significantly different than typical utility applications. Still, they provide useful comparisons.

TABLE 7.8

Service-Time Overhead Component Failure Rates for PG & E

Component	Failure Rates per Year	
	Rural	Urban
Transformers	0.0271%	0.0614%
Switches	0.126%	0.0775%
Fuses	0.45%	0.374%
Capacitors	1.05%	0.85%
Reclosers	1.50%	1.44%
Voltage regulators	2.88%	n/a
Conductors	1.22/100 mi	1.98/100 mi

Source: Horton, W. F., Golberg, S., and Volkmann, C. A., "The Failure Rates of Overhead Distribution System Components," IEEE Power Engineering Society Transmission and Distribution Conference, 1991. With permission. ©1991 IEEE.

TABLE 7.9

Overhead Component Failure Rates in the IEEE Gold Book

Component	Failure Rate per Year
Transformers (all)	0.62%
Transformers (300 to 10,000 kVA)	0.59%
Transformers (>10,000 kVA)	1.53%
Switchgear bus (insulated)	0.113%[a]
Switchgear bus (bare)	0.192%[a]

[a] For each circuit breaker and connected switch.

Source: From IEEE Std. 493-1997. Copyright 1998 IEEE. All rights reserved.

7.8 Faults in Equipment

Failures in equipment pose special hazards with important safety ramifications. Transformers deserve extra attention because they are so common. One utility has reported one violent distribution transformer failure for every 270 transformers containing an internal fault, and 20% of re-energizations had internal faults, which is one violent failure every 1350 reenergizations (CEA 149 D 491A, 1997). Cuk (2001) estimated that between 2 and 5% of overhead transformers are re-fused every year (based on analyzing fuse purchases; this number varies with fusing practices). This section discusses failure mechanisms and the consequences of an internal failure.

winding faults may increase pressure faster than the pressure-relief valve can dissipate (Lunsford and Tobin, 1997). As an interwinding fault arcs and causes damage and melts additional insulation, the fault current will increase; usually current jumps sharply to the bolted fault condition (not a slow escalation of current).

Overhead completely self protected transformers (CSPs) and padmounted transformers with under-oil fuses have less withstand capability than conventional transformers. The reason for this is that the under-oil fuse (called a weak-link fuse) provides another arcing location. When the weak-link fuse melts, an arc forms in place of the melted fuse element. This arc is in addition to whatever arc may exist within the tank that caused the fault in the first place. The arc across the fuse location is generally going to be longer than normal arcs that could occur inside a transformer. The length of an under-oil weak-link fuse is 2 to 3 in. (5.1 to 7.6 cm) for a 15-kV class fuse. Higher-voltage transformers have longer fuses — a 35-kV class fuse has a length of about 5 in. (12.7 cm). Also, the voltage gradient along an arc in a fuse tube under oil is greater than a "free" arc in oil [the fuse tube increases the pressure of the arc, which increases the voltage drop (Barkan et al., 1976)].

Transformers with under-oil arresters have a special vulnerability (Henning et al., 1989). The under-oil arrester provides another possible failure mode which can lead to very high energy in the transformer if the arrester fails. If the arrester blocks fail, a relatively long arc results. A 10-kV duty-cycle rated arrester has a total block length of about 4.5 in. (11.4 cm). With such a long arc, the energy in the transformer will be very high. Industry tests and ratings do not directly address this issue. To be conservative, consider using a current-limiting fuse upstream of the transformer if the line-to-ground fault current exceeds 1 or 2 kA if under-oil arresters are used.

Stand-alone arresters are another piece of equipment where failure is a concern. If an arrester fails, a long internal arc may cause the arrester to explode, sending pieces of the housing along with pieces of the metal oxide. The move from porcelain-housed arresters to polymer-housed arresters was motivated primarily by the fact that the polymer-housing is less dangerous if the arrester fails. With a porcelain-housed arrester, the thermal shock from the arc can shatter the housing and forcefully expel the "shrapnel." Polymer-housed arresters are safer because the fault arc splits the polymer housing, which relieves the pressure buildup (although if the arc originates inside the blocks, the pressure can expel bits of the metal oxide). Arresters have caused accidents, and they got a bad name when metal-oxide arresters were first introduced because they occasionally failed upon installation.

Distribution arresters may specify a fault current withstand which is governed by IEEE standards (ANSI/IEEE C62.11-1987). To pass the test, an arrester must withstand an internal failure of the given fault current and all components of the arrester must be confined within the enclosure. The duration of the test is a minimum of 0.1 sec (manufacturers often specify other times as well) which is a typical circuit breaker clearing time when the instantaneous relay element operates. If the available fault current is higher than the rated withstand, then current-limiting fuses should be considered. With polymer-housed arresters, the fault-current withstand is usually sufficient with manufacturers specifying withstand values of 10 to 20 kA for 0.1 sec.

The failure of arresters (especially porcelain-housed arresters) is also a consideration for fusing. If an arrester is downstream of a transformer fuse and the arrester fails, the relatively small transformer fuse will blow. If an arrester is upstream of the transformer fuse then a larger tap fuse or the substation circuit breaker operates, which allows a much longer duration fault current. Arresters have isolators that disconnect the arresters in case of failure. Isolators do *not* clear fault current. After the fuse or circuit breaker operates, the disconnect provides enough separation to allow the circuit to reclose successfully. If the next upstream device is a circuit breaker and an instantaneous element is not used, fault currents could be much longer than the tested 0.1 sec, so consider adding a fuse upstream of the arrester on porcelain-housed units.

7.9 Targeted Reduction of Faults

Since faults are the *root cause* of interruptions and voltage sags, obviously, if faults are reduced, the incidence of interruptions and sags will be reduced (power quality and reliability improve) (Mo and Taylor, 1993). This can be done in several ways including

- Tree trimming
- Animal guards
- Arrester protection
- Tree wire (covered wire)
- Aerial or underground cable
- Identifying and replacing poorly performing hardware
- Line patrols including infrared thermography

One way to improve the fault-rate performance is to track the location and type of faults. It is relatively straightforward to track permanent faults since the failed equipment is obvious. They can be classified by category to help determine what types of maintenance need to be performed. The effort in tracking faults is paid back by targeting maintenance efforts to the most important sections.

Temporary faults are harder to pinpoint. If a circuit breaker or recloser operates and successfully recloses, the utility may be unaware of it (unless they get complaints of blinking clocks). A fuse may be blown by a temporary fault (especially if fuse blowing is used). If a fuse blows, the area narrows considerably. For areas with repeat fuse operations, careful patrols may identify areas where repeated faults occur. Still, if a tap fuse operates but is re-fused successfully, the cause may have been a squirrel across a bushing, a tree branch that fell onto then burned off of a line, or wind pushing two conductors together. It often takes a trained eye to determine the cause.

The most important sections are not necessarily the locations with the most faults per mile. The number of customers on a circuit and the type of customers on a circuit are important considerations. For example, a suburban circuit with many high-tech commercial customers should warrant different treatment from a rural circuit with fewer, mostly residential and agricultural customers. How this is weighted depends on the utility's philosophy.

On-site investigations can help reduce faults. Faults tend to repeat at the same locations and follow patterns. For example, one particular type and brand of connector may have a high failure rate. If these are identified, replacement strategies can be implemented. Another example is animal faults. One particular pole, which happens to be a good travel path for squirrels, may have a transformer with no animal guards. The same location may have repeated outages. These may be difficult to find if they cause temporary faults.

Faults are not evenly distributed along lines. Faults are not inevitable. Not all faults are "acts of God." Most are from specific deficiencies at specific structures. On overhead circuits, most faults result from inadequate clearances, inadequate insulation, old equipment, or from trees or branches extending into a line.

Consider faults as *preventable*, then go look for them. Crews can be trained to spot pole structures where faults might be likely. During restoration, crews can identify several common causes of faults including

- Poor jumper clearances
- Old equipment (such as expulsion arresters)
- Bushings or cable terminations unprotected against animals
- Poor clearances with polymer arresters
- Damaged insulators
- Damaged covered wire
- Bad cutout placement
- Danger trees/branches present

Crews can fix problems identified during the outage restoration or target them for future repair.

Implement training out in the field for best results; show examples of fault sources (Taylor, 1995). Walk the line and use binoculars; this is more effective than "riding the line." Some fault sources are not obvious and require looking at a structure from different angles. Taylor provides another good piece of advice: never assume that the lines are built according to specifications.

Characteristic fault patterns are different at different locations. Factors that influence these differences include types of trees, weather conditions, construction practices, and types of equipment purchased. Know the local trends; they can help identify ways to reduce faults.

The best way to reduce faults over time is to "institutionalize" fault-reduction practices. After identifying the most common fault sources, implement programs to address these so performance improves continually. Start with a good design that eliminates fault sources, especially at equipment poles; use sufficient electrical clearance. Separate grounded objects from phase conductors as much as possible. Then employ procedures to ensure that more fault-resistant designs are implemented. Train linemen and field engineers to do it right. If possible, implement programs to bring old construction up to specifications; replace old arresters, increase poor clearances, add covered jumper wires, add animal guards, and so on. An opportune time to clean up poor construction is when crews are already doing work on a structure. Perform quality audits during work and after work is done. Give crews and field engineers feedback.

References

ABB, *Distribution Transformer Guide*, 1995.

Ackermann, R. H. and Baker, P., "Performance of Distribution Lines in Severe Lightning Areas," International Conference on Lightning and Power Systems, London, U.K., June 1984.

ANSI/IEEE C62.11-1987, *IEEE Standard for Metal-Oxide Surge Arresters for AC Power Circuits*, American National Standards Institute, Institute of Electrical and Electronics Engineers, Inc.

ANSI/IEEE Std. 100-1992, *Standard Dictionary of Electrical and Electronics Terms.*

ASTM F1506-94, *Textile Materials for Wearing Apparel for Use by Electrical Workers Exposed to Momentary Electric Arc and Related Thermal Hazards*, American Society for Testing and Materials, 1994.

Barkan, P., Damsky, B. L., Ettlinger, L. F., and Kotski, E. J., "Overpressure Phenomena in Distribution Transformers with Low Impedance Faults Experiment and Theory," *IEEE Transactions on Power Apparatus and Systems*, vol. PAS-95, no. 1, pp. 37–48, 1976.

Benton, R. E., "Energy Absorption Capabilities of Pad Mounted Distribution Transformers with Internal Faults," IEEE/PES Transmission and Distribution Conference, 1979.

Bernis, R. A. O. and de Minas Gerais, C. E., "CEMIG Addresses Urban Dilemma," *Transmission and Distribution World*, vol. 53, no. 3, pp. 56–61, March 2001.

Burke, J. J. and Lawrence, D. J., "Characteristics of Fault Currents on Distribution Systems," *IEEE Transactions on Power Apparatus and Systems*, vol. PAS-103, no. 1, pp. 1–6, January 1984.

CEA 149 D 491A, *Distribution Transformer Internal Pressure Withstand Test*, Canadian Electrical Association, 1997. As cited by Cuk (2001).

CEA 160 D 597, *Effect of Lightning on the Operating Reliability of Distribution Systems*, Canadian Electrical Association, Montreal, Quebec, 1998.

Cobine, J. D., *Gaseous Conductors*, McGraw-Hill, New York, 1941.

Cuk, N., "Seminar on Detection of Internal Arcing Faults in Distribution Transformers," Presentation at the IEEE Transformers Committee Meeting, Orlando, Florida, October 14–17, 2001.

Drouet, M. G. and Nadeau, F., "Pressure Waves Due to Arcing Faults in a Substation," *IEEE Transactions on Power Apparatus and Systems*, vol. PAS-98, no. 5, pp. 1632–5, 1979.

EPRI 1209-1, *Distribution Fault Current Analysis*, Electric Power Research Institute, Palo Alto, CA, 1983.

EPRI EL-6903, *Study of Fault-Current-Limiting Techniques*, Electric Power Research Institute, Palo Alto, CA, 1990.

EPRI TE-114915, *Mitigation of Animal-Caused Outages for Distribution Lines and Substations*, Electric Power Research Institute, Palo Alto, CA, 1999.

EPRI TR-106294-V2, *An Assessment of Distribution System Power Quality. Volume 2:* Statistical Summary Report, Electric Power Research Institute, Palo Alto, CA, 1996.

EPRI TR-106294-V3, *An Assessment of Distribution System Power Quality. Volume 3:* Library of Distribution System Power Quality Monitoring Case Studies, Electric Power Research Institute, Palo Alto, CA, 1996.

EPRI TR-109178, *Distribution Cost Structure — Methodology and Generic Data*, Electric Power Research Institute, Palo Alto, CA, 1998.

Finch, K., "Understanding Tree Outages," EEI Vegetation Managers Meeting, Palm Springs, CA, May 1, 2001.

Frazier, S. D. and Bonham, C., "Suggested Practices for Reducing Animal-Caused Outages," *IEEE Industry Applications Magazine*, vol. 2, no. 4, pp. 25–31, August 1996.

Girgis, A. A., Fallon, C. M., and Lubkeman, D. L., "A Fault Location Technique for Rural Distribution Feeders," *IEEE Transactions on Industry Applications*, vol. 29, no. 6, pp. 1170–5, November/December 1993.

Goda, Y., Iwata, M., Ikeda, K., and Tanaka, S., "Arc Voltage Characteristics of High Current Fault Arcs in Long Gaps," *IEEE Transactions on Power Delivery*, vol. 15, no. 2, pp. 791–5, April 2000.

Goedde, G. L., Kojovic, L. A., and Knabe, E. S., "Overvoltage Protection for Distribution and Low-Voltage Equipment Experiencing Sustained Overvoltages," Electric Council of New England (ECNE) Fall Engineering Conference, Killington, VT, 2002.

Goodfellow, J., "Understanding the Way Trees Cause Outages," 2000. http://www.eci-consulting.com/images/wnew/trees.pdf.

Goodman, E. A. and Zupon, L., "Static Pressures Developed in Distribution Transformers Due to Internal Arcing Under Oil," *IEEE Transactions on Power Apparatus and Systems*, vol. PAS-95, no. 5, pp. 1689–98, 1976.

Hart, B., "HV Overhead Line — the Scandinavian Experience," *Power Engineering Journal*, pp. 119–23, June 1994.

Hendrix, "Reliability of Overhead Distribution Circuits," Hendrix Wire & Cable, Inc., August 1998.

Henning, W. R., Hernandez, A. D., and Lien, W. W., "Fault Current Capability of Distribution Transformers with Under-Oil Arresters," *IEEE Transactions on Power Delivery*, vol. 4, no. 1, pp. 405–12, January 1989.

Horton, W. F., Golberg, S., and Volkmann, C. A., "The Failure Rates of Overhead Distribution System Components," IEEE Power Engineering Society Transmission and Distribution Conference, 1991.

IEEE C2-1997, *National Electrical Safety Code.*

IEEE Std. 493-1997, *IEEE Recommended Practice for the Design of Reliable Industrial and Commercial Power Systems (Gold Book).*

IEEE Std. 1024-1988, *IEEE Recommended Practice for Specifying Distribution Composite Insulators (Suspension Type).*

IEEE Tutorial Course 90EH0310-3-PWR, "Detection of Downed Conductors on Utility Distribution Systems," 1990.

IEEE Working Group on Distribution Protection, "Distribution Line Protection Practices Industry Survey Results," *IEEE Transactions on Power Delivery,* vol. 10, no. 1, pp. 176–86, January 1995.

Johnston, L., Tweed, N. B., Ward, D. J., and Burke, J. J., "An Analysis of Vepco's 34.5 kV Distribution Feeder Faults as Related to Through Fault Failures of Substation Transformers," *IEEE Transactions on Power Apparatus and Systems,* vol. PAS-97, no. 5, pp. 1876–84, 1978.

Kaufmann, G. H., "Impulse Testing of Distribution Transformers Under Load," *IEEE Transactions on Power Apparatus and Systems,* vol. PAS-96, no. 5, pp. 1583–95, September/October 1977.

Kaufmann, G. H. and McMillen, C. J., "Gas Bubble Studies and Impulse Tests on Distribution Transformers During Loading Above Nameplate Rating," *IEEE Transactions on Power Apparatus and Systems,* vol. PAS-102, no. 8, pp. 2531–42, August 1983.

Kersting, W. H. and Phillips, W. H., "Modeling and Analysis of Unsymmetrical Transformer Banks Serving Unbalanced Loads," *IEEE Transactions on Industry Applications,* vol. 32, no. 3, pp. 720–5, May-June 1996.

Kuntz, P. A., "Optimal Reliability Centered Vegetation Maintenance Scheduling in Electric Power Distribution Systems," Ph.D. thesis, University of Washington, 1999. As cited by Brown, R. E., *Electric Power Distribution Reliability,* Marcel Dekker, 2002.

Lampley, G. C., "Fault Detection and Location on Electrical Distribution System Case Study," IEEE Rural Electric Power Conference, 2002.

Lasseter, J. A., "Burndown Tests on Bare Conductor," *Electric Light & Power,* pp. 94–100, December 1956.

Lasseter, J. A., "Point Way to Reduce Lightning Outages," *Electrical World,* pp. 93–5, October 1965.

Laverick, D. G., "Performance of the Manweb Rural 11-kV Network During Lightning Activity," International Conference on Lightning and Power Systems, London, June 1984.

Littler, G. E., "High Speed Reclosing and Protection Techniques Applied to Wood Pole Lines," *Elec. Eng. Trans., Inst. Eng. Aust.,* vol. EE1, pp. 107–16, June 1965.

Lunsford, J. M. and Tobin, T. J., "Detection of and Protection for Internal Low-Current Winding Faults in Overhead Distribution Transformers," *IEEE Transactions on Power Delivery,* vol. 12, no. 3, pp. 1241–9, July 1997.

Mahieu, W. R., "Prevention of High-fault Rupture of Pole-Type Distribution Transformers," *IEEE Transactions on Power Apparatus and Systems,* vol. PAS-94, no. 5, pp. 1698–707, 1975.

Mazurek, B., Maczka, T., and Gmitrzak, A., "Effect of Arc Current on Properties of Composite Insulators for Overhead Lines," *Proceedings of the 6th International Conference on Properties and Applications of Dielectric Materials,* June 2000.

Mo, Y. C. and Taylor, L. S., "A Novel Approach for Distribution Fault Analysis," *IEEE Transactions on Power Delivery,* vol. 8, no. 4, pp. 1882–9, October 1993.

Mo, Y. C. and Taylor, L. S., "Analysis and Prevention of Animal-Caused Faults in Power Distribution Systems," *IEEE Transactions on Power Delivery*, vol. 10, no. 2, pp. 995–1001, April 1995.

NFPA 70E-2000, *Standard for Electrical Safety Requirements for Employee Workplaces*, National Fire Protection Association.

Parrish, D. E., "Lightning-Caused Distribution Circuit Breaker Operations," *IEEE Transactions on Power Delivery*, vol. 6, no. 4, pp. 1395–401, October 1991.

Peele, S., "Faults Per Feeder Mile," PQA Conference 2001.

PTI, "Distribution Transformer Application Course Notes," Power Technologies, Inc., Schenectady, NY, 1999.

Rizk, F. A. and Nguyen, D. H., "AC Source-Insulator Interaction in HV Pollution Tests," *IEEE Transactions on Power Apparatus and Systems*, vol. PAS-103, no. 4, pp. 723–32, April 1984.

Santoso, S., Dugan, R. C., Lamoree, J., and Sundaram, A., "Distance Estimation Technique for Single Line-to-Ground Faults in a Radial Distribution System," IEEE Power Engineering Society Winter Meeting, 2000.

Schweitzer, E. O., "A Review of Impedance-Based Fault Locating Experience," Fourteenth annual Iowa-Nebraska system protection seminar, Omaha, NE, October 16, 1990. Downloaded from www.schweitzer.com.

Short, T. A., "Distribution Fault Rate Survey," EPRI PEAC, 2001.

Simpson, P., "EUA's Dual Approach Reduces Tree-Caused Outages," *Transmission & Distribution World*, pp. 22–8, August 1997.

Slepian, J., "Extinction of a Long AC Arc," *AIEE Transactions*, vol. 49, pp. 421–30, April 1930.

Smith, D. R., "System Considerations — Impedance and Fault Current Calculations," *IEEE Tutorial Course on Application and Coordination of Reclosers, Sectionalizers, and Fuses, 1980.* Publication 80 EHO157-8-PWR.

St. Pierre, C., *A Practical Guide to Short-Circuit Calculations*, Electric Power Consultants, Schenectady, NY, 2001.

Strom, A. P., "Long 60-Cycle Arcs in Air," *AIEE Transactions*, vol. 65, pp. 113–8, March 1946.

Taylor, L., "Elimination of Faults on Overhead Distribution Systems," IEEE/PES Winter Power Meeting, 1995. Presentation to the Working Group on System Design, 1995 IEEE Winter Power Meeting, New York, NY.

Williams, C., "Tree Fault Testing on a 12 kV Distribution System," IEEE/PES Winter Power Meeting presentation to the Distribution Subcommittee, 1999.

You begin to think that trees possibly grow faster toward your power line just out of spite.

Powerlineman law #47, By CD Thayer and other Power Linemen
http://www.cdthayer.com/lineman.htm

8

Short-Circuit Protection

Overcurrent protection or short-circuit protection is very important on any electrical power system, and the distribution system is no exception. Circuit breakers and reclosers, expulsion fuses and current-limiting fuses — these protective devices interrupt fault current, a vital function. Short-circuit protection is the selection of equipment, placement of equipment, selection of settings, and coordination of devices to efficiently isolate and clear faults with as little impact on customers as possible.

Of top priority, good fault protection clears faults quickly to prevent:

- *Fires and explosions*
- *Further damage to utility equipment such as transformers and cables*

Secondary goals of protection include practices that help reduce the impact of faults on

- *Reliability* (long-duration interruptions) — In order to reduce the impact on customers, reclosing of circuit breakers and reclosers automatically restores service to customers. Having more protective devices that are properly coordinated assures that the fewest customers possible are interrupted and makes fault-finding easier.
- *Power quality* (voltage sags and momentary interruptions) — Faster tripping reduces the duration of voltage sags. Coordination practices and reclosing practices impact the number and severity of momentary interruptions.

8.1 Basics of Distribution Protection

Circuit interrupters should only operate for faults, not for inrush, not for cold load pickup, and not for transients. Additionally, protective devices should coordinate to interrupt as few customers as possible.

The philosophies of distribution protection differ from transmission-system protection and industrial protection. In distribution systems, protection is not normally designed to have backup. If a protective device fails to operate, the fault may burn and burn (until an upstream device is manually opened). Of course protection coverage should overlap, so that if a protective device fails due to an internal short circuit (which is different from failing to open), an upstream device operates for the internal fault in the downstream protector. Backup is not a mandatory design constraint (and is impractical to achieve in all cases).

Most often, we base distribution protection on standardized settings, standardized equipment, and standardized procedures. Standardization makes operating a distribution company easier if designs are consistent. The engineering effort to do a coordination study on every circuit reduces considerably.

Another characteristic of distribution protection is that it is not always possible to fully coordinate all devices. Take fuses. With high fault currents, it is impossible to coordinate two fuses in series because the high current can melt and open both fuses in approximately the same time. Therefore, close to the substation, fuse coordination is nonexistent. There are several other situations where coordination is not possible. Some low-level faults are very difficult — some would say impossible — to detect. A conductor in contact with the ground may draw very little current. The "high-impedance" fault of most concern (because of danger to the public) is an energized, downed wire.

8.1.1 Reach

A protective device must clear all faults in its protective zone. This "zone" is defined by

- *Reach* — The reach of a protective device is the maximum distance from a protective device to a fault for which the protective device will operate.

Lowering a relay pickup setting or using a smaller fuse increases the reach of the protective device (increasing the device's *sensitivity*). Sensitivity has limits; if the setting or size is too small, the device trips unnecessarily from overloads, from inrush, from cold-load pickup.

We have several generic or specific methods to determine the reach of a protective device. Commonly, we estimate the minimum fault current for faults along the line and choose the reach of the device as the point where the minimum fault current equals the magnitude where the device will operate. Some common methods of calculating the reach are

- Percentage of a bolted line-to-ground fault: The minimum ground fault current is some percentage (usually 25 to 75%) of a bolted fault.

- Fault resistance: Assume a maximum value of fault resistance when calculating the current for a single line-to-ground fault. Common values of fault resistance used are 1 to 2, 20, 30, and 40 Ω. Rural Electrification Administration (REA) standards use 40 Ω.

Other options for determining the reach are

- Point based on a maximum operating time of a device: Define the reach as the point giving the current necessary to operate a protective device in a given time (with or without assuming any fault impedance). Example: The REA has recommended taking the reach of a fuse as the point where the fuse will just melt for a single line-to-ground fault in 20 sec with a fault resistance of 30 or 40 Ω.
- Point based on a multiplier of the device setting: Choose the point where the fault current is some multiple of the device rating or setting. Example: The reach of a fuse is the point where the bolted fault current is six times the fuse rating.

None of these methods are exact. Some faults will always remain undetectable (high-impedance faults). The trick is to try to clear all high-current faults without being overly conservative.

Assuming a high value of fault resistance (20 to 40 Ω) is overly conservative, so avoid it. For a 12.47-kV system (7.2 kV line to ground), the fault current with a 40 Ω fault impedance is less than 180 A (this ignores the system impedance — additional system impedance reduces the calculated current even more). Using typical relay/recloser setting philosophies, which say that the rating of the recloser must be less than half of the minimum fault current, a recloser must be less than 90 A, which effectively limits the load current to an unreasonably low value. In many (even most, for some utilities) situations this is unworkable. As discussed in Chapter 7, faults with arc impedances greater than 2 Ω are not common, so take the approach that the minimum fault is close to the bolted fault current. On the other hand, high-impedance faults (common during downed conductors) generally draw less than 50 A and have impedances of over 100 Ω. The 40 Ω rule does not guarantee that a protective device will clear high-impedance faults, and in most cases would not improve high-impedance fault detection.

8.1.2 Inrush and Cold-Load Pickup

When an electrical distribution system energizes, components draw a high, short-lived inrush; the largest component magnetizes the magnetic material in distribution transformers (in most cases, it is more accurate to say *remagnetizes* since the core is likely magnetized in a different polarity if the circuit is energized following a short-duration interruption). The transformer inrush characteristics important for protection are:

- At a distribution transformer, inrush can reach peak magnitudes of 30 times the transformer's full-load rating.
- Relative to the transformer rating, inrush has higher peak magnitudes for smaller transformers, but the time constant is longer for larger transformers. Of course, on an absolute basis (amperes), a larger transformer draws more inrush.
- Sometimes inrush occurs, and sometimes it does not, depending on the point on the voltage waveform at which the reclosing occurs.
- System impedance limits the peak inrush.

The system impedance relative to the transformer size is an important concept since it limits the peak inrush for larger transformers and larger numbers of transformers. If one distribution transformer is energized by itself, the transformer is small relative to the source impedance, so the peak inrush maximizes. If a tap with several transformers is energized, the equivalent connected transformer is larger relative to the system impedance, so the peak inrush is lower (but the duration is extended). Several transformers energizing at once pull the system voltage down. This reduction in voltage causes less inrush current to be drawn from each transformer. For a whole feeder, the equivalent transformer is even larger, so less inrush is observed. Some guidelines for estimating inrush are:

- *One distribution transformer* — 30 times the crest of the full-load current.
- *One lateral tap* — 12 times the crest of the full-load current of the total connected kVA.
- *Feeder* — 5 times the connected kVA up to about half of the crest of the system available fault current.

At the feeder level, inrush was only reported to cause tripping by 15% of responders to an IEEE survey (IEEE Working Group on Distribution Protection, 1995). When a three-phase circuit is reclosed, the ground relay is most likely to operate since the inrush seen by a ground relay can be as high as the peak inrush on the phases (and it is usually set lower than the phase settings). An instantaneous relay element is most sensitive to inrush, but the instantaneous element is almost always disabled for the first reclose attempt. The ground instantaneous element could operate if a significant single-phase lateral is reconnected.

Transformers are not the only elements that draw inrush; others include resistive lighting and heating elements and motors. Incandescent filaments can draw eight times normal load current. The time constant for the incandescent filaments is usually very short; the inrush is usually finished after a half cycle. Motor starting peak currents are on the order of six times the motor rating. The duration is longer than transformer inrush with durations typically from 3 to 10 sec.

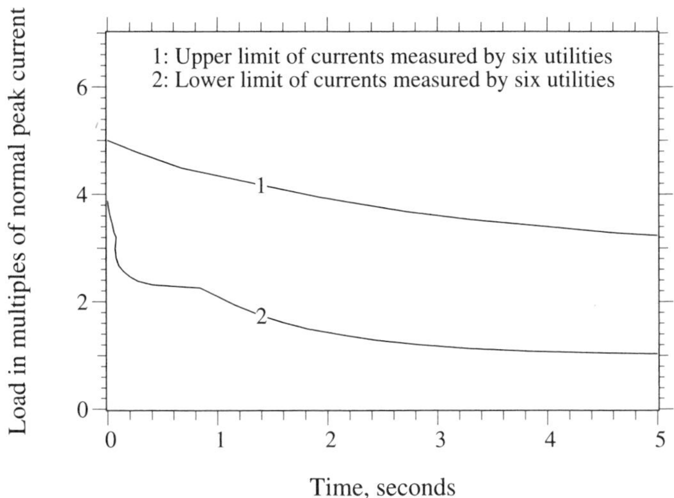

FIGURE 8.1
Ranges of cold-load pickup current from tests by six utilities. (Data from [Smithley, 1959].)

Cold-load pickup is the extra load following an extended interruption due to loss of the normal diversity between customers. Following an interruption, the water in water heaters cools down and refrigerators warm up. When the power is restored, all appliances that need to catch up energize at once. In cold weather, following an extended interruption, heaters all come on at once (so it is especially bad with high concentrations of resistance heating). In hot weather, houses warm up, so all air conditioners start following an interruption.

Cold-load pickup can be over three times the load prior to the interruption. As diversity is regained, the load slowly drops back to normal. This time constant varies depending on the types of loads and the duration of the interruption. Cold-load pickup is often divided into transformer inrush which last a few cycles, motor starting and accelerating currents which last a few seconds, and finally just the load due to loss of diversity which can last many minutes. Figure 8.1 shows the middle-range time-frame with motor starting and accelerating currents.

It is important to select relay settings and fuse sizes high enough to avoid operations due to cold-load pickup. Even so, cold-load pickup problems are hard to avoid in some situations. A survey of utilities reported 75% having experienced cold-load pickup problems (IEEE Working Group on Distribution Protection, 1995). When a cold-load pickup problem occurs at the substation level, the most common way to reconnect is to sectionalize and pick the load up in smaller pieces. For this reason, cold-load pickup problems are not widespread — after a long interruption, utilities usually sectionalize anyway to get customers back on more quickly. Two other ways that are sometimes used to energize a circuit are to raise relay settings or even to block tripping (not recommended unless as a very last resort).

In order to pick relays, recloser settings, and fuses, we often plot a cold-load pickup curve on a time-current coordination graph along with the protection equipment characteristics. Points can be taken from the curves in Figure 8.1. It is also common to choose one or two points to represent cold-load pickup. Three-hundred percent of full-load current at 5 sec is a common point.

Distribution protective devices tend to have steep time-overcurrent characteristics, meaning that they operate much faster for higher currents. K-link fuses and extremely-inverse relays are most commonly used, and these happen to have the steepest characteristics. This is no coincidence; a distribution protective device must operate fast for high currents (most faults) and slow for lower currents. This characteristic gives a protective device a better chance to ride through inrush and cold-load pickup.

8.2 Protection Equipment

8.2.1 Circuit Interrupters

All circuit interrupters — including circuit breakers, fuses, and reclosers — operate on some basic principles. All devices interrupt fault current during a zero-crossing. To do this, the interrupter creates an arc. In a fuse, an arc is created when the fuse element melts, and in a circuit breaker or recloser, an arc is created when the contacts mechanically separate. An arc conducts by ionizing gasses, which leads to a relatively low-impedance path.

After the arc is created, the trick is to increase the dielectric strength across the arc so that the arc clears at a current zero. Each half cycle, the ac current momentarily stops as the current is reversing directions. During this period when the current is reversing, the arc is not conducting and is starting to de-ionize, and in a sense, the circuit is interrupted (at least temporarily). Just after the arc is interrupted, the voltage across the now-interrupted arc path builds up. This is the *recovery voltage*. If the dielectric strength builds up faster than the recovery voltage, then the circuit stays interrupted. If the recovery voltage builds up faster than the dielectric strength, the arc breaks down again. Several methods used to increase the dielectric strength of the arc are discussed in the following paragraphs. The general methods are:

- *Cooling the arc* — The ionization rate decreases with lower temperature.
- *Pressurizing the arc* — Dielectric strength increases with pressure.
- *Stretching the arc* — The ionized-particle density is reduced by stretching the arc stream.
- *Introducing fresh air* — Introducing de-ionized gas into the arc stream helps the dielectric strength to recover.

An air blast breaker blasts the arc stream into chutes that quickly lengthen and cool the arc. Blowout coils can move the arc by magnetically inducing motion. Compressed air blasts can blow the arc away from the contacts.

The arc in the interrupter has enough resistance to make it very hot. This can wear contact terminals, which have to be replaced after a given number of operations. If the interrupter fails to clear with the contacts open, the heat from the arc builds high pressure that can breach the enclosure, possibly in an explosive manner.

In an oil device, the heat of an arc decomposes the oil and creates gasses that are then ionized. This process takes heat and energy out of the arc. To enhance the chances of arc extinction in oil, fresh oil can be forced across the path of the arc. Lengthening the arc also helps improve the dielectric recovery. In an oil circuit breaker, the contact parting time is long enough that there may be several restrikes before the dielectric strength builds up enough to interrupt the circuit.

Vacuum devices work because the dielectric strength increases rapidly at very low pressures (because there are very few gas molecules to ionize). Normally, when approaching atmospheric pressures, the dielectric breakdown of air decreases as pressure decreases, but for very low pressures, the dielectric breakdown goes back up. The pressure in vacuum bottles is 10^{-6} to 10^{-8} torr. A vacuum device only needs a very short separation distance (about 8 to 10 mm for a 15-kV circuit breaker). Interruption is quick since the mechanical travel time is small. The separating contacts draw an arc (it still takes a current zero to clear). Sometimes, vacuum circuit breakers chop the current, causing voltage spikes. The arc is a metal vapor consisting of particles melted from each side. Contact erosion is low, so vacuum devices are low maintenance and have a long life. Restrikes are uncommon.

SF_6 is a gas that is a very good electrical insulator, so it has rapid dielectric recovery. At atmospheric pressures, the dielectric strength is 2.5 times that of air, and at higher pressures, the performance is even better. SF_6 is very stable, does not react with other elements, and has good temperature characteristics. One type of device blows compressed SF_6 across the arc stream to increase the dielectric strength. Another type of SF_6 interrupter used in circuit breakers and reclosers has an arc spinner which is a setup that uses the magnetic field from a coil to cause the arc to spin rapidly (bringing it in contact with un-ionized gas). SF_6 can be used as the insulating medium as well as the interrupting medium. SF_6 devices are low maintenance, have short opening times, and most do not have restrikes.

Since interrupters work on the principle of the dielectric strength increasing faster than the recovery voltage, the *X/R* ratio can make a significant difference in the clearing capability of a device. In an inductive circuit, the recovery voltage rises very quickly since the system voltage is near its peak when the current crosses through zero. Asymmetry increases the peak magnitude of the fault current. For this reason, the capability of most interrupters decreases with higher *X/R* ratios. Some interrupting equipment is rated based on a symmetrical current basis while other equipment is based on asymmetrical current. Whether based on a symmetrical or asymmetrical basis, the interrupter has asymmetric interrupting capability.

8.2.2 Circuit Breakers

Circuit breakers are often used in the substation on the bus and on each feeder. Circuit breakers are available with very high interrupting and continuous current ratings. The interrupting medium in circuit breakers can be any of vacuum, oil, air, or SF_6. Oil and vacuum breakers are most common on distribution stations with newer units being mainly vacuum with some SF_6.

Circuit breakers are tripped with external relays. The relays provide the brains to control the opening of the circuit breaker, so the breaker coordinates with other devices. The relays also perform reclosing functions.

Circuit breakers are historically rated as constant MVA devices. A symmetrical short-circuit rating is specified at the maximum rated voltage [for more ratings information, see (ANSI C37.06-1997; IEEE Std. C37.04-1999; IEEE Std. C37.010-1999)]. Below the maximum rated voltage (down to a specified minimum value), the circuit breaker has more interrupting capability. The minimum value where the circuit breaker is a constant-MVA device is specified by the constant K:

$$\text{Symmetrical interrupting capability} = \begin{cases} V_R \cdot I_R & \text{for } \dfrac{V_R}{K} < V \leq V_R \\ K \cdot I_R & \text{for } V \leq \dfrac{V_R}{K} \end{cases}$$

where

I_R = Rated symmetrical rms short-circuit current operating at V_R

V_R = Maximum rms line-to-line rated voltage

V = Operating voltage (also rms line-to-line)

K = Voltage range factor = ratio of the maximum rated voltage to the lower limit in which the circuit breaker is a constant MVA device. Newer circuit breakers are rated as constant current devices (K = 1).

Consider a 15-kV class breaker application on a 12.47-kV system where the maximum voltage will be assumed to be 13.1 kV (105%). For an ANSI-rated 500-MVA class breaker with V_R = 15 kV, K = 1.3, and I_R = 18 kA, the symmetrical interrupting capability would be 20.6 kA (15/13.1 × 18). Circuit breakers are often referred to by their MVA class designation (1000-MVA class for example). Typical circuit breaker ratings are shown in Table 8.1.

Circuit breakers must also be derated if the reclose cycle could cause more than two operations and if the operations occur within less than 15 sec. The percent reduction is given by

TABLE 8.1

15-kV Class Circuit Breaker Short-Circuit Ratings

	500 MVA	750 MVA	1000 MVA
Rated voltage, kV	15	15	15
K, voltage range factor	1.3	1.3	1.3
Short circuit at max voltage rating	18	28	37
Maximum symmetrical interrupting, kA	23	36	48
Close and latch rating			
1.6 K × rated short-circuit current, kA (asym)	37	58	77
2.7 K × rated short-circuit current, kA (peak)	62	97	130

$$D = d_1(n-2) + d_1(15-t_1)/15 + d_1(15-t_2)/15 + \cdots$$

where

D = Total reduction factor, %

$$d_1 = \text{Calculating factor, \%} = \begin{cases} 3\% & \text{for } I_R < 18\text{kV} \\ \dfrac{I_R}{6} & \text{for } I_R > 18\text{kA} \end{cases}$$

n = Total number of openings

t_n = nth time interval less than 15 sec

The interrupting rating is then $(100-D)I_R$. The permissible tripping delay is also a standard. For the given delay period, the circuit breaker must withstand K times the rated short-circuit current between closing and interrupting. A typical delay is 2 sec.

Continuous current ratings are independent of interrupting ratings (although higher continuous ratings usually go along with higher interrupting ratings). Standard continuous ratings include 600, 1200, 2000, and 3000 A (the 600 and 1200-A circuit breakers are most common for distribution substations).

A circuit breaker also has a *momentary* or *close and latch* short-circuit rating (also called the first-cycle capability). During the first cycle of fault current, a circuit breaker must be able to withstand any current up to a multiple of the short-circuit rating. The rms current should not exceed $1.6K \times I_R$ and the peak (crest) current should not exceed $2.7K \times I_R$.

The circuit breaker interrupting time is defined as the interval between energizing the trip circuit and the interruption of all phases. Most distribution circuit breakers are five-cycle breakers. Older breakers interrupt in eight cycles.

Distribution circuit breakers are three-phase devices. When the trip signal is received, all three phases are tripped. All three will not clear simultaneously because the phase current zero crossings are separated. The degree of separation between phases is usually one-half to one cycle.

8.2.3 Circuit Breaker Relays

Several types of relays are used to control distribution circuit breakers. Distribution circuits are almost always protected by overcurrent relays that use inverse time overcurrent characteristics. An inverse time-current characteristic means that the relay will operate faster with increased current.

The main types of relays are

- *Electromechanical relays* — The induction disk relay has long been the main relay used for distribution overcurrent protection. The relay is like an induction motor with contacts. Current through the CT leads induces flux in the relay magnetic circuit. These flux linkages cause the relay disk to turn. A larger current turns the disk faster. When the disk travels a certain distance, the contacts on the disk meet stationary contacts to complete the relay trip circuit. An instantaneous relay functionality can be provided: a plunger surrounded by a coil or a disk cup design operates quickly if the current is above the relay pickup. Most electromechanical relays are single phase.
- *Static relays* — Analog electronic circuitry (like op-amps) provide the means to perform a time-current characteristic that approximates that of the electromechanical relay.
- *Digital relays* — The most modern relay technology is fully digital based on microprocessor components.

Electromechanical relays have reliably served their function and will continue to be used for many years. The main characteristics that should be noted as it affects coordination are overtravel and reset time. *Overtravel* occurs because of the inertia in the disk. The disk will keep turning for a short distance even after the short circuit is interrupted. A typical overtravel of 0.1 sec is assumed when applying induction relays. An induction disk cannot instantly turn back to the neutral position. This *reset time* should be considered when applying reclosing sequences. It is not desirable to reclose before the relay resets or *ratcheting* to a trip can occur.

Digital relays are slowly replacing electromechanical relays. The main advantages of digital relays are

- *More relay functions* — One relay performs the functions of several electromechanical relays. One relay can provide both instantaneous and time-overcurrent relay protection for three phases, plus the ground, and perform reclosing functions. This can result in considerable space and cost savings. Some backup is lost with this scheme if a relay fails. One option to provide relay backup is to use two digital relays, each with the same settings.
- *New protection schemes* — Advanced protection schemes are possible that provide more sensitive protection and better coordination with

TABLE 8.2

Relay Designations

	Westinghouse/ABB Designation	General Electric Designation
Moderately inverse	CO-7	
Inverse time	CO-8	IAC-51
Very inverse	CO-9	IAC-53
Extremely inverse	CO-11	IAC-77

other devices. Two good examples for distribution protection are negative-sequence relaying and sequence coordination. Advanced algorithms for high-impedance fault detection are also possible.

- *Other auxiliary functions* — Fault location algorithms, fault recording, and power quality recording functions.

Digital relays have another advantage: internal diagnostics with ability to self-test. With digital technology, the relay is less prone to drift over time from mechanical movements or vibrations. Digital relays also avoid relay overtravel and ratcheting that are constraints with electromechanical relays (though some digital relays do reset like an electromagnetic relay).

Digital relays do have disadvantages. They are a relatively new technology. Computer technology has a poor reputation as far as reliability. If digital relays were as unreliable as a typical personal computer, we would have many more interruptions and many fires caused by uncleared faults. Given that, most digital relays have proven to be reliable and are gaining more and more acceptance by utilities.

Just as computer technology continues to advance rapidly, digital relays are also advancing. While it is nice to have new features, technical evolution can also mean that relay support becomes more difficult. Each relay within a certain family has to have its own supporting infrastructure for adjusting the relay settings, uploading and downloading data, and testing the relay. Each relay requires a certain amount of crew learning and training. As relays evolve, it becomes more difficult to maintain a variety of digital relays. The physical form and connections of digital relays are not standardized. As a contrast, electromechanical relays change very little and require a relatively stable support infrastructure. Equipment standardization helps minimize the support infrastructure required.

The time-current characteristics are based on the historically dominant manufacturers of relays. Westinghouse relays have a CO family of relays, and the General Electric relays are IAC (see Table 8.2). Most relays (digital and electromechanical) follow the characteristics of the GE or Westinghouse relays. For distribution overcurrent protection, the extremely inverse relays are most often used (CO-11 or IAC-77).

The time-current curves for induction relays can be approximated by the following equation (Benmouyal and Zocholl, 1994):

TABLE 8.3

IEEE Standardized Relay Curve Equation Constants

	A	B	p
Moderately inverse	0.0515	0.114	0.02
Very inverse	19.61	0.491	2.0
Extremely inverse	28.2	0.1217	2.0

Source: IEEE Std. C37.112-1996. Copyright 1997 IEEE. All rights reserved.

$$t = TD\left(\frac{A}{M^p - 1} + B\right)$$

where

t = trip time, sec
M = multiple of pickup current ($M > 1$)
TD = time dial setting
A, B, p = curve shaping constants

With actual induction disk relays, the constants A and B change with the time dial setting, but with digital relays, they stay constant.

Standardized characteristics of relays have been defined by IEEE (IEEE Std. C37.112-1996). This is an attempt to make relay characteristics consistent (since the relay curve can be adjusted to almost anything in a digital relay). The equations for the standardized inverse relay characteristics are shown in Table 8.3. Figure 8.2 compares the shapes of these curves. The standard allows relays to have tripping times within 15% of the curves. The standard also specifies the relay reset time for $0 < M < 1$ as

$$t = TD\left(\frac{t_r}{M^2 - 1}\right)$$

where t_r = reset time, sec for $M = 0$

8.2.4 Reclosers

A recloser is a specialty distribution protective device capable of interrupting fault current and automatically reclosing. The official definition of a recloser is:

Automatic Circuit Recloser — A self-controlled device for automatically interrupting and reclosing an alternating-current circuit, with a predetermined sequence of opening and reclosing followed by resetting, hold closed, or lockout.

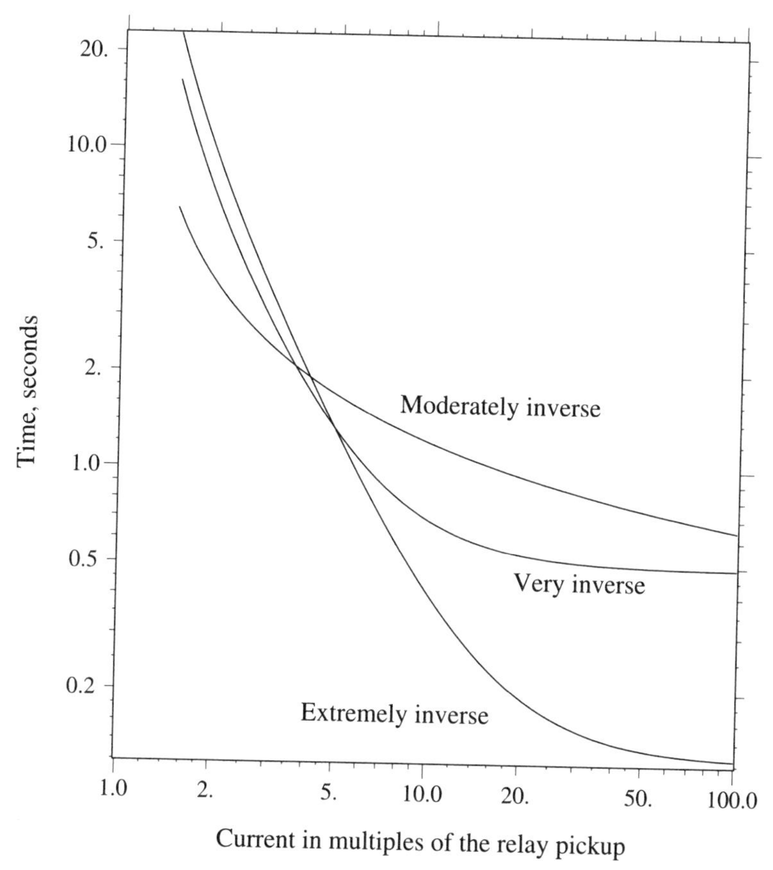

FIGURE 8.2
Relay curves following the IEEE standardized characteristics for a time dial = 5.

Like a circuit breaker, interruption occurs at a natural current zero. The interrupting medium of a recloser is most commonly vacuum or oil. The insulating medium is generally oil, air, a solid dielectric, or SF_6. The recloser control can be electronic, electromechanical (the relay for tripping is electromechanical, and the reclosing control is electronic) or hydraulic. A hydraulic recloser uses springs and hydraulic systems for timing and actuation.

The interrupting rating of a recloser is based on a symmetrical current rating. The interrupting current rating does not change with voltage. There is an exception that some reclosers have a higher interrupting current if operated at a significantly lower voltage than the rating. Smaller reclosers with a 50- to 200-A continuous rating typically have interrupting ratings of 2 to 5 kA (these would normally be feeder reclosers). Larger reclosers that could be used in substations have continuous current ratings as high as 1120 A and interrupting ratings of 6 to 16 kA. Historically, reclosers with series coil types had coil ratings of 25, 35, 50, 70, 100, 140, 200, 280, 400, and 560 A (each rating is approximately 1.4 times higher than the next lower rating).

Reclosers are tested at a specified *X/R* ratio as specified in ANSI/IEEE C37.60-1981. A typical test value is *X/R* = 16. While a lower *X/R* ratio at the point of application does not mean you can increase the rating of a recloser, the recloser must be derated if the *X/R* ratio is larger than that specified.

There are some other differences with recloser ratings vs. circuit breaker ratings (Cooper Power Systems, 1994). Reclosers do not have to be derated for multiple operations. Reclosers do not have a separate closing and latching (or first-cycle) rating. The symmetrical current rating is sufficient to handle the asymmetry during the first cycle as long as the circuit *X/R* ratio is lower than the tested value.

Reclosers have many distribution applications. We find reclosers in the substation as feeder interrupters instead of circuit breakers. An IEEE survey found that 51% of station feeder interrupting devices were reclosers (IEEE Working Group on Distribution Protection, 1995). Reclosers are used more in smaller stations and circuit breakers more in larger stations. Three-phase reclosers can be used on the main feeder to provide necessary protection coverage on longer circuits, along with improved reliability. Overhead units and padmounted units are available. Reclosers are available as single-phase units, so they can be used on single-phase taps instead of fuses. Another common application is in autoloop automation schemes to automatically sectionalize customers after a fault.

Since reclosers are devices built for distribution circuits, some have features that are targeted to distribution circuit needs. Three-phase units are available that can operate each phase independently (so a single-phase fault will only open one phase). Some reclosers have a feature called sequence coordination to enhance coordination between multiple devices.

The time-current characteristics of hydraulic reclosers have letter designations: A, B, and C. The A is a fast curve that is used similarly to an instantaneous relay element, and the B and C curves have extra delay ("delayed" and "extra delayed"). For a hydraulically-controlled recloser, the minimum trip threshold is twice the full-load rating of the trip coil of the recloser and is normally not adjustable. On electronically-controlled reclosers, the minimum trip threshold is adjustable independently of the rating (analogous to setting the pickup of a time-overcurrent relay).

8.2.5 Expulsion Fuses

Expulsion fuses are the most common protective device on distribution circuits. Fuses are low-cost interrupters that are easily replaced (when in cutouts). Interruption is relatively fast and can occur in a half of a cycle for large currents. An expulsion fuse is a simple concept: a fusible element made of tin or silver melts under high current. Expulsion fuses are most often applied in a fuse cutout. In a fuse tube, after the fuse element melts, an arc

FIGURE 8.3
Example operation of an expulsion fuse during a fault. (Courtesy of the Long Island Power Authority.)

remains. The arc, which has considerable energy, causes a rapid pressure buildup. This forces much of the ionized gas out of the bottom of the cutout (see Figure 8.3), which helps to prevent the arc from reigniting at a current zero. The extreme pressure, the stretching of the arc, and the turbulence help increase the dielectric strength of the air and clear the arc at a current zero. A fuse tube also has an organic fiber liner that melts under the heat of the arc and emits fresh, non-ionized gases. At high currents, the expulsion action predominates, while at lower currents, the deionizing gases increase the dielectric strength the most.

The "expulsion" characteristics should be considered by crews when placing a cutout on a structure. Avoid placement where a blast of hot; ionized gas blown out the bottom of the cutout could cause a flashover on another phase or other energized equipment. Implement and enforce safety procedures whenever a cutout is switched in (because it could be switching into a fault), including eye protection, arc resistant clothing, and, of course, avoiding the bottom of the cutout.

The *speed ratio* of a fuse quantifies how steep the fuse curve is. The speed ratio is defined differently depending on the size of the fuse (IEEE Std. C37.40-1993):

$$\text{Speed ratio for fuse ratings of 100 A and under} = \frac{\text{melting current at 0.1 seconds}}{\text{melting current at 300 seconds}}$$

$$\text{Speed ratio for ratings above 100 A} = \frac{\text{melting current at 0.1 seconds}}{\text{melting current at 600 seconds}}$$

Industry standards specify two types of expulsion fuses, the most commonly used fuses. The "K" link is a relatively fast fuse, and the "T" is somewhat slower. K links have a speed ratio of 6 to 8. T links have a speed ratio of 10 to 13. The K link is the most commonly used fuse for transformers and for line taps. The K and T fuse links are standardized well enough that they are interchangeable among manufacturers for most applications.

Two time-current curves are published for expulsion fuses: the minimum melt curve and the maximum total clear curve. The minimum-melt time is 90% of the average melt time to account for manufacturing tolerances. The total clearing time is the average melting time plus the arcing time plus manufacturing tolerances. Figure 8.4 shows the two published curves for 50-A K and T fuse links. The manufacturer's minimum melt curves for fuses less than or equal to 100 A normally start at 300 sec, and those over 100 A start at 600 sec.

The time-current characteristics for K and T links are standardized at three points (ANSI C37.42-1989). The minimum and maximum allowed melting current is specified for durations of 0.1 sec, 10 sec, and either 300 sec (for fuses rated 100 A or less) or 600 sec (for larger fuses).

Published fuse curves are for no loading and an ambient temperature of 25°C. Both loading and ambient temperature change the fuse melting characteristic. Load current causes the most dramatic difference, especially when a fuse is overloaded. Figure 8.5 shows the effect of loading on fuse melting time. Figure 8.6 through Figure 8.9 show time-current curves for K and T links.

For operation outside of this ambient range, the fuse melting time changes. The melting characteristic of tin fuse links changes 3.16% for each 10°C above or below 25°C, so a fuse operating in a 50°C ambient will operate in 92% of the published time ($100\% - \frac{25}{10}\,3.16\%$). Silver links are less sensitive to temperature (0.9% melting change for each 10°C above or below 25°C).

The I^2t of a fuse is often needed to coordinate between fuses. Table 8.4 shows the minimum melt I^2t of K and T links estimated from the time-current curves at 0.01 sec. This number is also useful to estimate melting characteristics for high currents below the published time-current characteristics, which generally have a minimum time of 0.01 sec.

The 6, 10, 15, 25, 40, 65, 100, 140, and 200-A fuses are standard ratings that are referred to as *preferred* fuses. The 8, 12, 20, 30, 50, and 80-A fuses are *intermediate* fuses. The designations are provided because two adjacent fuses (for those below 100 A) will not normally coordinate. A 40 and a 30-A fuse will not coordinate, but a 40 and a 25-A fuse will coordinate up to some maximum current. Most utilities pick a standard set of fuses to limit the number of fuses stocked.

K or T links with tin fuse elements can carry 150% of the nominal current rating indefinitely. It is slightly confusing that a 100-A fuse can operate continuously up to 150 A. Overloaded fuses, although they can be safely overloaded, operate significantly faster when overloaded, which could cause miscoordination. In contrast to tin links, silver links have no continuous overload capability.

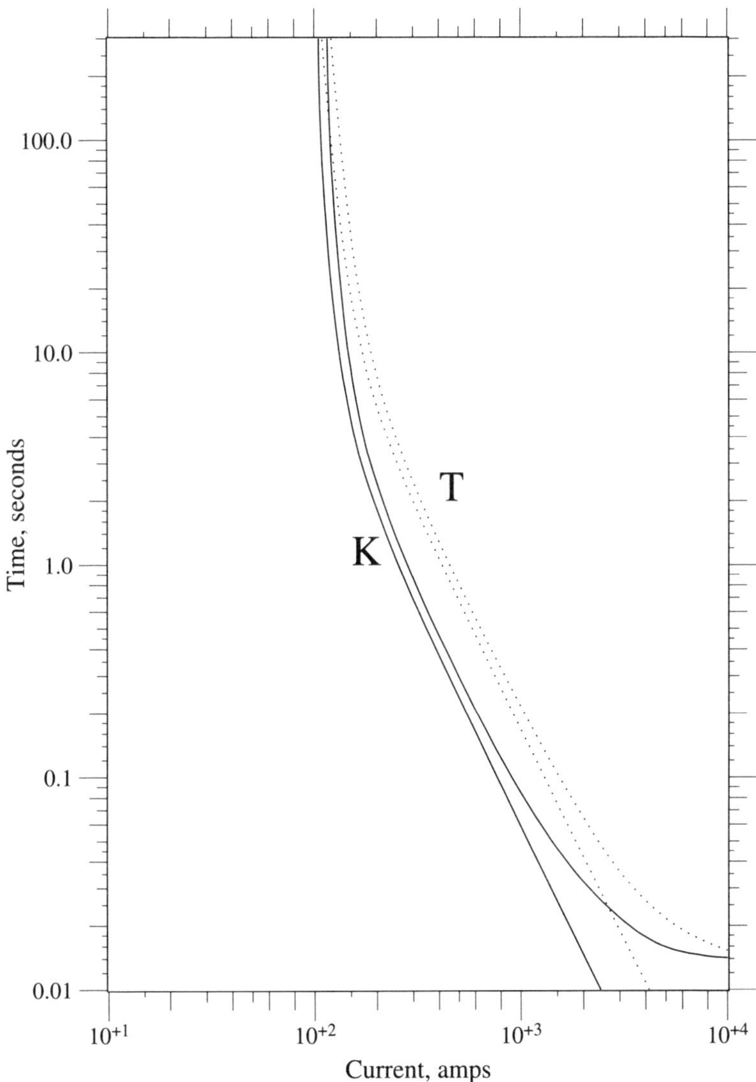

FIGURE 8.4
Minimum melt and total clearing curves for a T and K link (50 A).

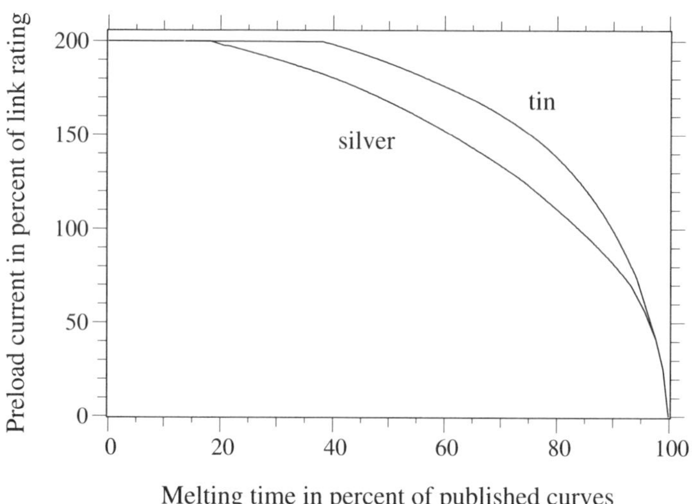

FIGURE 8.5
Effect of loading on fuse melting time. (Adapted from Cooper Power Systems, *Electrical Distribution — System Protection,* 3rd ed., 1990. With permission from Cooper Industries, Inc.)

Other nonstandard fuses are available from manufacturers for special purposes. One type of specialty fuse is a fuse even slower than a T link that is used to provide better coordination with upstream circuit breakers or reclosers in a fuse saving scheme. Another notable type of specialty fuse is a *surge-resistant* fuse that responds slowly to fast currents (such as surges) but faster to lower currents. These achieve better protection on transformers for secondary faults and faster operation for internal transformer failures while at the same time reducing nuisance fuse operations due to lightning.

Expulsion fuses under oil are another fuse variation. These "weak-links" are used on CSP (completely self-protected) transformers and some pad-mounted and vault transformers. Since they are not easily replaced, they have very high ratings — at least 2.5 times the transformer full load current and much higher if a secondary circuit breaker is used.

Transformers on underground circuits use a variety of fuses. For pad-mounted transformers, a common fuse is the replaceable Bay-O-Net style expulsion fuse. The time-current characteristics of this fuse do not follow one of the industry standard designations.

8.2.5.1 Fuse Cutouts

The cutout is an important part of the fuse interrupter. The cutout determines the maximum interrupting capability, the continuous current capability, the load-break capability, the basic lightning impulse insulation level (BIL), and the maximum voltage. Cutouts are typically available in 100, 200, and 300-A continuous ratings [ANSI standard sizes (ANSI C37.42-1989)].

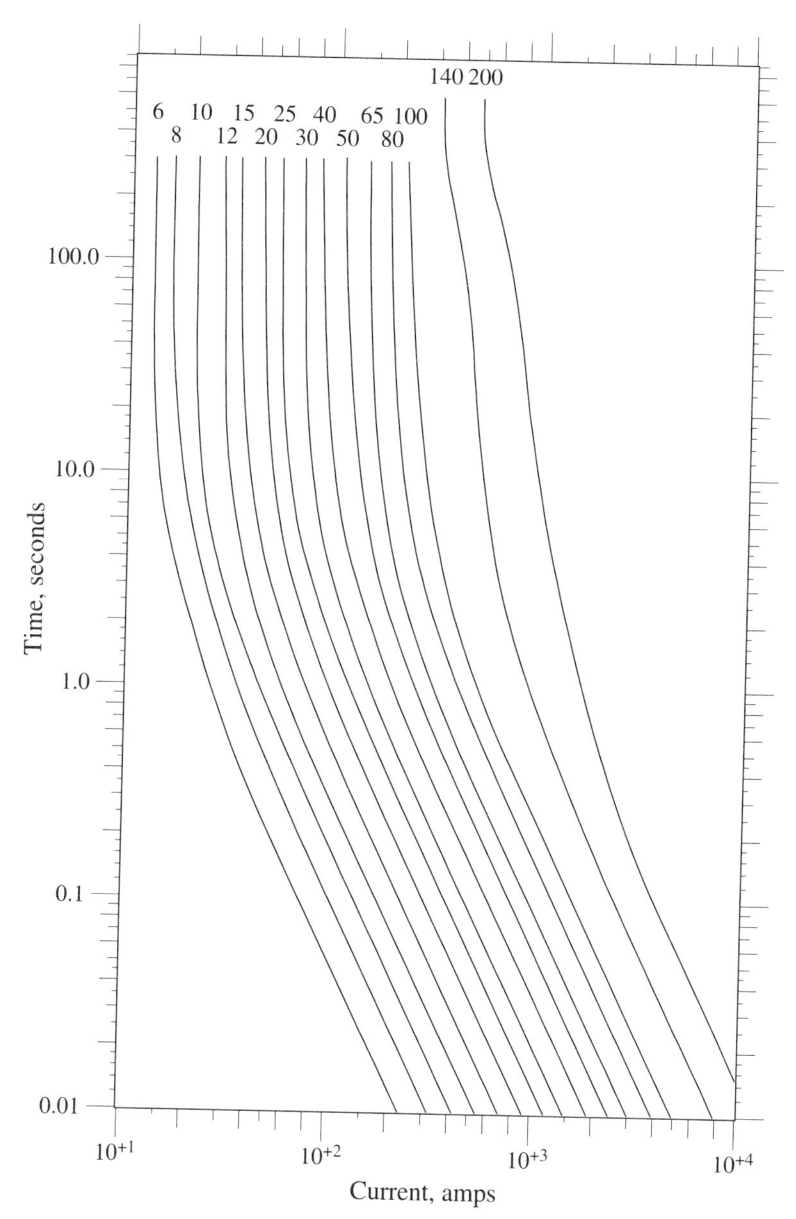

FIGURE 8.6
Minimum melt curves for K links. (S&C Electric Company silver links.)

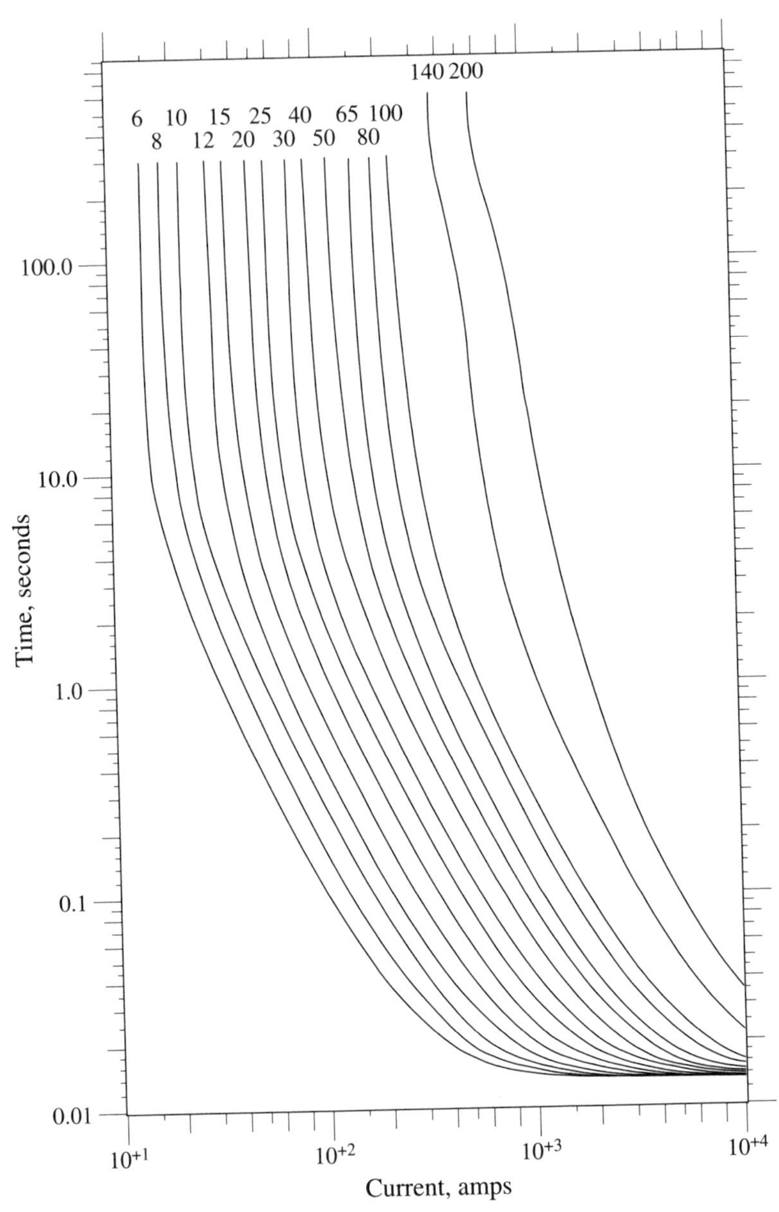

FIGURE 8.7
Maximum total clear curves for K links. (S&C Electric Company silver links.)

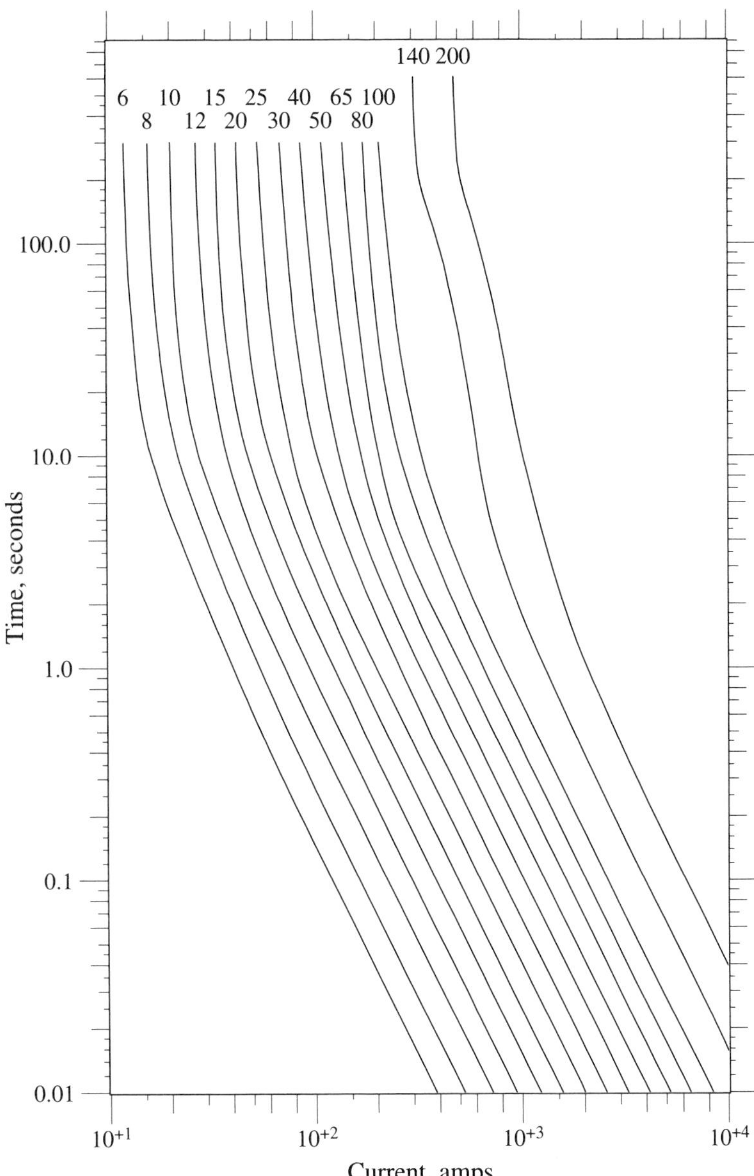

FIGURE 8.8
Minimum melt curves for T links. (S&C Electric Company silver links.)

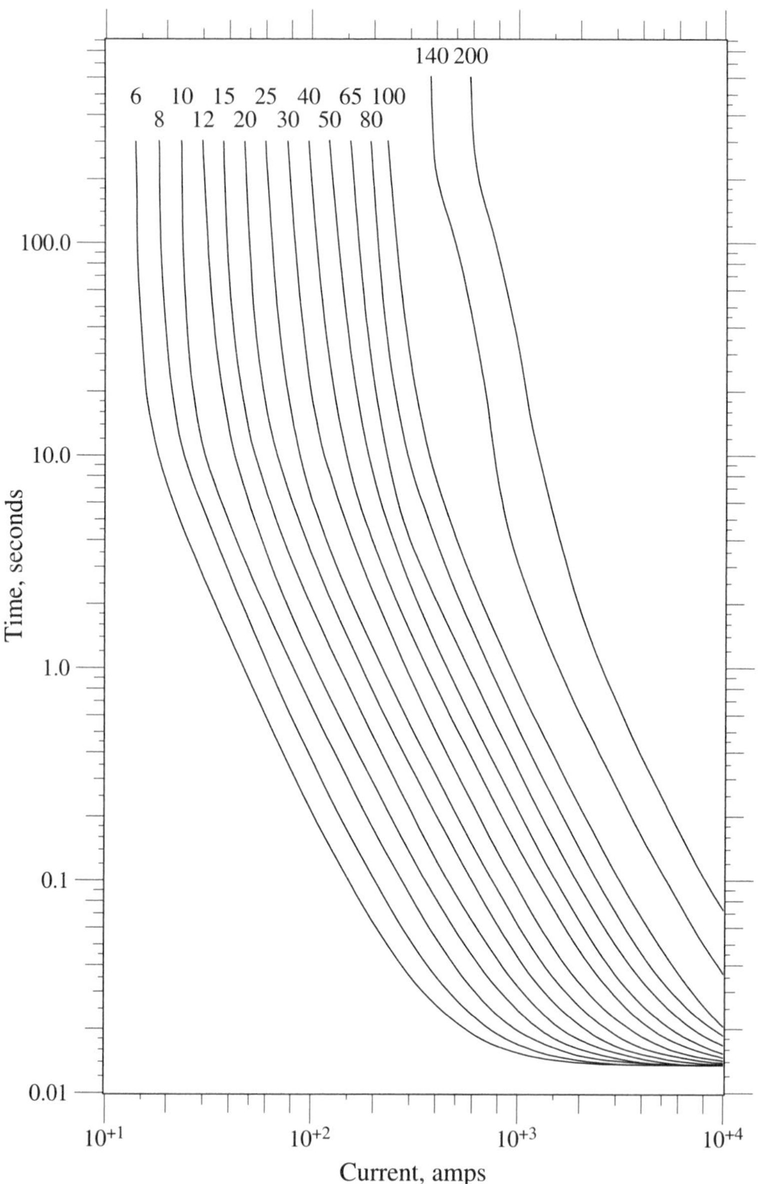

FIGURE 8.9
Maximum total clear curves for T links. (S&C Electric Company silver links.)

TABLE 8.4

Fuse Minimum-Melt I^2t (A^2-s)

Rating, A	K Links	T Links
6	534	1,490
8	1,030	2,770
10	1,790	5,190
12	3,000	8,810
15	5,020	15,100
20	8,500	24,500
25	13,800	40,200
30	21,200	65,500
40	36,200	107,000
50	58,700	173,000
65	90,000	271,000
80	155,000	425,000
100	243,000	699,000
140	614,000	1,570,000
200	1,490,000	3,960,000

Cutouts are rated on a symmetrical basis. Cutouts are tested at *X/R* ratios of 8 to 12, so if the *X/R* ratio at the application point is higher than the test value, the cutout should be derated. The fuse line holder determines the interrupting capability, not the fuse link.

Most cutouts are of the open variety with a removable fuse holder that is placed in a cutout with a porcelain-bushing type support. We also find enclosed cutouts and open-link cutouts. Open links have a fuse link suspended between contacts. Open links have a much lower interrupting capability (1.2 kA symmetrical).

Many cutouts available are full-rated cutouts that can be used on any type of system where the maximum line-to-line voltage is less than the cutout rating. Cutouts are also available that have *slant* voltage ratings, which provide two ratings such as 7.8/15 kV that are meant for application on grounded circuits (IEEE Std. C37.48-1997). One cutout will interrupt any current up to its interrupting rating and up to the lower voltage rating. On a grounded distribution system, in most situations, any cutout can be applied that has the lower slant rating voltage greater than the maximum line-to-ground voltage. On a 12.47Y/7.2-kV grounded distribution system, a 7.8/15-kV cutout could be used. If the system were ungrounded, a full-rated 15-kV cutout must be used. For three-phase grounded circuits, the recovery voltage is the line-to-line voltage for a line-to-line fault rather than the line-to-ground voltage (requiring a higher voltage rating). In this case, the slant rated cutouts are designed and tested so that two cutouts in series will interrupt a current up to its interrupting rating and up to the higher voltage rating. The two cutouts share the recovery voltage (even considering the differences in the melting times of the two fuses). On grounded systems, there are cases where slant-rated cutouts are "under-rated" — any time that a phase-to-phase fault could happen that would only be cleared by one cutout. This includes constructions where multiple circuits share a pole or cases where cutouts are applied on different poles.

Most cutouts used on distribution systems do not have load break capability. If the cutout is opened under load, it can draw an arc that will not clear. It is not an uncommon practice for crews to open cutouts under load (if it draws an arc, they slam it back in). Cutouts with load-break capability are available, usually capable of interrupting 100 to 300 A. Cutouts with load-break capability usually use an arc chute. A spring pulls the arc quickly through the arc chute where the arc is stretched, cooled, and interrupted. A load-break tool is available that can open standard cutouts (with no load break capability of its own) under load up to 600 to 900 A. Utilities also sometimes use solid blades in cutouts instead of a fuse holders; then crews can use the cutout as a switch.

8.2.6 Current-Limiting Fuses

Current limiting fuses (CLFs) are another interrupter having the unique ability to reduce the magnitude of the fault current. CLFs consist of fusible elements in silicon sand (see Figure 8.10). When fault current melts the fusible elements, the sand melts and creates a narrow tube of glass called a *fulgerite*. The voltage across the arc in the fulgerite greatly increases. The fulgerite constricts the arc. The sand helps cool the arc (which means it takes energy from the arc). The sand does not give off ionizable gas when it melts, and it absorbs electrons, so the arc has very little ionizable air to use as a conductor. Without ionizable air, the arc is choked off, and the arc resistance becomes very high. This causes a back voltage that quickly reduces the current. The increase in resistance also lowers the *X/R* ratio of the circuit, causing a premature current zero. At the current zero, the arc extinguishes. Since the *X/R* ratio is low, the voltage zero and current zero occur very close together, so there will be very little transient recovery voltage (the high arc voltage comes just after the element melts). Because the current-limiting fuse

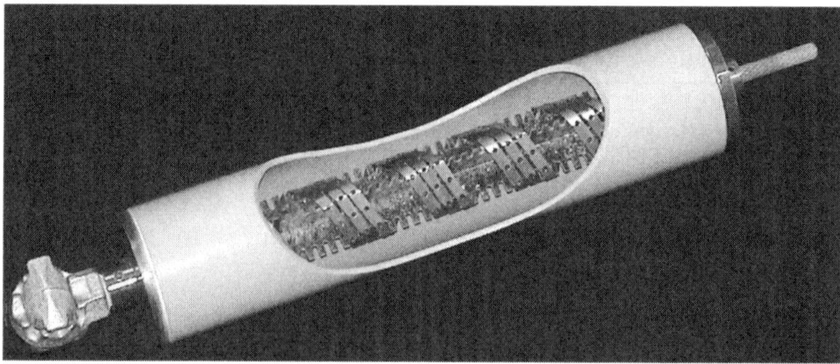

FIGURE 8.10
Example backup current-limiting fuse. (From Hi-Tech Fuses, Inc. With permission.)

TABLE 8.5

Use of Current-Limiting Fuses as Reported in a 1995 IEEE Survey

	5 kV	15 kV	25 kV	35 kV
General Purpose	15%	29%	30%	18%
Backup	15%	38%	43%	30%
On OH Line Laterals	5%	6%	9%	3%
On UG Line Laterals	7%	18%	20%	18%

Source: IEEE Working Group on Distribution Protection, "Distribution Line Protection Practices — Industry Survey Results," *IEEE Transactions on Power Delivery*, vol. 10, no. 1, pp. 176–86, January 1995.

forces an early current zero, the fuse can clear the short circuit in much less than one half of a cycle.

Current-limiting fuses are noted for their very high fault-clearing capability. CLFs have symmetrical maximum interrupt ratings to 50 kA; contrast that to expulsion fuses which may have typical maximum interrupt ratings of 3.5 kA in oil and 13 kA in a cutout. Current-limiting fuses also completely contain the arc during operation and are noiseless with no pressure buildup.

Current-limiting fuses are widely used for protection of equipment in high fault current areas. Table 8.5 shows the percentages of utilities that use CLFs. The major reason given for the use of CLFs is safety, and the second most common reason is high fault currents in excess of expulsion fuse ratings.

There are three types of current-limiting fuses (IEEE Std. C37.40-1993):

- *Backup:* A fuse capable of interrupting all currents from the maximum rated interrupting current down to the rated minimum interrupting current.
- *General Purpose:* A fuse capable of interrupting all currents from the maximum rated interrupting current down to the current that causes melting of the fusible element in one hour.
- *Full Range:* A fuse capable of interrupting all currents from the rated interrupting current down to the minimum continuous current that causes melting of the fusible element(s), with the fuse applied at the maximum ambient temperature specified by the manufacturer.

Current-limiting fuses are very good at clearing high-current faults. They have a much harder time with low-current faults or overloads. For a low-level fault, the fusible element will not melt, but it will get very hot and can melt the fuse hardware resulting in failure. This is why the most common CLF application is as a backup in series with an expulsion fuse. The expulsion fuse clears low-level faults, and the CLF clears high-current faults. Current-limiting fuses have very steep melting and clearing curves, much steeper than expulsion links. Many current-limiting fuses have steeper characteristics than I^2t. At low currents, heat from the notches transfers to the unnotched portion; at high currents, the element melts faster because heat cannot escape from the notched areas fast enough to delay melting.

General-purpose fuses usually use two elements in series — one for the high-current faults and one for the low-current faults. General-purpose fuses could fail for overloads, so restrict their application to situations where overloads are not present or are protected by some other device (such as a secondary circuit breaker on a transformer).

Full-range fuses provide even better low-current capability and can handle overloads and low-level faults without failing (as long as the temperature is within rating).

Current-limiting fuses can be applied in several ways including:

- Backup current-limiting fuse in series with an expulsion fuse in a cutout
- Full-range current-limiting fuse in a cutout
- Backup CLF under oil
- Full-range (or general-purpose) fuse under oil
- CLF in a dry-well canister or insulator

The best locations for use on distribution systems are close to the substation. This is where they are most appropriate for limiting damage due to high fault currents and where they are most useful for reducing the magnitude and duration of a voltage sag.

Some of the drawbacks of current-limiting fuses are summarized as

- *Voltage kick* — When a CLF operates, the rapidly changing current causes a voltage spike ($V = Ldi/dt$). Usually, this is not severe enough to cause problems for the fuse or for customer equipment.
- *Limited overload capability* — A backup or general-purpose fuse does not do well for overloads or low-current faults. A full-range fuse performs better but could still have problems with a transient overcurrent that partially melts the fuse.
- *Coordination issues* — A current-limiting fuse may be difficult to coordinate with expulsion fuses or reclosers or other distribution protective devices. CLFs are fast enough that they almost have to be used in a fuse-blowing scheme (fuse saving will not work because the fuse will be faster than the circuit breaker).
- *Cost* — High cost relative to an expulsion fuse.

Current-limiting fuses limit the energy at the location of the fault. This provides safety to workers and the public. Arc damage to life and property occurs in several ways:

- *Pressure wave* — The fault arc pressure wave damages equipment and personnel.
- *Heat* — The fault arc heat burns personnel and can start fires.
- *Pressure buildup in equipment* — An arc in oil causes pressure buildup that can rupture equipment.

All of these effects are related to the arc energy and all are greatly reduced with current-limiting fuses. Distribution transformers are a common application of current-limiting fuses to prevent them from failing violently due to internal failures.

8.3 Transformer Fusing

The primary purpose of a transformer fuse is to disconnect the transformer from the circuit if it fails. Some argue that the fuse should also protect for secondary faults. The fuse cannot effectively protect the transformer against overloads.

Engineers most commonly pick fuse sizes for distribution transformers from a fusing table developed by the utility, transformer manufacturer, or fuse manufacturer. These tables are developed based on criteria for applying a fuse such that the fuse should not have false operations from inrush and cold-load pickup.

One way to pick a fuse is to plot cold-load pickup and inrush points on a time-current coordination graph and pick a fuse with a minimum melt or damage curve that is above the cold-load and inrush points. Most fusing tables are developed this way. A fuse should withstand the cold-load and inrush points given in Table 8.6. The inrush points are almost universal, but the cold-load pickup points are more variable (and they should be since cold-load pickup characteristics change with predominant load types). An example application of the points given in Table 8.6 for a 50-kVA, 7.2-kV single-phase transformer which has a full-load current of 6.94 A is shown in Figure 8.11. The cold-load pickup and inrush points are plotted along with K links. The minimum melt time and the damage time (75% of the minimum melt time) are shown. Use the damage curve to coordinate. For this example, a 12-A K link would be selected; the 1-sec cold-load pickup point determines the fuse size. Since this point lies between the damage and minimum melt time of the 10K link, some engineers would pick the 10K link (not recommended).

Some utilities have major problems from nuisance fuse operations (especially utilities in high-lightning areas). A nuisance operation means that the fuse must be replaced, but the transformer was not permanently damaged. Nuisance fuse operations can be over 1% annually. Some utilities have thou-

TABLE 8.6

Inrush and Cold-Load Pickup Withstand Points for Transformer Fusing

	Full-Load Current Multiplier	Duration, sec
Cold-load pickup	2	100
	3	10
	6	1
Inrush points	12	0.1
	25	0.01

Source: Amundson, R. H., "High Voltage Fuse Protection Theory & Considerations," IEEE Tutorial Course on Application and Coordination of Reclosers, Sectionalizers, and Fuses, 1980. Publication 80 EHO157–8-PWR; Cook, C. J. and Niemira, J. K., "Overcurrent Protection of Transformers—Traditional and New Philosophies for Small and Large Transformers," IEEE/PES Transmission & Distribution Conference and Exposition, 1996. Presented at the training session on "Distribution Overcurrent Protection Philosophies."

sands of nuisance fuse operations per year. A utility in Florida had a region with 57% of total service interruptions due to transformer interruptions, and 63% of the storm-related interruptions required only re-fusing (Plummer et al., 1995). During a storm, multiple transformer fuses can operate on the same circuit. There are differences of opinion as to what is causing the nuisance operations. Some of the possibilities are

- *Inrush* — Transformer inrush may cause fuse operations even though the inrush points are used in the fuse selection criteria. Reclosing sequences during storms can cause multiple inrush events that can heat up the fuse. In addition, voltage sags can cause inrush (any sudden change in the voltage magnitude or phase angle can cause the transformer to draw inrush).
- *Cold-load pickup* — This is the obvious culprit after an extended interruption (many of the nuisance fuse operations have occurred when there is not an extended interruption).
- *Secondary-side transformer faults* — Secondary-side faults that self-clear can cause some nuisance fuse events.
- *Lightning current* — Lightning current itself can melt small fuses. Arrester placement is key here since the lightning current flows to the low-impedance provided by a conducting arrester. If the fuse is upstream of the arrester (which would be the case on a tank-mounted arrester), the lightning surge current flows through the fuse link. If the fuse is downstream, then little current should flow through the fuse.

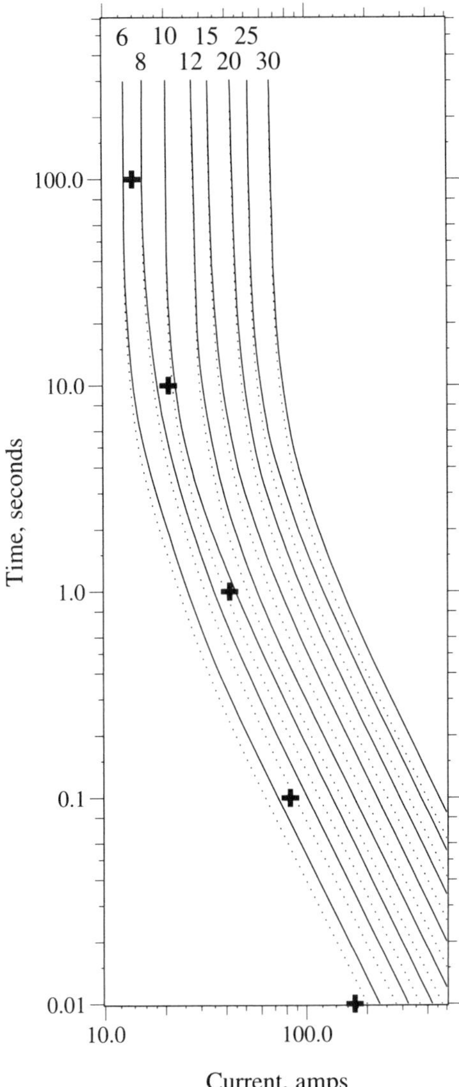

FIGURE 8.11
Transformer inrush and cold-load pickup points for a single-phase, 50-kVA, 7.2-kV transformer. The minimum-melt curves and damage curves (dotted lines) for K-link fuses are also shown.

- *Power-follow current through gapped arresters* — Following operation of a gapped arrester, a few hundred amps of power follow current flows in a gapped silicon carbide arrester until the gap clears (usually just for a half cycle if the gap is in good shape).
- *Transformer saturation from lightning currents* — Lightning can contain multiple strokes and long-duration components that last from 0.1 to

2 sec. These currents can saturate distribution transformers. Following saturation, the transformer becomes a low impedance and draws high current from the system through the fuse (Hamel et al., 1990).
- *Animal faults* — Across transformer bushings or arresters.

Several of these causes may add to the total. Nuisance fuse operations have occurred when circuits were out of service. This means that lightning is the cause since any type of inrush would require the system to be energized. Detroit Edison found that 70 to 80% of fuse operations were due to lightning (Gabrois et al., 1973). Lightning and inrush events are the most likely cause of nuisance fuse operations. Heavily loaded transformers are more susceptible to nuisance fuse operations because of the preheating of the fuse (a heavily loaded transformer is more susceptible to cold-load pickup as well).

Another method of choosing the transformer fuse size that gives "looser" fusing is the ×2 method (Burke, 1996). Choose a fuse size larger than twice the transformer full load current. A 50-kVA, 7.2-kV single-phase transformer, which has a full-load current of 6.94 A, should have a fuse bigger than 14 A (the next biggest standard size is a 15-A fuse). This applies for any type of fuse (K, T, or other). The factor of two provides a safety margin so that transformer fuses do not operate for inrush or cold-load pickup, and it helps with lightning.

The *fusing ratio* is the ratio of the fuse minimum melt current to the transformer full-load current (some sources also define a fusing ratio as the ratio of fuse rated current to transformer rated current which is different from this definition by a factor of two). Tight fusing means the fuse ratio is low. Relatively low fusing ratios have been historically used which has led to the nuisance fuse problems. The tighter fusing given using the Table 8.6 approach results in fusing ratios of 2 to 4. The looser ×2 method gives a fusing ratio of at least 4 (since the fuse rating is multiplied by two, and the minimum melting current at 300 sec is twice the fuse rating). The fusing ratio for the 50-kVA, 7.2-kV transformer with the 12-K fuse is 3.46, and it is 4.32 with the 15-K fuse.

Another strategy that is especially useful in high lightning areas: use a standard fuse size for all transformers up to a certain size. This also helps ensure that the wrong fuse is not applied on a given transformer. A standard fuse size of at least 15T or 20K results in few nuisance fuse operations (IEEE Std. C62.22-1997). At 12.5 kV, a 20K fuse should protect a 5-kVA transformer almost as well as it protects a 50-kVA transformer. It may lose some secondary protection relative to a smaller fuse, and a small portion of evolving faults will not be detected as soon; but other than that, there should not be much difference. If fuses get too big, they may start to bump up against tap fuse sizes and limit fuse options for lateral taps.

If looser fusing is used, some argue that overload protection of transformers is lost. Countering that argument, overload protection with fuses is not really possible if the transformer is used for its most economic performance (which means overloading a transformer at peak periods). To avoid nuisance fuse operations from load, we must use a fuse big enough so that thermal overload protection is impossible. It is also argued that most transformer failures start as failures between turns or layers and that a smaller, faster fuse detects this more quickly. Tests have indicated that a smaller fuse might not be much better than a larger fuse at detecting interwinding failures (Lunsford and Tobin, 1997) (pressure-relief valves limit tank pressures very well during this type of failure). All together, the arguments for a smaller fuse are not enough to overcome the concerns with nuisance fuse operations. If overload protection must be used, use a surge resistant fuse (it has a slower characteristic for high-magnitude, short-duration currents).

A few utilities practice group fusing where a lateral fuse provides protection to all of the transformers on the tap. If the transformer failure rate (including bushing faults) is low enough, then this practice will not degrade the overall frequency of interruptions significantly. One of the major disadvantages of this approach is that an internal transformer failure on a tap may be *very* hard to find. This drives up repair time (so the duration reliability numbers suffer but not necessarily the frequency indices). Also, the beneficial feature of being able to switch the transformer with the fused cutout is lost if group fusing is used.

Widely used, completely self-protected transformers (CSPs) have an internal weak-link fuse; an external fuse is not needed (although they may need an external current-limiting fuse to supplement the weak link).

Transformer bushing faults often caused by animals can have different impacts depending on fusing practices. A fault across a primary bushing operates an external transformer fuse. If the transformer is a CSP or group fusing is used, the upstream tap fuse operates (so more customers are affected).

Current-limiting fuses are regularly used on transformers in high fault-current areas to provide protection against violent transformer failure. NEMA established tests which were later adopted by ANSI (ANSI C57.12.20-1988) for distribution transformers to be able to withstand internal arcs. Transformers with external fuses are subjected to a test where an internal arcing fault with an arc length of 1 in. (2.54 cm) is maintained for 1/2 to 1 cycle. It was thought that 1 in. (2.54 cm) was representative of the length that arcs could typically achieve. The current is 8000 A. Under this fault condition, the transformer must not rupture or expel excessive oil. Note that this test does not include all of the possible failure modes and is no guarantee that a transformer will not fail with lower current. For example, a failure with an arc longer than 1 in. has more energy and ruptures the transformer at a lower level of current.

Table 8.7 shows rupture limits for several types of transformers based on tests for the Canadian Electrical Association. If fault current values exceed those given in this table, consider using current-limiting fuses to reduce the chance of violent failures (the CEA report considers the limits provisional

TABLE 8.7

Transformer Rupture Limits for Internal Faults

Transformer Type		$I \cdot t$, A-s, or Coulombs	Current Limit for a 1 Cycle Clearing Time, kA
Pole mounted	1ϕ	41	2.5
Pad mounted	1ϕ	150	9
Pad mounted/subway	3ϕ	180	11
Network with switch compartment	3ϕ	90	5.4
Submersible and vault	1ϕ	41	2.5

Source: CEA 288 D 747, *Application Guide for Distribution Fusing*, Canadian Electrical Association, 1998.

and suggests that more tests are needed). At arc energies within this range, the failure probability is on the order of 15 to 35%. Note that the 2.5-kA limit for pole-mounted transformers is much less than the ANSI test limit of 8 kA. Based on a series of tests with internal 2-in. (5 cm) arcs, Hamel et al. (2003) recommend considering current-limiting fuses for pole-type transformers when the short-circuit current exceeds 1.7 kA.

For transformers with an internal fuse, completely self-protected (CSP) transformers or padmounted transformers, the arcing test is done to the rating of the fuse which is generally much lower than 8000 A. Table 8.8 shows the maximum fault current ratings based on the ANSI tests. If the available line-to-ground fault current exceeds these values, then consider current-limiting fuses to reduce the possibility of violent failures. Not all utilities use current-limiting fuses in these situations, and in such instances, internal faults have failed transformers violently, blowing the cover.

If a transformer is applied in a location where the available line-to-ground fault current is higher than shown in Table 8.8, use current-limiting fuses.

TABLE 8.8

Maximum 1/2- to 1-Cycle Fault Current Rating on Distribution Transformers Based on the Test in ANSI C57.12.20-1988

Transformer	Maximum Tested Symmetrical Current	$\int Idt$ in the ANSI Test
Overhead transformer	8000 A	66.7 A-s
Under-oil expulsion fuse		
(based on typical fuse ratings)		
Up to 8.3 kV_{LG}	3500 A	29.2 A-s
Up to 14.4 kV_{LG}	2500 A	20.8 A-s
Up to 25 kV_{LG}	1000 A	8.3 A-s

Source: ANSI C57.12.20-1988, *American National Standard Requirements for Overhead-Type Distribution Transformers, 500 kVA and Smaller: High-Voltage, 67 000 Volts and Below; Low-Voltage, 15 000 Volts and Below,* American National Standards Institute.

8.4 Lateral Tap Fusing and Fuse Coordination

Utilities use two main philosophies to apply tap fuses: fusing based on load and standardized fusing schedules. With fusing based on load, we pick a fuse based on some multiplier of peak load current. The fuse should not operate for cold-load pickup or inrush to prevent nuisance operations. As an example, one utility sizes fuses based on 1.5 times the current from the phase with the highest connected kVA. With standardized fuse sizes, a typical strategy is to apply 100K links at all taps off of the mains (even if a tap only has one 15-kVA transformer). If using second-level fusing, use 65K links for these and 40K fuses for the third level. There is no clear winner; each has advantages and disadvantages:

- *Fusing based on load* — This tends to fuse more tightly. High-impedance faults are somewhat more likely to be detected. Nuisance fuse operations are more likely, especially with utilities that tightly fuse laterals. We are more likely to have load growth cause branch loadings to increase to the point of causing nuisance fuse operations. Fusing based on load helps on circuits that have covered wire because a smaller fuse helps protect against conductor burndown (taps that are more heavily loaded usually have a larger wire, which resists burndown).
- *Standardized fuse sizes* — It is simple: we spend less time coordinating fuses, we do not constantly check loadings, and utilities have less inventory. There is also less chance that the wrong fuse is installed at a location. A disadvantage of this approach is that larger fuses than needed are used at many locations, resulting in higher fault damage at the arc location, longer voltage sags, and more stress on in-line equipment.

Coordinating lateral tap fuses is generally straightforward. The fuse must coordinate with the station recloser or circuit breaker relays. Station ground relays are usually set to coordinate with the largest tap fuse. On the downstream side, a tap fuse should coordinate with the largest transformer fuse. This usually is not a problem.

In addition to sizing a fuse to avoid nuisance operations and coordinating with upstream and downstream protectors, we size fuses to ensure that the fuses provide protection to the line section that they are protecting. The reach of the fuse must exceed the length of the line section. Several methods are used to quantify the reach of a fuse:

- Where the fuse will clear a bolted single line-to-ground fault in 3 sec
- Where the bolted single line-to-ground fault current is six times the fuse rating

- Where the fuse will clear a single line-to-ground fault with a 30 Ω resistance in 5 sec

In most situations, typical fuse sizes provide sufficient reach by any of these methods. The first two methods are the best; the 30 Ω resistance is overly conservative and difficult to apply.

Reliability needs dictate the number of fuses used. The most common application for line fuses is at tap points. Occasionally, utilities fuse three-phase mains, but a recloser is more commonly used for this purpose. In the southwest U.S., in areas with few trees and little lightning, fuses may be rarely used. This is the exception, not the rule. Most utilities fuse most taps off the main line. Some go further and provide several levels of fusing, especially utilities with heavy tree coverage. Returns diminish: too many fuses leads to situations where fuses do not coordinate, and the extra fusing does not increase reliability significantly. Cutouts themselves contribute to causing faults by providing an easy location where animals, trees, and lightning cause faults, especially if they are poorly installed.

8.5 Station Relay and Recloser Settings

The main feeder circuit breaker relays (or recloser) must be set so that the circuit breaker coordinates with downstream devices, coordinates with upstream devices, and does not have trips from inrush or cold-load pickup. Station relays almost always use phase and ground time-overcurrent relays.

Table 8.9 shows typical settings used by several utilities. Many utilities try to use standardized relay settings at all distribution stations. This has the

TABLE 8.9

Time-Overcurrent and Instantaneous Station Relay Pickup Settings in Amperes on the Primary at Several Utilities

	Phase		Ground		
Utility	**TOC**	**Inst.**	**TOC**	**Inst.**	**Notes**
A	720	4000	480	4000	Assumes peak current = 400 A
B	720	1200	360	1200	Full load = 300 to 400 A
C	600	none	300	530	
D	960	1300	480	600	
E	800	none	340	none	
F	960	2880	240	1920	
G	≥2.25 × current rating	same as TOC	≥ 0.6 × current rating	same as TOC	
	600	600	160	160	Typical settings

advantage that relays are less likely to be set wrong, and there is less engineering effort in a coordination study. Some other utilities set each relay based on a coordination study.

Differences exist about the meaning of "peak load." Some utilities base it on the maximum design emergency load (which is typically something like 600 A). Others use the designed normal load (typically 400 A). Others may base it on some percentage of the total connected distribution transformer kVA.

Instantaneous relay settings vary more than phase relay settings. Several utilities also either disable or do not use an instantaneous relay setting. The instantaneous relay pickup ranges from one to almost ten times the phase relay pickup.

One reasonable set of base pickup settings is:

- *Phase TOC (time-overcurrent) relay* — Use two times the normal designed peak load on the circuit.
- *Ground TOC relay* — Use 0.75 times the normal designed peak load on the circuit.
- *Instantaneous phase and ground relays* — Use two times the TOC relay pickups.

Settings any less than this are prone to false trips from cold-load pickup and inrush.

In addition to avoiding nuisance trips, the relays (or recloser) must provide protection to its line section (to the end of the line or to the next protective device in series). Ensure that the relay has sufficient reach at the minimum operating current of the relay.

For a phase relay, sufficient reach is achieved by ensuring that 75% of the bolted line-to-line fault current at the end of the circuit is greater than the relay's pickup (its minimum operating current). So, if the line-to-line fault current at the end of the circuit is 1000 A, the pickup of the relay should be no more than 750 A. We use the line-to-line fault current because the two types of faults not seen by the ground relay are the three phase and the line to line. Of these, the line-to-line fault has the lower magnitude. The 75% factor provides a safety margin and allows some fault impedance. Another approach is to ensure that the line-to-ground fault current at the end of the circuit is less than the minimum operating current. The line-to-ground fault current is less than the line-to-line fault current, which provides the safety margin.

For a ground relay, ensure that the relay pickup is less than 75% of the line-to-ground fault current at the end of the line or to the next protective device. The ground relay must also coordinate with the largest lateral fuse.

Feeders dedicated to supplying secondary networks, either grid or spot networks, have similar settings as feeders supplying radial loads. Two main differences are related to loading and the ground relay setting. The pickup settings of station circuit breakers may have to account for higher peak loads.

Some utilities have phase relay pickups that are similar to radial circuits, from 600 to 800 A, but some have settings above 900 A with one utility having a 1680-A setting (Smith, 1999). Also, if feeders are supplying only network load, and all network transformers are connected delta – grounded wye, the unbalanced current seen by the ground relay is small. Utilities can set a low ground relay setting; some have settings ranging from 40 to 80 A (Smith, 1999). The main limitation on lowering the setting further is that during line-to-ground faults, the unfaulted circuits will backfeed the fault through the zero-sequence capacitance of that circuit. Lower ground-relay settings also help detect turn-to-turn or layer-to-layer faults within the primary windings of network transformers.

8.6 Coordinating Devices

Several details often arise when coordinating specific devices. Normally, we want to ensure that the downstream device clears before the upstream device operates over the range fault currents available at the downstream device. Time-current characteristics of both device normally show us how well two devices coordinate. Because of device differences, some combinations require slightly different approaches. We discuss some of the common combinations in the sections that follow.

8.6.1 Expulsion Fuse–Expulsion Fuse Coordination

When coordinating two fuses, the downstream fuse (referred to as the *protecting* fuse) should operate before the upstream fuse (the *protected* fuse). To achieve this goal, ensure that the total clear time of the protecting fuse is less than the damage time of the protected fuse. The damage time is 75% of the minimum melt time. An example for coordinating a 100K link with a 65K link is shown in Figure 8.12. Above a certain current, the two fuses do not coordinate; the protected fuse could suffer damage or melt before the protecting fuse can clear. For high fault currents, coordination is impossible because both fuses can open. The example shows that above 2310 A, the total clear curve of the 65K is above the damage curve of the 100K link. Utilities live with this common miscoordination. Table 8.10 lists the maximum coordination currents between K links. In cases where fuses do not coordinate, why have the second fuse? The second fuse still has some value; it adds another sectionalizing point (for a fuse in a cutout), and for a downstream fault, it identifies the fault location to a smaller area. Also, the downstream fuse may operate without damaging the upstream fuse. The amount of damage to the upstream fuse depends on the point of the waveform where the fault occurs (the extra 1/2 + cycle waiting for a current zero causes the extra heating to the protected fuse).

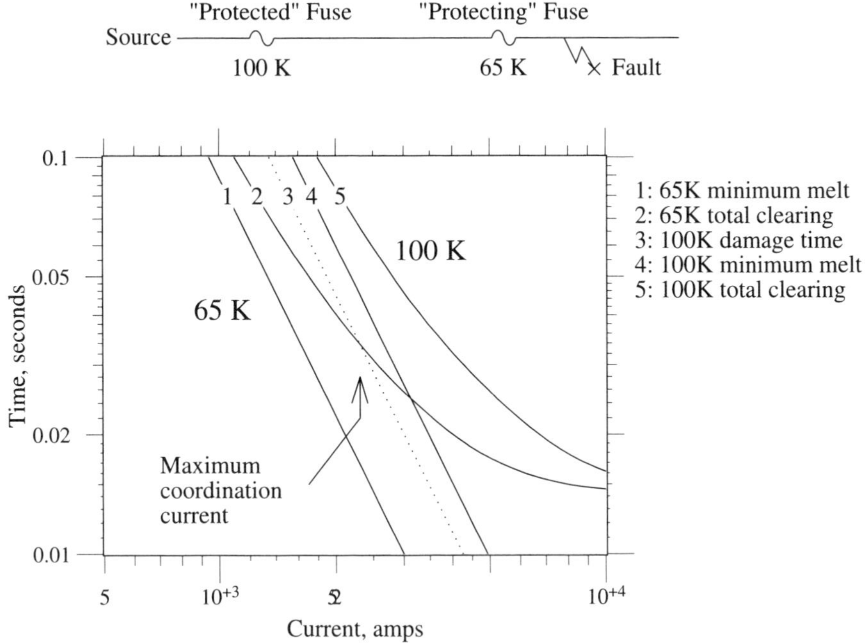

FIGURE 8.12
Example of fuse coordination between a 100-K (the protected fuse) and a 65-K link (the protecting link).

TABLE 8.10
Maximum Fault Currents for Coordination between the Given K Fuse Links

	10K	12K	15K	20K	25K	30K	40K	50K	65K	80K	100K	140K	200K
6K	170	310	460	640	840	1060	1410	1800	2230	2930	3670	5890	9190
8K	20	230	410	610	810	1040	1400	1790	2230	2930	3670	5890	9190
10K		40	300	550	780	1000	1370	1770	2220	2930	3670	5890	9190
12K			80	420	690	950	1330	1730	2190	2910	3650	5880	9190
15K				90	530	840	1250	1670	2120	2870	3640	5870	9190
20K					100	610	1120	1570	2040	2800	3590	5870	9190
25K						120	840	1380	1920	2710	3510	5830	9150
30K							240	1090	1690	2570	3380	5740	9110
40K								300	1240	2260	3210	5630	9010
50K									240	1750	2800	5500	8910
65K										970	2310	5210	8740
80K											420	4460	8430
100K												3550	7950
140K													4210

8.6.2 Current-Limiting Fuse Coordination

Coordinating two current-limiting fuses is similar to coordinating two expulsion links. Plot the time-current characteristics and ensure that the maximum clearing time of the load-side fuse is less than 75% of the minimum-melting time of the source-side fuse over the range of fault currents available at the load-side fuse. The 75% factor accounts for damage to the source-side fuse. Unlike expulsion links, current-limiting fuses can coordinate to very high currents. For coordination at higher currents than are shown on published time-current characteristics (operations faster than 0.01 sec), ensure that the maximum clearing I^2t of the load-side fuse is less than 75% of the minimum-melt I^2t of the source-side fuse. Manufacturers provide both of these I^2t values for current-limiting fuses.

Coordinating an expulsion link with a current-limiting fuse follows similar principles. Because the melting and clearing characteristics of current-limiting fuses are so much steeper than those of expulsion links, coordination is sometimes difficult; the operating characteristic curves are more likely to cross over. A load-side current-limiting fuse coordinates over a wide range of fault current. For a source-side current-limiting fuse, the clearing-time limitations of expulsion links (to about 0.8 cycles) prevent coordination at high currents. For currents above this value, either both will operate, or just the current-limiting fuse will operate.

Backup current-limiting fuse coordination requires special attention. To ensure that the CLF does not try to operate for currents below its minimum interrupting rating, the intersection of the expulsion fuse's total-clearing curve and the backup fuse's minimum-melting curve must be greater than the maximum interrupting rating of the backup fuse. Normally, we select backup current-limiting fuses for use with expulsion links based on *matched-melt* coordination. Select a backup current-limiting fuse that has a maximum melting I^2t below the maximum clear I^2t of the expulsion element. Also, check the time-current curves of the devices. The expulsion link should always clear for fault currents in the low-current operating region, especially below the minimum interrupting current of the current-limiting fuse.

With matched-melt coordination, the expulsion fuse always operates, including when the backup current-limiting fuse operates. In overhead applications with an expulsion fuse in a cutout, the dropout of the expulsion fuse provides a visible indication when the fuse(s) operate. Also, the backup fuse is unlikely to have full voltage across it.

The maximum melting I^2t of expulsion links is not provided from curves or data. To estimate this, take the minimum melting I^2t calculated from the minimum-melt curve at 0.0125 sec, and multiply by 1.2 for tin links or 1.1 for silver links. The multiplier allows for conservatism in minimum-melt curves and for manufacturing tolerances.

Somewhat less conservatively, experience has shown that fuses coordinate well if the maximum melt I^2t of the expulsion link does not exceed twice the minimum melt I^2t of the backup fuse (IEEE Std. C37.48-1997). We can tighten up the backup fuse because, under most practical situations, the backup fuse lets through significantly more I^2t than its minimum-melt value.

Manufacturers of backup current-limiting fuses normally provide coordination recommendations for their fuses, but review of the coordination approach is sometimes appropriate. Backup fuses often use a "K" nomenclature signifying the K link that it coordinates with. For example, a "25 K" backup link coordinates with a K link rated at 25 A or less. Figure 8.13 shows the time-current curves of a 40K expulsion link and the curves of one manufacturer's 40K backup current-limiting fuse. This graph extends below the normal cutoff time of 0.01 sec to show how the fuses coordinate at high currents. This example shows that the backup fuse does not coordinate using the strict matched-melt criteria (the maximum melting time of the expulsion link is more than the minimum melting time of the backup fuse). The minimum-melt I^2t of the backup fuse is 1.6 times the minimum melt I^2t of the backup fuse, so it meets the relaxed matched-melt criteria since this ratio is less than two.

The *time-current curve crossover coordination* allows a smaller backup current-limiting fuse. As before, the intersection of the expulsion fuse's total-clearing curve and the backup fuse's minimum-melting curve must be greater than the maximum interrupting rating of the backup fuse. We do not try to ensure that the backup fuse always melts. We can use a smaller fuse, which reduces the I^2t let through and reduces energy to faults. The backup CLF operates for a wider range of short-circuit currents. With a smaller fuse, the backup fuse can operate before the expulsion link melts for high fault currents. Utilities often use time-current curve crossover coordination for under-oil backup current-limiting fuses. In addition to lowering energy to faults, crossover coordination extends the range of current-limiting fuse protection to larger transformers.

For transformer protection, overload and secondary faults are also considerations for backup current-limiting fuses. Secondary faults at the terminals of the transformer should not damage or melt the backup CLF. One way to do this is to ensure that at the total clearing time of the expulsion link with the bolted secondary fault, the backup fuse's minimum melting current is at least 125% of the secondary fault current (Hi-Tech Fuses, 2002). Also, overload should not damage or melt the backup CLF.

8.6.3 Recloser–Expulsion Fuse Coordination

Normally, we want the recloser's fast curve (A curve) to clear before downstream fuses operate. This saves the fuse for temporary faults (we discuss fuse saving in more detail later in this chapter). Select the delayed curve (B, C, ...) to be above the clearing time of downstream fuses. A permanent fault downstream of a fuse should blow the fuse, not lockout the recloser.

To open the recloser before the fuse blows, Cooper (1990) recommends adjusting the A curve by multiplying the time by a factor of 1.25 for one fast

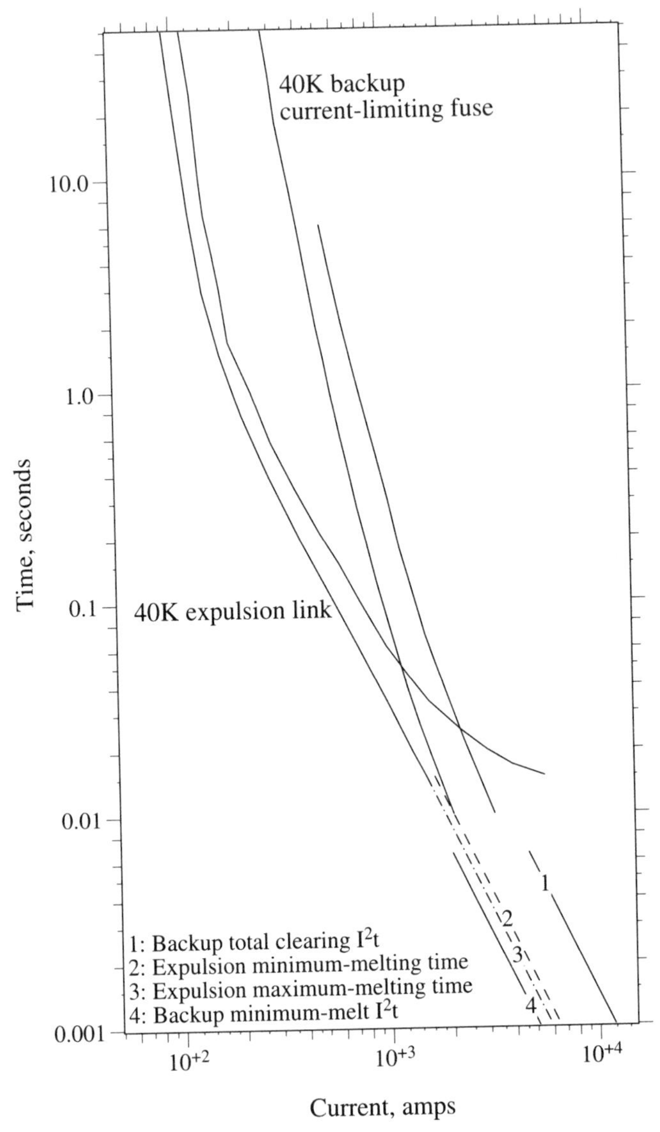

FIGURE 8.13
Coordination between a 40-K expulsion link and a 40-K backup current-limiting fuse using the relaxed matched-melt criteria.

operation, a factor of 1.35 for two fast operations with a reclosing time greater than or equal to 1 sec, and a factor of 1.8 for two fast operations with a reclosing time from 25 to 30 cycles. For applications with two or more delayed operations, the fast curve coordinates for fault currents up to the point where the adjusted A curve crosses the expulsion fuse's minimum-melting curve.

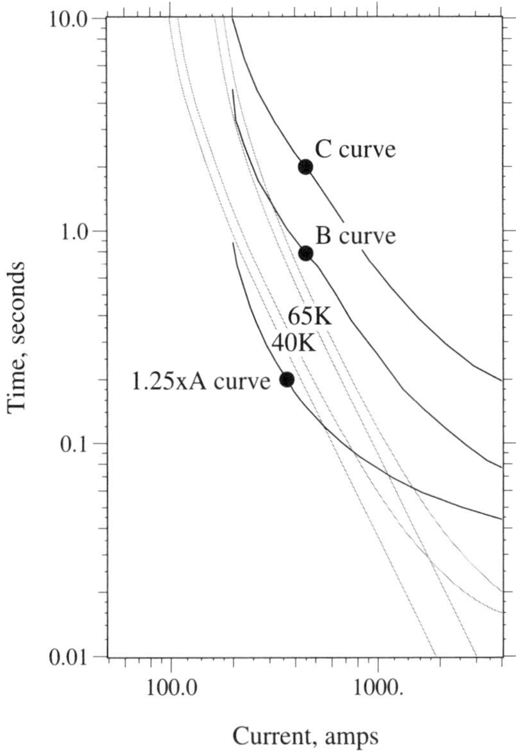

FIGURE 8.14
Example of coordination between K links and a Cooper 4E single-phase hydraulic recloser with a 100-A trip coil (A, B, and C curves shown).

On hydraulically controlled reclosers, the trip-coil rating determines the recloser's "pickup." Beyond that, hydraulically controlled reclosers have limited curve selections and no adjustments. Figure 8.14 shows average clearing curves for a single-phase Cooper 4E hydraulically controlled recloser overlayed on top of two K fuse links. For this example, only a limited range of fuses coordinate for low and high fault currents. Fuses larger than a 65K have significant overlap in the low-current area, leaving more chance that the recloser could lock out for a fault on a lateral tap. The slower delayed curves, such as the C curve shown, reduce the chance of miscoordination for lower fault currents. For the fast-trip A curve, the 40K link only coordinates with the fast curve for fault currents up to 360 A; smaller links are worse. Since K links are significantly steeper than these recloser curves, we must expect limited coordination for certain combinations. In this instance, T links coordinate over a wider current range because their time-current characteristics match the slope of the recloser characteristics more accurately. Miscoordination is more problematic in the low-current region. If the recloser locks out for faults downstream of a fuse, more customers are interrupted, and crews have a harder time finding the fault (more area to patrol).

Reclosers with electronic controls and relayed circuit breakers offer more flexibility. We can tailor tripping characteristics to coordinate over a wider range of currents. Three-phase reclosers have a ground-trip element that can increase the sensitivity of the recloser and also coordinate better with downstream fuses.

8.6.4 Recloser–Recloser Coordination

For coordinating two reclosers, the curve separation we need depends on the type of recloser. For hydraulically-controlled reclosers that are series coil operated, both operate if there is less than a 2-cycle separation; both *may* operate for a separation of 2 to 12 cycles, and both coordinate properly if there are more than 12 cycles of separation. For hydraulically-controlled reclosers that use high-voltage solenoid closing (larger reclosers), we need 8 cycles of separation for coordination (if it is less than 2 cycles, both devices operate). This data is for Cooper reclosers (Cooper Power Systems, 1990).

8.6.5 Coordinating Instantaneous Elements

Coordinating instantaneous relay elements or recloser fast curves is difficult. By the nature of an instantaneous element, two in series will both operate if the short-circuit current is above the pickup of both relays.

The most common way to coordinate two instantaneous elements is to raise the pickup of the upstream element. Find a setting where the instantaneous relay will not operate for faults downstream of the second protective device. The upstream relay cannot operate if its pickup is above the available fault current at the location of the downstream element. For this strategy, the instantaneous pickup on the element must be higher than its time-overcurrent pickup. This rules out hydraulic reclosers, which have the same pickup for the fast (A) curve and the delayed curves (B & C), but is not a concern with electronic reclosers because they have the same flexibility as relayed circuit breakers.

Rather than using an instantaneous relay element, we can perform the "fast trip" function with a time-overcurrent relay with a fast characteristic. Now, we might be able to coordinate the fast curve of a line recloser with the substation circuit breaker or recloser.

As another way to coordinate two instantaneous elements, use a time delay on the upstream instantaneous element. Choose enough time delay, 6 to 10 cycles, to allow the downstream device to clear before the station device operates.

Even with coordinated fast curves (either using a delay or using a fast TOC curve), nuisance momentary interruptions occur for faults cleared by a downstream line recloser. Consider a station recloser R1 and a downstream line recloser R2 each with one fast curve (A) and two delayed curves (B). If a permanent fault occurs downstream of R2, R2 will first operate on its A curve. If the fast curves of R1 and R2 are coordinated, R1 will not operate. After a delay, R2 recloses. The fault is still there, so R2 operates on its delayed curve (its B curve). Now, R1 *does* operate because it is on its A curve which operates before R2's B curve. After R1 recloses, R2 should then clear the fault on its B curve, which should operate before R1's B curve. The fault is still cleared properly, but customers upstream of R2 have extra momentary interruptions.

A more advanced form of coordination called *sequence coordination* removes this problem. Sequence coordination is available on electronic reclosers and also on digital relays controlling circuit breakers. With sequence coordination, the station device detects and counts faults — but does not open — for a fault cleared by a downstream protector on the fast trip. If the fault current occurs again (usually because the fault is permanent), the station device switches to the time-overcurrent element because it counted the first as an operation. Using this form of coordination eliminates the momentary interruption for the entire feeder for permanent faults downstream of a feeder recloser. On a relay or recloser that has sequence coordination, if the device senses current above some minimum trip setting and the current does not last long enough to trip based on the device's fast curve, the device advances its control-sequence counter as if the unit had operated on its fast curve. So when the downstream device moves to its delayed curve, the upstream device with sequence coordination also is operating on its delayed curve. With sequence coordination, for the fast curves, the response curve of the upstream device must still be slower than the clearing curve of the downstream device.

8.7 Fuse Saving vs. Fuse Blowing

Fuse saving is a protection scheme where a circuit breaker or recloser is used to operate before a lateral tap fuse. A fuse does not have reclosing capability; a circuit breaker (or recloser) does. Fuse saving is usually implemented with an instantaneous relay on a breaker (or the fast curve on a recloser). The instantaneous trip is disabled after the first fault, so after the breaker recloses, if the fault is still there, the system is time coordinated, so the fuse blows. Because most faults are temporary, fuse saving prevents a number of lateral fuse operations.

The main disadvantage of fuse saving is that all customers on the circuit see a momentary interruption for lateral faults. Because of this, many utilities are switching to a fuse blowing scheme. The instantaneous relay trip is disabled, and the fuse is always allowed to blow. The fuse blowing scheme is also called trip saving or breaker saving. Figure 8.15 shows a comparison of the sequence of events of each mode of operation. Fuse saving is primarily directed at reducing sustained interruptions, and fuse blowing is primarily aimed at reducing the number of momentary interruptions.

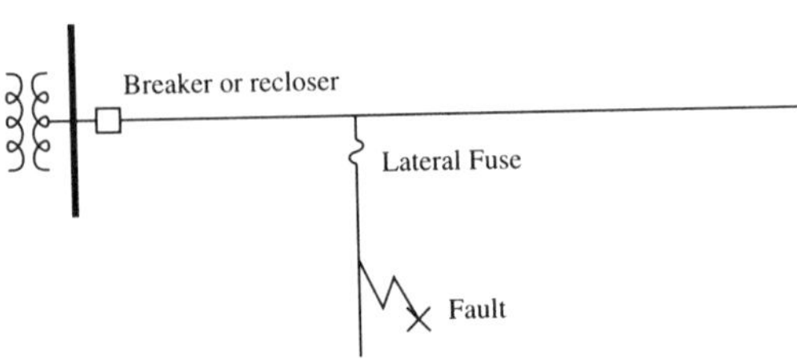

Fuse Saving

Temporary fault

1. The circuit breaker operates on the instantaneous relay trip (before the fuse operates).
2. The breaker recloses.
3. The fault is gone, so no other action is necessary.

Permanent fault

1. The circuit breaker operates on the instantaneous relay trip (before the fuse operates).
2. The breaker recloses.
3. The fault is still there.
4. The instantaneous relay is disabled, so the fuse operates.
5. Crews must be sent out to fix the fault and replace the fuse.

Fuse Blowing

Temporary fault

1. The fuse operates.
2. Crews must be sent out to replace the fuse.

Permanent fault

1. The fuse operates.
2. Crews must be sent out to fix the fault and replace the fuse.

FIGURE 8.15
Comparison of the sequence of events for fuse saving and fuse blowing for a fault on a lateral.

TABLE 8.11

IEEE Survey Results on the Percentage of Utilities that Use Fuse Saving

Survey Year	Percent of Utilities Using Fuse Saving
1988	91
1994	71
2000	66

Sources: IEEE Working Group on Distribution Protection, "Distribution Line Protection Practices — Industry Survey Results," *IEEE Transactions on Power Delivery*, vol. 3, no. 2, pp. 514–24, April 1988; IEEE Working Group on Distribution Protection, "Distribution Line Protection Practices — Industry Survey Results," *IEEE Transactions on Power Delivery*, vol. 10, no. 1, pp. 176–86, January 1995; Report to the IEEE working group on system performance, 2002.

TABLE 8.12

1996 Survey on the Usage of Fuse Saving

Use fuse saving	40%
Use a mixture	33%
Use fuse blowing	27%

Source: Short, T. A., "Fuse Saving and Its Effect on Power Quality," *EEI Distribution Committee Meeting*, 1999.

8.7.1 Industry Usage

Until the late 1980s, fuse saving was almost universally used. As power quality concerns grew, some utilities switched to a fuse blowing mode. An IEEE survey on distribution protection practices that is done periodically has shown a decrease in the use of fuse saving as shown in Table 8.11.

Another survey done by Power Technologies, Inc., in 1996 showed a mixture of practices at utilities as shown in Table 8.12. A few used fuse blowing because they indicated that fuse saving was not successful. Many of the "mixed practices" utilities decided on a case-by-case basis. Many of these normally used fuse saving but switched to fuse blowing if too many power quality complaints were received.

8.7.2 Effects on Momentary and Sustained Interruptions

The change in the number of momentary interruptions can be estimated simply by using the ratio of the length of the mains to the total length of the circuit including all laterals. For example, if a circuit has 5 mi of mains and 10 mi of laterals, the number of momentaries after switching to fuse blowing

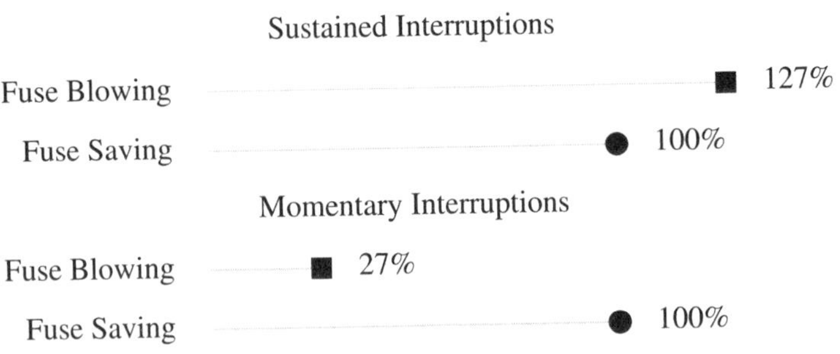

FIGURE 8.16
Comparison of fuse saving and fuse blowing on a hypothetical circuit. Mains: 10 miles (16.1 km), fault rate = 0.5/mile/year (0.8/km/year), 75% temporary. Taps: 10 miles (16.1 km) total, 20 laterals, fault rate = 2/mile/year (3.2/km/year), 75% temporary. It also assumes that fuse saving is 100% successful.

would be 1/3 of the number of momentaries with fuse saving (5/(5+10) = 1/3). This assumes that the mains and laterals have the same fault rate; if the fault rate on laterals is higher (which it often is because of less tree trimming, etc.), the number of momentaries is even less. Note how dramatically we can reduce momentaries by using fuse blowing. No other methods can so easily eliminate 30 to 70% of momentaries. The effect on reliability of going to a fuse blowing scheme is more difficult to estimate. Fuse blowing increases the number of fuse operations by 40 to 500% (Dugan et al., 1996; Short, 1999; Short and Ammon, 1997). This will increase the average frequency of sustained interruptions by 10 to 60%. Note that there are many variables that can change the ratios. One example is given in Figure 8.16.

Note that the effect on sustained interruptions is not equally distributed. Customers on the mains see no difference in the number of permanent interruptions. Customers on long laterals may have many more sustained interruptions with a fuse-blowing scheme.

8.7.3 Coordination Limits of Fuse Saving

One of the main reasons that utilities have decided not to use fuse saving is that it is difficult to make it work. Fuses clear quickly relative to circuit breakers, so where fault currents are high, the fuse blows before the breaker trips. This results in a fuse operation and a momentary interruption for all customers on the circuit. K links, the most common lateral fuses, are fast fuses. Most distribution circuit breakers take 5 cycles to clear. For fuse saving to work, the breaker must open before the fuse blows, so the fuse needs to survive for the time it takes the instantaneous relay to operate (about one cycle) plus the 5 cycles for the breaker. As an illustration, Figure 8.17 shows the limit of coordination of a 5-cycle breaker and a 100 K fuse. Fuse saving only coordinates for faults below 1354 A. Smaller fuses have lower current limits. Note that the breaker time is coordinated with the damage time of the fuse.

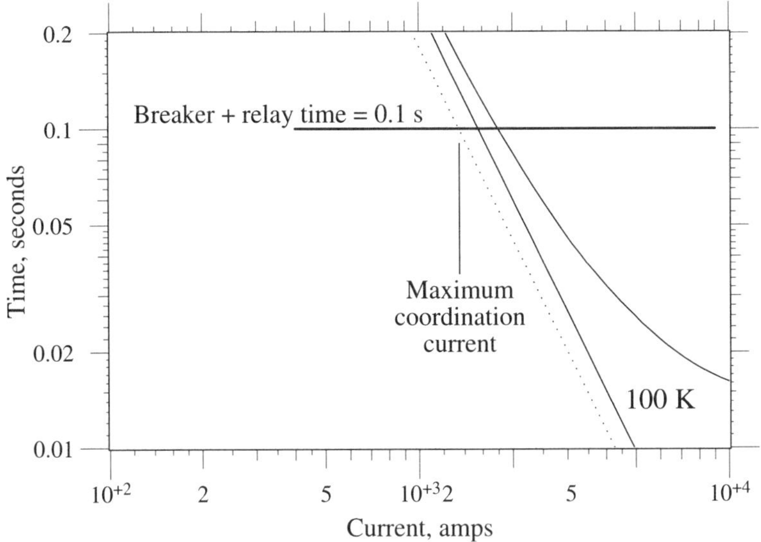

FIGURE 8.17
Coordination of a 100K lateral fuse with a 5-cycle circuit breaker.

Table 8.13 and Table 8.14 show the limits of coordination of several common lateral fuses for a standard circuit breaker (5 cycle) and for a fast breaker/relay combination (3-cycle circuit breaker and 1-cycle relay). Also shown are translations of these fault currents into distances from the substation at 12.47 kV (assuming an 8-kA fault level at the substation). Note

TABLE 8.13

Maximum Fault Currents and Critical Distances for Fuse Saving Coordination for Several Common Fuse Links for a 5-Cycle Circuit Breaker and a 1-Cycle Relay Time

Fuse	I_c, A	d_c, mi	d_c, km	Fuse	I_c, A	d_c, mi	d_c, km
20 K	254	26.5	42.6	20 T	433	15.5	25.0
25 K	323	20.8	33.5	25 T	552	12.2	19.6
30 K	398	16.9	27.2	30 T	699	9.6	15.5
40 K	520	12.9	20.8	40 T	896	7.5	12.1
50 K	665	10.1	16.3	50 T	1125	6.0	9.7
65 K	816	8.3	13.3	65 T	1428	4.8	7.7
80 K	1078	6.3	10.1	80 T	1790	3.8	6.2
100 K	1354	5.0	8.1	100 T	2277	3.1	4.9
140 K	2162	3.2	5.2	140 T	3447	2.1	3.4
200 K	3401	2.1	3.5	200 T	5436	1.5	2.4

Note: I_c = Maximum current where fuse saving works. d_c = Distance from the substation where fuse saving starts to work for 12.47-kV, 500-kcmil overhead line.

Source: Copyright © 2003. Electric Power Research Institute. 1001665. *Power Quality Improvement Methodology for Wire Companies.* Reprinted with permission.

TABLE 8.14

Maximum Fault Currents and Critical Distances for Fuse Saving Coordination for Several Common Fuse Links for a 3-Cycle Circuit Breaker and a 1-Cycle Relay Time

Fuse	I_c, A	d_c, mi	d_c, km	Fuse	I_c, A	d_c, mi	d_c, km
20 K	332	20.3	32.6	20 T	565	11.9	19.2
25 K	424	15.9	25.5	25 T	723	9.3	15.0
30 K	522	12.9	20.8	30 T	920	7.4	11.8
40 K	682	9.9	15.9	40 T	1175	5.8	9.3
50 K	875	7.7	12.4	50 T	1479	4.6	7.4
65 K	1070	6.3	10.2	65 T	1878	3.7	5.9
80 K	1407	4.8	7.8	80 T	2346	3.0	4.8
100 K	1763	3.9	6.3	100 T	2975	2.4	3.9
140 K	2823	2.5	4.1	140 T	4522	1.7	2.7
200 K	4409	1.7	2.8	200 T	7122	1.3	2.0

Note: I_c = Maximum current where fuse saving works. d_c = Distance from the substation where fuse saving starts to work for 12.47-kV, 500-kcmil overhead line.

Source: Copyright © 2003. 1001665. *Power Quality Improvement Methodology for Wire Companies.* Reprinted with permission.

that only the larger fuses shown (greater than 100 A) will coordinate for significant portions of the feeder. Smaller fuses used as second and third level fuses do not coordinate over the length of most feeders. The situation is even worse at higher voltages. At 24.94 kV, the distances in Table 8.13 and Table 8.14 are doubled, so fuse saving is more difficult to achieve at higher voltages. Reclosers are faster than standard 5-cycle breakers — the 4-cycle total operating time in Table 8.14 is representative of many reclosers.

If smaller K links are used such as 100K and 65K fuses (the most common lateral fuses), then fuse saving is not going to work very well. In that case, why use it? There is no sense in having a momentary every time a fuse blows (which is what will happen since the circuit breaker is not fast enough to save the fuse).

8.7.4 Long-Duration Faults and Damage with Fuse Blowing

Fuse blowing has drawbacks: faults on the mains can last a long time. With fuse saving, main-line faults normally clear in 5 to 7 cycles (0.1 sec) on the first shot with the instantaneous element. With fuse blowing, this same fault may last for 0.5 to 1 sec. Much more damage at the fault location occurs during this extra time. Some of the problems that have been identified are

- *Conductor burndowns* — At the fault, the heat from the fault current arc burns the conductor enough to break it, dropping it to the ground.
- *Damage of inline equipment* — The most common problem has been with inline hot-line clamps. If the connection is not good, the high-current fault arc across the contact can burn the connection apart.

- *Station transformers* — Extra duty on substation transformers
- *Evolving faults* — Ground faults are more likely to become two- or three-phase faults.
- *Underbuilt* — Faults on underbuilt distribution are more likely to cause faults on the transmission circuit above due to rising arc gases.

A fault current arc will expand after it is initiated. It has been found that the growth of the arc is generally in the vertical direction, and the growth is primarily a function of time and not of current or voltage [(Drouet and Nadeau, 1979) and Chapter 7]. The growth of the arc means that a 0.1-sec fault on the instantaneous trip (with fuse saving) is less likely to involve other phases or other circuits than a 0.2 to 1-sec fault on the time-delay trip (with fuse blowing).

8.7.5 Long-Duration Voltage Sags with Fuse Blowing

With fuse blowing, voltage sags last longer, especially for faults on the three-phase mains, which have to be cleared by phase or ground time-overcurrent elements. An example is shown in Figure 8.18 where voltage sag magnitudes and durations are shown for faults at various distances from a substation using fuse blowing. For the same circuit with fuse saving, all of the faults would have cleared in 0.1 sec. For a fault at the substation, the duration triples. For a fault one mile (1.6 km) from the substation, the duration qua-

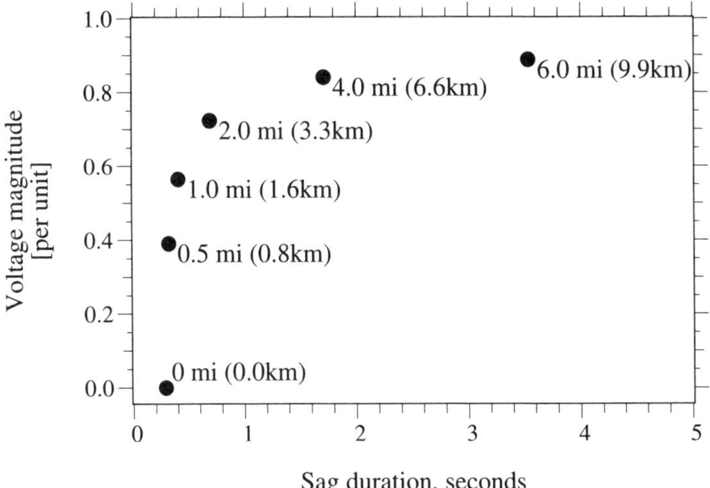

FIGURE 8.18
Magnitudes and durations of substation bus voltage sags for ground faults applied at the given distance with a fuse-blowing scheme. For the same circuit with fuse saving, all of the faults would clear in 0.1 sec. Assumptions: 12.47 kV, 500 kcmil, all-aluminum conductors. Ground relay: CO-11, TD = 5, pickup = 300 A.

druples. The situation is worse for phase-to-phase faults and three-phase faults because they must be cleared by the phase relays which are generally slower.

8.7.6 Optimal Implementation of Fuse Saving

In order to get a fuse saving scheme to work, it is necessary to get the substation protective device to open before fuses operate. We can achieve this in several ways:

- *Slow down the fuse* — Use big, slow fuses (such as a 140 or 200 T) near the substation to ensure proper coordination.
- *Faster breakers or reclosers* — If three-cycle circuit breakers are used instead of the normal 5-cycle breakers, fuse saving coordination is more likely. Some reclosers are even faster than 3-cycle breakers.
- *Limit fault currents*:
  - Open station bus ties: An open bus tie will reduce the fault current on each feeder and make fuse saving easier. This is the normal operating mode for most utilities.
  - Use a transformer neutral reactor: A neutral reactor reduces the fault current for single-phase faults (all faults on single-phase taps).
  - Use line reactors: This reduces the fault current for all types of faults. This has been an uncommon practice. An added advantage, reactors reduce the impact of voltage sags for faults on adjacent feeders.
  - Specify higher impedance transformers.

We can employ other strategies to limit the impact of momentary interruptions:

- *More downstream reclosers* — Extra downstream devices will reduce the number of momentaries for customers near the substation. It is important to coordinate reclosers with the upstream device (including sequence coordination).
- *Single-phase reclosers*
- *Immediate reclose*
- *Switch to a fuse blowing scheme on poor feeders* — For a feeder with many momentaries, disable the instantaneous relay for a time period. Identify poorly performing parts of the circuit during this time. The blown branch fuses provide a convenient fault location method. Once the poor performing sections are identified and improved, switch the circuit back to fuse saving.

8.7.7 Optimal Implementation of Fuse Blowing

Several strategies can optimize a fuse blowing scheme:

- *Fast fuses (or current-limiting fuses)* — If smaller or faster fuses are used, faults clear faster, so voltage sag durations are shorter. Current-limiting fuses also limit the magnitude and duration of the sag. Be careful not to fuse too small, or fuses will operate unnecessarily due to loading, inrush, and cold load pickup. Note that if smaller fuses are used, it is difficult to switch back to a fuse-saving scheme.
- *Covered wire or small wire* — Watch burndowns on circuits with covered wire or small wire that is protected by the station circuit breaker or recloser. If either of these cases exists, use a modified fuse blowing scheme with a time-delayed instantaneous element (see the next section).
- *Use single-phase reclosers on longer laterals* — A good way of maintaining some of the reliability of a fuse-saving scheme is to use single-phase reclosers instead of fuses on longer taps. Then, temporary faults on these laterals do not cause permanent interruptions to those customers.
- *More fuses* — Add more second and third level fuses to segment the circuit more.
- *Track lateral operations* — Temporary faults on fused laterals cause sustained interruptions. In order to minimize the impacts on lateral customers, track interruptions by lateral. Identify poorly performing laterals, patrol poor sections, then add tree trimming, animal guards, etc.

8.8 Other Protection Schemes

8.8.1 Time Delay on the Instantaneous Element (Fuse Blowing)

An alternative implementation of a fuse blowing scheme is to use a time delay on the instantaneous trip (rather than removing the instantaneous trip; a definite-time overcurrent relay also could do the same function) (Engelman, 1990). Faults do not last as long as they would if the relay went to a time-overcurrent element; there is less chance of wire burndowns, and voltage sags are of shorter duration for faults on the mains. A common delay is 0.1 sec.

An example implementation is shown in Figure 8.19 where a 0.1-sec delay is added to the instantaneous. A 100K fuse link is also shown. For the 100K link, the scheme is actually a mixture of fuse saving and fuse blowing. For fault currents above roughly 1700 A, it is a fuse blowing scheme (the fuse

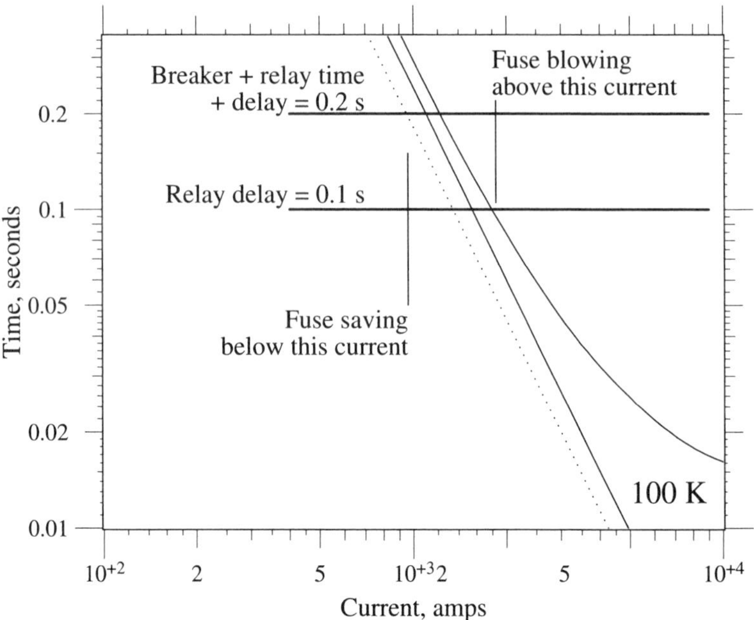

FIGURE 8.19
Example of a delayed instantaneous element used for fuse blowing.

clears before the instantaneous relay operates). For currents below 1000 A, the scheme is a fuse saving scheme (the circuit breaker trips before the fuse is damaged). Between 1000 and 1700 A, one or both devices operate.

Another option sometimes used with this scheme is a high-set instantaneous. The high-set instantaneous has no time delay and is set to clear faults close to the station. This removes the most damaging faults quickly (because they are the most likely to cause damage and cause the most severe voltage sags).

Using a time delay is a better fuse saving scheme than just removing the instantaneous relay. The disadvantage, and the reason that it is not implemented as much, is that it is usually more difficult and costly to implement. For electromechanical relays, another timer relay must be added, and the relay scheme must be engineered. Many digital relays ease the implementation since they have this time-delay option available.

8.8.2 High-Low Combination Scheme

Another option is to use fuse blowing at the substation and fuse saving at downstream reclosers (Burke, 1996).

- *Substation fuse blowing* — Fault currents are high near the substation, so it is difficult to get fuse saving to work here.

- *Recloser fuse saving* — Fault currents are lower downstream, and reclosers are faster, so fuse saving should work well here.

The high-low scheme is easy to implement. The station instantaneous trip is eliminated. Reclosers are operated with a fast trip (the A curve). Most are already in this mode, so no changes are necessary here.

8.3.3 SCADA Control of the Protection Scheme

Another option is to use SCADA to change back and forth between fuse saving and fuse blowing, getting some of the benefits of both schemes. Fuse blowing is the normal operating mode, but operators could switch to a fuse saving scheme during storms. This avoids clear-sky momentaries while at the same time improving storm restoration. Several factors make fuse saving better during storms:

- Faults are more likely to be temporary during storms (lightning, wind).
- Customers are more forgiving about momentaries during storms.
- Interruptions due to fuse operations last longer during storms (because crews have many repairs to perform). If fuses are blown due to temporary faults, this increases the number of repair locations. Saving fuses reduces the number of interruptions crews will have to address.

In order for SCADA control of fuse saving to work best, we must design the system for fuse saving to work: larger, slower fuses for laterals close to the substation, faster circuit breakers (or use substation reclosers), or possibly even using grounding reactors in the substation to limit fault currents. Likewise, we must design for fuse blowing, so avoid using tree wire (or go to delayed instantaneous relaying rather than removing the instantaneous trip).

Control is more readily available in the substation because the SCADA infrastructure may already be in place. If so, the cost of the SCADA system has already been justified, and this added functionality could be piggybacked on the existing system if there are free channels available. It is feasible to use automation technology to implement remote control of feeder reclosers, but the cost of the communication equipment may not justify having this functionality.

For SCADA control, microprocessor-controlled relays are not needed. A SCADA channel can be used to control a blocking relay on the instantaneous elements of the feeder relays. Alternatively, the SCADA channel could control the delay on the instantaneous relay element (no delay: fuse saving, with delay: fuse blowing). One SCADA channel could control the fuse saving/blowing status of all of the distribution feeders in a station. Alternatively, we could control each feeder independently.

8.8.4 Adaptive Control by Phases

Various protection schemes are classified as adaptive. An adaptive approach to a fuse blowing mode is to adjust the scheme depending on how many phases are faulted:

- *2- or 3-phase fault* — Use the instantaneous; the fault is assumed to be on the 3-phase mains. Tripping quickly reduces the duration of voltage sags for faults on the mains.
- *Single-phase fault* — Use fuse blowing (time delay curves or delayed instantaneous relay).

Adaptive control requires microprocessor-based relays. This is not a common scheme, and the expense and complexity are difficult to justify unless the chosen relay comes with this functionality.

8.9 Reclosing Practices

Automatic reclosing is a universally accepted practice on most overhead distribution feeders. On overhead circuits, 50 to 80% of faults are temporary, so if a circuit breaker or recloser clears a fault and it *recloses*, most of the time the fault is gone, and customers do not lose power for an extended period of time.

On underground circuits, since virtually all faults are permanent, we do not reclose. A circuit might be considered underground if something like 60 to 80% of the circuit is underground. Utility practices vary considerably relative to the exact percentage (IEEE Working Group on Distribution Protection, 1995). A significant number of utilities treat a circuit as underground if as little as 20% is underground while some others put the threshold over 80%.

The first reclose usually happens with a very short delay, either an immediate reclose which means a 1/3- to 1/2-sec dead time (discussed later) or with a 1- to 5-sec delay. Subsequent reclose attempts follow longer delays. The nomenclature is usually stated as 0–15–30 meaning there are three reclose attempts: the first reclose indicated by the "0" is made after no intentional delay (this is an immediate reclose), the second attempt is made following a 15-sec dead time, and the final attempt is made after a 30-sec dead time. If the fault is still present, the circuit opens and locks open. We also find this specified using circuit breaker terminology as O-0 sec-CO-15 sec-CO-30 sec-CO where "C" means close and "O" means open. Other common cycles that utilities use are 0–30–60–90 and 5–45.

With reclosers and reclosing relays on circuit breakers, the reclosing sequence is reset after an interval that is normally adjustable. This interval

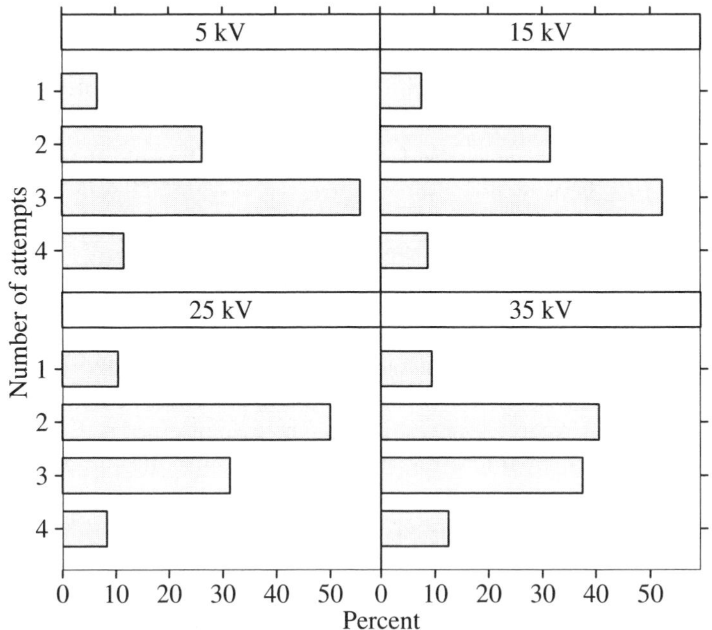

FIGURE 8.20
IEEE survey results on the number of reclose attempts for each voltage class. (Data from [IEEE Working Group on Distribution Protection, 1995].)

is generally set somewhere in the range of 10 sec to 2 min. Only a few utilities have reported excessive operations without lockout (IEEE Working Group on Distribution Protection, 1995).

8.9.1 Reclose Attempts and Dead Times

Three reclose attempts is most common as shown in Figure 8.20. More reclose attempts give the fault more chance to clear or burn free. Returns diminish; the chance that the third or fourth reclose attempt is successful is usually small. Additional reclose attempts have the following negative impacts on the system:

- *Additional damage at the fault location* — With each reclose into a fault, arcing does additional damage at the fault location. Faults in equipment do more damage. Cable faults are harder to splice, wire burndowns are more likely, and oil-filled equipment is more likely to rupture. Arcs can start fires. Faults (and the damage the arcs cause) can propagate from one phase to other phases.
- *Voltage sags* — With each reclose into a fault, customers on adjacent circuits are hit with another voltage sag. It can be argued that the

magnitude and duration of the sag should be about the same, so depending on the type of device, if the customer equipment survived the first sag, it will probably ride through subsequent sags of the same severity. If additional phases become involved in the fault, the voltage sag is more severe.

- *Through-fault damage to transformers* — Each fault subjects transformers to mechanical and thermal stresses. Virginia Power changed their reclosing practices because of excessive transformer failures on their 34.5-kV station transformers due to through faults (Johnston et al., 1978).
- *Through-fault damage to other equipment* — Cables, wires, and especially connectors suffer the thermal and mechanical stresses of the fault.
- *Interrupt ratings of breakers* — Circuit breakers must be derated if the reclose cycle involves more than one reclose attempt within 15 sec. This may be a consideration if fault currents are high and breakers are near their ratings. Reclosers do not have to be derated for a complete four-sequence operation. Extra reclose attempts increase the number of operations, which means more frequent breakers and reclosers maintenance.
- *Ratcheting of overcurrent relays* — An induction relay disc turns in response to fault current. After the fault is over, it takes time for the disc to spin back to the neutral position. If this reset is not completed, and another fault occurs, the disc starts spinning from its existing condition, making the relay operate faster than it should. The most common problem area is miscoordination of a substation feeder relay with a downstream feeder recloser. If a fault occurs downstream of the recloser, the induction relay will spin due to the current (but not operate if it is properly coordinated). Multiple recloses by the recloser could ratchet the station relay enough to falsely trip the relay. The normal solution is to take the ratcheting into account when coordinating the relay and recloser, but in some cases modification of the reclosing cycle of the recloser is an option. Another option is to use digital relays, which do not ratchet in this manner.

Given these concerns, the trend has been to decrease the number of reclose attempts. We try to balance the loss in reliability against the problems caused by extra reclose attempts. A major question is how often are the extra reclose attempts successful. Table 8.15 shows the success rate for one utility in a high-lightning area. Table 8.16 shows a second utility with similar reclosing practices but quite different reclose success (more lockouts and lower success rates for the first two reclose attempts). Reclose success rates change based on the types of faults most commonly seen in a region. Another data point with a broader distribution of utilities is obtained in the EPRI distribution power quality study. Table 8.17 shows the number of momentary interruptions (reclose attempts) that do not lead to sustained interruptions (lockouts).

TABLE 8.15

Reclose Success Rates for a Utility in a High Lightning Area

Reclosure	Success Rate	Cumulative Success
1st shot (immediate)	83.25%	83.25%
2nd shot (15 to 45 sec)	10.05%	93.30%
3rd shot (120 sec)	1.42%	94.72%
Locked out	5.28%	

Source: Westinghouse Electric Corporation, *Applied Protective Relaying*, 1982.

TABLE 8.16

Reclose Success Rates for One 34.5-kV Utility

Reclosure	Success Rate	Cumulative Success
1st shot (immediate)	25.3%	25.3%
2nd shot (15 sec)	42.1%	67.4%
3rd shot (80 sec)	11.6%	79.0%
Locked out	21.0%	

Source: Johnston, L., Tweed, N. B., Ward, D. J., and Burke, J. J., "An Analysis of Vepco's 34.5 kV Distribution Feeder Faults as Related to Through Fault Failures of Substation Transformers," *IEEE Transactions on Power Apparatus and Systems*, vol. PAS-97, no. 5, pp. 1876–84, 1978. With permission. ©1978 IEEE.

TABLE 8.17

Number of Interruptions per One Minute Aggregate Period that Do Not Lead to Sustained Interruptions

Number	Percentage
1	87%
2	9%
3	2%
4 or more	2%

Source: EPRI TR-106294-V2, *An Assessment of Distribution System Power Quality: Volume 2: Statistical Summary Report*, Electric Power Research Institute, Palo Alto, CA, 1996.

The key point is that it is relatively uncommon (but not rare) for the third or fourth reclose attempt to be successful.

We may block reclosing in some cases. It is common to block all reclose attempts when workers are doing maintenance on a circuit to provide an extra level of protection (an instantaneous relay element is also commonly enabled in this situation). Another situation is for very high-current faults. A high-set instantaneous relay covering just the first few hundred feet of circuit detects faults on the substation exit cables. If it operates, reclosing is disabled. This practice is done to reduce the damage for a failure of one of the station exit cables.

The duration of the open interval — the dead time between reclose attempts — is also a consideration. For a smaller number of reclose attempts, use longer delays to give tree branches and other material more time to clear.

Operator practices must also be considered as part of the reclosing scheme. Not uncommonly, an operator manually recloses the circuit breaker after a feeder lockout (especially during a storm). This sends the breaker or recloser through its whole reclosing cycle along with all of the bad effects (like more equipment damage and more voltage sags) with very little chance of success.

Some engineers and field personnel believe that the purpose of the extra reclose attempts is to *burn* the fault clear. This is dangerous. Faults regularly burn clear on low-voltage systems (<480 V), rarely at distribution primary voltages. Faults can burn clear on primary systems. The most common example is that tree branches or animals can be burned loose. The problem with this concept is that, just as easily, the fault burns the primary conductor, which falls to the ground causing a high-impedance fault. Fires and equipment damage are also more likely with the "burn clear" philosophy.

To reduce the impacts of subsequent reclose attempts, we could switch back to an instantaneous operation after the first time-overcurrent relay operation. If the fault does not clear after the first time-overcurrent relay operation, it means the fault is not downstream of a fuse (or a recloser). The reason to use a time overcurrent relay is to coordinate with the fuse. Since the fuse is out of the picture, why not use a faster trip for subsequent reclose attempts? While not commonly done, we could implement this with digital relays. The setting of the "subsequent reclose" instantaneous relay element should be different than the first-shot instantaneous. Set the pickup at the pickup of the time-overcurrent relay. Because of inrush on subsequent attempts, we may use a fast time-overcurrent curve or an instantaneous element with a short delay (something like 5 cycles).

As an example, if a utility uses a 0–15–30–90 sec reclosing cycle, the system is subjected to five faults if the system goes through its complete cycle. With the instantaneous operation enabled on the first attempt and disabled on subsequent attempts, we have a very high total duration of the fault current. For a CO-11 ground relay with a time-dial of 3, a 2-kA fault clears in roughly 1 sec. For the reclosing cycle to lock out, the system has a total fault time of 4.1 sec (one 0.1-sec fault followed by four 1-sec faults). If the instantaneous operation is enabled for reclose attempts 2 through 4, the total fault duration is 1.4 sec (one 0.1-sec fault followed by a 1-sec fault and three 0.1-sec faults). This greatly reduces the damage done by certain faults.

8.9.2 Immediate Reclose

An immediate reclose (also called an instantaneous or fast reclose) means having no intentional time delay (or a very short time delay) on the first reclose attempt on circuit breakers and reclosers. Many residential devices such as digital clocks, VCRs, and microwaves can ride through a 1/2-sec interruption but not a 5-sec interruption, so a fast reclose helps reduce residential complaints.

From a power quality point of view, a faster reclose is better. Some customers may not notice anything more than a quick blink of the lights. Many residential devices such as the digital clocks on alarm clocks, microwaves, and VCRs can ride through a 1/2-sec interruption where they usually cannot ride through a 5-sec interruption (a first reclose delay used by several utilities).

8.9.2.1 Effect on Sensitive Residential Devices

The most common power quality recorder in the world is the digital clock. Many complaints are due to the "blinking clocks." Using an immediate reclose reduces complaints. Florida Power has reported that a reclosing time of 18 to 20 cycles nearly eliminates complaints (Dugan et al., May/June, 1996). Another utility that has successfully used the immediate reclose is Long Island Lighting Company (now Keyspan) (Short and Ammon, 1997). According to an IEEE survey, a time to first reclose of less than 1 sec is the most common practice although the fast reclose practice tends to decline with increasing voltage (see Figure 8.21).

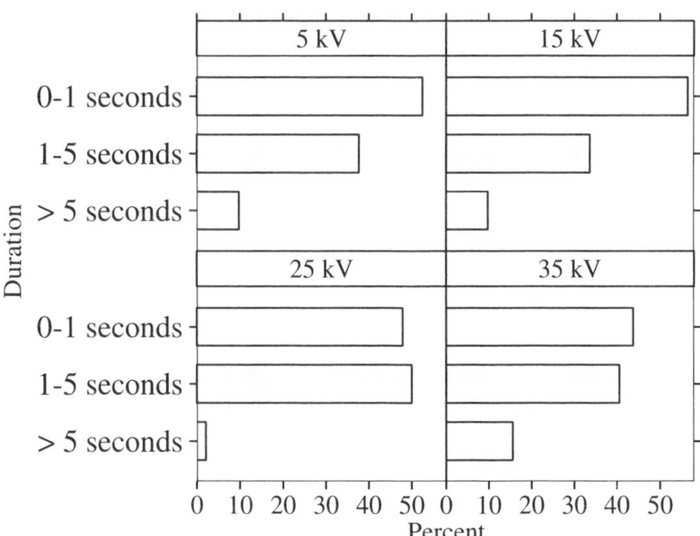

FIGURE 8.21
IEEE survey results of the intervals used before the first reclose attempt for each voltage class. (Data from [IEEE Working Group on Distribution Protection, 1995].)

Clocks have a wide range of voltage sensitivity, but most digital clocks will not lose memory for a complete interruption that is less than 0.5 sec. So, an immediate reclose helps residential customers ride through momentary interruptions without resetting many devices. Given the wide variation, some customers are sensitive to a 0.5-sec interruption. Note that the immediate reclose helps with digital clock-type devices whether it be on radio alarm clocks, VCRs, or microwaves. Fast reclosing does not help with most computers or other computer-based equipment, limiting the power quality improvement of using the immediate reclose to residential customers (no help for commercial or industrial customers).

8.9.2.2 Delay Necessary to Avoid Retriggering Faults

Sometimes a delayed reclose is necessary if there is not enough time to clear the fault. A fault arc needs time to cool, or the reclose could retrigger the arc. Whether the arc strikes again is a function of voltage and structure spacings. A 34.5-kV utility (Vepco) added a delay to the first reclose because the probability of success of the first reclose was much less than normal for distribution circuits (Johnston et al., 1978). The success rate for the first attempt after an instantaneous reclose was 25% which is much less than the 70 to 80% experienced by most utilities. Another item that added to the low success rate of Vepco's 34.5-kV system is that they used a lot of armless design, and the combination of higher voltage and tighter spacings requires a longer time delay for the arc to clear.

With the following equation, we can find the minimum de-ionization time of an arc based on the line-to-line voltage (Westinghouse Electric Corporation, 1982):

$$t = 10.5 + V / 34.5$$

where

t = minimum de-ionization time, 60-Hz cycles

V = rated line-to-line voltage, kV

The de-ionization time increases only moderately with voltage. Even for a 34.5-kV system, the de-ionization time is 11.5 cycles. This equation is a simplification (separation distances are not included) but does show that arcs rapidly de-ionize. Many high-voltage transmission lines successfully use a fast reclose. The reclose time for distribution circuit breakers and reclosers varies by design. A typical time is 0.4 to 0.6 sec for an immediate reclose (meaning no intentional delay). The fastest devices (newer vacuum or SF_6 devices) may reclose in as little as 11 cycles. This may prove to be too fast for some applications, so consider adding a small delay of 0.1 to 0.4 sec (especially at 25 or 35 kV).

On distribution circuits, other things affect the time to clear a fault besides the de-ionization of the arc stream. If a temporary fault is caused by a tree limb or animal, time may be needed for the "debris" to fall off the conductors or insulators. Because of this, with an immediate reclose use at least two reclose attempts before lockout. For example, use a 0–15–30 sec cycle (three reclose attempts), or if you wish to use two reclose attempts, use a 0–30 or 0–45 sec cycle (use a long delay before the last reclose attempt).

8.9.2.3 Reclose Impacts on Motors

Industrial customers with large motors have concerns about a fast reclose and damage to motors and their driven equipment. The major problem with reclosing is that the voltage on a motor will not drop instantly to zero when the utility circuit breaker (or recloser) is opened. The motor has residual voltage, where the magnitude and frequency decay with time. When the utility recloses, the utility voltage can be out-of-phase with the motor residual voltage, severely stressing the motor windings and shaft and its driven load. The decay time is a function of the size of the motor and the inertia of the motor and its load.

Motors in the 200 to 2000 hp range typically have open-circuit time constants of 0.5 to 2 sec (Bottrell, 1993). The time constant is the time it takes for the residual voltage to decay to 36.8% of its initial value. Reclose impacts are worse with

- Larger motors.
- Capacitor banks — excitation from the capacitor banks can greatly increase the motor decay time.
- Synchronous motors and generators — much larger time constants makes synchronous machines more vulnerable to damage than induction machines.
- Motors on contactors will drop out. Also, larger motors and synchronous motors normally have an undervoltage relay to trip when voltage is lost.

On the vast majority of distribution circuits, reclosing impacts will not be a concern because:

- Most utility feeders do not have individual motor loads larger than 500 hp.
- Even with feeders with large industrial customers, the non-motor load will be large enough to pull the voltage down to a safe level within the time it takes to do a normal immediate reclose (0.4 to 0.6 sec).

Because of this, we can safely implement an immediate reclose on almost all distribution circuits. One exception is a feeder with an industrial customer that is a majority of the feeder load, and the industrial customer has several large induction or (especially) synchronous motors. Another exception is a feeder with a large rotating distributed generator. In both of these cases, delay the first reclose or, alternatively, use line-side voltage supervision (if voltage is detected downstream of the breaker, reclosing is blocked to prevent an out-of-phase reclosing situation).

8.10 Single-Phase Protective Devices

Many distribution protective devices are single phase or are available in single-phase versions including reclosers, fuses, and sectionalizers. Single-phase protective devices are used widely on distribution systems; taps are almost universally fused. On long single-phase taps, single-phase reclosers are sometimes used. Most utilities also use fuses for three-phase taps. The utilities that do not fuse three-phase taps most often cite the problem of single-phasing motors of three-phase customers. Some utilities use single-phase reclosers that protect three-phase circuits (even in the substation).

Single-phase protective devices on single-phase laterals are widely used, and the benefits are universally accepted. The fuse provides an inexpensive way of isolating faulted circuit sections. The fuse also aids in finding the fault.

Using single-phase interrupters helps on three-phase circuits — only one phase is interrupted for line-to-ground faults. We can easily estimate the effect on individual customers using the number of phases that are faulted on average as shown in Table 8.18. Overall, using single-phase protective devices cuts the average number of interruptions in half. This assumes that all customers are single phase and that the customers are evenly split between phases.

Service to three-phase customers downstream of single-phase interrupters generally improves, too. Three-phase customers have many single-phase loads, and the loads on the unfaulted phases are unaffected by the fault. Three-phase devices may also ride-through an event caused by a single-phase fault (although motors may heat up because of the voltage unbalance

TABLE 8.18

Effect on Interruptions When Using Single-Phase Protective Devices on Three-Phase Circuits

Fault Type	Percent of Faults	Portion Affected	Weighted Effect
Single phase	70%	33%	23%
Two phase	20%	67%	13%
Three phase	10%	100%	10%
		TOTAL	47%

as discussed in the next section). Single-phase protective devices do have some drawbacks. The main concerns are

- Ferroresonance
- Single-phasing of motors
- Backfeeds

Ferroresonance usually occurs during manual switching of single-pole switching devices (where the load is usually an unloaded transformer). It is less common for ferroresonance to occur downstream of a single-phase protective device that is operating due to a fault. The reason for this is that if there is a fault on the opened phase, the fault prevents an overvoltage on the opened phase. Also, any load on the opened section helps prevent ferroresonant overvoltages. Because ferroresonance will be uncommon with single-phase protective devices, it is usually not a major factor in protective device selection. Still, caution is warranted on small three-phase transformers that may be switched unloaded (especially at 24.94 or 34.5 kV).

With single-phase protective devices, backfeeds can create hazards. During a line-to-ground fault where a single-phase device opens, backfeed through a three-phase load can cause voltage on the load side of an opened protective device. Backfeeds can happen with most types of three-phase distribution transformer connections (even with a grounded-wye – grounded-wye connection). The important points to note are that

- The backfeed voltage is enough to be a safety hazard to workers or the public (for example in a wire down situation).
- The available backfeed is a stiff enough source to maintain an arc of significant length. The arc can continue causing damage at the fault location during a backfeed condition. It may also be a low-level sparking and sputtering fault.

Based on these points, single-phasing can cause problems from backfeeding. Whether this constrains use of single-phase protective devices is debatable. Most utilities do use single-phase protective devices, usually with fuses, on three-phase circuits.

Under single-phasing, motors can overheat and fail. Motors have relatively low impedance to negative-sequence voltage; therefore, a small negative-sequence component of the voltage produces a relatively large negative-sequence current. Consequently the effect magnifies; a small negative-sequence voltage appears as a significantly larger percentage of unbalanced current than the percentage of unbalanced voltage.

Loss of one or two phases is a large unbalance. For one phase open, the phase-to-phase voltages become 0.57, 1.0, 0.57 for a wye – wye transformer and 0.88, 0.88, 0.33 for a delta – wye transformer. In either case, the negative-sequence voltage is 0.66 per unit. With such high unbalance, a motor overheats quickly. The negative-sequence impedance of a motor is roughly 15%, so for a 66% negative-sequence voltage, the motor draws a negative-sequence current of 440%.

Most utility service agreements with customers state that it is the customer's responsibility to protect their equipment against single phasing. The best way to protect motors is with a phase-loss relay. Nevertheless, some utilities take measures to reduce the possibility of single-phasing customers' motors, and one way to do that is to limit the use of single-phase protective devices. Other utilities are more aggressive in their use of single-phase protective equipment and leave it up to customers to protect their equipment.

8.10.1 Single-Phase Reclosers with Three-Phase Lockout

Many single-phase reclosers and recloser controls come with a controller option for a single-phase trip and three-phase lockout. Three-phase reclosers that can operate each phase independently are also available. For single-phase faults, only the faulted phase opens. For temporary faults, the recloser successfully clears the fault and closes back in, so there will only be a momentary interruption on the faulted phase. If the fault is still present after the final reclose attempt (a permanent fault), the recloser trips all three phases and will not attempt additional reclosing operations.

Problems of single-phasing motors, backfeeds, and ferroresonance disappear. Single-phasing motors and ferroresonance cause heating, and heating usually takes many minutes for damage to occur. Short-duration single-phasing occurring during a typical reclose cycle does not cause enough heat to do damage. If the fault is permanent, all three phases trip and lock out, so there is no long-term single phasing. A three-phase lockout also reduces the chance of backfeed to a downed wire for a prolonged period.

Single-phase reclosers are available that have high enough continuous and interrupting ratings that utilities can use them in almost all feeder applications and many substation applications.

Another consideration with single-phase reclosers vs. three-phase devices is that a ground relay is often not available on single-phase reclosers. A ground relay provides extra sensitivity for line-to-ground faults. Not having the ground relay is a tradeoff to using single-phase devices. Even if a ground relay is available on a unit with single-phase tripping, if the ground relay operates, it trips all three phases (which defeats the purpose of single-phase tripping).

References

Amundson, R. H., "High Voltage Fuse Protection Theory & Considerations," IEEE Tutorial Course on Application and Coordination of Reclosers, Sectionalizers, and Fuses, 1980. Publication 80 EHO157–8-PWR.

ANSI C37.06-1997, *AC High-Voltage Circuit Breakers Rated on a Symmetrical Current Basis — Preferred Ratings and Related Required Capabilities.*

ANSI C37.42-1989, *American National Standard Specifications for Distribution Cutouts and Fuse Links.*

ANSI C57.12.20-1988, *American National Standard Requirements for Overhead-Type Distribution Transformers, 500 kVA and Smaller: High-Voltage, 67 000 Volts and Below; Low-Voltage, 15 000 Volts and Below,* American National Standards Institute.

ANSI/IEEE C37.60-1981, *IEEE Standard Requirements for Overhead, Pad Mounted, Dry Vault, and Submersible Automatic Circuit Reclosers and Fault Interrupters for AC Systems.*

Benmouyal, G. and Zocholl, S. E., "Time-Current Coordination Concepts," Western Protective Relay Conference, 1994.

Bottrell, G. W., "Hazards and Benefits of Utility Reclosing," Proceedings, Power Quality '93, 1993.

Burke, J. J., "Philosophies of Distribution System Overcurrent Protection," IEEE/PES Transmission and Distribution Conference, 1996.

CEA 288 D 747, *Application Guide for Distribution Fusing,* Canadian Electrical Association, 1998.

Cook, C. J. and Niemira, J. K., "Overcurrent Protection of Transformers—Traditional and New Philosophies for Small and Large Transformers," IEEE/PES Transmission & Distribution Conference and Exposition, 1996. Presented at the training session on "Distribution Overcurrent Protection Philosophies."

Cooper Power Systems, *Electrical Distribution — System Protection,* 3rd ed., 1990.

Cooper Power Systems, "Comparison of Recloser and Breaker Standards," Reference data R280-90-5, February 1994.

Drouet, M. G. and Nadeau, F., "Pressure Waves Due to Arcing Faults in a Substation," *IEEE Transactions on Power Apparatus and Systems,* vol. PAS-98, no. 5, pp. 1632–5, 1979.

Dugan, R. C., Ray, L. A., Sabin, D. D., Baker, G., Gilker, C., and Sundaram, A., "Impact of Fast Tripping of Utility Breakers on Industrial Load Interruptions," *IEEE Industry Applications Magazine,* vol. 2, no. 3, pp. 55–64, May/June 1996.

Engelman, N. G., "Relaying Changes Improve Distribution Power Quality," *Transmission and Distribution,* pp. 72–6, May 1990.

EPRI 1001665, *Power Quality Improvement Methodology for Wire Companies,* Electric Power Research Institute, Palo Alto, CA, 2003.

EPRI TR-106294-V2, *An Assessment of Distribution System Power Quality: Volume 2: Statistical Summary Report,* Electric Power Research Institute, Palo Alto, CA, 1996.

Gabrois, G. L., Huber, W. J., and Stoelting, H. O., "Blowing of Distribution Transformer Fuses by Lightning," in *IEEE/PES Summer Meeting, Vancouver, Canada,* 1973. Paper C73-421-5.

Hamel, A., Dastous, J. B., and Foata, M., "Estimating Overpressures in Pole-Type Distribution Transformers Part I: Tank Withstand Evaluation," *IEEE Transactions on Power Delivery,* vol. 18, no. 1, pp. 113–9, 2003.

Hamel, A., St. Jean, G., and Paquette, M., "Nuisance Fuse Operation on MV Transformers During Storms," *IEEE Transactions on Power Delivery,* vol. 5, no. 4, pp. 1866–74, October 1990.

Hi-Tech Fuses, "Bulletin FS-10: Application of Trans-Guard EXT and OS Fuses," Hi-Tech Fuses, Hickory, NC, 2002.

IEEE Std. C37.04-1999, *IEEE Standard Rating Structure for AC High-Voltage Circuit Breakers.*

IEEE Std. C37.010-1999, *IEEE Application Guide for AC High-Voltage Circuit Breakers Rated on a Symmetrical Current Basis.*

IEEE Std. C37.40-1993, *IEEE Standard Service Conditions and Definitions for High-Voltage Fuses, Distribution Enclosed Single-Pole Air Switches, Fuse Disconnecting Switches, and Accessories.*

IEEE Std. C37.48-1997, *IEEE Guide for the Application, Operation, and Maintenance of High-Voltage Fuses, Distribution Enclosed Single-Pole Air Switches, Fuse Disconnecting Switches, and Accessories.*

IEEE Std. C37.112-1996, *IEEE Standard Inverse-Time Characteristic Equations for Overcurrent Relays.*

IEEE Std. C62.22-1997, *IEEE Guide for the Application of Metal-Oxide Surge Arresters for Alternating-Current Systems.*

IEEE Working Group on Distribution Protection, "Distribution Line Protection Practices — Industry Survey Results," *IEEE Transactions on Power Delivery*, vol. 3, no. 2, pp. 514–24, April 1988.

IEEE Working Group on Distribution Protection, "Distribution Line Protection Practices — Industry Survey Results," *IEEE Transactions on Power Delivery*, vol. 10, no. 1, pp. 176–86, January 1995.

Johnston, L., Tweed, N. B., Ward, D. J., and Burke, J. J., "An Analysis of Vepco's 34.5 kV Distribution Feeder Faults as Related to Through Fault Failures of Substation Transformers," *IEEE Transactions on Power Apparatus and Systems*, vol. PAS-97, no. 5, pp. 1876–84, 1978.

Lunsford, J. M. and Tobin, T. J., "Detection of and Protection for Internal Low-Current Winding Faults in Overhead Distribution Transformers," *IEEE Transactions on Power Delivery*, vol. 12, no. 3, pp. 1241–9, July 1997.

Plummer, C. W., Goedde, G. L., Pettit, E. L. J., Godbee, J. S., and Hennessey, M. G., "Reduction in Distribution Transformer Failure Rates and Nuisance Outages Using Improved Lightning Protection Concepts," *IEEE Transactions on Power Delivery*, vol. 10, no. 2, pp. 768–77, April 1995.

Report to the IEEE working group on system performance, 2002.

Short, T. A., "Fuse Saving and Its Effect on Power Quality," in *EEI Distribution Committee Meeting*, 1999.

Short, T. A. and Ammon, R. A., "Instantaneous Trip Relay: Examining Its Role," *Transmission and Distribution World*, vol. 49, no. 2, February 1997.

Smith, D. R., "Network Primary Feeder Grounding and Protection," in Draft 3 of Chapter 10 of the IEEE Network Tutorial, 1999.

Smithley, R. S., "Normal Relay Settings Handle Cold Load," *Electrical World*, pp. 52–4, June 15, 1959.

Westinghouse Electric Corporation, *Applied Protective Relaying*, 1982.

I feel that there is not a single manager that respects the work of a lineman, until they really need us. Hell ours could not even tell you what a shotgun stick is.

In response to: Do linemen feel they are respected by management and coworkers for the jobs they are doing, do management and coworkers understand what you do?

www.powerlineman.com

9

Reliability

Power outages disrupt more businesses than any other factor (see Figure 9.1). I lose two hours of work on the computer; Jane Doe gets stuck in an elevator; Intel loses a million dollars worth of computer chips; a refinery flames out, stopping production and spewing pollution into the air. End users expect good reliability, and expectations keep rising. Interruptions and voltage sags cause most disruptions. In this chapter we study "sustained" interruptions, long-duration interruptions generally defined as lasting longer than 1 to 5 min. We investigate momentary interruptions and voltage sags in the next chapter. Reliability statistics, based on long-duration interruptions, are the primary benchmark used by utilities and regulators to identify service quality. Faults on the distribution system cause most long-duration interruptions; a fuse, breaker, recloser, or sectionalizer locks out the faulted section.

Many utilities use reliability indices to track the performance of the utility or a region or a circuit. Regulators require most investor-owned utilities to report their reliability indices. The regulatory trend is moving to performance-based rates where performance is penalized or rewarded based on quantification by reliability indices. Some utilities also pay bonuses to managers or others based in part on indices. Some commercial and industrial customers ask utilities for their reliability indices when locating a facility.

9.1 Reliability Indices

9.1.1 Customer-Based Indices

Utilities most commonly use two indices, SAIFI and SAIDI, to benchmark reliability. These characterize the frequency and duration of interruptions during the reporting period (usually years) (IEEE Std. 1366-2000).

SAIFI, System average interruption frequency index

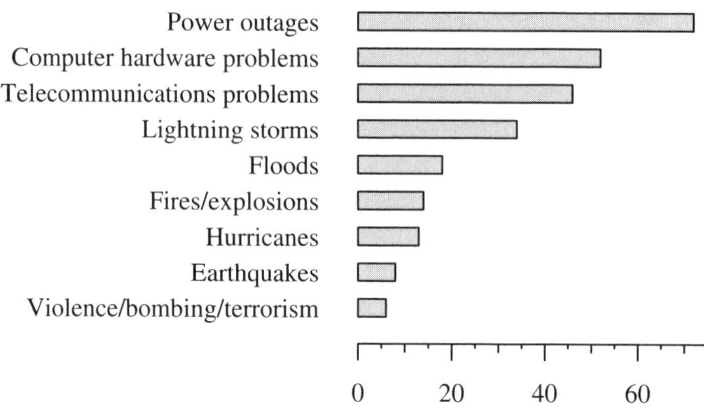

FIGURE 9.1
Percent of U.S. businesses disrupted by the given problem. (Data from [Rodentis, 1999].)

$$\text{SAIFI} = \frac{\text{Total number of customer interruptions}}{\text{Total number of customers served}}$$

Typically, a utility's customers average between one and two sustained interruptions per year. SAIFI is also the average failure rate, which is often labeled λ. Another useful measure is the mean time between failure (MTBF), which is the reciprocal of the failure rate: MTBF in years = $1/\lambda$.

SAIDI, System average interruption duration frequency index

$$\text{SAIDI} = \frac{\text{Sum of all customer interruption durations}}{\text{Total number of customers served}}$$

SAIDI quantifies the average total duration of interruptions. SAIDI is cited in units of hours or minutes per year. Other common names for SAIDI are CMI and CMO, standing for customer minutes of interruption or outage.

SAIFI and SAIDI are the most-used pair out of many reliability indices, which look like a wash of acronyms — most importantly, *D* for duration and *F* for frequency. Another related index is CAIDI:

CAIDI, Customer average interruption duration frequency index

$$\text{CAIDI} = \frac{\text{SAIDI}}{\text{SAIFI}} = \frac{\text{Sum of all customer interruption durations}}{\text{Total number of customer interruptions}}$$

CAIDI is the "apparent" repair time (from the customers' perspective). It is generally much shorter than the actual repair time because utilities nor-

TABLE 9.1

Reliability Indices Found by Industry Surveys

	SAIFI, No. of Interruptions/Year			SAIDI, h of Interruption/Year		
	25%	50%	75%	25%	50%	75%
IEEE Std. 1366-2000	0.90	1.10	1.45	0.89	1.50	2.30
EEI (1999) [excludes storms]	0.92	1.32	1.71	1.16	1.74	2.23
EEI (1999) [with storms]	1.11	1.33	2.15	1.36	3.00	4.38
CEA (2001) [with storms]	1.03	1.95	3.16	0.73	2.26	3.28
PA Consulting (2001) [with storms]				1.55	3.05	8.35
IP&L Large City Comparison (Indianapolis Power & Light, 2000)	0.72	0.95	1.15	1.02	1.64	2.41

Note: 25%, 50%, and 75% represent the lower quartile, the median, and the upper quartile of utilities surveyed.

mally sectionalize circuits to reenergize as many customers as possible before crews fix the actual damage.

Also used in many other industries, the availability is quantified as

ASAI, Average service availability index

$$\text{ASAI} = \frac{\text{SAIDI}}{\text{SAIFI}} = \frac{\text{Customer hours service availability}}{\text{Customer hours service demanded}}$$

We can find ASIFI from SAIDI specified in hours as

$$\text{ASAI} = \frac{8760 - \text{SAIDI}}{8760}$$

(Use 8784 h/year for a leap year.)

Survey results for SAIFI and SAIDI are shown in Table 9.1. Figure 9.2 shows the distribution of utility indices from the CEA survey. Much of this reliability index data is from Short (2002). Utility indices vary widely because of many differing factors, mainly

- Weather
- Physical environment (mainly the amount of tree coverage)
- Load density
- Distribution voltage
- Age
- Percent underground
- Methods of recording interruptions

Within a utility, performance of circuits varies widely for many of the same reasons causing the spread in utility indices: circuits have different lengths

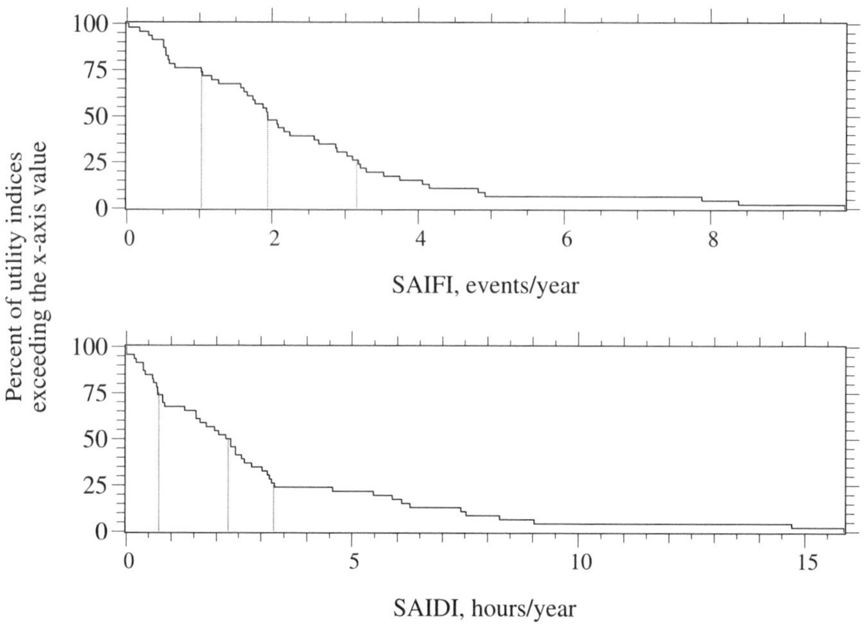

FIGURE 9.2
Distribution of utility indices in Canada (CEA survey, 36 utilities, two-year average). (Data from [CEA, 2000].)

necessary to feed different areas of load density, some are older than others, and some areas may have less tree coverage. Figure 9.3 shows the spread of reliability on individual feeders at two utilities for two years worth of data. Even though these two utilities are within the same state, SAIFI differs dramatically.

Customer reliability is not normally distributed. A skewed distribution such as the log-normal distribution is more appropriate and has been used in several reliability applications (Brown and Burke, 2000; Christie, 2002). A log-normal distribution is appropriate for data that is bounded on the lower side by zero. The skewed distribution has several ramifications:

- The average is higher than the median. The median is a better representation of the "typical" customer.
- Poor performing customers and circuits dominate the indices (which are averages).
- Storms and other outliers easily skew the indices.

Realize that SAIFI and SAIDI are weighted performance indices. They stress the performance of the worst-performing circuits and the performance during storms. SAIFI and SAIDI are not necessarily good indicators of the typical performance that customers have.

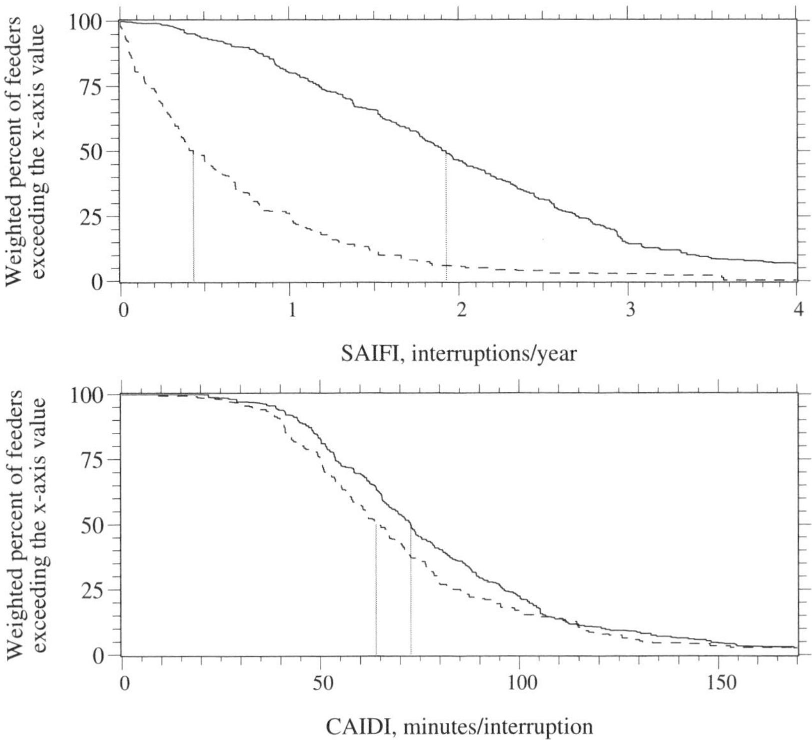

FIGURE 9.3
Reliability indices by feeder for two utilities. Forced events only—major events, scheduled events, and outside causes (substation or transmission) are excluded. The total SAIFI including all events was 0.79 for Utility A and 3.4 for Utility B.

9.1.2 Load-Based Indices

Residential customers dominate SAIFI and SAIDI since these indices treat each customer the same. Even though residential customers make up 80% of a typical utility's customer count, they may only have 40% of the utility's load. To more fairly weight larger customers, load-based indices are available; the equivalent of SAIFI and SAIDI, but scaled by load, are ASIFI and ASIDI:

ASIFI, Average system interruption frequency index

$$\text{ASIFI} = \frac{\text{Connected kVA interrupted}}{\text{Total connected kVA served}}(\text{Average number of interruptions})$$

ASIDI, Average system interruption frequency index

$$\text{ASIDI} = \frac{\text{Connected kVA interruption duration}}{\text{Total connected kVA served}}$$

Fewer than 8% of utilities track ASIFI and ASIDI, mainly since they are hard to track (knowing load interrupted is more difficult than knowing number of customers interrupted). Utilities also feel that commercial and industrial customers have enough clout that their problems are given due attention.

9.2 Storms and Weather

Much of the reliability data reported to regulators excludes major storm or major event interruptions. There are pros and cons to excluding storm interruptions. The argument for excluding storms is that storm interruptions significantly alter the duration indices to the extent that restoration performance dominates the index. Further, a utility's performance during storms does not necessarily represent the true performance of the distribution system. Including storms also adds considerable year-to-year variation in results. On the other hand, from the customer point of view an interruption is still an interruption. Also, the performance of a distribution system is reflected in the storm performance; for example, if a utility does more tree trimming and puts more circuits underground, their circuits will have fewer interruptions when a storm hits.

If storms are not excluded, the numbers go up as shown in an EEI survey in Figure 9.4. The interruption duration (CAIDI) and the average total interruption time (SAIDI) increase the most if storm data is included. Storms only moderately impact SAIFI. During storms, crew resources are fully used. Downed trees and wires plus traffic makes even getting to faults difficult. Add difficult working conditions, and it is easy to understand the great increase in repair times.

During severe storms, foreign crews, crews from other service territories, and general mayhem add large roadblocks preventing utilities from keeping records needed for tracking indices. Expedience rules — should I get the lights back on or do paperwork?

Utilities use various methods to classify storms. The two common categories are

- *Statistical method* — A common definition is 10% of customers affected within an operating area.
- *Weather based definition* — Common definitions are "interruptions caused by storms named by the national weather service" and "interruptions caused during storms that lead to a declaration of a state of emergency."

Some utilities exclude other interruptions including those scheduled or those from other parts of the utility system (normally substation or trans-

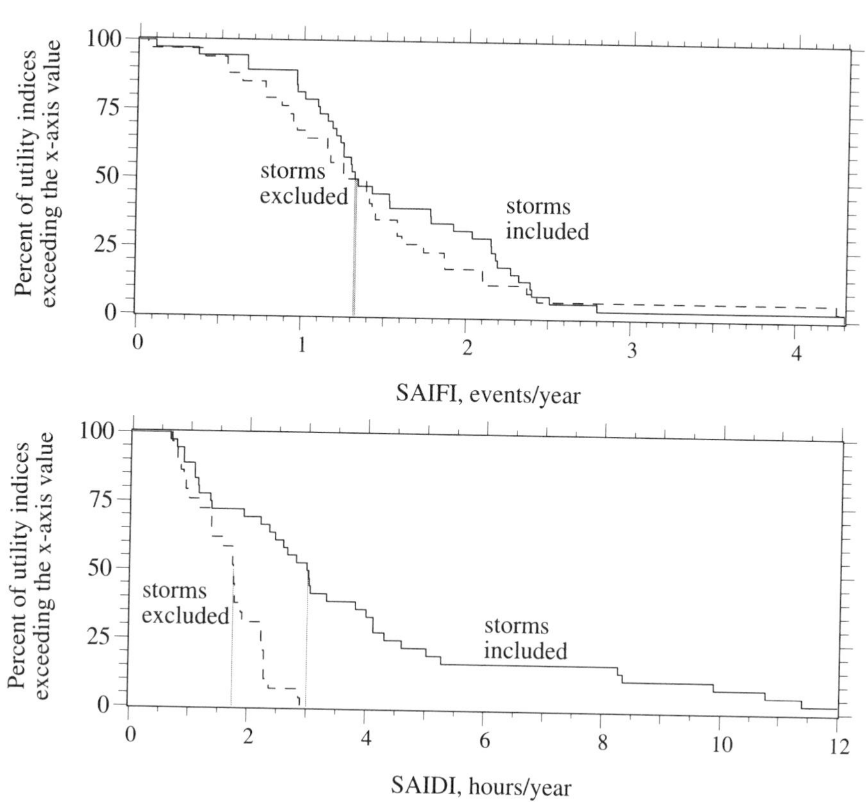

FIGURE 9.4
Distribution of utility indices with and without excluding storms. (Data from [EEI, 1999].)

mission-caused interruptions). Both are done for the same reasons as storm exclusions: neither scheduled interruptions nor transmission-caused interruptions reflect the normal operating performance of the distribution system.

The IEEE working group appears to favor a statistical approach to classifying major events (Christie, 2002). An argument against this approach is that major substation or transmission outages can be "major events" and get excluded from indices. From the customer point of view, major event or no major event, an interruption is still a loss of production or a spoiled inventory or a loss of productivity or a missed football game. For this reason, some regulators hesitate to allow exclusions.

During storm days, the interruption durations increase exponentially. Figure 9.5 shows probability distributions of the daily SAIDI based on data from four utilities. The plot is on a log-normal scale: the x-axis shows SAIDI for each day on a log scale, and the y-axis shows the probability on a normal-distribution scale. On this plot, data with a log-normal distribution comes out as a straight line. Most of the utility data fits a log-normal distribution. But, two of the utilities are even more skewed than a log-normal distribution indicates — at these utilities, storm days have even more customer-minutes

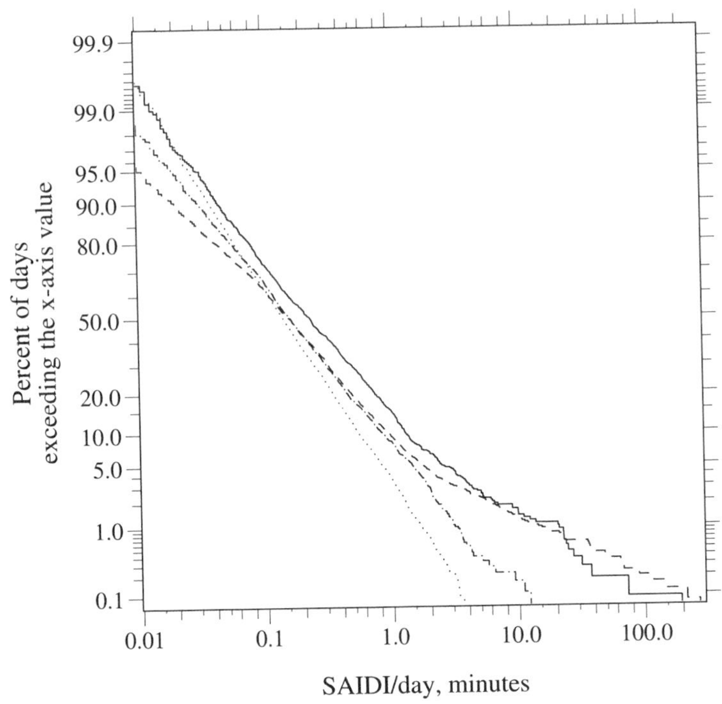

FIGURE 9.5
SAIDI per day probability distributions for four utilities.

interrupted. At one of these utilities, 0.2% of the days contributed 40% of the SAIDI index (over a 7-year period). During the worst year at this utility, 70% of SAIDI for that year happened because of three storms (impacting 5 days total). If we have one day with a SAIDI of over 100 min (the value for a whole year at some utilities), it is going to be a long year. Figure 9.6 again emphasizes how skewed the probability distribution of reliability data is. The average is much higher than the median, and the extreme days heavily influence the average.

Weather, even if it does not reach the level of "major event," plays a major role in reliability. Weather varies considerably from year to year — these weather variations directly affect reliability indices. Bad lightning years or excessively hot years worsen the indices.

Monitoring by Ontario Hydro Technologies in Ontario, Canada, gives some insight into storm durations and failure rates (CEA 160 D 597, 1998). In a mild to moderate lightning area, 20% of the interruptions occurred during storm periods and 15% in the 24 h following a storm. The study area had an average of 25 storm days per year and 73 storm hours per year. Therefore, about 35% of outages occur in 7% of the time in a year (and 20% in 0.8% of the time). The study found an interruption rate during storms of 10 to 20 times the non-storm rate.

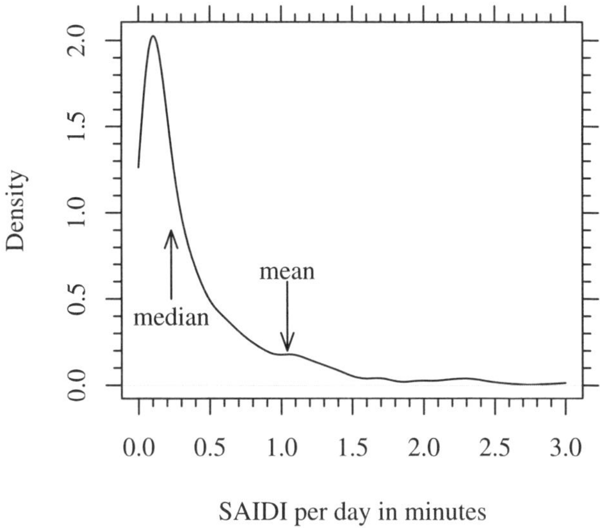

FIGURE 9.6
SAIDI per day probability density at one utility.

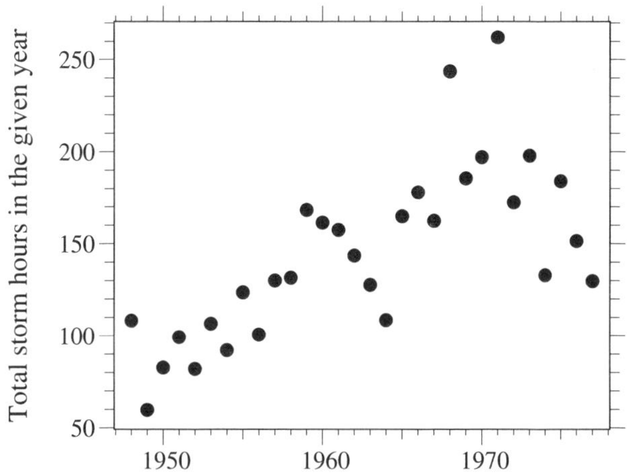

FIGURE 9.7
Thunderstorm duration by year for Tampa, Florida. (Data from [MacGorman et al., 1984].)

Faults and interruptions have significant year-to-year variation because weather conditions vary significantly. Just as severe storm patterns vary, normal storm frequencies and durations vary. Consider the thunderstorm duration plot in Figure 9.7. Over this 30-year period in Tampa, FL (a very high lightning area), some years had less than 80 h of storms, and a couple of years had more than 240 h. These are not "severe events," just variations in the normal weather patterns. These storm variations translate into variations in the number of faults and in the reliability indices. Even in areas with lower storm activity, significant variation is possible. Consider these variations if reliability "baselines" are going to be set for performance-based rates. Wind, icing, and temperature extremes all have significant year-to-year variations that directly impact reliability indices. Watch out for a few years of consistent weather; if the data from 1950–1955 of Figure 9.7 were used for performance-based rates, we would be in trouble in following years.

The first step in quantifying the effect of weather on interruptions is to track weather statistics along with interruption statistics. Lightning, wind, temperature, and other important weather statistics are available from national weather services as well as private groups, and many statistics have long historical records. Correlations between weather statistics and interruptions can help quantify the variations. Brown et al. (1997) show an example for a feeder in Washington state where wind-dependent failures were analyzed. For this case, they found 0.0065 failures/mi/year/mph of wind speed.

After correlating interruptions with weather data, we can extrapolate how much reliability indices could vary using historical weather data. One could even use weather statistics to come up with a normalized interruption index that tried to smooth out the weather variations.

9.3 Variables Affecting Reliability Indices

9.3.1 Circuit Exposure and Load Density

Longer circuits lead to more interruptions. This is difficult to avoid on normal radial circuits, even though we can somewhat compensate by adding reclosers, fuses, extra switching points, or automation. Most of the change is in SAIFI; the interruption duration (CAIDI) is less dependent on load circuit lengths. Figure 9.8 shows the effect on SAIFI at one utility in the southwest U.S.

It is easier to provide higher reliability in urban areas: circuit lengths are shorter, and more reliable distribution systems (such as a grid network) are more economical. The Indianapolis Power and Light survey results shown in Figure 9.9 only included performance of utilities in large cities. As expected, the urban results are better than other general utility surveys. Another comparison is shown in Figure 9.10 — in all states, utilities with higher load densities tend to have better SAIFIs.

Figure 9.11 and Figure 9.12 show reliability for different distribution services in several Commonwealth countries. The delineations used for this comparison for Victoria are

- Central business district: used map boundaries
- Urban: greater than 0.48 MVA/mi (0.3 MVA/km)

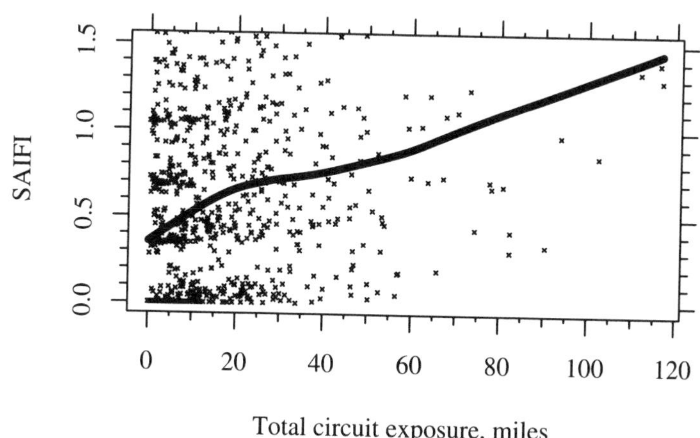

FIGURE 9.8
Effect of circuit length on SAIFI for one utility in the southwest U.S.

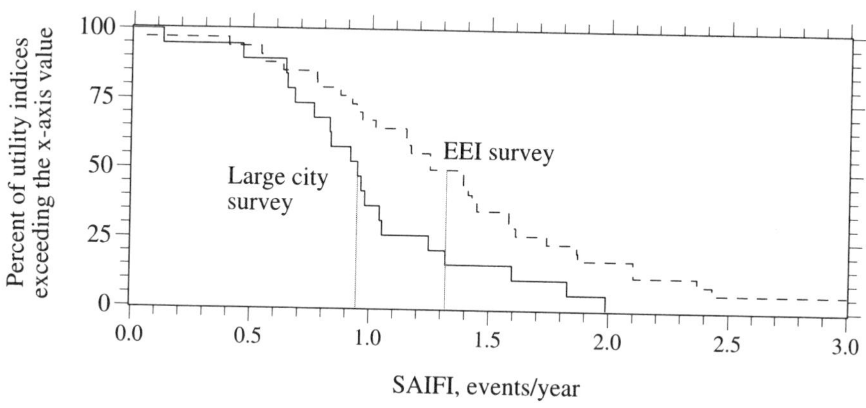

FIGURE 9.9
Comparison of the Indianapolis Power & Light Large City Survey of SAIFI to the general EEI survey results (with storms excluded). (Data from [Indianapolis Power & Light, 2000].)

- Short rural: less than 124 mi (200 km)
- Long rural: greater than 124 mi (200 km)

9.3.2 Supply Configuration

The distribution supply greatly impacts reliability. Long radial circuits provide the poorest service; grid networks provide exceptionally reliable service. Table 9.2 gives estimates of the reliability of several common distribution supply types developed by New York City's Consolidated Edison. Massive redundancy for grid and spot networks leads to fantastic reliability — 50 plus years between interruptions. Note that the interruption duration

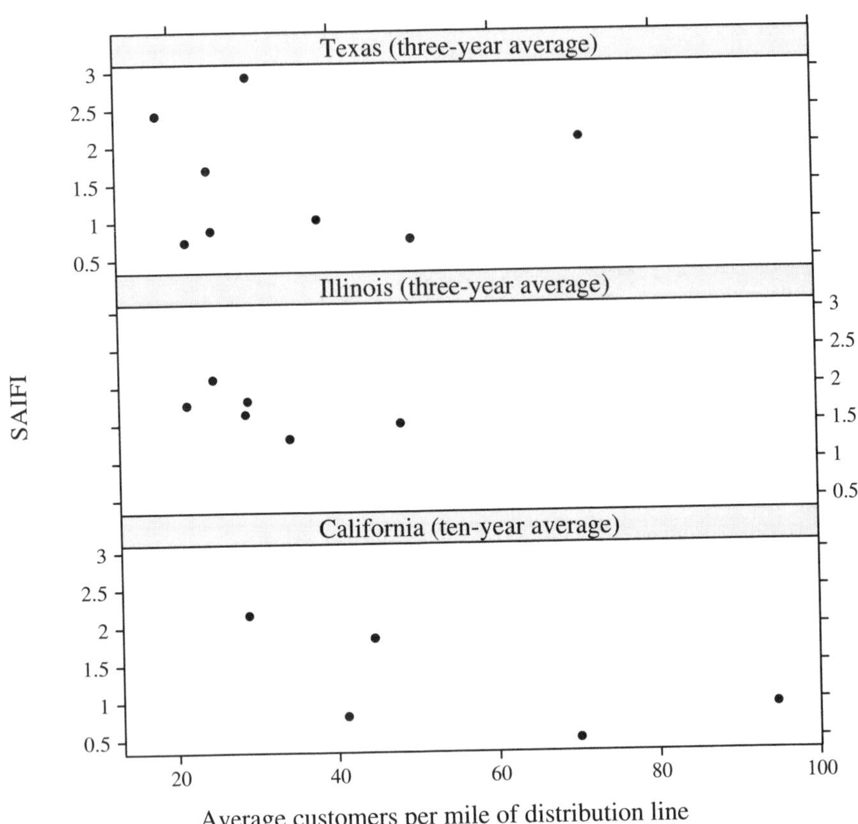

FIGURE 9.10
Effect of customer density on SAIFI.

(CAIDI) increases for the more urban configurations. Being underground and dealing with traffic increase the time for repairs.

9.3.3 Voltage

Higher primary voltages tend to be more unreliable, mainly because of longer lines. Figure 9.13 shows an example for one utility that is typical of many utilities: higher voltage circuits have more interruptions.

On higher-voltage primary circuits, we need to make more of an effort to achieve the same reliability as for lower voltage circuits: more reclosers, more sectionalizing switches, more tree trimming, and so forth. With the ability to build much longer lines and serve more customers, it is difficult to overcome the increased exposure. Keeping reliability in mind when planning higher-voltage systems helps. On higher-voltage circuits, wider is better than longer. Burke's analysis (1994) of the service length and width for a generalized feeder shows that for the best reliability, higher-voltage circuits should

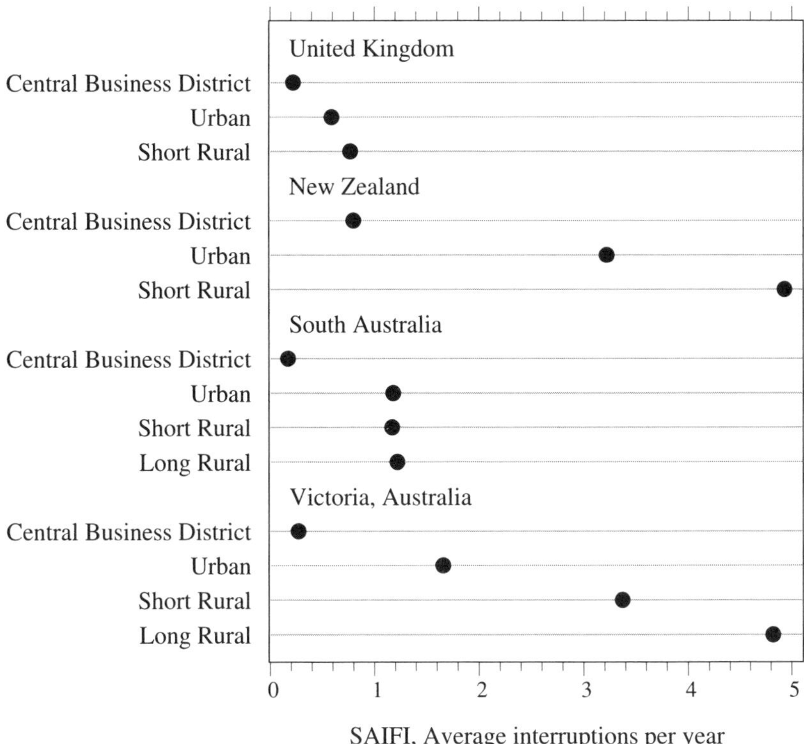

FIGURE 9.11
Comparison of SAIFI by load density for several former British Empire colonies. (Data from [Coulter, 1999].)

be longer and wider, not just longer (see Table 9.3). Usually, higher-voltage circuits are just made longer, which leads to poor reliability. Having a long skinny main feeder with short taps off of the mainline results in poor reliability performance.

9.3.4 Long-Term Reliability Trends

Utilities rarely have very long-term data covering decades. The Canadian Electrical Association has tracked reliability data for many years. Figure 9.14 shows SAIDI over a 40-year period for Canada. Significant variation exists from year to year. Part of this is due to the changing nature of the survey (the utility base was not consistent for the whole time period). Much of the variation is due to weather, even though the survey covers a huge geographic area (we expect more variations for smaller geographic areas). The data includes storms. Extreme years stand out. The worst year was 1998, which was dominated by the ice storm that hit Ontario and Quebec. Over 1.6 million customers lost power; Hydro Quebec's SAIDI for the year was almost 42 h when it is normally less than 4 h.

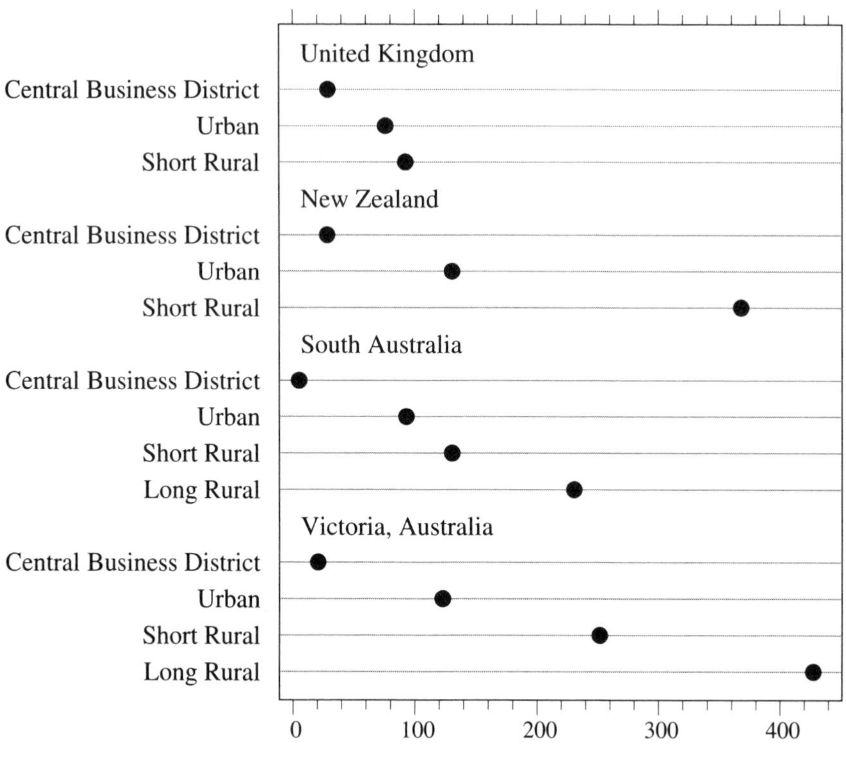

FIGURE 9.12
Comparison of SAIDI by load density for several former British Empire colonies. (Data from [Coulter, 1999].)

TABLE 9.2
Comparison of the Reliability of Different Distribution Configurations

	SAIFI Interruptions/Year	CAIDI min/Interruption	MAIFI Momentary Interruptions/Year
Simple radial	0.3 to 1.3	90	5 to 10
Primary auto-loop	0.4 to 0.7	65	10 to 15
Underground residential	0.4 to 0.7	60	4 to 8
Primary selective	0.1 to 0.5	180	4 to 8
Secondary selective	0.1 to 0.5	180	2 to 4
Spot network	0.02 to 0.1	180	0 to 1
Grid network	0.005 to 0.02	135	0

Source: Settembrini, R. C., Fisher, J. R., and Hudak, N. E., "Reliability and Quality Comparisons of Electric Power Distribution Systems," IEEE Power Engineering Society Transmission and Distribution Conference, 1991. With permission. ©1991 IEEE.

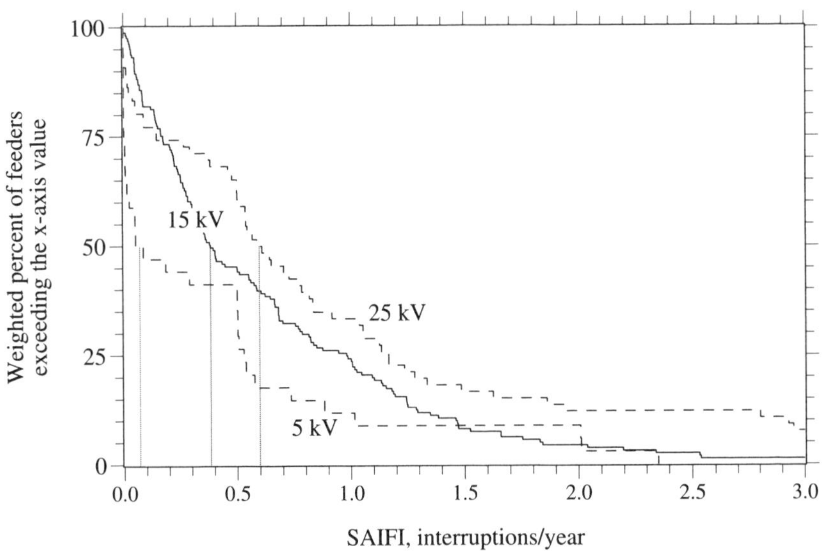

FIGURE 9.13
Effect of circuit voltage on SAIFI for one utility in the southern U.S.

TABLE 9.3
Mainline Lengths and Lateral Lengths for Optimal Reliability (Assuming a Constant Load Density)

Voltage, kV	Main Feeder Length, mi	Lateral Length, mi	Ratio of Main Feeder to Lateral Tap Length
13.8	1.51	0.95	1.59
23	1.81	1.32	1.37
34.5	2.09	1.71	1.22

Source: Burke, J. J., *Power Distribution Engineering: Fundamentals and Applications*, Marcel Dekker, New York, 1994.

Overall, the reliability trend is somewhat worsening. The main factor is probably the gradual move to higher-voltage distribution circuits and suburbanization. These trends lead to longer circuits and more exposure. Although, better record keeping (outage management systems) may be making SAIDI appear worse relative to earlier approaches because interruptions are recorded more accurately.

9.4 Modeling Radial Distribution Circuits

On purely radial circuits, the customers at the ends of the circuits unavoidably have the poorest reliability. On radial circuits, we can analyze the

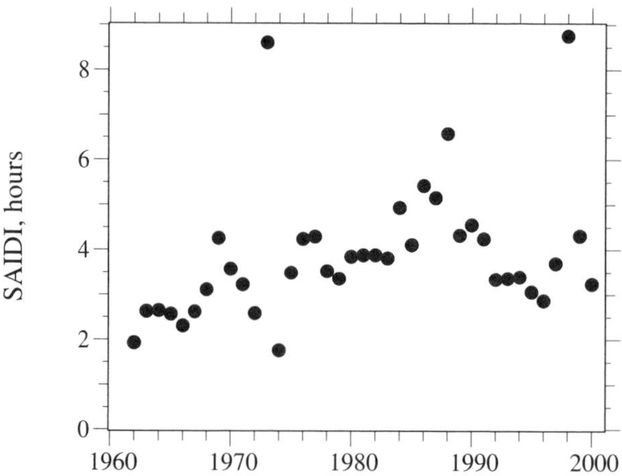

FIGURE 9.14
Yearly SAIDI for Canada. (Data from [Billinton, 1981, 2002; CEA, 2001].)

reliability using series combinations of individual elements. If any series component between the station and the customer fails, the customer loses power.

Series elements can be combined as

$$\lambda_S = \lambda_1 + \lambda_2 + \cdots + \lambda_n$$

$$U_S = U_1 + U_2 + \cdots + U_n = \lambda_1 r_1 + \lambda_2 r_2 + \cdots + \lambda_n r_n$$

$$r_S = \frac{U_S}{\lambda_S}$$

where

λ = failure rate, normally in interruptions per year

U = unavailability (total interruption time), normally in per unit, %, or h or min per year

r = average repair time per failure normally in per unit/year, %/year, or h or min

The subscript S is the total of the series combination, and the subscripts 1, 2, … n indicate the parameters of the individual elements.

The failure rate λ is analogous to SAIFI, U is analogous to SAIDI, and r is analogous to CAIDI.

We can use these basic reliability predictions to estimate reliability indices for radial circuits. Calculations quickly become complex if we try to account for sectionalizing or have circuits with parallel elements or backfeeds. Reliability analysis programs are available to model circuits with inputs similar to a load-flow program, except that switch characteristics are included as well as fault and equipment failure rates. Fault rates are the inputs most difficult to estimate accurately. These vary widely based on local conditions and construction practices.

With a given circuit configuration and SAIFI and MAIFI records for the circuit, Brown and Ochoa (1998) provide a way to back-estimate fault rates. For a given circuit configuration, the temporary and permanent fault rates are varied until the reliability prediction for a given circuit matches historical records. Once the failure rates are established, we can more accurately evaluate circuit changes such as automated switches.

9.5 Parallel Distribution Systems

To dramatically improve reliability for customers, parallel distribution supplies are needed. We see many forms of redundant distribution systems: autolooped primary distribution circuits with redundant paths, primary or secondary selective schemes with alternate supplies from two feeders, and spot or grid networks of several supply feeders with secondaries tied together. Analyzing the reliability of these interconnected systems is difficult. Several analytical techniques are available, and some are quite complicated.

With several components in series and parallel, we can find the failure rates and durations by reducing the network using the series or parallel combination of elements.

Parallel elements are combined with

$$\lambda_P = \frac{U_P}{r_p}$$

$$U_P = U_1 \times U_2 \times \cdots \times U_n = \lambda_1 \times \lambda_2 \times \cdots \times \lambda_n \times r_1 \times r_2 \times \cdots \times r_n$$

$$r_P = \frac{1}{1/r_1 + 1/r_2 + \cdots + 1/r_n}$$

for $n = 2$,

$$\lambda_P = \frac{U_P}{r_p} = \lambda_1 U_2 + \lambda_2 U_1 = \lambda_1 \lambda_2 (r_1 + r_2)$$

The subscript P is the total of the parallel combination. Note that the units must be kept the same: λ has units of 1/years, so the repair time, r, must be in units of years. Normally, this means dividing r by 8,760 if r is in hours or 525,600 if r is in minutes.

Including parallel elements is more complicated than series elements. The above equations are actually approximations that are valid only if the repair time is much less than the mean time between failure. This is generally true of distribution reliability applications (and more so for high-reliability applications).

The main problem with the equations representing the parallel combination of elements is that they are wrong for real-life electric supply reliability with multiple sources. A good illustration of this is from the data of the reliability of the utility supply found in a survey published in the *Gold Book* (IEEE Std. 493-1997). The average reliability of single-circuit supplies in the *Gold Book* has the following failure rate and repair time:

$$\lambda = 1.956 \text{ failures/year}$$

$$r = 79 \text{ min}$$

If a system were supplied with two parallel sources with the above failure rate characteristics, one would expect the following failure rates according to the ideal equations:

$$\lambda_P = (1.956)(1.956)(79{+}79)/525600 = 0.00115 \text{ failures/year}$$

$$r_P = 1/(1/79 + 1/79) = 39.5 \text{ min}$$

The actual surveyed reliability of circuits with multiple supplies is:

$$\lambda = 0.538 \text{ failures/year}$$

$$r = 22 \text{ min}$$

Another set of data for industrial supplies is shown in Figure 9.15 for the reliability of transmission supplies of Alberta Power. In this case, interruptions were defined as taking place for longer than 1 min. As with the *Gold Book* data, the multi-circuit Alberta Power supplies had better reliability than single-circuit supplies, but not by orders of magnitude. Single-circuit supplies had a 5-year average SAIFI of 0.9 interruptions per year (and SAIDI = 70 min of interruption/year). Multi-circuit supplies had a 5-year SAIFI of 0.42 interruptions per year (and SAIDI = 35 min of interruption/year). This information also helps show distribution engineers the number of transmission failures.

The failure rates with multiple circuits are reduced, but they are nowhere near the predicted value that is orders of magnitude lower. The reason the calculations are wrong is that the equations assume that the failures are totally independent. In reality, failures can have dependencies. The major factors are that

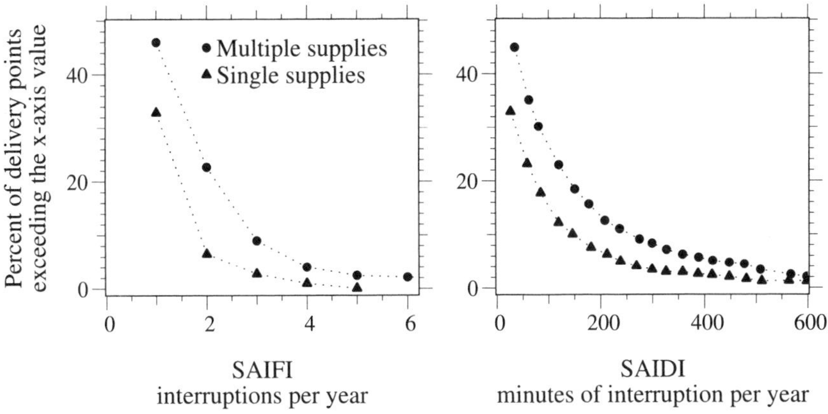

FIGURE 9.15
Comparison of single and multiple circuits for transmission supply. (Data from [Chowdhury and Koval, 1996].)

- Facilities share common space (utilities run two circuits on one structure).
- Separate supplies contain a common point upstream.
- Failures bunch together during storms.
- Maintenance must be considered.
- Hidden failures can be present.

Also, parallel supplies in many cases contain endpoint equipment that is not paralleled, including transformers, buswork, breakers, and cables.

It is possible to analytically model each of these effects. The problem is that much of the necessary input data is unknown, so many "educated" guesses are needed. For example, to analytically handle storm failures, one needs to find a storm failure rate and the duration of storms (both of these numbers are hard to come by).

It is common for multiple transmission and distribution circuits to be run on the same structures (and to be in proximity in underground facilities). If a car knocks down a pole with two circuits, both are lost. Lightning often causes multiple interruptions on structures with multiple circuits. There is also common space at the endpoint where the circuits are brought together at the customer. If there is a fire in the basement electrical room of a building with a spot network, the building will lose power.

Cable circuits are susceptible to a different type of outage bunching caused by overload failures. Cables are more sensitive to thermal overloads than are overhead lines. During high load, multiple cables can fail from overloads at the same time (and if one cable fails, the load on the others increases to make up the difference).

If a customer is supplied with two utility feeds off different feeders, the common point may be the distribution substation transformer. If the transformer or subtransmission circuit fails, both distribution feeders are lost simultaneously.

We can model common-mode failures of parallel supplies with the following equations:

$$\lambda_P = \lambda_1 \lambda_2 (r_1 + r_2) + \lambda_{12}$$

$$r_P = \frac{\lambda_1 \lambda_2 r_1 r_2 + \lambda_{12} r_{12}}{\lambda_P}$$

where λ_{12} and r_{12} are the frequency and repair time of the common-mode failures. The effect of common-mode failures is dramatic. If they are 10% of the individual failure rates, the common-mode failures dominate and degrade the parallel-combination failure rate.

One of the main problems is that there is little data on the common-mode failure rate of utility circuits (especially distribution circuits; some data is available for transmission circuits).

Failures requiring long-duration repair times are a special case with significant impacts. If something fails and takes two months to repair or replace, the system is much more vulnerable to failures during the maintenance. Long-duration repairs also violate the approximation that the repair time is much less than the MTBF, so the normal parallel and series combination equations are in error.

Also, on systems with redundancy, human nature acts to reduce the redundancy by increasing the repair time. Repairs generally take longer because people do not feel the urgency to make the repair. If sites are without power because of a failure, there are direct, immediate consequences. With redundancy, if a failure in one component occurs, the consequences are indirect (an increase in the likelihood of failure), so repair is not as urgent. If a network primary cable fails, customers are not without power, so crews have less urgency to repair the cable.

Failures due to lightning, wind, and rain occur during storms, and the failure rate during storms is much higher than normal. Since storms are a small portion of the total time, storm interruptions are bunched together, which dramatically raises the possibility of overlapping outages. Billinton and Allan (1984) provide ways to model the effects of storm bunching, and some reliability programs model these effects.

Hidden failures do not immediately show up or show up only when a failure occurs. Some examples might include the following:

- If a customer is served with a primary selective scheme, but the transfer switch is not working, the failure remains hidden until the switch is called upon to operate.

- A five-feeder grid network was originally designed to be able to lose two feeders and still serve the load. Subsequent load growth has reduced the redundancy so that four feeders must be on-line to handle the load. If more than one feeder is lost due to a failure and/or maintenance, the load could have an interruption.

Hidden failures are difficult to track down. In a parallel distribution system, the redundancy can mask failures. Protective equipment and diagnostics equipment that can isolate or identify failures is especially helpful in reducing hidden failures.

Hidden failures are difficult to model because they are obviously hidden. Hidden failures violate the assumption that the repair time is much shorter than the MTBF. This makes it difficult to use the equations listed above for parallel and series connections to analyze hidden failures. Finally, data on hidden failures is very limited.

When designing a redundant distribution system to one or more customers, consider these strategies that help reduce the possibility of overlapping failures:

- *Common space* — Limit sharing of physical space as much as possible. For services with multiple sources — primary or secondary selective schemes — try to use circuits that do not share the same poles or even right of ways. Use circuits that originate out of different distribution substations.
- *Storms* — Using underground equipment helps reduce the possibility of overlapping storm outages (although cables have their own bunched failures from overloads).
- *Maintenance* — Coordinate maintenance as much as possible to limit the loss of redundancy during maintenance. Try to avoid maintenance during stormy weather including heat waves for cables.
- *Testing* — Test switches, protective devices, and other equipment where hidden failures may lurk.
- *Loadings* — Review loadings periodically to ensure that overloads are not reducing designed redundancy.

9.6 Improving Reliability

We have many different methods of reducing long-duration interruptions including

- *Reduce faults* — tree trimming, tree wire, animal guards, arresters, circuit patrols

- *Find and repair faults faster* — faulted circuit indicators, outage management system, crew staffing, better cable fault finding
- *Limit the number of customers interrupted* — more fuses, reclosers, sectionalizers
- *Only interrupt customers for permanent faults* — reclosers instead of fuses, fuse saving schemes

Whether we are trying to improve the reliability on one particular circuit or trying to raise the reliability system wide, the main steps are

1. Identify possible projects.
2. Estimate the cost of each configuration or option.
3. Estimate the improvement in reliability with each option.
4. Rank the projects based on a cost-benefit ratio.

Prediction of costs is generally straightforward; predicting improvement is not. Some projects are difficult to attach a number to.

An important step in improving reliability is defining what measure to optimize: is it SAIFI, SAIDI, some combination, or something else entirely? The ranked projects change with the goal. Surveys have shown that frequency of interruptions is most important to customers (until you get to very long interruptions). Regulators tend to favor duration indicators since they are more of an indicator of utility responsiveness, and excessive cost cutting might first appear as a longer response time to interruptions.

Detailed analysis and ranking of projects can be done on a large scale. Brown et. al. (2001) provide an interesting example of applying reliability modeling to Commonwealth Edison's entire distribution system in Illinois to rank configuration improvements. Normally, large-scale projects require simplification (and often a good bit of guesswork).

Adding reclosers, putting in more fusing points, automating switches — these configuration changes are predictable. Many computer programs will quantify these improvements. Projects aimed at reducing the rates of faults, such as trimming more trees, adding more arresters, installing squirrel guards, are difficult to quantify. Improving fault-finding and repair are also more difficult to quantify. A sensitivity analysis helps when deciding on these projects. In the simplest form, rather than using one performance number, use a low, a best guess, and a high estimate. Pinpointing fault causes also helps frame how much benefit these targeted solutions can have (if there are few lightning-caused faults, additional arresters will provide little benefit).

9.6.1 Identify and Target Fault Causes

Tracking and targeting fault types helps identify where to focus improvements. If animals are not causing faults, we do not need animal guards.

Many utilities tag interruptions with identifying codes. The system-wide database of fault identifications is a treasure of information that we can use to help improve future reliability.

Different fault causes affect different reliability indices. Figure 9.16 shows the impact of several interruption causes on different reliability parameters for Canadian utilities. Relative impacts vary widely; for example, trees had a high repair time but impacted fewer customers.

Tracking this type of data for a utility operating region helps identify the most common problems for that service area. These numbers change by region depending on weather, construction practices, load densities, and other factors.

9.6.2 Identify and Target Circuits

Do not treat all circuits the same. The most important sections are usually not the locations with the most faults per mile. The number of customers on a circuit and the type of customers on a circuit are important considerations. For example, a suburban circuit with many high-tech commercial customers should warrant different treatment than a rural circuit with fewer, mostly residential and agricultural customers. How this is weighted depends on the utility's philosophy.

On radial distribution circuits, the three-phase mainline is critical. Sustained interruptions on the mains locks out all customers on the circuit until crews repair the damage. Feeders with extra-long mainlines have more interruptions. To reduce the impact of mainline exposure, trim trees more often on the mains and patrol the mains more often. Mainline sectionalizing switches help by allowing quick restoration of customers upstream of the fault; automated switches are even better. Another improvement is using normally open tie switches to other feeders, enabling crews to move load to other feeders during sectionalizing.

Lateral taps are another target. We can rank these by historical performance, taking into account their length and number of customers. Some longer laterals are good candidates for single-phase reclosers instead of fuses.

9.6.3 Switching and Protection Equipment

Fuses, sectionalizing switches, reclosers, sectionalizers — more, more, more — the more we have, the more we isolate faults to smaller chunks of circuitry, the fewer customers we interrupt.

Taps are almost universally fused, primarily for reliability. Fuses make cheap fault finders. Planners should also try to design to have tap exposure, not too much and not too little. We want to have a high percentage of a circuit's exposure on fused taps, so when permanent faults occur on those sections, only a small number of customers are interrupted.

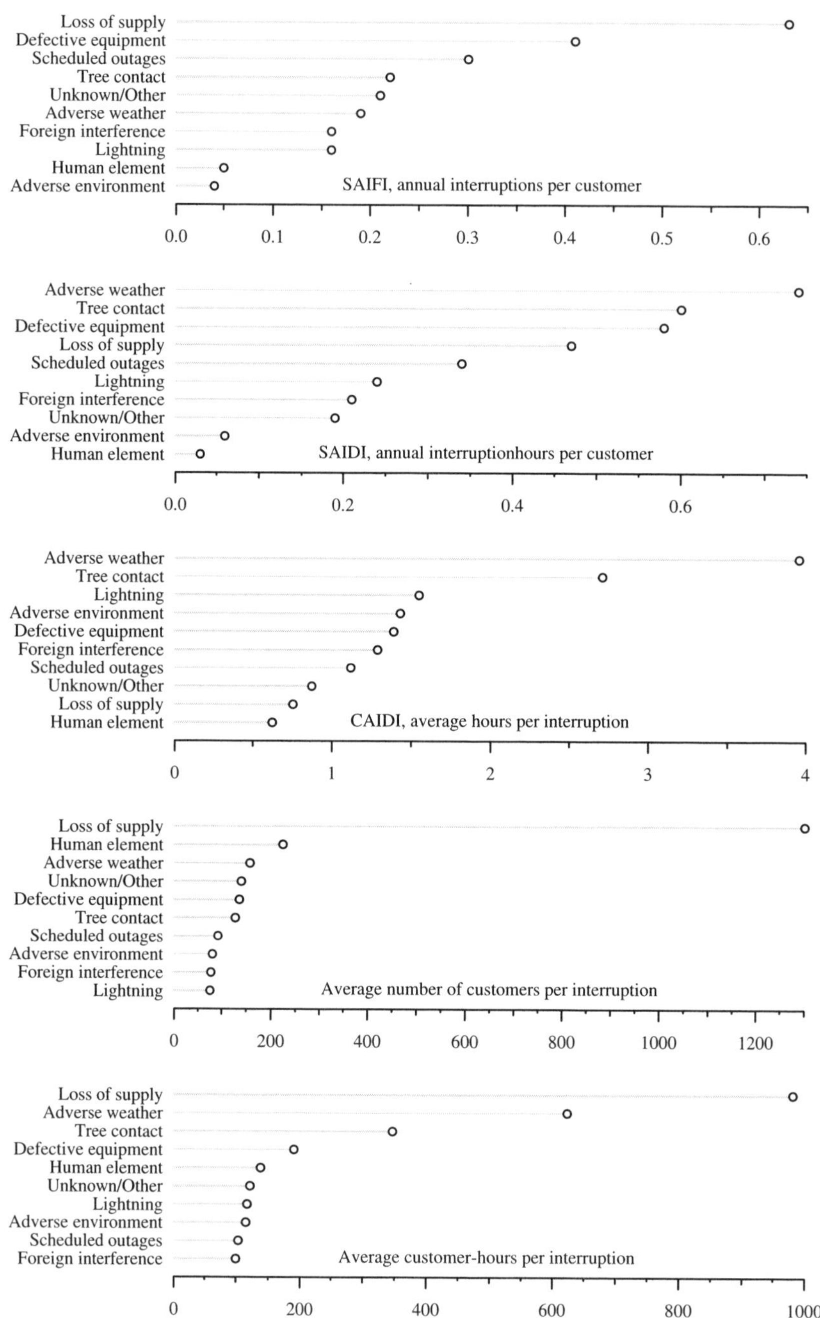

FIGURE 9.16
Root-cause contributors to different reliability parameters. (Data from [CEA, 2001].)

TABLE 9.4

Example Reliability Improvement Calculations

	SAIFI	MAIFI
Base case, fuse saving	5(0.3)+0.5(0.3) = 1.65	5(0.5)+10(0.9) = 12
Base case, fuse blowing	5(0.3)+0.5(0.9) = 1.95	5(0.6) = 3
One recloser, fuse blowing	(1.95+1.95/2)/2 = 1.46	(3+3/2)/2 = 2.25
Three-recloser auto-loop, fuse blowing	1.95/2 = 0.98	(3+3/2)/2 = 2.25
Five-recloser auto-loop, fuse blowing	1.95/3 = 0.65	(3+2+1)/3 = 2

Note: 5-mi mains, 15 total mi of exposure, 0.3 permanent faults/mi/year, 0.6 temporary faults/mi/year, fused laterals average 0.5 mi, customers are evenly distributed along the mainline and taps.

If taps become too long, use reclosers instead of fuses. Especially for circuits that fan out into two or three main sections (really like having two or three mainlines), reclosers on each of the main sections help improve reliability.

How circuits are protected and coordinated impacts reliability (see Chapter 8). Fuse saving, where the station breaker trips before tap fuses to try to clear temporary faults, helps long-duration interruptions most (but causes more momentary interruptions). Fuse blowing causes more long-duration interruptions because the fuse always blows, even for temporary faults.

Mainline reclosers also help improve reliability. Table 9.4 compares different scenarios for a common feeder with the following assumptions: 5-mi mains, 15 total mi of exposure, 0.3 permanent faults/mi/year, 0.6 temporary faults/mi/year, fused laterals average 0.5 mi, and customers are evenly distributed along the main line and taps. Mainline faults contribute most to SAIFI (1.5 interruptions per year). If the system were not fused at all, it would have 4.5 interruptions per year, pointing out the great benefit of the fuses. Branch-line faults only contribute an average of 0.15 interruptions with fuse saving (assuming fuse saving works right). This is an average; some customers on long taps have many more interruptions due to branch-line faults (these are good candidates for reclosers instead of fuses). On a purely radial system, reclosers do not help customers at the end of the line. An auto-loop scheme helps the customers at the ends of the line and significantly improves the feeder reliability indices.

This example only includes distribution primary interruptions. Supply-side interruptions and secondary interruptions should also be added as appropriate.

Sectionalizing switches can significantly improve SAIDI and CAIDI (but not SAIFI, unless the switches are automated). Such switches enable crews to easily reenergize significant numbers of customers well before they fix the actual damage. As Brown's analysis (2002) shows, the biggest gains are with the first few switches. Both SAIDI and CAIDI for mainline faults reduce in proportion to the difference between the mean time to repair (t_{repair}) and the mean time to switch (t_{switch}). With evenly spaced switches and customers on a radial circuit with no backfeeds, both CAIDI and SAIDI reduce with increasing numbers of sectionalizing switches as

$$R = \frac{\sum_{i=1}^{n} i}{(n+1)^2}(t_{repair} - t_{switch})$$

where

R = reduction in CAIDI and SAIDI
n = number of sectionalizing switches on the mainline
t_{repair} = time to repair the damage
t_{switch} = time to operate the sectionalizing switch

On a radial circuit, the improvement is not distributed equally; switches do not help customers at the end of the line at all. If a circuit has two evenly spaced sectionalizing circuits, the customers on the last part of the circuit see no improvement, the middle third see improvement for faults on the last third of the circuit, and the third at the front see improvement for faults on the last two-thirds of the circuit.

If the circuit has a backfeed from another circuit, the improvement is equal for all customers, so the overall circuit CAIDI improves. For evenly spaced switches and customers, the reduction is

$$R = \frac{n}{n+1}(t_{repair} - t_{switch})$$

Figure 9.17 shows how much sectionalizing switches can reduce CAIDI and SAIDI for the portion due to faults on the main line. The first switches provide the biggest bang for the buck. Beyond five switches, improvement is marginal. For application on real feeders (where the loads are not evenly placed), for biggest gains on radial circuits, place switches just downstream of large blocks of customers. If a circuit has a branch line with many customers, a mainline sectionalizing switch just downstream of the tap point allows crews to sectionalize that big block of customers for faults downstream of the switch.

The mean time to switch (t_{switch}) includes the time for the crew to get to the circuit, the time they need to find the fault, the time to find and open the appropriate sectionalizing switch, and finally the time to close the tripped breaker or recloser. Typically, t_{switch} is about 1 h. The mean time to repair (t_{repair}) is the time to travel to the circuit, find the fault, repair the damage, and close in the appropriate switching device to reconnect customers. The repair time varies widely — 4 h is a good estimate; actual repairs regularly range from 2 to 8 h.

Crews should decide whether to sectionalize based on local conditions. What is damaged? How long will it take to fix? How many customers would

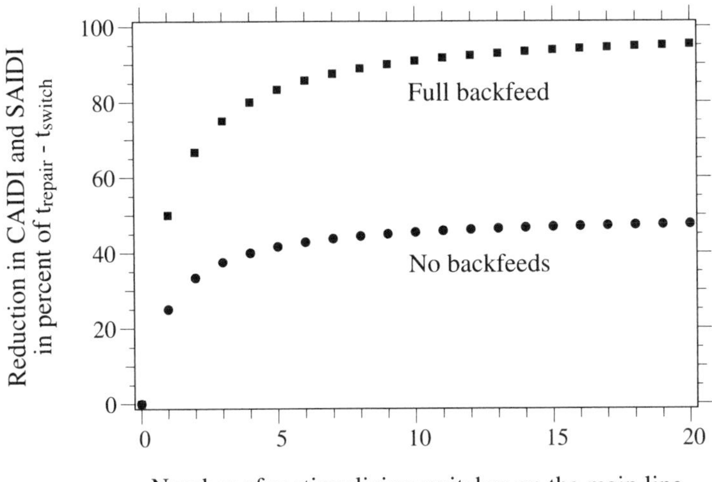

FIGURE 9.17
Reducing interruption durations with normally closed sectionalizing switches on the mainline.

sectionalizing bring back? Where are the sectionalizing switches? How long will it take to sectionalize the circuit? Sometimes, crews can start sectionalizing before the fault is found. If crews patrol a circuit section and do not see damage, they can open a sectionalizing switch and reclose the station breaker before they continue looking for the fault downstream.

9.6.4 Automation

Automation provides options for improving the reliability of the distribution supply. An auto-loop automated distribution configuration is a popular way to improve reliability on a normally radial circuit. These systems automatically reconfigure a distribution system: we do not need outside intervention or communications. In the three-recloser loop example in Figure 9.18, a normal sequence of operations for a fault upstream of recloser R1 is: (1) breaker B1 senses the fault and goes through its normal reclosing cycle and locks open; (2) recloser R1 senses loss of voltage and opens; and (3) recloser R2, the tie recloser, senses loss of voltage on the feeder and closes in. Since R2 can be switching into a fault, normally it is set for one shot; if the fault is there, it trips and stays open.

We can add more reclosers to divide the loop into more sections, but coordination of all of the reclosers is harder. Consider a five-recloser loop (Figure 9.19). Each feeder has two normally closed reclosers, and there is a normally open tie-point recloser. If feeder #1 is faulted close to the substation, breaker B1 locks out, recloser R1 opens, and the tie recloser closes. Now, we have a long radial circuit with the station breaker in series with four reclosers — that is a lot to try to coordinate. To ease the coordination, some reclosers can lower their tripping characteristics when operating in reverse mode. So,

B1
B2
R1 Normally closed recloser
R3 Normally closed recloser
R2
Normally open recloser

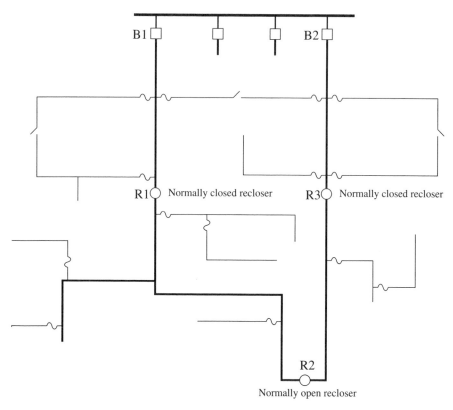

FIGURE 9.18
Example of an automated distribution feeder.

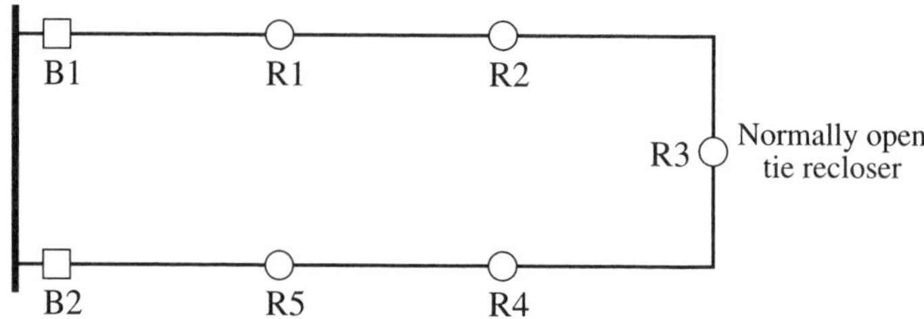

FIGURE 9.19
Five-recloser automated loop.

in this example, recloser R2 would drop its pickup setting. R2 sees much lower fault currents than it usually does, and we want it to trip before recloser R3 or R4.

For a fault between B1 and R1, the five-recloser loop responds similarly to a three-recloser loop: (1) breaker B1 locks out, (2) R1 opens on loss of voltage, (3) recloser R2 drops its trip setting, and (4) R3 senses loss of voltage on feeder 1 and closes in. For a fault between R1 and R2, the sequence is more complicated: (1) recloser R1 locks out, (2) recloser R2 drops its trip setting and goes to one shot until lockout, (3) R3 senses loss of voltage on feeder 1 and closes in (and closes in on the fault), and (4) R2 trips in one shot due to its lower setting. In a variation of this scheme, utilities use sectionalizers instead of reclosers at positions R2 and R4. Sectionalizers are easier to coordinate with several devices in series.

Remotely controlled switches are another option for automating a distribution circuit. The preferred communication is radio. Remotely controlled switches are more flexible than auto-loop schemes because it is easier to apply more tie points and we do not have to worry about coordinating protective equipment. Most commonly, operators decide how to reconfigure a circuit.

Even if a circuit is automated, doing another step of sectionalizing within the isolated section can squeeze out better reliability. Brown and Hanson (2001) show that manually sectionalizing after automated switches have operated can reduce SAIDI by several percent. As with other sectionalizing, crews should decide on a case-by-case basis whether to sectionalize.

Auto-loops will not necessarily help with momentary interruptions. Automation turns long-duration interruptions into momentary interruptions. To help with momentary interruptions, consider the following enhancements to automation schemes:

- *Line reclosers* — As part of an automated loop, line reclosers significantly improve momentaries; automated switches do not. Using single-phase reclosers helps interrupt fewer customers.
- *Tap reclosers* — Use reclosers on long lateral taps. Consider single-phase reclosers on three-phase taps. These will interrupt fewer customers.

9.6.5 Maintenance and Inspections

For many utilities, the best maintenance is trimming trees and then trimming some more trees; tree trimming is by far the largest maintenance expense for these utilities. Beyond tree trimming, distribution circuit maintenance practices vary widely. Distribution transformers, capacitors, insulators, wires, cables — most distribution equipment — do not need maintenance. Oil-filled switches, reclosers, and regulators need only occasional maintenance.

Most maintenance involves identifying old and failing equipment and targeting it for replacement. Equipment deteriorates over time. Several utilities have increasingly older infrastructure. Equipment fails at varying rates over its lifetime. Typically, it is a "bathtub curve": high failure rates initially during the break-in period (mostly due to manufacturing defects), a period of "normal" failure rates that increases over the equipment's lifetime. Some equipment sees more acceleration in failure rates than others. Data for distribution equipment is difficult to find. Early plastic cables — high-molecular weight polyethylene and cross-linked polyethylene — had dramatically increasing failure rates. Duckett and McDonough (1990) found dramatically increasing failure rates with age on Carolina Power & Light's 14.4-kV, 125-kV BIL transformers based on failures recorded from 1984 through 1988. The failure rate shot up when units reached 15 to 20 years old. An earlier study of CP&L's 7.2-kV, 95-kV BIL transformers did not show an increasing failure rate with age, staying at about 0.2 to 0.4% annually (Albrecht and Campbell, 1972). Aged transformers are more susceptible to failure, but we cannot justify replacement based on age; cables are the only equipment that utilities routinely replace solely based on age.

Storms trigger much "maintenance" — storms knock lines and equipment down, and crews put them back up (this is really restoration).

From birth to death, tracking equipment quality and failures helps improve equipment reliability. On most overhead circuits, most failures are external causes, not equipment failures (usually about 10 to 20% are equipment failures). Still, tracking equipment failures and targeting "bad apples" helps improve reliability. Many utilities do not track equipment failures at all. But, some utilities have implemented programs for tracking equipment and their failures. Failures occur in clusters: particular manufacturers, particular models, particular manufacturing years. Whether it is a certain type of connector or a brand of standoff insulator, some equipment has much higher than expected failure rates.

Proper application of equipment also helps, especially not overloading equipment excessively and applying good surge protection.

On underground circuits, equipment failures cause most interruptions. Tracking cable failures (usually by year of installation and type of insulation) and accessory failures and then replacing poor performers helps improve reliability. Monitoring loadings helps identify circuits that may fail thermally.

Quality acceptance testing of new equipment, especially cables, can identify poor equipment before it enters the field. For cables, tests can include microscopic evaluation of slices of cables to identify voids and impurities in samples. A high-pot test can also identify bad batches of cable.

On underground circuits, since workmanship plays a key role in quality of splices, tracking can also help. If a splice fails 6 months after it is installed and if we know who did the splice and who made the splice, we can work to correct the problem, whether it was due to workmanship or poor manufacturing quality.

Utilities use a variety of inspection programs to improve reliability. Of North American utilities surveyed (CEA 290 D 975, 1995), slightly more than half have regular inspection programs, and fewer than 5% have no inspections. Efforts varied widely: 27% spent less than 2% of operations and maintenance budgets on inspections, while 16% of utilities spent 10 to 30% of O&M on inspections.

Some distribution line inspection techniques used are

- *Visual inspections* — Most often, crews find gross problems, especially with drive-bys: severely degraded poles, broken conductor strands, and broken insulators. Some utilities do regular visual inspections; but more commonly, utilities have crews inspect circuits during other activities or have targeted inspections based on circuit performance. The most effective inspections are those geared towards finding fault sources — these may be subtle; crews need to be trained to identify them (see chapter 7).

- *Infrared thermography* — Roughly 40% of utilities surveyed use infrared inspections for overhead and underground circuits. Normally, crews watch a 20°C rise and initiate repair for more than a 30°C rise. Infrared scanning primarily identifies poor connectors. Some utilities surveyed rejected infrared monitoring and did not find it cost effective. Other utilities found significant benefit.
- *Wood pole tests* — Visual inspections are most common for identifying weak poles. A few utilities use more accurate measures to identify the mechanical strength left in poles. A hammer test, whacking the pole with a sledge, is slightly more sophisticated; a rotted pole sounds different when compared to a solid pole. Sonic testing machines are available that determine density and detect voids.
- *Operation counts* — Most utilities periodically read recloser operation and regulator tap changer counters to identify when they need maintenance.
- *Oil tests* — A few utilities perform oil tests on distribution transformers, reclosers, and/or regulators. While these tests can detect deterioration through the presence of water or dissolved gasses, the expense is difficult to justify for most distribution equipment.

Substation inspections and maintenance are more universally accepted. Most utilities track operation counts or station breakers and regulators, and most also sample and test station transformer oil periodically.

9.6.6 Restoration

Restoration affects SAIDI and CAIDI. Repair times vary considerably as shown in the example in Figure 9.20. Response time degrades quickly during storms as all crew resources are locked up. Even if "major events" are excluded, the responsiveness during bad weather still greatly influences restoration time.

The main way to improve restoration time is to sectionalize the circuit to bring as many customers back in as quickly as possible. Other methods that help reduce the repair time include the following:

- *Prepare* — Use weather information including lightning detection networks to track storms. Call out crews before the interruptions hit. Coordinate crews to distribute them as efficiently as possible.

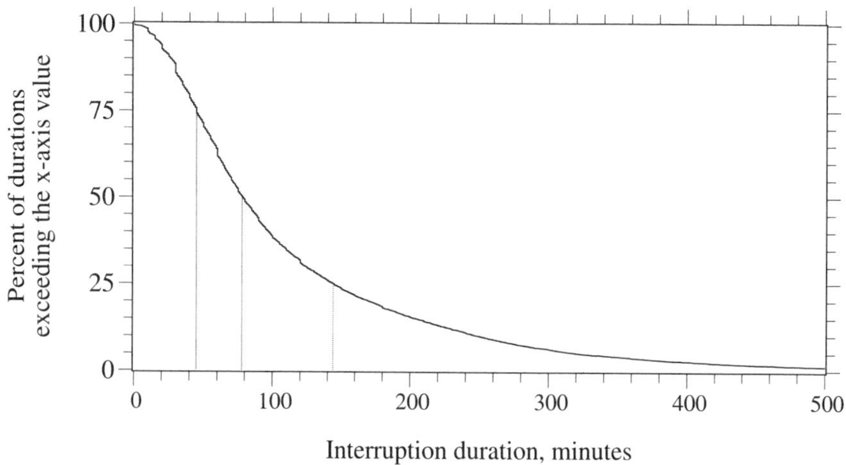

FIGURE 9.20
Distribution of interruption durations at one utility. (Data from [IEEE Working Group on System Design, 2001].)

- *Train* — Storm response training and other crew training help improve responsiveness.
- *Locate* — Use faulted circuit indicators and better cable locating equipment; have better system maps available to crews patrolling circuits. Use more fuses — a fuse is a cheap fault locator; with a smaller area downstream of a fuse, less length needs patrolling. Better communication between the call center and crews helps send crews to the right location.
- *Prioritize* — During storms, prioritize efforts based on those that get the most customers back in service quickly. Many of the first efforts will be sectionalizing; next will be efforts to target the repairs affecting most customers (faults on the distribution system mainline). Downed secondary and other failures affecting small numbers of customers have to wait. When prioritizing, safety implications should override reliability concerns; make sure downed wire cases are deenergized before other repairs.
- *Target* — Apply maintenance to address the faults that require long repair times. Tree faults have long repair times, so tree trimming reduces the repair time.

An outage management system helps with restoration and gives utilities information to help improve performance. But, be aware that implementing an outage management system will normally make reliability indices worse — just the indices. The actual effect on customers is not worse; in fact it should improve as utilities use the outage management system to improve responsiveness. Unfortunately, better record keeping translates into higher reliability indices. Several utilities have reported that SAIFI and SAIDI increase between 20 and 50% after implementing an outage management system. Outage management systems do help improve reliability and efficiency. Responsiveness improves as outage information is relayed more directly to crews. Outage management systems also calculate the reliability indices for utilities and can generate reports that utilities can use to target certain circuits for inspections or tree trimming. In addition to reliability, customer satisfaction improves, as call centers (either automated or people operated) are able to give customers better information on restoration times.

Knowing when most storms tend to occur and when most interruptions occur helps for scheduling crews. Typically, summer months are the busiest. Of course, each area has somewhat different patterns. Figure 9.21 shows SAIDI data from four U.S. utilities. Both a median and an average are shown — the median represents a typical day; the average counts towards the yearly index. One or two severe storm days can appreciably raise the average for the given month. Some utilities are hit much more by storms (those with high ratios of average to median).

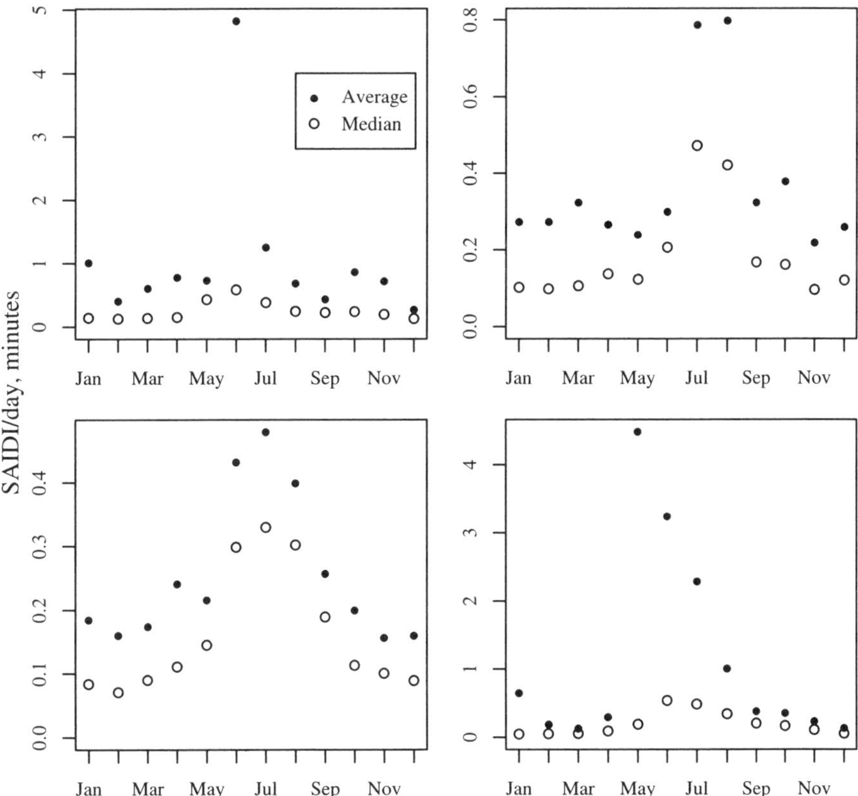

FIGURE 9.21
SAIDI per day by month of the year for four utilities.

Safety: Remember safety. Always. Reliability is important, but not worth dying for. Do not push repairs so quickly that crews take shortcuts that might create dangerous situations. Tired crews and rushed crews make more mistakes. Do not work during active lightning storms or other dangerous conditions. Make sure that the right people are doing the job; make sure they use the right tools, take enough breaks, and follow normal safety precautions.

9.6.7 Fault Reduction

An obvious approach to reliability improvement is to reduce the number of faults. In addition to long-duration interruptions, this strategy reduces the number of voltage sags and momentary interruptions and makes the system safer for workers and the public.

On-site investigations of specific faults can help reduce subsequent faults. Faults tend to repeat at the same locations and follow patterns. For example, one particular type and brand of connector may have a high failure rate. If these are identified, replacement strategies can be implemented. Another example is animal faults — one particular pole that happens to be a good travel path for squirrels may have a transformer with no animal guards. The same location may have repeated outages. These may be difficult to find at first, but crews can be trained to spot pole structures where faults might be likely.

9.7 Interruption Costs

Damaged equipment, overtime pay, lost sales, damage claims from customers — interruptions cost utilities money, plenty of money. An EPRI survey found that an average of 10% of annual distribution costs are for service restoration with ranges at different utilities from 7.6 to 14.8% (EPRI TR-109178, 1998). Restoration averaged \$14 per customer and \$0.20 per customer minute of interruption. The \$14 is per customer, not per customer interrupted, but it is close for a typical SAIFI of 1 to 1.5; assuming SAIFI equals 1.4, the average restoration cost is \$10 per customer interruption. Table 9.5 shows the restoration costs scaled by several factors. Not surprisingly, most of the cost is the actual construction to fix the problem as shown in Figure 9.22. Also, labor is the biggest portion of restoration costs, more than 70% in the survey. Note that the costs reported in the survey are costs directly associated with the restoration; lost kWh sales and damage claims are not included.

Costs escalate for major storms that severely damage distribution infrastructure. Table 9.6 shows the Duke Power Company's costs for several major storms. Many of these storms had much higher than normal costs per customer interrupted (as well as very high absolute costs).

TABLE 9.5

Surveyed Utility Restoration Costs in U.S. Dollars

	Average	Range
Per customer	14	12–17
Per customer minute of interruption	0.2	0.16–0.27
Per mile of overhead circuit	1000	300–1850
Per mile of underground circuit	3100	1700–5500

Source: EPRI TR-109178, *Distribution Cost Structure — Methodology and Generic Data,* Electric Power Research Institute, Palo Alto, CA, 1998.

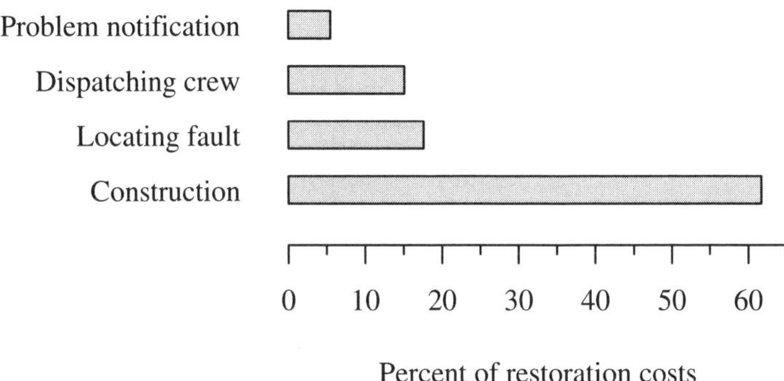

FIGURE 9.22
Breakdown of utility restoration costs. (Data from [EPRI TR-109178, 1998].)

TABLE 9.6

Restoration Costs During a Typical Year and During Major Storms for the Duke Power Company

Date	Storm Type	Customers Interrupted	Cost, $k	Cost Per Customer Interrupted, $
May-89	Tornadoes	228,341	15,190	67
Sep-89	Hurricane Hugo	568,445	64,671	114
Mar-93	Wind, Ice, and Snow	146,436	9,176	63
Oct-95	Hurricane Opal	116,271	1,655	14
Jan-96	Western NC Snow	88,076	873	10
Feb-96	Ice Storm	660,000	22,906	35
Sep-96	Hurricane Fran	409,935	17,472	43

Source: Keener, R. N., "The Estimated Impact of Weather on Daily Electric Utility Operations," Social and Economic Impacts of Weather, Proceedings of a workshop at the University Corporation for Atmospheric Research, Boulder, CO, 1997. Available at http://sciencepolicy.colorado.edu/socasp/weather1/keener.html.

TABLE 9.7

Survey of Interruption Costs to 299 Large Commercial and Industrial Customers

	4-h Interruption, No Notice	1-h Interruption, No Notice	1-h Interruption with Notice	Momentary Interruption	Voltage Sag
Production time lost, hours	6.67	2.96	2.26	0.70	0.36
Percent of work stopped	91%	91%	91%	57%	37%
Average total costs	$74,835	$39,459	$22,973	$11,027	$7,694
Costs per monthly kWh	0.2981	0.0182	0.0438	0.0506	0.0492

Source: Sullivan, M. J., Vardell, T., and Johnson, M., "Power Interruption Costs to Industrial and Commercial Consumers of Electricity," *IEEE Transactions on Industry Applications*, vol. 33, no. 6, pp. 1448–58, November/December 1997.

TABLE 9.8

Survey of Interruption Costs to Puget Sound Energy Customers

	12-h Interruption	4-h Interruption	1-h Interruption	Momentary Interruption
Commercial and industrial	$5144	$2300	$1008	$109
Residential	$25.95	$12.73	$8.32	$3.64

Source: Sullivan, M. and Sheehan, M., "Observed Changes in Residential and Commercial Customer Interruption Costs in the Pacific Northwest Between 1989 and 1999," IEEE Power Engineering Society Summer Meeting, 2000.

Some utilities also consider the costs to customers when planning for reliability. Costs of interruptions for customers vary widely, depending on the type of customer, the size of customer, the duration of the interruption, and the time of day and day of the week. Costs are highest for large commercial and industrial customers — Table 9.7 shows averages of customer costs for large commercial and industrial customers for various interruptions and short-duration events. Costs rise for longer-duration interruptions. Table 9.8 shows surveyed interruption costs for one utility's customers. We have to be careful of surveyed results of reliability surveys; utility customers often will not actually pay for solutions to eliminate the interruptions, even if the solution has a very short payback assuming their claimed costs of losses.

References

Albrecht, P. F. and Campbell, H. E., "Reliability Analysis of Distribution Equipment Failure Data," EEI T&D Committee, New Orleans, LA, January 20, 1972. As cited by Duckett and McDonough (1990).

Billinton, R., "Comprehensive Indices for Assessing Distribution System Reliability," IEEE International Electrical, Electronics Conference and Exposition, 1981.

Billinton, R., 2002. Personal communication.

Billinton, R. and Allan, R. N., *Reliability Evaluation of Power Systems*, Pitman Advanced Publishing Program, 1984.

Brown, R. E., *Electric Power Distribution Reliability*, Marcel Dekker, New York, 2002.

Brown, R. E. and Burke, J. J., "Managing the Risk of Performance Based Rates," *IEEE Transactions on Power Systems*, vol. 15, no. 2, pp. 893–8, May 2000.

Brown, R. E., Gupta, S., Christie, R. D., Venkata, S. S., and Fletcher, R., "Distribution System Reliability Assessment: Momentary Interruptions and Storms," *IEEE Transactions on Power Delivery*, vol. 12, no. 4, pp. 1569–75, October 1997.

Brown, R. E. and Hanson, A. P., "Impact of Two-Stage Service Restoration on Distribution Reliability," *IEEE Transactions on Power Systems*, vol. 16, no. 4, pp. 624–9, November 2001.

Brown, R. E., Hanson, A. P., Willis, H. L., Luedtke, F. A., and Born, M. F., "Assessing the Reliability of Distribution Systems," *IEEE Computer Applications in Power*, vol. 14, no. 1, pp. 44–49, 2001.

Brown, R. E. and Ochoa, J. R., "Distribution System Reliability: Default Data and Model Validation," *IEEE Transactions on Power Systems*, vol. 13, no. 2, pp. 704–9, May 1998.

Burke, J. J., *Power Distribution Engineering: Fundamentals and Applications*, Marcel Dekker, New York, 1994.

CEA 160 D 597, *Effect of Lightning on the Operating Reliability of Distribution Systems*, Canadian Electrical Association, Montreal, Quebec, 1998.

CEA 290 D 975, *Assessing the Effectiveness of Existing Distribution Monitoring Techniques*, Canadian Electrical Association, 1995.

CEA, *CEA 2000 Annual Service Continuity Report on Distribution System Performance in Electric Utilities*, Canadian Electrical Association, 2001.

Chowdhury, A. A. and Koval, D. O., "Delivery Point Reliability Measurement," *IEEE Transactions on Industry Applications*, vol. 32, no. 6, pp. 1440–8, November/December 1996.

Christie, R. D., "Statistical methods of classifying major event days in distribution systems," IEEE Power Engineering Society Summer Meeting, 2002.

Coulter, R. T., "2001 Electricity Distribution Price Review Reliability Service Standards," Prepared for the Office of the Regulator-General, Victoria, Australia and Service Standards Working Group, 1999.

Duckett, D. A. and McDonough, C. M., "A guide for transformer replacement based on reliability and economics," Rural Electric Power Conference, 1990. Papers Presented at the 34th Annual Conference, General Electric Co., Hickory, NC, 1990.

EEI, "EEI Reliability Survey," Minutes of the 8th Meeting of the Distribution Committee, March 28–31, 1999.

EPRI TR-109178, *Distribution Cost Structure — Methodology and Generic Data*, Electric Power Research Institute, Palo Alto, CA, 1998.

IEEE Std. 493-1997, *IEEE Recommended Practice for the Design of Reliable Industrial and Commercial Power Systems (Gold Book).*

IEEE Std. 1366-2000, *IEEE Guide for Electric Power Distribution Reliability Indices.*

IEEE Working Group on System Design, http://grouper.ieee.org/groups/td/dist/sd/utility2.xls, 2001.

Indianapolis Power & Light, "Comments of Indianapolis Power & Light Company to Proposed Discussion Topic, Session 7, Service Quality Issues," submission to the Indiana Regulatory Commission, 2000.

Keener, R. N., "The Estimated Impact of Weather on Daily Electric Utility Operations," Social and Economic Impacts of Weather, Proceedings of a workshop at the University Corporation for Atmospheric Research, Boulder, CO, 1997. Available at http://sciencepolicy.colorado.edu/socasp/weather1/keener.html.

MacGorman, D. R., Maier, M. W., and Rust, W. D., "Lightning Strike Density for the Contiguous United States from Thunderstorm Duration Records." Report to the U.S. Nuclear Regulatory Commission, # NUREG/CR-3759, 1984.

PA Consulting, "Evaluating Utility Operations and Customer Service," FMEA-FMPA Annual Conference, Boca Raton, FL, July 23–26, 2001.

Settembrini, R. C., Fisher, J. R., and Hudak, N. E., "Reliability and Quality Comparisons of Electric Power Distribution Systems," IEEE Power Engineering Society Transmission and Distribution Conference, 1991.

Short, T. A., "Reliability Indices," T&D World Expo, Indianapolis, IN, 2002.

Sullivan, M. and Sheehan, M., "Observed Changes in Residential and Commercial Customer Interruption Costs in the Pacific Northwest Between 1989 and 1999," IEEE Power Engineering Society Summer Meeting, 2000.

Sullivan, M. J., Vardell, T., and Johnson, M., "Power Interruption Costs to Industrial and Commercial Consumers of Electricity," *IEEE Transactions on Industry Applications*, vol. 33, no. 6, pp. 1448-58, November/December 1997.

Selected responses to: Close In Or Patrol?

Our company policy is to try to close and then patrol.

A line fuse is blown. Primary is on the ground. Children are around it. And you want to try it first! You better patrol the line!

I think you should patrol first. We once killed a farmer's cow because primary was down on his fence. Lucky it was just a cow.

Our Company Policy is to Close in on the line on arrival, if the phone center has not recieved any wire down calls. I'm not sure this is a good policy, because Mr and Mrs Customer don't know a powerline from a washline......I personally would ride the line and then when I was convinced that I have made a good call, close in.................Be CAREFUL............

www.powerlineman.com

10

Voltage Sags and Momentary Interruptions

The three most significant power quality concerns for most customers are

- Voltage sags
- Momentary interruptions
- Sustained interruptions

Different customers are affected differently. Most residential customers are affected by sustained interruptions and momentary interruptions. For commercial and industrial customers, sags and momentaries are the most common problems. Each circuit is different, and each customer responds differently to power quality disturbances. These three power quality problems are caused by faults on the utility power system, with most of them on the distribution system. Faults can never be completely eliminated, but we have several ways to minimize the impact on customers.

Of course, several other types of power quality (PQ) problems can occur, but these three are the most common; sags and momentary interruptions are addressed in this chapter (other power quality disturbances are discussed in the next chapter).

"The lights are blinking" is the most common customer complaint to utilities. Other common complaints are "flickering," "clocks blinking," or "power out." The first step to improving power quality is identifying the actual problem. Sustained interruptions are the easiest to classify since the power is usually out when the customer calls. The "blinking" is harder to classify:

- Is it momentary interruptions caused by faults on the feeder serving the customer?
- Is it voltage sags caused by faults on lateral taps or adjacent feeders?
- Is it periodic voltage flicker caused by an arc welder or some other fluctuating load on the same circuit?

Some strategies for identifying the problem are:

- For commercial or industrial customers, does the customer lose all computers or just some of them? Losing all indicates the problem is momentaries; losing some indicates the problem is sags.
- Is it just the lights flickering? Do any computers or other electronic equipment reboot or reset? If it is just the lights, the problem is likely to be voltage flicker caused by some fluctuating load, which could be in the facility that is having problems.
- If approximate times of events are available from the customer, we can compare these times against the times of utility protective device operations. Of course to do this, the utility times must be recorded by a SCADA system or a digital relay or recloser controller. If these are available, it is often possible to correlate a customer outage to a utility protective device. If the protective device is a circuit breaker or recloser upstream of the customer, the cause was probably a momentary interruption. If the protective device is on an adjacent circuit or the subtransmission system, the likely cause was a voltage sag.
- A review of the number of operations of the protective devices on the circuit, if these records are kept, can reveal whether the customer is seeing an abnormal number of momentary interruptions or possibly sags from faults on adjacent feeders.
- Does the flickering occur because of changes in the customer load? For example in a house, does sump-pump starting cause the lights to dim in another room? If so, look for a local problem. A likely candidate — a loose neutral connection — causes a reference shift when load is turned on or off.
- Are other customers on the circuit having problems? If so, then the problem is probably due to momentary interruptions and not just a customer that is very sensitive to sags. Momentary interruptions affect most end users; voltage sags only impact the more sensitive end users.

10.1 Location

Fault location is the primary factor that determines the disturbance severity to customers. Figure 10.1 shows several fault locations and how they impact a specific customer differently. A fault on the mains causes an interruption for the customer. If the fault is permanent, the customer has a long-duration interruption, but if the fault is temporary, the interruption is short as the protective device recloses successfully. A fault on a lateral tap causes a voltage sag unless fuse saving is used. With fuse saving, the fault on the tap

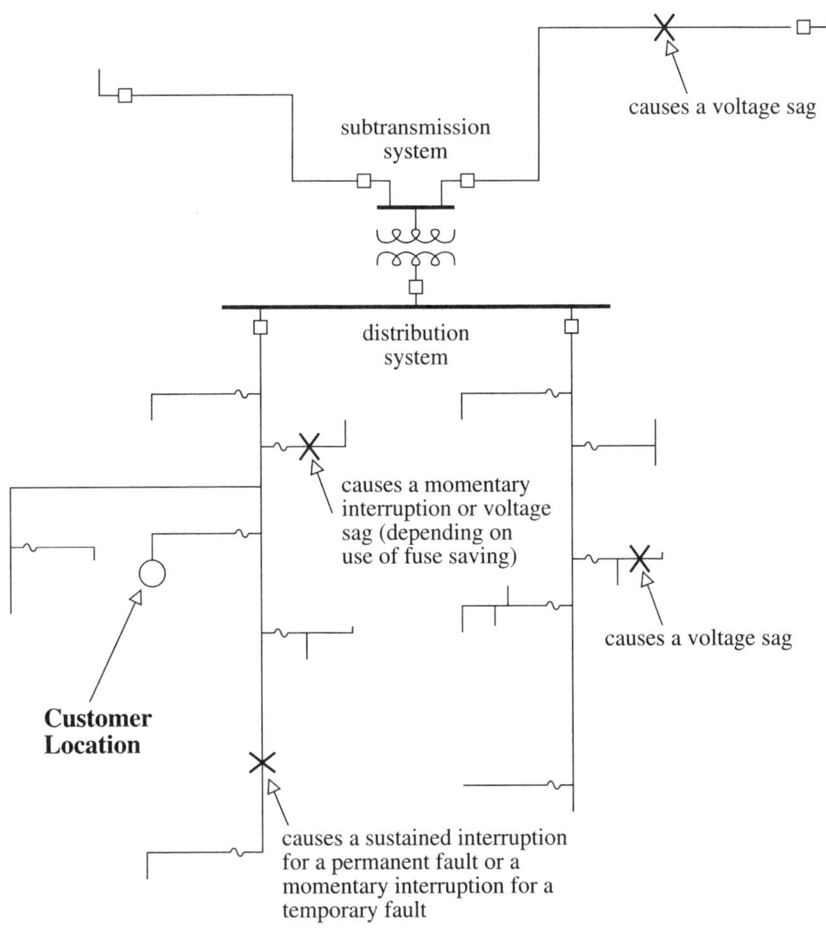

FIGURE 10.1
Example distribution system showing fault locations and their impact on one customer.

causes a momentary interruption as the substation breaker or recloser tries to prevent the fuse from blowing.

Faults on adjacent feeders cause voltage sags, the duration of which depends on the clearing time of the protective device. The depth of the sag depends on how close the customer is to the fault and the available fault current. Faults on the transmission system cause sags to all customers off of nearby distribution substations. We can depict all of the possible fault locations by areas of exposure or areas of vulnerability as shown in Figure 10.2. Each exposure area defines the vulnerability for the specific customer. For sags, we have different areas of vulnerability based on the severity of the sag. An outline of the area that causes sags to below 50% is tighter than the area of vulnerability for sags to below 70%. We can use the area of vulner-

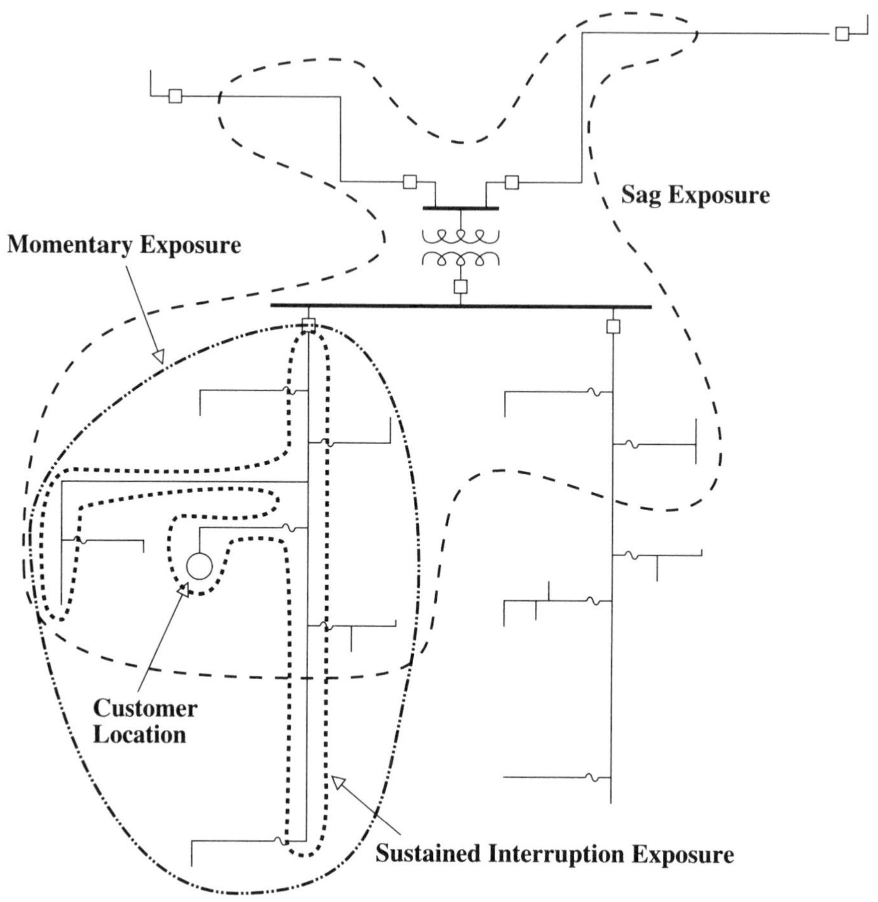

FIGURE 10.2
Example distribution system showing outlines of circuit exposure that cause a voltage sag, a momentary interruption, and a sustained interruption for one customer location.

ability curves to help target maintenance and improvements for important sensitive customers.

10.2 Momentary Interruptions

Momentary interruptions primarily result from reclosers or reclosing circuit breakers attempting to clear temporary faults, first opening and then reclosing after a short delay. The devices are usually on the distribution system; but at some locations, momentary interruptions also occur for faults on the subtransmission system. Terms for short-duration interruptions include *short interruptions*, *momentary interruptions*, *instantaneous interruptions*, and

TABLE 10.1
Surveys of MAIFI

Survey	Median
1995 IEEE (IEEE Std. 1366-2000)	5.42
1998 EEI (EEI, 1999)	5.36
2000 CEA (CEA, 2001)	4.0

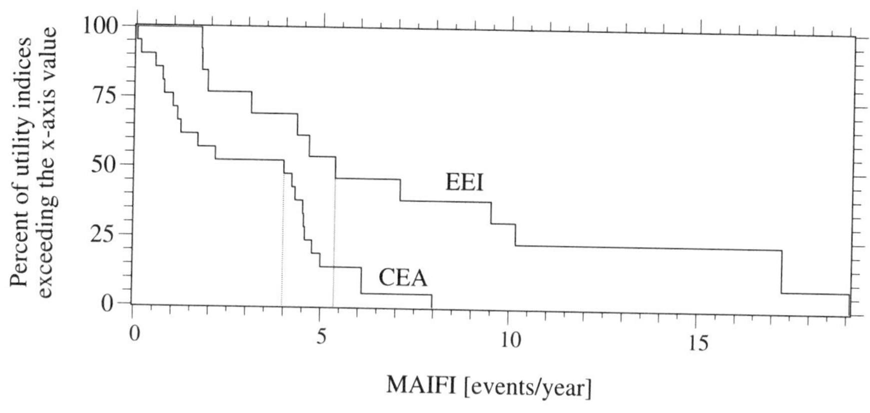

FIGURE 10.3
Distribution of utility MAIFI indices based on industry surveys by EEI and CEA. (Data from [CEA, 2001; EEI, 1999].)

transient interruptions, all of which are used with more or less the same meaning. The dividing line for duration between sustained and momentary interruptions is most commonly thought of as 5 min (1 min is also a common definition).

Table 10.1 shows the number of momentary interruptions based on surveys of the reliability index MAIFI. MAIFI is the same as SAIFI, but it is for short-duration rather than long-duration interruptions.

The number of momentary interruptions varies considerably from circuit to circuit and utility to utility. For example, in the EEI survey, the median of the utility averages is 5.4, but MAIFI ranged from 1.4 at the "best" utility to 19.1 at the "worst." Weather is obviously an important factor but so are exposure and utility practices. See Figure 10.3 for distributions of utility survey results.

There is a difference between the reliability definition and the power quality definition of a momentary interruption. The reliability definition (IEEE Std. 1366-2000) is

> A single operation of an interrupting device that results in zero voltage.

In addition, there is a distinction (IEEE Std. 1366-2000) between momentary interruptions and momentary interruption *events*:

> An interruption of duration limited to the period required to restore service by an interrupting device. Note: Such switching operations must be completed in a specified time not to exceed 5 min. This definition includes all reclosing operations which occur within 5 min of the first interruption. For example, if a recloser or breaker operates two, three, or four times and then holds, the event shall be considered one momentary interruption event.

Momentary interruption events and the associated index $MAIFI_E$ (E for event) better represent the impact on customers. Since we expect the first momentary disrupts the device or process, subsequent interruptions are unimportant. Momentary interruptions are most commonly tracked by using breaker and recloser counts, which implies that most counts of momentaries are based on MAIFI and not $MAIFI_E$. To accurately count $MAIFI_E$, a utility must have a SCADA system or other time-tagging recording equipment.

The power quality definition of a momentary interruption (IEEE Std. 1159-1995) is based on the voltage characteristics rather than the cause:

> A type of *short duration variation*. The complete loss of voltage (<0.1 pu) on one or more phases for a time period between 0.5 cycles and 3 sec.

Several extra events fall under the power quality definition of a momentary interruption. The power quality definition includes both operations of interrupting devices as well as very deep voltage sags. For this book, the reliability definition of a momentary interruption is used. The difference is worth remembering. Momentary interruptions that are tracked by using breaker and recloser counts are different from momentary interruptions recorded by power quality recorders. Table 10.2 shows momentary interruptions as recorded by several power quality studies using the power quality definition and an estimate of the reliability definition where very short events are excluded.

TABLE 10.2

Average Annual Number of Momentary Interruptions from Monitoring Studies

Study	Power Quality Definition[a] 1 cycle–10 sec	Reliability Definition[a] 20 cycles–10 sec
EPRI feeder sites (5-min filter)	6.4	4.5
NPL (5-min filter)	7.9	6.8
CEA primary (no filter)	3.2	1.3
CEA secondary (no filter)	6.5	2.8

[a] These are not industry standard definitions, just arbitrary time windows chosen to illustrate that the power quality definition of momentary interruptions has more events than a reliability definition.

Source: Dorr, D. S., Hughes, M. B., Gruzs, T. M., Juewicz, R. E., and McClaine, J. L., "Interpreting Recent Power Quality Surveys to Define the Electrical Environment," *IEEE Transactions on Industry Applications*, vol. 33, no. 6, pp. 1480–7, November 1997.

Momentaries can be improved in several ways including the following:

- *Reduce faults* — Tree trimming, tree wire, animal guards, arresters, circuit patrols, etc.
- *Reclose faster*
- *Limit the number of customers interrupted* — Single-phase reclosers, extra downstream reclosers, not using fuse saving, etc.

10.3 Voltage Sags

Voltage sags cause some of the most common and hard-to-solve power quality problems. Sags can be caused by faults some distance from a customer's location. The same voltage sag affects different customers and different equipment differently. Solutions include improving the ride-through capability of equipment, adding additional protective equipment (such as an uninterruptible power supply), or making improvements or changes in the power system.

A voltage sag is defined as an rms reduction in the ac voltage, at the power frequency, for durations from a half cycle to a few seconds (IEEE Std. 1159-1995). Sags are also called *dips* (the preferred European term). Faults in the utility transmission or distribution system cause most sags. Utility system protective devices clear most faults, so the duration of the voltage sag is the clearing time of the protective device.

Voltage-sag problems are a contentious issue between customers and utilities. Customers report that the problems are due to events on the power system (true), and that they are the utility's responsibility. The utility responds that the customer has overly-sensitive equipment, and the power system can never be designed to be disturbance free. Utilities, customers, and the manufacturers of equipment all share some of the responsibility for voltage sag problems. There are almost no industry standards or regulations to govern these disputes, and most are worked out in negotiations between a customer and the utility.

Terminology is a source of confusion. A 30% voltage sag can be interpreted as the voltage dropping to 70% of nominal or to 30% of nominal. Be more precise and say a "sag *to* X (volts or percent)." There is also some difference between a sag to 60% of nominal and a sag to 60% of the prefault voltage. Since most (but not all) equipment is sensitive to the actual voltage, generally refer to sags based on the percentage of nominal voltage.

Figure 10.4 shows a voltage sag that caused the system voltage to fall to approximately 45% of nominal voltage for 4.5 cycles.

Voltage sags can be improved with several methods on the utility system:

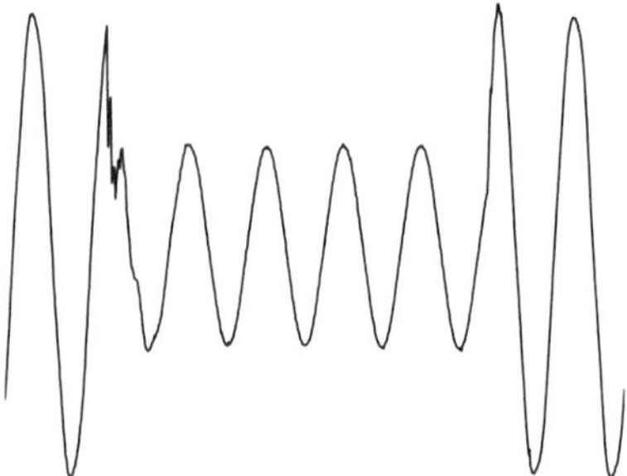

FIGURE 10.4
Example voltage sag caused by a fault.

- *Reduce faults* — Tree trimming, tree wire, animal guards, arresters, circuit patrols
- *Trip faster* — Smaller fuses, instantaneous trip, faster transmission relays
- *Support voltage during faults* — Raising the nominal voltage, current-limiting fuses, larger station transformers, line reactors

The voltage during the fault at the substation bus is given by the voltage-divider expression in Figure 10.5 based on the source impedance (Z_s), the feeder line impedance (Z_f), and the prefault voltage (V).

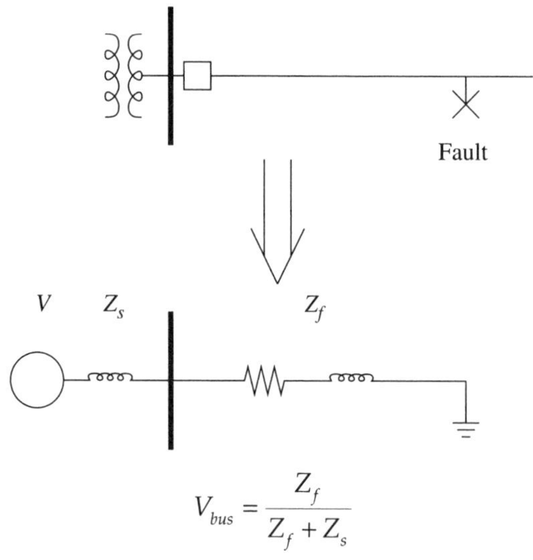

FIGURE 10.5
Voltage divider equation giving the voltage at the bus for a fault downstream. (This can be the substation bus or another location on the power system.)

The voltage sags deeper for faults electrically closer to the bus (smaller Z_f). Also, as the available fault current decreases (larger Z_s), the sag becomes deeper. The source impedance includes the transformer impedance plus the subtransmission source impedance (often, subtransmission impedance is small enough to be ignored). The impedances used in the equation depend on the type of fault it is. For a three-phase fault (giving the most severe voltage sag), use the positive-sequence impedance ($Z_f = Z_{f1}$). For a line-to-ground fault (the least severe voltage sag), use the loop impedance, which is $Z_f = (2Z_{f1} + Z_{f0})/3$. A good approximation is one ohm for the substation transformer (which represents a 7 to 8-kA bus fault current) and 1 Ω/mi (0.6 Ω/km) of overhead line for ground faults. For accuracy, use complex division since the impedances are complex, but for back-of-the-envelope, first-approximation calculations, use the impedance magnitude.

Another way to approximate the voltage divider equation is to use the available short-circuit current at the substation bus and the available short-circuit current at the fault location:

$$V_{bus} = 1 - \frac{I_f}{I_s}$$

where

V_{bus} = per unit voltage at the substation
I_f = the available fault current on the feeder at the fault location
I_s = the available fault current at the substation bus

Note that this can be used for any type of fault as long as the appropriate fault values are used in the equation. If the angles are ignored, the equation is an approximation (which is usually acceptable). Figure 10.6 shows a profile of the substation bus voltage for faults at the given distance along the line for 12.47, 24.94, and 34.5 kV. The higher-voltage systems have more severe voltage sags for faults at a given distance. The graph also shows that three-phase faults cause more severe sags. Figure 10.7 compares sags on underground and overhead systems.

The effect of feeder faults on voltage sags at the substation bus can be estimated with the following equation:

$$S(V_{sag}) = n_f \lambda \frac{V_{sag}}{1 - V_{sag}} \left(\frac{Z_s}{Z_f} \right)$$

where

S = annual number of sags per year where the voltage sags below V_{sag}
V_{sag} = per unit voltage sag level of interest (in the range of 0 to 1, e.g., 0.7)
n_f = number of feeders off of the bus

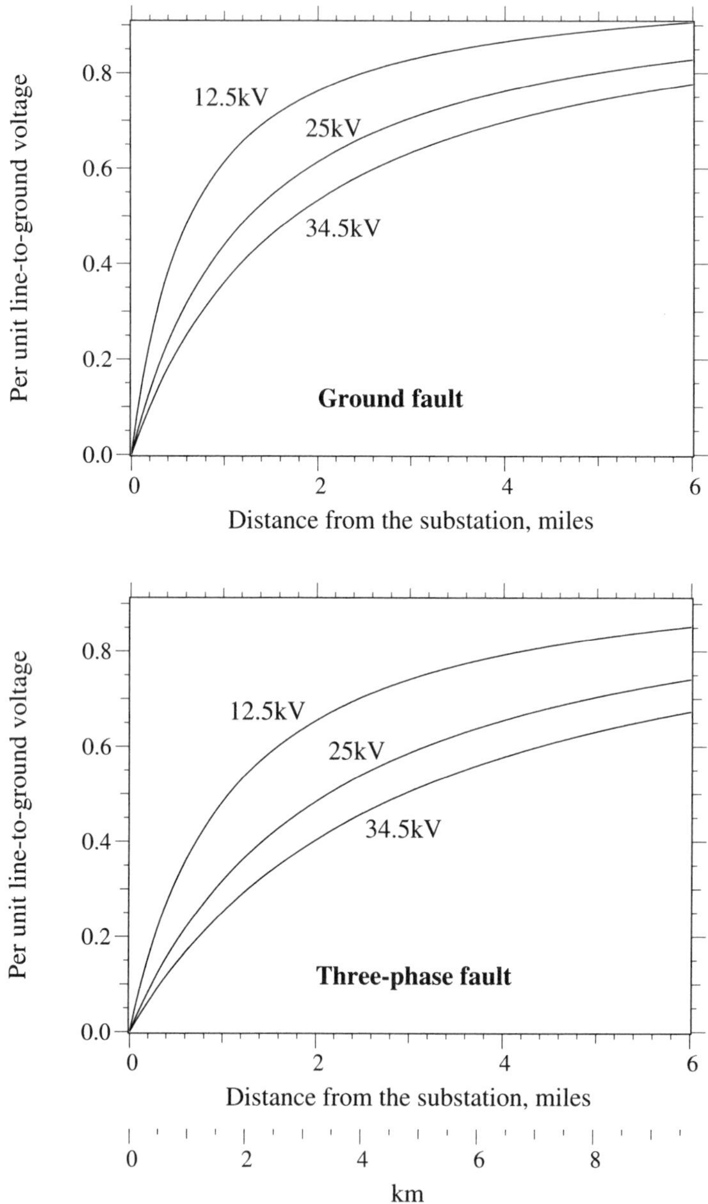

FIGURE 10.6
Substation voltage profile for faults at the given distance (single-phase and three-phase faults are shown for each voltage — the circuit parameters for the 500-kcmil circuit are the same as those in Figure 7.11).

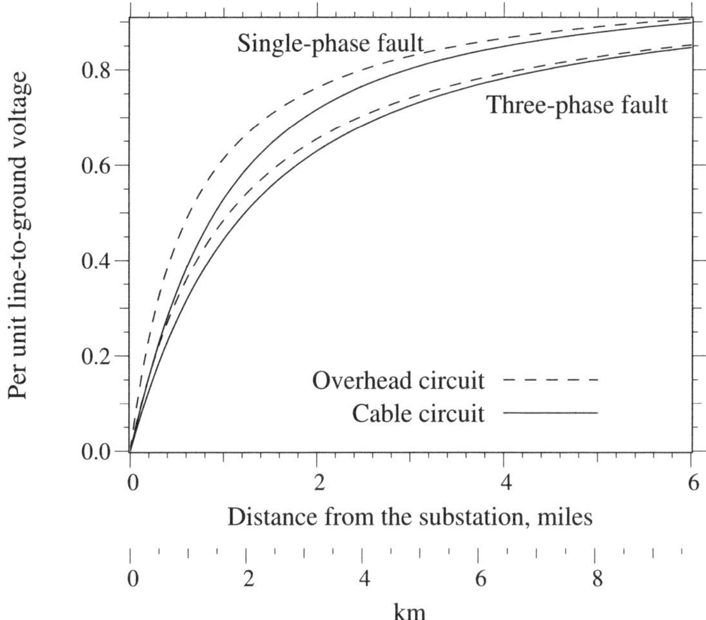

FIGURE 10.7
Comparison of substation voltage for faults on overhead circuits and cable circuits at the given distance (single-phase and three-phase faults are shown; the circuit parameters are the same as those in Figure 7.11 and Figure 7.12).

λ = feeder mains fault rate per mile (or other unit of distance) per phase including faults on laterals and including both temporary and permanent faults

Z_f = feeder impedance, Ω/mi (or other unit of distance); usually use $Z_f = (2Z_1+Z_0)/3$ for ground faults

Z_s = source impedance, Ω

The distribution of voltage sags based on this equation is shown in Figure 10.8 for some common parameters. Several points are noted from this analysis on voltage sags:

- *Exposure* — For 15-kV circuits, we can ignore exposure beyond the first 2 or 3 mi (4 or 5 km) for sags to the bus voltage. The first mile or two is most important as far as circuit improvement, maintenance, or application of current-limiting fuses.
- *System voltage* — Sags are more severe on higher voltage distribution systems (especially at 34.5 kV). A fault 4 mi from the substation sags the voltage much more on a 25-kV system than on a 12-kV system because the substation transformer is a higher impedance relative to the line impedance at higher system voltages. For 24.94 kV, exposure as far as 5 mi from the station is significant.

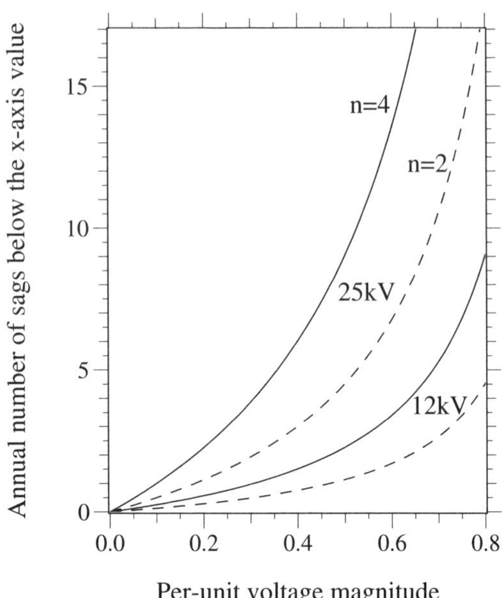

FIGURE 10.8
Cumulative distribution of substation bus voltage sags per year for the given (25-MVA, 10% transformer, 500-kcmil feeder, n = 2 or 4 feeders off of the bus, λ = 1 faults/phase/mile of mains/year, assumes line-to-ground faults only). Copyright © 2003. Electric Power Research Institute. 1001665. *Power Quality Improvement Methodology for Wires Companies.* Reprinted with permission.

- *Single vs. three-phase faults* — Three-phase faults cause more severe sags than single-line-to-ground faults. Three-phase faults farther away can pull the voltage down.
- *Underground vs. overhead* — All-underground circuits have more exposure to sags because cables have lower impedance than overhead lines.
- *Number of feeders* — The number of sags on the station bus is directly proportional to the number of feeders off the bus.
- *Transformer impedance* — A lower station transformer impedance (a bigger transformer or lower percent impedance) improves voltage sags.
- *Bus tie* — It does not matter whether a substation bus tie is open or closed. If it is open, a fault only affects half of the feeders. A fault that does occur forces a deeper sag because of a higher effective source impedance. These two effects tend to cancel each other.
- *Voltage regulation* — Raising the nominal voltage improves the voltage seen by customers during a fault. Say that a fault drops the voltage to 0.8 per unit, and the prefault voltage was 1.0 per unit. If the prefault voltage were 1.1 per unit, the voltage during the sag is 0.88 per unit. This is not a big difference, but for equipment sensitive

TABLE 10.3

Line-to-Ground and Line-to-Line Voltages on the Low-Voltage Side of a Transformer with One Phase on the High-Voltage Side Sagged to Zero

Voltages	Primary Voltages			Voltages Downstream of a Delta–Wye Transformer		
Line-ground	0.00	1.00	1.00	0.58	0.58	1.00
Line-line	0.58	0.58	1.00	0.33	0.88	0.88

to sags to 0.7 to 0.85 per unit, higher voltages appreciably reduce the number of tripouts.

Customers at the end of a circuit have more severe voltage sags because almost all faults upstream appear as little or no voltage (most actually fit the power quality definition of an interruption, a voltage to below 10%).

10.3.1 Effect of Phases

Three-phase loads are often controlled by single-phase devices (the controls are often the most sensitive element). The effect on three-phase customers depends on how loads are connected and depends on the transformer connection as shown in Table 10.3. In general, if the transformer causes more phases to be affected, the voltage drop is less severe. One situation is not always better than the other. Severity depends on which phases the sensitive devices are located. The type and design of the device and its controls are also factors.

For facility equipment connected *line-to-line*, the wye – wye transformer connection provides the best performance. For facility equipment connected *line-to-ground*, the delta – wye facility transformer is best.

Single-phase sags on distribution systems are more common than two- or three-phase sags. This is expected since most faults on distribution systems are single phase. For example, in EPRI's Distribution Power Quality Study (EPRI TR-106294-V2, 1996), about 64% of voltage sags to below 70% were single phase, while three-phase sags made up 25%, and two-phase sags, 10%. For severe sags below 30% voltage, three-phase events are more common; more than half are three-phase events (see Figure 10.9). This includes momentary interruptions, most of which are three-phase.

10.3.2 Load Response

During a voltage sag, rotating machinery supports the voltage by feeding current back into the system. Synchronous motors and generators provide the largest boost. Induction motors also provide benefit, but the support decays quickly.

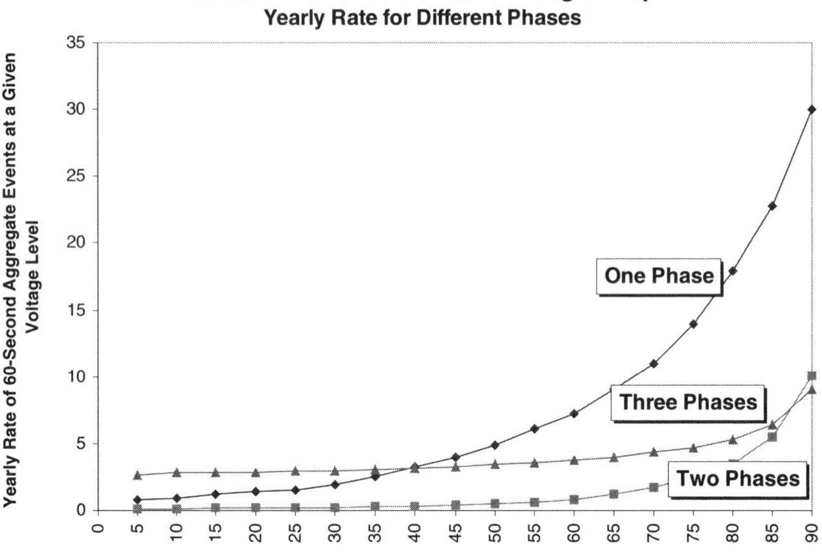

FIGURE 10.9
Rate of number of phases with a voltage drop in the EPRI DPQ study. (Copyright © 1996. Electric Power Research Institute. TR-106294-V2. *An Assessment of Distribution System Power Quality: Volume 2: Statistical Summary Report.* Reprinted with permission.

Increasing a motor's inertia is one way to increase the ride through of the motor, which also increases the support to other loads in the facility.

Following a sag, however, the response of loads — particularly motors — may further disturb the voltage. During a sag, motors slow down. After the sag, the motors draw inrush current to speed up. If motors are a large enough portion of the load, this inrush pulls the voltage down, delaying the recovery of voltage. Motors with small slip and those with large inertia draw the most inrush following a sag. These effects are more severe for customers or areas with a large percentage of motor loads and for longer fault clearing times (Bollen, 2000; IEEE Std. 493-1997).

An extreme case of motor inrush sometimes happens with air conditioners. Single-phase air conditioner compressors are prone to stall during voltage sags; during which, the compressor draws locked rotor current, about five or six times normal. Tests by Williams et al. (1992) found that voltages below 60% of nominal for five cycles stalled single-phase air conditioners. Longer-duration sags also stall compressors for less severe sags (in the range of 60 to 70% of nominal). The compressor stays stalled long after the system voltage has returned to normal. It keeps drawing current until thermal overload devices trip the unit, which can take one half of a second. On the distribution system, this extra current may trip breakers or blow fuses in addition to aggravating the voltage sag.

Adjustable-speed drives and other loads with capacitors (mainly rectifiers) also draw inrush following a voltage sag. During the sag, rectifiers stop drawing current until the dc voltage on the rectifier drops to the sagged voltage. After the sag, the rectifier draws inrush to charge the capacitor. This spikes to several times normal, but the duration is short relative to motor inrush. The inrush may blow fuses or damage sensitive electronics in the rectifier. For severe sags, much of the rectifier-based load trips off, which reduces the inrush.

Normally, we neglect the load response for voltage sag evaluations, but occasionally, we must consider the response of the load, either for its direct impact on voltage sags, or for the impact of the inrush.

10.3.3 Analysis of Voltage Sags

The calculation of the voltage magnitude at various points on a system during a fault at a given location is easily done with any short-circuit program. We make the fairly accurate assumption that the fault impedance is zero. The engineer or computer program finds the duration of the sag using the time-current characteristics of the protective device that should operate along with the fault current through it.

Based on a short-circuit program, the *fault positions method* repeatedly applies faults at various locations and tallies the voltages at specified locations during the faults. The procedures, which may apply thousands of fault locations, result in predictions of the number of voltage sags below a given magnitude at the specified locations. This procedure is well documented in the *Gold Book* (IEEE Std. 493-1997) [see also (Conrad et al., 1991)].

The faults are applied along each line in a system. The end results are scaled by the fault rate on the line, which can be based on historical results or typical values for the voltage and construction.

We need considerable detail for the fault-positions analysis, especially a complete system model including proper zero-sequence impedances and transformer connections (these are left out of many transmission system load-flow models).

Another simpler method for voltage sags is the *method of critical distances* (Bollen, 2000). The approach is to find the farthest distance, the *critical* distance, to a fault that causes a sag of a given magnitude. Pick a sag voltage of interest, 0.7 per unit for example. Find the critical distance for the chosen voltage. Using a feeder map, add up the circuit lengths within the critical distance. Multiply the total exposed length by the fault rate — this is the number of events expected. This method is not as accurate as the fault positions method, but is much simpler: we can calculate the results by hand, and the process of doing the calculations provides insight on the portions of distribution and transmission system that can cause sags to the given customer. We can also target this *area of vulnerability* for inspection or additional maintenance or apply faster protection schemes covering those circuits (to clear faults and sags more quickly).

10.4 Characterizing Sags and Momentaries

10.4.1 Industry Standards

The most commonly cited industry standard for ride through was developed by the Information Technology Industry Council (ITI) (Figure 10.10). The ITI curve updates the CBEMA curve (Computer Business Equipment Manufacturers' Association, which became ITI) and is often referred to as the new CBEMA curve. The ITI curve is not an actual tested standard — computers do not have to be certified to pass some test. The ITI curve is used as a

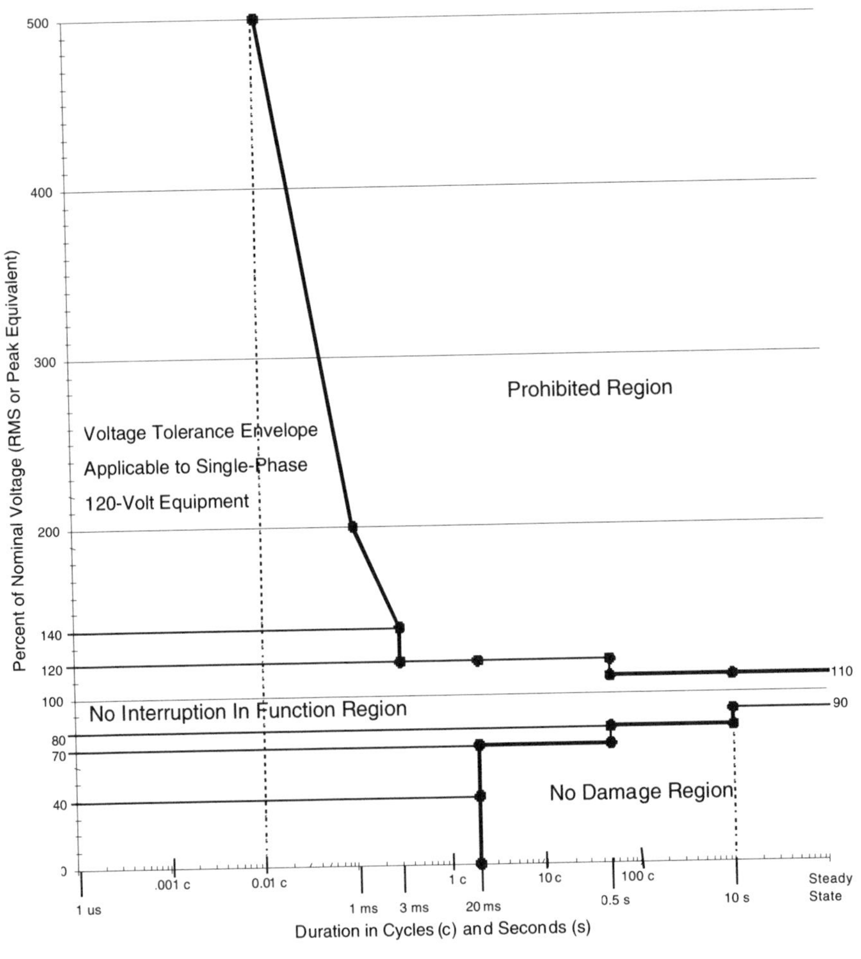

FIGURE 10.10
ITI curve that shows the *typical* voltage sensitivity of information technology equipment. (From [Information Technology Industry Council (ITI), 2000]. With permission.)

benchmark indicator for comparison of power quality between sites and to track performance over time. Because the ITI curve somewhat represents the ride through of computers, we can single out events below the ITI curve as "suspects," which may trip sensitive equipment.

Another major equipment standard has been produced by the semiconductor industry (SEMI F47-0200, 1999). The major advance of the SEMI set of standards is that there is an actual test standard for the equipment. To meet the SEMI standard, equipment must pass a series of voltage sag tests (SEMI F42-0600, 1999). The standard defines many factors, including sag generator and other test apparatus requirements, sampling specimens, test procedure, and reporting of test results. The SEMI standard is only for single-phase sags; for three-phase equipment with a neutral, six tests are done: each phase-to-neutral voltage is "sagged," and each phase-to-phase voltage is sagged in turn. For three-phase equipment without a neutral, each phase-to-phase voltage is tested with a sag generator.

The SEMI curve focuses exclusively on voltage sags. In some cases, the SEMI curve is more strict than the ITI curve, and it appears that way when the two curves are graphed together as in Figure 10.11. The SEMI curve has a deeper voltage sag characteristic. The most severe point on the SEMI curve is the 0.2-sec sag for a voltage to 50% of nominal. However, some equipment could meet the SEMI requirement but not pass the ITI curve points. The main types of equipment that fall into this category are relays and contactors. The ITI curve has a 0.02-sec interruption that is enough to disengage many relays and contactors that may survive a 0.2-sec sag to 50% of nominal voltage.

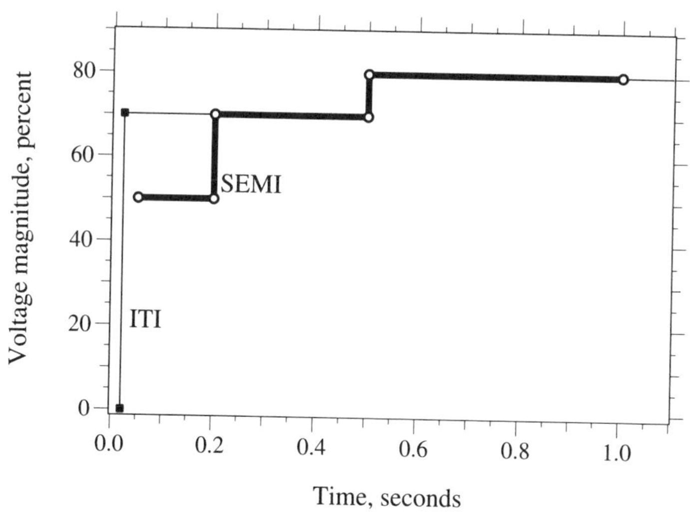

FIGURE 10.11
SEMI voltage sag ride-through requirement compared against the ITI curve. (SEMI curve from [SEMI F47-0200, 1999].)

Several power quality indices have been introduced that are similar to the reliability indices (EPRI TP-113781, 1999). Utilities can use these for some of the same purposes as reliability indices: targeting areas for maintenance and circuit upgrades, tracking the performance of regions, and documenting performance to regulators. The most widely used index is SARFI (EPRI TP-113781, 1999; Sabin et al., 1999) defined as

> $SARFI_X$, System Average RMS (Variation) Frequency Index: $SARFI_X$ represents the average number of specified rms variation measurement events that occurred over the assessment period per customer served, where the specified disturbances are those with a magnitude less than X for sags or a magnitude greater than X for swells.

$$SARFI_X = \frac{\sum N_i}{N_T}$$

where

- X = rms voltage threshold; possible values — 140, 120, 110, 90, 80, 70, 50, and 10
- N_i = number of customers experiencing short-duration voltage deviations with magnitudes above X% for X > 100 or below X% for X < 100 due to measurement event i
- N_T = number of customers served from the section of the system to be assessed

The breakpoints were not chosen arbitrarily. The 90, 80, and 70% thresholds are boundaries of the ITI curve, the 50% threshold is a typical breakpoint for motor contactors, and 10% is the dividing line between a sag and an interruption. Two special variations of SARFI have also been defined. $SARFI_{ITIC}$ is the number of events below the lower ITI curve. In similar fashion, $SARFI_{SEMI}$ is the number of events below the SEMI curve. SARFI can be applied for one monitor (and one customer) or for several monitored locations. It is difficult to extend this concept to make SARFI a system-wide performance indicator like SAIFI — what is straightforward for reliability indices becomes much more complicated for sags because a fault causes different voltages at different locations on the distribution system. It is difficult to find a system-wide average without a vast number of monitors. Approximations must be used to estimate the effects at different customers based on a small number of monitored points.

10.4.2 Characterization Details

Several disturbances often occur within a short time of each other. Commonly, a breaker or recloser goes through several reclosing attempts. The customer sees a sequence of voltage sags. If one of these events causes an end-use disruption, from their point of view, it does not matter if additional events follow within the next few minutes, as the customer is already disturbed. To account for this, we can aggregate events within a rolling time window. Commonly, time windows are 1 and 5 min for calculating $SARFI_X$ or other power quality benchmarks.

Since voltage sags can have different impacts on each phase, how do we account for the differences between a three-phase sag and a single-phase sag? We can tabulate sags in two different ways:

- *Per phase* — Each phase is tracked independently. A three-phase sag counts three times that of a single-phase sag. Single-phase recorders automatically calculate the number of sags per phase.
- *Minimum phase* — A sag event is recorded as the lowest of the three phase voltages. A three-phase sag counts the same as a single-phase sag. $SARFI_X$ uses this approach.

Both approaches are useful depending on the customer and load characteristics. The per-phase method is better for single-phase customers and for customers with three-phase load that is more sensitive to multiple-phase sags. The minimum-phase method is better for facilities where sags on any of the three phases could trip a process. At a three-phase location, the minimum-phase method gives higher numbers of voltage sags.

The line-to-ground and line-to-line voltages may be significantly different during a voltage sag. Ideally, we want to record and benchmark what the critical load sees, but sometimes that is unknown (and, some facilities may have critical loads connected line to line and line to ground). Normally, SARFI is tracked based on however the recorders are connected.

Most voltage sags have a simple shape — the voltage drops in magnitude and stays at a constant value until the fault clears. After that, the voltage returns to its pre-sag value. The rms change is approximately a rectangular wave. The rectangular shape makes classification easy — only a magnitude and a duration are needed. Sometimes, sags do not follow the rectangular shape. If the fault current is not constant, the voltage will not be constant. If the fault evolves from a single line-to-ground fault into a multiple-phase fault, the voltage will change. These types of events are hard to classify, but most of the time, we can ignore them for the purposes of monitoring and collecting statistics at a site. For analysis of specific events that disrupted equipment, review of the rms shape may provide additional meaning beyond just having a magnitude and a duration.

10.5 Occurrences of Voltage Sags

Several power quality monitoring studies have characterized the frequency of voltage sags. The two most widely quoted studies are EPRI's Distribu-

TABLE 10.4

Average Annual Number of Voltage Sags below the Given Magnitude for Longer than the Given Duration from the NPL Data with a Five-Minute Filter

	Duration							
Magnitude	**1 Cycle**	**6 Cycles**	**10 Cycles**	**20 Cycles**	**0.5 sec**	**1 sec**	**2 sec**	**10 sec**
87%	126.4	56.8	36.4	27.0	23.0	18.1	14.5	5.2
80%	44.8	23.7	17.0	13.9	12.2	10.0	8.0	4.3
70%	23.1	17.3	14.5	12.8	11.5	9.7	7.9	4.3
50%	15.9	14.1	12.9	11.8	10.6	9.4	7.8	4.3
10%	12.2	12.0	11.7	11.0	10.2	9.0	7.5	4.2

Source: Dorr, D. S., Hughes, M. B., Gruzs, T. M., Juewicz, R. E., and McClaine, J. L., "Interpreting Recent Power Quality Surveys to Define the Electrical Environment," *IEEE Transactions on Industry Applications*, vol. 33, no. 6, pp. 1480–7, November 1997.

tion Power Quality (DPQ) study and the National Power Laboratory's end-use study.

NPL's end-use study recorded power quality at the point of use at residential, commercial, and industrial customers. At 130 sites within the continental U.S. and Canada, single-phase line-to-neutral monitors were connected at standard wall receptacles (Dorr, 1995). The survey resulted in a total of 1200-monitor months of data. Table 10.4 shows the average number of voltage sags that dropped below the given magnitude for longer than the given duration.

EPRI's Distribution Power Quality (DPQ) project recorded power quality in distribution substations and on distribution feeders, measured on the primary at voltages from 4.16 to 34.5 kV (EPRI TR-106294-V2, 1996; EPRI TR-106294-V3, 1996). It was seen that 277 sites resulted in 5691 monitor-months of data. In most cases three monitors were installed for each randomly selected feeder, one at the substation and two at randomly selected places along the feeder. Table 10.5 shows average numbers of voltage sags for a given magnitude and duration for the DPQ data.

TABLE 10.5

Average Annual Number of Voltage Sags Below the Given Magnitude for Longer than the Given Duration from the EPRI Feeder Data with a Five-Minute Filter

	Duration							
Magnitude	**1 Cycle**	**6 Cycles**	**10 Cycles**	**20 Cycles**	**0.5 sec**	**1 sec**	**2 sec**	**10 sec**
90%	77.7	31.2	19.7	13.5	10.7	7.4	5.4	1.8
80%	36.3	17.4	12.4	9.3	7.9	6.4	4.9	1.7
70%	23.9	13.1	10.3	8.3	7.2	6.2	4.8	1.7
50%	14.6	9.5	8.4	7.5	6.6	5.9	4.6	1.7
10%	8.1	6.5	6.4	6.2	5.6	5.1	4.0	1.7

Source: Dorr, D. S., Hughes, M. B., Gruzs, T. M., Juewicz, R. E., and McClaine, J. L., "Interpreting Recent Power Quality Surveys to Define the Electrical Environment," *IEEE Transactions on Industry Applications*, vol. 33, no. 6, pp. 1480–7, November 1997.

TABLE 10.6

Annual Number of Power Quality Events (Upper Quartile, Median, and Lower Quartile) for the EPRI DPQ Feeder Sites with a One-Minute Filter

	Duration, seconds						
Voltage	0	0.02	0.05	0.1	0.2	0.5	1
0.9	32.8 57.5 104.8	30.8 49.0 95.1	24.4 35.3 65.6	13.6 22.7 38.7	7.6 13.2 24.0	3.3 7.3 14.2	1.4 3.2 8.9
0.8	16.4 31.6 54.1	14.8 26.0 50.1	12.1 20.9 37.9	8.1 15.0 25.1	4.9 9.6 16.9	2.4 5.3 11.0	0.9 2.7 7.5
0.7	10.1 20.5 33.8	8.6 18.8 32.7	8.1 15.3 27.6	5.8 11.3 18.8	4.0 7.8 13.5	1.8 4.5 9.3	0.9 2.5 7.0
0.5	4.7 9.7 19.2	4.5 9.0 17.4	4.2 7.7 14.3	3.5 5.9 11.2	2.3 5.0 9.6	1.4 3.3 7.7	0.8 2.2 5.7
0.3	2.1 4.8 12.8	1.8 4.5 11.0	1.6 4.2 9.5	1.4 3.6 8.6	1.1 3.5 8.3	0.8 2.8 6.6	0.5 1.6 5.1
0.1	0.9 3.2 8.3	0.9 2.9 7.8	0.8 2.8 7.8	0.8 2.7 7.8	0.7 2.7 7.8	0.5 2.2 6.1	0.3 1.6 4.9

Note: A *B* C represent the lower quartile *A*, the median *B*, and the upper quartile *C* of the total number of events below the given magnitude and longer than the given duration (up to 1 min).

As expected, the number of voltage sags is higher for the end-use NPL study than for the primary-level DPQ study. At the point of use, the nominal voltage is lower, which picks up more voltage sags, especially minor sags. End-use monitoring also picks up events caused internally, mainly voltage sags.

Table 10.6 shows cumulative numbers of voltage sags measured at sites during the DPQ study. Table 10.4 and Table 10.5 presented results based on averages — Table 10.6 shows the data based on the median, upper, and lower quartiles. One use of it is to estimate the number of times a year disturbances will affect a device — for example, if a device is sensitive to any event below a voltage of 50% of nominal for longer than 0.1 sec, then Table 10.6 predicts that at half of the sites in the U.S. distribution system, the device misoperates more than 5.9 times per year.

As an indicator, the average misrepresents the typical site power quality. The median represents site data better; here, by definition, 50% of sites have values higher than the median, and 50% have values lower. With balanced distributions such as the normal distribution, the average equals the median. In a skewed distribution, the average is higher than the median. Additionally, poor sites and anomalies such as a severe storm skew the average upward. In the DPQ data, the average is 31 to 115% higher than the median depending on the quality indicator as shown in Table 10.7.

10.5.1 Site Power Quality Variations

EPRI's Distribution Power Quality (DPQ) project allows us the opportunity to explore how power quality varies at different sites. Completed in 1995, the DPQ project collected data from 24 utility systems at a total of 277 locations on 100 distribution system feeders over a 27-month period. Site and circuit descriptors help us analyze the causes for site variations. Some

TABLE 10.7

Ratio of Median and Average for DPQ Site Statistics at Feeder Sites

	Median	Average	Ratio of Average to Median
$SARFI_{ITIC}$	21.27	27.86	131%
$SARFI_{SEMI}$	15.28	18.92	124%
$SARFI_{10}$	2.52	5.42	215%

notable details about the DPQ measurements and our analysis (Short et al., 2002):

- All measurements were on the distribution primary. Of course, most customers connect to the distribution secondary. Normally, this means that a customer's equipment sees more events below a given threshold. Also note that for three-phase customers, a delta – wye transformer distorts the secondary voltages relative to the primary voltages.
- All data was measured at three-phase points on the distribution circuit (single-phase locations were not monitored).
- We present all data based on the worst of the three phases, which is conservative because most faults are single phase. Single-phase customers see fewer sags. In addition, some three-phase equipment is less sensitive to single-phase sags than to three-phase sags.
- Most of the measurements are from phase to ground (the monitors on the ungrounded circuits show phase-to-phase measurements).
- We only used sites with at least 200 days of monitoring.

Power quality varies widely by site. Figure 10.12 shows cumulative distributions of different power quality indices along with statistics and a fit to a log-normal distribution. The left column (SARFI 70, 50, and 10) gives the average annual number of voltage sags below 70, 50, and 10%, which are most applicable for relays, contactors, and other devices that drop out quickly. $SARFI_X$ considers only short-duration rms events, defined as 1/2 cycle to one minute (IEEE Std. 1159-1995). The right column of Figure 10.12 shows data similar to the left column but for criteria that disregards very short events. The ITI curve (Information Technology Industry Council, 2000) disregards sags less than 0.02 sec, and the SEMI curve (SEMI F47-0200, 1999) disregards sags less than 0.05 sec. The indices that exclude short events are more appropriate for computer power supplies and other devices that ride through short-duration events. $SARFI_{10\ (>0.4sec)}$ is for momentary interruptions greater than 0.4 sec, which differentiates between deep sags and total loss of voltage due to operation of a breaker or recloser.

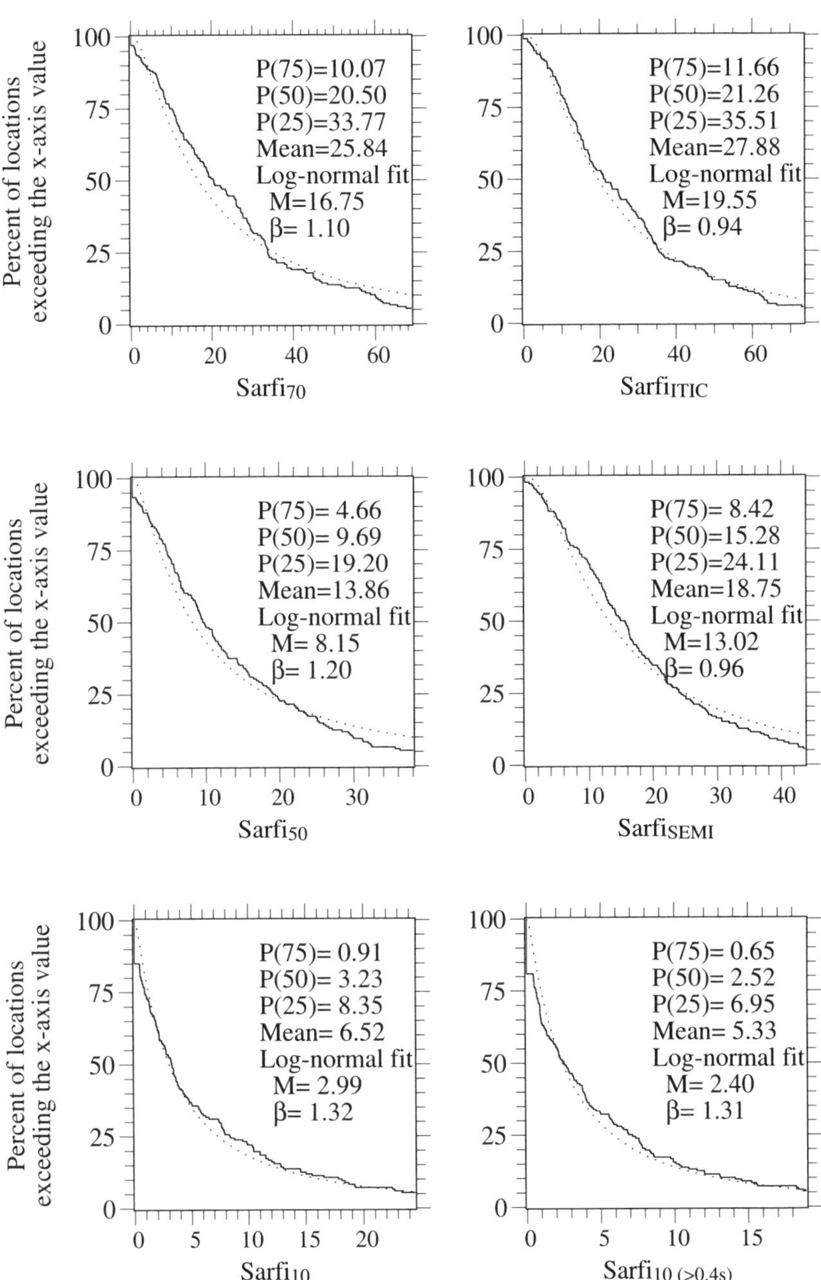

FIGURE 10.12
Cumulative distributions of DPQ feeder data along with statistics for various indices. SARFI 70, 50, and 10 gives the number of voltage sags below 70, 50, and 10%. $SARFI_{ITIC}$ and $SARFI_{SEMI}$ are events below the ITI curve and the SEMI curve, respectively. The dotted line fits a log-normal distribution.

TABLE 10.8

Statistics for Power Quality from the SEMI Monitoring Study, Which Are Primarily Transmission Service

	Average	Median		
		P(75%)	P(50%)	P(25%)
$SARFI_{ITIC}$	4.60	2.05	3.80	5.10
$SARFI_{SEMI}$	2.05	0.00	1.90	3.64
$SARFI_{70}$	4.40	2.05	3.50	5.10
$SARFI_{50}$	0.97	0.00	0.69	1.20
$SARFI_{10}$	0.24	0.00	0.00	0.23

Source: Stephens, M., Johnson, D., Soward, J., and Ammenheuser, J., *Guide for the Design of Semiconductor Equipment to Meet Voltage Sag Immunity Standards*, International SEMATECH, 1999. Technology Transfer #99063760B-TR, available at http://www.sematech.org/public/docubase/document/3760btr.pdf.

The site data is not normally distributed. The site indices are nonnegative, and the distribution skews upward; therefore, we need another distribution, the log-normal, the Gamma, or the Weibull. Figure 10.12 includes fits to log-normal distributions. The median (M) of the log-normal distribution equals the mean of the natural log of the values (x_i) raised to e: $M = e^{\mathrm{mean}(\ln(x_i))}$. The log standard deviation is $\beta = \mathrm{sd}[\ln(x_i)]$.

10.5.2 Transmission-Level Power Quality

Large industrial customers, utility's prize customers, are primarily fed with transmission-level service and expect high-quality power. Several semiconductor manufacturing sites provided a basis for developing the SEMI F47 standard for semiconductor tools (Stephens et al., 1999). These sites were primarily served from transmission lines; not all were direct transmission services, but distribution exposure was minimal. While not as extensive as the DPQ study, the monitoring provides good data on the number of events that are primarily from the transmission exposure. Table 10.8 shows summary statistics from the SEMI dataset of 16 sites with 30 total monitor-years of data. Figure 10.13 compares distributions of SEMI data with the DPQ substation data. As expected, the semiconductor manufacturing sites experience fewer events compared to the typical DPQ site. This comparison provides some guidance on the portion of distribution events that are caused on the transmission system. Use caution though since these are two independent data sets.

10.6 Correlations of Sags and Momentaries

Figure 10.14 shows the number of momentary interruptions at a site plotted against the number of voltage sags. We see that sites with high numbers of

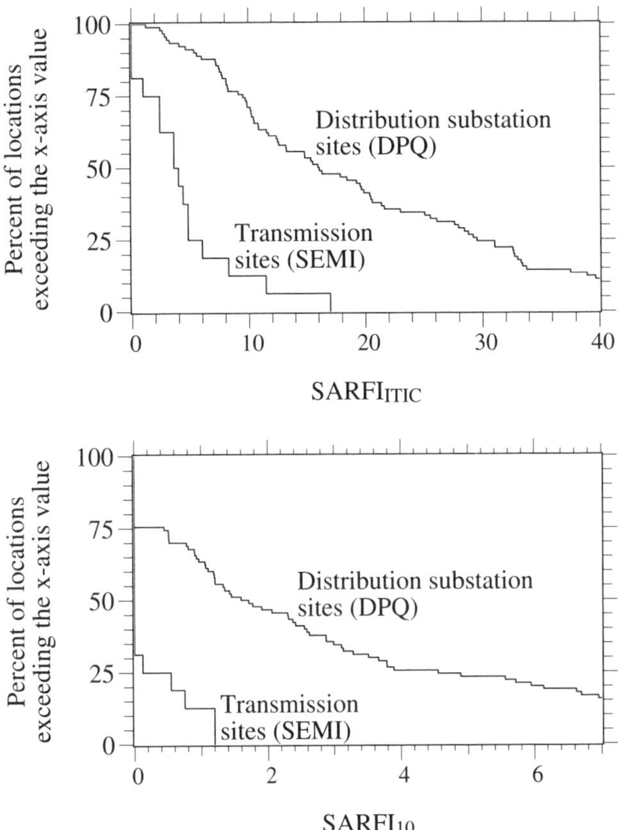

FIGURE 10.13
Comparison of the 16 SEMI sites with the DPQ substation sites.

momentary interruptions probably also have high numbers of voltage sags. Sites with low numbers of momentary interruptions may have high or low numbers of voltage sags. The correlation coefficient between sags and momentaries for the DPQ sites is 44.8%.

Correlations between deep voltage sags and shallow sags are more pronounced. $SARFI_{90}$ and $SARFI_{50}$ have a 56.9% correlation coefficient. If we break the sites down by load density, the correlation coefficients improve to 90, 84, and 74% for urban, suburban, and rural sites.

10.7 Factors That Influence Sag and Momentary Rates

Power system faults cause voltage sags and momentary interruptions. The frequency of faults depends on many factors including weather, maintenance, and age of equipment. The protection schemes and location of circuit

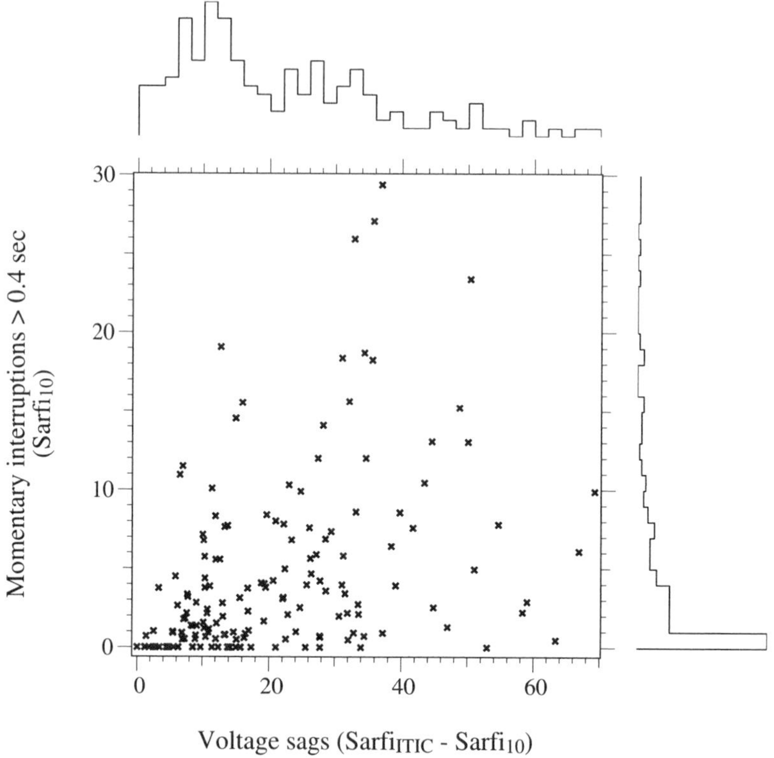

FIGURE 10.14
Relationship between voltage sags and momentary interruptions (greater than 0.4 sec). Each point gives the average voltage sags and momentary interruptions at a site (n = 158).

interrupters determine whether a fault causes a voltage sag or an interruption, and the protection system determines the event duration. The following sections describe work using EPRI's DPQ data to investigate what factors influence sags and momentaries (Short et al., 2002).

10.7.1 Location

Three monitors were used on each circuit in the DPQ study. One was always at the substation, and two were on the feeder, named "feeder middle" and "feeder end." The feeder sites were randomly picked on the circuits, so the naming is somewhat misleading; "feeder end" does not mean the most distant point from the substation (it just means the most distant of the two monitors randomly placed on the circuit). Since one third of the monitors are at the substation, the set is biased to "near-substation" customers since most customers are not located near the substation. Although there is some difference between measurement locations, it turns out that it is not drastic.

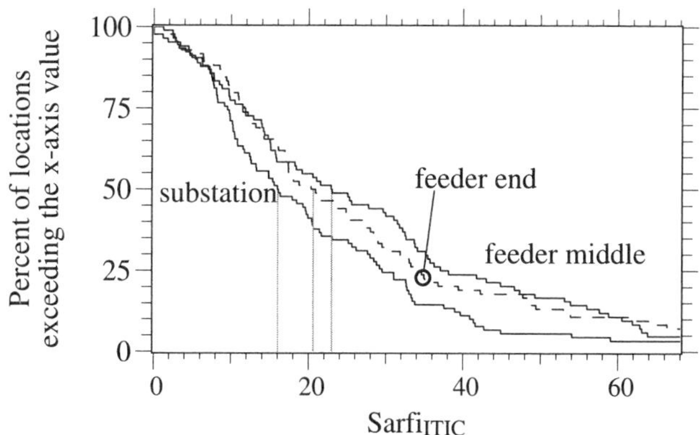

FIGURE 10.15
Comparison of feeder sites and substation sites in the DPQ data for $SARFI_{ITIC}$.

There is surprisingly little difference between the distributions of monitoring locations (see Figure 10.15 for $SARFI_{ITIC}$).

Figure 10.16 shows a more specific comparison of the substation's performance plotted against its two feeder sites. As expected, most feeder sites have more sags than their substation site, especially rural sites. A significant number of feeder sites were better than the substation. Measurement anomalies could produce this (the substation recorder is down for part of a bad storm season), or it could be real (downstream regulation devices keep the nominal voltage higher or the connected load "pushes" back on the source

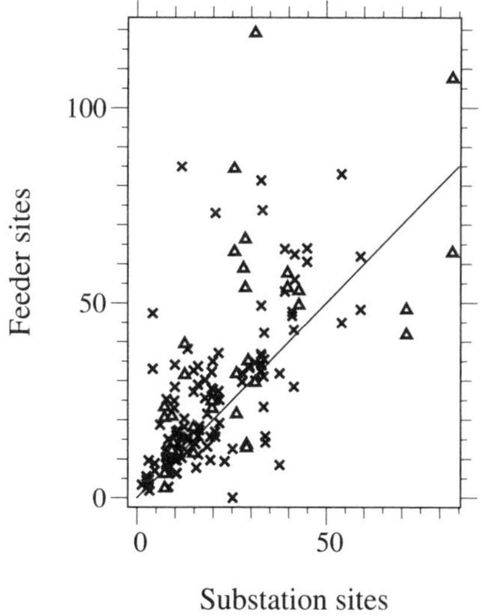

FIGURE 10.16
$SARFI_{ITIC}$ at substation sites plotted against $SARFI_{ITIC}$ at that substation's feeder sites (triangles indicate rural sites).

TABLE 10.9

Statistics for Momentary Interruptions Longer than 0.4 sec

	P(75%)	Median P(50%)	P(25%)
Rural	2.37	8.56	18.31
Suburban	0.23	2.39	6.71
Urban	0.00	1.37	2.82

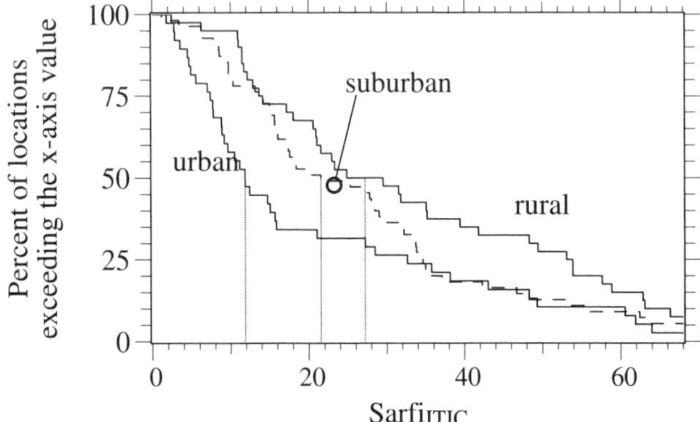

FIGURE 10.17
Comparison of urban, suburban, and rural sites for $SARFI_{ITIC}$ (feeder sites only).

impedance during bus faults). For most of our analysis, we excluded substation sites, thinking that the feeder sites better represent a random feeder location where customers are fed.

10.7.2 Load Density

Rural sites have more voltage sags and momentary interruptions (see Table 10.9 and Figure 10.17). This is not surprising given the extra lengths of line needed to serve load in low-density areas. Interruptions showed the most dramatic difference.

Why do urban sites not have even more profoundly lower voltage sag rates than suburban and rural sites? After all, urban sites are shorter and mostly underground with fewer faults per mile. The main answer is that urban sites have many more feeders off of a bus. In addition, even though urban circuits are shorter, most of the exposure is close to the substation. So, while many of the faults on rural and suburban circuits are too far away to pull down the substation voltage, almost every fault on an urban circuit causes a significant voltage sag for all customers off of that substation bus.

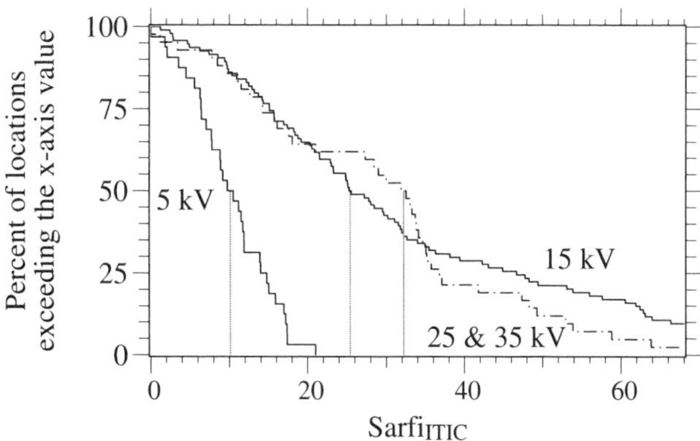

FIGURE 10.18
Comparison of feeder sites by voltage class in the DPQ data.

10.7.3 Voltage Class

Figure 10.18 shows that 5-kV systems have much lower numbers of voltage sags and interruptions. Lower voltage systems have less feeder exposure, and higher line impedance relative to the station transformer. Fault rates are often lower on 5-kV systems. Somewhat surprisingly, the 25 and 35-kV systems were not worse than the 15-kV systems.

10.7.4 Comparison and Ranking of Factors

We analyzed data available on the DPQ site characteristics to determine what parameters most affected power quality events. Figure 10.19 shows the variations of $SARFI_{ITIC}$ with site characteristics.

The three most significant predictors of excursions below the lower ITI curve are:

1. *Circuit exposure* — The total exposure on the circuit including three-phase and single-phase portions is a good predictor of voltage sags. Any fault on the circuit sags the voltage.
2. *Lightning* — Lightning causes many faults on distribution systems, and lightning strongly correlates with voltage sags (based on the 10-year average, 1988–98, from the U.S. National Lightning Detection Network). In addition, lightning predicts weather patterns — areas with high lightning tend to have more storms and more wind and tree-related faults.
3. *Transformer impedance and number of feeders* — The $n_f \cdot kV^2/MVA_{xfrmr}$ term in Figure 10.19 contains the number of feeders off of the transformer bus along with an estimate of the transformer impedance. The transformer impedance is $Z_{\%}kV^2/MVA$; but since the per-unit

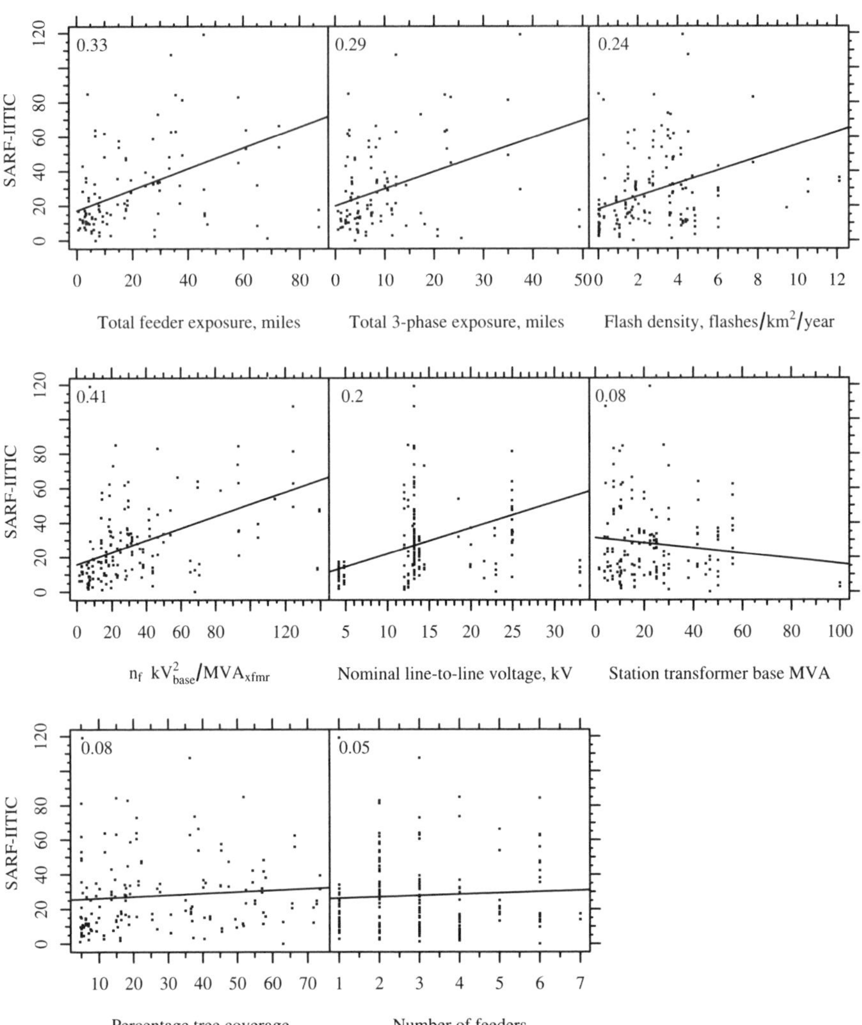

FIGURE 10.19
Variations in the number of excursions below the lower ITI curve (which are mainly voltage sags) vs. various site parameters. The correlation coefficients (r) are given in the upper-left corner of each plot.

impedance of station transformers is roughly constant (7 to 10%), we use kV^2/MVA.

This last term requires a bit more explanation. The number of bus sags is directly proportional to n_f, the number of feeders off the bus and to Z_S, the source impedance (a lower station transformer impedance — a bigger transformer or lower percent impedance — improves voltage sags at the station bus). We approximate these two terms as $n_f \cdot kV^2/MVA_{xfrmr}$.

Other variables have much less impact on the number of voltage sags than the three main parameters given.

10.8 Prediction of Quality Indicators Based on Site Characteristics

We derive a formula for predicting the number of events for a quality indicator based on a few of the characteristics of the site. If no measurement or historical data is available, this is useful in estimating the utility-side quality.

Regression techniques are commonly used to find a model prediction formula. A generalized linear model is a least-squares fit to an equation of the following form:

$$y = a_1 x_1 + a_2 x_2 + \cdots + a_n x_n + \varepsilon$$

The x's are site characteristics (such as base voltage or lightning flash density), and the a's are coefficients fitted to the model. The generalized linear model is somewhat different from a standard linear model; we used a generalization where the distribution of the error ε is assumed to be a gamma distribution rather than a normal distribution in a strictly linear model. A gamma distribution skews to the right, like the log-normal distribution.

A model for estimating $\text{SARFI}_{\text{ITIC}}$ is

$$N_{ITIC} = 4.74 + 0.76l + 2.47N_g + 0.192\frac{n_f \cdot kV^2}{MVA_{xfmr}}$$

$$+ 8.2 \text{ if moderate to heavy tree coverage}$$

where

N_{ITIC} = predicted annual number of events which fall under the lower ITI curve

l = total exposure (including three-phase and single-phase portions) on the circuit, mi (multiply kilometers by 1.609)

N_g = lightning ground flash density, flashes/km^2/year

kV = base line-to-line voltage, kV

n_f = total number of feeders off of the substation bus

MVA_{xfmr} = station transformer base rating (open-air rating), MVA

If any of the circuit characteristics are unknown, we could use the following medians from the DPQ data:

where

$$l = 14.5 \text{ mi } (23.4 \text{ km})$$

$$N_g = 2.57 \text{ flashes/km}^2\text{/year}$$

$$\frac{n_f \cdot kV^2}{MVA_{xfmr}} = 25$$

All three variable terms in the linear regression are significant to at least 99% (there is less than a 1% chance that the terms of the model do not influence the prediction). The tree coverage term is less certain; there is a 9% chance that the term is not significant. We based the tree coverage term on the University of Maryland's Global Land Cover Facility data from the Advanced Very High Resolution Radiometer (AVHRR). Half of the DPQ sites had more than 19% of the land area covered by trees, which we defined as "moderate to heavy tree cover."

How good is the model? It is decent given all the factors that affect sags and momentary interruptions and inherent variability. Given the variability of power quality events, it is surprising that the model is this good: 34% of the values are within 25% of the prediction, and 60% of the values are within 50% of the prediction. See Figure 10.20 for the prediction scatter.

For an example 12.47-kV case with three feeders, a 25-MVA transformer, a flash density of 4 flashes/km²/year, moderate tree coverage, and a total exposure of 32 km, the model predicts 29.8 events per year. For this case, the data shows a prediction interval with a 50% confidence level of between 15.6 and 34.2 events per year (the 90% confidence prediction interval is between 0 and 68.3). The data is so dispersed that the model is not good enough to use for precision estimates (such as in a contract for premium power).

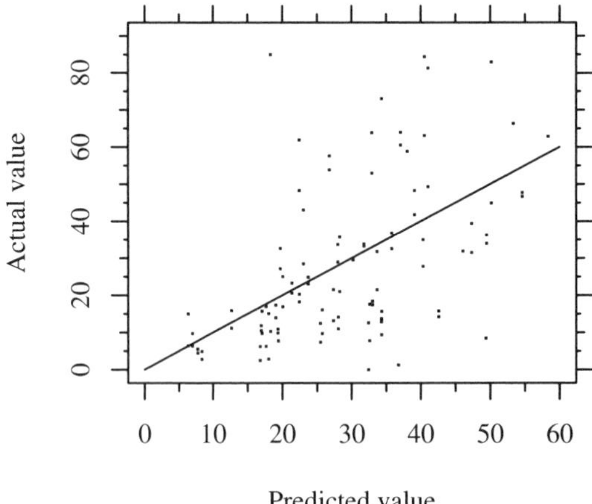

FIGURE 10.20
Actual values vs. predicted values for the model predicting the annual average number of events below the lower ITI curve.

The site characteristics most affecting sags but not included in this model (because no information was available) are (1) subtransmission exposure and characteristics and (2) percentage of the circuit that was underground.

A reasonable model for predicting momentary interruptions is

$$N_{10} = \begin{pmatrix} 5.52 \text{ if Rural} \\ 0.29 \text{ if Suburban} \\ -1.61 \text{ if Urban} \end{pmatrix} + 0.30 l_3 + 0.27 N_g + 1.24 \frac{n_f \cdot kV}{MVA_{xfmr}}$$

where

N_{10} = the predicted annual number of events with voltage less than 10% of nominal for more than 0.4 sec

l_3 = the three-phase circuit exposure, mi

The parameters differ somewhat from $\text{SARFI}_{\text{ITIC}}$ predictors. Two of the strongest indicators of momentary interruptions are load density and three-phase circuit exposure. Other significant parameters are the lightning activity and a term with voltage, number of feeders, and transformer MVA. The model is not as good as the ITI model, but all parameters have more than a 95% probability of affecting the result. The site characteristic most affecting momentaries that is not included in the model for lack of information is whether fuse saving is used.

10.9 Equipment Sensitivities

10.9.1 Computers and Electronic Power Supplies

Computers and other equipment with electronic power supplies are the most widely found equipment that is sensitive to power quality disturbances. The power supply is typically a switched mode power supply as shown in Figure 10.21. Computers have a wide range of sensitivities. The ride-through capability for interruptions of several computers is summarized in Figure 10.22. Many of the computers had ride through of more than 0.1 sec (0.28 sec was the best of this set of studies), and some could not even ride through a 0.01-sec interruption.

The ride-through capability of computers is close to rectangular. Two points describe the characteristic on a volt-time curve: the interruption ride through time and the steady-state ride-through point. There is usually a steep transition between the interruption ride-through point and the steady-state ride-through point. Other characteristics of computer ride through are:

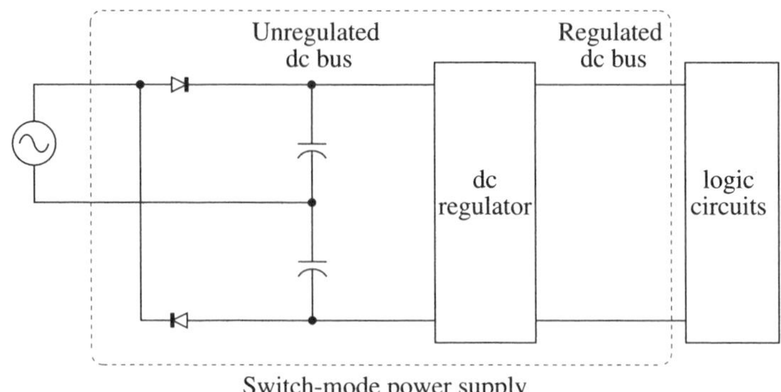

FIGURE 10.21
Switch-mode power supply used in most computers.

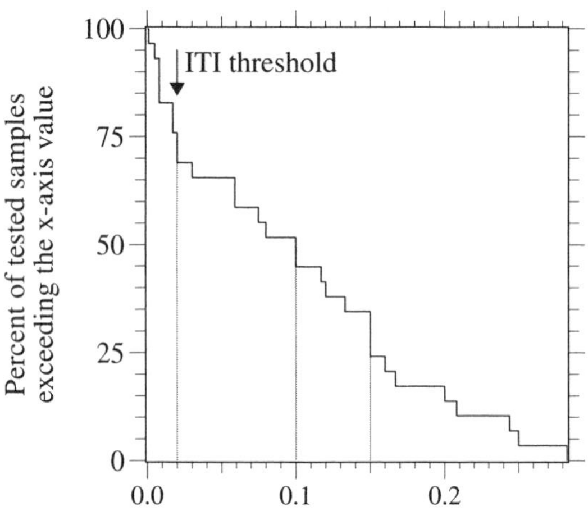

FIGURE 10.22
Capability of computers to ride through an interruption (n = 27). (Data from [Bowes, 1990; Chong, 2000; Courtois, 2001; Courtois and Deslauriers, 1997; EPRI PEAC Brief No. 7, 1992].)

- There is little difference between the performance when the computer is processing or accessing disk and when the computer is idle.
- The point on the waveform when the disruption occurs does not matter.

Figure 10.23 shows volt-time sensitivities of computers from several studies. The most sensitive units, those that violate the ITI curve, were made and tested before the ITI curve was created. That is not much of an excuse as the most sensitive computers also violated the CBEMA curve which was available to the manufacturers at that time (remember, there is no standard that requires testing computers to meet the ITI/CBEMA curve).

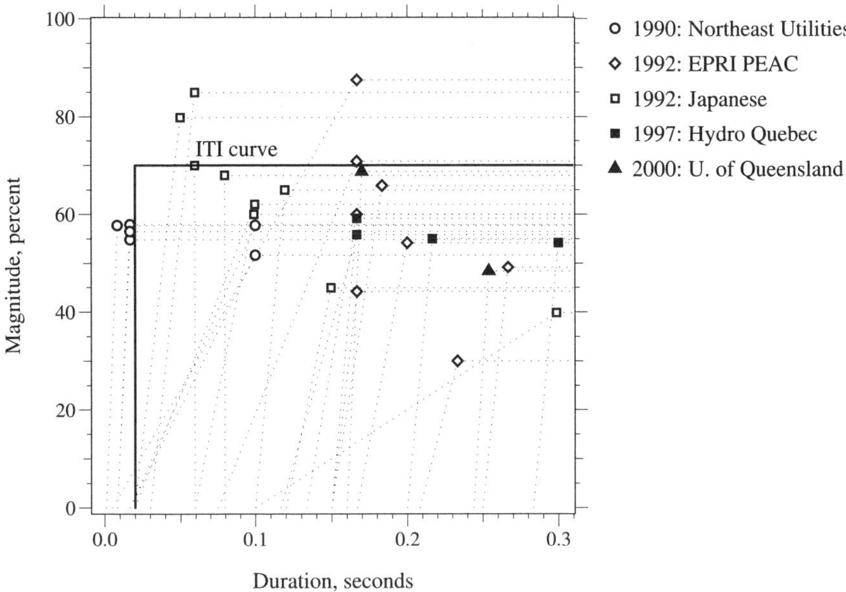

FIGURE 10.23
Volt-time characteristics of several computers tested in different studies. (Data from [Bowes, 1990; Chong, 2000; Courtois, 2001; Courtois and Deslauriers, 1997; EPRI PEAC Brief No. 7, 1992; Sekine et al., 1992]. Figure copyright © 2002. Electric Power Research Institute. 1007281. *Analysis of Extremely Reliable Power Delivery Systems.* Reprinted with permission.)

An important factor regarding the ride-through capability of computers is that it varies significantly depending on the voltage just before the interruption. The energy storage in a switch-mode power supply is from the front-end rectifier capacitors. The energy stored in a capacitor is $1/2CV^2$. Power supplies typically have two 470-μF capacitors in series, and the voltage across the two capacitors in series is $V_p = 2\sqrt{2} \cdot 120 = 339.4$ V. We can estimate the ride-through capability of a computer as:

$$t = \frac{C(V_p^2 + V_d^2)}{4 \times 10^6 P}$$

where

t = ride-through duration for an interruption, sec
P = load on the computer, W
C = capacitance on one half of the bridge rectifier, μF (470 is common)
V_p = peak of the ac voltage, V (339.4 V for 120 V nominal)
V_d = voltage on the unregulated dc bus where the computer will drop out (use half of V_p if unknown or 0 for the maximum ride through)

Since the energy is a function of V^2, a voltage of 90% of nominal means the capacitor stores only 81% of the energy that it would at nominal voltage.

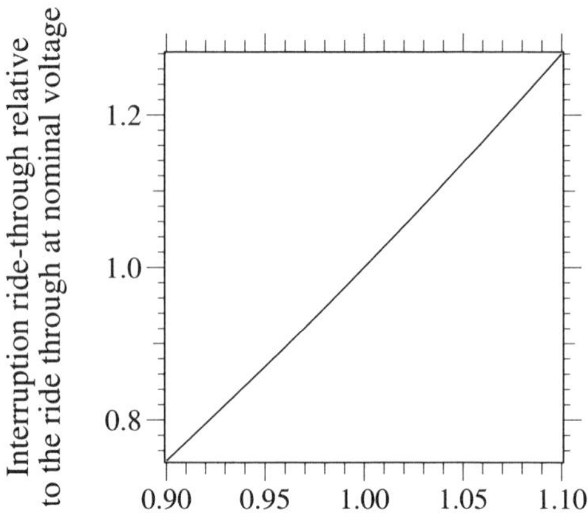

FIGURE 10.24
Change in the ride-through capability of computers vs. the voltage prior to the interruption.

Even worse, the computer drops out before all of the energy in the capacitor is used. Figure 10.24 shows the relative ride through as a function of the voltage prior to the interruption, assuming the computer drops out when the unregulated dc bus voltage reaches half of nominal.

The pre-disturbance voltage affects ride through for any device that has capacitance for energy storage, including most computer power supplies, programmable-logic controllers, digital clocks, and adjustable-speed drives. So, either on the utility side or the customer's side, raise voltages to inexpensively increase ride through of devices.

Even more ride-through capability is possible with computers. EPRI PEAC has done tests of a computer power supply modified with extra ride-through capability (EPRI PEAC Brief No. 12, 1993). The enhanced supply, developed by the New England Electric Company, had an extra 4500 μF of capacitors installed in parallel with the existing capacitors, which increased the ride through from 0.175 sec to 1.8 sec. In the near future, ultracapacitors may supply an even more economical ride-through enhancement.

Intelligent power management might also increase ride through. Laptops and most desktops have sophisticated ways of managing power to conserve energy. We could apply similar techniques to short-duration power interruptions. The processor, disk, and other power-hungry equipment could be "quick suspended" during a power interruption to increase the normal 0.05 to 0.2-sec ride-through capability. Just suspending the processor (30–50 W typically in fast, hot chips) would extend the ride through considerably. This enhancement requires very little extra hardware; a sensor to measure the incoming ac power or the unregulated dc bus voltage would be needed — no microprocessor-level changes are required.

Industrial dc power supplies share the same characteristics as the computer power supply. Heavily loaded power supplies are more susceptible to voltage sags and interruptions. Use a supply rated at twice the load on the supply to increase ride through.

A power supply with a universal input operates over a wide range of voltages (85 to 264 V typically), but the ride-through capability changes dramatically with operating voltage. Operation as close to the upper end as possible improves ride through. For this reason, prefer a line-to-line connection (208 V) over a line-to-ground connection (120 V). The low-voltage limit of 85 V is 71% of nominal at 120 V, but we obtain much better ride through when applied at 208 V (now the lower limit is 41% of nominal). The difference in $1/2CV^2$ is dramatic in the two cases. By the same token, if the power supply has alternate settings, use the setting that positions the actual voltage near the high end of the range. Consider a power supply with a 95 to 250-V range designed for Japanese and European loads and a 110 to 270-V range designed for America. The range with an upper limit of 250 V for a 208-V line-to-line connection results in the maximum ride through (McEachern, 2001).

Another option for some industrial supplies and large computer systems is a three-phase power supply instead of a single-phase supply. A three-phase supply is less sensitive to voltage sags. Single-phase sags only slightly depress the dc bus voltage of a three-phase rectifier because the remaining two phases can fully supply the load. Even a two-phase sag is significantly less severe than a three-phase sag.

Linear power supplies have much less ride-through capability than switch-mode power supplies (switch-mode supplies may have 100 times the capacitance). Fortunately, most power supplies are switch-mode supplies (primarily because they are lighter, more efficient, and cost less).

10.9.2 Industrial Processes and Equipment

A variety of industrial equipment is sensitive to voltage sags. Some of the main sensitive equipment used in industrial facilities are

- Programmable logic controllers (PLCs)
- Adjustable-speed drives (ASDs), also called variable-speed drives (VFDs)
- Contactors
- Relays
- Control equipment

Depending on the process and load, any number of devices can be the weak link. Table 10.10 shows the breakdown of weak links for semiconductor tools serving the semiconductor manufacturing industry.

TABLE 10.10

Breakdown of Semiconductor-Tool Voltage Sag Sensitivities (n = 33)

Weak Link	Overall Percentage
Emergency off (EMO) circuit: pilot relay (33%), main contactor (14%)	47%
dc power supplies: PC (7%), controller (7%), I/O (5%)	19%
3-phase power supplies: magnetron (5%), rf (5%), ion (2%)	12%
Vacuum pumps	12%
Turbo pumps	7%
ac adjustable-speed drives	2%

Source: Stephens, M., Johnson, D., Soward, J., and Ammenheuser, J., *Guide for the Design of Semiconductor Equipment to Meet Voltage Sag Immunity Standards,* International SEMATECH, 1999. Technology Transfer #99063760B-TR, available at http://www.sematech.org/public/docubase/document/3760btr.pdf.

10.9.2.1 Relays and Contactors

Contactors are electromechanical switches used for a variety of power and control applications. A contactor uses a solenoid to engage when an appropriate voltage is applied. More voltage is required to close the contactor than is required to keep it closed.

Relays and contactors can drop out very quickly. Figure 10.25 shows the ride-through duration for an interruption for several relays and contactors, and Figure 10.26 shows the dropout levels for voltage sags. The devices are somewhat dependent on the point on the wave where the voltage sag starts. Ride through is longest for sags starting at the voltage zero crossing; but unfortunately, faults tend to occur when the voltage is near its peak. The fast dropout of contactors limits some of the utility-side solution options — faster relaying, smaller fuses, or 1.5-cycle transfer switches may provide good improvement to computers but offer little help for many relays and contactors. Because they trip very quickly, voltage mainly dominates, not the duration.

The volt-time capability of relays and contactors approximates a rectangular shape. Contactors can have the unusual property that the ride-through capability improves at lower voltages. An example volt-time ride-through characteristic is shown in Figure 10.27. The reason for this property relates to the fact that current, and not voltage, holds a contactor in. A contactor contains shading rings, which are analogous to damper windings in a rotating machine. A shading ring is a shorted winding around the magnetic core. In response to a voltage transient, the shading ring produces a back emf that opposes the transient. A larger transient (deeper sag) creates more current that holds the contactor in (Collins Jr. and Bridgwood, 1997).

A larger relay generally has more ride through; a contactor usually has more ride through than a relay. Some of the most sensitive relays are small industrial relays with clear plastic cases referred to as *ice-cube* relays.

Several options are available to help hold in contactors and relays (St. Pierre, 1999):

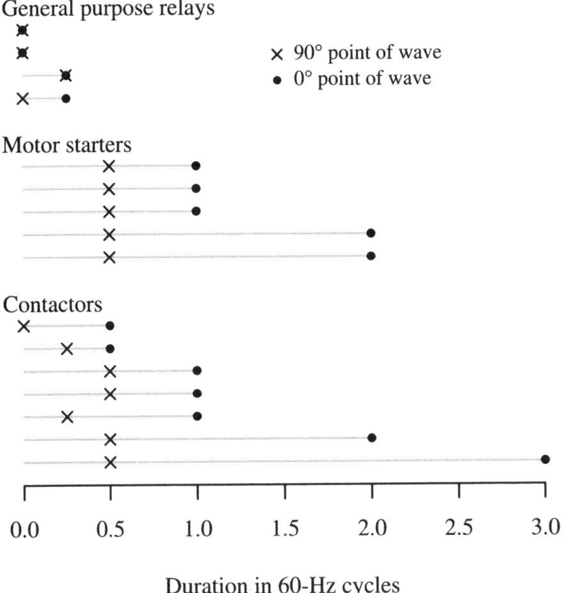

FIGURE 10.25
Ride-through duration for an interruption to several relays and contactors. (Data from [EPRI PEAC Brief No. 44, 1998].)

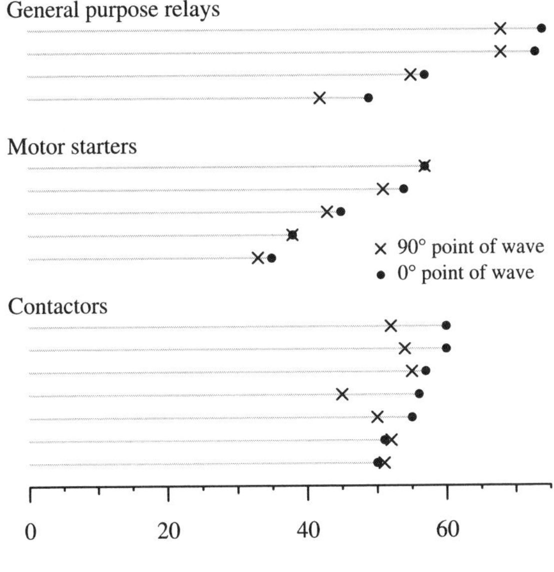

FIGURE 10.26
Voltage magnitude for dropout of several relays and contactors for a five-cycle voltage sag. (Data from [EPRI PEAC Brief No. 44, 1998].)

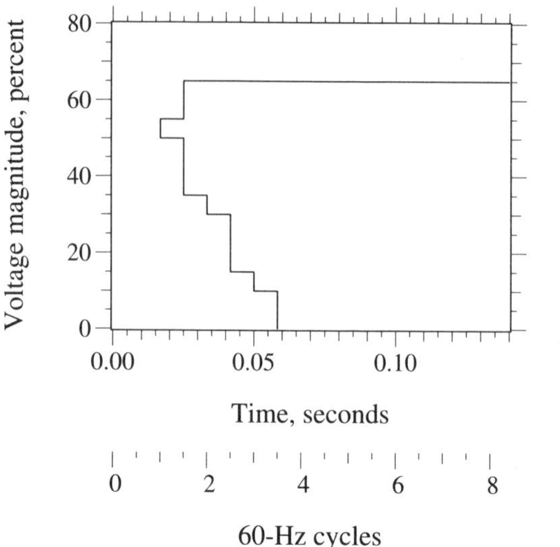

FIGURE 10.27
Ride-through capability of a NEMA size 1 contactor rated 0 to 50 hp at 480 V. (Data from [EPRI PEAC Brief No. 10, 1993].)

- *Coil hold in* — Use a coil hold-in device. Coil hold-in devices supply current to keep a relay or contactor coil held in during a voltage sag to about 25%.
- *dc* — Rectify the ac voltage and use a dc contactor. A dc contactor generally has a longer ride through than an ac contactor. A capacitor added in parallel with the contactor coil can extend the ride-through time.
- *Time-delay relay* — Add a time-delay relay to the control circuit in parallel with the contactor. If the contactor drops out because of a voltage sag, the time-delay relay keeps the control circuit energized (but the motor still drops out). If the sag finishes before the time-delay setting has elapsed, the contactor pulls back in, and the motor reconnects. This is not the best solution because the motor is disconnected and reconnected. The reconnection draws inrush current, which can itself cause local disruptions, especially if multiple motors are energized together. The high current may trip facility relays and possibly damage motors. Additionally, the disconnection of many motors within a facility removes the voltage support provided by the motors feeding back into the utility system. Furthermore, when the motors reconnect, the inrush creates a voltage sag.
- *Power conditioner* — Apply a constant voltage transformer or other power conditioning device such as a dip-proofing inverter to the control circuit. This provides enough ride through for all but the deepest voltage sags.

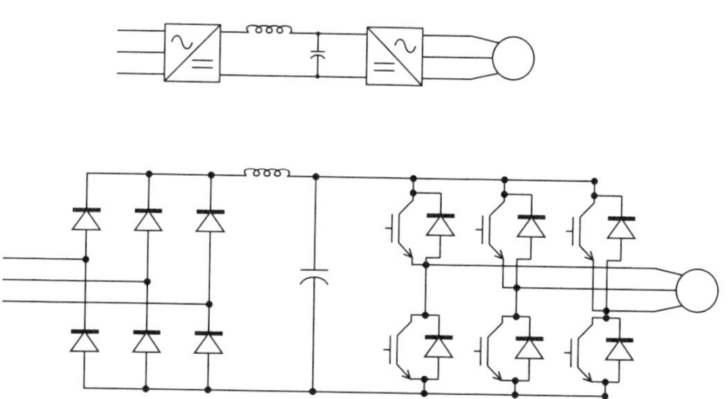

FIGURE 10.28
Common adjustable-speed drive topology.

10.9.2.2 Adjustable-Speed Drives

Adjustable-speed drives (ASDs) are very common industrial tools used to perform a variety of tasks. Figure 10.28 shows the most common drive topology: the three-phase incoming supply is rectified to dc, and a pulse-width modulation (PWM) inverter converts the dc to a variable frequency three-phase ac voltage that drives an induction motor at variable speeds.

The capacitor on the dc bus provides some energy storage but not much. Adjustable-speed drives are sensitive to voltage sags and almost always drop out for momentary interruptions. A common cause for shutdown is that the dc bus drops too low, and the drive shuts down on an undervoltage trip (normally 70 to 85% of the nominal dc bus voltage). Even if a drive does not actually shut down, the motor may stall and not be able to start without removing the mechanical load. External factors can also cause shutdown: if the drive is wired through a contactor and the contactor trips, the drive shuts down (and the contactor may be more sensitive than the drive) (EPRI PEAC commentary #3, 1998). Also, if programmable logic controllers provide stop, start, or other signals, loss of the controller can shut the drive down. Following a voltage sag, the drive can draw large inrush, as much as three or four times normal current. This can blow fuses or damage the input diodes and trip the drive.

Drives have significant variation. Of small drives tested by EPRI PEAC (Brief No. 9, 1993) at full load, two could ride through five-cycle three-phase sags down to zero volts; another two tripped at about 80% of rating. At half load, three drives survived a five-cycle sag down to zero volts, and the other tripped at 70% of rating. Some of the most sensitive drives are:

- *Older drives* — Many older drives power their electronics from the ac system, which makes the controller more sensitive. Modern drives power the controls from the dc bus.
- *Higher horsepower drives* — The front-end circuit normally uses thyristors instead of diodes in a current-source topology. To prevent

commutation failure, the dc undervoltage relay is set more sensitively, often 85 to 90%.

- *dc drives* — A thyristor-bridge feeds directly into the dc motor armature. dc drives have no built-in energy, and sags can disturb the timing circuitry for firing the thyristors. Regenerative converters may be very sensitive, especially during regeneration (reverse power flow); a voltage sag can prevent a thyristor from shutting off, which puts a short on the system that will blow a fuse.

Adjustable-speed drives are less sensitive to single-phase sags than three-phase sags because all three phases are rectified (Mansoor et al., 1997). A three-phase sag sags the dc bus voltage down in similar proportions. With a single-phase sag, the two "unsagged" phases can support the drive's dc bus. Whether the drive trips or not depends on how heavily it is loaded and how the undervoltage detection circuitry works. Figure 10.29 shows an example ride-through capability for a 60-kW drive.

Configuration adjustments can sometimes improve ride through; reducing the undervoltage trip setting significantly improves ride through. Increasing the overcurrent trip setting and setting appropriate restart parameters can also help. Also, some models have firmware upgrades providing additional ride through. Drives with a flying restart feature (the drive can restart while the motor is spinning) are better for critical loads. A drive with a synchronous flying restart following a sag to 50% voltage for five cycles allowed only a 5% decrease in motor speed and was fully restored in 1/2 sec (a nonsynchronous flying restart is not nearly as good) (EPRI PEAC Brief No. 30, 1995).

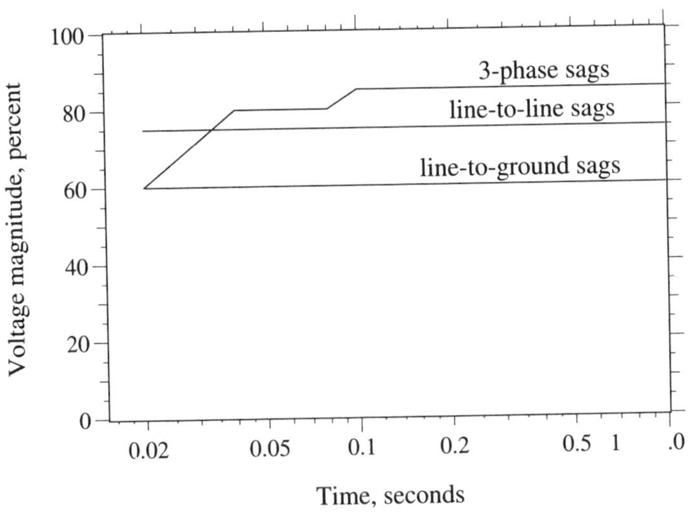

FIGURE 10.29
Ride-through capability of a 60-kW adjustable-speed drive under different voltage sag conditions. (Data from [Abrahams et al., 1999].)

10.9.2.3 Programmable-Logic Controllers

The performance of programmable-logic controllers (PLCs) varies widely. A PLC is a hardened electronic controller used to control many types of industrial processes and equipment. PLCs have multiple input and output channels (I/O racks) used to measure and control equipment. A dc power supply powers the cpu and the I/O racks.

In most cases, the PLC power supply is the same as other computer power supplies, a switched-mode power supply usually capable of riding through an interruption of several cycles. The problem is that the power supply is often not the weakest link in the system. A power supply monitoring circuit that senses the input voltage may initiate a shutdown during a voltage sag. PLCs are more sensitive as a result.

Another important concern is that sags or interruptions can not only cause a shutdown of the PLC, a sag can produce faulty outputs on some PLCs. Faulty outputs can cause more havoc with some processes than if the PLC actually shut down. Figure 10.30 shows the sensitivity of several programmable-logic controllers.

10.9.3 Residential Equipment

The digital clock has been quoted as being "the world's best-selling power quality recorder." The "blinking clocks" are a nuisance for customers and generate many phone calls for utilities. That said, clocks have a wide range of sensitivities, and many actually have very good ride-through capability.

Figure 10.31 shows ride-through capabilities for several digital clocks tested by EPRI PEAC and Hydro Quebec. There is a wide range of voltage sensitivity, but few of the digital clocks tested (25%) lose memory for a complete interruption that is less than 0.5 sec. The main consideration for distribution circuits is the dead time on the first reclose attempt. This is usually about 0.3 to 5 sec depending on the delay on the reclosing relay. An immediate reclose attempt has a dead time of 0.3 to 0.5 sec, making it a good option for reducing blinking clocks. Table 10.11 shows the ride through of different residential devices from a study by Northeast Utilities that also shows most devices have good ride through for events less than 0.5 sec.

Most residential devices have a fairly rectangular volt-time characteristic. Figure 10.32 shows the characteristic of several residential devices. Only longer-duration voltage sags affect most residential devices.

10.10 Solution Options

10.10.1 Utility Options for Momentary Interruptions

We can reduce momentary interruptions in several ways:

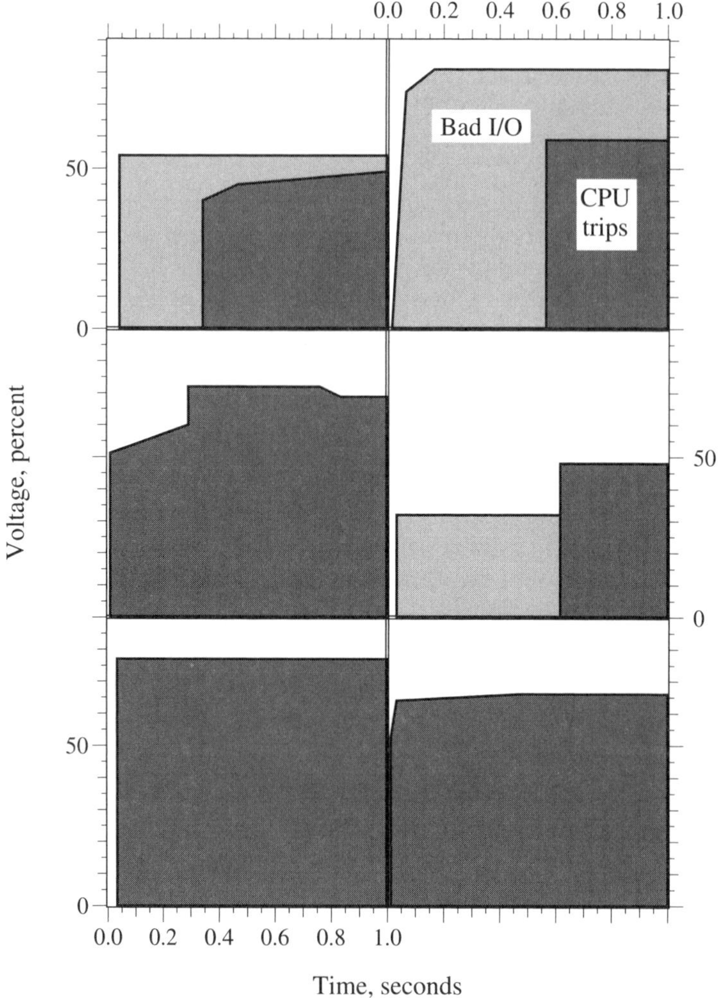

FIGURE 10.30
Sensitivity of six PLCs. (Adapted from [EPRI PEAC Brief No. 39, 1996].)

- Immediate reclose
- Use of fuse blowing
- Single-phase reclosers
- Extra downstream devices (fuses or reclosers)
- Sequence coordination with downstream devices
- Reduce faults

Reducing faults is a universally good approach to improving power quality. Other approaches target specific disturbances. For momentaries, the single

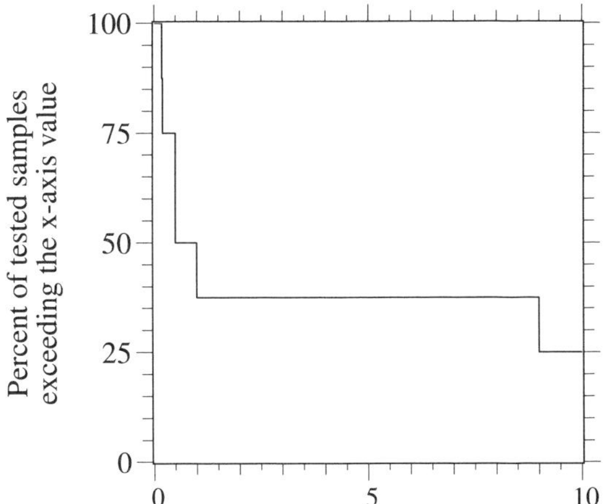

FIGURE 10.31
Capability of digital clocks to ride through an interruption (n = 8). (Data from [Courtois, 2001; Courtois and Deslauriers, 1997; EPRI PEAC Brief No. 17, 1994].)

TABLE 10.11
Percentage of Devices That Were Able to Successfully Ride Through a Momentary Interruption of the Given Duration

Device	0.5 sec	2 sec	16.7 sec
Digital clock	70	60	0
Microwave oven	60	0	0
VCR	50	37.5	0
Computer	0	0	0

Source: Bowes, K. B., "Effects of Power Line Disturbances on Electronic Products," *Power Quality Assurance Magazine*, vol. Premier V, pp. 296–310, 1990.

biggest improvement is to use fuse blowing instead of fuse saving (see Chapter 8). Using an immediate reclose, while not reducing the number of momentaries counted, reduces complaints from residential customers. Both of these changes for improving momentaries are relatively easy to implement.

Other methods of reducing momentaries involve better application of protection equipment. Extra protection devices that segment the circuit into smaller sections help improve momentary interruptions (and long-duration interruptions). Improving coordination between devices (including use of sequence coordination to improve coordination between reclosers) helps eliminate some unnecessary blinks. Single-phase reclosers instead of three-phase reclosers or breakers helps reduce the number of phases interrupted for single line-to-ground faults.

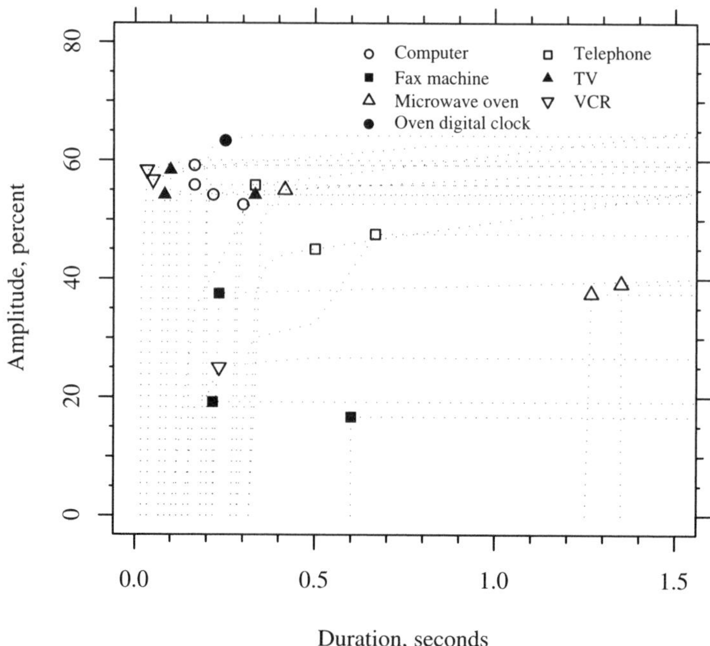

FIGURE 10.32
Ride-through capability of various residential devices. (Data from [Courtois, 2001; Courtois and Deslauriers, 1997].)

10.10.2 Utility Options for Voltage Sags

Utility-side options for reducing voltage sags are limited and are rarely done (at least solely for the purposes of voltage sags). Some of the strategies and equipment that help are:

- Use of fuse saving
- Current limiting fuses
- Smaller lateral fuses
- Faster breakers or reclosers
- Raise the nominal voltage
- Reduce faults

Faster relaying or faster interrupters — any changes in protection schemes that clear faults faster — help reduce the sag's duration. The next few sections address some of the options for reducing the impact of power-system faults on the voltage.

10.10.2.1 Raising the Nominal Voltage

Raising the nominal voltage helps the ride through of many types of equipment. Computers, adjustable-speed drives, and other equipment with capacitors benefit if the voltage is regulated near the upper end of the ANSI range A, or at least avoid the lower end. On the utility system, we use LTCs, regulators, and switched capacitors. This approach is the opposite of demand-side management programs which are designed to deliver low voltages in order to reduce peak demand or customer energy usage.

10.10.2.2 Line Reactors

Series line reactors provide electrical separation between feeders off the substation bus. Figure 10.33 shows the effect of line reactors on the station bus voltage for different configurations. Reactors have the added benefit of limiting the fault current. The reactors provide good protection against some

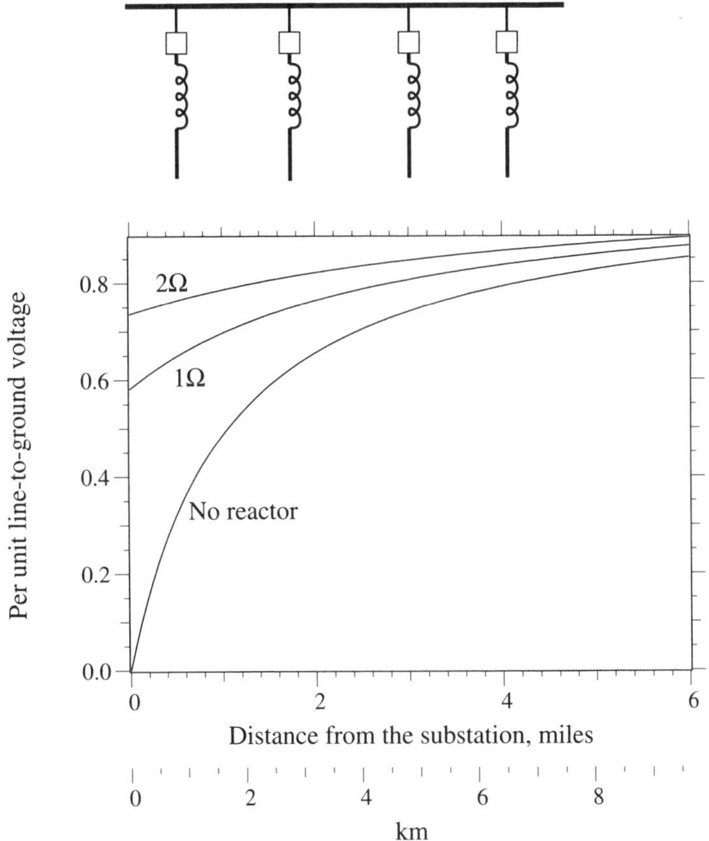

FIGURE 10.33
Substation bus voltage for a three-phase fault at the given distance for different line reactor configurations (for a 12.5-kV circuit with the same parameters as Figure 7.11).

voltage sags but do not help with interruptions or sags caused by transmission-level faults. On the faulted feeder, line reactors make the voltage sag worse, but much of the feeder may have very low voltage anyway (downstream of the fault), or the breaker may open and give all of the customers an interruption.

Utilities have used line reactors in this application, mainly at urban stations to reduce high fault currents. Line reactors do not have the best reputation; they are expensive, take up substation space, and increase voltage drop. The reactor must be designed to withstand the fault currents that it will regularly see.

10.10.3.2 Neutral Reactors

Phase reactors effectively isolate feeders and limit the voltage sag, so how about the reactor sometimes used between the substation transformer neutral and ground? The answer depends on the transformer connection serving the customer and the connection of the load. Neutral reactors generally make voltage sags worse for loads connected line-to-ground on the distribution circuit because:

- *Duration* — A neutral reactor lowers the fault current, so tap fuses and time overcurrent relays take slightly longer to operate, increasing the sag duration.
- *Magnitude* — A neutral reactor increases the zero-sequence impedance. The weaker ground source cannot hold up the station line-to-ground voltage as well for a downstream line-to-ground fault.

For line-to-line connected loads, the neutral reactor significantly improves voltages to end-use equipment, because of

- *Neutral shift* — The reactor adds impedance which shifts the neutral point. This raises the voltage on the unfaulted phases, but supports the line-to-line voltages. (A very large reactor in a high-impedance grounded system would have almost no drop in the line-to-line voltages during a single line-to-ground fault.)

With a single line-to-ground fault, the line-to-line voltage in per unit is (assuming that the circuit resistances are zero for simplification)

$$V = \frac{\sqrt{X_0^2 + X_0X_1 + X_1^2}}{2X_1 + X_0}$$

The neutral reactor of X ohms adds $3jX$ ohms to the zero-sequence impedance and raises the X_0/X_1 ratio. Table 10.12 and Figure 10.34 show that a

TABLE 10.12

Line-to-Ground and Line-to-Line Voltages on the Low-Voltage Side of a Transformer with One Phase on the High-Voltage Side Sagged To Zero

Voltages	Primary Voltages			Voltages Downstream of a Delta – Wye Transformer		
No neutral reactor ($X_0/X_1 = 1$)						
Line-ground	0.00	1.00	1.00	0.58	0.58	1.00
Line-line	0.58	0.58	1.00	0.33	0.88	0.88
Neutral reactor that gives $X_0/X_1 = 3$						
Line-ground	0.00	1.25	1.25	0.72	0.72	1.00
Line-line	0.72	0.72	1.00	0.60	0.92	0.92

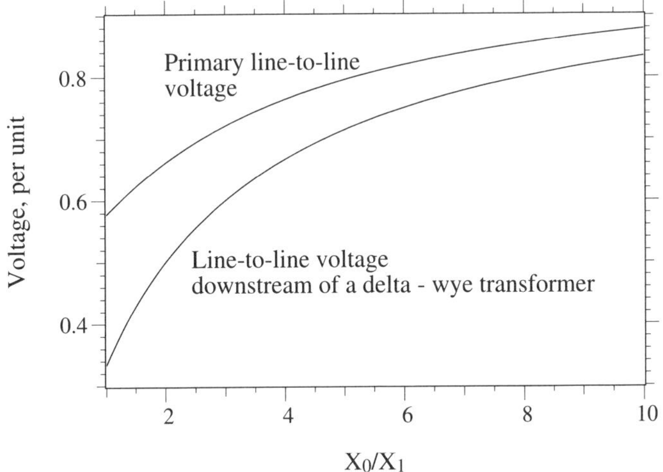

FIGURE 10.34
Impact of a neutral reactor (higher X_0) on voltage sags.

modest neutral reactor significantly helps line-to-line loads during line-to-ground faults. Also, with delta – wye distribution transformers, the line-to-ground secondary voltages see equivalent sags as the line-to-line voltage on the primary. The neutral reactor provides benefit only for line-to-ground faults and no help for three-phase or line-to-line faults. A disadvantage of the neutral reactor is that it increases the voltage rise (the swell) on the unfaulted phases (see Chapter 13).

10.10.2.4 Current-Limiting Fuses

Current-limiting fuses reduce the fault current and force an early zero crossing. In the process, the fuse reduces the severity of the magnitude and

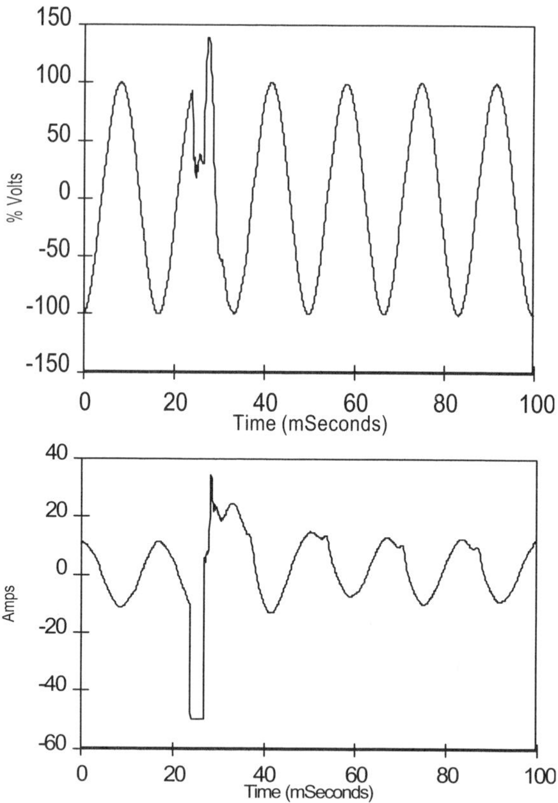

FIGURE 10.35
Example of a current-limiting fuse operation recorded in the EPRI DPQ study. (Copyright © 1996. Electric Power Research Institute. TR-106294-V3. *An Assessment of Distribution System Power Quality: Volume 3: Library of Distribution System Power Quality Monitoring Case Studies.* Reprinted with permission.)

reduces the duration. Figure 10.35 shows an example of a fault cleared by a current-limiting fuse. The duration of the sag is very short, and the depth is minimal. These types of results have been verified by other measurements and computer models (Kojovic and Hassler, 1997; Kojovic et al., 1998). Close to the substation is the best location for use, where they are most useful for reducing the magnitude and duration of voltage sags and are most appropriate for limiting damage due to high fault currents.

10.10.3 Utility Options with Nontraditional Equipment

10.10.3.1 Fast Transfer Switches

Medium-voltage static transfer switches are a utility-side option for providing improved power quality and protection from voltage sags and momentary interruptions. Thyristors are used to obtain a 1/2-cycle transfer between two sources. The switches can be configured as preferred/alternate or as a split bus. The normal configuration employs static transfer switches for sensitive customers in a primary selective scheme. After the switch has operated, it may or may not switch back to the original feeder.

We should also consider the impacts on the distribution system. Switching a large load between weak locations on circuits could cause objectionable voltage changes. Possible interaction with other voltage-regulating equipment warrants investigation. This includes other custom-power devices that may be trying to dynamically correct voltage.

Another fast-transfer technology can be used in the same applications as static switches. High-speed mechanical source transfer switches use high-speed vacuum switches and a sophisticated microprocessor based control to provide "break-before-make" transfers in approximately 25 msec or 1.5 cycles. The two sources are not paralleled during the transfer; therefore, the load experiences an interruption of approximately 1.5 cycles. This level of protection may be acceptable for some equipment (but the 1.5-cycle interruption may trip some sensitive equipment). These switches have the advantage of being very efficient (99%), inexpensive (one-fifth to one-tenth the cost of a static switch), and small. Both pole-mounted and padmounted versions are available. For most loads, the fast transfer switch provides significant benefit. Relays and contactors, though, can drop out for a 1.5-cycle interruption.

Normally, fast transfer switches have been applied at individual customers in a primary-selective scheme. The technology could nicely apply to feeder-level application as shown in Figure 10.36. This arrangement provides improved performance for voltage sags, momentary interruptions, and long-duration interruptions for the customers at the end of the circuit (these are customers that usually get the worst power quality).

10.10.3.2 DVRs and Other Custom-Power Devices

In addition to the static transfer switch, medium-voltage power electronics have enabled a wide variety of utility-level solutions to power-quality problems: series injection devices, static regulators, shunt devices, and medium-voltage uninterruptible power supply (UPS) systems with a variety of energy storage options. These power-electronics solutions have been coined "custom power." The advantage of utility-side approaches is that the whole facility is supported (we do not have to track down, test, and fix every possible piece of sensitive equipment).

Figure 10.37 shows the configuration of several custom-power configurations. Most provide support during voltage sags. Single-phase sags are more easily corrected (since devices can use energy from the unfaulted phases). For momentary interruptions, some sort of energy storage is necessary such as batteries, ultracapacitors, flywheel, or superconducting coil. A summary of the cost and capability of the most common devices is shown in Table 10.13.

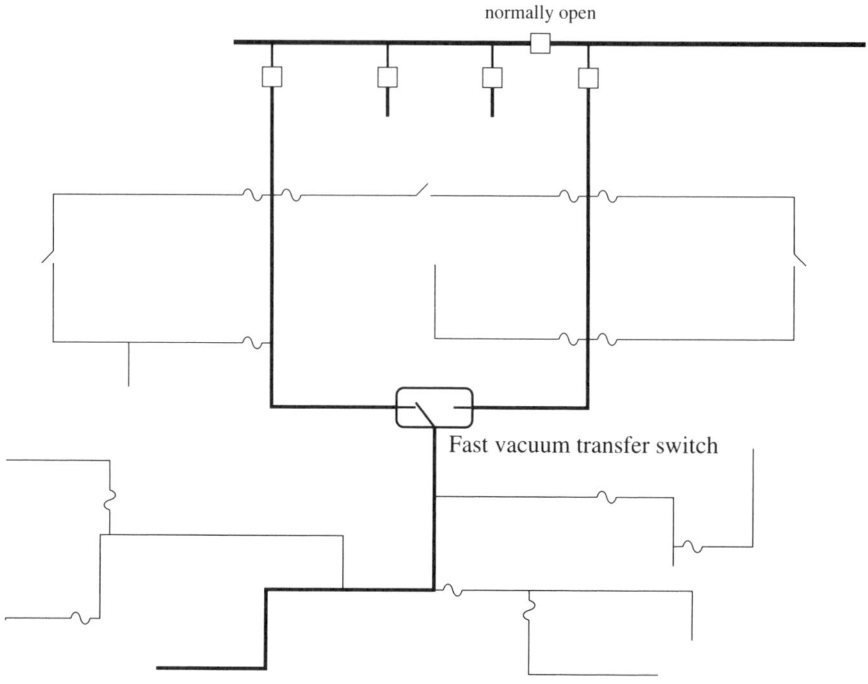

FIGURE 10.36
Using a fast transfer switch to enhance power quality to a downstream section of circuit. (Copyright © 2002. Electric Power Research Institute. 1007281. *Analysis of Extremely Reliable Power Delivery Systems.* Reprinted with permission.)

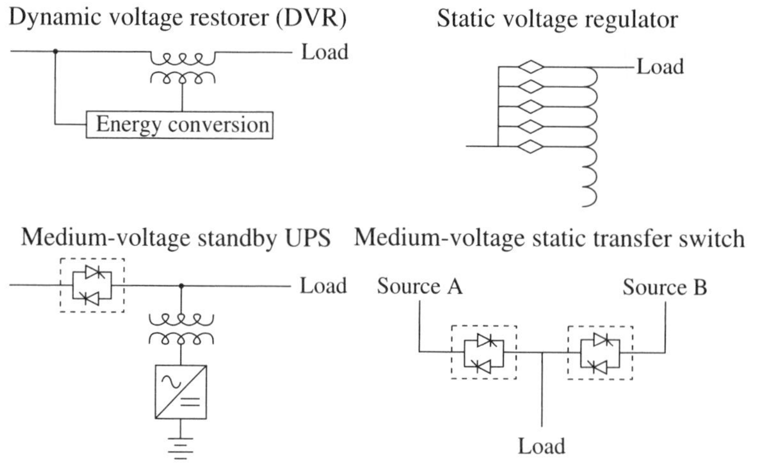

FIGURE 10.37
Common custom-power devices.

TABLE 10.13

Comparison of Custom-Power Equipment for Correcting Sags and Momentary Interruptions

	Cost, U.S. $	Capability
Static shunt compensation	50-200/kvar	sags to 70%
Source transfer switch	500-1000/A	
DVR	150-250/kVA	sags to 50%
Static voltage regulator	80-125/kVA	sags to 50%
MV UPS	750-1500/kVA	

Source: EPRI 1000340, *Guidebook on Custom Power Devices*, EPRI, Palo Alto, CA, 2000.

TABLE 10.14

Comparison of Selected Power Conditioning Equipment

	Sags to 80%	Sags to 50%	Sags to 25%	Below 25%	Outage
Dip-proofing inverter	Solves	Solves	Solves	Solves	To 1 sec
Constant-voltage transformer	Solves	Size[a]	Size[a]	No	No
DySC	Solves	Solves	To 0.33 sec	To 0.26 sec	To 0.15 sec
Uninterruptible power supply	Solves	Solves	Solves	Solves	Solves
Coil hold-in devices	Solves	Solves	Solves	No	No

[a] Size = capability depends on the size of the device

Source: Stephens, M., Johnson, D., Soward, J., and Ammenheuser, J., *Guide for the Design of Semiconductor Equipment to Meet Voltage Sag Immunity Standards*, International SEMATECH, 1999. Technology Transfer #99063760B-TR, available at http://www.sematech.org/public/docubase/document/3760btr.pdf.

The dynamic-voltage restorer (DVR) is one of the devices most suited for correcting for voltage sags (Woodley et al., 1999). During a voltage sag, the DVR adds voltage through an in-line transformer to offset the missing voltage. A DVR can correct for any sag to 50%, and optional energy storage increases the range of performance.

10.10.4 Customer/Equipment Solutions

The best place to provide solutions for customers is often in the facility. Some of the customer solution options are shown in Table 10.14.

Other tips (McEachern, 2001; Stephens et al., 1999) include

- Wire devices in a phase-to-phase configuration where possible.
- Avoid mismatched equipment voltages.
- Avoid the use of ac "ice cube" general-purpose relays.
- Do not use phase-monitoring relays in interlocking circuits.
- Utilize a non-volatile memory.

- Do not overload dc power supplies.
- Use a targeted voltage conditioning approach.
- Use robust inverter drives.

Make sure that computer-controlled equipment can recover from interruptions. This sounds like an obvious solution, but user complaints reveal that a lot of computer-controlled equipment gets completely confused by a disturbance. State machine programming is a promising approach to help make sure that processes recover from disturbances. Another step that helps is to use battery-backed up memory or a disk to store process step information.

10.11 Power Quality Monitoring

Power quality monitoring helps utilities and their customers diagnose and fix power quality problems. Many locations are suitable for monitoring. The distribution substation is a good place because it can monitor the voltage and currents feeding a large number of customers. For a specific customer having a problem, nothing is better than recording right at or near the sensitive process or piece of equipment. Some suggestions for monitoring are as follows:

- *Voltage and current* — Record both voltage and current, and for three-phase installations, record all three phases plus the ground current. For substation monitoring, currents identify which circuit the fault was on or whether it was caused on the transmission system. With the current, we can estimate the distance to the fault. For facility monitoring, the current shows us if a sag is caused by something internal to the facility or on the utility system. More importantly, the current may show what tripped off within the facility and when it tripped.
- *Triggers* — Do not set the triggers too sensitively. Even with larger onboard memories and disk capabilities, there is usually no reason for highly sensitive settings. The commonly used ± 5% is too sensitive. Sensitive triggers create too much data for downloading, storage, and most importantly, for analysis. When recording sags and momentaries, we can safely ignore sags that do not drop below 90%, and a trigger setting of 85% is often appropriate (except for very sensitive loads).
- *Output contacts* — For monitoring at sensitive equipment, record output contacts of any equipment that may indicate its operational status or the process status. Also, be aware that some equipment (like many adjustable-speed drives and UPSs) keeps an event log that may help track down problems. At the substation, record the output contacts of the pertinent relays to determine which relays tripped.

- *Timing* — Synchronize the monitoring clocks to a GPS time reference to help correlate with other monitoring data or event records. If this is not feasible, establish a protocol of regularly setting the recorder clocks (and logging the errors to be able to adjust for drift) to keep them close to a time standard. If a process has internal data recording, synchronize those clocks. The same goes for the utility SCADA system — accurate timing increases the certainty with which we can diagnose power quality problems.
- *Trend data* — Record trend data (say 1-min maximum, minimum, and average values) as well as triggered data (sags, swells, etc.). This helps pinpoint problems related to the steady-state voltage. The ride through of some devices depends on the pre-disturbance voltage.

Fault-recorder type devices with power quality capability are good for substation application; the large number of channels and ability to monitor relay contacts is very beneficial. Also, many types of devices such as relays, recloser controllers, and customer meters can record power quality disturbances.

A drawback to monitoring is that it requires expert manpower to operate. It takes considerable work to install the monitoring, set up the downloading, and (most significantly) analyze the data. Good power quality software helps by highlighting or paring down the data and creating summary reports. Sometimes power quality recorders are used in "stealth" mode: they are installed and largely forgotten until there are customer or utility problems.

An important question is, what do we do with all of the data from power quality recorders? Oftentimes, the best answer is nothing, until power quality problems are reported. A good use of the data is benchmarking. Pick an indicator such as $\text{SARFI}_{\text{ITIC}}$, and collect data from each recording site. We can compare the benchmark on a site-by-site basis to help determine which circuits need more maintenance attention. Further analysis could reveal what practices are better (for example, circuits with a three-year tree-trimming cycle could be compared with those having a 5-year trim cycle). Benchmarking can also be used as an advertising tool to try and attract sensitive, high-tech customers.

For high accuracy at a given site, we need long monitoring periods. Events that happen less often require longer monitoring for good accuracy. Table 10.15 shows an approximation by Bollen (2000) based on a Poisson distribution. The length of monitoring needed is at least

$$t = \frac{4}{\mu\varepsilon^2}$$

where

ε = accuracy needed, per unit

μ = number of events expected per time period

TABLE 10.15

Monitoring Duration Needed to Achieve the Given Level of Accuracy

Event Frequency	50% Accuracy	10% Accuracy
1 per day	2 weeks	1 year
1 per week	4 months	7 years
1 per month	1 year	30 years
1 per year	16 years	400 years

Source: Bollen, M. H. J., *Understanding Power Quality Problems: Voltage Sags and Interruptions*, IEEE Press, New York, 2000. With permission. ©2000 IEEE.

Treat these as minimum numbers; Bollen's analysis assumes that events randomly appear, while in reality they appear in clusters, during stormy weather or during overloads. This clustering means that even longer monitoring is needed for a given level of accuracy. Another concern is that power systems change over time, which adds further uncertainty to historical monitoring. Bollen also suggests a method of adjusting sag and momentary measurements by the fault rate measured over a longer time period. A better estimate of the actual event rate is

$$\overline{N}_{sags} = N_{sags} \frac{\overline{N}_{faults}}{N_{faults}}$$

where

N_{sags} and N_{faults} = the number of sags and the number of faults over the recording period

$\overline{N}_{sags}$ and $\overline{N}_{faults}$ = the number of sags and the number of faults over a longer period of time

For sites where monitoring cannot meet accuracy needs, prediction methods can help produce an estimate. This requires developing a model of the system and using a stochastic approach to the fault positions analysis method.

References

Abrahams, R., Keus, A. K., Koch, R. G., and van Coller, J. M., "Results of Comprehensive Testing of a 120kW CSI Variable Speed Drive at Half Rating," PQA Conference, 1999.

Bollen, M. H. J., *Understanding Power Quality Problems: Voltage Sags and Interruptions*, IEEE Press, New York, 2000.

Bowes, K. B., "Effects of Power Line Disturbances on Electronic Products," *Power Quality Assurance Magazine*, vol. Premier V, pp. 296–310, 1990.

CEA, *CEA 2000 Annual Service Continuity Report on Distribution System Performance in Electric Utilities*, Canadian Electrical Association, 2001.

Chong, W. Y., "Effects of Power Quality on Personal Computer," Masters thesis, University of Queensland, 2000.

Collins Jr., E. R. and Bridgwood, M. A., "The Impact of Power System Disturbances on AC-Coil Contactors," Textile, Fiber, and Film Industry Technical Conference, 1997.

Conrad, L., Kevin, L., and Cliff, G., "Predicting and Preventing Problems Associated with Remote Fault-Clearing Voltage Dips," *IEEE Transactions on Industry Applications*, vol. 27, pp. 167–72, 1991.

Courtois, E. L., 2001. Personal communication.

Courtois, E. L. and Deslauriers, D., "Voltage Variations Susceptibility of Electronic Residential Equipment," PQA Conference, 1997.

Dorr, D. S., "Point of Utilization Power Quality Study Results," *IEEE Transactions on Industry Applications*, vol. 31, no. 4, pp. 658–66, July/August 1995.

Dorr, D. S., Hughes, M. B., Gruzs, T. M., Juewicz, R. E., and McClaine, J. L., "Interpreting Recent Power Quality Surveys to Define the Electrical Environment," *IEEE Transactions on Industry Applications*, vol. 33, no. 6, pp. 1480–7, November 1997.

EEI, "EEI Reliability Survey," Minutes of the 8th Meeting of the Distribution Committee, March. 28–31, 1999.

EPRI 1000340, *Guidebook on Custom Power Devices*, EPRI, Palo Alto, CA, 2000.

EPRI 1001665, *Power Quality Improvement Methodology for Wires Companies*, Electric Power Research Institute, Palo Alto, CA, 2003.

EPRI 1007281, *Analysis of Extremely Reliable Power Delivery Systems: A Proposal for Development and Application of Security, Quality, Reliability, and Availability (SQRA) Modeling for Optimizing Power System Configurations for the Digital Economy*, Electric Power Research Institute, Palo Alto, CA, 2002.

EPRI PEAC Brief No. 7, *Undervoltage Ride-Through Performance of Off-the-Shelf Personal Computers*, EPRI PEAC, Knoxville, TN, 1992.

EPRI PEAC Brief No. 9, *Low-Voltage Ride-Through Performance of 5-hp Adjustable-Speed Drives*, EPRI PEAC, Knoxville, TN, 1993.

EPRI PEAC Brief No. 10, *Low-Voltage Ride-Through Performance of AC Contactor Motor Starters*, EPRI PEAC, Knoxville, TN, 1993.

EPRI PEAC Brief No. 12, *Low-Voltage Ride-Through Performance of a Modified Personal Computer Power Supply*, EPRI PEAC, Knoxville, TN, 1993.

EPRI PEAC Brief No. 17, *Electronic Digital Clock Performance During Steady-State and Dynamic Power Disturbances*, EPRI PEAC, Knoxville, TN, 1994.

EPRI PEAC Brief No. 30, *Ride-Through Performance of Adjustable- Speed Drives with Flying Restart*, EPRI PEAC, Knoxville, TN, 1995.

EPRI PEAC Brief No. 39, *Ride-Through Performance of Programmable Logic Controllers*, EPRI PEAC, Knoxville, TN, 1996.

EPRI PEAC Brief No. 44, *The Effects of Point-on-Wave on Low-Voltage Tolerance of Industrial Process Devices*, EPRI PEAC, Knoxville, TN, 1998.

EPRI PEAC commentary #3, *Performance of AC Motor Drives During Voltage Sags and Momentary Interruptions*, 1998.

EPRI TP-113781, *Reliability Benchmarking Application Guide for Utility/Customer PQ Indices*, Electric Power Research Institute, Palo Alto, CA, 1999.

EPRI TR-106294-V2, *An Assessment of Distribution System Power Quality: Volume 2: Statistical Summary Report*, Electric Power Research Institute, Palo Alto, CA, 1996.

EPRI TR-106294-V3, *An Assessment of Distribution System Power Quality: Volume 3: Library of Distribution System Power Quality Monitoring Case Studies*, Electric Power Research Institute, Palo Alto, CA, 1996.

IEEE Std. 493-1997, *IEEE Recommended Practice for the Design of Reliable Industrial and Commercial Power Systems (Gold Book).*

IEEE Std. 1159-1995, *IEEE Recommended Practice for Monitoring Electric Power Quality.*

IEEE Std. 1366-2000, *IEEE Guide for Electric Power Distribution Reliability Indices.*

Information Technology Industry Council (ITI), "ITI (CBEMA) Curve Application Note," 2000. Available at http://www.itic.org.

Kojovic, L. A. and Hassler, S. P., "Application of Current Limiting Fuses in Distribution Systems for Improved Power Quality and Protection," *IEEE Transactions on Power Delivery*, vol. 12, no. 3, pp. 791–800, April 1997.

Kojovic, L. A., Hassler, S. P., Leix, K. L., Williams, C. W., and Baker, E. E., "Comparative Analysis of Expulsion and Current-Limiting Fuse Operation in Distribution Systems for Improved Power Quality and Protection," *IEEE Transactions on Power Delivery*, vol. 13, no. 3, pp. 863–9, July 1998.

Mansoor, A., Collins, E. R., Bollen, M. H. J., and Lahaie, S., "Effects of Unsymmetrical Voltage Sags on Adjustable Speed Drives," PQA-97, Stockholm, Sweden, June 1997.

McEachern, A., "How to Increase Voltage Sag Immunity," 2001. Available at http://powerstandards.com/tutorials/immunity.htm.

Sabin, D. D., Grebe, T. E., and Sundaram, A., "RMS Voltage Variation Statistical Analysis for a Survey of Distribution System Power Quality Performance," IEEE/PES Winter Meeting Power, February 1999.

Sekine, Y., Yamamoto, T., Mori, S., Saito, N., and Kurokawa, H., "Present State of Momentary Voltage Dip Interferences and the Countermeasures in Japan," International Conference on Large Electric Networks (CIGRE), September 1992.

SEMI F42-0600, *Test Method for Semiconductor Processing Equipment Voltage Sag Immunity*, Semiconductor Equipment and Materials International, 1999.

SEMI F47-0200, *Specification for Semiconductor Processing Equipment Voltage Sag Immunity*, Semiconductor Equipment and Materials International, 1999.

Short, T. A., Mansoor, A., Sunderman, W., and Sundaram, A., "Site Variation and Prediction of Power Quality," IEEE Transactions on Power Delivery, 2002. Preprint submission.

St. Pierre, C., "Don't Let Power Sags Stop Your Motors," *Plant Engineering*, pp. 76–80, September 1999.

Stephens, M., Johnson, D., Soward, J., and Ammenheuser, J., *Guide for the Design of Semiconductor Equipment to Meet Voltage Sag Immunity Standards*, International SEMATECH, 1999. Technology Transfer #99063760B-TR, available at http://www.sematech.org/public/docubase/document/3760btr.pdf.

Williams, B. R., Schmus, W. R., and Dawson, D. C., "Transmission Voltage Recovery Delayed by Stalled Air Conditioner Compressors," *IEEE Transactions on Power Systems*, vol. 7, no. 3, pp. 1173–81, August 1992.

Woodley, N. H., Morgan, L., and Sundaram, A., "Experience with an Inverter-Based Dynamic Voltage Restorer," *IEEE Transactions on Power Delivery*, vol. 14, no. 3, pp. 1181–6, July 1999.

All I know is that when shit hits the fan, management calls us we don't call them.

In response to: Do linemen feel they are respected by management and coworkers for the jobs they are doing, do management and coworkers understand what you do?

www.powerlineman.com

11

Other Power Quality Issues

While voltage sags and momentary interruptions cause the most widespread power quality problems, several other power quality disturbances can damage equipment, overheat equipment, disrupt processes, cause data loss, and annoy and upset customers. In this chapter, we explore several of these, including transients, harmonics, voltage flicker, and unbalance.

11.1 Overvoltages and Customer Equipment Failures

Often, customers complain of equipment failures, especially following power interruptions. Is it lightning? Voltage swells during faults? Some sort of switching transient? Sometimes explanations are obvious, sometimes not. Several events can fail equipment during a fault/interruption, either from the disturbance that caused a fault, the voltage sag during the fault, a voltage swell during the fault, or the inrush while the system is recovering. Some possibilities are

- *Overvoltages* — Lightning and other system primary-side overvoltages can enter the facility and damage equipment.
- *Grounding* — Poor facility grounding practices can introduce overvoltages at equipment from fault current.
- *Capacitive coupling* — Reclose operations and other switching transients can create fast-rising voltage on the primary that capacitively couples through the transformer, causing a short pulse on the secondary.
- *Inrush current* — While recovering from a voltage sag or momentary interruption, the inrush current into some electronic equipment can blow fuses or fail semiconductor devices.
- *Unbalanced sags* — Three-phase electronic equipment like adjustable-speed drives can draw excessive current during a single-phase sag

or other unbalanced sag. The current can blow fuses or fail the front-end power electronics.

- *Equipment aging* — Some equipment is prone to failure during turn on, even without a voltage transient. The most obvious example is an incandescent light bulb. Over time, the filament weakens, and the bulb eventually fails, usually when turned on. At turn on, the rapid temperature rise and mechanical stress from the inrush can break the filament.

Lightning can cause severe overvoltages, both on the primary and on the secondary. Damaging surges can enter from strikes to the primary, strikes to the secondary, strikes to the facility, strikes to plumbing, and strikes to cable-television or telephone wires. Poor grounding practices can make lightning-caused failures more likely.

Another source of severe overvoltages is primary or secondary conductors contacting higher voltage lines. Other overvoltages are possible; normally these are not severe enough to damage most equipment, except for sensitive electronics:

- Voltage swells — Peaks at about 1.3 per unit on most distribution circuits.
- Switching surges — Normally peaks at less than 2 per unit and decays quickly.
- Ferroresonance — Normally peaks at less than 2 per unit.

Just as arresters on distribution lines are sensitive to overvoltages, arresters inside of electronic equipment often are the first thing to fail. The power supply on most computers and other electronics contains small surge arresters (surge suppressors) that can fail quickly while trying to clamp down on overvoltages, especially longer-duration overvoltages. These small suppressors have limited energy absorption capability.

In addition to proper grounding, surge arresters are the primary defense against lightning and other transients. For best protection, use surge protection at the service entrance and surge protection at each sensitive load.

Surge arresters work well against short-duration overvoltages — lightning and switching transients. But arresters have trouble conducting temporary power-frequency overvoltages; they absorb considerable energy trying to clamp the overvoltage and can fail. Small arresters often are the first component to fail in equipment. Using a higher voltage rating helps give more protection to the surge arrester during temporary overvoltages (for example, end users should not use arresters with a maximum continuous operating voltage below 150 V). Surge arresters should be coordinated; the large surge arrester at the service entrance should have the lowest protective level of all of the arresters within the facility. Because arresters are so nonlinear, the unit with the lowest protective level will conduct almost all of the current.

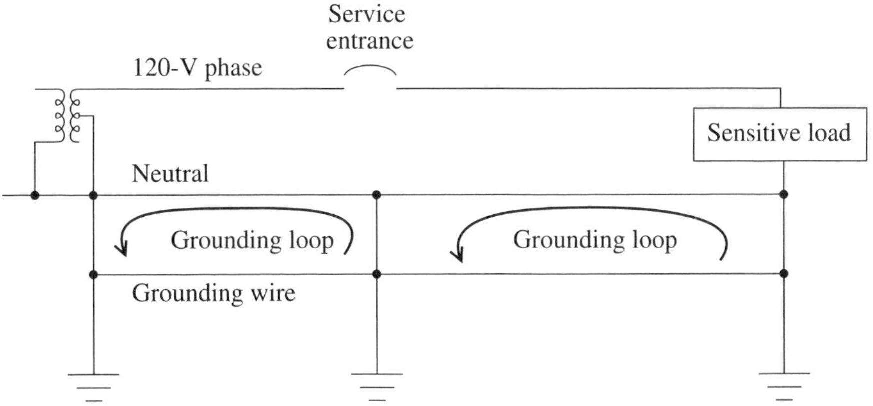

FIGURE 11.1
Ground loops within a facility.

So, we want the arrester with the most energy capability to absorb most of the energy.

11.1.1 Secondary/Facility Grounding

Grounding problems within a facility can lead to equipment failures or malfunctions. A common problem is ground loops. If the secondary neutral has multiple connections within the facility, external ground currents from lightning or from faults can induce voltages along the neutral. Figure 11.1 shows an example. The neutral voltage shifts can impose overvoltages on sensitive equipment. For more information, see IEEE Std. 1100-1999 or IEEE Std. 142-1991. Single-point grounding of the neutral breaks the loops and reduces noise and possible overvoltages.

Grounding loops between multiple "ports" can create damaging scenarios. Any significant wired connection to equipment counts as a port — power, telephone, cable TV, printer cables, and networking cables. Figure 11.2 shows an example of a television where the cable enters via a different route and has a different ground than its electric service. A fault or lightning strike on the utility side can elevate the potential of the electric-supply ground relative to the television cable ground. Even if both the electric and cable television cables have independent surge protection, surges still create voltage differences between components. To avoid multiple-port difficulties, arrange to have all power and communications enter a building at one location, provide a common ground, and apply surge protection to all ports. To overcome this problem, users can install a "surge reference equalizer" rather than independent surge protection on each port (EPRI PEAC Solution No. 1, 1993). Both the electric supply and the other port (telephone, cable TV, ethernet, etc.) plug into the surge reference equalizer. The equalizer provides surge protection for both incoming ports and a single common grounding connection.

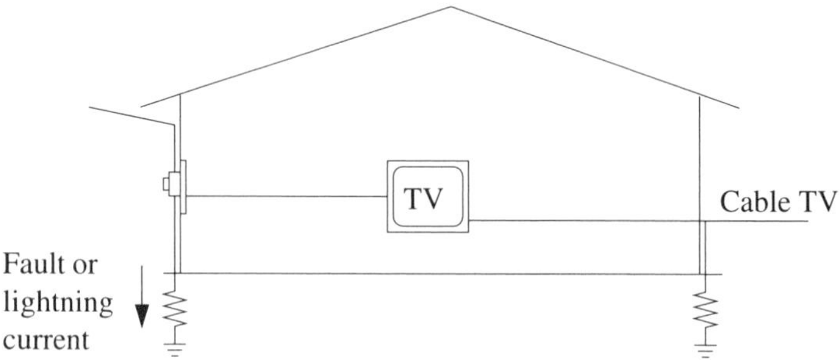

FIGURE 11.2
Example of a two-port voltage stress at a television between its power source and the cable TV cable.

For equipment, overvoltages between the phases and neutral are most important. While it is commonly believed that neutral-to-ground overvoltages cause problems for sensitive electronics, EPRI PEAC found that computers are quite tolerant of neutral-to-ground overvoltages. In tests, neither a continuous 50-V neutral-to-ground voltage nor a 3-kV, 100-kHz ring wave upset the operation of a computer (EPRI PEAC Brief No. 21, 1994).

Pole-mounted controllers for regulators, reclosers, capacitors, and switches face particularly severe environments. Because of their location, they have significant exposure to transients from lightning strikes to nearby poles. Additionally, they are right in the path of fault current returning from faults downstream of the controller. Some utilities have had reliability problems with controllers; some problems may stem from poor powering and grounding arrangements. Figure 11.3 shows how significant voltages can develop on the low-voltage supply when the power is supplied at a remote pole and a fault occurs downstream. Lightning current following this same path can create very severe voltages. Additionally, two-port vulnerabilities can arise

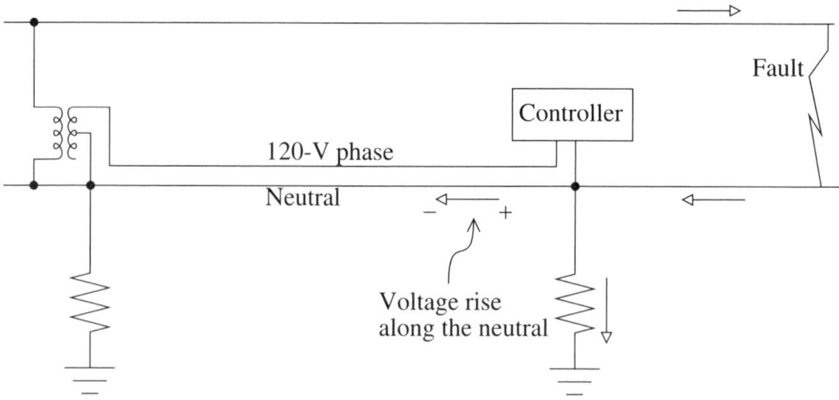

FIGURE 11.3
Pole-mounted controller exposure to faults along the circuit.

between the controller power and potential transformers (PTs). Other sources of two-port problems are current transformers and communication lines. Consider the following strategies to better protect these controllers:

- *Surge protection* — On the power-supply inputs, apply extra low-voltage surge protection, especially in high lightning areas.
- *Power source* — To help avoid two-port problems, try to power the controller from a PT or small transformer on the same structure as the controller.

11.1.2 Reclose Transients

One source of possibly damaging surges is from capacitive coupling through transformers. The very short-duration pulses come from electrostatic coupling through the transformer. Much analysis of this has been done for larger power transformers at generating stations (Abetti et al., 1952). At high frequencies, the transformer acts like a capacitor. A steep surge can pass from the primary to the secondary. How much of the surge gets from the primary to the secondary depends on the capacitances of the transformer and the secondary load — not the transformer turns ratio. A transformer has capacitance from the high-voltage winding to the low-voltage winding (C_1) and capacitance from the secondary winding to ground (C_2). As a first approximation, the voltage on the secondary as a function of the voltage on the primary is a capacitive voltage divider (Greenwood, 1991):

$$V_S = \frac{C_1}{C_1 + C_2} V_P$$

Line energization and capacitor switching and other transients can create surges that pass through the transformer. It does not even have to be an overvoltage on the primary, just a fast rise to the nominal peak — a normal reclose operation — can create a surge on the secondary. When a line is energized near its peak voltage, right when the switch engages, a traveling wave with a very steep front rushes down the line. At a transformer, this front can be steep enough to couple capacitively through to the secondary at voltages much higher than the turns ratio of the transformer. Figure 11.4 shows a 1-μsec wide transient to just over 2000 V following a reclose operation.

These capacitively coupled surges are worse with

- Higher voltage transformations from the primary to the secondary (34.5 kV to 480 V is worse than 12.5 kV to 480 V).
- Loads close to the substation or recloser (the wave front flattens out with distance).
- Minimal resistive load on the secondary.

Millisecond-scale view of the transient:

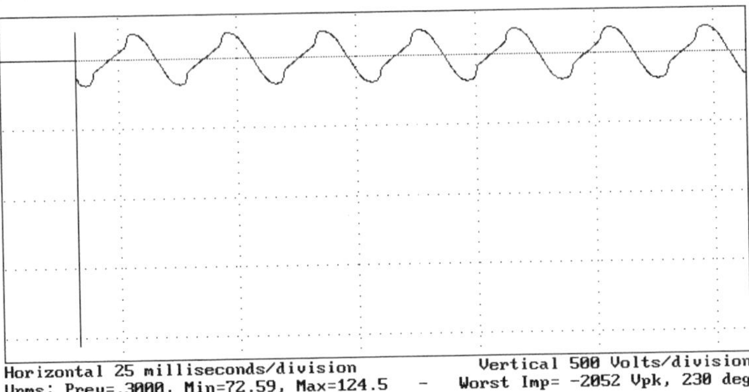

Microsecond-scale view:

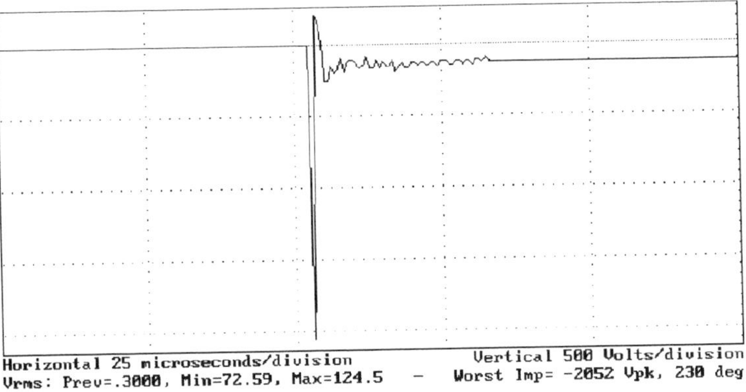

FIGURE 11.4
Measurements of a transient captured during a circuit reclose. (Recordings courtesy of Francisco Ferrandis Mauriz, Iberdrola Distribucion Electrica.)

We can protect against these surges with:

- Line-to-ground capacitor banks in the facility (either power-factor correction or surge capacitors)
- Surge arresters

11.2 Switching Surges

Transients are triggered from capacitor switching, from line energization, and from faults. Capacitor switching transients normally cause the highest peak magnitudes. If a capacitor is switched just when the system voltage is near its peak, the capacitor pulls the system voltage down (as current rushes into the capacitor to charge it up). The system rebounds; the voltage overshoots and oscillates about the fundamental-frequency waveform (theoretically rising to two per unit, but normally less than that). This transient normally decays quickly. The oscillation occurs at the natural resonance frequency between the capacitor bank and the system, usually in the hundreds of hertz:

$$f = \frac{1}{2\pi\sqrt{L_S C}} = 60\sqrt{\frac{X_C}{X_S}} = 60\sqrt{\frac{MVA_{sc}}{Mvar}}$$

where

f = frequency, Hz
C = capacitance, F
L_S = system inductance up to the capacitor bank at nominal frequency, H
X_C = line-to-ground impedance of one phase of the capacitor bank at nominal frequency
X_L = system impedance up to the capacitor bank at nominal frequency
MVA_{SC} = three-phase short-circuit MVA at the point where the capacitor is applied
$Mvar$ = three-phase Mvar rating of the capacitor

Capacitor switching transients tend to be more severe when the capacitor is at a weaker point on the system. Other capacitors on the circuit normally help, but if two capacitors are very close together, switching in the second capacitor can create a higher voltage transient. Capacitors with significant separation can magnify the switching surges.

Normally, switching surges are not particularly severe on distribution systems. The voltages decay quickly, and magnitudes are normally not severe enough to fail line equipment (Figure 11.5 shows a typical example). Switching transients can be large enough to affect sensitive end-use loads, particularly adjustable-speed drives and uninterruptible-power supplies. The oscillation frequency is in the hundreds of hertz, low enough for the transient to pass right through distribution transformers and into customers' facilities.

EPRI's distribution power quality (DPQ) study found regular but mild transient overvoltages, most presumably due to switching operations. Transients measured on the distribution primary above 1.6 per unit are rare, averaging less than two per year (where per unit is relative to the peak of the nominal sinusoidal voltage wave). Figure 11.6 shows the average distribution of transients measured during the study (EPRI TR-106294-V2, 1996). This graph shows occurrences of transients with a peak magnitude between 1.05 and 1.9 per unit with a principal oscillation frequency between 240 and 3000 Hz (excluding transients associated with faults).

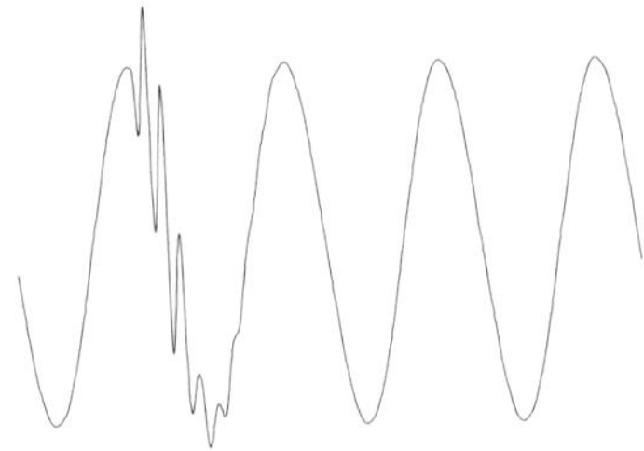

FIGURE 11.5
Capacitor switching transient. (Copyright © 1996. Electric Power Research Institute. TR-106294-V2. *An Assessment of Distribution System Power Quality: Volume 2: Statistical Summary Report.* Reprinted with permission.)

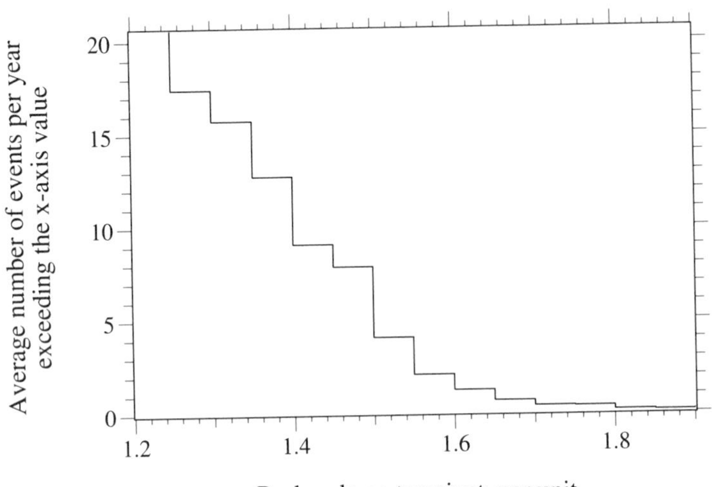

FIGURE 11.6
Average magnitudes of voltage transients measured on the distribution primary in EPRI's DPQ study. (Data from [EPRI TR-106294-V2, 1996; Sabin et al., 1999b].)

During EPRI's DPQ study, most of the transients measured occurred during the morning hours, from 5 a.m. to 10 a.m., when most switched capacitors are coming on line. The wide majority of oscillating transients measured in the study had dominant frequencies between 250 and 800 Hz.

AEP staged various switching tests on a 12.47-kV circuit and found that most capacitor switching and line energization transients were less than two per unit on the primary (Kolcio et al., 1992). In their tests, switching a 450-kvar capacitor bank produced the highest transient, just under 2.3 per unit on closing — vacuum switches produced slightly higher transients than oil switches. Line energizations caused peak primary voltage of just under 1.9 per unit. The transients decayed in less than 5 msec.

Switching transients are normally more problematic on higher voltage distribution circuits, such as 34.5 kV. Several instances of equipment failures have been reported on these higher voltage systems, usually from switching a large (10 plus Mvar) capacitor bank (Shankland et al., 1990; Troedsson et al., 1983). Capacitors tend to be larger, and sources are weaker. Plus, at higher voltages, primary level insulation capability is lower on a relative basis compared to lower voltage systems.

In capacitive circuits, switch *restrike* or *prestrike* generates more severe transients. A restrike can occur when switching a capacitor off; the switch opens at a current zero, trapping the peak voltage on the capacitor (see Figure 11.7). As the system voltage wave decreases from the peak, the voltage across the switch rapidly increases. By the time the system voltage reaches the opposite system peak, the voltage across the switch is 2 per unit (this extra voltage is the main reason that switches have a different rating for breaking capacitive circuits). If the switch restrikes because of this overvoltage, the voltage swings about the new voltage with a peak-to-peak voltage of 4 per unit, forcing the line-to-ground voltage to a peak of 3 per unit (theoretically). The interrupter can clear at another current zero and trap even more voltage on the capacitor, which can cause another more severe restrike; such multiple restrikes can escalate the voltage. Fortunately, such voltage escalation is rare. Restrikes are most likely if the interrupting contacts have not fully separated when the current first interrupts. Restrikes can fail switches and cause higher voltages on the system. Most switching technologies, including vacuum, can restrike under some conditions.

Prestriking occurs when a switch closes into a capacitor. If the gap between a switch's contacts breaks down before the contacts have fully closed, the system voltage charges up the capacitor. Because the contacts are not closed, the switch can clear at a current zero, leaving the capacitor charged. We are

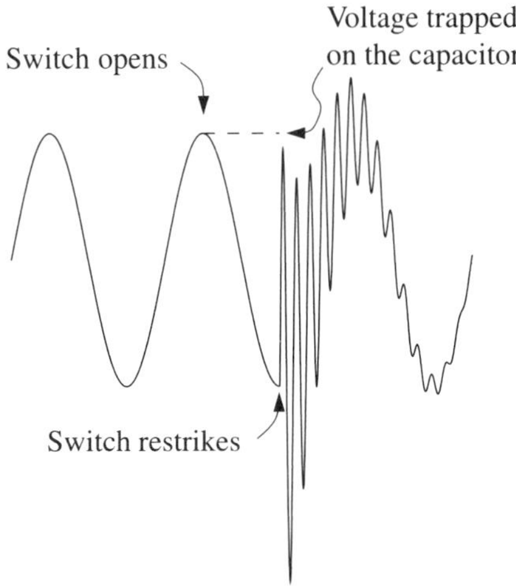

FIGURE 11.7
Restrike of a capacitor bank.

back to the restriking scenario — when the gap breaks down again, the voltage transient can oscillate around the system voltage at a much higher level. Capacitor switches should not and normally do not regularly restrike or prestrike, and voltages rarely escalate, but some failures can be traced to these switching problems.

Capacitor switching is not the only switching transient that can occur on a distribution circuit. Line energization can cause transients, as can faults. Both of these are normally benign. Walling et al. (1995) tested load-break elbows and switching-compartment disconnects. Such switching results in repeated arc restrikes and prestrikes, but the overvoltages are not particularly severe because the oscillations dampen out without clearing at a transient zero crossing (no escalation of the voltage). With a maximum overvoltage of 2.44 per unit, Walling et al. do not expect consequential effects on underground cable insulation.

11.2.1 Voltage Magnification

Certain circuit resonances can *magnify* capacitor switching transients at locations away from the switched capacitor bank (Schultz et al., 1959). Magnification often increases a transient's peak magnitude by 50%, sometimes even by 100%. Figure 11.8 shows the classic voltage magnification circuit; two sets of resonant circuits interact to amplify the transient that occurs when C_1 is switched. The most severe magnification occurs when:

- C_1 is much larger than C_2.
- The series combination of L_1 and C_1 resonates at about the same frequency as L_2 and C_2.
- There is little series or shunt resistance to dampen the transient.

If C_1 is much larger than the series impedance $L_2 + C_2$, switching C_1 in will initiate a transient that oscillates at a frequency of $\omega_1 = 1/\sqrt{L_1C_1}$. If the downstream pair, L_2 and C_2, happen to have a natural resonant frequency

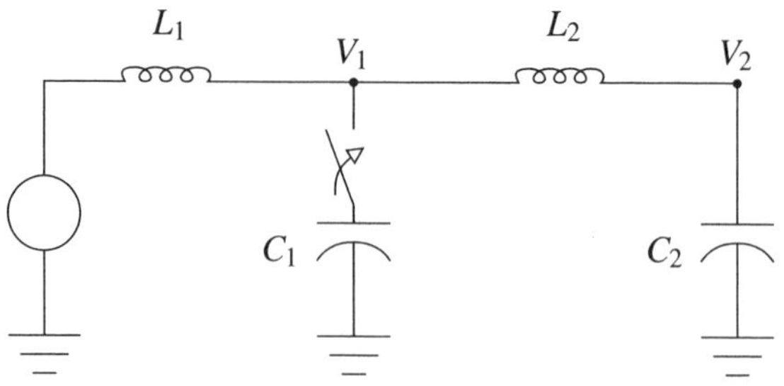

FIGURE 11.8
Voltage magnification circuit.

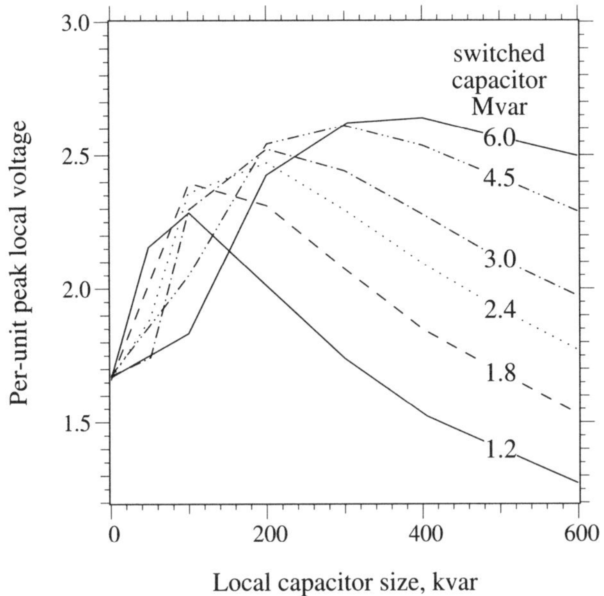

FIGURE 11.9
Transient magnitude for various sizes of switched distribution feeder capacitors and low-voltage capacitors. (From McGranaghan, M. F., Zavadil, R. M., Hensley, G., Singh, T., and Samotyj, M., "Impact of Utility Switched Capacitors on Customer Systems — Magnification at Low Voltage Capacitors," *IEEE Trans. Power Delivery*, 7(2), 862-868, April 1992. With permission. ©1992 IEEE.)

near that of ω_1, oscillation between L_1 and C_1 will pump the natural oscillation between L_2 and C_2, causing higher voltages at V_2.

Consider a 10-Mvar substation capacitor and a 1200-kvar line capacitor on a 12.5-kV circuit with a substation source impedance of 1 Ω at 60 Hz. Under these conditions, the loop resonances match when X_{L2} is 8.3 Ω, or when the capacitor is about 13 mi from the substation (given typical line impedances). For maximum magnification, the ratio of the inductive impedances equals the inverse of the capacitor kvar ratios ($kvar_1/kvar_2 = X_{L2}/X_{L1}$).

In another common scenario, low-voltage power factor correction capacitors magnify transients from switched utility capacitors (McGranaghan et al., 1992). In this scenario, L_2 is primarily determined by the customer's transformer, and L_1 is determined by the system impedance to the capacitor bank. Their parametric study found that a wide combination of local and switched utility capacitors can magnify transients as shown in Figure 11.9. Larger utility banks are more likely to generate transients that are magnified on the low-voltage side. Resistive load helps dampen these transients; facilities with primarily motor loads have higher transients. Magnified transients can also inject considerably more energy into low-voltage metal-oxide arresters than they are typically rated to handle (especially for switching large utility banks).

Transients from switched capacitor banks on the high-side of the substation bus can magnify at distribution capacitors or customer-owned capacitors (Bellei et al., 1996). Again, a wide combination of capacitor ranges can magnify surges.

11.2.2 Tripping of Adjustable-Speed Drives

The main power quality symptoms related to distribution capacitor switching is shutdown of adjustable-speed drives (ASD) and other sensitive process equipment. The front-end of an adjustable-speed drive rectifies the incoming line-to-line ac voltages to a dc voltage. The rectifier peak-tracks the three incoming ac phases, so a switching surge on any phase charges up the dc bus. Because the electronics fed by the dc bus are sensitive to overvoltages, drives normally have sensitive dc bus overvoltage trip settings, typically 1.2 per unit with some as low as 1.17 per unit. It does not take much of a transient to reach 1.2 per unit. At such low sensitivities, the system and customer voltage regulation plays an important role. If the voltage at the drive is at 1.05 per unit when the capacitor switches, the transient does not have to oscillate much to reach 1.17 or 1.2 per unit. These voltages are on the dc bus, which is fed by the line-to-line voltage. The actual line-to-ground transients on the distribution system normally need to be higher than 1.2 per unit to raise the drive dc link voltage to 1.2 per unit (and this depends on the customer transformer connection).

McGranaghan et al. (1991) used transient simulations to analyze several variables impacting the tripping of drives:

- *dc capacitor* — If the drive's rectifier has a dc capacitor, it impacts the drive's sensitivity to switching transients. Drives with larger dc capacitor take longer to charge up, which makes them less likely to trip on overvoltage.
- *Switch closing* — The worst transients are when the capacitor switches all switch at the peak of the waveform.
- *Capacitor size* — Larger capacitor banks cause a more severe transient. For the case McGranaghan et al. (1991) analyzed, capacitor banks less than 1200-kvar banks were less likely to cause drive tripouts (see Figure 11.10).
- *Local power-factor correction capacitors* — If the facility has local capacitors, magnification of the incoming transient makes the drive more likely to trip. End users can avoid this problem by configuring their power-factor correction capacitors as tuned filters (McGranaghan et al., 1992).

Locally, the addition of chokes (series inductors) connected to the input terminals of the adjustable-speed drive reduces the voltage rise on the drive's

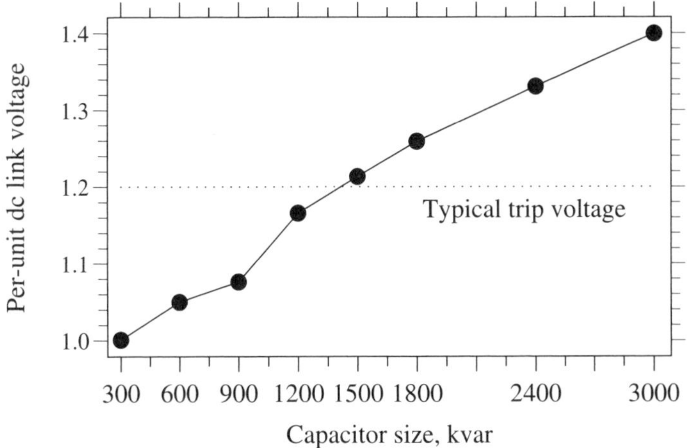

FIGURE 11.10
dc voltage on an adjustable-speed drive from switching of the given size capacitor bank (with no local power-factor correction capacitors). (From McGranaghan, M. F., Grebe, T. E., Hensley, G., Singh, T., and Samotyj, M., "Impact of Utility Switched Capacitors on Customer Systems. II. Adjustable-Speed Drive Concerns," *IEEE Trans. Power Delivery*, 6(4), 1623-1628, October 1991. With permission. ©1991 IEEE.)

dc bus. Standard sizes are 1.5, 3, and 5% impedance on the drive's kVA rating. A parametric simulation study of transmission-switched capacitors found that a 3% inductor greatly reduced the number of drive trips, but a number of transients can still trip some drives (depending on the size capacitor switched and many other variables) (Bellei et al., 1996).

11.2.3 Prevention of Capacitor Transients

Switched capacitors cause the most severe switching transients. Utility-side solutions to prevent the transient include zero-voltage closing switches or pre-insertion resistors or reactors. Zero-voltage closing switches time their closing to minimize transients. Switches with pre-insertion resistors place a resistor across the switch gap before fully closing contacts; this reduces the inrush and dampens the transient. Any of these options are available for substation banks, and utilities use them often, especially in areas serving sensitive commercial or industrial customers.

For distribution feeder banks, the zero-crossing switch controllers are the only available option for reducing transients. These controllers time the closing of each switch contact to engage the capacitor at the instant where the system voltage is zero, or very close to zero. An uncharged capacitor closing in at zero volts causes no transient.

The switches time the initiation of the switching so that by the time the switch engages, the system voltage is at zero. Since the capacitor should have no voltage, and it is switched into the system when it has no voltage, the switching does not create a transient. Timing and repeatability are important. One popular 200-A vacuum switch takes 0.015 sec to close and has a repeatability of ±3 degrees (±0.14 msec). Each phase is controlled separately.

Other claimed benefits of zero-crossing switches include increased capacitor and switch life, reduction of induced voltages into the low voltage control wiring, and reduction of ground transients. Zero-crossing switches also eliminate capacitor inrush, including the much more severe inrush found when switching a capacitor with a nearby capacitor already energized.

11.3 Harmonics

Distortions in voltage and current waveshapes can upset end-use equipment and cause other problems. Harmonics are a particularly common type of distortion that repeats every cycle. Harmonically distorted waves contain components at integer multiples of the base or *fundamental* frequency (60 Hz in North America). Resistive loads like incandescent lights, capacitors, and motors do not create harmonics — these are passive elements — when applied to 60-Hz voltage; they draw 60-Hz current. Electronic loads, which create much of the harmonics, do not draw sinusoidal currents in response to sinusoidal voltage.

A very common harmonic producer is the power supply for a computer, a switched-mode power supply that rectifies the incoming ac voltage to dc (see Figure 10.21). The bridge rectifier has diodes that conduct to charge up capacitors on the dc bus. The diodes only conduct when the ac supply voltage is above the dc voltage for just a portion of each half cycle. So, the power supply draws current in short pulses, one each half cycle. Each pulse charges up the capacitor on the power supply. The current is heavily distorted compared to a sine wave, containing the odd harmonics, 3, 5, 7, 9, etc. The third harmonic may be 80% of the fundamental, and the fifth may be 60% of the fundamental. Figure 11.11 shows examples of the harmonic current drawn by switched-mode power supplies in computers along with other sources of harmonics.

Other very common sources of harmonics are adjustable-speed drives and other three-phase dc power supplies (see Figure 10.28). These rectify the incoming ac waveshape from each of the three phases. In doing so, drives create current with harmonics of order 5, 7, 11, 13, etc. — all of the odd harmonics except multiples of three. In theory, drives create a fifth harmonic that is one-fifth of the fundamental, a seventh harmonic that is one-seventh of the fundamental, and so forth. Other harmonic-producing loads include arc furnaces, arc welders, fluorescent lights (with magnetic and especially with electronic ballasts), battery chargers, and cycloconverters.

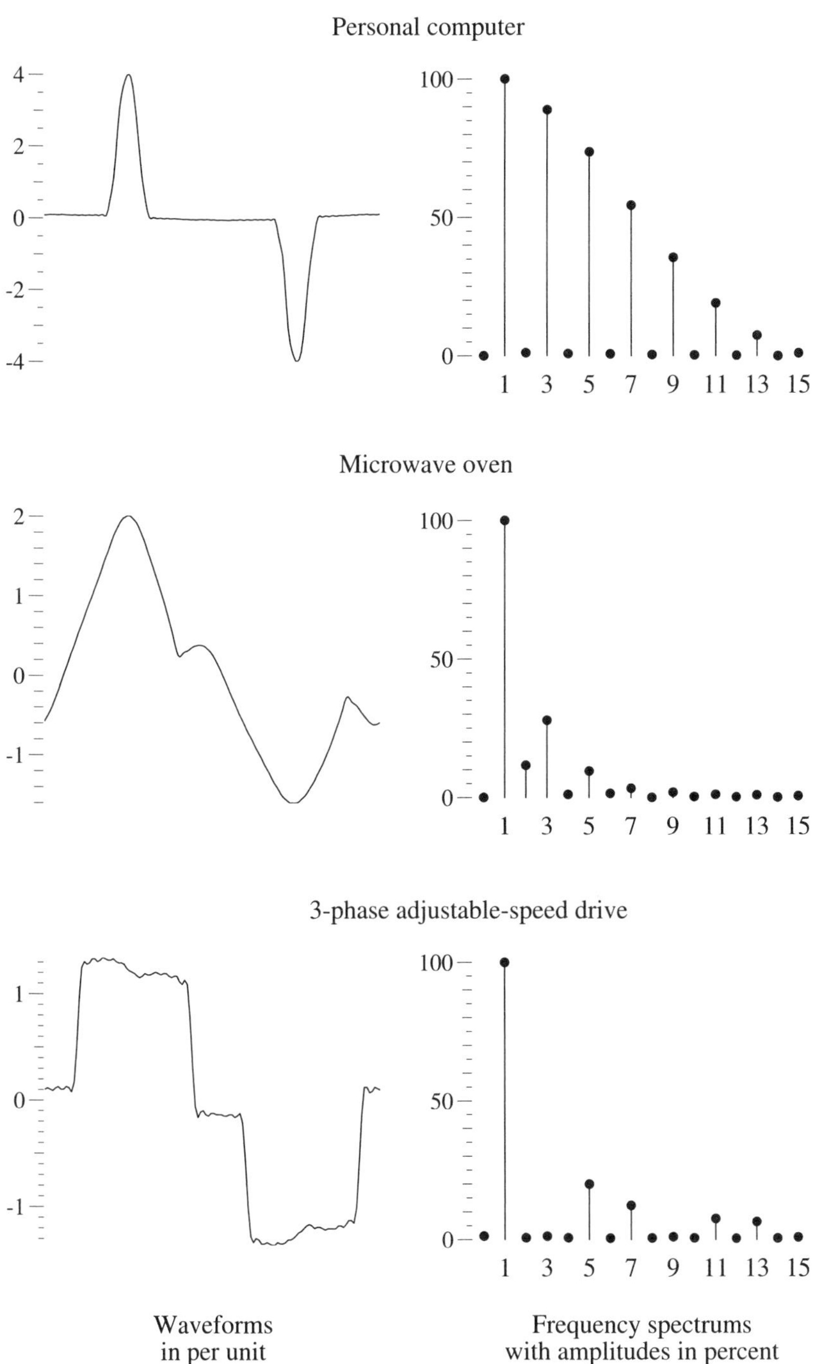

FIGURE 11.11
Examples of harmonic-current sources. (Data from EPRI PEAC, various measurements.)

Harmonic currents and harmonic voltages are interrelated. Commonly, nonlinear loads are modeled as ideal current injections ("ideal" meaning they inject the same current regardless of the voltage). An *n*th-harmonic source driving current through an impedance $R+jX$ raises the *n*th voltage to

$$V_n = R \cdot I_n + jn \cdot X \cdot I_n$$

Utility distribution systems are normally stiff enough (low Z) that they can absorb significant current distortions without voltage distortion.

Harmonics have been heavily researched and studied, probably more so than any other power quality problem. Despite the attention, harmonic problems are uncommon on distribution circuits. Most harmonic problems are isolated to industrial facilities; end-use equipment produces harmonics, which cause other problems within the facility.

We can characterize harmonics by the magnitude and phase angle of each individual harmonic. Normally, we specify each harmonic in percent or per unit of the fundamental-frequency magnitude. Another common measure is the *total harmonic distortion*, THD, which is the square-root of the sum of the squares of each individual harmonic. Voltage THD is

$$V_{THD} = \frac{\sqrt{\sum_{h=2}^{\infty} V_h^2}}{V_1}$$

where V_1 is the rms magnitude of the fundamental component, and V_h is the rms magnitude of component h. The Fourier transform is used to convert one or more waveshape cycles to the frequency domain and V_h values.

IEEE provides a recommended practice for harmonics, IEEE Std. 519-1992, which gives limits for utilities and for end users. Utilities are expected to maintain reasonably distortion-free voltage to customers. For suppliers at 69 kV and below, IEEE voltage limits are 3% on each individual harmonic and 5% on the total harmonic distortion (Table 11.1). Both of these percentages are referenced to the nominal voltage.

Current distortion limits (Table 11.2) were developed with the objective of matching the voltage distortion limits. If users limit their current injections according to the guidelines, the voltage limits will remain under the guidelines imposed on the utility (as long as no major circuit resonance exists). The individual harmonics and the total harmonics (total demand distortion, TDD) are referenced to the maximum demand at that point, either the 15- or 30-min demand. Harmonic current limits depend on the facility size relative to the utility at the interconnection point. Larger facilities — relative to the strength of the utility at the interconnection point — are more restricted. Likewise, a facility at a stronger point in the system is allowed to

TABLE 11.1

Voltage Distortion Limits

	Voltage Distortion Limit
Individual harmonics	3%
Total harmonic distortion (THD)	5%

Note: For a bus limit at the point of common coupling at and below 69 kV. For conditions lasting less than 1 h, the limits may be exceeded by 50%.

Source: IEEE Std. 519-1992.

TABLE 11.2

Current Distortion Limits for Distribution Systems (120 V through 69 kV)

I_{sc}/I_L	$h < 11$	$11 \le h < 17$	$17 \le h < 23$	$23 \le h < 35$	$34 \le h$	TDD
< 20	4.0	2.0	1.5	0.6	0.3	5.0
20–50	7.0	3.5	2.5	1.0	0.5	8.0
50–100	10.0	4.5	4.0	1.5	0.7	12.0
100–1000	12.0	5.5	5.0	2.0	1.0	15.0
> 1000	15.0	7.0	6.0	2.5	1.4	20.0

Note: Even harmonics are limited to 25% of the odd harmonic limits above. Current distortions that result in a dc offset are not allowed. For conditions lasting less than 1 h, the limits may be exceeded by 50%.

Source: IEEE Std. 519-1992.

inject more than the same facility at a weaker point because the same injection at a weaker point creates more voltage distortion. The ratio I_{sc}/I_L determines which set of limits applies. I_{sc} is the maximum short-circuit current; I_L is the maximum demand for the previous year.

Both the voltage and current limits apply at the *point of common coupling*, the point on the system where another customer can be served. Either side of the distribution transformer can be the point of common coupling, depending on whether the transformer can supply other customers. The point of common coupling is often taken as the metering point.

Harmonics can impact a variety of end-use equipment (IEEE Task Force, 1985; IEEE Task Force, 1993; Rice, 1986). Some of the most common end-use effects are as follows:

- *Transformers* — Current distortion increases transformer losses and heating, enough so that transformers may have to be derated [see (IEEE Std. C57.110-1998)]. Dry-type transformers are particularly sensitive.
- *Motors* — Voltage distortion induces heating on the stator and on the rotor in motors and other rotating machinery. The rotor presents a relatively low impedance to harmonics, like a shorted transformer

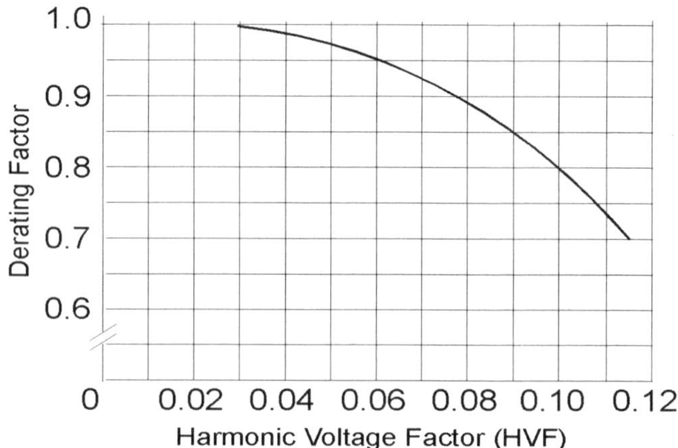

FIGURE 11.12
Motor derating due to harmonics. NEMA's harmonic voltage factor is slightly different from THD; it includes only the odd harmonics that are not multiples of three. (From NEMA MG-1, *Standard for Motors and Generators*, 1998. With permission.)

winding. The effects are similar to motors operating with voltage unbalance, although not quite as severe since the impedance to the harmonics increases with frequency. Figure 11.12 shows the National Electrical Manufacturers Association's (NEMAs) recommended derating factor for motors running with distorted voltage.

- *Conductors and Cables* — Resistance increases with frequency due to skin effect and proximity effect, particularly on larger conductors. As an example, at the seventh harmonic (420 Hz), a 500-kcmil cable has a resistance that is 2.36 times its dc resistance (Rice, 1986). Neutral conductors in facilities are especially prone to problems; third-harmonic currents from single-phase loads add in the neutral, which can actually increase neutral current above that in the phase conductors (EPRI PEAC Application #6, 1996).
- *Sensitive electronics* — Ironically, some of the main harmonic producers are also sensitive to harmonics. If an uninterruptible power supply (UPS) detects excessive distortion, it may switch to its batteries; after the batteries run out, the critical load may drop out. Harmonics can impact digital devices that depend on a voltage zero crossing for timing (digital clocks being the simplest example). Excessive harmonics can introduce additional zero crossings that can disrupt controllers.

Most harmonic problems occur in industrial or commercial facilities. Distribution equipment is fairly immune to problems from harmonic voltages or currents. Heating from harmonics is not normally a problem for overhead conductors. Underground cables are more sensitive to heating. Harmonic currents can appreciably increase cable temperatures.

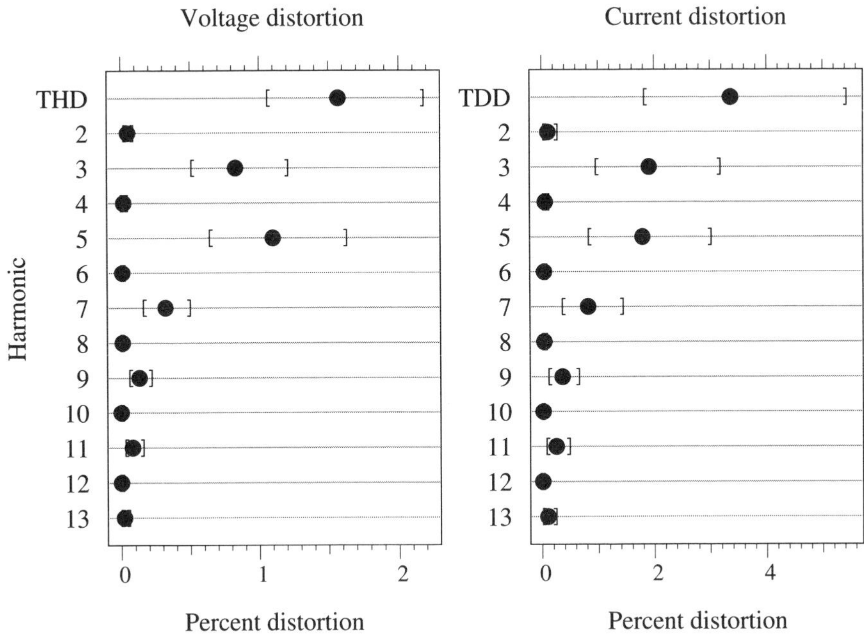

FIGURE 11.13
Voltage and current distortion measured in EPRI's DPQ study. The dot marks the mean distortion for the given harmonic, and the brackets mark the 5th and 95th percentiles. (Data from [EPRI TR-106294-V2, 1996; Sabin et al., 1999a].)

Many harmonic problems relate to heating, either in transformers, motors, neutral conductors, or capacitors. Because of the thermal time constants of equipment, heat-related problems occur for sustained levels of harmonics (usually along with heavy power-frequency loading). Heating degrades insulation and leads to premature failure. Higher peak voltages can also cause insulation breakdowns, with capacitors being highly sensitive.

Most distribution circuits have very little distortion, having neither voltage distortion nor current distortion. Figure 11.13 shows average distortion measured during EPRI's DPQ study (EPRI TR-106294-V2, 1996; Sabin et al., 1999a). None of the DPQ sites had an average voltage THD above the IEEE 519 limit of 5% (the worst couple of sites averaged just under 4.8% THD). And, only 4% of the DPQ sites exceeded this IEEE 519 limit more than 5% of the time. At the DPQ sites, 95% of the time, the voltage THD was less than 2.2%.

Substation sites had slightly lower voltage and current distortions than those recorded at feeder sites: substation average THD = 1.37%; overall average THD = 1.57%. Industrial feeders (defined as having at least 40% of the load coming from industrial loads) showed only slightly higher voltage distortion than the average for all sites.

Other monitoring studies also show fairly low harmonics. Surveys of 76 sites on the Southwestern Electric Power Company's system found that none exceeded a voltage THD of 5% (less than 5% of exceeded 4%; 87% had THD less than 2.5%) (Govindarajan et al., 1991). A few of the sites had current distortion exceeding IEEE 519 standards. In a large survey of locations on the Sierra Pacific Power Company, voltage THD averaged 1.59%, and current THD averaged 5.3% (Etezadi-Amoli and Florence, 1990).

Some work has been done to investigate the impact of new harmonic loads on the harmonic distortion. Compact fluorescent lights with electronic ballasts, heat pumps and air conditioners with adjustable-speed drives, computers, and electric vehicle battery chargers are some of the loads that could drive up harmonics. Pileggi et al. (1993) predicted that if every home used two or three electronically ballasted fluorescent lights, voltage distortion may exceed 5%. This is hard to believe, but it does make us consider the impact of future electronic loads. Dwyer et al. (1995) also came to similar conclusions: modest numbers of electronic compact fluorescent lights (50 W per house) can raise the voltage distortion above 5%, especially if feeder capacitors create resonances that amplify the harmonics. It helps that multiple single-phase loads have some phase-angle cancellation; harmonic loads do not necessarily sum linearly (Mansoor et al., 1995).

In the northeastern U.S., Emanuel et al. (1995) predict rising levels of harmonic distortion based on predictions of increased use of adjustable-speed drives, electronic fluorescent lights, computer loads, and other nonlinear loads. They predict that more nonlinear loads can add as much as 0.3% per year to the voltage THD under the worst-case scenario. A limited number of measurements by the same group of researchers over a 10-year time period found that harmonic voltages have increased at a rate of about 0.1% per year (it takes 10 years for the THD to increase by a percentage point) (Nejdawi et al., 1999).

Penetrations of air conditioners driven by adjustable-speed drives to 10 to 20% can raise voltage THD levels above 5% (Gorgette et al., 2000; Thallam et al., 1992). In another study, Bohn (1996) predicted that electric-vehicle battery charger penetration levels as low as 5% could increase voltage THD above 5%, especially on weak systems. In both of these cases, investigators found wide differences in equipment designs. Some generate much more harmonics than others. For example, some adjustable-speed drives with more sophisticated rectifier sections cause much less harmonics than standard rectifiers. As electronic loads grow, utilities must involve themselves in setting guidelines to limit the harmonics introduced by end-use equipment.

Finding the source of harmonics can be tricky — IEEE 519A provides two ways to track down harmonics (IEEE P519A draft 7, 2000):

1. *Time variations* — If harmonics are intermittent, correlating the time variations of the voltage or current harmonics with the operation time of facilities or of specific equipment can identify the harmonic producer.
2. *Monitoring with capacitor banks off* — In a radial circuit with no capacitors, harmonic sources inject current that flows back to the power

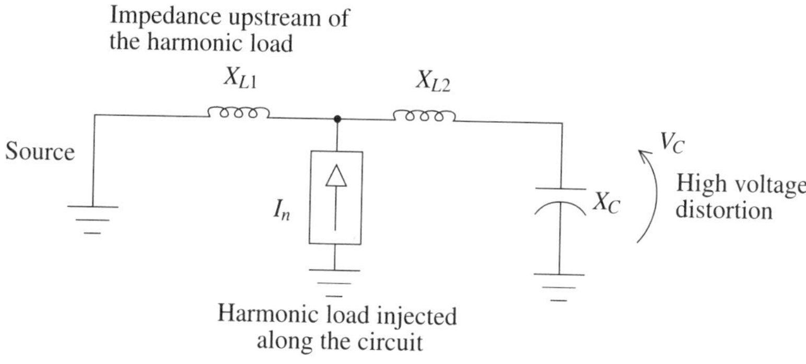

FIGURE 11.14
Harmonic resonance.

source; following the current leads to the source of the harmonics. If capacitors are present, resonances can mask the true source of harmonics.

11.3.1 Resonances

When harmonics cause problems, capacitors are often a contributing factor. Capacitors can cause resonances that amplify harmonics. Harmonic problems often show up first at capacitor banks, where harmonics cause fuse operations or even capacitor failure. Most problems with severe harmonics are found in industrial facilities where capacitors resonate against the system impedance. Utility resonances are infrequent but sometimes cause problems.

A capacitor on a distribution circuit will resonate against the inductance back to the system source (including the line impedances and transformer impedances). At the resonant frequency, the circuit amplifies harmonics injected into the circuit (see Figure 11.14).

The resonance point between the capacitor and the system for the scenario in Figure 11.14 (this is the same frequency that the system will ring at during a switching surge) is

$$n = \sqrt{\frac{X_C}{X_L}} = \sqrt{\frac{MVA_{sc}}{Mvar}}$$

where

n = order of the harmonic (multiples of the fundamental frequency)
X_C = line-to-ground reactance of one phase of the capacitor bank at nominal frequency
$X_L = X_{L1} + X_{L2}$ = system reactance at nominal frequency
MVA_{SC} = three-phase short-circuit MVA at the point where the capacitor is applied
$Mvar$ = three-phase Mvar rating of the capacitor

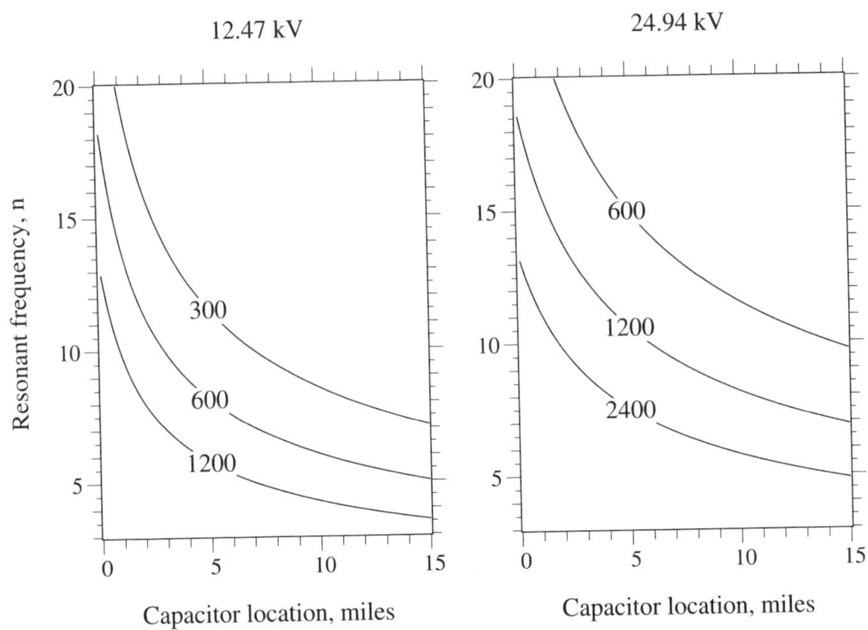

FIGURE 11.15
Resonant frequency vs. capacitor size (in kvar) and location for balanced, positive-sequence resonances (500-kcmil AAC).

If a nonlinear load injects a harmonic frequency equal to the system's resonant point, it can create significant harmonic voltages. Common problem frequencies for resonances are n = 5, 7, 11, and 13. Larger capacitors lower the resonant point to where it is more likely to cause problems. For 15-kV class systems, resonances are possible at common capacitor locations for 600 to 1200-kvar banks (see Figure 11.15). For the single-capacitor circuit in Figure 11.14, the voltage distortion at the capacitor is:

$$V_C = \frac{X_C X_{L1}}{nX_{L1} + nX_{L2} - \frac{1}{n}X_C} I_n$$

where I_n is the nth harmonic current. The worst conditions are when the harmonic source is right at the capacitor or downstream of the capacitor (X_{L2} is small). Harmonics injected further upstream are amplified less. This simplification ignores the resistance of the line and the resistance of the loads. Both reduce the peak magnitude.

Multiple capacitors on a circuit create multiple resonant points that can require more sophisticated analysis. For complicated circuits, especially with capacitor banks, a harmonic analysis program helps identify problems and try out solution options. A harmonic program models multiple harmonic sources at different locations; the solution shows the harmonic current flows

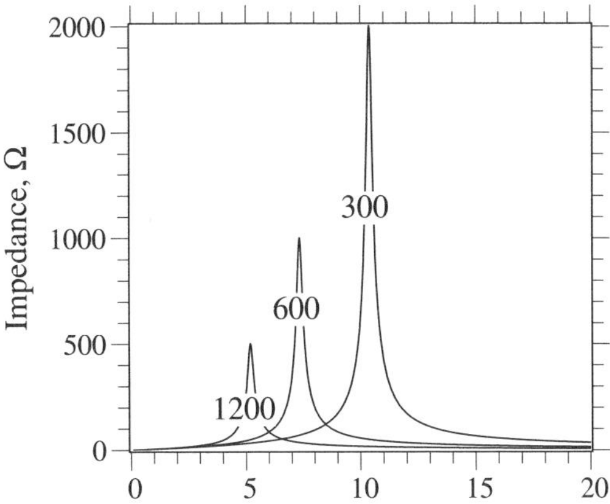

FIGURE 11.16
Harmonic scans for various size capacitors (shown in kvar) applied on a 12.47-kV circuit 6 miles from the substation.

and the harmonic voltages. Another tool, the frequency scan — a plot of the impedance at a location vs. frequency — helps identify system resonances that appear as peaks in the scan (see Figure 11.16).

Distribution systems rarely need solutions to harmonics because typical voltage distortion levels are so low. When problems arise, specific harmonic producers or specific resonant conditions are often to blame. For offending producers, we can enforce harmonic standards. Where resonances occur, the solutions include:

- *Relocate or remove capacitors* — The easiest solution is to disconnect the capacitors that form the resonant circuit. Relocating capacitors or changing their size can also shift the resonant point enough to ease problems.
- *Harmonic filters* — While rarely done, tuned filters, consisting of a capacitor in series with an inductor, absorb harmonics near their tuned frequency. The best location for filters is closest to the harmonic producers. If harmonic producers are distributed, filters located near the center of the line can absorb some of the harmonics. Applying filters requires study to ensure that additional resonances are not created.

11.3.2 Telephone Interference

Current flowing through distribution conductors induces voltage along parallel conductors, both power conductors and communication conductors. On low-voltage communication lines, induced voltages can interfere with

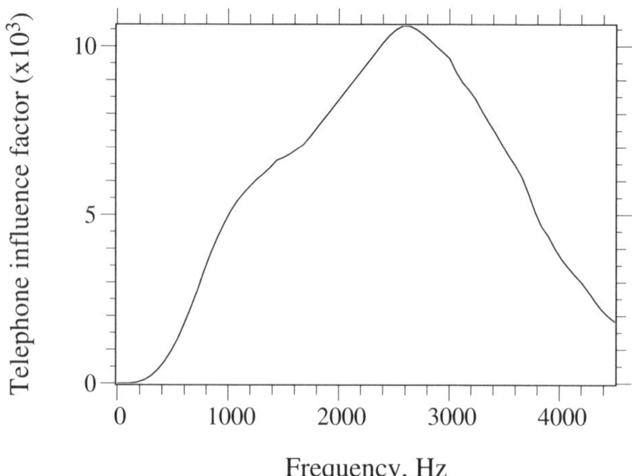

FIGURE 11.17
Telephone interference factor weights (W_f) with frequency.

communication signals, particularly on telephone circuits. Telephone interference is less common than in the past, primarily because telephone companies use more immune circuits with shielded cables or fiber optics. Open-wire communication lines are the most susceptible to interference. Interference depends on the offending current magnitude, the distance between conductors, and the length that the circuits run in parallel.

Harmonics increase the likelihood of telephone interference. Higher frequencies introduce noise that transfers more readily through telephone equipment and is more easily heard. *Telephone influence factors (TIF)* weight harmonics differently, depending on frequency. Figure 11.17 shows TIF weighting factors with frequency. This curve is based on the response of the telephone equipment (mainly the handset) and human perception of different frequencies. The most sensitive frequency is near 2600 Hz (n = 43). Lower-order harmonics scaling as far down as the third harmonic can interfere with telephone and other communication circuits. The total TIF for either voltage or current is a weighted factor at each harmonic relative to the total rms current:

$$TIF = \sqrt{\sum \left(\frac{X_f W_f}{X_t} \right)^2}$$

where

X_t = total rms voltage or current
X_f = single-frequency rms current or voltage at frequency f
W_f = single-frequency TIF weighting at frequency f

TABLE 11.3

Balanced $I \cdot T$ Guidelines between Overhead Power Circuits and Overhead Telephone Circuits

Classification	$I \cdot T$
Levels most unlikely to cause interference	$\leq$ 10,000[a]
Levels that might cause interference	10,000–25,000
Levels that probably will cause interference	> 25,000

[a] For some areas that use a ground return for either telephone or power circuits, this value may be as low as 1500.

Source: IEEE Std. 519-1992.

TIF is a factor that indicates the harmonic content weighted by frequency; TIF is like THD but scales each harmonic according to its relative interference capability. Because interference depends on current magnitudes as well as harmonic content, interference potential is often described in terms of the product of current and the telephone influence factor, $I \cdot T$, where I is the rms current in amperes, and T is the TIF. Table 11.3 shows guidelines for $I \cdot T$ from IEEE 519.

Harmonics in the ground return path are much more likely to interfere with communication circuits. Balanced third and ninth harmonics as well as other multiples of three (the *triplens* or *triples*) are zero-sequence frequencies. These currents generally add in the neutral. Because zero-sequence circuit impedances are higher than positive-sequence impedances, zero-sequence resonances can arise with smaller capacitor banks than for positive-sequence resonances. Figure 11.18 shows capacitor locations that cause zero-sequence resonances that force more current into the ground-return path. Table 11.4 shows locations where capacitors resonate for zero-sequence third harmonics.

While the triplen harmonics are most likely to interfere with communications circuits, other common harmonics such as the fifth or seventh can cause trouble. Although the fifth and seventh harmonics are not naturally zero-sequence currents, if the harmonic components on each of the three phases are unbalanced, the unbalanced portion of harmonics will flow in the ground.

Telephone interference problems are usually solved by the telephone company in cooperation with the electric utility involved. Solutions are often trial-and-error; harmonic load-flow models can help test out solutions. The normal utility-side solutions are (IEEE Std. 776-1992; IEEE Std. 1137-1991; IEEE Working Group, 1985):

- *Change size* — For problems involving a resonance, increasing or decreasing the size of a bank can shift the resonant point enough to ease interference problems. The easiest solution is to disconnect the bank that is contributing to the problem. This is also a good first step to quantify the role of the capacitor bank in the interference.

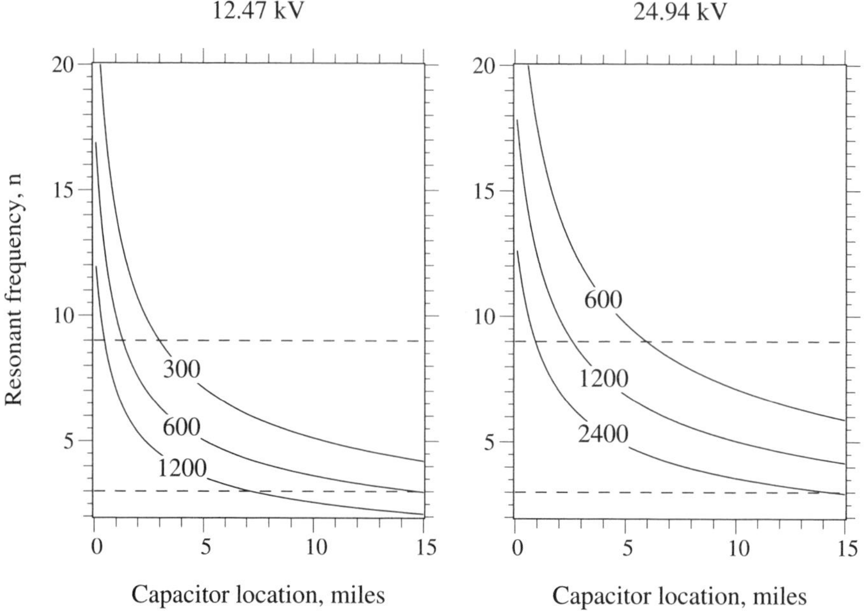

FIGURE 11.18
Resonant frequency vs. capacitor size (in kvar) and location for zero-sequence resonances (500-kcmil AAC, $X_0 = 1.9\Omega$/mile).

Table 11.4

Capacitor Location in Miles from the Substation Where a 3rd-Harmonic Zero-Sequence Resonance Occurs for the Given Size Capacitor (for 500-kcmil AAC with $X_0 = 0.3605\Omega/1000$ ft)

	System Voltage, kV			
kvar	**4.8**	**12.47**	**24.94**	**34.5**
600	2.1	14.8	59.8	
900	1.3	9.7	39.6	76.2
1200	1.0	7.2	29.5	56.9
1800		4.7	19.4	37.6
2400			14.4	27.9
4800			6.8	13.4

- *Balancing* — Balancing harmonic loads between phases reduces the ground return current. In cases of long single-phase runs, we can upgrade the line to a two-phase line to reduce ground currents. Note that this will not help with the third-harmonic currents.
- *Move the bank* — Moving a capacitor can change a resonant point enough to stop interference problems. On some circuits, one can also move the capacitor away from the telephone circuits having problems.

- *Unground the bank* — A floating-wye connection has no connection to the ground, so the connection blocks zero-sequence harmonic currents. Two-bushing capacitor units are necessary for floating the wye point (unless the utility floats the capacitor tanks and deals with the safety issues that accompany that practice).
- *Add a neutral filter* — While not a common solution, a tuned reactor in the ground path of a wye-connected capacitor bank is invisible to positive and negative sequences, but it changes the zero-sequence resonant frequency of a distribution feeder and often eliminates the resonant problem. As another approach, on a bank close to the harmonic source, one could also tune the filter to drain the offending harmonics (so they do not flow back along the electric/phone lines).
- *Add a grounding bank* — Grounding banks can change zero-sequence impedances to shift resonances or shift current flows away from parallel utility/communication runs to avoid interference. They also provide a low-impedance path for zero-sequence characteristic currents to flow. A zig-zag bank can be added, or existing floating-wye – delta connections can be converted to grounded wye – delta.

On the communications side, several solutions are possible: conversion to more immune circuits such as fiber optics or shielded conductors. 60-Hz rejection filters, drainage coils between conductor pairs to reduce induced voltages, and neutralizing transformers which add a voltage opposing the induced voltages. Finally, if the interference can be traced to major harmonic producers, the offending producer can be turned off or harmonics can be filtered at the source.

11.4 Flicker

Light flicker is due to rapidly changing loads that cause fluctuation in the customer's voltage. Even a small change in voltage can cause noticeable lamp flicker. Flicker is an irritation issue. Flicker does not fail equipment; flicker does not disrupt sensitive equipment; flicker does not disrupt processes. Flickering lights or televisions or computer monitors annoy end users.

Annoying lamp flicker occurs when rapid changes in load current cause the power system voltage to fluctuate. Sawmills, irrigation pumps, arc welders, spot welders, elevators, laser printers — all can rapidly change their current draw. In arc furnaces, arcs fluctuate wildly from cycle to cycle, continuously flickering the voltage. When starting, motors draw significant inrush, five or six times their normal current, possibly depressing the voltage for tens of seconds. Some loads like elevators turn on and off repeatedly. All of these fluctuating load changes can cause flicker.

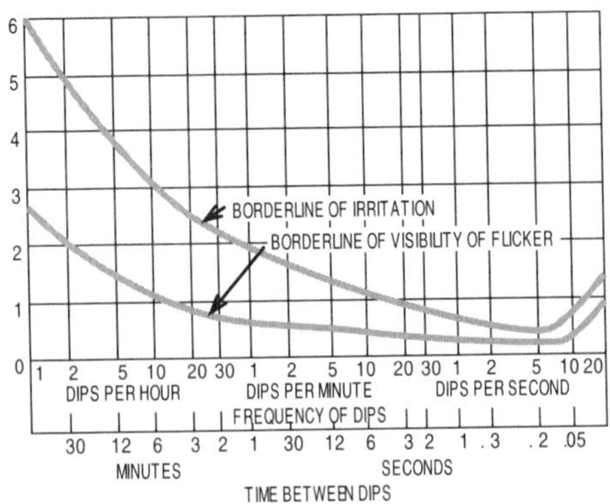

FIGURE 11.19
GE flicker curve. (From IEEE Std. 141-1993. Copyright 1994 IEEE. All rights reserved.)

Susceptibility to flicker depends on the stiffness of the supply system. So, flicker is more common on lower-voltage systems and at the ends of long circuits.

Both magnitude and frequency of fluctuations affect flicker perceptions. People are most sensitive to flicker that changes from two to ten times per second, and flicker is visible up to about 35 Hz. The most common flicker reference curve that has been developed is the GE flicker curve based on tests by General Electric and several utilities around 1930 (General Electric GET-1008L) and republished in the IEEE Red Book (IEEE Std. 141-1993). [See Walker (1979) for comparison with other flicker curves.] An IEEE survey found that 69% of utilities were using the GE curve (Seebald et al., 1985). The GE curve shows both a "threshold of perception" and a "threshold of irritation" (see Figure 11.19). The GE flicker curve is based on square-wave changes to the supply voltage at the frequencies indicated. Load changes that are more gradual than a stepped square wave result in less noticeable flicker; the eye–brain response is more sensitive to rapid light changes (up to a point). The flicker curve is based on a change in voltage (ΔV) relative to the steady-state voltage (V) (see Figure 11.20). Both ΔV and V are best represented by rms quantities; mixing rms, peak, or peak-to-peak quantities gives errors of 2 or $\sqrt{2}$. The frequency of the changes is also confusing, even for changes that are regular and periodic. The GE flicker curve is based on the number of *dips* per minute or second or hours. Some other curves or criteria are based on the number of *changes* per unit of time (frequency of changes are twice the frequency of dips).

The flicker tendency of different lights varies significantly:

- *Smaller incandescent bulbs* — Smaller filaments cool more quickly, so their light output changes more for a given fluctuation. Higher-voltage bulbs also have smaller filaments, so they flicker less.

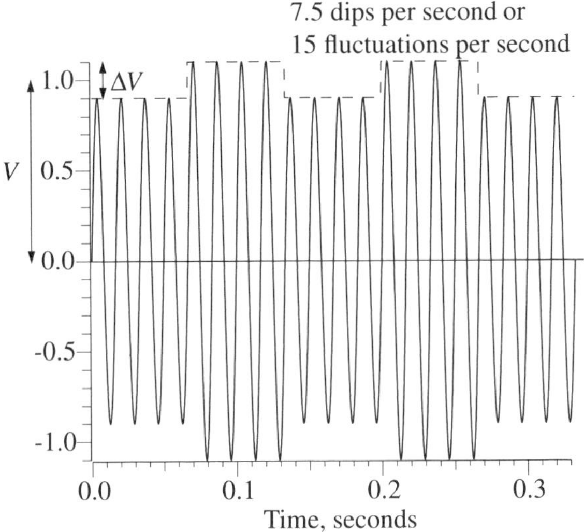

FIGURE 11.20
Characterizing the magnitude and frequency of voltage changes.

- *Dimmers* — Some electronic dimmers greatly increase voltage flicker, especially at low-light settings (EPRI PEAC Brief No. 25, 1995). Flicker on some dimmed lights is equivalent to the same flicker on undimmed lights with double the voltage fluctuation.
- *Fluorescent lights* — Whether magnetic or electronically ballasted, fluorescent lights normally flicker less for a given voltage input, some by a factor of over five (voltage changes on a fluorescent light need to be five times greater to cause the same level of flicker perception on an incandescent light). This is not always the case though; some fluorescent lights flicker as much as incandescent lights, and there is no general rule to identify whether a fluorescent light is sensitive. Fluorescent light ouput depends on the point on wave where the phosphors ignites, so peak voltage affects fluorescent lights more than rms-voltage changes.

While lights receive the most attention, voltage fluctuations can also cause televisions and computer monitors to waver.

Because flicker is based on human perception, it is an inexact science. Some people are more sensitive than others. Younger people are more sensitive. Flicker is more noticeable in quiet, low-light settings. In addition, the steady-state voltage during the time of the fluctuation impacts flicker — lower nominal voltages lead to less flicker; a 5% reduction in voltage reduces perceived flicker by 5% (EPRI PEAC Brief No. 36, 1996). Flicker is more noticeable with side vision (peripheral vision) than with straight-ahead vision. The perception process involves the thermal characteristics of the filament and the human eye–brain response.

The European community has derived sophisticated methods of characterizing and analyzing flicker, developing a flickermeter (IEC 868, 1986) and a comprehensive set of standards. The IEEE is moving towards adopting the European flickermeter approach to quantify flicker (Halpin et al.). The flickermeter models the complex lamp–eye–brain interaction to uniformly quantify flicker from a variety of sources, whether the offending load is an arc furnace or repetitive motor starts. The flickermeter measures voltage and produces these outputs:

- *Instantaneous flicker sensation* — This output is in "perception units," a unit-less quantity. One per unit is defined as the threshold of perception for 50% of the population.
- *Short-term flicker indicator* (P_{st}) — The short-term flicker indicator is a weighted factor based on probabilities of the instantaneous flicker sensation over a ten-minute monitoring period. A P_{st} = 1 is the threshold of irritation, the level that the majority of the population finds annoying. P_{st} values as low as 0.7 have been found to be visible under some conditions.
- *Long-term flicker indicator* (P_{lt}) — Long-term flicker is based on 2 h of flicker measurement and combines 12 consecutive P_{st} values (using the cubed root of the sum of the cubes of the 12 P_{st} values).

The flickermeter produces comparable results to the GE flicker curve when the flickermeter has a square-wave input as shown in Figure 11.21. Both the GE curve and the IEC curve (calibrated for 120-V lamps) produce comparable results. A flicker curve analysis is easier for loads that fluctuate periodically in a square-wave fashion. For intermittent or chaotic loads like welders, a flicker curve is impossible to use. Converting a measured waveform to the frequency domain with a Fourier transform and plotting the results on a flicker curve is imprecise. While the exercise reveals predominant frequencies, the Fourier transform does not mimic the brain's response to flickering light. The flickermeter is much more appropriate for cases where loads fluctuate irregularly.

The P_{st} is an rms quantity that we can treat like a per-unit voltage. If the current draw from the fluctuating load doubles (and the waveshape stays the same), the voltage deviation (ΔV) will double, and P_{st} will double. Just as we can use voltage dividers, we can use impedance dividers to estimate flicker propagation and estimate flicker severity at other locations on a circuit (see Figure 11.22). Flicker decreases as one moves upstream of the fluctuating load on a radial circuit. Customers at or downstream of the fluctuating load experience the worst flicker.

Multiple sources of harmonics can influence flicker differently, depending on the characteristics of the fluctuating loads. Multiple sources combine as

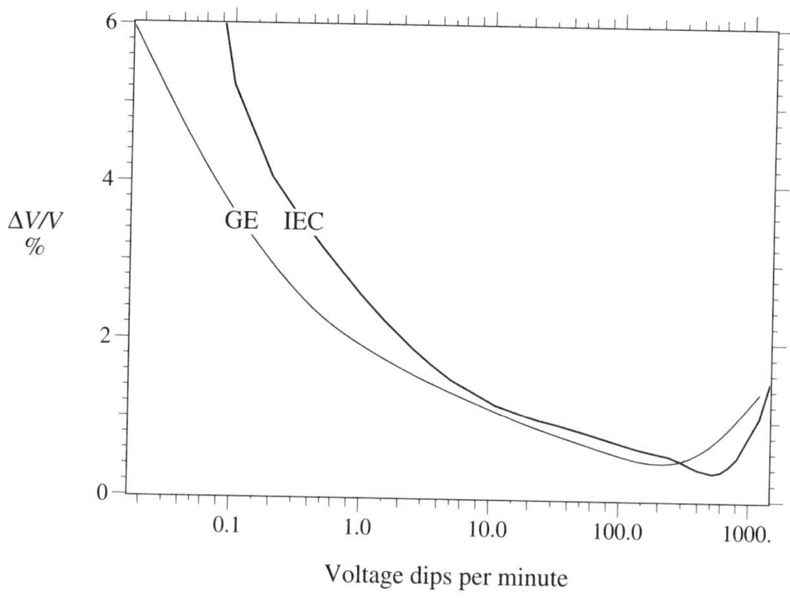

FIGURE 11.21
GE borderline of irritation curve compared to the voltage change necessary to produce $P_{st} = 1$ from the IEC flickermeter with a square-wave input.

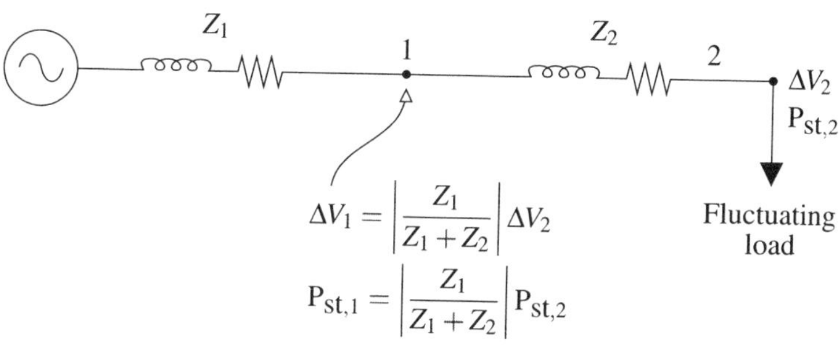

$$\Delta V_1 = \left|\frac{Z_1}{Z_1 + Z_2}\right| \Delta V_2$$

$$P_{st,1} = \left|\frac{Z_1}{Z_1 + Z_2}\right| P_{st,2}$$

FIGURE 11.22
Propagation of voltage flicker.

$$P_{st,TOTAL} = \left(\sum P_{st,i}{}^{m}\right)^{\frac{1}{m}}$$

UIE, the organization that developed many of the European flicker standards (UIEPQ-9715, 1999), suggests the following values of m for combining multiple fluctuating loads (smaller m is more conservative):

$m = 4$: for arc furnaces operated to avoid simultaneous operations
$m = 3$: when the risk of simultaneous voltage variations is minimal (Most unrelated loads are in this category.)

$m = 2$: simultaneous operation of devices producing random-noise type fluctuations, such as arc furnaces operating at the same time
$m = 1$: when there is a high probability of simultaneous operation

The flickermeter can effectively quantify flicker for existing loads; it does not help much for predicting the flicker at new installations. It is possible to model the time variations of fluctuating loads and run a software "flicker-meter" on the results. If a similar load exists on the system, another approach is to measure P_{st} at the existing location and scale it by the relative differences between impedances at the existing location and the new location:

$$P_{st,B} = \frac{Z_B}{Z_A} P_{st,A}$$

where
$P_{st,A}$ = the short-term flicker indicator measured at a location with an impedance back to the source of Z_A (the three-phase short-circuit impedance)
$P_{st,B}$ = the short-term flicker indicator for the same fluctuating load moved to a point B which has a short-circuit impedance of Z_B

A basic screening criteria based on the stiffness of the distribution circuit and the amount of fluctuation also helps determine if flicker could be a problem at new installations. The Europeans have adopted a set of criteria depending on the rate of change of the load and the fluctuation relative to the stiffness of the supply (Table 11.5). The change in load (ΔI) is referenced to the short-circuit current at the nearest customer (I_{SC}). At less than ten changes per minute, with a short-circuit level on the primary of 1000 A, a load could fluctuate by 4 A (as measured on the primary) without causing flicker if the fluctuations are fewer than 10/min. North American limits could be less conservative than this because the lower-voltage North American light bulb is less sensitive to flicker than filaments at higher European voltages (120-V vs. 230-V bulbs).

Motor starts are a special case of voltage flicker. Most motors start only a few times per day, which is possibly off the flicker curve. While motors may

TABLE 11.5

Thresholds Where Fluctuating Loads Will Not Cause Flicker Problems per (IEC 61000-3-7, 1995; UIEPQ-9715, 1999)

Number of Voltage changes per minute, r	ΔI / I_{SC}, %
$r > 200$	0.1
$10 \leq r \leq 200$	0.2
$r < 10$	0.4

TABLE 11.6

Voltage Flicker Limits at Several Utilities during Motor Starting

Utility	Voltage Criteria
Dense urban area	3%
Urban/suburban	3%
Suburban and rural	3%
Urban and rural	3%
Rural, mountainous	4 V (on a 120-V base)
Rural, mountainous	none

Source: Willis, H. L., *Power Distribution Planning Reference Book*, Marcel Dekker, New York, 1997.

TABLE 11.7

One Utility's Allowable Motor Starting Currents

System	Maximum Allowable Starting Current, A
Single Phase	
120 V	100
208 V	160
240 V	200
Three Phase	
208 V	1554
240 V	1346
480 V	673
2400 V	135

Note: Automatically controlled motors are limited to half of the allowable starting currents in the table.

start infrequently, when they do start, the voltage change is sharp, and light change can be deep and easily visible. Motors normally draw five to six times full-load current during starting. Normally, the voltage drops suddenly and then gradually recovers over several seconds as the motor comes up to speed.

Utilities normally have a criteria for motor starts based on the change of voltage and how often the customer starts the motor. Willis (1997) reported that many utilities use a criteria of 3% during motor starting (Table 11.6), but some utilities vary on how they define the percentage (either relative to the nominal voltage, the minimum voltage, or the voltage at the time of the motor start). Also, utilities often limit the size of motors or the starting current allowed by end users, depending on the voltage (see Table 11.7 for one utility's limits).

The source impedance at the customer with the motor (or other fluctuating load) is an important component. As with harmonics, we are most concerned

TABLE 11.8

Reference Impedances as Seen from the Electrical Panel Board

Location	Impedance, Ω
120-V residential, phase to neutral	$0.19 + j0.062$
240-V residential, phase to return phase	$0.20 + j0.080$
600-V industrial, phase to neutral	$0.58 + j0.107$
600-V industrial, phase to phase	$0.57 + j0.135$

Source: Ba, A. O., Bergeron, R., and Laperriere, A., "Source impedances of the Canadian distribution systems," CIRED 97, 1997.

about the voltage drop at the point of common coupling, the point where other customers can be tapped off. For customers fed from the same transformer, the point of common coupling is at the transformer. The impedance at the transformer is dominated by the transformer's impedance; the primary-side impedance is normally small. If customers share secondary circuits, also add the impedance of the shared secondary. For larger customers with their own dedicated transformer, the point of common coupling is on the primary. To evaluate the flicker to the customer that has the fluctuating load, consider the secondary and service drop to the meter. Hydro Quebec estimated reference impedances for Canadian residential and industrial facilities (Ba et al., 1997). At the panel, they estimated that 95% of customers have lower impedances than the values in Table 11.8, which we could use as a first approximation for motor starting and other fluctuation limits.

Interharmonics — harmonic distortions that are not integer multiples of the fundamental — can cause voltage flicker (Gunther, 2001). Noninteger harmonics are less common than integer harmonics, so this problem is not particularly widespread. Cycloconverters, arc furnaces, arc welders, and induction furnaces inject interharmonics. Interharmonics also come from loads like ovens or furnaces with integral cycle control, where the load controls the average voltage by either giving the heater full voltage or no voltage on a cycle-by-cycle basis (60% average voltage could come from six cycles on, then four cycles off). Standard six-pulse or twelve-pulse power converters do not normally create noninteger harmonics, but a converter can create noninteger harmonics if its electronic switches fire at the wrong time. Misfiring can come from incorrect control settings or a variety of hardware problems. The Wisconsin Electric Power Company had such a problem with a dc arc furnace that caused flicker (Tang et al., 1994).

Two superimposed frequencies beat against each other at this frequency:

$$f = |f_{ih} - f_0|$$

where

f_{ih} = frequency of the interharmonic

f_0 = integer multiple of the fundamental frequency closest to f_{ih}

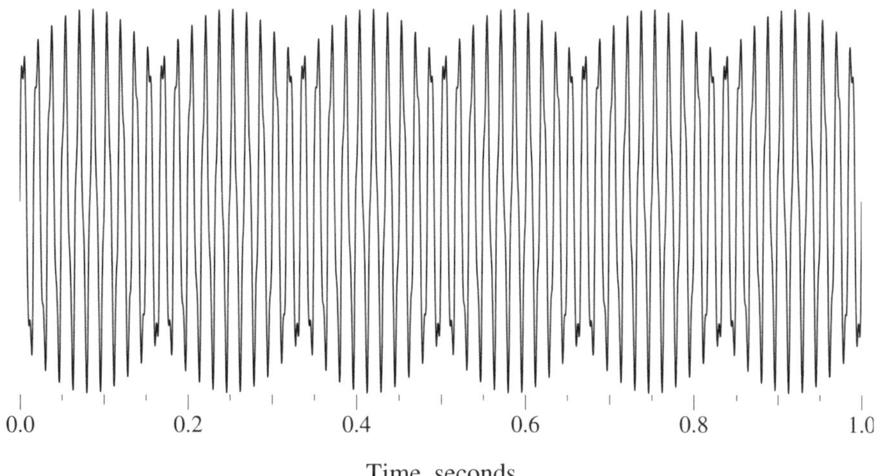

FIGURE 11.23
A 20% interharmonic frequency of n = 3.1 (186 Hz) causing a beat frequency of 6 Hz.

So, for n = 3.1, the beat frequency is |3.1–3| = 0.1 per unit, or |186 Hz–180 Hz| = 6 Hz (see Figure 11.23), right at the most sensitive flicker frequency. Frequencies of n = 1.9, 2.1, 2.9, 3.1, 3.9, and 4.1 per unit each beat at the same frequency: 0.1 per unit. Lower-frequency interharmonics — those near the first through third harmonic — are most likely to cause flicker. The IEC flickermeter detects flicker due to rms changes. Low-frequency interharmonics below the second harmonic change the rms, especially near the fundamental. But interharmonics above the second harmonic only modulate the peak; so at higher-frequency interharmonics, where the rms stays constant, the IEC flickermeter does not detect flicker (even though the fluctuations in the peak can result in noticeable flicker on fluorescent lights).

Incandescent lamps respond to rms voltage. Interharmonics below the second harmonic can cause flicker; higher frequencies do not. Fluorescent lights with electronic ballasts are particularly sensitive to the waveform peak; as the peak fluctuates, light output fluctuates. Electronic ballasts rectify the ac, which tracks the waveform peaks.

Just as capacitors can amplify integer harmonics, resonances can amplify noninteger harmonics. Solutions to flicker from interharmonics can include some of the solutions to regular flicker discussed in the next section. Also, harmonic solutions are appropriate in some situations, especially removing resonances by resizing or moving capacitors.

11.4.1 Flicker Solutions

11.4.1.1 Load Changes

Often, flicker is most economically solved at the source of the problem, the fluctuating load.

- *Welders* — For single large welders, changing the electrode firing sequence to draw smaller but more frequent current pulses can reduce flicker (EPRI PEAC Case Study No. 1, 1997). For multiple welders, control equipment can add a short delay to some welders to prevent simultaneous firing of several units. Also, on some welders, users can lower the weld "heat" (and current draw) without sacrificing the quality of the weld.
- *Motors* — For motor-starting flicker, one of several reduced-voltage starters will reduce the current draw and voltage deviation during starting. Common ways to reduce the voltage include an autotransformer start, a reactor start, and electronically switching the voltage (like a light dimmer) during starting. Reducing voltage to the motor during starting reduces the current inrush from 5 to 6 times normal current to under 2 per unit in some cases. Electronic motor starters can ramp the current to a preset maximum level. However, reduced-voltage starts also reduce the starting torque, so mechanical load issues can limit how much we can reduce the voltage. For motors susceptible to stalling (such as wood grinders), more careful operations on the mechanical side can reduce flicker (such as controlling the stock fed to the grinder).
- *Resistive heaters* — Heaters with on–off type controls flicker from repeated on and off cycling. Often, more precise control with less fluctuation is possible by upgrading the heater controls — regulating the voltage to the heating elements or splitting one heater into separately controlled elements.
- *Operational limits* — A simple option requiring no physical changes is to limit operation of offending loads to times when neighboring customers are unlikely to be bothered by flickering lights. Midnight to 6 a.m. is normally safe. Daytime operation might be tolerated. Operation from 6 p.m. to 11 p.m. is most likely to cause complaints. For facilities with multiple fluctuating loads, facilities can stagger operations to limit overlapping operation of multiple units.

Another option is to provide solutions where the customers are complaining. Review the conditions when flicker occurs and what lighting fixtures are flickering. A review of the lighting may reveal especially sensitive light fixtures such as some types of dimmers. These light fixtures can be replaced by fixtures that flicker less. One option is to convert from incandescent to fluorescent lights, which do not flicker as much.

If the facility causing the fluctuations is the only one experiencing the flicker, one solution is to isolate the lighting circuits from the circuits with fluctuating loads. Run separate circuits to the lighting loads. For single-phase fluctuating loads, put them on one phase and the lighting on the other two phases.

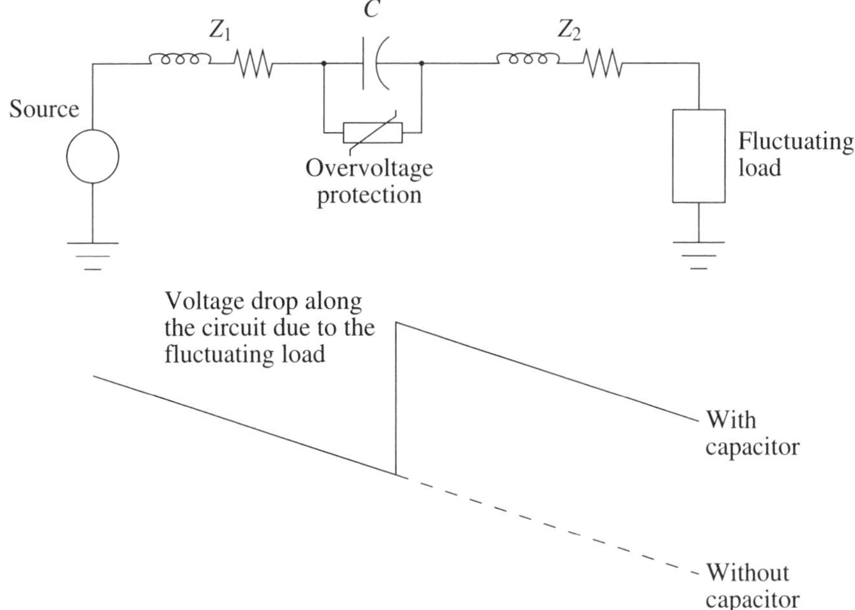

FIGURE 11.24
Series capacitor.

11.4.1.2 Series Capacitor

On distribution lines, much of the voltage drop is due to a circuit's inductance. If we add a capacitor in series with the inductance, the capacitor cancels out a portion of the inductance (see Figure 11.24). Now, the circuit has less total inductance, so fluctuations in load cause less voltage drop. A series capacitor is a passive circuit element. It responds instantaneously and automatically. Series capacitors help with voltage regulation as well as improving voltage flicker.

The best place for a series capacitor depends on where the complaining customers are relative to the offending load. Customers upstream of the series capacitor see no difference in flicker. Voltage improves downstream of the series capacitor. Sizing and placement involves tradeoffs. Ideally, we want to size the capacitor such that its impedance equals the circuit impedance to the fluctuating load (smaller series capacitor impedances are more expensive: more kvars, more units in parallel). For placement, a good rule of thumb is electrically half way between the source and the fluctuating load — the point where the voltage drop is half of the total voltage drop due to the fluctuating load. Locations closer to the source reduce flicker for more customers but increase fault duties. An IEEE working group suggests a rule-of-thumb location where the voltage drop is one-third to one-half of the total drop (Miske, 2001).

Series capacitors can superbly compensate flicker from fluctuating inductive loads. For resistive loads, series capacitors provide much less benefit. The voltage drop through the distribution system is approximately $I_R R + I_X X$. Reducing X with a series capacitor helps reduce the $I_X X$ term for fluctuating I_X; but if just I_R fluctuates, series capacitors provide little benefit.

Electrically, series capacitors elegantly solve voltage flicker; in practice, they are not widely used, mainly because of

- *Reliability of short-circuit protection* — Historically, spark gaps were used to protect the capacitors during downstream faults. Utilities have had many problems with failures.
- *Cost* — Series capacitors are nonstandard and must be custom engineered.
- *Unusual* — Line crews and field engineers find series capacitors strange. For example, if crews switch in a shunt capacitor downstream of a series capacitor, the voltage may go down instead of up. Increased fault currents downstream of the capacitor can make coordination of protective devices more difficult.
- *Ferroresonance* — Series capacitors may also ferroresonate with downstream transformers under the right conditions.

A key design issue is the voltage across the capacitor during faults downstream of the unit. During a fault downstream of the series capacitor, the voltage across the capacitor units is the short-circuit current times the capacitor impedance. Depending on the impedances upstream of the capacitor relative to the capacitor size, the voltage across the series capacitor can exceed the system's nominal voltage. When calculating the available fault current downstream of the capacitor, include the impedance reduction caused by the capacitor. The fault current normally can rise past the capacitor (depending on the amount of capacitance relative to the system impedance and the line X/R ratio). The highest fault current can be some distance from the capacitor. Figure 11.25 shows an example on a 12.5-kV system for a series capacitor placed 3 mi from the substation to compensate for a fluctuating load at 8 mi from the substation. The series capacitor significantly alters the fault-current profile, and the fault current impresses significant voltage across the capacitor. Some sort of protection is needed to prevent this.

The increased fault current raises the primary voltage and the customers' voltage during the fault. Figure 11.26 shows a profile of the line voltages for a fault at 4.75 mi for the example shown in Figure 11.25. Under this scenario, the fault current is leading (capacitive), so voltages rise along the line until the series capacitor. These overvoltages are unacceptable, so some overvoltage protection across the capacitor is vital to prevent them. This example has a rather poor choice of location and impedances, but it illustrates how important it is to consider these applications carefully. Reducing the impedance of the capacitor reduces the overvoltages and reduces the fault currents.

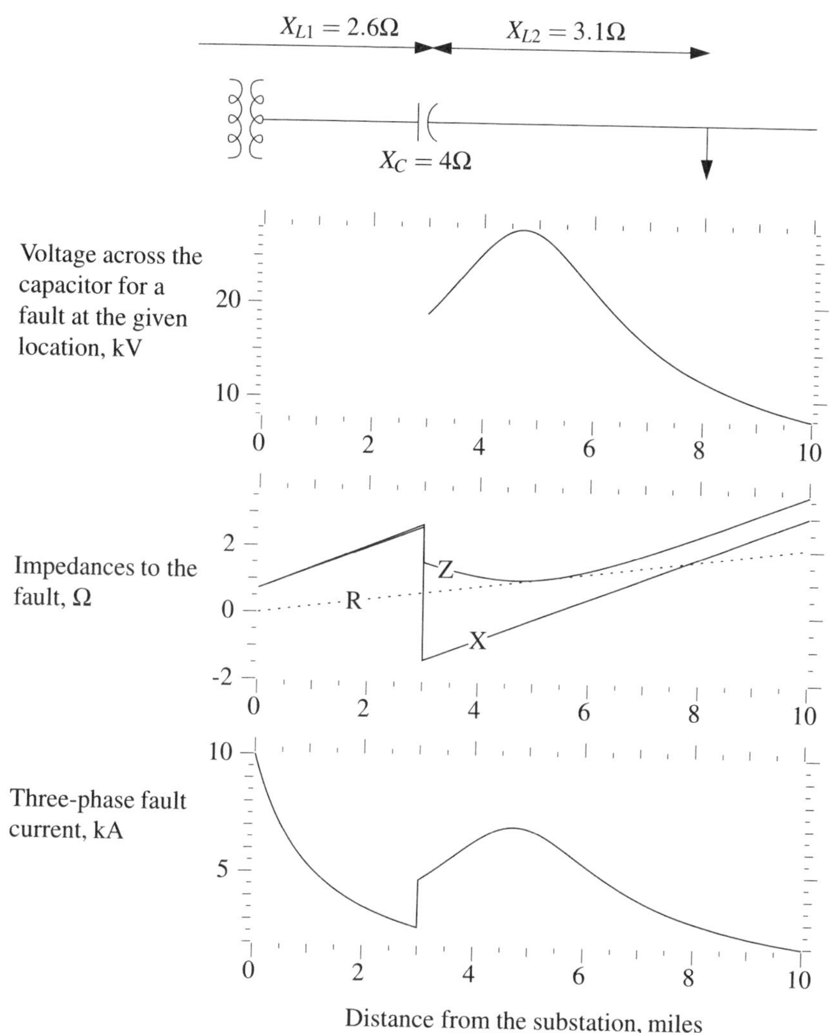

FIGURE 11.25
Fault-current profile on a 12.5-kV circuit with a series capacitor located 3 miles from the substation (the same line parameters as Figure 7.11, 500-kcmil AAC conductors). The series capacitor has no short-circuit protection.

Lowering the capacitor's impedance to maintain a total line impedance that is always reactive (never capacitive) eliminates the problem of overvoltages. But we still have higher fault currents downstream of the capacitor, and lowering the capacitor's impedance reduces the flicker-prevention capability.

Several protection arrangements have been used to protect capacitors from overvoltages. A spark gap is the simplest device but has been problematic. Duke Power reported that many of the gaps on their units kept arcing and started fires (Morgan et al., 1993). Gap erosion also reduced protection and

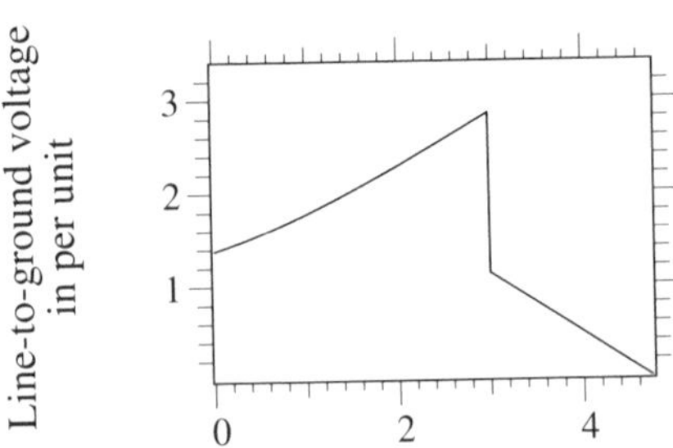

FIGURE 11.26
Voltages along the circuit for a three-phase fault at 4.75 miles for the application shown in Figure 11.25.

led to capacitor failures. Lat et al. (1990) used an improved gap that reduced erosion. The most modern units use metal-oxide arresters as protection against overvoltages; several stacks of metal-oxide blocks in parallel are needed to absorb the energy during a fault. Series capacitors normally have bypass switches to short the unit out if needed. Some designs close the bypass as soon as a fault is detected to reduce the duty on the capacitors and the arresters. Another benefit of using arresters for overvoltage protection is that the arresters short out the capacitors so that the series capacitor does not significantly increase the downstream fault current or cause overvoltages.

Marshall (1997) dealt with the problem of short-circuit protection by designing installations so that they did not need extra protection. This is possible at locations with low fault currents and where we can use a capacitor with a reasonably low voltage rating, especially if we do not try to compensate too aggressively. This is more likely on long rural circuits where the impedances are high, and we want a high capacitive impedance (which means we need less kvar in our series bank). Picking a higher voltage rating for a capacitor unit requires using several more units in parallel. For a given kvar rating, doubling the voltage rating raises the unit's impedance by a factor of four; now, we need four times the kvar for the same impedance.

11.4.1.3 Static Var Compensator

Static var compensators (SVCs) dynamically change their reactance, injecting or absorbing reactive power to regulate the voltage in real time. They are often used at large arc furnaces to control voltage flicker. Utilities sometimes use them to reduce voltage flicker on distribution circuits (Jatskevich et al., 1998; Kemerer and Berkebile, 1998; Wong et al., 1990).

Several styles of var compensators are available:

- *Switched inductor* — A thyristor controls how much reactive power is drawn from the system, depending on where on the voltage waveshape the thyristor turns on. Full vars are drawn if the thyristor turns on when the voltage is at its peak (a full half cycle of current is drawn). Less vars are drawn if the controller waits to turn on the thyristor. Once fired, the thyristor stays on until the current goes through zero and the thyristor shuts off. A drawback to this configuration is that the current blips are not sinusoidal and create harmonics. The inductor may be used in parallel with a shunt capacitor to control vars negatively and positively.

- *Switched capacitors* — In this configuration, several shunt capacitors each have thyristors to control whether the capacitor is connected. The number of capacitor units switched in determines the amount of vars injected. Having more individual blocks of capacitors helps keep the control smooth. The switched capacitors may be paralleled with a fixed inductor to allow the unit to control vars in both directions.
- *STATCOM* — The most advanced reactive power controller, this custom power device has a power electronics interface (normally a pulse-width-modulated inverter using IGBTs or other fast-switching electronics) that injects or absorbs reactive power that is temporarily stored in capacitors on a dc bus. The STATCOM (static compensator) can smoothly vary the reactive power with little harmonics. The STATCOM may be extended to inject or absorb real power with auxiliary energy storage.

None of these devices can instantly counter changes from fluctuating loads; some delay is always present. The delay is normally on the order of one-half to two cycles, depending on the compensator and controller technology.

Reactive power is more efficient than real power at countering voltage changes. Reactive power is more efficient because the equivalent source impedance is normally more reactive than resistive, and the voltage drop is approximately

$$V_{drop} \approx R \cdot I_R + X \cdot I_X$$

where R and X are the source resistance and reactance, I_R is the resistive component of the load, and I_X is the reactive component of the load.

11.4.1.4 Other Solutions

Other utility-side solutions to flicker include

- *Dedicated circuit* — Running a separate load to the customer with a fluctuating load isolates the flicker to the customer causing the flicker. We can tap the dedicated circuit where the voltage change due to the fluctuating load is tolerable.
- *Reconductoring* — Larger conductors have lower resistance and slightly lower reactance. If the voltage fluctuation is due to changes

in the resistive load, larger conductors will help; but if the voltage fluctuation is due to reactive load, larger conductors will not help. Tighter phase spacings reduce the reactance somewhat, but not dramatically. Converting to aerial or underground cable does dramatically reduce the reactance.

- *Upgrading voltage* — On a percentage basis, voltage drop decreases as the square of system voltage. Of course, upgrading the system voltage requires new insulators, arresters, switches, and transformers.

Unfortunately, all of these are normally expensive options. Jenner and Brockhurst (1990) reported success using vacuum-switched shunt capacitor banks to counteract the voltage drop from large motor starts, although the measurements shown still left a significant voltage change.

11.5 Voltage Unbalance

Three-phase end-use equipment expects balanced voltages, each with the same magnitude. If the voltages are not balanced, some equipment can perform poorly, mainly induction motors. Voltage unbalance (or imbalance, an equivalent term used by many) is defined as:

$$\text{Percent Unbalance} = \frac{\text{Maximum deviation from the average}}{\text{Average of the three phase - to - phase voltages}} \times 100\%$$

Normally, phase-to-phase voltages are used for voltage unbalance rather than phase-to-ground voltages.

ANSI C84.1-1995 states that utilities should limit the maximum voltage unbalance to 3%, measured under no-load conditions. Most distribution systems have modest unbalance, well under the ANSI 3% limit. Field surveys sited by ANSI C84.1 report that only 2% of electric supply systems have unbalance that exceeds 3%.

Motors are most efficient with balanced voltages; unbalanced voltage heats motors significantly due to negative-sequence currents. When applied to unbalanced voltages, motors should be derated. Figure 11.27 shows a derating factor that ANSI and NEMA provides for motors.

Motors have relatively low impedance to negative-sequence voltage (approximately equal to the motor's locked rotor impedance); therefore, a small negative-sequence component of the voltage produces a relatively large negative-sequence current. The negative-sequence voltage creates a rotating field that rotates in the opposite direction that the motor is spinning. The rotor is a shorted winding to this counter-rotating field.

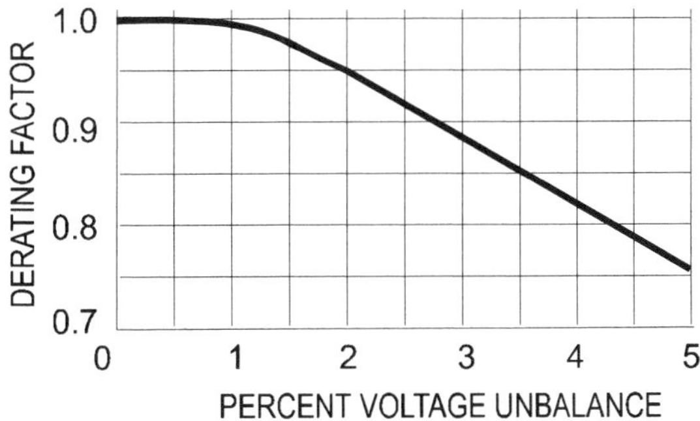

FIGURE 11.27
Motor derating curve based on voltage unbalance. (From NEMA MG-1, *Standard for Motors and Generators*, 1998. With permission.)

NEMA standard MG-1 allows a 1% voltage unbalance without derating; operation above 5% unbalance is not recommended (NEMA MG-1, 1998).

Because motors are really influenced by the negative-sequence voltage, another more-precise indicator is the unbalance factor, defined in European standards:

$$\%VUF = \frac{V_2}{V_1} \times 100\%$$

where
V_1 = positive-sequence voltage
V_2 = negative-sequence voltage

The percentage negative-sequence voltage is close to unbalance as defined by NEMA and ANSI using phase-to-phase voltage measurements, with a difference of at most 13% between the two terms (UIEPQ-9715, 1996). The EPRI DPQ study found that 98% of the time, including all sites monitored, the unbalance factor (V_2/V_1) was less than 1.5% with an average just above 0.8% (EPRI TR-106294-V2, 1996). In addition, only 2 out of 277 sites averaged above a 3% unbalance factor.

While negative-sequence voltages impact motors and other line-to-line connected loads the most, zero-sequence unbalance can affect line-to-ground connected three-phase loads. Most sites have little zero-sequence voltage unbalance (although the current unbalance may be significant). The DPQ study found similar levels of V_0/V_1 as V_2/V_1: V_0/V_1 averaged just over 0.8%; only 3 of 277 sites averaged above 3% zero-sequence voltage. Whereas transformer connections do not change the ratio of V_2/V_1, zero-sequence voltages are blocked by ungrounded transformer connections like the delta – grounded wye.

While motors receive much of the attention, unbalance also affects other loads, particularly electronic loads like adjustable-speed drives. With voltage unbalance, adjustable-speed drives draw significantly unbalanced currents, currents so unbalanced that overload protection circuits can trip (EPRI PEAC Brief No. 28, 1995). An adjustable-speed drive rectifies the three incoming phases. Each phase-to-phase input takes its turn charging up the capacitor on the drive's dc bus; but if one of these three inputs has higher voltage than the others, that phase conducts significantly more current than the other two. Even a 3% voltage unbalance can cause some drives to trip. Voltage unbalance also increases the harmonics that drives create, including the third harmonic, which balanced three-phase drives do not normally produce. Voltage unbalance also increases the ripple on the dc bus voltage.

Some causes of voltage unbalance in the order of most to least likely include:

- *Current unbalance* — Voltage unbalance normally comes from excessive current unbalance — more heavily loaded phases have more voltage drop than the lightly loaded phases. Large single-phase laterals are often culprits. Normally, crews can change the phase connections of single-phase laterals to balance the voltages. Also consider the possibility of shifting load to another circuit.
- *Capacitor banks with blown fuses* — Banks with blown fuses inject unbalanced vars, which can severely unbalance voltages.
- *Malfunctioning regulators* — Single-phase line voltage regulators adjust each phase independently. If one phase is stuck or is otherwise working improperly or has inappropriate settings, the regulator will unbalance the voltages. Note that because regulators adjust each phase independently, they perform well for improving a circuit's voltage balance.
- *Open-wye–open-delta transformers* — These banks can supply three-phase load from a two-phase supply. But the connection is prone to causing voltage unbalance to the end user on the transformer. Also, because it injects significant current into the ground, it can contribute to voltage unbalance elsewhere on the circuit.
- *Untransposed circuit* — Transposing distribution circuits is almost unheard of; it is just not needed. Rarely, balanced load can cause unbalanced voltages because the line impedances are unbalanced. As an example, on a flat configuration with an 8-ft crossarm with 336-kcmil ACSR conductors at 12.47 kV, a balanced, unity-power factor current of 100 A through 10 mi of line causes 1% voltage unbalance at the end.

Normally, we can solve unbalanced voltages by correcting the current balance or fixing a problem regulator or capacitor. After that, adding voltage regulators is an option.

References

Abetti, P. A., Johnson, I. B., and Schultz, A. J., "Surge Phenomena in Large Unit-Connected Steam Turbine Generators," *AIEE Transactions*, vol. 71, part III, pp. 1035–47, December 1952.

Ba, A. O., Bergeron, R., and Laperriere, A., "Source impedances of the Canadian distribution systems," CIRED 97, 1997.

Bellei, T. A., O'Leary, R. P., and Camm, E. H., "Evaluating Capacitor-Switching Devices for Preventing Nuisance Tripping of Adjustable-Speed Drives Due to Voltage Magnification," *IEEE Transactions on Power Delivery*, vol. 11, no. 3, pp. 1373–8, July 1996.

Bohn, J., "Utility Concerns for Electric Vehicle Battery Charger Penetration," Panel on Utility Application of Harmonic Limits, IEEE PES Summer Power Meeting, 1996.

Dwyer, R., Khan, A. K., McGranaghan, M., Tang, L., McCluskey, R. K., Sung, R., and Houy, T., "Evaluation of Harmonic Impacts from Compact Fluorescent Lights on Distribution Systems," IEEE/PES Winter Power Meeting, 1995.

Emanuel, A. E., Janczak, J., Pileggi, D. J., Gulachenski, E. M., Breen, M., Gentile, T. J., and Sorensen, D., "Distribution Feeders with Nonlinear Loads in the Northeast USA. I. Voltage Distortion Forecast," *IEEE Transactions on Power Delivery*, vol. 10, no. 1, pp. 340–7, January 1995.

EPRI PEAC Application #6, *Avoiding Harmonic-Related Overload of Shared-Neutral Conductors*, 1996.

EPRI PEAC Brief No. 21, *The Effects of Neutral-to-Ground Voltage Differences on PC Performance*, EPRI PEAC, Knoxville, TN, 1994.

EPRI PEAC Brief No. 25, *The Effect of Lamp Dimmers on Light Flicker*, EPRI PEAC, Knoxville, TN, 1995.

EPRI PEAC Brief No. 28, *Input Performance of ASDs During Supply Voltage Unbalance*, EPRI PEAC, Knoxville, TN, 1995.

EPRI PEAC Brief No. 36, *Lamp Flicker Predicted by Gain-Factor Measurements*, EPRI PEAC, Knoxville, TN, 1996.

EPRI PEAC Case Study No. 1, *Light Flicker Caused By Resistive Spot Welder*, EPRI PEAC, Knoxville, TN, 1997.

EPRI PEAC Solution No. 1, *Equalizing Potential Differences*, EPRI PEAC, Knoxville, TN, 1993.

EPRI TR-106294-V2, *An Assessment of Distribution System Power Quality. Volume 2: Statistical Summary Report*, Electric Power Research Institute, Palo Alto, CA, 1996.

Etezadi-Amoli, M. and Florence, T., "Voltage and Current Harmonic Content of a Utility System — A Summary of 1120 Test Measurements," *IEEE Transactions on Power Delivery*, vol. 5, no. 3, pp. 1552–7, July 1990.

General Electric GET-1008L, "Distribution Data Book."

Gorgette, F. A., Lachaume, J., and Grady, W. M., "Statistical Summation of the Harmonic Currents Produced by a Large Number of Single Phase Variable Speed Air Conditioners: a Study of Three Specific Designs," *IEEE Transactions on Power Delivery*, vol. 15, no. 3, pp. 953–9, July 2000.

Govindarajan, S. N., Cox, M. D., and Berry, F. C., "Survey of Harmonic Levels on the Southwestern Electric Power Company System," *IEEE Transactions on Power Delivery*, vol. 6, no. 4, pp. 1869–75, October 1991.

Greenwood, A., *Electrical Transients in Power Systems*, 2nd ed., Wiley Interscience, New York, 1991.

Gunther, E. W., "Interharmonics in Power Systems," IEEE Power Engineering Society Summer Meeting, 2001.

Halpin, S. M., Bergeron, R., Blooming, T., Burch, R. F., Conrad, L. E., and Key, T., "Voltage and Lamp Flicker Issues: Should the IEEE Adopt the IEC Approach?" IEEE P1453 working group paper. Downloaded from http://www.manta.ieee.org/groups/1453/drpaper.html.

IEC 868, *Flickermeter: Functional and Design Specifications*, International Electrical Commission (IEC), 1986.

IEC 61000-3-7, *Limitation of Voltage Fluctuation and Flicker for Equipments Connected to Medium and High Voltage Power Supply Systems*, International Electrical Commission (IEC), 1995.

IEEE P519A draft 7, *Guide for Applying Harmonic Limits on Power Systems*, 2000.

IEEE Std. 141-1993, *IEEE Recommended Practice for Electric Power Distribution for Industrial Plants.*

IEEE Std. 142-1991, *IEEE Recommended Practice for Grounding of Industrial and Commercial Power Systems.*

IEEE Std. 519-1992, *IEEE Recommended Practices and Requirements for Harmonic Control in Electrical Power Systems.*

IEEE Std. 776-1992, *IEEE Recommended Practice for Inductive Coordination of Electric Supply and Communication Lines.*

IEEE Std. 1100-1999, *IEEE Recommended Practice for Powering and Grounding Electronic Equipment.*

IEEE Std. 1137-1991, *IEEE Guide for the Implementation of Inductive Coordination Mitigation Techniques and Application.*

IEEE Std. C57.110-1998, *IEEE Recommended Practice for Establishing Transformer Capability When Supplying Nonsinusoidal Load Currents.*

IEEE Task Force, "The Effects of Power System Harmonics on Power System Equipment and Loads," *IEEE Transactions on Power Apparatus and Systems*, vol. PAS-104, no. 9, pp. 2555–63, September, 1985.

IEEE Task Force, "Effects of Harmonics on Equipment," *IEEE Transactions on Power Delivery*, vol. 8, no. 2, pp. 672–80, April 1993.

IEEE Working Group, "Power Line Harmonic Effects on Communication Line Interference," *IEEE Transactions on Power Apparatus and Systems*, vol. PAS-104, no. 9, pp. 2578–87, September 1985.

Jatskevich, J., Wasynczuk, O., and Conrad, L., "A Method of Evaluating Flicker and Flicker-Reduction Strategies in Power Systems," *IEEE Transactions on Power Delivery*, vol. 13, no. 4, pp. 1481–7, October 1998.

Jenner, M. and Brockhurst, F. C., "Vacuum Switched Capacitors to Reduce Induction Motor Caused Voltage Flicker on a 12.47 kV Rural Distribution Line," IEEE Rural Electric Power Conference, 1990.

Kemerer, R. S. and Berkebile, L. E., "Directly Connected Static VAr Compensation in Distribution System Applications," IEEE Rural Electric Power Conference, 1998.

Kolcio, N., Halladay, J. A., Allen, G. D., and Fromholtz, E. N., "Transient Overvoltages and Overcurrents on 12.47 kV Distribution Lines: Field Test Results," *IEEE Transactions on Power Delivery*, vol. 7, no. 3, pp. 1359–70, July 1992.

Lat, M. V., Kundu, D., and Bonadie, G., "Overvoltage Protection Scheme for Series Capacitor Banks on High Voltage Distribution Systems," *IEEE Transactions on Power Delivery,* vol. 5, no. 3, pp. 1459–65, July 1990.

Mansoor, A., Grady, W. M., Staats, P. T., Thallam, R. S., Doyle, M. T., and Samotyj, M. J., "Predicting the Net Harmonic Currents Produced by Large Numbers of Distributed Single-Phase Computer Loads," *IEEE Transactions on Power Delivery,* vol. 10, no. 4, pp. 2001–6, October 1995.

Marshall, M. W., "Using Series Capacitors to Mitigate Voltage Flicker Problems," IEEE Rural Electric Power Conference, 1997.

McGranaghan, M. F., Grebe, T. E., Hensley, G., Singh, T., and Samotyj, M., "Impact of Utility Switched Capacitors on Customer Systems. II. Adjustable-Speed Drive Concerns," *IEEE Transactions on Power Delivery,* vol. 6, no. 4, pp. 1623–8, October 1991.

McGranaghan, M. F., Zavadil, R. M., Hensley, G., Singh, T., and Samotyj, M., "Impact of Utility Switched Capacitors on Customer Systems — Magnification at Low Voltage Capacitors," *IEEE Transactions on Power Delivery,* vol. 7, no. 2, pp. 862–8, April 1992.

Miske, S. A., "Considerations for the Application of Series Capacitors to Radial Power Distribution Circuits," *IEEE Transactions on Power Delivery,* vol. 16, no. 2, pp. 306–18, April 2001.

Morgan, L., Barcus, J. M., and Ihara, S., "Distribution Series Capacitor with High-Energy Varistor Protection," *IEEE Transactions on Power Delivery,* vol. 8, no. 3, pp. 1413–19, July 1993.

Nejdawi, I. M., Emanuel, A. E., Pileggi, D. J., Corridori, M. J., and Archambeault, R. D., "Harmonics Trend in NE USA: A Preliminary Survey," *IEEE Transactions on Power Delivery,* vol. 14, no. 4, pp. 1488–94, October 1999.

NEMA MG-1, *Standard for Motors and Generators*, 1998.

Pileggi, D. J., Gulachenski, E. M., Root, C. E., Gentile, T. J., and Emanuel, A. E., "The Effect of Modern Compact Fluorescent Lights on Voltage Distortion," *IEEE Transactions on Power Delivery,* vol. 8, no. 3, pp. 1451–9, July 1993.

Rice, D. E., "Adjustable Speed Drive and Power Rectifier Harmonics — Their Effect on Power Systems Components," *IEEE Transactions on Industry Applications,* vol. IA-22, no. 1, pp. 161–77, January/February 1986.

Sabin, D. D., Brooks, D. L., and Sundaram, A., "Indices for Assessing Harmonic Distortion from Power Quality Measurements: Definitions and Benchmark Data," *IEEE Transactions on Power Delivery,* vol. 14, no. 2, pp. 489–96, April 1999a.

Sabin, D. D., Grebe, T. E., and Sundaram, A., "Assessing Distribution System Transient Overvoltages Due to Capacitor Switching," International Conference on Power Systems Transients, Technical University of Budapest, Budapest, Hungary, 1999b.

Schultz, A. J., Johnson, J. B., and Schultz, N. R., "Magnification of Switching Surges," *AIEE Transactions on Power Apparatus and Systems*, pp. 1418–26, February 1959.

Seebald, R. C., Buch, J. F., and Ward, D. J., "Flicker Limitations of Electric Utilities," *IEEE Transactions on Power Apparatus and Systems*, vol. PAS-104, no. 9, pp. 2627–31, September 1985.

Shankland, L. A., Feltes, J. W., and Burke, J. J., "The Effect of Switching Surges on 34.5 kV System Design and Equipment," *IEEE Transactions on Power Delivery,* vol. 5, no. 2, pp. 1106–12, April 1990.

Tang, L., Mueller, D., Hall, D., Samotyj, M., and Randolph, J., "Analysis of DC Arc Furnace Operation and Flicker Caused by 187 Hz Voltage Distortion," *IEEE Transactions on Power Delivery*, vol. 9, no. 2, pp. 1098–107, April 1994.

Thallam, R. S., Grady, W. M., and Samotyj, M. J., "Estimating Future Harmonic Distortion Levels In Distribution Systems Due To Single-Phase Adjustable-Speed-Drive Air Conditioners: A Case Study," ICHPS V International Conference on Harmonics in Power Systems, 1992.

Troedsson, C. G., Gramlich, E. F., Gustin, R. F., and McGranaghan, M. F., "Magnification of Switching Surges as a Result of Capacitor Switching on a 34.5 kV Distribution System," Proceedings of the American Power Conference, 1983.

UIEPQ-9715, *Guide to Quality of Electrical Supply for Industrial Installation — Part 4: Voltage Unbalance*, Union Internationale de Electrothermie (UIE), 1996.

UIEPQ-9715, *Guide to Quality of Electrical Supply in Industrial Installations — Part 5: Flicker and Voltage Fluctuations*, Union Internationale de Electrothermie (UIE), 1999.

Walker, M. K., "Electric Utility Flicker Limitations," *IEEE Transactions on Industry Applications*, vol. IA-15, no. 6, pp. 644–55, November/December 1979.

Walling, R. A., Melchior, R. D., and McDermott, B. A., "Measurement of Cable Switching Transients in Underground Distribution Systems," *IEEE Transactions on Power Delivery*, vol. 10, no. 1, pp. 534–9, January 1995.

Willis, H. L., *Power Distribution Planning Reference Book*, Marcel Dekker, New York, 1997.

Wong, W. K., Osborn, D. L., and McAvoy, J. L., "Application of Compact Static VAR Compensators to Distribution Systems," *IEEE Transactions on Power Delivery*, vol. 5, no. 2, pp. 1113–20, April 1990.

Although they try hard, bless their hearts, consumers that diagnose their own powerline problems tend to miss a thing or two at times.

Powerlineman law #18, By CD Thayer and other Power Linemen, http://www.cdthayer.com/lineman.htm

12

Lightning Protection

Protection of power systems from lightning-related damage and faults is crucial to maintaining adequate power quality, reliability, and controlling damage costs to the utility system. In the U.S., the most severe lightning activity occurs in the southeast and Gulf coast states. However, lightning is also a major cause of faults and interruptions in areas with just modest lightning activity, such as New England. There are many facets in the design of lightning protection, including surge arrester sizing and placement, grounding issues, and the selection of appropriate power system equipment and insulation ratings. This chapter focuses on the factors that impact the lightning performance of power distribution systems, providing background on the characteristics of lightning, methods for calculating lightning flashover rates, and guidelines in the application of lightning protection equipment.

In the majority of cases, lightning causes temporary faults on distribution circuits; the lightning arcs externally across insulation, but neither the lightning nor the fault arc permanently damages any equipment. Normally, less than 20% of lightning strikes cause permanent damage. In Florida (high lightning), EPRI monitoring found permanent damage in 11% of circuit breaker operations that were coincident with lightning (EPRI TR-100218, 1991; Parrish, 1991). Where equipment protection is done poorly, more equipment failures happen. Any failure in transformers, reclosers, cables, or other enclosed equipment causes permanent damage. One lightning flash may cause multiple flashovers and equipment failures. Even if a lightning-caused fault does no damage, a long-duration interruption occurs if the fault blows a fuse.

The protection strategy at most utilities is

1. Use surge arresters to protect transformers, cables, and other equipment susceptible to permanent damage from lightning.
2. Use reclosing circuit breakers or reclosers to reenergize the circuit after a lightning-caused fault.

Lightning causes most damage by directly striking an overhead phase wire and injecting an enormous current surge that creates a very large voltage.

The voltage impulse easily breaks down most distribution-class insulation unless it is protected with a surge arrester. Almost all direct lightning strokes cause flashovers. In addition, the lightning current may start a pole fire or burn through conductors. Also, nearby lightning strokes that do not hit the line may couple damaging voltages to the line. These induced voltages may fail equipment or cause flashovers.

Figure 12.1(a) shows an example of a lightning flash composed of several individual strokes hitting a distribution circuit (the movement of the lightning channel and camera shaking makes it easier to distinguish the individual strokes). Two phases initially flashed over, and the third eventually flashed over (either due to a subsequent stroke or the gasses from the initial fault). On three-phase circuits, lightning normally causes two- or three-phase faults. The lightning-caused fault in Figure 12.1(a) resulted in a downed-wire high-impedance fault that was not cleared by the station protection (the circuit was tripped manually). Taken moments after the lightning flash, Figure 12.1(b) shows the glow of an arcing downed conductor. In this case, the 60-Hz fault arc, not the original lightning impulse, burned the wires down. Two characteristics made this circuit prone to burndowns: covered wire and relaying that allowed long-duration faults (the instantaneous relay in the substation did not cover this portion of the line). The second bright blob in the lower right of the lightning strike picture is another flashover further down the line on the same circuit. These supplementary flashovers — away from the fault point and on the same circuit — are common, especially in areas with high ground impedances (the ground potential rise pushes the surge to other locations on the circuit). These pictures were taken by a lightning-activated camera developed by the Niagara Mohawk Power Corporation (now part of National Grid) (Barker and Burns, 1993).

Locations with higher lightning activity have higher fault rates on overhead circuits. Figure 12.2 shows fault rates reported by different utilities against estimated ground flash density. Utilities in higher lightning areas have more faults. Not all of these are due to lightning; many are due to wind and other storm-related faults. Lightning is a good indicator of storm activity at a location.

The linear curve fit to the data in Figure 12.2 shows line fault rates per 100 mi per year varying with the ground flash density, N_g, in flashes/km^2/year as

$$f = 20.6N_g + 43$$

12.1 Characteristics

Driven by the need to better protect transmission and distribution lines and equipment, the electrical industry performed much of the early research on

(a) Lightning strike and flashovers (b) Arcing downed conductor

FIGURE 12.1
Example of a lightning flash to a 13.2-kV distribution line and the downed wire that resulted. (Copyright 1991, Niagara Mohawk Power Corporation. Reprinted with permission.)

lightning and electrical breakdowns. Charles P. Steinmetz, Charles F. Wagner, Walter W. Lewis, Karl B. McEachron, Charles L. Fortescue, Basil F. J. Schonland — the list of early contributors is long — increased our understanding of the physics of lightning and its electrical characteristics.

Lightning is the electric breakdown of the air from high electric fields generated when electric charge separates within a cloud. Lightning may flash within a cloud, from one cloud to another, or from the cloud to the ground. Distribution lines are only affected by cloud-to-ground lightning. In the normal scenario, charge separates within a thundercloud. The upper portion becomes positively charged and the lower portion becomes negatively charged. The ground just underneath the cloud becomes positively charged

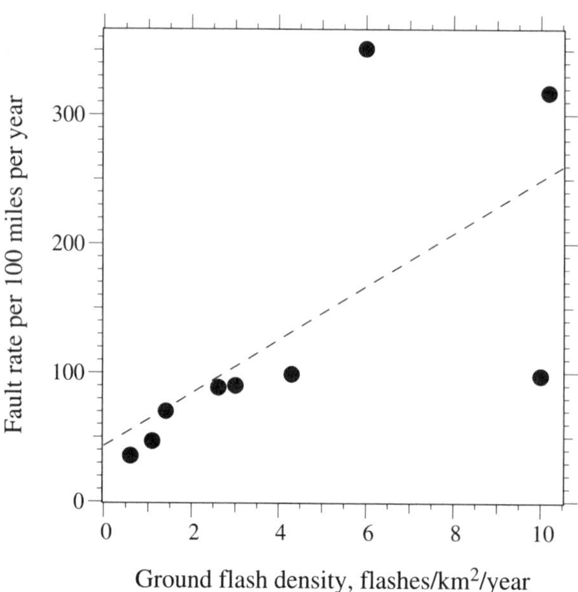

FIGURE 12.2
Distribution line fault rate vs. ground flash density. (Based on the data in Figure 7.2.)

(being attracted to the negatively charged lower portion of the cloud). The lightning breakdown begins in the lower portion of the cloud. The air breaks down in steps called *stepped* leaders. Each step is about 150 ft (50 m) with pauses of about 50 μsec between steps. The stepped leader may fork and form branches that each progress towards the ground. As the stepped leader progresses closer to the ground (see Figure 12.3), more charge is lowered closer to the ground. More positive charge collects on the earth in response — short upward leaders extend to meet the downward negative stepped leader.

When the downward leader meets the upward leader, a *return stroke* occurs. The negative charge held in the stepped leader rushes into the ground, brilliantly lighting the channel and creating a large pressure wave (thunder). The return stroke propagates up the channel at roughly 20% of the speed of light, releasing charge as it goes. The charge rushing into the ground creates a current of tens of thousands of amps peaking in a few microseconds. The current may extinguish in about 100 μsec, or lower-level continuing current in the range of hundreds of amperes may flow for several milliseconds (about 25% of the time, continuing currents flow following the return stroke).

Subsequent strokes may follow the first stroke. After the current extinguishes and the channel becomes dark, another pocket of charge may work its way down the same path. Fast-moving leaders called dart leaders break down the recently deionized path of the first stroke. Subsequent strokes typically have lower magnitudes of current and charge transferred, but subsequent stroke currents have higher rates of rise. Subsequent strokes have higher

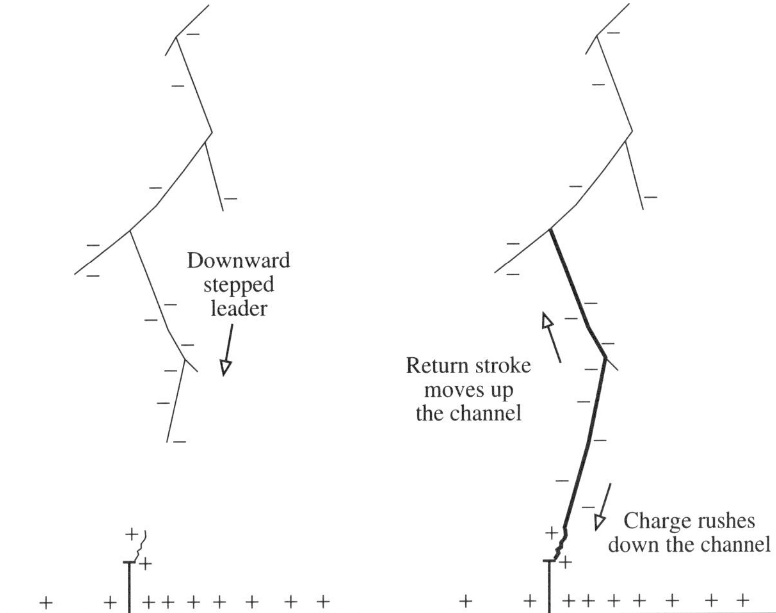

FIGURE 12.3
Cloud-to-ground lightning.

return-stroke velocities, often greater than 50% of the speed of light. The first stroke and subsequent strokes make up a lightning *flash*.

While the downward negative flash is the most common, other types of cloud-to-ground lightning occur. About 5 to 10% of cloud-to-ground flashes are positive. Downward positive lightning lowers positive charge from the cloud to the ground. Breakdown starts at a positive portion of the cloud usually near the top of the cloud; a positive downward stepped leader moves downward until it meets an upward negative leader close to the ground. Some positive flashes may have very large peak currents and charge. Positive flashes occur more often during winter storms, especially in some areas. Positive flashes usually only have one stroke. Cloud-to-ground lightning may also start at the ground and rise upward, with an upward stepped leader starting at the ground. These are common on tall objects like the Empire State Building, but rare from distribution lines.

Normally, the lightning current injection is considered an ideal current surge (it does not really matter what is struck; the electrical characteristics of the current stay the same). Table 12.1 shows characteristics of a downward negative current flash. Many of the characteristics fit a log-normal distribution, which is common for data bounded at zero. The log standard deviation, $\beta = \text{sd}(\ln(x_i))$, is shown for the characteristics that have a log-normal characteristic. The 5th and 95th percentiles are shown based on the log-normal fit. The first stroke peak current data does not fit a log-normal distribution, but Anderson and Eriksson found a good fit using two log-normal param-

TABLE 12.1

Lightning Current Parameters for Downward Negative Flashes

Parameter	Percent of Cases More than Value			β
	5%	50% (*M*)	95%	
First Strokes				
Peak current, kA	8	33.3	90	
Model for I≤20kA		61.1		1.33
Model for I>20kA		33.3		0.61
Time to peak, μsec (virtual front time based on the time from 30 to 90% of the peak = $T_{30\text{–}90\%}/0.6$)	1.5	3.83	10	0.553
Steepness, 30–90%, kA/μsec	2.6	7.2	20	0.921
Tail, time to half the peak, μsec	30	77.5	200	0.577
Charge, C	1.1	4.65	20	0.882
$\int I^2dt$, $(A)^2$sec ×10^3	6	57	546	1.373
Subsequent Strokes				
Peak current, kA	5.2	12.3	29.2	0.530
Time to peak, μsec (30–90% virtual front)	0.2	0.67	3.0	
Steepness, 30–90%, kA/μsec	4.1	20.1	99	0.967
Tail, μsec	6.5	30.2	140	0.933
Charge, C	0.2	0.938	4	0.882
$\int I^2dt$, $(A)^2$sec ×10^3	0.6	5.5	52	1.366
Flash				
Charge, C	1.3	7.5	40	1.02
Flash duration, sec	0.03	0.2	1	
Number of strokes	1	2–3	9	
Interval between strokes, msec	6	35	202	1.066

Sources: Anderson, R. B. and Eriksson, A. J., "Lightning Parameters for Engineering Applications," *Electra*, no. 69, pp. 65–102, March 1980a; Anderson, R. B. and Eriksson, A. J., "A Summary of Lightning Parameters for Engineering Applications," CIGRE Paper No. 33–06, 1980b; Berger, K., Anderson, R. B., and Kröninger, H., "Parameters of Lightning Flashes," *Electra*, no. 41, pp. 23–37, July 1975.

eters, one for low currents and one for high currents. Another common approximation to Berger's data for the probability of the peak magnitude of a first stroke is (EPRI, 1982):

$$P(I_0 \geq i_0) = \frac{1}{1+(i_0/31)^{2.6}}$$

Table 12.2 shows characteristics for positive strokes. This data is not as well defined since the data set is more limited.

Although most stroke and flash characteristics are independent of each other, there are some interdependencies. Cigre (1991) examined correlations

TABLE 12.2

Lightning Current Parameters for Downward Positive Flashes

Parameter	Percent of Cases More than Value 5%	50% (*M*)	95%	β
Peak current, kA	5	35	260	1.21
Time to peak, μsec (30–90% virtual front)	3	22	140	1.23
Stroke charge, C	2	16	84	1.36
Flash charge, C	18	80	350	0.90

Source: Berger, K., Anderson, R. B., and Kröninger, H., "Parameters of Lightning Flashes," *Electra*, no. 41, pp. 23–37, July 1975.

between various parameters. Larger first strokes tend to have longer rise times. For first strokes, the equivalent front risetime correlates some with the peak current; the average rate of rise does not. For subsequent strokes, the peak current is independent of the rise time, although the peak current partially correlates with the rate of rise. For both first and subsequent strokes, the peak current correlates to some degree with the maximum rate of rise. The correlations are not particularly strong in any of these cases. Cigre used these interdependencies to find derived distributions that are useful in some stochastic simulations.

More than half of cloud-to-ground lightning flashes are composed of more than one stroke (see Figure 12.4). A quarter of them have at least four strokes. The subsequent strokes usually have less current than first strokes, but the rate of rise of current is higher (important for the inductive voltage rise, Ldi/dt, in arrester leads, arrester spacings on lines, and traveling-wave issues in cables). Subsequent stroke characteristics are thought to be independent of the first stroke.

A lightning flash may last more than 1 sec; this impacts line reclosing practices. If a circuit breaker or recloser clears a lightning-caused fault and

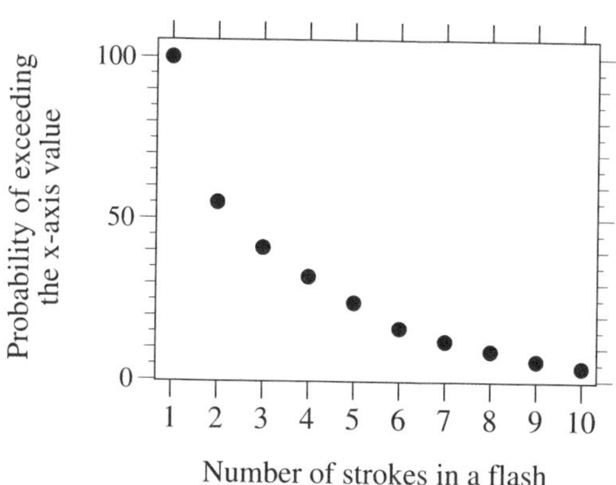

FIGURE 12.4
Number of strokes in a flash. (Data from [Anderson and Eriksson, 1980a,b].)

TABLE 12.3

Probability of a Successful Reclose Following a Lightning-Caused Fault

Duration of the Dead Time Before the Reclose, sec	Probability of a Successful Reclosure
0.3	83%
0.4	90%
0.5	95%

Source: Anderson, R. B. and Eriksson, A. J., "Lightning Parameters for Engineering Applications," *Electra*, no. 69, pp. 65–102, March 1980a.

immediately recloses, a subsequent stroke may flash the line over again (or continuing current in the flash may maintain the fault arc). About 5% of lightning flashes will last beyond a typical immediate reclosing time of 0.5 sec (see Table 12.3). Normally, the extra fault will not cause any more damage; the circuit breaker must open and reclose again. The practical effect of this is to extend a half-second interruption to a 10-sec interruption (or whatever the next delay time is in the reclosing sequence). It does show that it is important to have more than one reclose attempt if an immediate reclose is used.

A good percentage of multiple-stroke flashes have subsequent strokes to different points on the ground (Thottappillil et al., 1992). This implies that ground flash densities from flash counters and lightning detection networks may underestimate the number of lightning flash ground terminations.

The lightning activity in an area can be measured. Many areas of the world have lightning detection networks that measure the magnetic and/or electric field generated by a lightning stroke, determine if the stroke is from cloud to ground, and triangulate the stroke's position. Such systems help utilities prepare for storms; information on storm intensity, direction, and location helps determine the number of crews to call up and where to send them. Maps generated from lightning detection networks of *ground flash density* (GFD or N_g) are the primary measure of lightning activity. Figure 12.5 shows a 10-year ground flash density contour map of the U.S. from the U.S. National Lightning Detection Network (NLDN), which has been operating since before 1990.

Lightning detection networks are also useful for correlating faults with lightning. This data helps with forensics and is even used in real time to direct crews to damage locations. From personnal experience with correlating faults with the U.S. NLDN and with camera monitoring studies, the system successfully captures about 90% of strokes. The most important characteristic that allows accurate correlation of faults and lightning is accurate time tagging of power system event recorders including power quality recorders, SCADA, or fault recorders (GPS works well). Position accuracy of detection networks is not good enough to determine if strokes hit a line, but it is good enough to narrow the choices of strokes considerably — almost

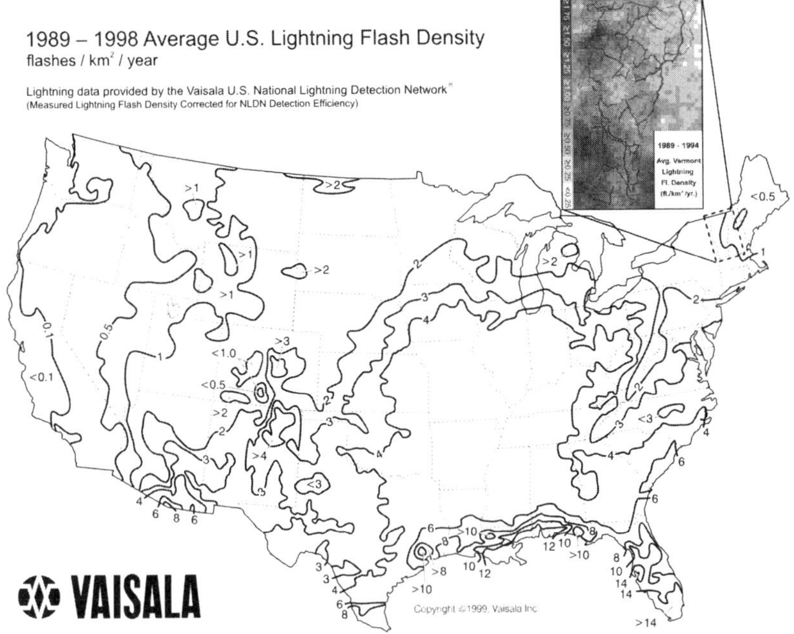

FIGURE 12.5
Ground flash density from the United States National Lightning Detection System. (Lightning map provided by Vaisala.)

all strokes found by the U.S. NLDN are accurate to within 1 mi (1.6 km), with most accurate to 2000 ft (0.5 km).

If direct measurements of ground flash density are unavailable, meteorological records of thunderstorms are available, most commonly the number of days with thunderstorms (or keraunic level). Thunderstorm days approximately relate to ground flash density as (Eriksson, 1987):

$$N_g = 0.04T_d^{1.25}$$

Records of the number of thunderstorm hours per year relate to N_g as (MacGorman et al., 1984):

$$N_g = 0.054T_h^{1.1}$$

where T_h is the number of thunderstorm hours per year.

Another crude estimate of the lightning level is from NASA's Optical Transient Detector, which measured worldwide lightning activity for 5 years (Chisholm et al., 1999).

Lightning is highly variable. It takes several hundred lightning flash counts to obtain modest accuracy for an estimate of the average flash density. A

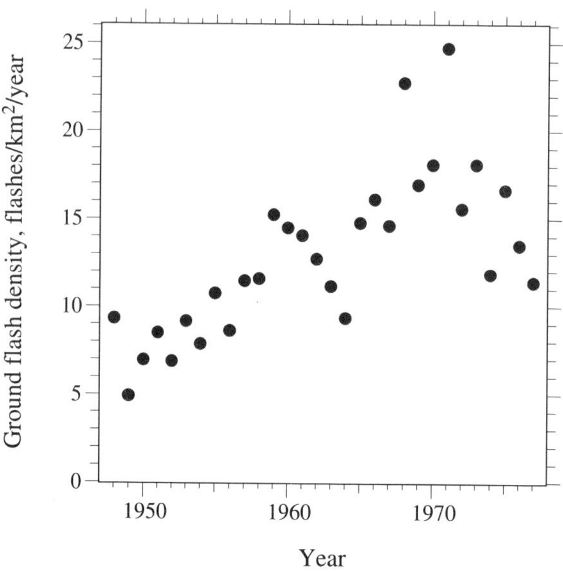

FIGURE 12.6
Estimated annual ground flash density for Tampa, Florida based on thunderstorm-hour measurements. (Data from [MacGorman et al., 1984].)

smaller geographic area requires more measurement time to arrive at a decent estimate. Similarly, a low-lightning area requires more measurement time to accurately estimate the lightning. Standard deviations for yearly measurements of lightning activity range from 20 to 50% of the mean (IEEE Std. 1410-1997). Figure 12.6 shows the variability of ground flash density in a high-lightning area. Lightning and storms have high variability. Lightning and weather patterns may have cycles that last many years.

Distribution lines cover small geographic areas, so lightning damage and lightning flashover rates show high variability on a circuit. One year a circuit may get nailed with damage coming from many storms; the next year it may have next to nothing.

The variability of lightning and the variability of storms is also important for utility planning regarding regulatory incentives for reliability and for performance guarantees for customers. Just a few years of data usually does not accurately depict the performance of weather-related events for a circuit or even for a whole system.

12.2 Incidence of Lightning

Lightning flashes hit typical-height distribution lines that are in the open at the rate of about 17 flashes/100 circuit mi/year for an area with N_g = 1 flash/km²/year (11 flashes/100 km/year). Flashes to lines vary directly as the ground flash density. Almost all of these flashes to lines cause faults, and some damage equipment. Taller lines attract more flashes. The most accepted model of lightning strikes to power lines in open ground is Eriksson's model, which finds the number of strikes as a function of the line height (Eriksson, 1987):

$$N = N_g\left(\frac{28h^{0.6} + b}{10}\right)$$

where

N = flashes/100 km/year to the line

N_g = ground flash density/km²/year

h_t = height of the conductor (or overhead groundwire) at the tower, m

b = overhead ground wire separation, m

For distribution lines, the b term can be ignored, which gives:

$$N = 2.8N_g \cdot h^{0.6}$$

for h in meters and N in flashes/100 km/year, or

$$N = 2.2N_g \cdot h^{0.6}$$

for h in feet and N in flashes/100 mi/year.

For a 30-ft (10-m) line, about 17 flashes/100 mi/year (10.6 flashes/100 km/year) hit a distribution line for N_g = 1 flash/km²/year. The main data point for distribution lines is a South African test line heavily monitored in the 1980s (Eriksson, 1987). The test line had 19 flashes/100 mi/year (12 flashes/100 km/year) normalized for N_g = 1 flash/km²/year for a line height of 28 ft (8.6 m).

The equivalent shadow width or total width that a distribution line attracts lightning is on the order of 360 ft (110 m). This is about 12 times the line height. For many years, a shadow width of four times the line height was used; this model underestimates the number of flashes hitting lines.

More exposed lines — those with no trees nearby or lines on the tops of ridges — attract more flashes. Eriksson's equation is for a line in open ground. Many lines have fewer hits because of shielding from nearby objects (mainly trees, but also buildings and other power lines). For lines in forested areas, the number of strikes is significantly reduced, so this equation should be an upper limit for most lines. Some very exposed lines on hills may have more hits than Eriksson's equation predicts. The number of hits to a line in a shielded area is:

$$N_S = N(1 - S_f)$$

where S_f is the shielding factor, which is between 0 and 1. In Florida, shielding factors of around 0.7 were found for circuits out of four substations (Parrish and Kvaltine, 1989). Lines in these areas had moderate to heavy shielding by nearby objects.

12.3 Traveling Waves

When lightning injects surge currents into power lines, they move down the conductors as traveling waves. Understanding the nature of traveling waves helps in predicting the voltage and current levels that occur on power systems during lightning strikes. Surges travel at 99% the speed of light on overhead circuits — about 1000 ft/μsec(300 m/μsec). The inductance L and capacitance C uniformly distributed along the line determine the velocity and the relationship between voltage and current. The velocity in a unit length per second is based on L and C per length unit as

$$v = \frac{1}{\sqrt{LC}}$$

The voltage and current are related by Z, the *surge impedance*,

$$V = ZI$$

where Z is a real value, typically 300 to 400Ω. The distributed inductance and capacitance determines the surge impedance:

$$Z = 60\sqrt{\frac{L}{C}} = 60\ln\frac{2h}{r}$$

where

h, average conductor height

r, conductor radius

When surge voltages of like polarity pass over each other, the voltages add as shown in Figure 12.7. The currents subtract; current is charge moving in a direction. Charge of the same polarity flowing in opposite directions cancels.

$$V = V_F + V_B = Z(I_F - I_B)$$

$$I = Z(I_F + I_B)$$

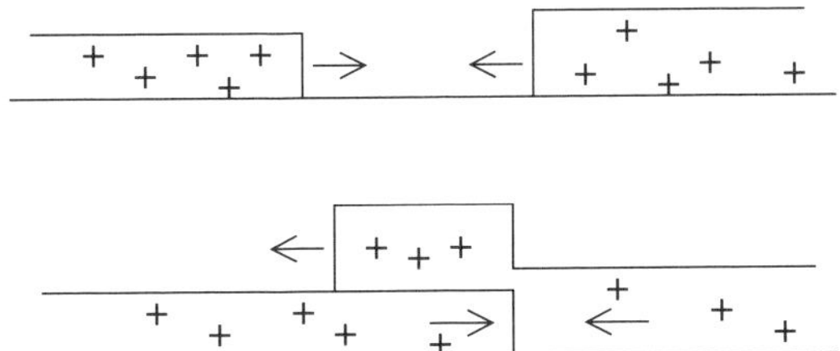

FIGURE 12.7
Forward and backward traveling waves passing each other.

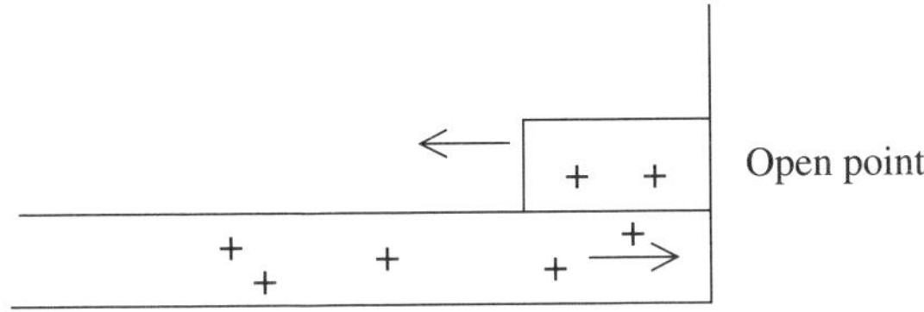

FIGURE 12.8
Traveling wave reflection at an open point.

A surge arriving at an open point in the circuit reflects, sending back a voltage wave equal to the incoming wave (see Figure 12.8). The voltage doubles. Think of the incoming wave as a stream of electrons. When these electrons hit the open point, they stop, but the pileup of electrons at the end forces the electrons back in the direction they came from (like charges repel each other). While the voltage doubles, the current cancels to zero as the return wave counters the incoming wave.

Voltage doubling at open points is an important consideration for the protection of distribution insulation. It is a very important consideration when protecting underground cables. It also comes into play when placing arresters to protect a substation.

When a surge hits a short circuit, the voltage drops to zero as we expect. The ground releases opposing charge that cancels the voltage, creating a reverse wave traveling back toward the source. Now the current doubles. Two cases are very close to a short circuit: a conducting surge arrester and a fault arc. Both of these act like a short. They generate a voltage canceling wave.

Line taps, capacitors, open points, underground taps — any impedance discontinuities — cause reflections. In the general case, the voltage wave reflected from a discontinuity is

$$V_R = aV_I$$

where a is the reflection coefficient:

$$a = \frac{Z_B - Z_A}{Z_B + Z_A}$$

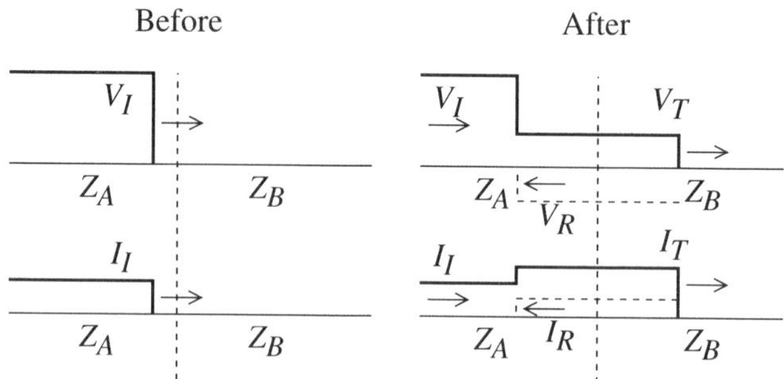

FIGURE 12.9
Traveling wave reflection at a surge-impedance discontinuity where $Z_B < Z_A$.

For an open circuit ($Z_B = \infty$), the reflection coefficient, $a = 1$, and for a short circuit ($Z_B = 0$), $a = -1$.

The voltage at the discontinuity is the sum of the incoming wave and the reflected wave:

$$V_T = V_I + V_R = V_I + aV_I = bV_I$$

where b is the refraction coefficient:

$$b = 1 + a = \frac{2Z_B}{Z_B + Z_A}$$

The current waves are found from the voltages as $I_F = V_F / Z$ for forward waves and $I_B = -V_B / Z$ for backward waves. See Figure 12.9.

When a traveling wave hits a split in the line, a portion of the wave goes down each path and another wave reflects back toward the source as shown in Figures 12.10. Use $V_T = bV_I$ to calculate the voltage at the split, just use a new Z_B' as the parallel combination of Z_B and Z_C, $Z_B' = Z_B || Z_C = Z_B Z_C / (Z_B + Z_C)$. For equal surge impedances ($Z_A = Z_B = Z_C$), the total voltage is 2/3 of the incoming wave.

To a traveling wave, a capacitor initially appears as a very small impedance, almost a short circuit. Then, as the capacitor charges, the effective

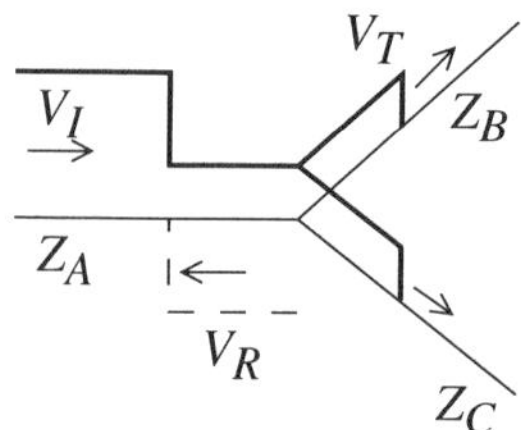

FIGURE 12.10
Traveling wave reflections at a split.

impedance rises until it becomes an open circuit with the capacitor fully charged. For a traveling wave section terminated in a capacitor, the voltage ramps up to double the voltage as

$$V_T = 2V_I\left(1 - e^{-\frac{t}{ZC}}\right)$$

An inductance initially appears almost as an open circuit. As the inductor allows more current to flow, its impedance drops until it is equivalent to a short circuit. For a traveling wave section terminated in an inductor, the voltage initially doubles then decays down to zero as

$$V_T = 2V_I\left(1 - e^{-\frac{Z}{L}t}\right)$$

A cable section attached to an overhead circuit behaves similarly to a capacitor. It is like an open-circuited traveling wave section with a low surge impedance. In fact, a capacitor can be modeled in a traveling wave program such as EMTP as a traveling wave section with a surge impedance of $\Delta t/2C$ and a travel time of $\Delta t/2$, where Δt is small (usually the time step of the simulation) (Dommel, 1986). Likewise, we may model an inductor as a traveling wave section *shorted* at the end with a surge impedance equal to $2L/\Delta t$ and a travel time of $\Delta t/2$.

Corona impacts traveling waves by modifying the surge impedance. When lightning strikes an overhead distribution circuit, tremendous voltages are created. The voltage stress on the conductor surface breaks down the air surrounding the conductor, releasing charge into the air in many small streamers. A high-voltage transmission line may have a corona envelope of several feet (1 m), which increases the capacitance. This decreases the surge impedance and increases coupling to other conductors and slopes off the front of the wave. Normally, distribution line insulation flashes over before significant corona develops, so we can neglect corona in most cases.

Predischarge currents, an effect similar to corona, may also form between conductors. When the voltage between two parallel conductors reaches the flashover strength of air (about 180 kV/ft or 600 kV/m), a sheet of many small streamers containing predischarge currents flows between the conductors. This is a precursor to a complete breakdown. The predischarge current delays the breakdown and relieves voltage stress between the conductors. While the predischarge currents are flowing, the resistance across the gap is 400 Ω/ft or 1310 Ω/m of line section (Wagner, 1964; Wagner and Hileman, 1963; Wagner and Hileman, 1964). Normally, on overhead distribution circuits, the weakest insulation is at poles, so in most cases, the circuit flashes over before the circuit enters the predischarge state. Predischarge currents do make midspan flashovers less likely and may help with using arresters to protect lines.

Normally, for modeling lightning in circuits or cables, we can ignore corona and predischarge and use lossless line sections (wire resistance and damping is ignored). This is normally sufficient since only short sections need to be modeled for most situations.

A multiconductor circuit has several modes of propagation; each has different velocity and different surge impedance. In most cases, these multiconductor effects are ignored, and we assume each wave travels on each conductor at 95 to 100% of the speed of light.

For multiconductor circuits, the coupling between circuits is important. The mutual surge impedance determines the voltage on one conductor for a current flowing in another as $V_2 = Z_{12}I_1$. The mutual surge impedance is

$$Z_{ij} = 60 \ln \frac{a_{ij}}{b_{ij}}$$

where

a_{ij}, distance between conductor i and the image conductor j
b_{ij}, distance between conductor i and conductor j

Conductors that are closer together couple more tightly. This generally helps by reducing the voltage stress between the conductors.

12.4 Surge Arresters

Modern surge arresters have metal-oxide blocks, highly nonlinear resistors. Under normal voltages, a metal-oxide surge arrester is almost an open circuit, drawing much less than a milliampere (and the watts losses are on the order of 0.05 W/kV of MCOV rating). Responding to an overvoltage almost instantly, the impedance on the metal-oxide blocks drops to a few ohms or less for severe surges. After the surge is done, the metal oxide immediately returns to its normal high impedance. Metal-oxide blocks are primarily zinc oxide with other materials added. Boundaries separate conductive zinc-oxide grains. These boundaries form semiconductor junctions like a semiconducting diode. A metal-oxide block is equivalent to millions of series and parallel combinations of semiconducting diodes. With low voltage across the boundaries, the resistance is very high, but with high voltage, the resistance of the boundaries becomes very low.

Metal-oxide surge arresters overcame many drawbacks of earlier designs. The spark gap was one of the earliest devices used to protect insulation against lightning damage. A spark gap has many of the desirable benefits of a surge arrester; the gap is an open circuit under normal conditions and when it sparks over, the gap is virtually a short circuit. The problem is that after the surge, the gap is still shorted out, so a fuse or circuit breaker must

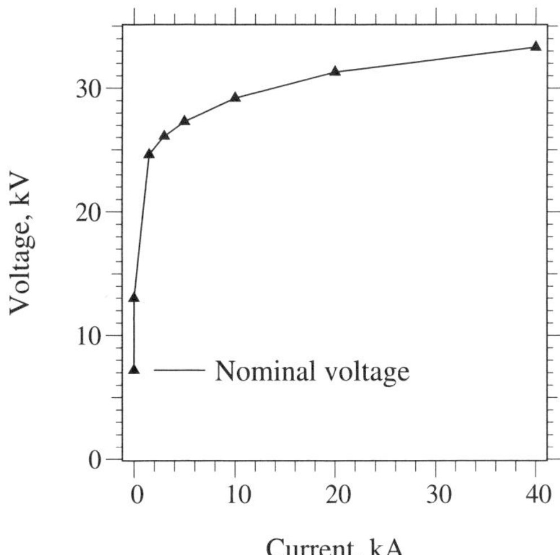

FIGURE 12.11
Typical characteristics of an 8.4-kV MCOV arrester (typically used on a system with a nominal voltage of 7.2 or 7.62 kV from line to ground).

operate. In some parts of the world, notably Europe, spark gaps are still widely used to protect transformers. Spark gaps were improved by adding nonlinear resistors in series. The nonlinear material would clear the power follow current flowing through the spark gap after the surge was done. Gapped arresters with silicon carbide blocks were developed in the 1940s and used for many years. Metal oxide is so nonlinear that a gap is not required. This allows an arrester design, which is simple and highly reliable, with excellent "fast-front" response characteristics. Figure 12.11 shows the voltage-current relationships of a typical metal-oxide arrester. At distribution voltages, metal oxide was first used to provide better protection at riser poles feeding underground cables (Burke and Sakshaug, 1981).

Distribution surge arresters are primarily for protection against lightning (not switching surges or other overvoltages). In areas without lightning they would not be needed. That said, most utilities in low-lightning areas use arresters even if the cost of arresters exceeds the cost of occasional failures. If a utility is hit with a 1-in-40-year storm, it can wipe out a significant portion of unprotected transformers, enough that the utility does not have the inventory to replace them all, and some customers would have very long interruptions.

Arresters must be placed as close as possible to the equipment being protected. This means on the same pole because lightning has such high rates of rise, an arrester one pole span away provides little protection. At the pole structure, the best place for the arrester is right on the equipment tank to minimize the inductance of the arrester leads.

Arresters protect any equipment susceptible to permanent damage from lightning. Figure 12.12 shows examples of arrester applications. Transform-

FIGURE 12.12
Example arrester applications.

ers should have arresters. Two-terminal devices such as reclosers, regulators, and vacuum switches should have arresters on both the incoming and outgoing sides. Arresters on reclosers are especially important. When the recloser is open, a surge hitting the open point will double. Commonly, reclosers open to clear lightning-caused faults, and if the downstream side is not protected, a subsequent lightning stroke that followed the one that caused the fault could fail the open recloser. Cables need special attention to prevent lightning entry and damage. At equipment with predominantly air flashover paths — insulators, switches, cutouts — utilities do not normally use arresters.

Most utilities also protect capacitor banks with arresters. Some believe that capacitors protect themselves because capacitors have low impedance to a fast-changing surge. While this is somewhat true, nearby direct lightning flashes may still fail capacitors. About 1.9 C, which is a small first stroke, charges a 400-kvar, 7.2-kV capacitor unit to 95 kV (the BIL for 15-kV equipment).

Several classes of arresters are available for protecting distribution equipment:

- *Riser pole* — Designed for use at a riser pole (the junction between an overhead line and a cable); has better protective characteristics than heavy-duty arresters (because of voltage doubling, underground protection is more difficult)
- *Heavy duty* — Used in areas with average or above average lightning activity
- *Normal duty* — Higher protective levels and less energy capability. Used in areas with average or below average lightning activity
- *Light duty* — Light-duty arrester used for protection of underground equipment where most of the lightning current is discharged by another arrester at the junction between an overhead line and the cable

Two protective levels quantify an arrester's performance (IEEE Std. C62.22-1997):

- *Front-of-wave protective level (FOW)* — The crest voltage from a current wave causing the voltage to rise to crest in 0.5 μsec.
- *Lightning impulse protective level (LPL)* — The crest voltage for an 8/20-μsec current injection (8 μsec is the equivalent time to crest based on the time from 10 to 90% of the crest, and 20 μsec is the time between the origin of the 10 to 90% virtual front and the half-value point). These characteristics are produced for current peaks of 1.5, 3, 5, 10, and 20 kA.

Table 12.4 gives common ranges. Lower protective levels protect equipment better.

TABLE 12.4

Common Ranges of Protective Levels

Duty Cycle Rating kV rms	MCOV kV rms	Front-of-Wave Protective Level, kV 5 kA Normal Duty	10 kA Heavy Duty	10 kA Riser Pole	Maximum Discharge Voltage, kV 8/20 μsec Current Wave 5 kA Normal Duty	10 kA Heavy Duty	10 kA Riser Pole
3	2.55	11.2–17	13.5–17	10.4	10.2–16	9.1–16	8.2
6	5.1	22.3–25.5	26.5–35.3	17.4–18	20.3–24	18.2–25	16.2
9	7.65	33.5–36	26.5–35.3	22.5–36	30–33.5	21.7–31.5	20–24.9
10	8.4	36–37.2	29.4–39.2	26–36	31.5–33.8	24.5–35	22.5–26.6
12	10.2	44.7–50	35.3–50	34.8–37.5	40.6–44	32.1–44	30–32.4
15	12.7	54–58.5	42–59	39–54	50.7–52	35.9–52	33–40.2
18	15.3	63–67	51–68	47–63	58–60.9	43.3–61	40–48
21	17.0	73–80	57–81	52–63.1	64–75	47.8–75	44–56.1
24	19.5	89–92	68–93	63–72.5	81.1–83	57.6–83	53–64.7
27	22.0	94–100.5	77–102	71–81.9	87–91.1	65.1–91	60–72.1
30	24.4	107–180	85–109.5	78–85.1	94.5–99	71.8–99	66–79.5
36	29.0	125	99–136	91–102.8	116	83.7–125	77–96

Source: IEEE Std. C62.22-1997. Copyright 1998 IEEE. All rights reserved.

12.4.1 Ratings and Selection

The choice of arrester rating depends on the voltage under normal conditions and possible temporary overvoltages. Temporary overvoltages during line-to-ground faults on the unfaulted phases are the main concern, therefore, system grounding plays a large role in arrester application.

The most important rating of a metal-oxide arrester is the maximum continuous operating voltage (MCOV). Arresters also have another voltage rating called the duty-cycle rating (this rating is very commonly used to refer to a particular arrester). The duty cycle rating is about 20% higher than the MCOV rating. Within a given rating class, arresters may have different performance characteristics (such as the discharge voltage and the energy-handling capability). For example, one 8.4-kV MCOV normal-duty arrester may have a protective level for a 10-kA, 8/20-μsec current injection of 36 kV. A riser-pole arrester of the same rating may have a 27-kV protective level.

Operating voltage is the first criteria for applying arresters. The arrester MCOV rating must be above the normal upper limit on steady-state voltage. Some engineers use the upper (ANSI C84.1-1995) range B, which is slightly less than 6% above nominal. Use 10% above the nominal line-to-ground voltage to be more conservative.

Temporary overvoltages are the second application criteria. An arrester must withstand the temporary overvoltages that occur during line-to-ground faults on the unfaulted phases. The overvoltages are due to a neutral shift. On an ungrounded system, a ground fault shifts the neutral point to the faulted phase. The arresters connected to the unfaulted phases see line-to-line voltage. Four-wire multigrounded systems are grounded but still see

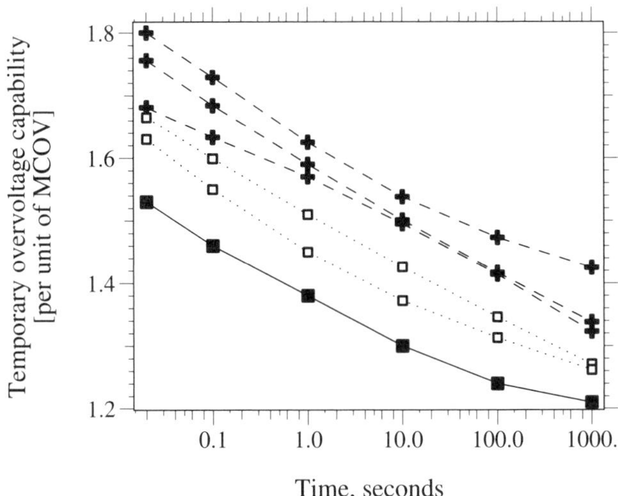

FIGURE 12.13
Temporary-overvoltage capabilities of several different arrester manufacturers and models (dashed-lines: heavy-duty arresters, dotted lines: riser-pole arresters, bottom line: lowest TOV capability expected as published in IEEE Std. C62.22-1997).

some neutral shift. (IEEE C62.92.4-1991) recommends using a 1.35 per unit overvoltage factor for a multigrounded system. Chapter 13 has more information on overvoltages during ground faults.

Arrester temporary overvoltage (TOV) capabilities are time-dependent; manufacturers publish temporary overvoltage curves for their arresters (see Figure 12.13). The duration of the overvoltage depends on the relaying and fusing. Normally, this varies, and since arrester applications are not individually engineered, a value is normally assumed.

On four-wire multigrounded systems, the MCOV criteria, not the TOV criteria, determines the arrester application. On ungrounded or impedance-grounded circuits, TOV capability determines the rating.

Riser-pole arresters have less temporary overvoltage capability than normal or heavy-duty arresters. Riser-pole arresters normally have the same arrester block size as heavy-duty arresters; the manufacturer chooses blocks carefully to find those with lower discharge voltages. A lower discharge characteristic means more current and energy for a given temporary overvoltage.

Utilities typically standardize arrester ratings for a given voltage and grounding configuration. Table 12.5 shows some common applications. For the four-wire systems, most are okay, but try to avoid the tight applications such as the 9-kV duty cycle arrester at 12.47 kV and the 10-kV duty cycle arrester at 13.8 kV.

For ungrounded circuits, the voltage on the unfaulted phases rises to the line-to-line voltage. This normally requires an arrester with an MCOV equal to the line-to-line voltage because the fault current is so low on an ungrounded system that single line-to-ground faults are difficult to detect.

TABLE 12.5

Commonly Applied Arrester Duty-Cycle Ratings

Nominal System Voltage (V)	Four-Wire Multigrounded Neutral Wye	Three-Wire Low Impedance Grounded	Three-Wire High Impedance Grounded
2400			3 (2.55)
4160Y/2400	3 (2.55)	6 (5.1)	6 (5.1)
4260			6 (5.1)
4800			6 (5.1)
6900			9 (7.65)
8320Y/4800	6 (5.1)	9 (7.65)	
12000Y/6930	9 (7.65)	12 (10.2)	
12470Y/7200	9 (7.65) or 10 (8.4)	15 (12.7)	
13200Y/7620	10 (8.4)	15 (12.7)	
13800Y/7970	10 (8.4) or 12 (10.1)	15 (12.7)	
13800			18 (15.3)
20780Y/12000	15 (12.7)	21 (17.0)	
22860Y/13200	18 (15.3)	24 (19.5)	
23000			30 (24.4)
24940Y/14400	18 (15.3)	27 (22.0)	
27600Y/15930	21 (17.0)	30 (24.4)	
34500Y/19920	27 (22.0)	36 (29.0)	

Note: MCOV ratings are shown in parenthesis.

Source: IEEE Std. C62.22-1997.

12.4.2 Housings

Polymer materials such as silicone or EPR compounds are used for arrester housings. Porcelain housings are available, and all arresters up until the 1980s were porcelain housed. The main advantage of polymer over the porcelain is a safer failure mode. An internal fault in a porcelain arrester can explode the housing and expel porcelain like shrapnel. The polymer housing will split from the pressures of an internal fault. Arc energies may still crack apart and eject portions of the blocks; but overall, the failure is less dangerous than in a porcelain housing. Polymer-housed arresters can pass the ANSI housing test where a specified fault current is applied for at least 0.1 sec, and all components of the arrester must remain confined within the enclosure. Polymer-housed arresters typically can pass these criteria for a 5-kA fault current for 1/2 s ($\int I dt = I \cdot t = 2500$ C). Porcelain-housed arresters fail violently for these levels of current. Lat and Kortschinski (1981) ruptured porcelain arresters with fault currents totaling as low as 75 C (6.8 kA for 11 msec, another failed at 2.7 kA for 50 msec). Expulsion fuses may prevent explosions in porcelain arresters at low short-circuit currents but not at high currents (current-limiting fuses are needed at high currents).

A polymer housing keeps water out better than a porcelain housing (water initiates many failure modes). Polymer-housed arresters — although better than porcelain-housed arresters — are not immune from moisture ingress. Moisture can enter through leaks in fittings or by diffusing through the polymer material. One utility in a high-lightning area has found polymer-housed arresters with moisture ingress degradation. These show up on thermal imaging systems where an arrester with moisture ingress may be 50°C hotter than adjacent arresters. The utility replaces any arresters with significant thermovision temperature discrepancies. Lahti et. al. (1998, 1999) performed high humidity tests and hot-water immersion tests on polymer-housed arresters. Arresters with the housing directly molded onto the arrester body performed well (no failures). Those with housings pressed on and fitted with caps at the ends often allowed moisture in through leaks in the end caps.

12.4.3 Other Technologies

Many older arrester technologies are still in place on distribution circuits. The most prevalent is the gapped silicon-carbide arrester. Silicon-carbide is a nonlinear resistive material, but it is not as nonlinear as metal oxide. It requires a gap to isolate the arrester under normal operating voltage. When an impulse sparks the gap, the resistance of the silicon-carbide drops, conducting the impulse current to ground. With the gap sparked over, the arrester continues to conduct 100 to 300 A of power follow current until the gap clears. If the gap fails to clear, the arrester will fail. Annual failure rates have been about 1% with moisture ingress into the housing causing most failures of these arresters [an Ontario-Hydro survey found 86% of failures were from moisture (Lat and Kortschinski, 1981)]. Moisture degrades the gap, failing it outright or preventing it from clearing a surge properly. Darveniza et al. (1996) recommended that gapped silicon-carbide arresters older than 15 years be progressively replaced with metal-oxide arresters. Their examinations and tests found a significant portion of silicon-carbide arresters had serious deterioration with a pronounced upturn after about 13 years.

Under-oil arresters are another protection option. Metal-oxide blocks, mounted inside of transformers, protect the windings with almost no lead length. Eliminating the external housing eliminates animal faults across arrester bushings; on padmounted transformers, connector space is freed. Since the arrester blocks are immersed in the oil, thermal characteristics are excellent, which improves temporary overvoltage characteristics [Walling (1994) shows tests where external arresters failed thermally during ferroresonance but under-oil arresters survived]. The main disadvantages of under-oil arresters are related to failures. If the blocks fail, either due to lightning or other reason, the energy in the long fault arc around or through the arrester blocks makes transformer tank failures more likely (Henning et al., 1989). Even if the transformer tank does not fail, since the arrester is a fundamental component of the transformer, an arrester failure is a transformer failure: a crew must take the whole transformer back to the shop.

Most metal-oxide arresters are gapless; but a gapped metal-oxide arrester is available. A portion of the metal-oxide blocks that would normally be used are replaced with a gap. This improves the protection. When the gap sparks over, the surge current creates less voltage because there is less metal oxide. Once the overvoltage clears, the metal oxide provides considerable impedance to clear the gap with very little power follow current. The gapped arrester also has good temporary overvoltage capabilities if the gap does not spark over. If a lightning stroke or ferroresonance sparks the gap, temporary overvoltages can fail the arrester more quickly. Another disadvantage is the gap itself, which was the most unreliable part of the gapped silicon-carbide arrester.

12.4.4 Isolators

Distribution arresters have isolators that remove failed arresters from the circuit. The isolator has an explosive cartridge that blows the end off of a failed arrester, which provides an external indication of failure. The isolator itself is not designed to clear the fault. An upstream protective device normally must clear the fault (although, in a few cases, the isolator may clear the fault on its own, depending on the available short-circuit current and other parameters).

Crews should take care with the end lead that attaches to the bottom of the arrester. It should not be mounted so the isolator could swing the lead into an energized conductor if the isolator operates. The proper lead size should also be used (make sure it is not too stiff, which might prevent the lead from dropping).

Occasionally, lightning can operate an isolator in cases where the arrester is still functional. One study found that 53% of arresters removed due to operation of the isolator were because the isolator operated improperly (Campos et al., 1997) (this ratio is much higher than normal).

Arrester isolators have a small explosive charge similar to a 0.22 caliber cartridge. The firing mechanism is a gap in parallel with a resistor. Under high currents, the gap flashes over. The arc through the gap attaches to the shell and generates heat that fires the shell. The heat from an arc is roughly a function of $\int I \, dt$, which is the total charge that passes. Most isolators operate for a total charge of between 5 and 30 C as shown in Figure 12.14. A significant portion of lightning flashes contains charge reaching these amounts. Even though only a portion of the lightning current normally flows into arresters, a significant number of events will still cause isolator operation.

Isolators should coordinate with upstream protective devices. Some small fuses can blow before the isolator clears (the melting time of fuses varies as $\int I^2 \, dt$, but the isolators vary as $\int I \, dt$, so there is some crossover point). Most small transformer fuses may clear in less than a half cycle for faults above 1000 A (the clearing time depends on the timing of the fault relative to the next zero crossing). A 1/2-cycle fault of 1 kA is 8.3 C, which may not fire an isolator. This is a safety issue since line crews may replace a blown fuse right next to a failed arrester that was not isolated properly.

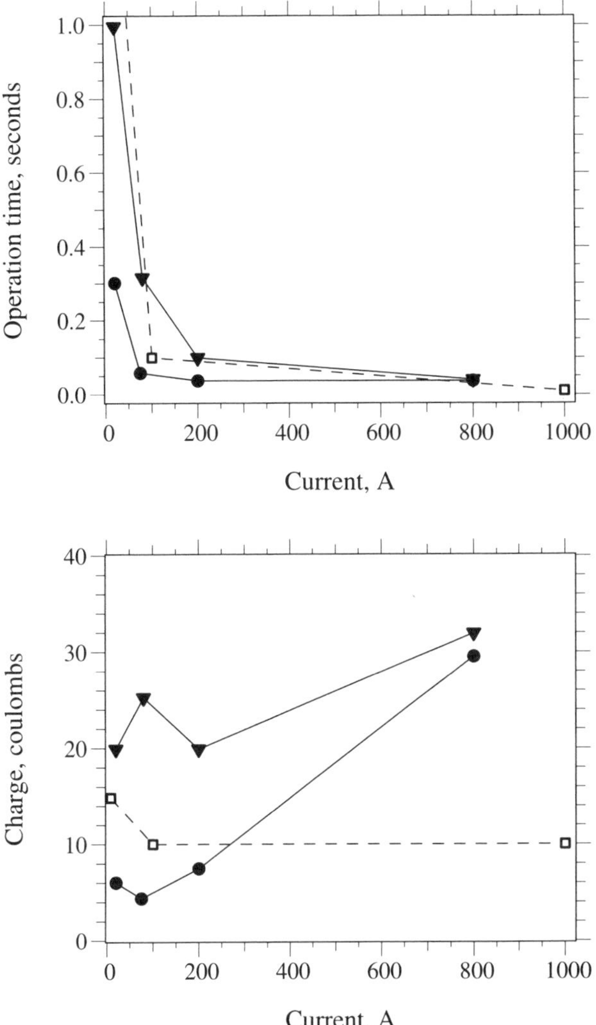

FIGURE 12.14
Operation time and charge required to operate isolators for different current levels (from three manufacturers).

12.4.5 Arrester Reliability and Failures

Modern arresters are reliable components with failure rates much less than 1% annually. They do have somewhat of a bad reputation though, probably because there were bad products produced by some manufacturers during the early years of metal-oxide arresters. And their tendency to fail violently did not help.

One manufacturer states a failure rate less than 0.05% for polymer-housed arresters. This is primarily based on returned arresters, which are a low estimate of the true failure rate; arresters are almost a throw-away commodity; not all failures are returned. Arresters may also fail without utilities noticing; if the isolator operates and a circuit breaker or recloser closes back in, the failed arrester is left with a dangling lead that may not be found for some time. Ontario Hydro (CEA 160 D 597, 1998) has recorded failure rates averaging about 0.15% in a moderately low-lightning area with about one flash/km^2/year. About 40% of their arrester failures occurred during storm periods.

Lightning arresters fail for a variety of reasons. Moisture ingress, failure due to lightning, and temporary overvoltages beyond arrester capability are some of the possibilities. Early in my career, I was involved with lab tests of arresters and EMTP modeling to evaluate system conditions such as ferroresonance, presence of distributed generation, and regulation voltages (Short et al., 1994). The main conclusion was that well-made arresters should perform well. More problems were likely for tightly applied arresters (such as using a 9-kV duty-cycle arrester with a 7.65-kV MCOV on a 12.47/7.2-kV system), so avoid tightly applying arresters under normal circumstances.

Darveniza et al. (2000) inspected several arresters damaged in service in Australia. Of the gapped metal-oxide arresters inspected, both polymer and porcelain-housed units, moisture ingress caused most failures. Of the gapless metal-oxide arresters inspected, several were damaged under conditions that were likely ferroresonance — arresters on riser poles with single-pole switching or with a cable fault. Another portion were likely from severe lightning, most likely from multiple strokes. Most of the failures occurred along the outer surface of the blocks. A small number of metal-oxide arresters, both polymer and porcelain, showed signs of moisture ingress damage, but overall, the metal-oxide arresters had fewer problems with moisture ingress than might be expected.

Lightning causes some arrester failures. Figure 12.15 shows an example. The standard test waves (4/10 μsec or 8/20 μsec) do not replicate lightning very well but are assumed to test an arrester well enough to verify field performance. The energy of standard test waves along with the charge is shown in Table 12.6. Charge corresponds well with arrester energy input since the arrester discharge voltage stays fairly constant with current.

Studies of surge currents through arresters show that individual arresters only conduct a portion of the lightning current. Normally, the lightning current takes more than one path to ground. Often that path is a flashover caused by the lightning; the flashover is a low-impedance path that "protects" the arrester. The largest stroke through an arrester measured during an EPRI study with more than 200 arrester years of monitoring using lightning transient recorders measured 28 kA (Barker et al., 1993). During the study, 2% of arresters discharged more than 20 kA annually. The largest energy event was 10.2 kJ/kV of MCOV rating (the arrester did not fail, but larger than normal 4.7-cm diameter blocks were used).

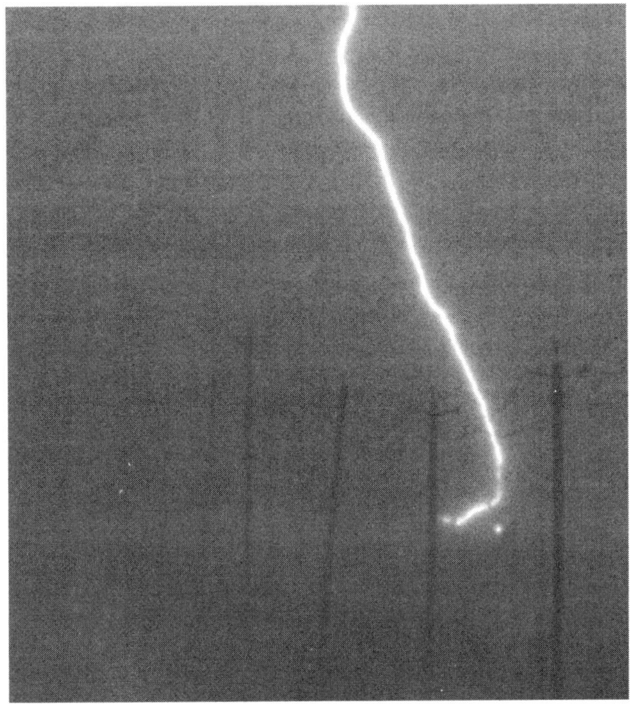

FIGURE 12.15
Lightning flash that failed an arrester. The circuit in the foreground curves around to the right about a mile beyond the visible poles. Flashovers occurred at poles on either side of the strike point. An arrester failed on one of the poles. (Copyright 1991, Niagara Mohawk Power Corporation. Reprinted with permission.)

TABLE 12.6
Approximate Energy and Charge in ANSI/IEEE Arrester Test Current Waves

(ANSI/IEEE C62.11-1987) Test Wave	Energy, kJ/kV of MCOV	Charge, C
100 kA, 4/10 μsec	4.5	1.0
40 kA, 8/20 μsec	3.6	0.8
10 kA, 8/20 μsec	0.9	0.2

Manufacturers often cite 2.2 kJ/kV of MCOV rating as the energy capability of heavy-duty arresters. The actual capability is probably higher than this. If test results from station-class arresters (Ringler et al., 1997) are translated to equivalent distribution size blocks, heavy-duty arresters should withstand 6 to 15 kJ/kV of MCOV (the 100-kA test produces about 4 to 7 kJ/kV of MCOV). Metal-oxide blocks exhibit high variability in energy capability, even in the same manufacturing batch.

Darveniza et al. (1994, 1997) found that multiple strokes are more damaging to arresters. Less energy is needed to fail the arrester. The failures occur as flashovers along the surface of the metal-oxide blocks. The block surface coating plays an important role; contamination and moisture increased the probability of failure. Five 8/20-μsec test waves were applied within an interval of 40 msec between each. A set of five 10-kA impulses applies about 4 kJ/kV of MCOV. Many of the arrester designs failed these tests. The best design had a housing directly molded onto the blocks.

Longer-duration surges also fail arresters at lower energy levels. Tests by Kannus et al. (1999) on polymer-housed arresters found many failures with 2.5/70-μsec test waves with a peak value of about 2 kA. As with multiple impulse tests, the outside of the blocks flashed over. Some arresters failed with as little as 0.5 kJ/kV of MCOV. Moisture ingress increased the failure probability for some arrester designs (Lahti et al., 2001).

This brings us to a common but difficult to answer question, should we use normal-duty or heavy-duty arresters? Heavy-duty arresters have more energy capability. This reduces the chance of failure due to severe lightning or temporary overvoltages such as ferroresonance. They also have lower discharge voltages (although the extra margin is not usually needed for protecting overhead equipment). Normal duty arresters cost less, so the question remains as to whether the extra capability of the larger arrester is worth it. For most locations, heavy-duty arresters are appropriate, except in low lightning areas with less than 0.5 flashes/km^2/year. However, this depends on the environmental exposure (how many trees are around), utility stocking and standardization considerations (for example, are transformer arresters the same as arresters at riser poles?), and cost considerations.

12.5 Equipment Protection

Insulation coordination of distribution systems involves some simple steps:

1. Choose the arrester rating based on the nominal system voltage and the grounding configuration.
2. Find the discharge voltage characteristics of the arrester.
3. Find the voltage that may be impressed on the insulation, considering any lead length and for cables, any traveling wave effects.
4. Finally, ensure that the impressed voltage is less than the equipment insulation capability, including a safety margin.

As with many aspects of distribution engineering, standards help simplify applications. Normally, utilities can standardize equipment protection for a particular voltage including arrester rating, class, and location, and excep-

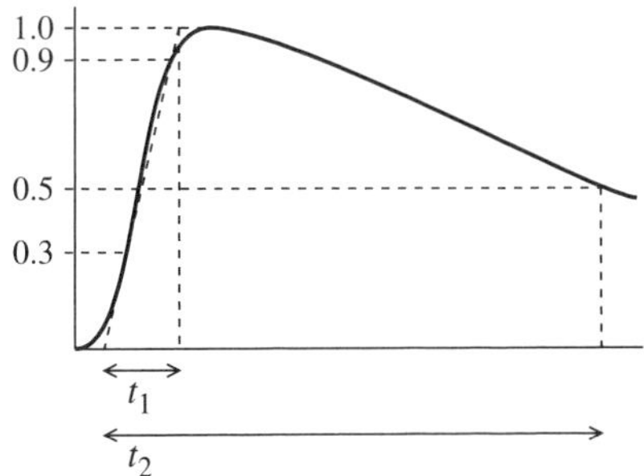

FIGURE 12.16
Standard impulse voltage test wave. (From IEEE Std. 4-1995. Copyright 1995 IEEE. All rights reserved.)

tions are rarely needed. Construction framing drawings specify how to apply arresters to minimize lead length.

12.5.1 Equipment Insulation

Distribution insulation performance is characterized with the *Basic Lightning Impulse Insulation Level* (BIL). BIL is the peak withstand voltage for a 1.2/50-µsec impulse wave. IEEE Std. 4-1995 defines voltage impulse test waves as t_1/t_2 µsec where t_1 is the equivalent time to crest based on the time taken to rise from 30 to 90% of the crest (see Figure 12.16). The time to half value t_2 is the time between the origin of the 30 to 90% virtual front and the point where it drops to half value. Equipment must withstand a certain number of applications of the test wave under specified conditions.

Standard BIL ratings for distribution equipment are 30, 45, 60, 75, 95, 125, 150, 200, 250, and 350 kV. Most equipment has the BILs given in Table 12.7.

Some insulation is also tested with a chopped wave. A chopped wave has the same characteristic as the 1.2/50-µsec wave, but the waveshape is chopped off after 2 or 3 µsec. Because the voltage stress does not last as long, with most equipment, the chopped wave withstand (CWW) is higher than the BIL. The CWW values given in Table 12.7 are for transformers (IEEE

TABLE 12.7

Distribution Equipment Insulation Impulse Levels (BIL and CWW)

Voltage Class, kV	BIL, kV	CWW, kV[a]
5	60	69
15	95	110
25	125	145
35	150	175

[a] For transformers: other equipment may have different CWW.

C57.12.00-2000). For transformers and other oil-filled equipment and insulators and other air mediums, assume CWW = 1.15 · BIL. For cable insulation, assume they are equal; CWW = BIL.

Insulation withstands higher voltages when the voltage is applied for shorter periods. The volt-time or time-lag characteristic of insulation shows this effect with the time to failure of a given crest magnitude. The volt-time characteristic of air insulation turns up significantly; oil-filled insulation turns up less, and solid insulation, very little. For solid insulations like EPR and XLPE, the small turnup is why the chopped wave withstand is the same as the BIL.

12.5.2 Protective Margin

Insulation coordination for distribution systems involves checking to see if there is enough margin between the voltage across the insulation and the insulation's capability. A protective margin quantifies the margin between the voltage the surge arrester allows (protective level) and the insulation withstand. For overhead transformers and other overhead equipment, we evaluate two protective margins both in percent, one for the chopped wave and one for the full wave (IEEE Std. C62.22-1997):

$$PM_{\text{CWW}} = \left[\frac{\text{CWW}}{\text{FOW} + Ldi/dt} - 1\right] \cdot 100$$

$$PM_{\text{BIL}} = \left[\frac{\text{BIL}}{\text{LPL}} - 1\right] \cdot 100$$

where

CWW = Chopped wave withstand, kV
FOW = Front-of-wave protective level, kV
BIL = Basic lightning impulse insulation level, kV
LPL = Lightning impulse protective level, kV
Ldi/dt = the lead wire voltage due to the rate of change of the current, kV

Both margins should be over 20% (IEEE Std. C62.22-1997); and for various reasons, we are safer having at least 50% margin. The chopped-wave protective margin includes voltage due to the rate of change of current through the arrester leads.

Protective margins for common voltages are shown in Table 12.8 and Table 12.9. The BIL margin is high in all cases; but with long lead lengths, the CWW margin is low enough to pose increased risk of failure.

The lead length component is very important; the lead voltage can contribute as much as the arrester protective level for long lengths. The arrester lead inductance is approximately 0.4 μH/ft (1.3 μH/m). Commonly, a rate of current change is assumed to be 20 kA/μsec. Together, this is 8 kV/ft of

TABLE 12.8

Typical BIL Protective Margins for Common Distribution Voltages (Using the 10-kA Lightning Protective Level)

System Voltage (kV)	Arrester MCOV (kV)	BIL (kV)	LPL (kV)	PM_{BIL}
12.47Y/7.2	8.4	95	32	197%
24.9Y/14.4	15.3	125	58	116%
34.5Y/19.9	22	150	87	72%

TABLE 12.9

Typical CWW Protective Margins for Common Distribution Voltages (Assuming 8 kV/ft, 26 kV/m of Lead Length)

System Voltage (kV)	Arrester MCOV (kV)	CWW (kV)	FOW (kV)	Ldi/dt (kV)	PM_{CWW}
No Lead Length					
12.47Y/7.2	8.4	110	36	0	206%
24.9Y/14.4	15.3	145	65	0	123%
34.5Y/19.9	22	175	98	0	79%
3-ft (0.9-m) Lead Length					
12.47Y/7.2	8.4	110	36	24	83%
24.9Y/14.4	15.3	145	65	24	63%
34.5Y/19.9	22	175	98	24	43%
6-ft (1.8-m) Lead Length					
12.47Y/7.2	8.4	110	36	48	31%
24.9Y/14.4	15.3	145	65	48	28%
34.5Y/19.9	22	175	98	48	20%

lead length (26 kV/m). This is not an unreasonable rate of rise to use in the calculation; 20 kA/μsec is about the median value for subsequent strokes during the rise from 30 to 90%.

Lead lengths less than 3 ft (1 m) are necessary to achieve a 50% margin for protecting overhead equipment. The easiest approach is to tank mount arresters. Pole or crossarm mounting makes it harder to keep reasonable lead lengths. It is important to remember that lead length includes the ground lead as well as the phase wire lead, and the lead-length path is along the path that the lightning current flows from the phase wire to ground (see Figure 12.17). Some important, but obvious directions for arrester application are:

1. *Do not coil leads* — While this may look tidy, the inductance is very high.

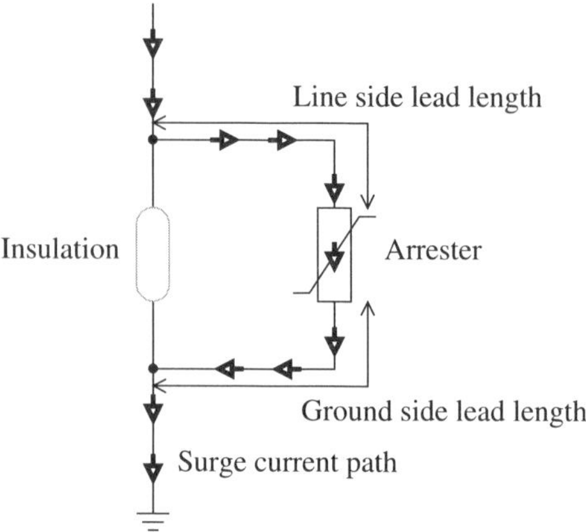

FIGURE 12.17
Lead length.

2. *Tie the ground lead to the tank* — The NESC (IEEE C2-1997) requires arrester ground leads to be tied to an appropriate ground. To achieve any protection, the ground lead must be tied to the tank of the equipment being protected. Without attaching the ground lead to the tank, the transformer or other equipment is left completely unprotected.

Arrester vs. fuse placement is an ongoing industry debate. Tank mounting an arrester protects the transformer best; but since the transformer is downstream of the fuse, the lightning surge current passes through the fuse. Lightning may blow the fuse unnecessarily, and many utilities have histories of nuisance fuse operations. Applying the arrester upstream of the fuse keeps the surge current out of the fuse, but usually results in long lead lengths. I prefer the tank mounted approach along with using larger fuses or surge resistant fuses to limit unnecessary fuse operations.

12.5.3 Secondary-Side Transformer Failures

Single-phase residential-type transformers (three-wire 120/240-V service) may also fail from surge entry into the low-voltage winding. This damage mode has been extensively discussed within the industry (sparking competing papers from manufacturers) and summarized by an IEEE Task Force (1992). Lightning current into the neutral winding of the transformer (usually X2) induces possibly damaging stresses in the high-voltage winding near the ground and line ends, both turn-to-turn and layer-to-layer. Lightning current can enter X2 from strikes to the secondary or strikes to the primary (where it gets to the neutral through a surge arrester or flashover, see Figure 12.18).

The most concern with secondary surge entry is on small (< 50 kVA) overhead transformers with a noninterlaced secondary winding (pad-

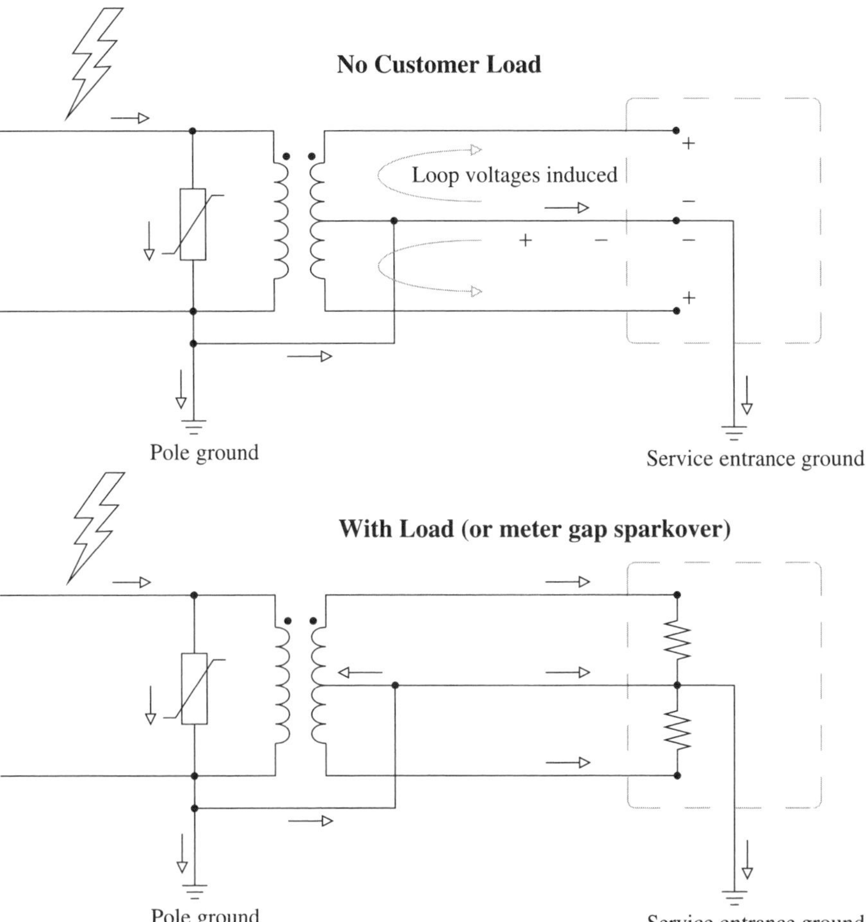

FIGURE 12.18
Surge entry via the secondary.

mounted transformers are also vulnerable). The primary arrester does not limit the voltage stresses on the ends of the primary winding. Currents couple less to the high-voltage winding on transformers with an interlaced secondary winding, which is used on core-type transformers and some shell-type transformers. Smaller kVA transformers are most prone to damage from this surge entry mode.

When the surge current gets to the transformer neutral, it has four places to go: down the pole ground, along the primary neutral, along the secondary neutral towards houses, or into the transformer winding. Current through the secondary neutral creates a voltage drop along the neutral. This voltage drop will push current through the transformer when load is connected at the customer (Figure 12.18). The voltage created by flow through the neutral is partially offset by the mutual inductance between the secondary neutral

TABLE 12.10

Typical Transformer Inductances

Interlaced Transformers		Noninterlaced Transformers	
10 kVA Core	17.4 μH	10 kVA Shell	204 μH
10 kVA Shell	18.1 μH		
15 kVA Shell	17.8 μH	15 kVA Shell	257 μH
25 kVA Shell	22.6 μH	25 kVA Shell	124 μH

Source: EPRI TR-000530, *Lightning Protection Design Workstation Seminar Notes*, Electric Power Research Institute, Palo Alto, CA, 1992.

and the phases (tighter coupling is better). The major factors impacting secondary-surge severity are:

- *Interlacing* — For a balanced surge (equal in both windings), the inductance provided by the transformer is approximately the hot leg to hot leg short-circuit reactance. Transformers with interlaced secondaries have significantly lower inductances (see Table 12.10). Higher transformer inductance induces a higher voltage on the primary-side winding.
- *Transformer size* — Smaller transformers have higher inductances, inducing a higher voltage on the primary-side winding.
- *Grounding* — Better grounding on the transformer helps reduce the current into the transformer. Poor grounding sends more current into the transformer and more current into houses connected to the transformer. Other nearby primary grounds help to reduce the current into the transformer.
- *Secondary wire* — Since triplex has tighter coupling than open-wire secondaries (and less voltage induced in the loop), a secondary of triplex has less current into the transformer neutral terminal.
- *Multiple secondaries* — The worst case is with one secondary drop from the transformer; several in parallel reduce the current by providing parallel paths.
- *Secondary length* — Longer secondaries are more prone to failure (more voltage induced in the loop).

In normal lightning areas, this surge failure mode is normally neglected. In high lightning areas, using interlaced transformers leads to a lower failure rate [by a factor of three for lightning-related failures according to Goedde et al. (1992)]. Utilities may use secondary arresters or spark gaps to reduce the transformer failure rate on noninterlaced units. Goedde also reported that transformer failure rates reduced to 0.05% (from their normal 0.2 to 1%) on 25,000 transformers that they made with secondary arresters. Either a metal-oxide secondary arrester or a secondary spark gap effectively protects transformers from secondary-side failure modes. Economically, secondary-side surge protection can make sense in higher lightning areas. An IEEE Task Force report (1992) (based on Ward, 1990) cites that about 1% of a transformer's first cost is saved for every 0.1% reduction in failure rate.

12.6 Underground Equipment Protection

Underground equipment is susceptible to lightning damage; in the normal scenario, lightning hits an overhead line near the cable, and a surge travels into the cable at the riser pole. The most important item that makes cable protection difficult is:

- *Voltage doubling* — when a surge voltage traveling down a cable hits an end point, the voltage wave reflects back, doubling the voltage.

Normally, for analysis of underground protection, we assume no attenuation in the surge and that the voltage doubles at open points. The surge impedance of cable is approximately 50 Ω, which is lower than the 400 Ω overhead line surge impedance. So, if a lightning current has a choice between the cable and an overhead line, most of it enters the cable. Normally though, the cable has a surge arrester at the riser pole, and a conducting surge arrester has an impedance of less than five ohms (for an arrester with LPL = 30 kV and a 10-kA stroke, LPL/I = 30 kV/10 kA = 3 Ω), so the surge arrester conducts most of the current. Surges travel more slowly on cables, roughly one half of the speed of light. Transformers along the cable have little effect, and a transformer termination is the same as an open circuit to the surge.

The chopped wave withstand equals the BIL for cable (so we do not have the extra 15% for the chopped wave protective margin). Since the BIL and CWW are the same, we only need one protective margin equation:

$$PM_{\text{CWW}} = \left[\frac{\text{BIL}}{V_t} - 1\right] \cdot 100$$

where V_t is the total voltage impressed across the insulation. With an arrester at the riser pole and no arresters in the underground portion, we must use twice the voltage at the riser pole, including the lead length:

$$V_t = 2(\text{FOW} + Ldi/dt)$$

so

$$PM_{\text{CWW}} = \left[\frac{\text{BIL}}{2(\text{FOW} + Ldi/dt)} - 1\right] \cdot 100$$

Table 12.11 shows protective margins for several voltages using a riser-pole arrester. All of the margins are unacceptable, except for the 12.47-kV case with no lead length (the 24.9-kV case is barely acceptable per the standards). With a heavy-duty arrester (FOW = 36 kV) for the 12.47-kV case, the protective margin drops to:

$$PM_{\text{CWW}} = \left[\frac{95}{2(36)} - 1\right] \cdot 100 = 32\%$$

While a 20% protective margin is considered acceptable per IEEE C62.22-1997, there are many reasons why it is prudent to design for larger margins. First, insulation in cables and transformers can degrade with time. Second, non-standard lightning waveshapes, 60-Hz cable charging effects, and extreme rates of rise may increase the voltage above what we have estimated. Ideally, a 50% margin or better is a good design objective. To have a 50% margin with the riser-pole arrester, the lead length must be less than 4 in. (10 cm). Figure 12.19 shows how to obtain the smallest lead length possible. To obtain minimum lead length, be sure to jumper to the arrester first and tie the ground lead to the cable neutral.

Table 12.11 shows that it was difficult to obtain reasonable protective margins with just an arrester at the riser pole. In the next section, we will discuss how to increase margins by adding an open-point arrester to prevent voltage doubling.

12.6.1 Open Point Arrester

For grounded 15-kV class systems, an arrester just at the riser pole is sufficient, but only if it is a riser-pole type arrester and only if the arrester is right across the pothead bushing (minimal lead length). For higher-voltage applications, we need additional arresters. Putting an arrester at the end of the cable prevents doubling. Although the voltage cannot double, reflections that occur before the endpoint arrester starts full conduction will raise the voltage in the middle of the cable (see Figure 12.20). To account for the reflected portion, use:

$$V_t = \text{LPL}_{\text{riser}} + \frac{1}{2}\text{LPL}_{\text{openpoint}}$$

where $\text{LPL}_{\text{riser}}$ is the 10-kA protective level of the riser-pole arrester, and $\text{LPL}_{\text{openpoint}}$ is the 1.5-kA protective level of the open-point arrester. The lower-

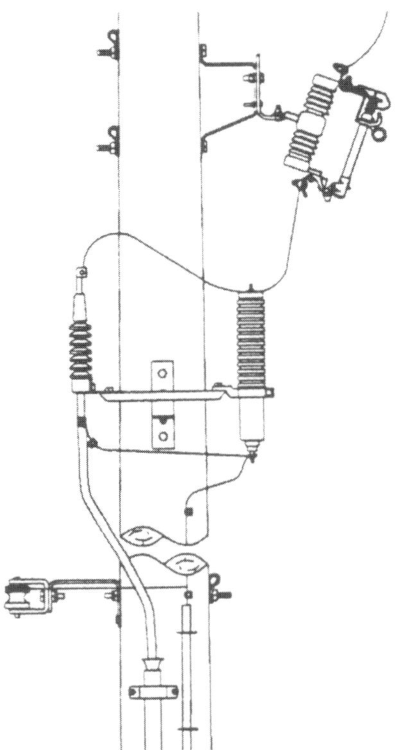

FIGURE 12.19
Minimum lead length on a riser pole. (From IEEE Std. 1299/C62.22.1-1996. Copyright 1997 IEEE. All rights reserved.)

TABLE 12.11

Typical Underground Protective Margins with a Riser-Pole Arrester, No Other Arresters, and Using 8 kV/ft (26 kV/m) of Lead Length

System Voltage (kV)	Arrester MCOV (kV)	CWW (= BIL) (kV)	FOW (kV)	Ldi/dt (kV)	V_t (kV)	PM_{CWW}
No Lead Length						
12.47Y/7.2	8.4	95	29	0	58	64%
24.9Y/14.4	15.3	125	51	0	102	23%
34.5Y/19.9	22	150	77	0	154	−3%
3-ft (0.9-m) Lead Length						
12.47Y/7.2	8.4	95	29	24	106	−10%
24.9Y/14.4	15.3	125	51	24	150	−17%
34.5Y/19.9	22	150	77	24	202	−26%

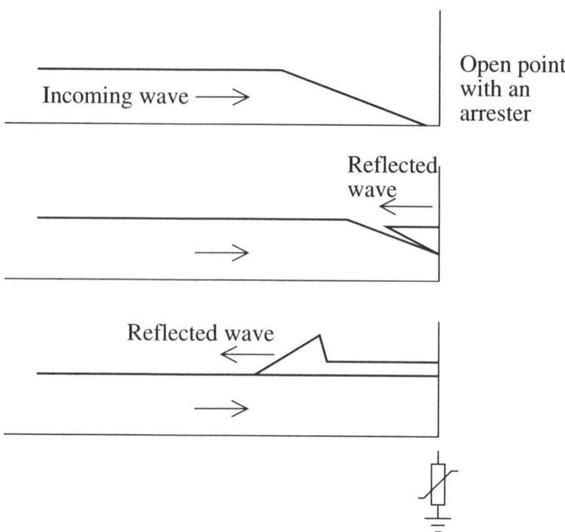

FIGURE 12.20
Voltage reflections with an open-point arrester.

current protective level of the open-point arrester is used since the riser pole arrester drains off most of the current leaving much less in the cable (with a surge impedance of 50 Ω, a 35-kV voltage surge corresponds to 0.7 kA of current). Note that the protective level of the open-point arrester is higher (it is a light-duty arrester).

The portion of the wave that is reflected before the open-point arrester starts to conduct adds to the incoming wave. Because of this, surges with low rise times cause the most severe overvoltage (contrary to most overvoltage scenarios). We have neglected the riser-pole lead length for the reflected voltage, primarily because lead length causes a short-duration voltage on the front of a fast-rising surge. This is just the sort of surge that the open-point arrester handles well because it immediately starts conducting.

With an open-point arrester, we still need to include the lead length in the chopped-wave margin calculation (we just do not add any factor for a reflection from the end of the cable). Protective margins in Table 12.12 and Table 12.13 show that we have more leeway with lead length if we have an open-point arrester.

Several options are available for arresters on cable systems. Elbow arresters attach to padmounted transformers, switching enclosures, or other equipment with 200-A loadbreak connectors (IEEE Std. 386-1995). A parking-stand arrester attaches to the enclosure and allows the open, energized cable to be "parked" on an arrester (with a 200-A elbow connector). A bushing arrester fits between a cable elbow and an elbow bushing. Under-oil arresters used in padmounted transformers are another option for protecting underground equipment; they have good thermal performance and have cost advantages.

TABLE 12.12

Typical Underground Protective Margins with a Riser-Pole Arrester and an Open Point Arrester

System Voltage (kV)	Arrester MCOV (kV)	BIL (kV)	LPL_{riser} (kV)	$LPL_{openpoint}$ (kV)	V_t (kV)	PM_{BIL}
12.47Y/7.2	8.4	95	27	32	43	121%
24.9Y/14.4	15.3	125	48	58	77	62%
34.5Y/19.9	22	150	72	90	117	28%

TABLE 12.13

Typical CWW Underground Protective Margins with a Riser-Pole Arrester and an Open Point Arrester

System Voltage (kV)	Arrester MCOV (kV)	CWW (= BIL) (kV)	FOW (kV)	Ldi/dt (kV)	V_t (kV)	PM_{CWW}
3-ft (0.9-m) Lead Length						
12.47Y/7.2	8.4	95	29	24	53	79%
24.9Y/14.4	15.3	125	51	24	75	67%
34.5Y/19.9	22	150	77	24	101	49%

Normally, we neglect attenuation in cables. This is the conservative approach. Owen and Clinkenbeard (1978) found that a 1/40-μsec wave injected into a cable attenuated less than 5% even at 6000 ft (1800 m). Attenuation increases significantly with frequency. Owen and Clinkenbeard found that a sparkover voltage wave attenuated by 40% after 6000 ft (1800 m) (the sparkover voltage is a narrow voltage spike — on a gapped arrester, it is the portion of voltage that occurs before the gap sparks over). Most significantly for cable protection, attenuation softens the lead length inductive kick. The attenuation depends on cable type. Zou and Boggs (2002) showed that EPR has an order of magnitude higher attenuation than TR-XLPE, mainly because of higher dielectric losses in the cable. Their EMTP simulations of cable voltages were significantly lower on EPR cable than on TR-XLPE cable, especially with long lead lengths, because of the difference in attenuation.

For cable protection, utilities are wise to have high protective margins. Most of the history verifying distribution overvoltage protection practices comes from arresters applied with over 50% protective margin. Effective lightning protection helps limit high failure rates that have been experienced in the industry (and riser poles expose a significant amount of equipment to overvoltages). Several other factors can cause margins to be less than they appear:

- *Quadrupling* — Barker (1990) showed that voltages can as much as quadruple for bipolar surges into the riser-pole arrester. While lightning currents are not known to have bipolarities, induced voltages from nearby strikes can induce bipolar waveshapes.
- *System voltage* — The system voltage or trapped charge can add to the voltage in the cable. If lightning hits a line at a time when the system voltage happens to be at a peak of the opposite polarity, the arrester will not go into complete conduction until the voltage impulse has compensated for system voltage. On a 12.47Y/7.2-kV system, this adds another $\sqrt{2}(7.2) = 10.18$ kV to the traveling wave.
- *Insulation deterioration* — Transformer and cable insulation degrades over time, which reduces the margin of protection provided by arresters.

Application of arresters is both a science and an art and should be applied based on local conditions. Higher lightning areas justify better protection (a higher protective margin), especially if a utility has high historical cable failure rates. As always, we must weigh the cost against the risk.

12.6.2 Scout Arresters

Another option for protecting cables is to use *scout* arresters, arresters applied on the overhead line on both sides of the riser pole (Kershaw, 1971). A scout arrester intercepts and diverts a lightning current that is heading towards the riser pole. Since most of the current conducts through the closest arrester, less voltage gets in the cable at the riser pole (unless lightning hits almost right at the cable). Virginia Power has applied scout arresters to riser poles to try to reduce high failure rates of certain types of cables at 34.5 kV (Marz et al., 1994). Transient simulations done by Marz et al. found improved protective margins with the scout arresters.

Scout arrester effectiveness depends on grounding the scout arresters well. Without good grounding, the ground potential rises at the scout arrester, causing high voltage on the phase and neutral wire (but little voltage difference between them). When the surge arrives at the riser pole, the low-impedance ground path offered by the cable drops the neutral potential (and increases the phase-to-neutral voltage). This pulls significant current through the riser-pole arrester (and sends a voltage wave down the cable), which reduces the effectiveness of the scout arresters. The lower the impedance of the scout arrester grounds, the less this effect occurs.

12.6.3 Tapped Cables

If cables are tapped, complex reflections can cause voltage to more than double, with voltage at an open point reaching over 2.8 times the peak voltage at the riser pole [see Figure 12.21 and (Hu and Mashikian, 1990)].

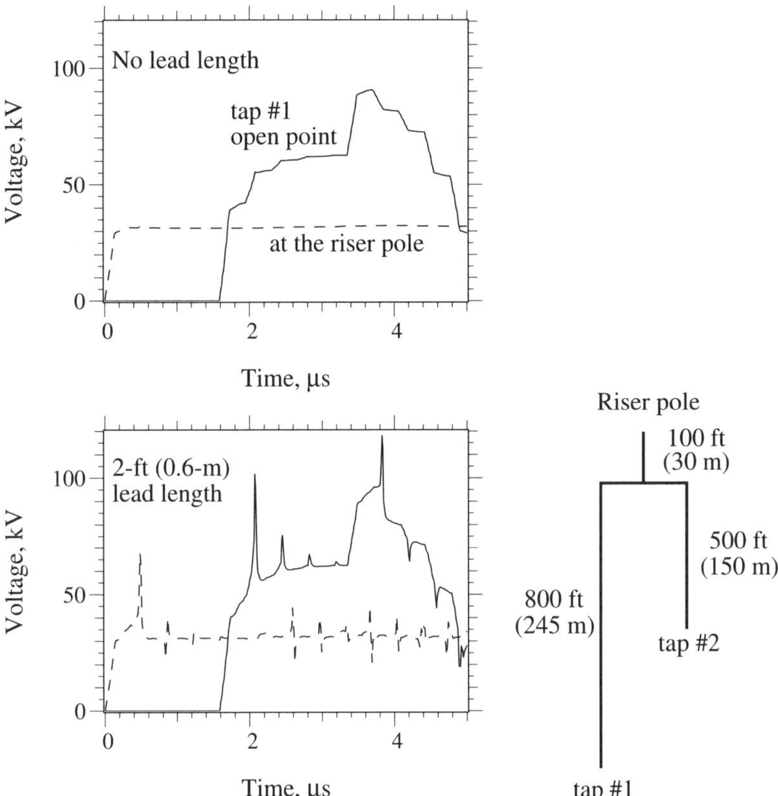

FIGURE 12.21
Transient simulation of traveling waves on a tapped cable (0.5-μs risetime).

The voltage varies depending on the lengths of each section beyond the tap point. For tapped cables with no open point arresters, find both the BIL and chopped protective margins and include a 2.6 multiplier in the BIL protective margin calculation (this is not the worst case but applies some extra safety):

$$PM_{BIL} = \left[\frac{BIL}{2.6(LPL)} - 1\right] \cdot 100$$

$$PM_{CWW} = \left[\frac{BIL}{2.6(FOW) + Ldi/dt} - 1\right] \cdot 100$$

The lead length voltage is not doubled in the chopped-wave protective margin since the inductive voltage spike from the lead length does not coincide with the reflections causing the highest overvoltage. On tapped cables, applying open point arresters to both open points provides the safest protection. Arresters at the tap points are also an option for protection.

FIGURE 12.22
Lightning flash triggered by a rocket trailing a wire. (Copyright © 1993. Electric Power Research Institute. Reprinted with permission.)

12.6.4 Other Cable Failure Modes

Lightning may also puncture the jackets in jacketed cable. Ros (1993) demonstrated that lightning current into a riser-pole ground created enough ground potential rise to puncture jackets and leave pin holes. Fifty-mil jackets punctured with 150 to 160 kV, and 80-mil jackets punctured with 155 to 170 kV. The voltage impressed across the cable jacket is the surge current times the ground resistance at the riser pole, which may be very high with poor grounds. Ros' tests found semiconducting jackets did not allow nearly the voltage to develop across the jacket.

Lightning can also damage cables from strikes above the ground. EPRI tests found that rocket-triggered lightning strikes to the ground above buried cables caused extensive damage (Barker and Short, 1996). Lightning normally breaks down the soil and arcs to the cable (see Figure 12.22 and Figure 12.23). We tested three cable configurations: unjacketed direct buried, jacketed direct buried, and jacketed cable in conduit. All were single-conductor 220-mil XLPE cables with a full copper neutral. Lightning above the ground damaged them all to varying degrees. Lightning melted most strands of the concentric neutral and punctured the jacket and conduit if present. In some cases, lightning also damaged the insulation shield. What made the damage particularly bad was that lightning hit the soil surface and continued an arcing path 3 ft (1 m) underground directly to the cable. Measurements

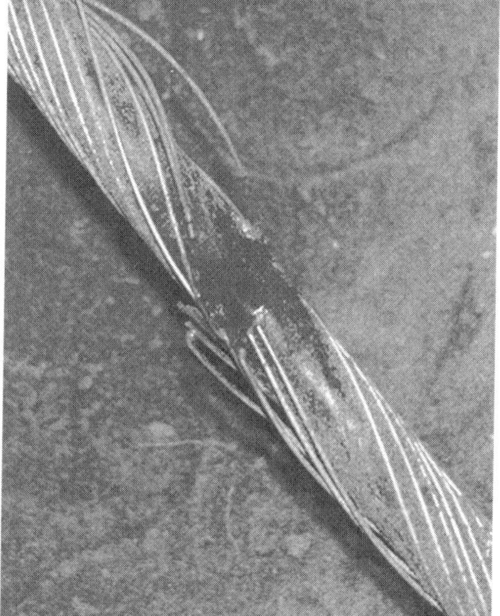

FIGURE 12.23
Cable and conduit damage observed from rocket-triggered lightning flashes above the ground. (Copyright © 1993. Electric Power Research Institute. Reprinted with permission.)

showed that 15 to 25% of the lightning currents reached the padmounted transformers on either side of the flash point. This type of cable damage would not show up immediately. More likely, cable failure would accelerate from increased water entry and localized neutral heating. Voltages we measured across the cable insulation were not a significant threat (they were less than 17 kV).

The tests also pointed to secondary voltages as a concern. We found nearly 4 kV on the closest transformer secondary. Surge entry into the secondary stresses the transformer insulation and sends possibly damaging surges into homes. Current entering into the secondary neutral terminal of the transformer induces voltages that stress the primary winding. Williams (1988) presented evidence that padmounted transformers fail from secondary-side surge entry in similar percentages to overhead transformers.

During the rocket-triggered lightning tests, strokes to the cable attached from as far away as 15 ft (5 m). Furrows from trees to underground cables as long as 300 ft (100 m) have been found (Sunde, 1968). Sunde derived a model for lightning attraction to buried cables based on the lightning current and the soil resistivity. The number of flashes attracted to cables depends on flash density, the peak current magnitude, and the ground resistivity. Higher peak lightning currents make longer jumps to cables. Cables in higher-resistivity soil attract flashes from farther away. The Sunde model predicts that cables attract strikes within the following radius:

$$r = \begin{cases} 0.079\sqrt{\rho \cdot I_p} & \rho \le 100\ \Omega \cdot \mathrm{m} \\ 0.046\sqrt{\rho \cdot I_p} & \rho \ge 1000\ \Omega \cdot \mathrm{m} \end{cases}$$

where

r = attractive radius, m

ρ = soil resistivity, Ω-m

I_p = peak lightning current, kA

Using the IEEE distribution of peak first strokes, this gives the following collection rates to cables:

$$N = \begin{cases} 0.092 N_g \sqrt{\rho} & \rho \le 100\ \Omega \cdot \mathrm{m} \\ 0.053 N_g \sqrt{\rho} & \rho \ge 1000\ \Omega \cdot \mathrm{m} \end{cases}$$

for N in flashes/100 km/year (multiply by 1.609 for results in flashes/100 mi/year) and N_g in flashes/km²/year. For resistivities in the region that Sunde left out (100 to 1000 Ω·m), either interpolate the results from each equation or use the closest equation. For 1000 Ω·m soil with N_g = 1, cables attract 2.7 flashes/100 mi/year (1.7 flashes/100 km/year). For areas in trees, cables collect more flashes; EPRI (1999) recommends doubling the number of strikes for areas in trees.

These flashes to cables puncture jackets and conduits. The amount of heating damage to the neutral and to the insulation is primarily a function of the total charge in the lightning flash (since arcs exhibit a fairly constant voltage drop at the point of attachment, the energy is a function of $\int I dt$). Burying shield wires above the cables is one way to offer protection against this type of damage. An AT&T handbook (1985) provides estimates on shield-wire effectiveness. Note that there is no current power industry practice for this type of protection, and the amount of damage (percent of cable faults) is unknown.

12.7 Line Protection

Line protection is the attempt to reduce the number of lightning-caused faults. Utilities have increased interest in line protection as one way to improve reliability and power quality. Since lightning can flash over a 230-kV or 500-kV transmission line, we should not be surprised that protecting a 13-kV distribution line is difficult. To protect against direct hits, we need either a shield wire or arresters to divert the stroke to ground without a flashover. Lightning strokes close to a line may induce enough voltage to flash over the line insulation. Induced voltages are easier to contain since induced voltages have much lower magnitudes than direct strike voltages. Maintaining enough insulation capability is the normal way to limit induced-voltage flashovers. Line arresters can also greatly reduce induced voltage flashovers from nearby strokes.

12.7.1 Induced Voltages

Lightning strikes near a distribution line will induce voltages into the line from the electric and magnetic fields produced by the lightning stroke. These *induced* voltages are much less severe than direct strikes, but close strikes can induce enough voltage to flash insulation and damage poorly protected equipment.

The charge and current flow through the lightning channel creates fields near the line. These fields induce voltages on the line. The vertical electric field is the major component that couples voltages into the line (magnetic fields also play a role). As the highly charged leader approaches the ground, the electric field increases greatly; and when the leader connects, the electric field collapses very quickly. The rapidly changing vertical electric field induces a voltage on a conductor, which is proportional to the height of the conductor above ground. The induced voltage waveform is usually a narrow pulse, less than 5 or 10 μsec wide, and it may be bipolar (negative then positive polarity).

Most measurements of induced voltages have been less than 300 kV, so the most common guideline for eliminating problems with induced voltages is to make sure that the line insulation capability (CFO) is higher than 300 kV. Lines with insulation capabilities less than 150 kV have many more flashovers due to induced voltages.

A simplified version of a model developed by Rusck (1958, 1977) approximates the peak voltage induced by nearby lightning. Rusck's model can be simplified to estimate the peak voltage developed on a conductor (IEEE Std. 1410-1997):

$$V = 36.5\frac{I \cdot h}{y}$$

where
V = peak induced voltage, kV
I = peak stroke current, kA
h = height of the line, usually ft or m
y = distance of the stroke from the line

This equation is for an ungrounded circuit. For a circuit with a grounded neutral or shield wire, the voltage from the phase to the neutral is less. For normal distribution line conductor spacings, multiply the answer by 0.75 for lines with a grounded neutral. So, for a 30-ft (10-m) distribution line, a 40-kA stroke 200 ft (60 m) from the line induces 165 kV on a line with a grounded neutral. Most lines are immune from strikes farther than 500 ft (150 m) from the line.

Models of attraction to distribution lines show that for lines out in the open, most flashes that do not hit the line are too far away to induce a particularly high voltage across the insulation. For environmentally shielded lines — those with nearby trees and buildings — fewer strikes hit the line, but the line should have higher induced voltages because lightning strokes could hit closer to the line.

EPRI sponsored rocket triggered lightning tests in 1993 that showed induced voltages could be higher than predicted by Rusck's model for some strikes (Barker et al., 1996). For strikes to the ground 475 ft (145 m) from the line, voltages were 63% higher than Rusck's model. Hydro Quebec and New York State Electric and Gas sponsored another round of tests in 1994 that were also led by P. P. Barker. Closer strokes, strokes 60 ft (18 m) from the line, induced less voltage than the Rusck model.

Table 12.14 compares the rocket-triggered lightning measurements with the Rusck model. High ground resistivity at the Florida test site probably explains why the 475-ft (145-m) measurements were higher than the Rusck prediction (Ishii, 1996; Ishii et al., 1994). Why the closer measurements are lower is not

TABLE 12.14

Comparison of Induced Voltage Measurements to Rusck Predictions

Distance	Number of Data Samples	Rusck Prediction V_{ind}/I_s	Rocket-Triggered Lightning Measurements V_{ind}/I_s
60 ft (18 m)	20	12.2	5.25
400 ft (125 m)	8	1.7	2.67
475 ft (145 m)	63	1.4	2.24

verified. My best guess is because those strokes were triggered from rockets launched on a 45-ft (14-m) tower. Since the tower is above ground, the positive charge on the tower shields some of the negative charge in the downward leader, so the electric field inducing voltages into the line is smaller. Environmentally shielded lines should act similarly to the tall tower. The charge collected on the tree should shield the line and reduce the induced voltage.

The three induced-voltage data points when normalized for a 30-ft (10-m) tall distribution line fit the following equation:

$$V = \frac{9.8I}{\left(1 + \frac{y}{121}\right)^{1.8}}$$

Figure 12.24 shows estimates of induced voltages with insulation level for both open ground and for lines that are environmentally shielded (usually by trees). One is based on Rusck's model using the approach from IEEE Std. 1410-1997. Another is based on the curve fit of the triggered lightning results, which shows more reasonable answers for environmentally shielded lines.

12.7.2 Insulation

High insulation levels on structures help prevent induced voltage flashovers. Insulation levels are also critical on some types of line protection configurations with shield wires or arresters. Note that for a normal distribution line, higher insulation levels may actually stress nearby cables and transformers more. With high insulation strengths, a higher voltage develops across the insulation before it flashes over. By allowing a larger magnitude surge on the line before flashover, damage to nearby equipment is more likely. The flashover of the insulation acts as an arrester and protects other equipment.

The critical flashover voltage (CFO) of self-restoring insulation (meaning no damage after a flashover) is the voltage where the insulation has a 50% probability of flashing over for a standard 1.2/50-μsec voltage wave. For insulators, manufacturers' catalogs specify the CFO. CFO and BIL are often used interchangeably, but they have slightly different definitions. A statistical BIL is the 10% probability of flashing over for a standard test wave. Normally, CFO and statistical BIL are within a few percent of each other.

Lightning may flash along several paths: directly between conductors across an air gap or along the surface of insulators and other hardware at poles (normally the easiest path to flashover). We need to consider phase-to-ground and phase-to-phase paths. At a pole structure, the flashover path may involve several insulating components, the insulator, wood pole or crossarm, and possibly fiberglass. Wood and fiberglass greatly increase the structure insulation.

Table 12.15 shows the critical flashover voltage of common components. When more than one insulator is in series, the total insulating capability is

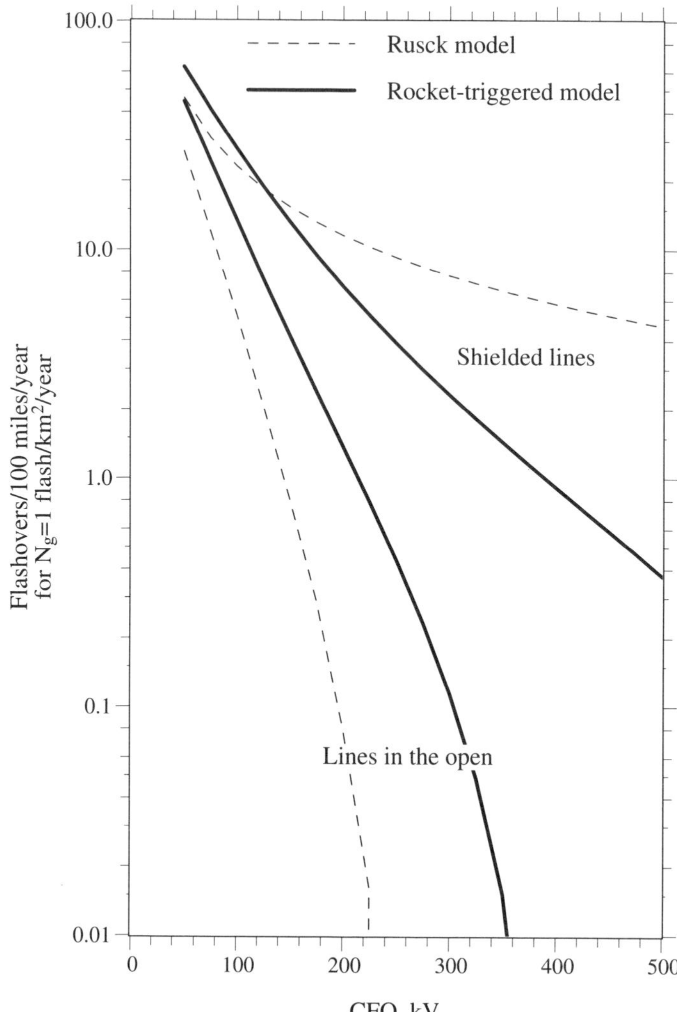

FIGURE 12.24
Induced voltage flashovers vs. insulation level for a line with a grounded neutral.

TABLE 12.15
CFO of Common Distribution Components (By Themselves)

Component	kV/ft	kV/m
Air	180	600
Wood pole or crossarm	100	350
Fiberglass standoff	150	500

less than the sum of the components. When insulation is subjected to a voltage surge, the voltage across each component splits based on the capacitance between each element. Normally, this voltage division does not split the voltage by the same ratio as their insulation capability, so one component flashes first leaving more voltage across the rest of the components.

The simplest way to estimate the insulation level of a structure is to take the CFO of the insulator (usually about 100 kV) and add the wood length at 75 kV/ft (250 kV/m). Often the wood provides more insulation than the insulator. Estimate the air-to-air gap using 180 kV/ft (600 kV/m). The air gap between conductors usually has higher insulation than the path along the insulator and wood. So, most flashovers occur at the poles, the weakest point. Typical distribution structures generally have CFOs between 150 and 300 kV. And, to eliminate induced voltage flashovers, we try to have 300 kV of CFO.

Another way to estimate the structure CFO of several components in series is to take the square root of the sum of the squares of each component. Another more precise way to estimate structure CFOs is with the extended CFO added method described by Jacob et al. (1991) and Ross and Grzybowski (1991) (or, see IEEE Std. 1410-1997 for a simplified version).

12.7.2.1 Practical Considerations*

Equipment and support hardware on distribution structures may severely reduce CFO. These "weak-link" structures may greatly increase flashovers from induced voltages. Several situations are described below.

Guy wires. Guy wires may be a major factor in reducing a structure's CFO. For mechanical advantage, guy wires are generally attached high on the pole in the general vicinity of the principal insulating elements. Because guy wires provide a path to the ground, their presence will generally reduce the configuration's CFO. The small porcelain guy-strain insulators that are often used provide very little in the way of extra insulation (generally less than 30 kV of the CFO).

A fiberglass-strain insulator may be used to gain considerable insulation strength. A 20-in. (50-cm) fiberglass-strain insulator has a CFO of approximately 250 kV.

Fuse cut-outs. The mounting of fuse cut-outs is a prime example of unprotected equipment which may lower a pole's CFO. For 15-kV class systems, a fuse cut-out may have a 95-kV BIL. Depending on how the cut-out is mounted, it may reduce the CFO of the entire structure to approximately 95 kV (approximately because the BIL of any insulating system is always less than the CFO of that system).

On wooden poles, the problem of fuse cut-outs may usually be improved by arranging the cut-out so that the attachment bracket is mounted on the pole away from any grounded conductors (guy wires, ground wires, and

* This section is from IEEE Std. 1410-1997. Copyright 1997. All rights reserved. The author chaired the IEEE working group on the Lightning Performance of Distribution Lines during the development and approval of this guide.

neutral wires). This is also a concern for switches and other pieces of equipment not protected by arresters.

Neutral wire height. On any given line, the neutral wire height may vary depending on equipment connected. On wooden poles, the closer the neutral wire is to the phase wires, the lower the CFO.

Conducting supports and structures. The use of concrete and steel structures on overhead distribution lines is increasing, which greatly reduces the CFO. Metal crossarms and metal hardware are also being used on wooden pole structures. If such hardware is grounded, the effect may be the same as that of an all-metal structure. On such structures, the total CFO is supplied by the insulator, and higher CFO insulators should be used to compensate for the loss of wooden insulation. Obviously, trade-offs should be made between lightning performance and other considerations such as mechanical design or economics. It is important to realize that trade-offs exist. The designer should be aware of the negative effects that metal hardware may have on lightning performance and attempt to minimize those effects. On wooden pole and crossarm designs, wooden or fiberglass brackets may be used to maintain good insulation levels.

Multiple circuits. Multiple circuits on a pole often cause reduced insulation. Tighter phase clearances and less wood in series usually reduces insulation levels. This is especially true for distribution circuits built underneath transmission circuits on wooden poles. Transmission circuits will often have a shield wire with a ground lead at each pole. The ground lead may cause reduced insulation. This may be improved by moving the ground lead away from the pole with fiberglass spacers.

Spacer-cable circuits. Spacer-cable circuits are overhead-distribution circuits with very close spacings. Covered wire and spacers [6 to 15 in. (15 to 40 cm)] hung from a messenger wire provide support and insulating capability. A spacer-cable configuration will have a fixed CFO, generally in the range of 150 to 200 kV. Because of its relatively low insulation level, its lightning performance may be lower than a more traditional open design (Powell et al., 1965). There is little that can be done to increase the CFO of a spacer-cable design.

A spacer-cable design has the advantage of a messenger wire which acts as a shield wire. This may reduce some direct-stroke flashovers. Back flashovers will likely occur because of the low insulation level. Improved grounding will improve lightning performance.

Spark gaps and insulator bonding. Bonding of insulators is sometimes done to prevent lightning-caused damage to wooden poles or crossarms, or it is done to prevent pole-top fires. Spark gaps are also used to prevent lightning damage to wooden material [this includes Rural Electrification Administration specified pole-protection assemblies (REA Bulletin 50-3, 1983)]. In some parts of the world, spark gaps are also used instead of arresters for equipment protection.

Spark gaps and insulator bonds will greatly reduce a structure's CFO. If possible, spark gaps, insulator bonds, and pole-protection assemblies should not be used to prevent wood damage. Better solutions for damage to wood and pole fires are local insulator-wood bonds at the base of the insulator.

12.7.3 Shield Wires

Shield wires are effective for transmission lines but are difficult to make work for distribution lines. A shield wire system works by intercepting all lightning strokes and providing a path to ground. If the path to ground is not good enough, a voltage develops on the ground with respect to the phases (called a *ground potential rise*). If this is high enough, the phase can flashover (it is called a *backflashover*).

Grounding and insulation are important. Good grounding reduces the ground potential rise. Extra insulation protects against backflashover. As an example, consider Figure 12.25 where a 22-kA stroke (which is on the small size for lightning) is hitting a distribution line. The ground potential rises to 400 kV relative to the phase conductor, enough voltage to flashover most distribution lines.

To keep the insulation high, use fiberglass standoffs to keep the ground wire away from the pole to maximize the wood length. Also, make sure guy wires and other hardware do not compromise insulation.

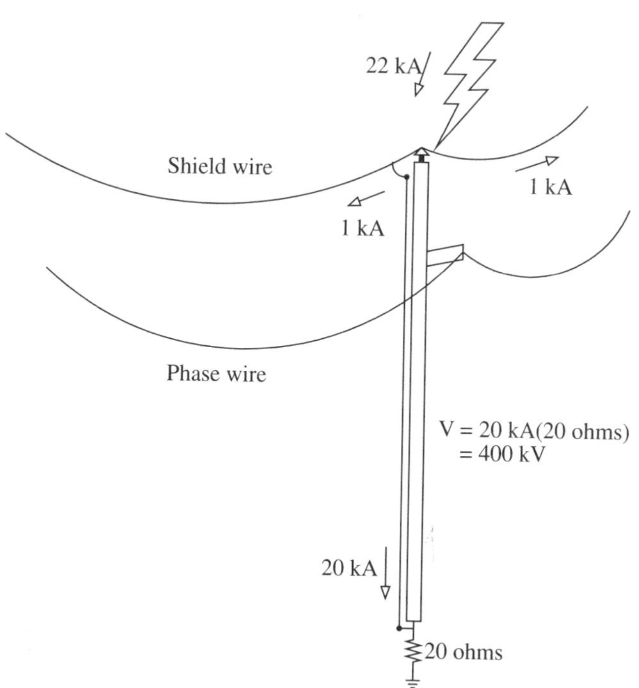

FIGURE 12.25
Shield-wire lightning protection system.

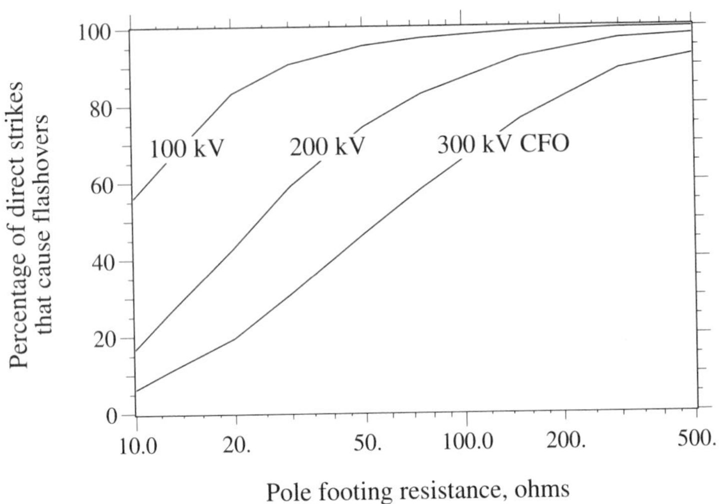

FIGURE 12.26
Performance of a shield wire depending on grounding and insulation level.

Ground the shield wire at each pole. Lightning has such fast rise times that if a pole is not grounded, and a lightning strike hits the shield wire it will flash over before the grounds at adjacent poles can provide any help in relieving the voltage stress.

At all poles, obtain a ground that is 20 Ω or less. Good grounding is vital! Exposed sections of circuit such as at the top of a hill or ridge should have the most attention. Getting adequate grounds may require:

- More than one ground rod. Make sure to keep them spaced further than one ground rod apart.
- Deeper ground rods
- Chemical soil treatments
- Counterpoise wires (buried lengths of wire)

Figure 12.26 shows estimates of performance versus grounding for several insulation levels based on the approach of IEEE Std. 1410-1997. In order to ensure that lightning hits the shield wire and not the phase conductors, maintain a shielding angle of 45° or less (as defined in Figure 12.27).

12.7.4 Line Protection Arresters

Arresters are normally used to protect equipment. Some utilities are using them to protect lines against faults, interruptions, and voltage sags. To do this, arresters are mounted on poles and attached to each phase. For protec-

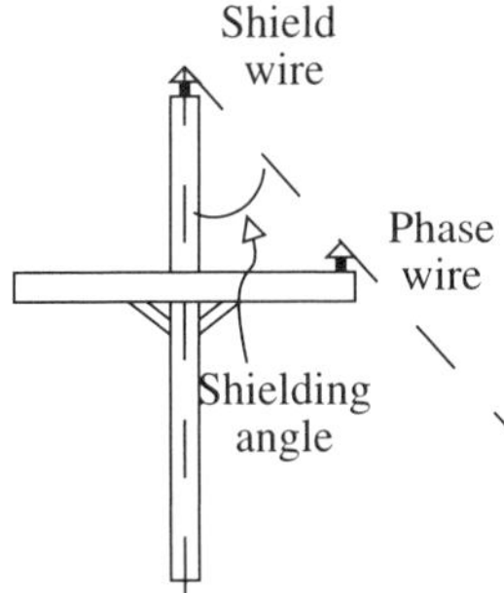

FIGURE 12.27
Shield wire shielding angle.

tion against direct strikes, arresters must be spaced at every pole (or possibly every other pole on structures with high insulation levels) (McDermott et al., 1994). This is a lot of arresters, and the cost prohibits widespread usage. The cost can only be justified for certain sections of line that affect important customers.

Arresters have been used at wider spacings such as every four to six poles by utilities in the southeast for several years. This grew out of some work done in the 1960s by a task force of eight utilities and the General Electric Company (1969a, 1969b). Anecdotal reports suggest improvement, but there is little hard evidence. Recent field monitoring and modeling suggest that this should not be effective for direct strikes. One of the reasons that this may provide some improvement is that arresters at wider spacings improves protection against induced voltages. Nevertheless, arresters applied at a given spacing is not recommended as the first option. Fixing insulation problems or selectively applying arresters at poles with poor insulation are better options for reducing induced-voltage flashovers.

For direct strike protection, arresters are needed on all poles and on all phases. The amount of protection quickly drops if wider spacings are used. Lead length is not as much of an issue as it is with equipment protection, but it is always good practice to keep lead lengths short. The arrester rating would normally be the same as the existing arresters.

Grounding is normally not an overriding concern if arresters are used on all phases. If grounds are poor, one effect is that surges tend to get pushed out away from the strike location (since there is no good path to ground). One of the implications is if just a section of line is protected (such as an exposed ridge crossing), then grounding at the ends is important. Good grounds at the end provide a path to drain off the surge.

One concern with arresters is that they may have a relatively high failure rate. Direct strikes can cause failures of nearby arresters. Something in the range of 5 to 30% of direct lightning strikes may cause an arrester failure. This is still an undecided (and controversial) subject within the industry. It is recommended that the largest block size available be used (heavy-duty or intermediate-class blocks) to reduce the probability of failure.

Field trials on Long Island, NY, of arresters at various spacings did not show particularly promising results for distribution line protection arresters (Short and Ammon, 1999). LILCO added line arresters to three circuits. One had spacings of 10 to 12 spans between arresters (1300 ft, 400 m), one had spacings of five to six spans (600 ft, 200 m), and one had arresters at every pole (130 ft, 40 m). Arresters were added on all three phases. We also monitored two other circuits for comparison. None of the three circuits with line arresters had dramatically better lightning-caused fault rates than the two circuits without arresters. Statistically, we cannot infer much more than this since the data is limited (always a problem with lightning studies). One of the most significant results was that the circuit with arresters on every pole had several lightning-caused interruptions, and theoretically it should have had none. Missing arresters is the most likely reason for most of the lightning-caused faults. One positive result was that the arrester failure rate was low on the circuit with arresters on every pole.

Automated camera systems captured a few direct flashes to the line during the LILCO study. Figure 12.28 shows a direct stroke almost right at a pole protected by arresters (arresters were every five to six spans on this circuit). Ideally, the arresters on that pole should divert the surge current to ground without a flashover. The arresters prevented a flashover on that pole, but two of the three insulators flashed over one pole span away. An arrester at the struck pole also failed.

Another field study showed more promise for line arresters. Commonwealth Edison added arresters to several rural, open feeders in Illinois (McDaniel, 2001). ComEd uses an arrester spacing of 1200 ft (360 m); as a trial, they tightened the arrester spacing to 600 ft (180 m) on 30 feeders (and all existing arresters that were not metal oxide were replaced). The 30 feeders with the new spacing were compared to 30 other feeders that were left with the old standard. Over three lightning seasons of evaluation, the upgraded circuits showed that circuit interruptions improved 16% (at a 95% confidence level). Note that most of the interruptions were transformer fuse operations.

Another way to apply arresters is to use them on the top phase only. The top phase is turned into a shield wire. When lightning hits the top phase, the arrester conducts and provides a low-impedance path to the pole ground. Just as with a shield-wire system, grounding and insulation are critical. A top-phase arrester application cannot be used on typical three-phase crossarm designs because there is no top phase.

In areas where grounding is poor and arrester failure is a concern, arresters can be used with a shield wire system. The shield wire takes away most of the energy concerns, and the arresters protect against backflashovers. This provides very good lightning protection (and is very expensive).

FIGURE 12.28
Lightning-caused fault on a Long Island Lighting Company 13-kV circuit. (From Short, T. A. and Ammon, R. H., "Monitoring Results of the Effectiveness of Surge Arrester Spacings on Distribution Line Protection," *IEEE Trans. Power Delivery*, 14(3), 1142-1150, July 1999. ©1999 IEEE. With permission.)

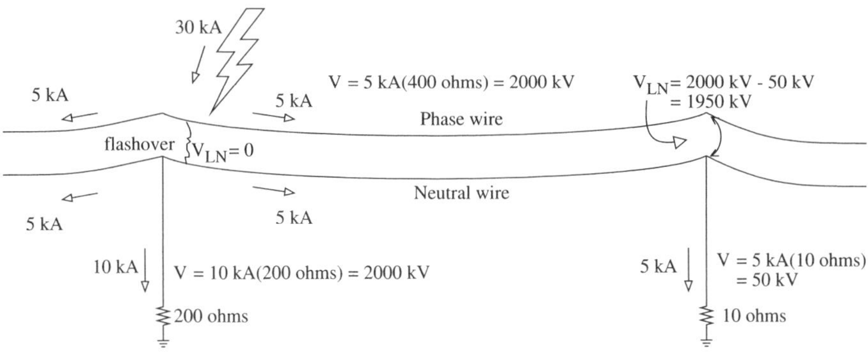

FIGURE 12.29
Impact of grounding.

12.8 Other Considerations

12.8.1 Role of Grounding

With poor grounding, lightning strikes to distribution lines subject more equipment to possibly damaging surges due to ground potential rise. Lightning current must flow to ground somewhere. If the pole ground near the strike point has high impedance, more of the surge flows in the line, exposing more equipment to possible overvoltages. At the point of a direct stroke to a distribution conductor, the huge voltages flash over the insulation, shorting the phase to the neutral. From there, the lightning travels to the closest ground. If there is no proper pole ground nearby, the next most likely path is down a guy wire. If the path to ground is poor, the phase conductor and neutral wire all rise up in voltage together. This surge (on both the phase and neutral) travels down the circuit. When the surge reaches another ground point, current drains off the neutral wire, which increases the voltage between the phase and the neutral. The voltage across the insulation grows until a surge arrester conducts more current to ground or another insulation flashover occurs. Figure 12.29 shows a drawing describing the ground potential effect, and Figure 12.1 shows a photograph of a direct strike that caused another flashover remote from the flash point.

Good grounding helps confine possible lightning damage to the immediate vicinity of the strike. If the distribution insulation flashes over, the short circuit acts as an arrester and helps protect other equipment on the circuit (as long as grounds are good).

Normally, the ground resistance right at a piece of equipment protected with an arrester does not impact the primary-side protection. The primary surge arrester is between the phase and ground. For a lightning hit to the primary, the arrester conducts the current to ground and limits the voltage across the equipment insulation (even though the potential of all conductors may rise with poor grounding).

Grounding does play a role for transformers, as they are vulnerable to surge current entering through the secondary. Poor grounding pushes more current into the transformer on the secondary side and increases the possibility of failure. Poor grounding also forces more current to flow into the secondary system to houses connected to the transformer. And, poor grounding may also push more current into telephone and cable television systems.

12.8.2 Burndowns

The lightning arc itself can pass enough charge to burn small primary and secondary wires although sometimes it is hard to tell whether it is the lightning or the fault arc that does most of the damage. Normally, it is the power frequency arc that does the most damage. Chisholm et al. (2001) cites a damage rate on Hydro-One's primary distribution system of 0.13/100 mi/year (0.08/100 km/year) for an area with a ground flash density of about 1 flash/km^2/year. This is about 1% of the lightning flashes hitting circuits.

The lightning arc causes damage the same way that a fault current arc (and also an arc welder) does: the heat of the arc melts conductor strands. Damage is mainly a function of the charge in the flash. Lightning tends to be more damaging than the equivalent ac fault arc. A negative direct current is about 50% more damaging than an alternating current passing equal coulombs. Negative flashes are more destructive than positive flashes; it takes about three times the charge for a positive stroke to cause the same damage as a negative stroke. Lightning also tends to stay in one place more than an arc because it does not have the motoring effect.

12.8.3 Crossarm and Pole Damage and Bonding

Although service experience indicates that lightning rarely damages poles or crossarms; in high-lightning areas, concern is warranted. Darveniza (1980) cites a survey by Zastrow published in 1966 that found poles failing from lightning in the range of 0.008 to 0.023% annually (versus an overall failure rate of 0.344 to 1.074% with more than half of these due to decay). Other surveys summarized by Darveniza also found minimal lightning-caused damage. Lightning overvoltages damage and shatter poles and crossarms when the wood breaks down internally rather than along the surface. If the wood is green and wet, internal breakdowns and damage are more likely. Damage within the first year of service is more likely.

If historical records show that wood damage is a problem, bonding the insulators (grounding the base of each insulator) protects the wood, but this shorts out the insulation capability provided by the wood. Better are surface electrodes fitted near the insulator pin, including wire-wraps, bands, or other metal extensions attached near the insulator in the likely direction of flashover. This local bonding encourages breakdown near the surface rather than internally.

Preventative measures for lightning damage to wood also reduce the likelihood of pole-top fires. Leakage current arcs at metal-to-wood interfaces start fires (Darveniza, 1980; Ross, 1947). Local bonding using wire bands or wraps helps prevent pole damage. Bonding bridges the metal-to-wood contacts where fires are most likely to start. Local bonding is better than completely bonding the insulators because the insulation level is not compromised.

12.8.4 Arc Quenching of Wood

Wood poles and crossarms can prevent faults from forming after lightning flashes across the wood (Armstrong et al., 1967; Darveniza et al., 1967). Whether this arc quenching occurs depends mainly on the power frequency voltage gradient along the arc at the time of the flashover. If the power-frequency voltage happens to be near a zero crossing when the lightning strikes, the lightning-caused arc is likely to extinguish rather than becoming a power-frequency fault.

The voltage gradient for an arc inside wood is much higher than an arc voltage gradient in the air. Darveniza (1980) found voltage gradients in wood on the order of 1000 V/in. (400 V/cm) through wood compared to about 30 V/in. (12 V/cm) for arcs in air. When dry wood insulation arcs, the path is often just barely below the surface (about a millimeter); and the arc is surrounded by wood fibers, so the arc-suppression action occurs. Wood fibers raise the arc voltage primarily by cooling it and absorbing ionized particles (similar to the arc-absorbing action of fuses). The arc voltage gradient is lower for wet wood (flashovers are no longer fully contained and tend to arc on the surface rather than just below the surface). There is considerable variability in the arc voltage, especially for wet wood. To transition from a lightning arc to a power-frequency fault, the power-frequency voltage must be higher than the arc voltage to keep the arc conducting.

Darveniza calculated several probabilities for arc quenching based on the arc voltage gradients (see Figure 12.30). Arc quenching is less likely on wet wood and when multiple phases flash over (even if the flashovers are on the same phase). To achieve significant benefit, voltage gradients must be kept less than 3 kV/ft (10 kV/m) of wood. For a 12.47Y/7.2-kV system, this is at least 2.4 ft (7.2/3 = 2.4 ft or 0.7 m) of wood in the phase-to-ground flashover paths and at least 4.1 ft (1.2 m) of wood in the phase-to-phase

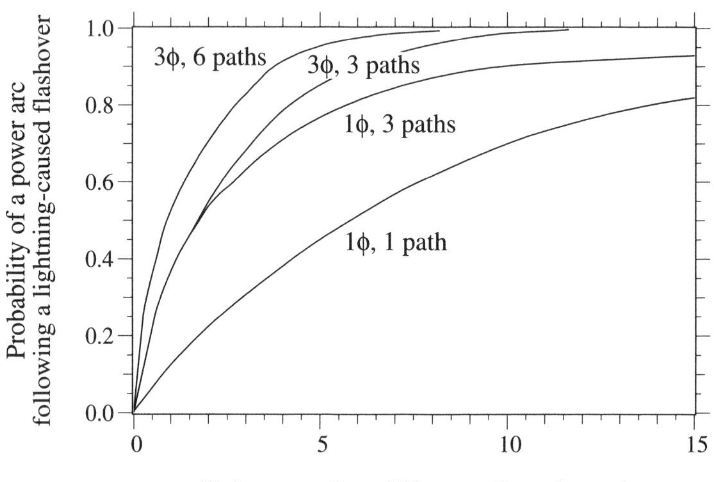

FIGURE 12.30
Probability of a power-frequency fault due to a lightning-caused flashover over wet wood crossarms with different numbers of flashover paths. (Adapted from Darveniza, M., *Electrical Properties of Wood and Line Design*, University of Queensland Press, 1980. With permission.)

flashover paths. And, if we really want to count on this effect, we should at least double these lengths.

References

Anderson, R. B. and Eriksson, A. J., "Lightning Parameters for Engineering Applications," *Electra*, no. 69, pp. 65–102, March 1980a.

Anderson, R. B. and Eriksson, A. J., "A Summary of Lightning Parameters for Engineering Applications," CIGRE Paper No. 33–06, 1980b.

ANSI C84.1-1995, *American National Standards for Electric Power Systems and Equipment — Voltage Ratings (60 Hz)*.

ANSI/IEEE C62.11-1987, *IEEE Standard for Metal-Oxide Surge Arresters for AC Power Circuits*, American National Standards Institute, Institute of Electrical and Electronics Engineers, Inc.

Armstrong, H. R., Stoetling, H. O., and Veverka, E. F., "Impulse Studies on Distribution Line Construction," *IEEE Transactions on Power Apparatus and Systems*, vol. 86, pp. 206–14, 1967.

AT&T Technologies Inc., "Telecommunication Electrical Protection." Select code 350-060, 1985.

Barker, P. P., "Voltage Quadrupling on a UD Cable," *IEEE Transactions on Power Delivery*, vol. 5, no. 1, pp. 498–501, January 1990.

Barker, P. P. and Burns, C. W., "Photography Helps Solve Distribution Lightning Problems," *IEEE Power Engineering Review*, vol. 13, no. 6, June 1993.

Barker, P. P., Mancao, R. T., Kvaltine, D. J., and Parrish, D. E., "Characteristics of Lightning Surges Measured at Metal Oxide Distribution Arresters," *IEEE Transactions on Power Delivery*, vol. 8, no. 1, pp. 301–10, January 1993.

Barker, P. P. and Short, T. A., "Lightning Effects Studied — The Underground Cable Program," *Transmission & Distribution World*, vol. 48, no. 5, May 1996.

Barker, P. P., Short, T. A., Eybert, B. A. R., and Berlandis, J. P., "Induced Voltage Measurements on an Experimental Distribution Line During Nearby Rocket Triggered Lightning Flashes," *IEEE Transactions on Power Delivery*, vol. 11, no. 2, pp. 980–95, April 1996.

Berger, K., Anderson, R. B., and Kröninger, H., "Parameters of Lightning Flashes," *Electra*, no. 41, pp. 23–37, July 1975.

Burke, J. J. and Sakshaug, E. C., "The Application of Gapless Arresters on Underground Distribution Systems," *IEEE Transactions on Power Apparatus and Systems*, vol. PAS-100, no. 3, pp. 1234–43, March 1981.

Campos, M. L. B., Coelho, V. L., et al., "Evaluation of the Sealing System and of the Electric Performance of the Distribution Lightning Arresters," CIRED, June 1997.

CEA 160 D 597, *Effect of Lightning on the Operating Reliability of Distribution Systems*, Canadian Electrical Association, Montreal, Quebec, 1998.

Chisholm, W. A., Beattie, J., and Janischewskyj, W., "Analysis of the Optical Transient Detector Measurements of Lightning over North and South America," Proceedings of V International Symposium on Lightning Protection (SIPDA), Sào Paulo, Brazil, May 1999.

Chisholm, W. A., Levine, J. P., and Pon, C., "Lightning Protection Aspects for Applications of Optical Fibre Ground Wire," International Symposium on Lightning Protection (SIPDA), Santos, Brazil, 2001.

Cigre, "Guide to Procedures for Estimating the Lightning Performance of Transmission Lines," Working group 01 (lightning) of study committee 33 (overvoltages and insulation co-ordination), 1991.

Darveniza, M., *Electrical Properties of Wood and Line Design*, University of Queensland Press, 1980.

Darveniza, M., Limbourn, G. J., and Prentice, S. A., "Line Design and Electrical Properties of Wood," *IEEE Transactions on Power Apparatus and Systems*, vol. PAS-86, pp. 1344–56, 1967.

Darveniza, M., Mercer, D. R., and Watson, R. M., "An Assessment of the Reliability of In-Service Gapped Silicon-Carbide Distribution Surge Arresters," *IEEE Transactions on Power Delivery*, vol. 11, no. 4, pp. 1789–97, October 1996.

Darveniza, M., Roby, D., and Tumma, L. R., "Laboratory and Analytical Studies of the Effects of Multipulse Lightning Current on Metal Oxide Arresters," *IEEE Transactions on Power Delivery*, vol. 9, no. 2, pp. 764–71, April 1994.

Darveniza, M., Saha, T. K., and Wright, S., "Comparisons of In-Service and Laboratory Failure Modes of Metal-Oxide Distribution Surge Arresters," IEEE/PES Winter Power Meeting, October 2000.

Darveniza, M., Tumma, L. R., Richter, B., and Roby, D. A., "Multipulse Lightning Currents and Metal-Oxide Arresters," *IEEE Transactions on Power Delivery*, vol. 12, no. 3, pp. 1168–75, July 1997.

Dommel, H. W., "Electromagnetic Transients Program Reference Manual (EMTP Theory Book)," prepared for Bonneville Power Administration, 1986.

EPRI, *Transmission Line Reference Book: 345 kV and Above*, 2nd ed., Electric Power Research Institute, Palo Alto, CA, 1982.

EPRI. Lightning Protection Design Workstation version 5.0 CFlash online help: Electric Power Research Institute, Palo Alto, CA, 1999.

EPRI TR-000530, *Lightning Protection Design Workstation Seminar Notes*, Electric Power Research Institute, Palo Alto, CA, 1992.

EPRI TR-100218, *Characteristics of Lightning Surges on Distribution Lines. Second Phase — Final Report*, Electric Power Research Institute, Palo Alto, CA, 1991.

Eriksson, A. J., "The Incidence of Lightning Strikes to Power Lines," *IEEE Transactions on Power Delivery*, vol. PWRD-2, no. 3, pp. 859–70, July 1987.

Goedde, G. L., Dugan, R. C., and Rowe, L. D., "Full Scale Lightning Surge Tests of Distribution Transformers and Secondary Systems," *IEEE Transactions on Power Delivery*, vol. 7, no. 3, pp. 1592–600, July 1992.

Henning, W. R., Hernandez, A. D., and Lien, W. W., "Fault Current Capability of Distribution Transformers with Under-Oil Arresters," *IEEE Transactions on Power Delivery*, vol. 4, no. 1, pp. 405–12, January 1989.

Hu, H. and Mashikian, M. S., "Modeling of Lightning Surge Protection in Branched Cable Distribution Network," *IEEE Transactions on Power Delivery*, vol. 5, no. 2, pp. 846–53, April 1990.

IEEE C2-1997, *National Electrical Safety Code*.

IEEE C57.12.00-2000, *IEEE Standard General Requirements for Liquid-Immersed Distribution, Power, and Regulating Transformers*.

IEEE C62.92.4-1991, *IEEE Guide for the Application of Neutral Grounding in Electrical Utility Systems, Part IV - Distribution*.

IEEE Std. 386-1995, *IEEE Standard for Separable Insulated Connector Systems for Power Distribution Systems Above 600 V.*

IEEE Std. 1410-1997, *IEEE Guide for Improving the Lightning Performance of Electric Power Overhead Distribution Lines.*

IEEE Std. C62.22-1997, *IEEE Guide for the Application of Metal-Oxide Surge Arresters for Alternating-Current Systems.*

IEEE Task Force Report, "Secondary (Low-Side) Surges in Distribution Transformers," *IEEE Transactions on Power Delivery,* vol. 7, no. 2, pp. 746–56, April 1992.

Ishii, M. Discussion to (Barker et al., 1996), 1996.

Ishii, M., Michishita, K., Hongo, Y., and Oguma, S., "Lightning-Induced Voltage on An Overhead Wire Dependent on Ground Conductivity," *IEEE Transactions on Power Delivery,* vol. 9, no. 1, pp. 109–18, January 1994.

Jacob, P. B., Grzybowski, S., and Ross, E. R. J., "An Estimation of Lightning Insulation Level of Overhead Distribution Lines," *IEEE Transactions on Power Delivery,* vol. 6, no. 1, pp. 384–90, January 1991.

Kannus, K., Lahti, K., and Nousiainen, K., "Effects of Impulse Current Stresses on the Durability and Protection Performance of Metal Oxide Surge Arresters," High Voltage Engineering Symposium, August 22–27, 1999.

Kershaw, S. S., Jr., "Surge Protection for High Voltage Underground Distribution Circuits," Conference on Underground Distribution, Detroit, MI, September 27–October 1, 1971.

Lahti, K., Kannus, K., and Nousiainen, K., "Behaviour of the DC Leakage Currents of Polymeric Metal Oxide Surge Arresters in Water Penetration Tests," *IEEE Transactions on Power Delivery,* vol. 13, no. 2, pp. 459–64, April 1998.

Lahti, K., Kannus, K., and Nousiainen, K., "A Comparison Between the DC Leakage Currents of Polymer Housed Metal Oxide Surge Arresters in Very Humid Ambient Conditions and in Water Immersion Tests," *IEEE Transactions on Power Delivery,* vol. 14, no. 1, pp. 163–8, January 1999.

Lahti, K., Kannus, K., and Nousiainen, K., "The Durability and Performance of Polymer Housed Metal Oxide Surge Arresters Under Impulse Current Stresses," CIRED, 2001.

Lat, M. V. and Kortschinski, J., "Distribution Arrester Research," *IEEE Transactions on Power Apparatus and Systems,* vol. PAS-100, no. 7, pp. 3496–505, July 1981.

MacGorman, D. R., Maier, M. W., and Rust, W. D., "Lightning Strike Density for the Contiguous United States from Thunderstorm Duration Records." Report to the U.S. Nuclear Regulatory Commission, # NUREG/CR-3759, 1984.

Marz, M. B., Royster, T. E., and Wahlgren, C. M., "A Utility's Approach to the Application of Scout Arresters for Overvoltage Protection of Underground Distribution Circuits," IEEE/PES Transmission and Distribution Conference, 1994.

McDaniel, J., "Line Arrester Application Field Study," IEEE/PES Transmission and Distribution Conference and Exposition, 2001.

McDermott, T. E., Short, T. A., and Anderson, J. G., "Lightning Protection of Distribution Lines," *IEEE Transactions on Power Delivery,* vol. 9, no. 1, pp. 138–52, January 1994.

Owen, R. E. and Clinkenbeard, C. R., "Surge Protection of UD Cable Systems. I. Cable Attenuation and Protection Constraints," *IEEE Transactions on Power Apparatus and Systems,* vol. PAS-97, no. 4, pp. 1319–27, 1978.

Parrish, D. E., "Lightning-Caused Distribution Circuit Breaker Operations," *IEEE Transactions on Power Delivery,* vol. 6, no. 4, pp. 1395–401, October 1991.

Parrish, D. E. and Kvaltine, D. J., "Lightning Faults on Distribution Lines," *IEEE Transactions on Power Delivery,* vol. 4, no. 4, pp. 2179–86, October 1989.

Powell, R. W., Thwaites, H. L., and Stys, R. D., "Estimating Lightning Performance of Spacer-Cable Systems," *IEEE Transactions on Power Apparatus and Systems,* vol. 84, pp. 315–9, April 1965.

REA Bulletin 50-3, *Specifications and Drawings for 12.5/7.2-kV Line Construction,* United States Department of Agriculture, Rural Electrification Administration, 1983.

Ringler, K. G., Kirkby, P., Erven, C. C., Lat, M. V., and Malkiewicz, T. A., "The Energy Absorption Capability and Time-to-Failure of Varistors Used in Station-Class Metal-Oxide Surge Arresters," *IEEE Transactions on Power Delivery,* vol. 12, no. 1, pp. 203–12, January 1997.

Ros, W. J., "Neutral-to-Ground Impulse Voltage Effects on URD Cable," IEEE Rural Electric Power Conference, 1993.

Ross, E. R. and Grzybowski, S., "Application of the Extended CFO Added Method to Overhead Distribution Configurations," *IEEE Transactions on Power Delivery,* vol. 6, no. 4, pp. 1573–8, October 1991.

Ross, P. M., "Burning of Wood Structures by Leakage Currents," *AIEE Transactions,* vol. 66, pp. 279–87, 1947.

Rusck, S., "Induced Lightning Overvoltages on Power Transmission Lines With Special Reference to the Overvoltage Protection of Low Voltage Networks," *Transactions of the Royal Institute of Technology,* no. 120, 1958.

Rusck, S., "Protection of Distribution Lines," in *Lightning,* R. H. Golde, ed., London, Academic Press, 1977.

Short, T. A. and Ammon, R. H., "Monitoring Results of the Effectiveness of Surge Arrester Spacings on Distribution Line Protection," *IEEE Transactions on Power Delivery,* vol. 14, no. 3, pp. 1142–50, July 1999.

Short, T. A., Burke, J. J., and Mancao, R. T., "Application of MOVs in the Distribution Environment," *IEEE Transactions on Power Delivery,* vol. 9, no. 1, pp. 293–305, January 1994.

Sunde, E. D., *Earth Conduction Effects in Transmission Systems,* Dover Publications, New York, 1968.

Task force of eight utility companies and the General Electric Company, "Investigation and Evaluation of Lightning Protective Methods for Distribution Circuit. I. Model Study and Analysis," *IEEE Transactions on Power Apparatus and Systems,* vol. PAS-88, no. 8, pp. 1232–8, August 1969a.

Task force of eight utility companies and the General Electric Company, "Investigation and Evaluation of Lightning Protective Methods for Distribution Circuits. II. Application and Evaluation," *IEEE Transactions on Power Apparatus and Systems,* vol. PAS-88, no. 8, pp. 1239–47, August 1969b.

Thottappillil, R., Rakov, V. A., Uman, M. A., Beasley, W. H., Master, M. J., and Shelukhin, D. V., "Lightning Subsequent-Stroke Electric Field Peak Greater Than the First Stroke Peak and Multiple Ground Terminations," *Journal of Geophysical Research,* vol. 97, no. D7, pp. 7503–9, May 20, 1992.

Wagner, C. F., "Application of Predischarge Currents of Parallel Electrode Gaps," *IEEE Transactions on Power Apparatus and Systems,* vol. 83, pp. 931–44, September 1964.

Wagner, C. F. and Hileman, A. R., "Effect of Predischarge Currents Upon Line Performance," *IEEE Transactions on Power Apparatus and Systems,* vol. 82, pp. 117–31, April 1963.

Wagner, C. F. and Hileman, A. R., "Predischarge Current Characteristics of Parallel Electrode Gaps," *IEEE Transactions on Power Apparatus and Systems*, vol. 83, pp. 1236–42, December 1964.

Walling, R. A., Hartana, R. K., Reckard, R. M., Sampat, M. P., and Balgie, T. R., "Performance of Metal-Oxide Arresters Exposed to Ferroresonance in Padmount Transformers," *IEEE Transactions on Power Delivery*, vol. 9, no. 2, pp. 788–95, April 1994.

Ward, D. J., "Evaluating Product Reliability Costs," *IEEE Transactions on Power Delivery*, vol. 5, no. 2, pp. 724–9, 1990.

Williams, C. W., "Findings of Low-Voltage Surge Effect on the Florida Power Corp. System," Presented to the Task Force on Low-Side Surge Requirements for Distribution Transformers, IEEE Transformers Committee, Washington, DC, April 1988. As cited by IEEE Task Force Report (1992).

Zhou, L.-M. and Boggs, S. A., "Effect of Shielded Distribution Cables on Lightning-Induced Overvoltages in a Distribution System," *IEEE Transactions on Power Delivery*, vol. 17, no. 2, pp. 569–574, April 2002.

Check the top of your arrester and make sure you have the proper size for the voltage. Some times they look alike. Don't assume, it ain't good for your Fruit of the Looms.

Anonymous poster, on a near miss when a 3-kV arrester was mistakenly installed instead of a 10-kV arrester (7.2 kV line to ground)

www.powerlineman.com

13

Grounding and Safety

Grounding is one of the main defenses against hazardous electric shocks and hazardous overvoltages. Good equipment grounding helps reduce the chances that line workers and the public get shocks from internal failures of the equipment. System grounding determines how loads are connected and how line-to-ground faults are cleared. Most North American distribution systems have effective grounding; they have a neutral that acts as a return conductor and as an equipment safety ground.

13.1 System Grounding Configurations

13.1.1 Four-Wire Multigrounded Systems

There are several grounding configurations for three-phase power distribution systems. On distribution systems in North America, the four-wire, multigrounded neutral system predominates. Figure 13.1 shows how loads normally connect to the four-wire system. One phase and the neutral supply single-phase loads. The neutral carries unbalanced current and provides a safety ground for equipment. Low cost for supplying single-phase loads is a major reason the four-wire system evolved in North America. More than half of most distribution systems consist of single-phase circuits, and most customers are single phase. On a multigrounded neutral system, the earth serves as a return conductor for part of the unbalanced current during normal and fault conditions.

Four-wire multigrounded systems have several advantages over three-wire systems. Four-wire systems provide low cost for serving single-phase loads:

- Single cables for underground single-phase load (and potheads and elbows and padmounted transformer bushings)
- Single-phase overhead lines are less expensive
- Single-bushing transformers
- One arrester and one fuse for a single-phase transformer

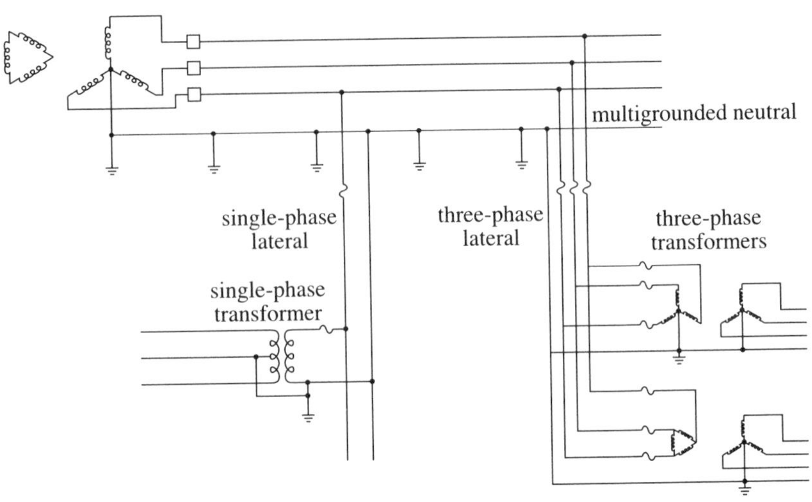

FIGURE 13.1
A four-wire multigrounded distribution system.

- The neutral can be located lower on the pole
- Lower rated arresters
- Lower insulation required; transformer insulation can be graded

Extensive use of single-phase lines provides significant cost savings. The primary neutral may be shared with the secondary neutral for further cost savings. And flexibility increases: upgrading a single-phase line to two- or three-phase service on overhead lines is a cost-effective way to upgrade a circuit following load growth.

Safety. The multigrounded neutral provides significant safety benefits for crews and end users. If a transformer pole-ground down lead gets broken, little safety margin is lost. With the transformer still connected to the multigrounded neutral, there is still a low-impedance path to earth, which protects against internal faults in the transformer.

The underbuilt neutral also partially blocks access to the phase conductors (to ladders or bucket trucks). It warns crews that high-voltage phase conductors are overhead. And, if a phase wire breaks, there is a good chance that it will contact the neutral as it falls (if the neutral does not catch it, it might result in a high-impedance fault).

The substation transformer normally provides the grounding source, the low-impedance path to zero sequence. On circuits converted from an ungrounded configuration to a grounded configuration, a grounding transformer, usually a zig-zag transformer, may provide the zero-sequence source, so a neutral can be added. A grounding transformer must not be disconnected during operation of the distribution line. If the grounding transformer becomes disconnected, a single line-to-ground fault causes high overvoltages

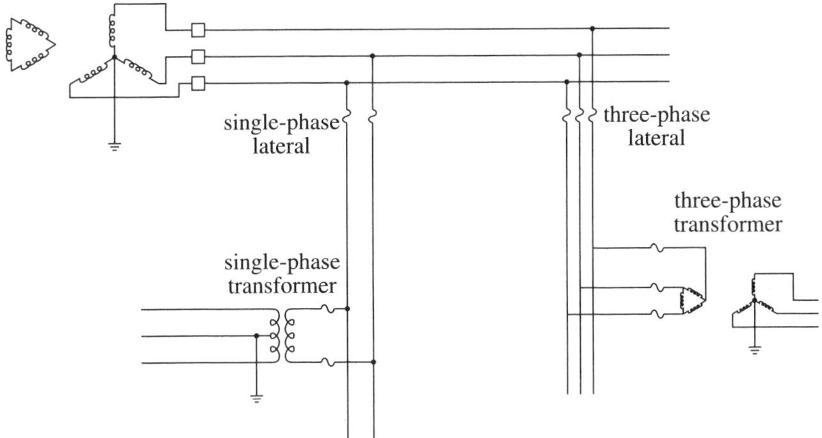

FIGURE 13.2
A three-wire unigrounded distribution system

on the unfaulted phases. Load unbalances may also create overvoltages if the grounding transformer is disconnected.

13.1.2 Other Grounding Configurations

Most other distribution configurations are three-wire systems, either grounded or ungrounded. Figure 13.2 shows a three-wire unigrounded service. A unigrounded is grounded at one place, normally the substation. Many older distribution systems are three-wire systems, including 2400-, 4160-, and 4800-V systems. Many utilities have portions of these circuits that are holdovers from installations in the 1920s to the 1940s. Whether unigrounded or ungrounded, three-wire systems have the following advantages:

- Lower cost for three-phase loads
- Lower stray voltage
- Easier detection of high-impedance faults
- Lower magnetic fields

European distribution designs are normally three-wire systems. All load is three phase or phase-to-phase connected, so there is no need for a neutral. But without the neutral, these systems must do without the multigrounded safety ground. Most European designs make up for the lack of a multigrounded primary ground by having a multigrounded secondary neutral, which is grounded at equipment and at each customer. Since European designs tend to have extensive secondary systems and many houses on each transformer, the secondary ground provides a safe grounding conductor.

Some parts of the world use a *single-wire earth-return* (SWER) distribution system. The earth can carry all of the return current. On earth-return systems, the resistance of ground electrodes is critical. Poor ground connections can cause low end-use voltage and create very dangerous step-and-touch potentials to the transformer tank. The single-wire system is used extensively in Australia to serve remote rural customers with single-phase service; a high primary voltage (33 kV) is normally used, which keeps currents low. Note that the National Electrical Safety Code (IEEE C2-1997) does not allow an earth-return system; it prohibits using the earth as the sole conductor for any part of the circuit.

Other types of grounding such as high-resistance grounding or high-reactance grounding are rarely used on distribution circuits. European systems occasionally use a three-wire *resonant*-grounded system, where the substation transformer neutral is grounded through a tapped reactor (Peterson coil). The coil is tuned to cancel the line-to-ground capacitance. It is an ungrounded system. The main advantage is that it can operate with one phase faulted. On overhead distribution circuits, the main advantage is that temporary faults clear by themselves without opening a breaker or recloser. Without the tuning reactor, the capacitance creates enough current to maintain the fault arc (especially with cables). In the past, it was hard to keep the reactor in tune with the system capacitance. Clerfeuille et al. (1997) describe a resonant grounding distribution system that uses an adaptively tuned reactor that continuously maintains proper tuning.

The five-wire design is a new approach that may reduce stray voltage and magnetic fields and also make high-impedance faults more easily detectable. The fifth wire is an isolated neutral that carries all of the unbalanced return current. Under normal conditions, the 5-wire design operates very similar to the 4-wire design with one major exception: all the return current is confined to the neutral that is isolated from ground. The multi-grounded ground wire continues to perform the safety functions associated with a multi-grounded system. New York State Electric and Gas (NYSEG) converted a section of circuit to a five-wire configuration as a research project, and monitoring results showed reduction in stray voltage and magnetic fields along with promising results that high-impedance faults could be detected (EPRI 1000074, 2000; Short et al., 2002). Challenges found during the project were: interfacing between the standard four-wire section and the five-wire section, how to insulate the fifth wire, surge protection on the fifth wire, how to protect against an open neutral, and crew acceptance. Increased cost of the five-wire system would also factor in any real installation.

13.2 System Grounding and Neutral Shifts During Ground Faults

The system grounding configuration determines the overvoltages that can occur during a line-to-ground fault. A single line-to-ground fault shifts the

Normal conditions

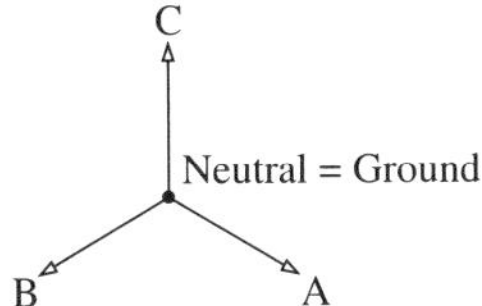

Ungrounded System

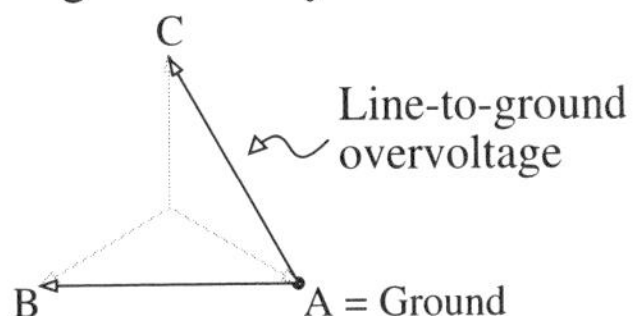

1.73 per unit overvoltage
Total ground reference shift for a fault from ground to phase A

Perfectly Grounded System

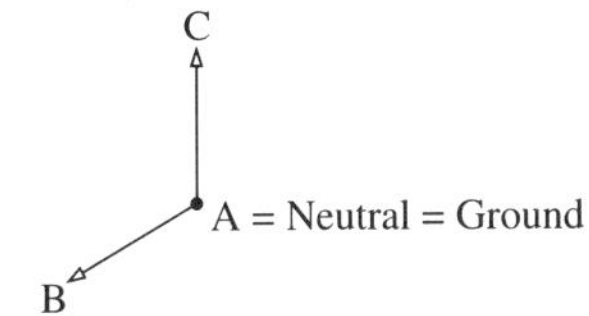

No overvoltage
Phase A collapses to the neutral (and ground) voltage

Typical Multigrounded System

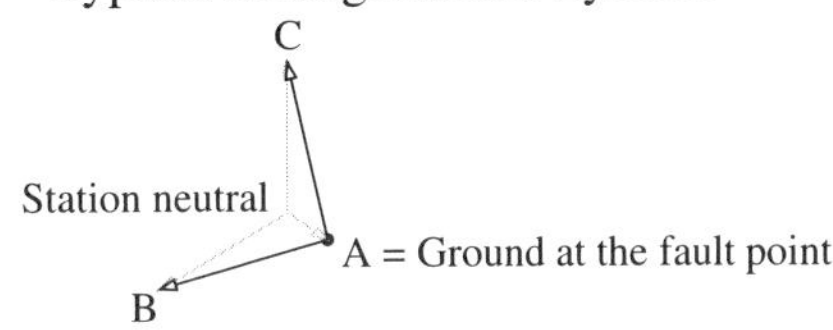

1.2 per unit overvoltage
Partial ground reference shift for a fault from ground to phase A

FIGURE 13.3
Ground potential shifts and overvoltages depending on the grounding configuration.

ground potential at the fault location. The severity of this ground reference shift depends on the grounding configuration (see Figure 13.3). On a solidly grounded system with a good return path to the grounding source, little reference shift occurs. On an ungrounded system, a full offset occurs — the line-to-ground voltage on the unfaulted phases rises to the line-to-line voltage, 1.73 per unit. On a multigrounded distribution system with a solidly grounded station transformer, overvoltages above 1.3 per unit are rare.

Two factors relate the overvoltage to the system voltage:

Coefficient of grounding:

$$\text{COG} = V'_{LN}/V_{LL}$$

Earth fault factor:

$$\text{EFF} = V'_{LN}/V_{LN}$$

where

V'_{LN} = maximum line-to-ground voltage on the unfaulted phases during a fault from one or more phases to ground

V_{LN}, V_{LL} = nominal line-to-neutral and line-to-line voltages

A system is "effectively grounded" if the coefficient of grounding is less than or equal to 80% (the earth-fault factor is less than 138%) (ANSI/IEEE C62.92-1987). This is met approximately with the following conditions:

- $X_0/X_1 < 3$
- $R_0/X_1 < 1$

For a single line-to-ground fault on phase A, the voltages on phases B and C are

$$V_b = \left(a^2 + \frac{Z_1 - Z_0}{2Z_1 + Z_0 + 3R_F}\right)E$$

$$V_c = \left(a + \frac{Z_1 - Z_0}{2Z_1 + Z_0 + 3R_F}\right)E$$

where
Z_1 = positive-sequence impedance
Z_0 = zero-sequence impedance
$a = 1\angle 120°$
R_F = fault resistance
E = line-to-neutral voltage magnitude prior to the fault

For a double line-to-ground fault, the voltage on the unfaulted phase is

$$V = \frac{3Z_0 + 6R_F}{Z_1 + 2Z_0 + 6R_F}E$$

In some cases, the double line-to-ground fault causes overvoltages that are slightly higher than the single line-to-ground fault. But since single line-to-ground faults are so much more common, we often design for the single line-to-ground fault. For single line-to-ground faults, the voltage is always worse with the fault impedance $R_F = 0$. For double line-to-ground faults, it may not always be worse with $R_F = 0$. Figure 13.4 shows overvoltage charts as a function of X_0/X_1 and R_0/X_1. This includes overvoltages due to single line-to-ground faults and for double line-to-ground faults (assuming that $R_F = 0$).

IEEE suggests the overvoltage multiplier factors for different systems shown in Table 13.1 (IEEE C62.92.4-1991). The multipliers include the neutral shift during line-to-ground faults at a voltage of 105% (so, the ungrounded system has an overvoltage of $1.73 \times 1.05 = 1.82$).

The higher overvoltage factor of 1.35 for multigrounded systems with metal-oxide arresters was identified as a more conservative factor for four-wire systems because of the reduced saturation of newer transformers and

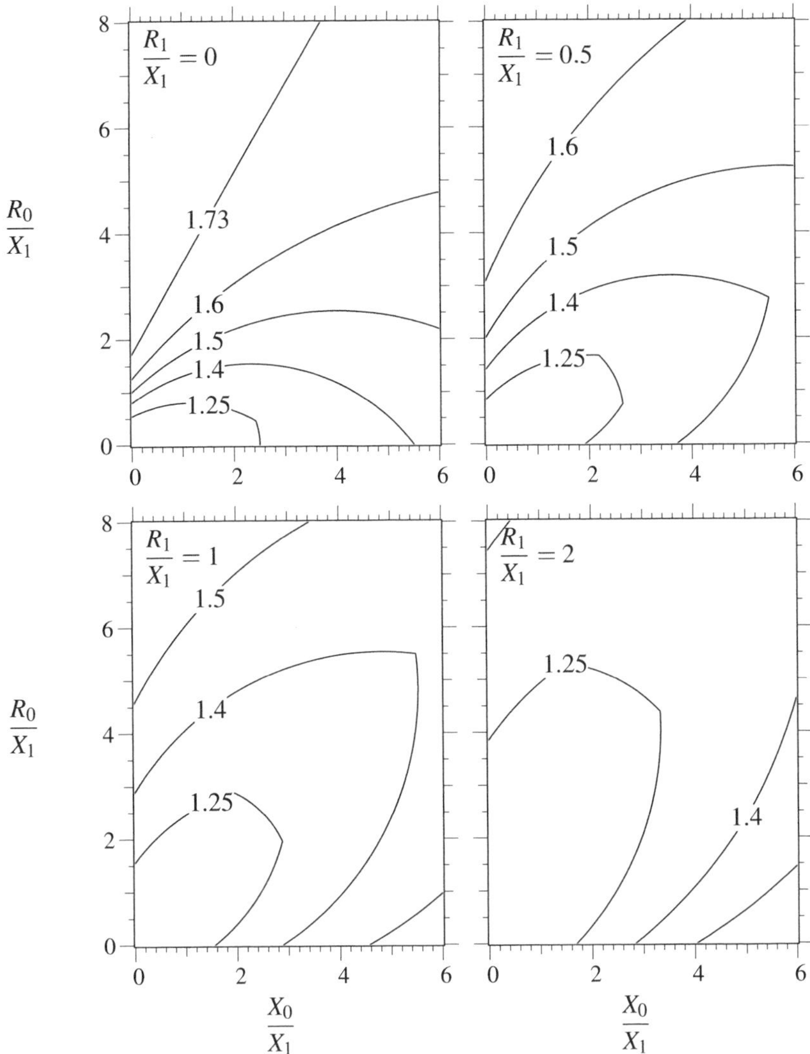

FIGURE 13.4
Maximum overvoltages in per unit for line-to-ground faults based on X_0/X_1 and R_0/X_1 (the contours mark the threshold of voltage).

use of metal-oxide arresters (they are always connected, so they are more sensitive to overvoltages than older arresters, which have an isolating gap).

13.2.1 Neutral Shifts on Multigrounded Systems

On four-wire systems with a multigrounded neutral, standard short-circuit calculation programs calculate overvoltages incorrectly. The problem is not symmetrical components; the problem is with the impedance of the return

TABLE 13.1

Overvoltage Factors for Different Grounding Systems

System	Overvoltage Factor
Ungrounded system	1.82
Four-wire multigrounded system (spacer cable)	1.5
Three- or four-wire unigrounded system (open wire)	1.4
Four-wire multigrounded system (open wire; gapped arrester)	1.25
Four-wire multigrounded system (open wire; metal-oxide arrester)	1.35

Source: IEEE C62.92.4-1991, *IEEE Guide for the Application of Neutral Grounding in Electrical Utility Systems, Part IV — Distribution.*

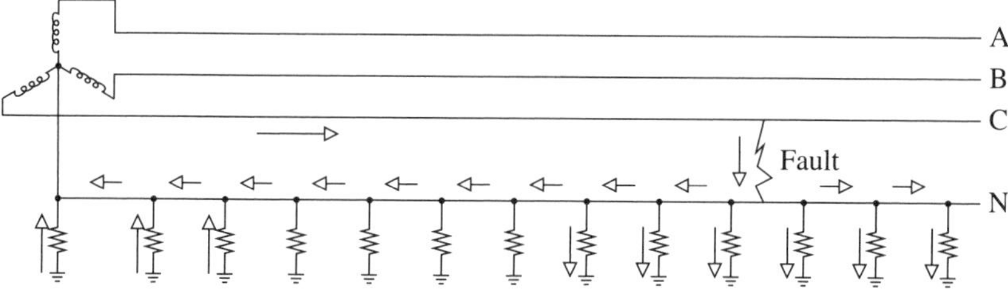

FIGURE 13.5
Fault current paths for a line-to-ground fault.

path, the zero-sequence impedance. When the impedance of the neutral is rolled into the zero-sequence impedance, the neutral is modeled as being perfectly grounded; ground resistances are ignored.

Assuming the neutral is perfectly grounded incorrectly models overvoltages. Lat (1990) analyzed overvoltages with several techniques and found that the traditional calculation method has serious shortcomings vs. modeling each ground and line section separately. We need to model each line section between grounds as shown in Figure 13.5. Standard short-circuit programs would not model this detail; we need a flexible steady-state analysis program such as EMTP.

Lat's analysis and additional work by Mancao et al. (1993) showed several interesting findings:

- *Fault location* — Overvoltages are highest for faults near the substation. This contradicts the common belief that the highest overvoltages are for faults at the end of the circuit (based on the standard symmetrical components method).
- *Feeder length* — Overvoltages are higher on shorter feeders (assuming the same number of grounds per unit length, see Figure 13.6). The ground return path has a higher impedance on shorter lines because there are fewer grounds beyond the fault location.
- *Neutral size* — A larger neutral provides a better return path.

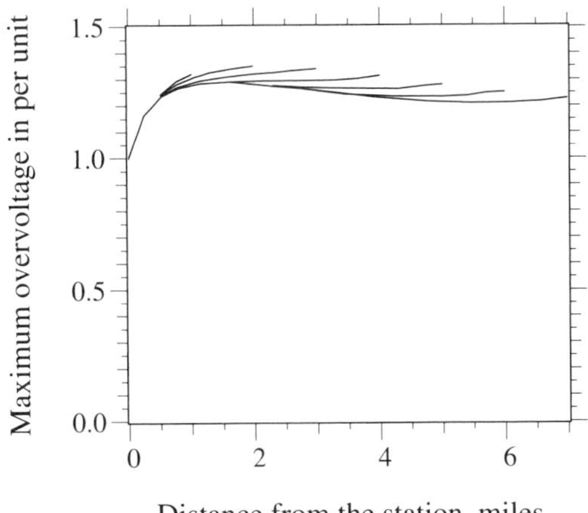

FIGURE 13.6
Overvoltages for different fault locations on feeders of different lengths. 3/0 ACSR phase conductors, 1/0 ACSR neutral, 4-foot (1.2-m) spacings, four 25-Ω grounds/mile (normally, there are many more grounds than this). (Adapted from Lat, M. V., "Determining Temporary Overvoltage Levels for Application of Metal Oxide Surge Arresters on Multigrounded Distribution Systems," *IEEE Trans. Power Delivery*, 5(2), 936-946, April 1990. ©1990 IEEE. With permission.)

- *Ground rod resistance and number of ground rods* — More ground rods and lower rod resistances reduce overvoltages (see Figure 13.7). But the results are not particularly sensitive to ground rod resistance values.
- *Neutral reactor* — A neutral reactor on the substation transformer causes a higher overvoltage, mainly for faults near the station.

Hydro Quebec staged faults on rural circuits and measured the neutral voltages and currents (Rajotte et al., 1995). The measured neutral offsets compared very well with simulations done with Lat's modeling approach. But, they compared well only if customer's grounds were modeled. On this circuit, the customer grounds contributed more to grounding the line than the utility ground rods. There were three times more customer ground rods, and their average resistance was almost eight times lower. It is not unusual for customer grounds to be better than utility grounds, as customer grounds are often tied to water pipes and wells. More extensive measurements of ground rods and customer grounds on 16 rural feeders found an equivalent of 47 Ω for ground rods and 16 Ω for customer ground connections (these equivalents are the inverse of the average admittance) (Rajotte et al., 1995).

If the neutral conductor breaks, overvoltages are much higher. Mancao et al. (1993) found overvoltages as high as 1.58 per unit on simulated systems for faults downstream of an open point in the neutral. The worst case is for a fault just past the break, but even a fault several miles from the open point causes higher than normal overvoltages.

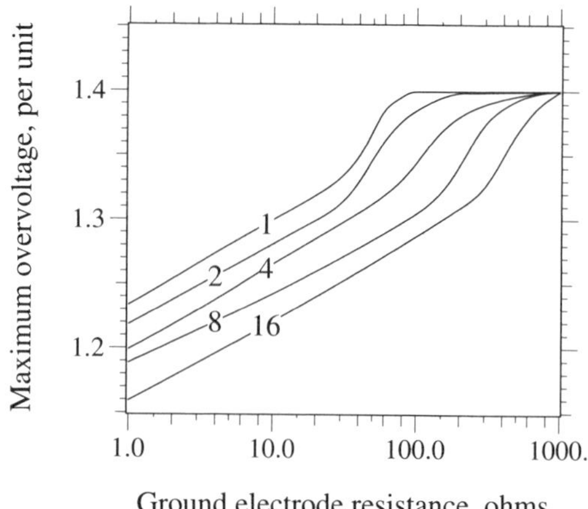

FIGURE 13.7
Effect of ground rod resistance and spacings. Each curve is labeled with the number of grounds per mile. (Adapted from Lat, M. V., "Determining Temporary Overvoltage Levels for Application of Metal Oxide Surge Arresters on Multigrounded Distribution Systems," *IEEE Trans. Power Delivery*, 5(2), 936-946, April 1990. ©1990 IEEE. With permission.)

Figure 13.8 shows simulated divisions of current for several neutral sizes for a fault near the end of the circuit. Near a line-to-ground fault, most of the current is in the neutral. In the middle, more flows through the earth. Near the station, additional current returns to the neutral. And, some current returns via grounds beyond the fault location. Larger neutrals carry more return current.

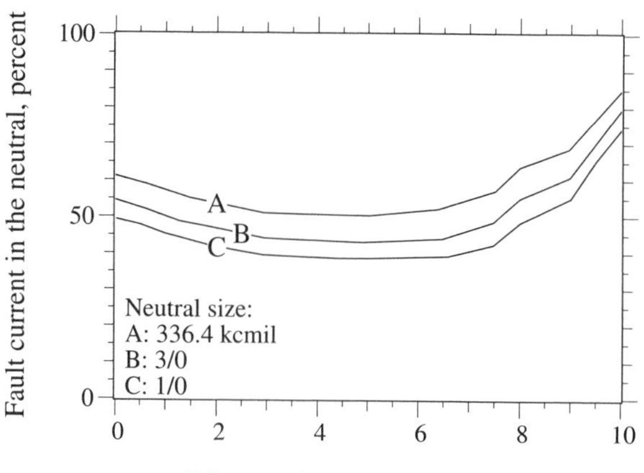

FIGURE 13.8
Division of fault current between the neutral and earth. Armless construction, 336-kcmil phase conductors, four 25-Ω grounds per mile. (Adapted from Mancao, R. T., Burke, J. J., and Myers, A., "The Effect of Distribution System Grounding on MOV Selection," *IEEE Trans. Power Delivery*, 8(1), 139-45, January 1993. ©1993 IEEE. With permission.)

An IEEE Working Group (1972) stated that transformer saturation holds down overvoltages based on transient network analyzer evaluations. But others disagree. Field tests found that saturation did not lower the peak voltage (Rajotte et al., 1995). A transformer core saturates when the flux is highest. The flux is 90° out of phase with the voltage. The peak flux occurs when the voltage crosses through zero. So, when a transformer saturates, it distorts the voltage but does not appreciably limit the peak voltage.

Normally, the highest overvoltages are for single line-to-ground faults, but in some cases slightly higher overvoltages occur for double line-to-ground faults. These cases normally involve large conductors (high X/R).

Spacer cable systems have higher overvoltages because of the messenger wire (not because of the tighter phase and neutral spacings). An IEEE Working Group (1972) found overvoltages between 1.36 and 1.45 per unit with spacer cable having a 5/16-in. or 3/8-in. copperweld messenger with a 30% conductivity. If a spacer cable uses a lower-resistivity messenger or has a separate neutral, overvoltages are similar to those of open-wire configurations. ESEERCO (1992) simulated different spacer cable configurations and found overvoltages of about 1.5 per unit on configurations with just a 3/8-in. copperweld messenger. With a 1/0 ACSR neutral added to the spacer cable, the overvoltages dropped to 1.3 per unit.

Underground cables normally have lower overvoltages. Rajotte et. al. (1990) simulated overhead and underground systems and found that the neutral shift was a factor of ten lower on underground systems. Urban locations, whether overhead or underground, have many more grounds, so overvoltages and other grounding issues such as stray voltage are less likely.

However, circuits of cables with tape shields or lead sheaths can have significantly higher overvoltages. On these cables, the tape shields have very high resistance. Figure 13.9 shows how the overvoltage on the unfaulted phases increases with distance of the fault from the substation for an example scenario. This example assumes that the sheath is perfectly grounded. If poorly grounded, more current will return through the high-resistance sheath and increase the overvoltage. Utilities normally use tape shield or lead sheathed power cables in urban areas where distances are fortunately short. The high resistance also contributes to transients; the X/R ratio is low enough that faults often spark and cause transients at each zero crossing. In a resistive circuit, when the arc clears at a zero crossing, the voltage builds up slowly, which delays the arc breakdown. When the arc does restrike, a transient can occur because of the trapped voltage just prior to the arc restrike.

13.2.2 Neutral Reactor

A reactor connected between the substation transformer neutral point and the substation ground reduces ground fault currents. This limits duty on equipment, but it reduces the effectiveness of the grounding system. During line-to-ground faults, the voltages on the unfaulted phases are higher. A

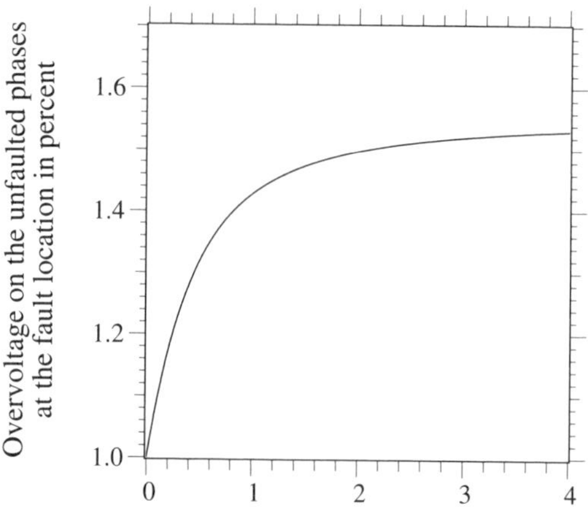

FIGURE 13.9
Voltage on the unfaulted phases for a system with 500-kcmil, triplex, copper-conductor, tape-shield cables. 12.47 kV, $Z_1 = 0.0292 + j0.0355\ \Omega/1000$ feet, $Z_0 = 0.3765 + j0.2882\ \Omega/1000$ feet, $Z_{src} = j0.7\ \Omega$.

neutral reactor of $X\ \Omega$ adds $3jX\ \Omega$ to the zero-sequence impedance. The overvoltage is most pronounced for faults near the substation. Beyond a few miles, the line impedance dominates.

Figure 13.10 shows an example of the tradeoff between fault current at the substation bus and neutral reactor impedance on a 12.47-kV system. A 0.5 Ω reactor increases the Z_0/Z_1 ratio to 3.1, which reduces the fault current to under 6 kA at the substation bus while still keeping the overvoltage below 130%. A reactor larger than this must be considered carefully to avoid subjecting metal-oxide arresters and customer's loads to undue overvoltages.

This graph was developed from a simplified equation for overvoltages. With all circuit resistances equal to zero, the voltage magnitude in per unit on the worst unfaulted phase for a bolted single line-to-ground fault simplifies (Westinghouse Electric Corporation, 1950) to

$$V = 1.732\frac{\sqrt{X_0^2 + X_0X_1 + X_1^2}}{2X_1 + X_0}$$

where

$X_0 = X_{0,circuit} + 3X_{reactor}$

13.2.3 Overvoltages on Ungrounded Systems

Many older distribution systems are ungrounded, typically operating at 2400 or 4800 V. Although ungrounded systems have no intentional ground, they

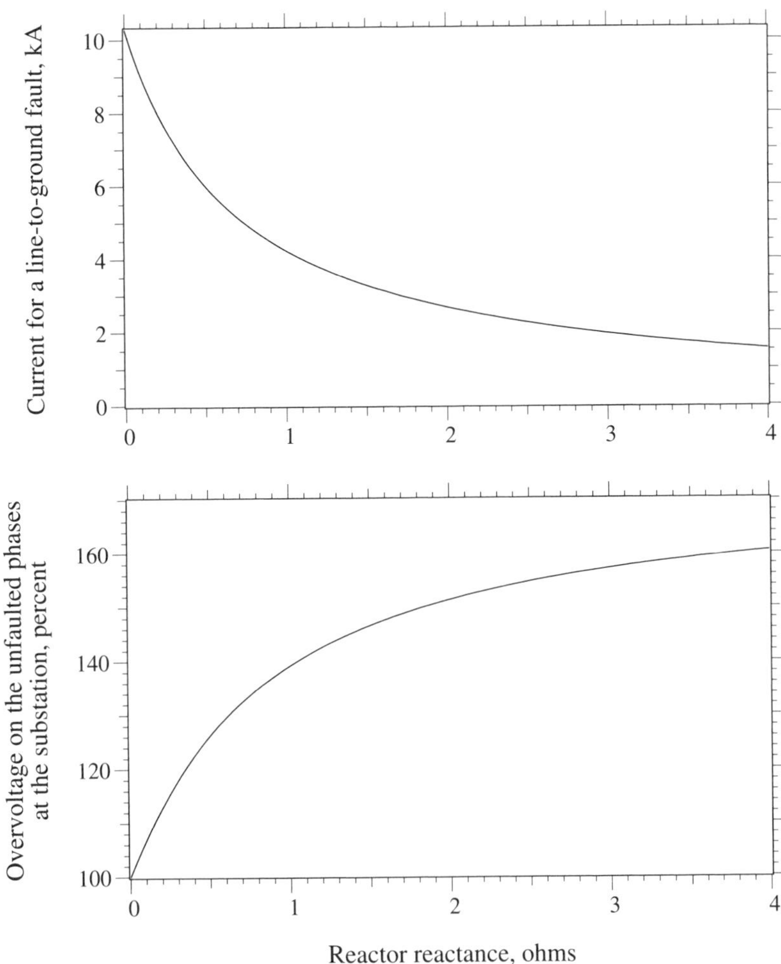

FIGURE 13.10
Tradeoff between overvoltage on the unfaulted phases and fault current for a single line-to-ground fault at the substation bus for different size neutral reactors. (12.47-kV circuit with a 20-MVA, 9% transformer, subtransmission impedance neglected)

are grounded through the line's capacitance. Under normal conditions, the phase-to-ground voltage equals the phase-to-phase voltage divided by $\sqrt{3}$ if the total line-to-ground capacitance on each phase is the same.

Ungrounded systems may have extra-high overvoltages under certain conditions:

- *Resonance* — The line-to-ground capacitance can resonate with the system reactance.
- *Arcing ground faults* — A restriking ground fault can create significant line-to-ground voltage.

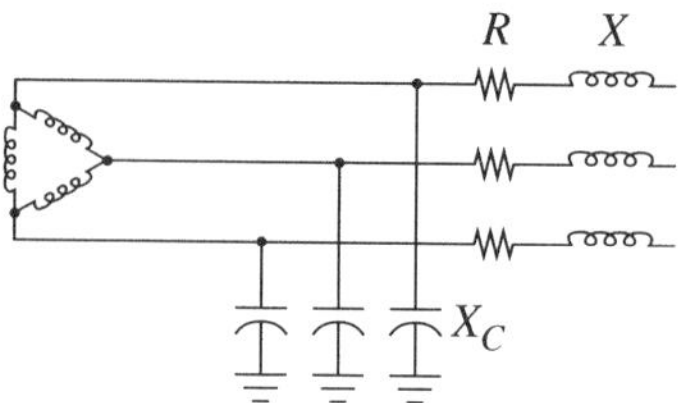

Equivalent circuit referred to one phase

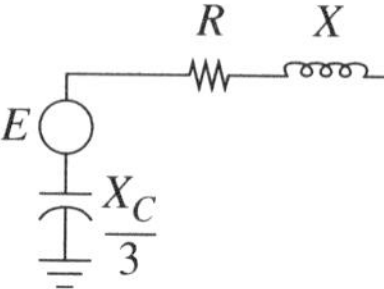

FIGURE 13.11
Ungrounded system schematic.

The main impact of these overvoltages is stress on insulation and impact on arresters (higher rated arresters are necessary). A bolted line-to-ground fault on one phase draws very little current on an ungrounded system. The current is normally much less than 10 A, so the system may continue supplying power. This is an operational advantage of the ungrounded system. Because it can operate through line-to-ground faults, ungrounded systems may have higher reliability. Most lower-voltage ungrounded systems (2400 or 4800 V) are very resistant to trees.

Figure 13.11 shows a simplification of an ungrounded system that shows how resonances can create overvoltages.* For a bolted single line-to-ground fault on a phase, the voltages on phases B and C can become extremely high if the system capacitance resonates with the inductance, which occurs when

$$X = X_C/3$$

where

X = reactance of the system, Ω

X_C = line-to-ground capacitance, Ω

The overvoltage reaches a maximum when at the resonant point and can reach significant voltages for ratios of X_C/X of up to 15. Normal line capacitances and reactances are well away from the resonant point; even cable capacitances are too low to cause resonance. For example, 10 mi (16 km) of cable with a capacitance of 0.5 μF/mi (0.9 μF/km) has an impedance of 530 Ω, and line impedances are less than 1 Ω/mi. Even with several circuits in

* The simplified circuit shows how resonance occurs, but the circuit is actually more complicated than shown. The capacitance is not necessarily lumped on the source side of the line impedance, it is distributed along the line (although circuit taps may concentrate the capacitance at points on the line). The figure also ignores the line-to-line capacitance and the system load.

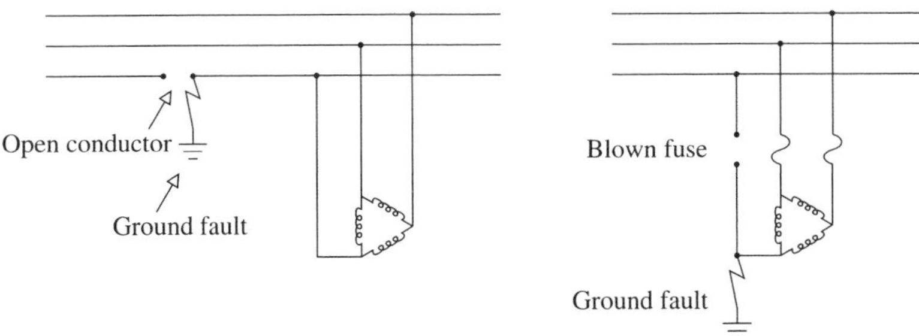

FIGURE 13.12
Accidental connections of reactances to ground that may cause high overvoltages on ungrounded systems.

parallel (more capacitance), such resonances are unlikely. For a more thorough analysis, see Peterson (1951).

Resonances can occur when high reactances are accidentally connected to ground. Figure 13.12 shows ground fault scenarios that may cause overvoltages when three-phase distribution transformers become connected to ground.

Ungrounded systems are also susceptible to arcing ground faults. An intermittent, restriking fault from ground to a phase conductor may spark high voltages, possibly as high as five per unit. Consider the equivalent single-phase circuit in Figure 13.11. This series circuit is a classic restriking circuit often studied for switching capacitive loads; but now, the restriking is in an arc to ground instead of in the circuit interrupter. The voltage can ratchet up as the restriking charges the system capacitance. Since the intermittent ground fault may last for hours, this overvoltage mode subjects equipment to continued overvoltages. The overvoltages are likely to be chaotic and random because of the random restriking times.

Fortunately, end users do not see the overvoltages from neutral shifts or from arcing ground faults. The overvoltage is line to ground, and all loads are connected line to line. But utility equipment insulation from the primary to the equipment case does see the voltage, especially surge arresters, which are sensitive to overvoltages.

13.3 Equipment/Customer Grounding

The National Electrical Safety Code (NESC) (IEEE C2-1997) places several requirements on grounding systems. The NESC requires effective grounds on

- Surge arresters
- Primary, secondary, and common neutrals
- Unguarded metal or metal-reinforced supporting structures

- Other intentionally grounded wires
- Instrument transformers
- Cable concentric neutrals
- Most equipment including transformers, capacitors, regulators, reclosers, and sectionalizers
- Uninsulated guy and messenger wires
- Customer meters

Some items that may be ungrounded on overhead circuits are insulators, cutouts, and some open-air switches. For multigrounded systems, the grounded conductor may serve as the ground for many applications. On a multigrounded system, the NESC requires ground electrodes at transformer locations and at customer meters. For an effectively grounded neutral (from an NESC safety perspective) there must be at least four grounds in each mile (1.6 km) of the entire line. Customer meter grounds do not count (they are not necessarily under the control of the utility). Also, the neutral must be continuous.

The NESC does not allow the earth as the sole conductor for any part of the circuit.

On systems without a multigrounded neutral, where a grounding electrode is required, it must be less than 25 Ω. If it is higher than that, another grounding electrode must be made. The second electrode must be at least 6 ft (1.8 m) from the first. Note that there is no requirement on the final resistance (it can still be greater than 25 Ω and meet this rule). It should be low enough so that a protective device will clear for a fault from the primary through the grounding electrode.

The NESC does not require a specific grounding resistance for multigrounded systems. Multigrounded systems achieve their performance mainly by having many grounds in parallel, so the resistance of individual electrodes is less significant. Still, many utilities use the ungrounded NESC approach for all grounding points: drive one rod, if it is more than 25 Ω, drive another.

At least 20 ft (6.1 m) must separate grounding electrodes of different systems (Clapp, 1997; IEEE C2-1997). The primary and secondary are "different systems." Grounding electrodes of different systems may not be bonded above or below ground.

The NESC requires grounding conductors for surge arresters to be at least #6 copper or #4 aluminum. Other grounding electrodes may be smaller, but they must be able to handle the fault current they will see. For multigrounded systems, ground down leads from the neutral to the grounding electrode will normally not carry significant current. The critical connection is between the equipment frame and the neutral. First and most importantly, make sure there is such a connection. The connection between the frame and the neutral must be designed to take the entire fault current for a line-to-ground fault from the line to the equipment case. Since the fault current duration plays a role in whether the wire will handle the fault, we must consider the fault interrupter. For most applications where normal-sized fuses are used, a #4 or #6 copper bonding conductor is sufficient. For situations with high fault currents and where a station breaker clears the fault, a larger conductor or parallel conductors may be needed.

Since the tank-to-neutral connection is critical on multigrounded systems, on transformers, utilities typically tie two connections from the transformer tank. One runs from the tank to the pole ground lead, which is attached to the primary neutral. Another runs from the secondary neutral terminal of the transformer (X2) to the neutral.

13.3.1 Special Considerations on Ungrounded Systems

Systems that do not have a multigrounded neutral — ungrounded or unigrounded systems — have special grounding considerations. Without the multigrounded neutral, the secondary neutral grounding system is the whole grounding system. If a transformer fails from the primary to the secondary (a high-to-low fault), the secondary grounding system must be effective enough to prevent unsafe voltages on customer wiring. Crews should make sure that distribution transformers are grounded as well as possible and that customer meters are grounded as well as possible.

On systems without a multigrounded neutral, the NESC requires the primary surge arrester to be separate from the secondary ground. The NESC does allow interconnection of the primary and secondary grounds through a spark gap as long as the 60-Hz breakdown value of the gap is at least twice the primary circuit voltage but not necessarily more than 10 kV. If a spark gap is used, an additional ground electrode must be attached, not including the customer's grounds; and the extra ground must be at least 20 ft (6.1 m) from the arrester's grounding electrode.

13.3.2 Secondary Grounding Problems

Many complaints are related to grounding problems on the secondary. Some common problems are shown in Figure 13.13. One of the most common power quality complaints is due to an open secondary neutral. If the neutral is open or has a poor connection, the unbalanced current must return through the grounding electrodes (these can have high impedance). The customer may see this as light flicker, and high voltage may occur on the less loaded of the two secondary legs.

Another common problem on secondary systems is when load is connected to the equipment safety ground (the green wire) instead of the neutral wire, or the ground wire is wrongly connected to the neutral wire somewhere within the facility. Both situations violate the National Electrical Code (NFPA

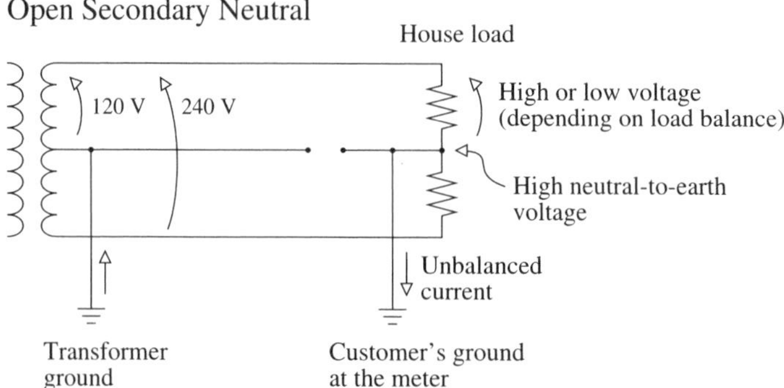

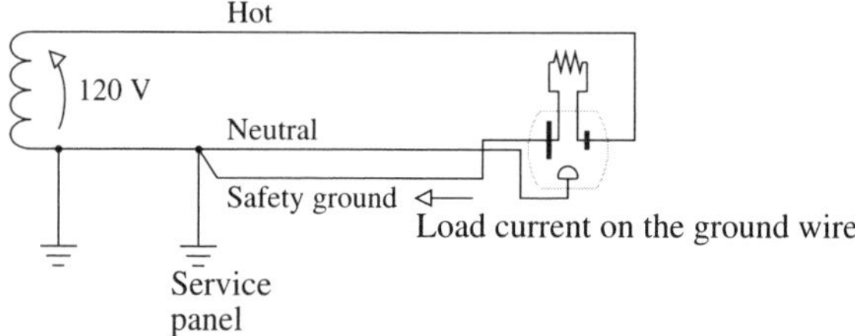

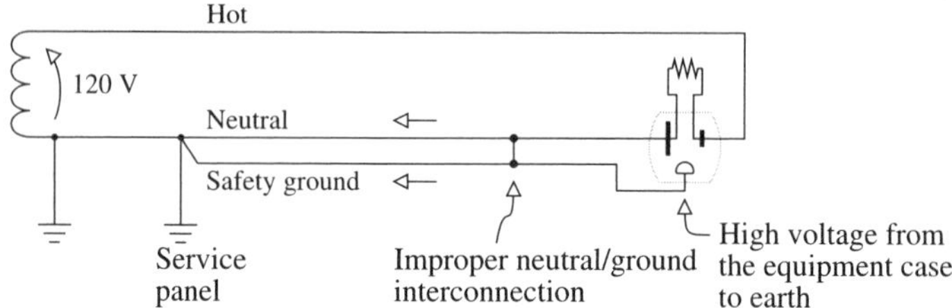

FIGURE 13.13
Common secondary problems.

70, 2002). Currents flowing in the grounding conductor wire increase shock hazards. Other secondary problems that can occur are isolated grounds, ground loops, and missing safety grounds.

13.4 Ground Rods and Other Grounding Electrodes

The earth, the soil, has very low resistance, primarily because there is so much of it. The trick is getting from the concentrated grounding electrode to the soil beneath that is almost infinite by comparison.

Ground rods are the most common grounding electrode. The NESC requires at least an 8-ft (2.45-m) rod; iron or steel rods must be at least 5/8 in. (15 mm) in diameter; copper-clad stainless steel, or stainless steel-clad rods must be at least 1/2-in. (12 mm). Rods should be driven to at least 8 ft (2.45 m). If a driven rod hits a rock bottom, the depth may be less than 8 ft (2.45 m). Strategies are to drive the rod at an angle or to use another type of ground such as a buried strip. NESC requires buried strips to be at least 10 ft (3 m) and have a total surface area of at least 5 ft^2 (0.47 m^2). Guy anchors are not allowed as grounding electrodes.

The resistivity of a ground rod is

$$R = \frac{\rho}{2\pi L}\left(\ln\frac{4L}{a} - 1\right)$$

where

ρ = soil resistivity, Ω-m
L = length of the ground rod, m
a = radius of the ground rod, m

As a good approximation for 10-ft (3.3-m) ground rods, the resistance in ohms equals the ground resistivity in Ω-m divided by three. Divide earth resistivity by 2.5 for an 8-ft rod.

Rod diameter has little impact on the electrode resistance; the diameter is needed mainly for mechanical strength and to ensure the rod has enough material to survive corrosion.

Twenty percent of the resistance of a ground rod occurs within 1.2 in. (3 cm) of the rod, and half of the resistance is in the first 6 in. (15 cm) (IEEE Std. 142-1991). These numbers vary with the ground resistivity.

The number of rods and the length are the main factors that influence the resistance. Rods can be driven deeper. Normally, crews drive a section and then thread on another rod section to pound it in deeper. Deeper rods are best when lower levels of the soil have lower resistivity, normally because of more moisture.

Multiple ground rods in parallel effectively reduce the overall resistance. But, if two ground rods are too close together; they act as one ground rod with a larger diameter, reducing much of the gain of using parallel rods. One common rule: separate the rods by a distance of at least the length of one of the ground rods. The NESC requires ground rods to be at least 6 ft (1.8 m) apart. Even with these separations, the resistance of n ground rods in parallel is more than $1/n$ times the resistance of a single rod. Table 13.2 shows multiplying factors and final resistances for rods placed symmetrically one rod length apart.

The NESC allows other grounding electrodes. Water piping systems make fantastic grounds if they are available; grounds less than 1 Ω are common. Unfortunately, conducting water pipes are not typically installed anymore,

TABLE 13.2

Resistance of Multiple Rods

Number of Rods	Multiplying Factor	Total Resistance as a Percentage of One Rod
2	1.16	58%
3	1.29	43%
4	1.36	34%
8	1.68	21%
12	1.80	15%
16	1.92	12%
20	2.00	10%
24	2.16	9%

Source: IEEE Std. 142-1991, *IEEE Recommended Practice for Grounding of Industrial and Commercial Power Systems.*

so water system grounds are less common. Gas piping systems are not allowed as grounding electrodes.

The NESC allows the concentric neutral of a cable to be a ground for direct-buried cables of at least 100 ft (30 m). The cable neutral must be bare or have a semi-conducting jacket with a radial resistivity not exceeding 10 Ω-m (meaning that 30 m of cable has a resistance of 0.33 Ω across the jacket).

The NESC allows other "made" electrodes other than ground rods including buried wires, strips, and plates. These are especially appropriate in areas with a rock bottom. Buried wires must have at least a 0.162 in. (4 mm) diameter, be at least 100 ft (30 m) long, and be at least 18 in. (0.45 m) deep (less is allowed if a rock bottom is hit). Strips must be at least 10 ft (3 m) long, buried at least 18 in. (0.45 m) deep, and have a total surface area on both sides of 5 ft^2 (0.47 m^2). Plates must be buried at least 5 ft (1.5 m) deep and have a total surface area of at least 2 ft^2 (0.185 m^2). Both plates and strips must have a thickness of at least 0.06 in. (1.5 mm) wide for nonferrous metals and at least 0.25 in. (6 mm) wide for ferrous metals. Table 13.3 shows resistance calculations for several different configurations.

Two-pole grounding techniques, butt plates and wire wraps (see Figure 13.14) are available but have several deficiencies (Clapp, 1997). A driven ground provides more than twice the surface area of a butt plate. Additionally, pole movement can reduce the contact between the butt plate or wire wrap and the earth, and the pole can draw moisture out of the soil enough to increase its resistivity. The NESC only considers butt plates and wire wraps as appropriate in areas with low ground resistivity (less than 30 Ω-m), and even then, two such grounds must be used, and they must not be used as the sole ground at transformer locations.

13.4.1 Soil Characteristics

Soil resistivity is the resistance of a certain volume of soil. Normally, resistivity is specified in Ω-m, the resistance between opposite faces of cube of

TABLE 13.3

Ground Electrode Resistance Calculations

Hemisphere radius a	$R = \frac{\rho}{2\pi a}$
One ground rod length L, radius a	$R = \frac{\rho}{2\pi L}\left(\ln\frac{4L}{a} - 1\right)$
Two ground rods $s > L$; spacing s	$R = \frac{\rho}{4\pi L}\left(\ln\frac{4L}{a} \quad 1 + \frac{\rho}{4\pi s} \quad 1 - \frac{L^2}{3s^2} + \frac{2L^4}{5s^4} \cdots\right)$
Two ground rods $s < L$; spacing s	$R = \frac{\rho}{4\pi L}\left(\ln\frac{4L}{a} + \ln\frac{4L}{s} - 2 + \frac{s}{2L} - \frac{s^2}{16L^2} + \frac{s^4}{512L^4} \cdots\right)$
Buried horizontal wire length $2L$, depth $s/2$	$R = \frac{\rho}{4\pi L}\left(\ln\frac{4L}{a} + \ln\frac{4L}{s} - 2 + \frac{s}{2L} - \frac{s^2}{16L^2} + \frac{s^4}{512L^4} \cdots\right)$
Right-angle turn of wire length of arm L, depth $s/2$	$R = \frac{\rho}{4\pi L}\left(\ln\frac{2L}{a} + \ln\frac{2L}{s} - 0.2373 + 0.2146\frac{s}{L} + 0.1035\frac{s^2}{L^2} - 0.0424\frac{s^4}{L^4} \cdots\right)$
Three-point star length of arm L, depth $s/2$	$R = \frac{\rho}{6\pi L}\left(\ln\frac{2L}{a} + \ln\frac{2L}{s} + 1.071 - 0.209\frac{s}{L} + 0.238\frac{s^2}{L^2} - 0.054\frac{s^4}{L^4} \cdots\right)$
Four-point star length of arm L, depth $s/2$	$R = \frac{\rho}{8\pi L}\left(\ln\frac{2L}{a} + \ln\frac{2L}{s} + 2.912 - 1.071\frac{s}{L} + 0.645\frac{s^2}{L^2} - 0.145\frac{s^4}{L^4} \cdots\right)$
Six-point star length of arm L, depth $s/2$	$R = \frac{\rho}{12\pi L}\left(\ln\frac{2L}{a} + \ln\frac{2L}{s} + 6.851 - 3.128\frac{s}{L} + 1.758\frac{s^2}{L^2} - 0.490\frac{s^4}{L^4} \cdots\right)$
Eight-point star length of arm L, depth $s/2$	$R = \frac{\rho}{16\pi L}\left(\ln\frac{2L}{a} + \ln\frac{2L}{s} + 10.98 - 5.51\frac{s}{L} + 3.26\frac{s^2}{L^2} - 1.17\frac{s^4}{L^4} \cdots\right)$
Ring of wire, depth $s/2$ diameter of ring D, diameter of wire d	$R = \frac{\rho}{2\pi^2 D}\left(\ln\frac{8D}{d} + \ln\frac{4D}{s}\right)$
Buried horizontal strip length $2L$, section a by b, depth $s/2$, $b < a/8$	$R = \frac{\rho}{4\pi L}\left(\ln\frac{4L}{a} + \frac{a^2 - \pi ab}{2(a+b)^2} + \ln\frac{4L}{s} - 1 + \frac{s}{2L} - \frac{s^2}{16L^2} + \frac{s^4}{512L^4} \cdots\right)$
Buried horizontal round plate, radius a, depth $s/2$	$R = \frac{\rho}{8a} + \frac{\rho}{4\pi s}\left(1 - \frac{7}{12}\frac{a^2}{s^2} + \frac{33}{40}\frac{a^4}{s^4} + \cdots\right)$
Buried vertical round plate, radius a, depth $s/2$	$R = \frac{\rho}{8a} + \frac{\rho}{4\pi s}\left(1 + \frac{7}{24}\frac{a^2}{s^2} + \frac{99}{320}\frac{a^4}{s^4} + \cdots\right)$

Sources: Dwight, H. B., "Calculation of Resistance to Ground," *AIEE Transactions*, vol. 55, pp. 1319–28, December 1936; IEEE Std. 142-1991.

soil with a volume of 1 m^3. Units of Ω-cm are also common — the resistance between faces of a cube with a volume of a cubic centimeter: Ω-cm = 100 × Ω-m.

Soil resistivities vary widely. Rich, moist organic soil may have a resistivity of 10 Ω-m, while bedrock may have resistivities greater than 10^4 Ω-m. Resistivity depends on the soil grain characteristics (size, variability, and density) as well as moisture and chemical content. See Table 13.4 for typical values.

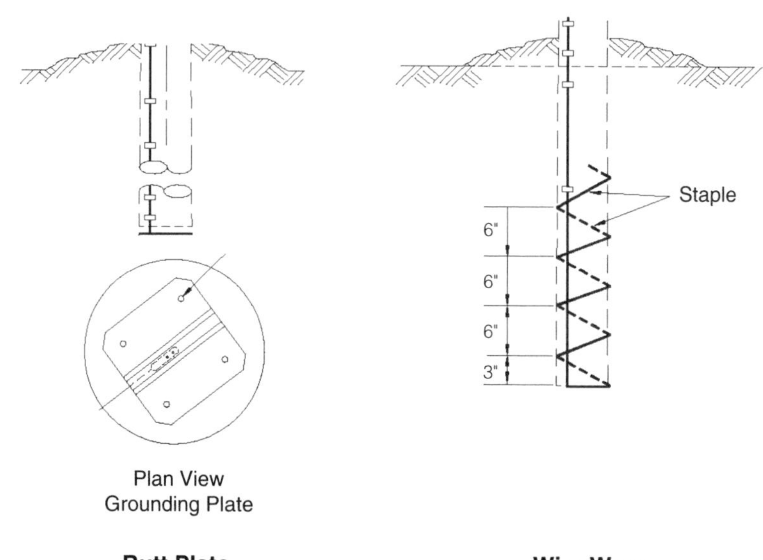

FIGURE 13.14
Butt plate and wire wrap grounds. (From RUS 1728F-803, *Specifications and Drawings for 24.9/14.4 kV Line Construction*, United States Department of Agriculture, Rural Utilities Service, 1998.)

TABLE 13.4
Resistivity of a Variety of Soils along with Resistances of a 10-Foot Ground Rod

Soil	Average Resistivity, Ω-m	Resistance of a 5/8 in. (16 mm) × 10 ft (3 m) Rod, Ω
Well graded gravel, gravel-sand mixtures, little or no fines	600–1000	180–300
Poorly graded gravels, gravel-sand mixtures, little or no fines	1000–2500	300–750
Clayey gravel, poorly graded gravel, sand-clay mixtures	200–400	60–120
Silty sands, poorly graded sand-silts mixtures	100–500	30–150
Clayey sands, poorly graded sand-clay mixtures	50–200	15–60
Silty or clayey fine sands with slight plasticity	30–80	9–24
Fine sandy or silty soils, elastic silts	80–300	24–90
Gravelly clays, sandy clays, silty clays, lean clays	25–60[a]	17–18[a]
Inorganic Clays of high plasticity	10–55[a]	3–16[a]

[a] Soils highly influenced by moisture.

Source: IEEE Std. 142-1991.

Soil resistivity often varies significantly with depth. In most areas, resistivity improves significantly with depth as moisture increases, and normally, grounds driven down to the water table provide a very good ground. In other areas, an underlying bedrock significantly increases resistivity at lower levels (and makes driving ground rods difficult). Several different layers of soil with different resistivities may be present. A two-layer soil model is often used for engineering analysis.

Soil resistivity is primarily affected by

- *Moisture* — Moisture is one of the main factors determining soil resistivity. The dryer the soil, the higher the resistivity. Changes in moisture level throughout the year are the biggest reason for the change in ground electrode resistance. Once the moisture exceeds 22%, resistivity changes very little (IEEE Std. 142-1991).
- *Temperature* — Above the freezing point of water, temperature does not impact resistivity significantly. Below freezing, resistivity rises significantly.
- *Salts* — The presence of soluble salts significantly impacts the resistivity.

Figure 13.15 shows these three main effects for different example soils. These effects are somewhat interrelated and depend on the soil's specific characteristics.

One option to reduce ground electrode resistivity is to use chemical treatments or backfills to reduce the soil resistivity near the electrode. Bentonite backfills or salt treatments such as sodium chloride or calcium chloride significantly reduce the soil resistivity. If soil treatments are used to enhance ground rods, make sure to consider the effect on corrosion. Some chemical treatments increase the metal corrosion rate. Also, consider the environmental impacts and regulations. Most normal distribution grounding needs outside of the substation do not warrant chemical enhancement.

If a grounding electrode sustains significant current, the current may dry out the surrounding soil and increase the resistance of the electrode. With a multigrounded system, several grounds share current, but on systems with isolated grounds, sustained current through a grounding electrode may degrade the electrode's resistance (another good reason to use multiple grounding electrodes in parallel whenever possible).

The resistivity of soil is normally very linear, but for very high voltage stresses, the soil breaks down; and arcing in the soil reduces the effective resistivity. For a ground rod, the soil breaks down only when high current flows into the rod, normally from lightning (a localized effect) or from faults with very little sharing between ground rods. The breakdowns improve the resistance of a grounding electrode.

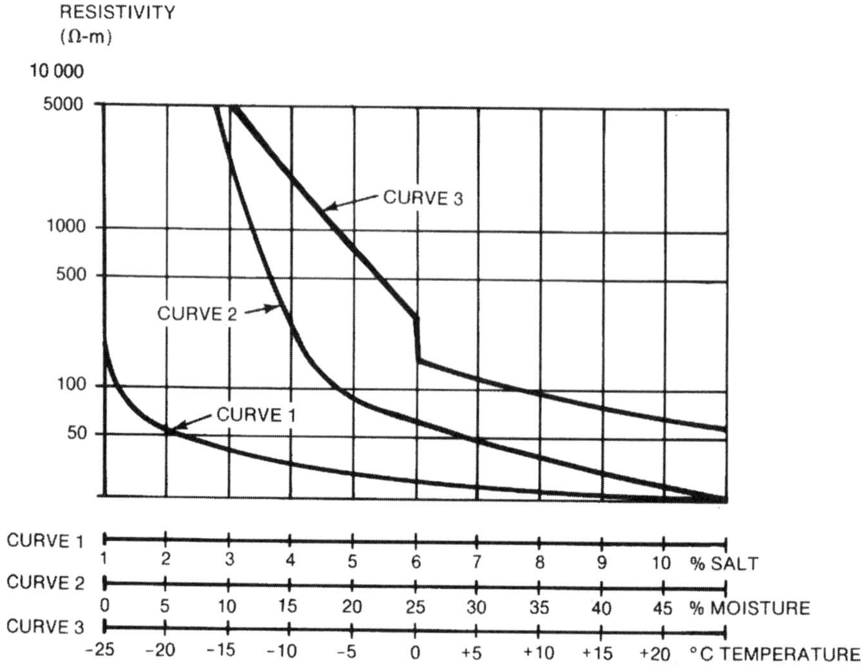

FIGURE 13.15
Effect of salt, moisture, and temperature on soil resistivity. (From IEEE Std. 80-2000. Copyright 2000 IEEE. All rights reserved.)

13.4.2 Corrosion and Grounding Electrodes

Corrosion is a major consideration for grounding electrodes. Ground rods are normally copper-clad steel or galvanized steel. Most grounding wires are copper. Steel and aluminum corrode too quickly for grounding use. One problem with copper is its attractiveness to thieves. In problem areas, copper-clad steel has reduced thefts since it has less commercial value.

Mixing metals significantly increases corrosion. The problem is galvanic corrosion. Different metals create an electrochemical battery in the soil. A voltage develops between grounding electrodes made of different metals (or between a grounding electrode and another metallic object in the ground). When these electrodes are connected, current flows in the earth between the metal objects, corroding one of the metals at a much higher-than-normal rate.

Galvanic corrosion between dissimilar metals can eat up grounding electrodes. Galvanic corrosion operates on the same principal as a battery (batteries release their energy with controlled corrosion). Pure metals are unstable; they are found in nature as ores and must be refined to a pure state. Metals corrode to return to their natural state. Iron dissolves in a corrosive electrolyte and gives up electrons, forming a positive ion. Some of the iron ions attach to free oxygen and become iron oxide (rust). At low corrosion rates, the metal reaches equilibrium with the surrounding electrolyte that prevents additional iron from releasing electrons. Two different metals separated by an electrolyte (a conducting medium) have a voltage difference between them as they give up electrons at different rates. If these different metals are electrically connected together, current flows between the metals and increases corrosion.

Consider a galvanized-steel ground rod next to a copper ground rod. The galvanized rod is coated with zinc. The zinc releases more electrons into the electrolyte, so it becomes positively charged relative to the copper rod. At the zinc surface (the positive terminal, the anode), the flow of this dc current dissolves the zinc at an increasing rate. At the copper rod (the negative terminal), the galvanic circuit decreases the rate of corrosion. Figure 13.16 shows the galvanic series for several different metals. When dissimilar metals form a galvanic circuit, the most anodic of the metal corrodes (corrosion slows down, stops, or even reverses on the cathodic metal).

In fact, most corrosion occurs by the same action. An anode and a cathode develop between different portions of a metal, and galvanic currents flow

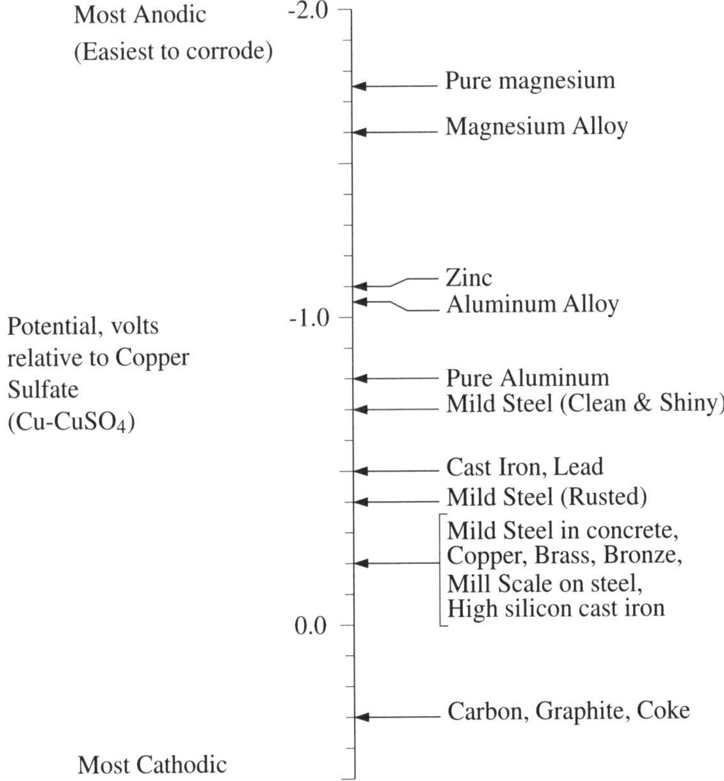

FIGURE 13.16
Relative potentials of various metals in an electrolyte.

between them, corroding the anode. One part of a metal may become anodic to another in several ways based on differences in impurities in the metal, mechanical stresses, and oxygen concentration.

Copper has naturally good properties that limit corrosion. When copper corrodes, it becomes copper oxide; the copper-oxide layer greatly reduces the corrosion rate and protects the underlying copper. Copper is normally the most cathodic element, so it rarely corrodes because of dissimilar metals, but it may help corrode other grounding electrodes.

A zinc coating on a steel ground rod (which is a galvanized steel rod) provides corrosion protection for the rod. Zinc by itself is resistant to corrosion because it develops a self-protecting zinc-carbonate film that retards further corrosion. In addition, the zinc provides galvanic protection to the steel core. If part of the rod has exposed steel (due to cuts, scratches, or corrosion of the zinc), the zinc acts as a very large anode to the small area of exposed steel. Galvanic action reduces the corrosion on the exposed steel (the cathode).

With lower soil resistivity, galvanic currents are higher, so grounding electrodes corrode faster. Resistivity is the main soil parameter that determines corrosion. Other factors that increase corrosion are very high or very low pH levels, high water content, and the presence of chlorides (salts) and sulfates.

To limit excessive corrosion caused by different metals, avoid using different grounding rods (either stick with galvanized steel or with copper-clad steel). Provide as much separation as possible between ground electrodes of different metals. Consider the presence of guy anchors and other possible buried wires including bare concentric neutral cables and water pipes. With galvanized steel rods, use steel or aluminum pole ground leads. With copper-clad steel rods, use copper down leads.

13.4.3 Resistance Measurements

The NESC does not specify how a ground measurement should be taken. IEEE Std. 81-1983 documents several of the methods of measuring grounds. Normally, on systems with a multigrounded neutral, the actual impedance is not critical, so neither is the measurement. (Just make sure ground rods are connected.) On systems without a multigrounded neutral, individual grounds (and ground resistance checks) are more important.

Accurate ground electrode measurements of a concentrated electrode (ground rods, anchors, or short strips of wire) are normally performed using a three-point method (AEMC Instruments, 1998; Biddle Instruments, 1982; IEEE Std. 81-1983). Two auxiliary ground probes are driven into the earth forming a straight line with the test electrode. The tester injects current I through the electrode under test and the outer auxiliary electrode. The tester measures voltage V between the test electrode and the inner auxiliary electrode. If the middle electrode is out of the influence of the outer two electrodes, then the resistance of the test electrode is V/I.

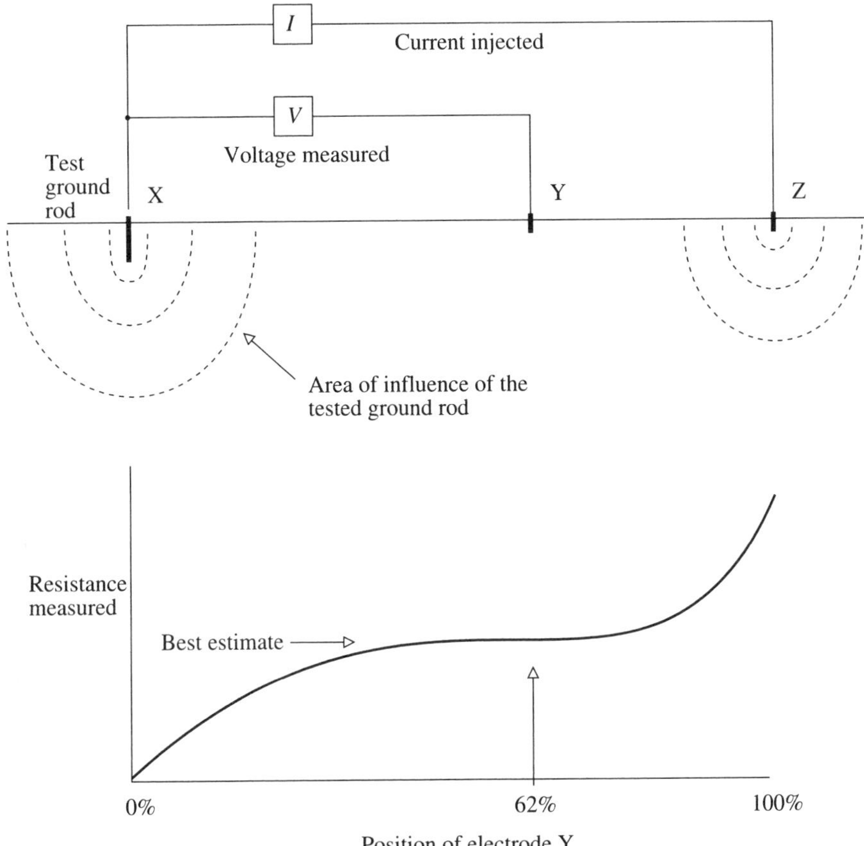

FIGURE 13.17
Three-point ground-electrode measurement.

The *fall of potential* method is the most common three-point technique (see Figure 13.17). It requires several measurements. Take several measurements with the middle electrode at different locations between the test electrode and the outer electrode. Plot the resistance vs. the distance, and pick the resistance from the portion of the curve where resistance flattens out. The point where the fall-of-potential curve flattens out is most likely near 62% of the distance from the test electrode to the outer electrode. A simplified version is just to measure near the 62% mark.

The three-point methods must be done with the ground electrode disconnected from the multigrounded neutral and connections to other grounds that may interfere (including cable TV and telephone grounding connections). When disconnecting a grounding conductor from the multigrounded neutral, crews should know that a significant potential can develop between the neutral and the disconnected ground if a line-to-ground fault occurs somewhere on the line.

To have adequate spacing with the thee-point method, normally the outer electrodes must be spaced at least 75 to 100 ft apart (23 to 30 m). Deeper grounds or wider grounds (strips or multiple ground rods) require more distance between the electrodes.

If an excellent ground is available nearby as a reference, the two-point measurement technique is suitable. Just like an ohmmeter, the tester injects current through the grounding electrode under test and the reference ground. The resistance reading is the sum of both grounds, so if the reference ground is very good (like a water piping system), the reading is predominantly the resistance of the test electrode. A multigrounded neutral can be used as the ground reference but only if the frequency of the injected signal is different from the standard power frequency (normal power-frequency current flows can foul the ground reading). As with the three-point method, the ground electrode must be disconnected from the multigrounded neutral.

A convenient ground electrode tester is a clamp-on device that tests a grounding electrode by clamping around the grounding lead. The clamp-on unit only works on electrodes connected to a very good ground like the multigrounded neutral (it relies on the good ground). It works just like the two-point test. The clamp-on is a current transformer that couples a voltage to the grounding lead; the voltage is in the lower kilohertz range. The resistance of the grounding electrode is the voltage divided by the current (if all of the parallel grounds provide a low-impedance path, the resulting impedance is only the grounding electrode).

The clamp-on ground-resistance tester more than adequately tests ground-rod connections on systems with a multigrounded neutral. It is fast and convenient, crews do not have to disconnect the ground rod, and it tests the connector continuity. Since values of individual grounds are not as critical on multigrounded-neutral systems, the clamp-on's accuracy is good enough for most grounding checks on distribution circuits.

Ground resistivity is normally measured using the four-point method. Four electrodes of the same size are driven to the same depth in a straight line with equal spacing between them. A tester injects current through the outer two electrodes and measures the voltage between the inner two electrodes. The distance between two electrodes determines the depth at which the average resistivity is measured. If the electrode spacing is at least 20 times the electrode depth, the resistivity is

$$\rho = 2\,\pi\,A\,R$$

where

ρ = earth resistivity, Ω-m
A = electrode spacing, m
R = resistance reading from the ground tester, Ω

Since the rod spacing determines the depth at which the resistivity is measured, testing with multiple electrode separations shows how the soil varies with depth. But keep in mind that other buried equipment, pipes or cables, may give false readings.

13.5 Shocks and Stray Voltages

Contact with high voltage utility circuits poses dangers to utility workers and the public. Electricity is dangerous. Just a small amount of current through a person's body can kill. Backhoes, kites, downed wires, trees into lines — direct and indirect contacts to utility circuits may occur in many ways. When equipment fails or external factors cause line-to-ground faults, current flow into the ground can pose step and touch potential hazards to the public and to utility workers. Many electric dangers are invisible. When a crane boom accidentally contacts an overhead primary wire, there may be no visible or audible indications from the ground to warn someone, to prevent someone from grabbing the crane's door handle and being electrocuted.

Even under normal operating conditions, the multigrounded neutral may be imperfectly grounded and may have a voltage above that of true ground. Under certain conditions, this voltage potential may shock people when they touch metallic equipment that is connected to the neutral. These same "stray voltages" also cause concern on dairy farms and other agricultural facilities.

Given all of the possible hazards in dealing with high voltage and given the great lengths of distribution line in operation, utilities do a very good job of safely operating their lines and minimizing accidents. Several components help maintain a safe system — designs with good clearances and ample mechanical strength, public education to avoid things like kites into overhead lines, proper overcurrent protection schemes, effective grounding, and rapid response and repair.

13.5.1 Biological Models

The body's resistance is an important part of the electrical circuit in possible shocking situations. The body's internal resistance is about 300 Ω. Resistance including the skin contact resistance is in the range of 500 to 5000 Ω [(IEEE Std. 80-2000) and many other publications] but varies widely as shown in Table 13.5. It is common to use a resistance of 1000 Ω and assume that the hand and foot contact resistances are zero (as in IEEE Std. 80-2000). The 1000 Ω is used for the path from one hand to the other, from hand to foot, and from foot to foot (all are roughly the same impedance). Much of this impedance is concentrated at the extremities, the forearm, and the ankle and shin. For more information on biological impacts and models, see Reilly (1998) and the IEEE Working Group on Electrostatic and Electromagnetic Effects et al. (1978).

TABLE 13.5

Human Resistance Values for Several Contact Conditions

	Resistance Range, Ω	
	Dry	Wet
Finger touch	40,000–1,000,000	4000–15,000
Hand holding wire	15,000–50,000	3000–6000
Finger-thumb grasp (interpolated)	10,000–30,000	2000–5000
Hand holding pliers	5000–10,000	1000–3000
Palm touch	3000–8000	1000–2000
Hand around 3.8 cm (1.5 in.) pipe (or drill handle)	1000–3000	500–1500
Two hands around 3.8 cm (1.5 in.) pipe	500–1500	250–750
Hand immersed in water		200–500
Foot immersed in water		100–300
Human body, internal, excluding skin	200–1000	

Source: Lee, R. H., "Electrical Safety in Industrial Plants," *IEEE Spectrum*, June 1971. With permission. ©1971 IEEE. (Also in IEEE Std. 902-1998, *IEEE Guide for Maintenance, Operation, and Safety of Industrial and Commercial Power Systems*.)

Electric current does the damage. Table 13.6 shows electrical effects on people for different ranges of currents. Currents as low as 0.1 mA are perceptible to some, and currents above 1 mA are perceptible to most. Above the "let-go" current, at about 5 to 25 mA, the hand and arm muscles involuntarily clench, and a person may be unable to release an electrified object. These current levels may be quite painful but do not cause permanent damage. A person may have severe breathing difficulties for currents above 30 mA, which in some cases may cause death.

Electrocution can occur for currents through the heart on the order of 60 to 100 mA; 250 mA through the heart is almost always fatal (without life-saving emergency care). These levels of current interfere with the heart's internal electrical pacemaker and force the heart to stop beating properly. During this ventricular fibrillation, the heart muscle fibers contract erratically and unpredictably, leading quickly to death from lack of oxygen to the brain. Above 5 A, current can burn tissue.

The destructive effect of current is a function of the energy into the body; assuming a constant resistance, the energy input is a function of I^2t. One commonly cited model, Dalziel's formula, describes the relationship between fibrillation and the duration of current (Dalziel, 1946; Dalziel, 1972) as

$$I = \frac{116}{\sqrt{t}}$$

where

I = minimum body current in mA necessary to cause ventricular fibrillation for at least 0.5% of the population for a 110-lb (50-kg) adult

t = duration of the current, sec

TABLE 13.6

Threshold Levels for 60-Hz Contact Currents

rms Current mA	Threshold Reaction/Sensation
Perception	
0.09	Touch perception for 1% of women
0.13	Touch perception for 1% of men
0.24	Touch perception for 50% of women
0.33	Grip perception for 1% of women
0.36	Touch perception for 50% of men
0.49	Grip perception for 1% of men
0.73	Grip perception for 50% of women
1.10	Grip perception for 50% of men
Startle	
2.2	Estimated borderline hazardous reaction, 50% probability for women (arm contact)
3.2	Estimated borderline hazardous reaction, 50% probability for women (pinched contacts)
Let-Go	
4.5	Estimated let-go for 0.5% of children
6.0	Let-go for 0.5% of women
9.0	Let-go for 0.5% of men
10.5	Let-go for 50% of women
16.0	Let-go for 50% of men
Respitory Tetanus	
15	Breathing difficulty for 50% of women
23	Breathing difficulty for 50% of men
Fibrillation	
35	Estimated 3-s fibrillating current for 0.5% of 44-lb (20-kg) children
100	Estimated 3-s fibrillating current for 0.5% of 150-lb (70-kg) adults

Source: IEEE Working Group on Electrostatic and Electromagnetic Effects, Delaplace, L. R., and Reilly, J. P., "Electric and Magnetic Field Coupling from High Voltage AC Power Transmission Lines — Classification of Short Term Effects on People," *IEEE Transactions on Power Apparatus and Systems*, vol. PAS-97, no. 6, pp. 2243–52, November/December 1978. ©1978 IEEE.

Larger people are able to withstand higher currents. A more general formulation of Dalziel's formula includes body weight as

$$I = \frac{\sqrt{S}}{\sqrt{t}}$$

where
 S = body weight, kg

Dalziel's formula applies for the time ranges of faults, from about one-half cycle to 3 sec (it may not apply for longer- or shorter-duration currents).

Electricity can kill at astonishingly low energies. A 100-mA current through a body resistance of 500 Ω is only 50 W of power.

If an electrical shock stops someone's heart, cardiopulmonary resuscitation (CPR) can supply some oxygen to the brain until emergency care arrives. In many cases, resuscitation after electrical shocks is possible with a defibrillator. The electric current causing ventricular fibrillation shuts down the heart's electrical system but may not significantly damage organs or nerves. Note that a victim of fibrillation may not immediately become unconscious and can deny needing help (death may be quick or take hours). Always call emergency help immediately, and apply CPR if appropriate (but be careful not to apply CPR to someone whose heart is not in fibrillation).

Currents above 5 A can permanently burn tissue. Fast fault clearing helps reduce the amount of tissue damage and increases the likelihood of successful emergency care.

While current does the damage, voltage plays an important role. Normally, 120-V secondary voltages are not as dangerous as higher voltages (but 120 V kills many people every year). In many shocking situations, 120 V may produce painful shocks but not kill (the circuit resistance including contacts is often high enough to limit the current). But 120 V can kill: remember bathtubs. Current in the path depends primarily upon what resistance is in the current's path — wet vs. dry skin, shoe resistance, glove resistance, and other impedances in the circuit path. To reach 100 mA, the total resistance must be less than 1200 Ω, which can happen, especially if contacted when wet. Higher secondary voltages such as 480 V are more likely to kill. Above 600 V, the skin offers no extra resistance; the voltage punctures right through, leaving only the internal body resistance or about 300 Ω. Above 2400 V, the voltage drives currents high enough for burning and tissue damage.

13.5.2 Step and Touch Potentials

Step and touch potentials are very important in substations because during ground faults, all of the ground current returns to the substation transformer. The current that returns through the earth can create a significant voltage gradient along the ground and between the ground and conducting objects. A touch potential is normally considered a hand-to-foot or a hand-to-hand contact; a step potential creates a path through the legs from one foot to the other (see Figure 13.18).

The approximate impedance of one foot-to-ground contact is

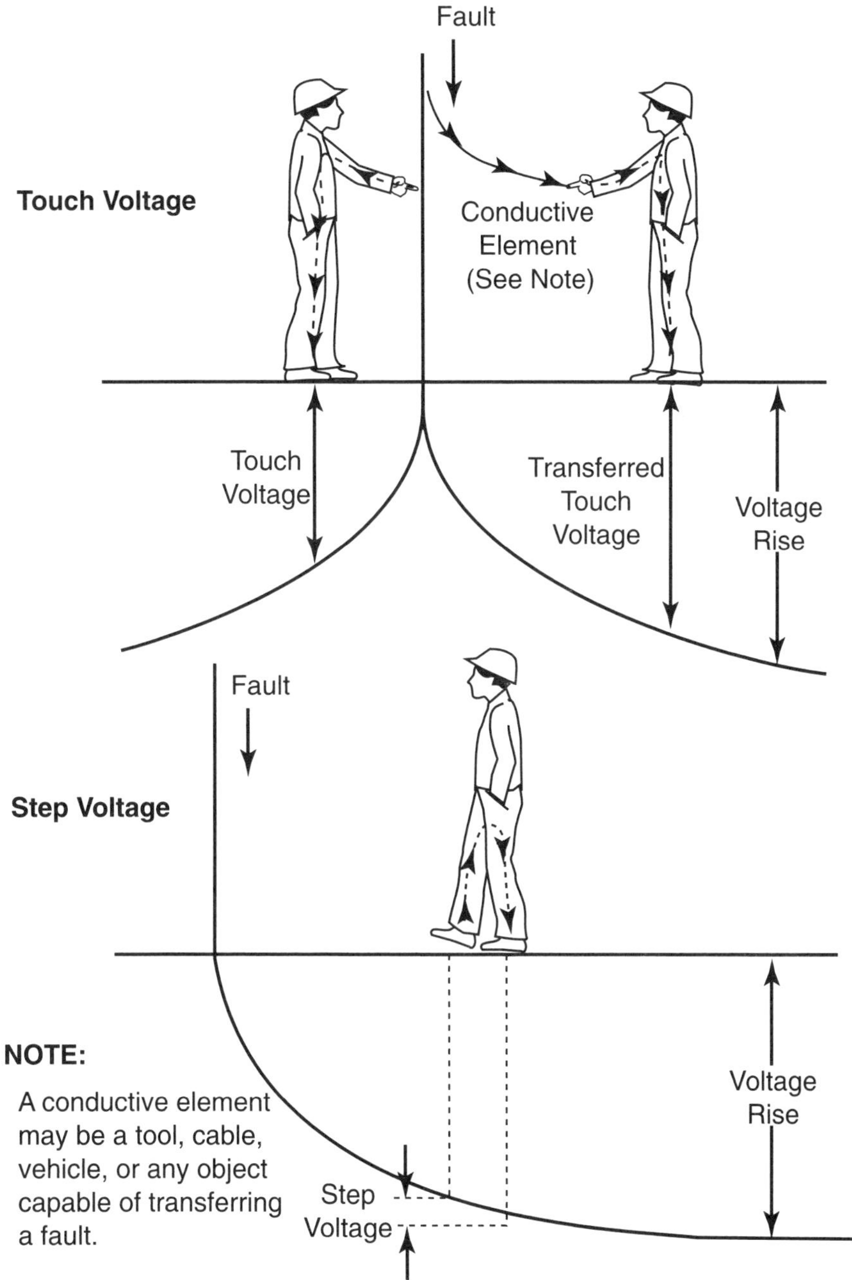

FIGURE 13.18
Paths for touch and step potentials. (From WAPA, *Power System Maintenance Manual, Chapter 1: Personal Protective and Vehicle Grounding*, Western Area Power Administration, U.S. Department of Energy, 1997.)

$$R = 3\rho$$

where

ρ = earth resistivity, Ω-m

For a step potential, two of the foot-contact resistances are in series with the body resistance R_B, so the body current I_B is

$$I_B = \frac{V_{step}}{R_B + 6\rho}$$

With a hand-and-feet touch potential, the circuit impedance is two foot-contact resistances in parallel plus the body resistance. For a touch potential, the body current is

$$I_B = \frac{V_{touch}}{R_B + 1.5\rho}$$

Step and touch potentials are of concern during normal conditions and during a ground fault. Under normal conditions, unbalanced current can raise the neutral-to-earth voltage. This is not normally dangerous, but it can cause annoying shocks and concerns on farms (see the next section on stray voltage). Step and touch potentials during faults are more dangerous. This is an important design consideration in substation grounding. Unfortunately, outside the substation we cannot control many situations nearly as well. The multigrounded neutral helps reduce the chance of dangerous step-and-touch potentials during line-to-ground faults. First, by creating a low-impedance path back to the source, faults are cleared quickly by fault interrupters. And, having multiple grounding electrodes tied together helps reduce touch potentials at the fault point. Away from the substation on systems with multigrounded neutrals, step potentials are usually not dangerous since fault current spreads between several grounding electrodes.

Simulations by Rajotte et. al. (1990) help define the touch potentials between the multigrounded neutral and the earth. Figure 13.19 shows neutral-to-earth voltages on overhead and underground systems for different fault locations and earth resistivities. Underground circuits have touch voltages an order of magnitude less than overhead circuits because less current flows in the earth due to the cable neutral being the preferred return path (the reason being the tight coupling with the phase conductor).

The highest ground potential rise is normally at the fault location, but Rajotte simulated cases where an underground section followed an overhead section: the maximum potential rise was sometimes at the pothead rather than the fault location.

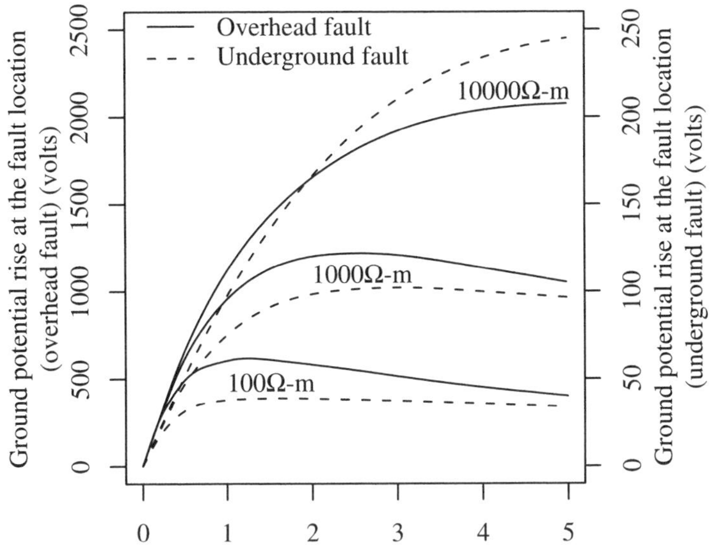

FIGURE 13.19
Neutral-to-earth voltages during line-to-ground faults on 25-kV overhead and underground circuits. (Adapted from Rajotte, Y., Bergeron, R., Chalifoux, A., and Gervais, Y., "Touch Voltages on Underground Distribution Systems During Fault Conditions," *IEEE Trans. Power Delivery*, 5(2), 1026-1033, April 1990. ©1990 IEEE. With permission.)

Using a reactor on the substation transformer neutral helps limit step-and-touch potentials. While utilities normally use these neutral reactors to limit fault currents, the reduction of the ground fault current also reduces step-and-touch potentials and reduces current in grounding and bonding connectors.

13.5.3 Stray Voltage

Stray voltage is a touch voltage that end-users experience. Stray voltage causes complaints from residential customers, but most concern has centered around dairy farms. On farms, milking cows are subjected to a voltage between the milking equipment and the earth they are standing on. Cows and people have somewhat similar discomfort responses. Currents above 1 mA may cause discomfort for cows. The Wisconsin Public Service Commission defines a "level of concern" of 1 V for the cow contact voltage, which corresponds to 2 mA of current through a 500 Ω resistor (Reines et al., 1995). Figure 13.20 shows a U.S. Department of Agriculture guideline on the effects of stray voltage on milk production.

The amount of current through an animal's body depends on the neutral-to-earth voltage, the animal's resistance, and the characteristics of the soil (especially its moisture) and/or concrete.

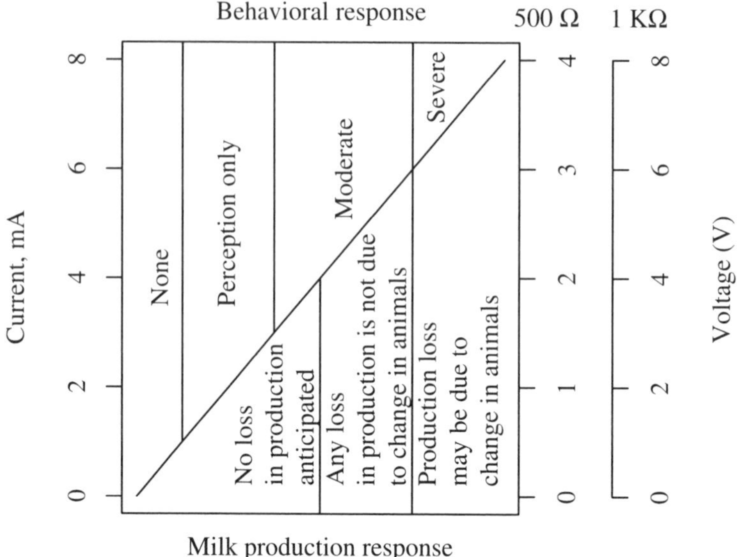

FIGURE 13.20
Milk production and behavioral responses to stray voltage based on current. (From [USDA, 1991].)

Stray voltage at barns is normally measured across a 500-Ω resistor, which represents the typical resistance of a cow.

Stray voltage has stirred significant public attention. The "danger" of stray voltage has been exaggerated in some articles in the press. Studies have found mixed results when trying to quantify the effect of stray voltage on milk production, reduced food or water intake, somatic cell count, and other important physiological effects. The Wisconsin Public Service Commission database of stray voltage field surveys found little correlation between milk production and electrical parameters such as cow contact current (Reines et al., 1995; Reines et al., 1998). A report of scientific advisors to the Minnesota Public Utilities Commission found some correlation between stray voltage and high-health herds and low-health herds, but other factors were more significant (Minnesota Public Utilities Commission, 1998). The CEA (1992) found no significant effect on cows' behavior or production for stray voltages of 1 or 2.5 V. Some change in water and feed consumption were noted for 5-V stray voltages.

Many stray voltage problems are due to improper facility grounding and bonding. The National Electrical Code (NEC) (NFPA 70, 2002) covers the issues that could cause problems. Many stray voltage problems originate on the secondary system. Improper bonding or other violations of the NEC can create excessive voltages between equipment frames and the earth. Another common problem is failed equipment insulation — a failure to a bonding frame (in a motor, for example) can create stray voltages but not draw enough current to blow a fuse or trip a breaker. So, identifying if the problems originate on the secondary is the first step in any stray voltage problem. The easiest way to determine if secondary problems are the culprit is to trip the main breaker or pull the meter. If the stray voltage exists after the secondary loads have been removed, then local problems are ruled out; the problems are likely due to a primary neutral-to-earth voltage or a secondary voltage due to other customers on the same secondary. Secondary-side fixes include using a larger neutral, fixing poor neutral connections, balancing 120-V load, and fixing NEC grounding or bonding violations.

Regardless of the origin of the stray voltage, localized bonding can eliminate touch potentials. In a barn, we can bury wires or sheets of metal just under the earth that a cow stands on during milking. Bonding the buried wires or sheets to the milking equipment creates an equipotential plane that eliminates shocks during milking. Another local solution option is an active voltage suppression device that reduces shocks by actively countering the stray voltage.

EPRI (1999) investigated several residential stray voltage problems, many related to swimming pools. Pools are often involved in shock incidents because wet skin has lower resistance and is more susceptible to shock. Many pool-related shocking sensations were found in cases where handrails, ladders or decks were not bonded. The NEC requires all metallic parts of a pool to be bonded together including ladders, concrete decks, and diving boards as well as lights and pumps. Voltage differences between bonded and unbonded metal or conducting elements cause shocking situations. The bonded parts of the pool are at the neutral potential (which may be at an elevated potential relative to the earth). Unbonded parts float to a different voltage. Decks in contact with the earth that are not bonded will be at earth potential. Bonding helps eliminate annoying shocks, but more importantly, it reduces the chance of electrocution if the unbonded part happens to become energized at 120 V. Hot tubs and outdoor water faucets are also involved in shocking incidents (faucets are grounded, the soil beneath is at earth potential).

Neutral-to-earth voltages cannot be analyzed with existing load-flow programs. To accurately model stray voltage, a more general steady-state circuit analysis program such as EMTP is needed. Accurate modeling involves modeling the neutral and the individual grounding electrodes.

Neutral-current harmonics can occasionally contribute to stray voltages, primarily from the third harmonic which adds in the neutral [see Tran et al. (1996) for a more complete analysis]. Leap et al. (2002) showed an example where the rms stray voltage at one location on Portland General Electric's system increased from 0.4 V to 8.3 V after adding a 1200-kvar capacitor bank. The capacitor resonated with the system impedance and drew third harmonic current into the neutral, enough to significantly raise the neutral voltage relative to remote earth. Reducing the size of the capacitor to 300 kvar reduced the stray voltage acceptably.

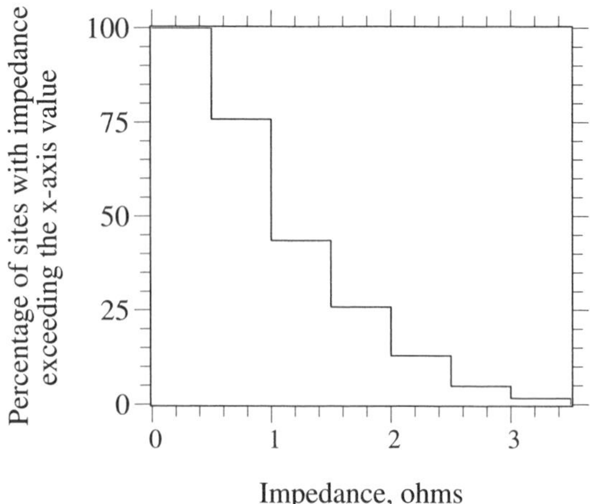

FIGURE 13.21
Distribution of measured impedances of the multigrounded neutral on rural circuits. (Data from [Rajotte et al., 1997].)

Figure 13.21 shows distributions of the neutral impedance measured at several locations on 16 rural feeders (Rajotte et al., 1997). Even though the ground resistances of these circuits varied over orders of magnitude, the neutral impedance stayed consistent and fairly low. This highlights one of the main advantages of the multigrounded neutral — many grounds overcome poor ground resistivity, providing an effective return path regardless of the performance of individual grounds. Even with such low impedances, neutral-to-earth voltages can reach several volts given enough current unbalance.

High neutral-to-earth voltages are more likely near the substation and far from the substation. Utility-side solutions to stray voltage include

- *Balancing* — Balancing load among the phases reduces the earth-return current (and thus the neutral-to-earth voltage).
- *Single-phase lines* — Converting single-phase sections to two or three phase reduces the neutral-to-earth voltage.
- *Neutral* — Upgrading the neutral size provides a lower-impedance path for earth-return current.
- *Capacitor banks* — Check capacitor banks for blown fuses. If a grounded-wye capacitor has a blown fuse, the unbalance can create excessive neutral current.

Another utility option is to use a neutral isolation device. Isolating the secondary neutral keeps utility neutral-to-earth voltages away from customer equipment. Keeping the primary neutral separate from the secondary is difficult. Make sure phone, cable television, and metal water pipes are all disconnected; this may require coordination with other utilities. The NESC allows using a spark gap as a neutral separation device: the 60-Hz breakdown voltage of the gap must not exceed 3 kV. Other neutral separation devices are possible although they may not be approved by the NESC (the NESC allows devices that "perform an equivalent function" as a spark gap). A saturating reactor has been proposed as an interconnecting device. Under normal conditions, the reactor is a high impedance (unsaturated), but under fault conditions, the reactor saturates and becomes a much lower impedance (but a saturating reactor may allow narrow voltage spikes through). The reactor has a surge arrester in parallel to conduct lightning-type surges. Another interconnecting device uses a solid-state switch — when a controller detects high voltage, it fires thyristors in parallel to short out the neutral connection. Use neutral isolation devices sparingly. Widespread usage reduces the effectiveness of the multigrounded neutral since the isolation devices disconnect the secondary grounds from the primary grounds.

13.5.4 Tree Contacts

Trees cause another possibly dangerous shocking hazard. A tree in contact with an energized phase conductor may create a touch potential hazard near the ground. For a tree in contact with one conductor, the resistance of the tree is high enough to remain a high-impedance connection; it will not draw enough current to operate a fuse or other protective device, but it may be enough to create significant touch potentials. St. Clair (1999) reported two electrocutions from tree contacts to 12.47-kV lines, one a tree trimmer and another at ground level where a man leaned against a tree that had fallen into conductors.

Daily (1999) analyzed the potentials involved and showed threatening step voltages. Daily's measurements of American elms had resistivities from 37 to 55 Ω-m. A resistivity of 50 Ω-m translates into 93 Ω/ft (305 Ω/m) for a tree trunk with a diameter of 18 in. (0.46 m). Using a tree resistivity of 50 Ω-m and a soil resistivity of 100 Ω-m, Daily calculated a touch voltage of 140 V along with a body current of 391 mA, definitely in the fatal defibrillating range. Lower soil resistivities caused higher body currents (even though the step potentials were lower).

The body current for a touch potential to a tree involves many factors including the resistance of the tree, the tree's resistance to ground in the earth, the conductor voltage, and a person's hand and foot contact resistance. The tree resistance is an important component, and it varies considerably, especially with the moisture content. Hoffmann et al. (1984) measured a poplar tree and found the distribution of impedances shown in Figure 13.22, with impedances of 1.8 kΩ/ft (6 kΩ/m) near the base of the tree rising to 25 kΩ/ft (80 kΩ/m) near the top of the tree. The total impedance of this tree was 320 kΩ, which would draw from 0.2 to 0.7 A from a 12.47-kV line. Hoffman also found resistances from the base of the tree to ground of about

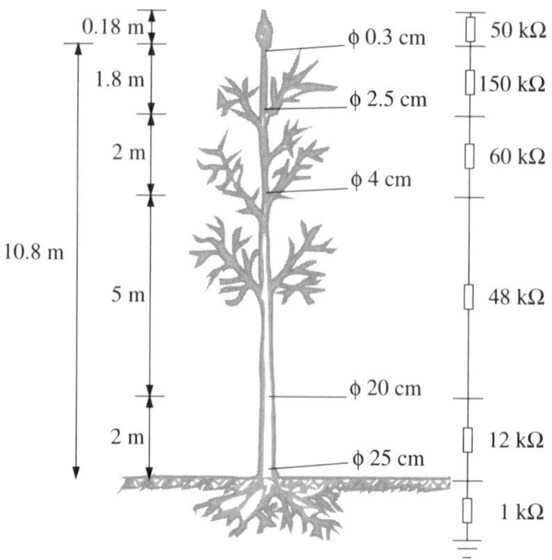

FIGURE 13.22
Resistance measurements on a live poplar tree. (From Hoffmann, E., Rheinbaben, H. V., and Stosser, B., "Trees in Electrical Contact with High Voltage Lines," CIGRE Session, 22-03, September 1984. With permission.)

300 Ω for soil resistivities near 600 Ω-m. Defandorf (1956) measured resistances to ground in the range of 10 to 40 kΩ for a tulip tree at typical distribution conductor heights, which would draw only 0.2 to 0.7 mA from a 12.47-kV line (7.2 kV line to ground). Hoffman et al. tested a tree contact to a 20-kV line, which drew 80 mA initially but rose to 600 mA with a potential at 6.6 ft (2 m) reaching 4 kV. Niagara Mohawk (Finch, 2001) tested several trees at 7.62 kV from line to ground — the highest was just under 0.5 A (see Table 13.7). Larger tree limbs have lower resistances, so the most

TABLE 13.7
Current Drawn by Several Trees in Contact with 7.62 kV from Line to Ground

Location	Specimen	Maximum Current, mA
Test 1	Black Gum	92
Test 2	Black Gum	69
Test 3	Black Cherry	110
Test 4	Black Cherry	100
Test 5	White Ash	37
Test 6	Aspen	484
Test 7	Red Maple	125.5

Source: Finch, K., "Understanding Tree Outages," EEI Vegetation Managers Meeting, Palm Springs, CA, May 1, 2001.

dangerous scenario is a primary conductor solidly contacting a large branch or the main trunk. Shoes and the insulation that they provide often limit such currents below the let-go threshold.

Williams (1999) found tree touch voltages for a tree contacting a 12-kV circuit at 7 ft varied between 90 and 130 V (open circuit). But he found that the current through a shorting jumper at 7 ft (2.1 m) tied to a ground rod 2 ft (0.6 m) from the tree drew less than a milliamp of current (imperceptible to some people).

Texas A&M researchers tested several scenarios involving tree contacts to a single-phase portion of a 12.47-kV circuit (Butler et al., 1999). At 7 ft (2.1 m) the step voltages ranged from 200 to 1000 V. Voltage gradients along the bottom portion of the tree reached as high as 118 V/ft (387 V/m) during the tests. The current into the tree during the tests was about 1 to 2 A. Although they did not make any current measurements through a resistor or jumper, we can verify that these voltages can produce dangerous currents. For an 800-V step potential across 6 ft (1.8 m) of the tree and 2 A of current flow, this length of tree represents a resistance of about 400 Ω. For a body in parallel with the tree resistance, the current through the body is determined by a current divider. For this scenario with 2 A of current from the utility primary through the tree, any body contact resistance less than 3600 Ω causes a possibly fatal 100 mA through the body. Trees passing this level of current hissed and crackled, which hopefully warns bystanders to stay away.

The most dangerous conditions are when the ground is wet and the tree is wet. Low earth resistivity "grounds" the tree better, so it draws more current from the line. Foot-to-earth contact resistances are also lower with drenched soil, which draws more current through your body. A wet tree — due to rain or humid weather — is more dangerous as the tree's resistance drops appreciably.

Strategies to reduce the possibility of tree-related shocks include the following:

- Educate the public on the dangers of cutting trees in the vicinity of power lines.
- Trim trees away from power lines.
- Train tree-trimming crews and ground support crews properly.
- Use covered wire instead of bare wire.

Wooden poles can also cause dangerous touch potentials if a phase conductor contacts the pole directly. The U.S. National Institute for Occupational Safety and Health has documented more than one case where an electrical worker died when helping guide the butt of a pole into its hole at ground level after the top of the pole contacted a live distribution circuit. Resistance of the pole plays an important role. Wood poles have a wide resistance, depending primarily on their internal moisture content and surface moisture (the resistance is largely independent of the pole species). Poles normally have higher resistance than live trees because they are drier. Stewart (1936) measured resistances of approximately 500 Ω/ft to 10 MΩ/ft (1600 Ω/m to 30 MΩ/m) on a variety of treated and untreated poles. The lowest resistances are for new poles with moisture contents above 30%. Surface moisture, due to humid weather or rain, decreases a pole's resistance. Seasoned dry poles normally have resistances above 2.5 MΩ/ft (8 MΩ/m) when dry and 50 kΩ/ft (160 kΩ/m) when wet.

13.6 Protective Grounding

Extensive repair or maintenance or installation work requires working the circuit deenergized. Remember the line worker's gospel:

If it is not grounded, it is not dead!

Or, better yet:

If it is not tested and grounded, it is not dead!

During deenergized work on a structure, protective grounds tie the phase conductors to a ground point. The main purpose of these protective jumpers is to protect workers in case the circuit is accidentally energized (another worker could close the disconnecting switch or another energized circuit could fall into the deenergized circuit). The jumpers also help protect against voltages from magnetic or capacitive coupling and backfeed through transformer or capacitor banks. With a properly grounded protection system, line workers do not need rubber gloves.

Utilities use two types of grounding systems: bracketed grounds and work-site grounds (IEEE Std. 1048-1990). Bracketed grounds are attached on structures on both sides of the work site. Work-site grounds are attached on the work-site structure. Work-site grounds protect workers (lower possible body currents) better than bracketed grounds. Work-site grounds provide the best protection, but keep them far enough away so that they do not interfere with the work on the structure and so the work does not disturb the jumpers. With single-point grounds, it is safest to have the grounding set on the source side of the work site. But, be careful, many lines may be energized from more than one direction. Bracketed grounds can effectively protect workers. Puget Sound Power & Light allows bracket grounds up to 250 ft from the worker (King, 1990). If grounds are farther than one span, rubber gloves or other work practices may be appropriate (based on a thorough review).

Normally, crews tie protective grounds to the multigrounded neutral. On circuits without a multigrounded neutral, crews tie bonds together at

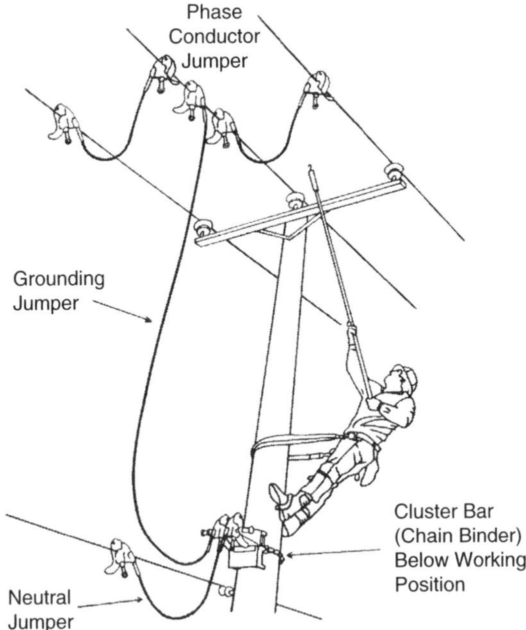

FIGURE 13.23
Work-site protective ground. (From Hubbell/Chance, "Tips of the Trade: Grounding," 2000. Bulletin 09-9001. With permission.)

a pole band, which may be bonded to a driven ground rod or guy anchor or other grounding point (see Figure 13.23). Crews may tie individual grounding jumpers from each phase to the neutral (or other bonding point), or they may jumper the three phases together and tie another jumper from a phase jumper to the pole ground. Excessive jumper lengths should be avoided.

Testing the line for voltage before installing jumpers is safer; attempting to jumper an energized line is extremely dangerous — a jumper that causes a fault creates an arc and the magnetic forces can swing the jumper. A variety of testers attached at the end of live-line tools are available.

When the work is done, the protective grounds should not be removed until everyone is clear of the work site. When removed, the phase end should be removed first with an appropriate hotline tool.

On three-phase circuits, grounding all three phases is generally preferred since all three conductors on the pole are grounded. If the circuit is accidentally energized, a three-phase fault has higher fault current and may trip protective equipment more quickly. A three-phase fault also keeps most of the fault current out of the ground, which reduces step and touch potentials.

Protective grounding jumpers must withstand the heat generated during a fault, and they must withstand the mechanical forces. ASTM F-855 (1997) specifies grounding cables, clamps, and ferrules.

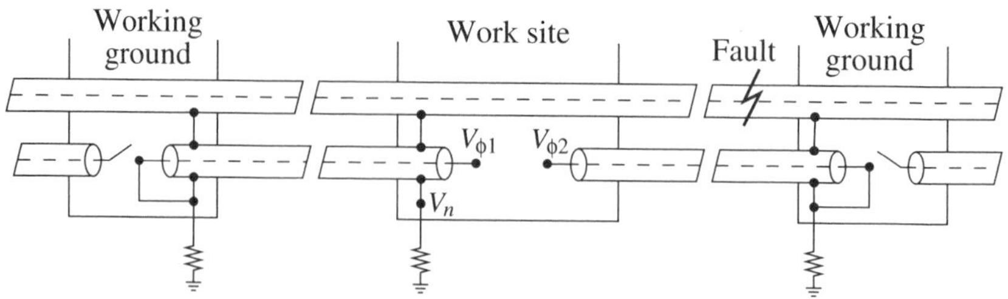

FIGURE 13.24
An underground cable under repair with a fault on an adjacent cable. (Adapted from Rajotte, Y., Bergeron, R., Chalifoux, A., and Gervais, Y., "Touch Voltages on Underground Distribution Systems During Fault Conditions," *IEEE Trans. Power Delivery*, 5(2), 1026-1033, April 1990. ©1990 IEEE. With permission.)

Maintenance and proper application of grounding jumpers is important. Always clean conductor surfaces and tighten all terminals. Utilities should test protective-grounding sets periodically.

Underground work has special protective grounding issues. With insulated cables, protective grounding at the work site is not always possible. Normally, the cable is opened and protective grounds are placed at the nearest switching point, but this may be hundreds of feet (meters) from the work site. In urban installations, duct banks often have several cables. Faults on adjacent cables may induce voltages at the work site (see Figure 13.24). The sheath covering the cable limits capacitive coupling between phases, but inductive coupling may induce voltages on the cable sheath. Since the cable sheaths are tied together, a portion of the fault current flows along the cable under repair; this current induces voltage in the cable phase conductors.

Field tests by Rajotte et. al. (1990) found very small neutral-to-earth voltages in the manhole, less than 20 V on a 25-kV circuit with a 4800-A fault current. The voltages were low because the circuit tested had low ground resistivity (50 Ω-m) and many connections to ground.

On cables, the highest touch voltages occur between the two phases of the separated cable, and the highest voltages occur when the protective grounds are far from the work site. The voltages on the phase conductor are induced by current flow in the cable's concentric sheath. Since the phase and the concentric neutral have tight inductive coupling, significant voltage develops on the phase conductor for current flow through the concentric neutral. Rajotte et al. (1990) measured 169 V on the phase conductor of the cable; the protective ground on that cable was 1.6 mi (2.5 km) from the work site. The voltage between the two cables ($V_{\phi1} - V_{\phi2}$ in Figure 13.24) reached 221 V. The two phase conductors have voltage of opposite polarity, so the voltage between the two is the sum of the voltages. The best way to reduce the voltages on the phase conductors is to keep the protective grounds close, preferably right at the work site.

The voltage induced between the phase conductor and the neutral depends on the fault current (I) and the distance to the protective ground (d):

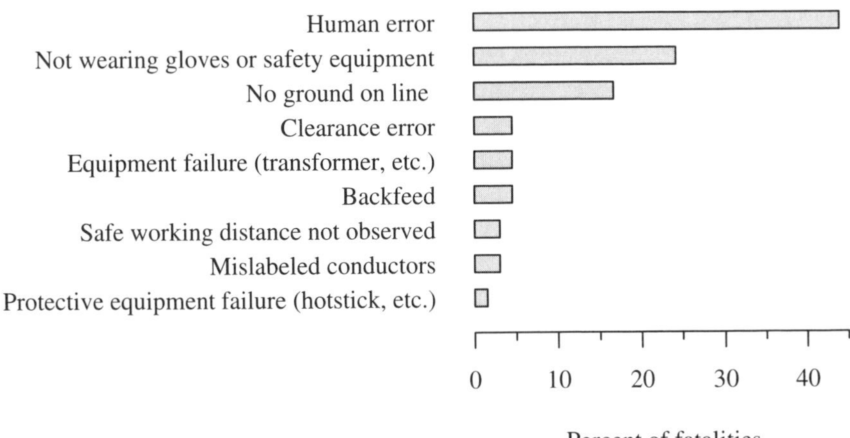

FIGURE 13.25
Transmission and distribution fatalities (4 – 46 kV) by cause from 1988 – 1990 from the International Brotherhood of Electrical Workers (IBEW) database (n = 66). (Data from [EPRI 1001917, 2002].)

$$V_{\phi\text{-N}} = k\, d\, I$$

where k is a factor in ohms per unit length that is the coupling factor to the phase conductor; k depends on the number of cables in the duct bank, their size, and their position. Rajotte et al. (1990) calculated $k = 0.016\ \Omega$/mi (0.01 Ω/km) for a configuration with five 750-kcmil cables and $k = 0.1\ \Omega$/mi (0.06 Ω/km) for a configuration with two 350-kcmil cables. The voltage between two cable phase conductors is based on the sum of the length to the protective grounds on each side since the voltages add.

The Occupational Safety and Health Administration (OSHA) regulates several aspects of utility operations and equipment related to safety in OSHA Standard 29 CFR parts 1910 (Occupational Safety and Health Standards) and 1926 (Safety and Health Regulations for Construction). The Lineman's and Cableman's Handbook (Kurtz et al., 1997) also has useful safety information.

Following proper procedures and using common sense are critical. Most accidents are not from equipment failure (see Figure 13.25); they are from human failure — carelessness, doing something stupid. The best piece of safety equipment is a brain — an alert, well-trained, and cautious brain.

References

AEMC Instruments, *Understanding Ground Resistance Testing*, 1998.

ANSI/IEEE C62.92-1987, *IEEE Guide for the Application of Neutral Grounding in Electrical Utility Systems. Part I — Introduction.*

ASTM F855-97e1, *Standard Specifications for Temporary Protective Grounds to Be Used on De-energized Electric Power Lines and Equipment*, American Society for Testing and Materials, 1997.

Biddle Instruments, *Getting Down to Earth*, 1982.

Butler, K. L., Russell, B. D., Benner, C., and Andoh, K., "Characterization of Electrical Incipient Fault Signature Resulting from Tree Contact with Electric Distribution Feeders," IEEE Power Engineering Society Summer Meeting, 1999.

CEA 178 D 477, *Stray Voltage Effects on Dairy Cattle*, Canadian Electrical Association, 1992.

Clapp, A. L., *National Electrical Safety Code Handbook*, The Institute of Electrical and Electronics Engineers, Inc., 1997.

Clerfeuille, J. P., Juston, P., and Clement, M., "Extinguishing Faults Without Disturbances," *Transmission & Distribution World*, pp. 52–9, August 1997.

Daily, W. K., "Engineering Justification for Tree Trimming (Power System Maintenance)," *IEEE Transactions on Power Delivery*, vol. 14, no. 4, pp. 1511–8, October 1999.

Dalziel, C. F., "Dangerous Electric Currents," *AIEE Transactions on Power Apparatus and Systems*, vol. 65, pp. 579–85, 1946.

Dalziel, C. F., "Electric Shock Hazard," *IEEE Spectrum*, pp. 41–50, February. 1972.

Defandorf, F. M., "Electrical Resistance to the Earth of a Live Tree," *AIEE Transactions*, vol. 75, pp. 936–41, October 1956.

Dwight, H. B., "Calculation of Resistance to Ground," *AIEE Transactions*, vol. 55, pp. 1319–28, December. 1936.

EPRI 1000074, *Five-Wire Distribution System Demonstration Project*, EPRI, Palo Alto, CA; ESEERCO, New York, NY; NYSERDA, Albany, NY, 2000.

EPRI 1001917, *Live Working Application Guide*, Electric Power Research Institute, Palo Alto, CA, 2002.

EPRI TR-113566, *Identifying, Diagnosing, and Resolving Residential Shocking Incidents*, Electric Power Research Institute, Palo Alto, CA, 1999.

ESEERCO, *Distribution System Arresters*, Empire State Electric Energy Research Corporation, 1992.

Hoffmann, E., Rheinbaben, H. v., and Stosser, B., "Trees in Electrical Contact with High Voltage Lines," CIGRE Session, 22-03, September 1984.

Hubbell/Chance, "Tips of the Trade: Grounding," 2000, Bulletin 09-9001.

IEEE C2-1997, *National Electrical Safety Code.*

IEEE C62.92.4-1991, *IEEE Guide for the Application of Neutral Grounding in Electrical Utility Systems, Part IV — Distribution.*

IEEE Std. 80-2000, *IEEE Guide for Safety in AC Substation Grounding.*

IEEE Std. 81-1983, *IEEE Guide for Measuring Earth Resistivity, Ground Impedance, and Earth Surface Potentials of a Ground System.*

IEEE Std. 142-1991, *IEEE Recommended Practice for Grounding of Industrial and Commercial Power Systems.*

IEEE Std. 902-1998, *IEEE Guide for Maintenance, Operation, and Safety of Industrial and Commercial Power Systems.*

IEEE Std. 1048-1990, *IEEE Guide for Protective Grounding of Power Lines.*

IEEE Working Group, "Voltage Rating Investigation for Application of Lightning Arresters on Distribution Systems," *IEEE Transactions on Power Apparatus and Systems*, vol. 91, no. 3, pp. 1067–74, May/June 1972.

IEEE Working Group on Electrostatic and Electromagnetic Effects, Delaplace, L. R., and Reilly, J. P., "Electric and Magnetic Field Coupling From High Voltage AC Power Transmission Lines — Classification of Short Term Effects on People," *IEEE Transactions on Power Apparatus and Systems*, vol. PAS-97, no. 6, pp. 2243–52, November/December 1978.

King, C. C., "Why Proper Grounding Is Vital for Worker Safety," *Electrical World*, November. 1990.

Kurtz, E. B., Shoemaker, T. M., and Mack, J. E., *The Lineman's and Cableman's Handbook*, McGraw Hill, New York, 1997.

Lat, M. V., "Determining Temporary Overvoltage Levels for Application of Metal Oxide Surge Arresters on Multigrounded Distribution Systems," *IEEE Transactions on Power Delivery*, vol. 5, no. 2, pp. 936–46, April 1990.

Leap, N., Harries, J., Siebel, J., and Lusky, C., "Power Quality Problems Associated With Distribution and Transmission Line Capacitors," PQA Conference, Portland, OR, 2002.

Mancao, R. T., Burke, J. J., and Myers, A., "The Effect of Distribution System Grounding on MOV Selection," *IEEE Transactions on Power Delivery*, vol. 8, no. 1, pp. 139–45, January 1993.

Minnesota Public Utilities Commission, "Final Report of the Science Advisors to the Minnesota Public Utilities Commission: Research Findings and Recommendations Regarding Claims of Possible Effects of Currents in the Earth on Dairy Cow Health and Milk Production," 1998.

NFPA 70, *National Electrical Code*, National Fire Protection Association, 2002.

Peterson, H. A., *Transients in Power Systems*, Dover Publications, New York, 1951.

Rajotte, Y., Bergeron, R., Chalifoux, A., and Gervais, Y., "Touch Voltages on Underground Distribution Systems During Fault Conditions," *IEEE Transactions on Power Delivery*, vol. 5, no. 2, pp. 1026–33, April 1990.

Rajotte, Y., Fortin, J., Cyr, B., and Raymond, G., "Characterization of the Ground Impedance of Rural MV Lines on Hydro-Quebec's System," CIRED 97, June 1997.

Rajotte, Y., Fortin, J., and Raymond, G., "Impedance of Multigrounded Neutrals on Rural Distribution Systems," *IEEE Transactions on Power Delivery*, vol. 10, no. 3, pp. 1453–9, July 1995.

Reilly, J. P., *Applied Bioelectricity: From Electrical Stimulation to Electropathology*, Springer Verlag, New York, 1998.

Reines, R., Cook, M., and Dasho, D., "PSC Staff Report: The Wisconsin Stray Voltage Experience," Public Service Commission of Wisconsin, 1995.

Reines, R., Cook, M., and Dasho, D., "PSC Staff Report: Wisconsin's Stray Voltage Experience — An Update," Public Service Commission of Wisconsin, 1998.

Short, T. A., Stewart, J. R., Smith, D. R., O'Brien, J., and Hampton, K., "5-Wire Distribution System Demonstration Project," *IEEE Transactions on Power Delivery*, vol. 17, no. 2, pp. 649–54, April 2002.

St. Clair, J. G., 1999. Discussion to Daily (1999).

Stewart, P. B., "The Electrical Resistance of Wood Poles," *Proceedings of the American Wood-Preservers' Association*, vol. 32, pp. 353–85, 1936. As cited by Darveniza, M., *Electrical Properties of Wood and Line Design*, University of Queensland Press, 1980.

Tran, T. Q., Conrad, L. E., and Stallman, B. K., "Electric Shock and Elevated EMF Levels Due to Triplen Harmonics," *IEEE Transactions on Power Delivery*, vol. 11, no. 2, pp. 1041–9, April 1996.

USDA, "Effects of Electrical Voltage/Current on Farm Animals: How to Detect and Remedy Problems," in *Agriculture Handbook No. 696*, U.S. Department of Agriculture, 1991.

Westinghouse Electric Corporation, *Electrical Transmission and Distribution Reference Book*, 1950.

Williams, C., "Tree Fault Testing on a 12 kV Distribution System," IEEE/PES Winter Power Meeting presentation to the Distribution Subcommittee, 1999.

Some crane and digger operators think that "up just a touch" means a bare minimum of 12 inches.

Powerlineman law #41, By CD Thayer and other Power Linemen, http://www.cdthayer.com/lineman.htm

14

Distributed Generation

Distributed generation (DG) or embedded generation (the European term) refers to generation applied at the distribution level. EPRI (1998) defines distributed generation as the "utilization of small (0 to 5 MW), modular power generation technologies dispersed throughout a utility's distribution system in order to reduce T&D loading or load growth and thereby defer the upgrade of T&D facilities, reduce system losses, improve power quality, and reliability."

No exact size or voltages are accepted as definitions of distributed generation. A draft IEEE standard (IEEE P1547-D11-2003) applies to generation under 10 MW. Distribution substation generation is normally considered as distributed generation; sometimes subtransmission-level generation is also considered as DG since many of the application issues are the same. Related terms — non-utility generator (NUG) and independent power producer (IPP) — refer to independent generation that may or may not be at the distribution level. A broader term, distributed resources (DR), encompasses distributed generation, backup generation, energy storage, and demand side management (DSM) technologies.

Smaller-sized generators continue to improve in cost and efficiency, moving closer and closer to the performance of large power plants. At the same time, utilities face significant obstacles when building large facilities — both power plants and transmission lines. Utilities or end users can install modular distributed generation quickly. This local generation reduces the need for large-scale utility projects. Distributed generation can allow utilities to defer transmission and distribution upgrades. Also, DG reduces losses and improves voltage. With the right configuration, distributed generation can also improve customer reliability and power quality.

Both utilities and end users can install distributed generators, usually for different reasons. Distributed generators are most cost effective if the customer has need for

- *Cogeneration* — Using the generator waste heat locally significantly improves the economics of many applications. Uses include process steam, heating water, facility heat, or running air conditioning.

- *Reliability* — Many locations need backup generation for reliability purposes. Once they are bought, running them full time or for peak shaving becomes more economical.

Utilities most need local generation with

- *Overloaded circuits* — Distributed generation can delay infrastructure upgrades.
- *Generation shortages* — If a utility needs generation anywhere it can get it, the utility or end users can quickly install distributed generation.

Applying generation closer to the load benefits the transmission and distribution infrastructure. Local generation can relieve overburdened transmission and distribution facilities as well as reduce losses and voltage drop.

While offering benefits and opportunities, generation is not always easy to integrate into existing distribution systems (Barker and De Mello, 2000; CEA 128 D 767, 1994; EPRI 1000419, 2000). Distribution systems were never designed to include generation; they were designed for one-way power flow, from the utility substation to the end users. Generators violate this basic assumption, and generators can disrupt distribution operations if they are not carefully applied. One of the most critical situations is that a distribution interrupter may isolate a section of circuit, and the generator might continue supplying the load on that section in an *island*. Islanding poses safety hazards, and islanded generators can cause overvoltages on the circuit. In addition to islanding, generators can disturb protection, upset voltage regulation, and cause other power quality problems. We will investigate many of these issues in this chapter, how to identify problems and options for fixing them.

In addition to the technical difficulties, distributed generation raises several other issues. How do we meter a generator? What is a fair price to pay a generator injecting power into the system? How can distributed generators be dispatched or controlled, especially if they are owned by end users? How can we apply the generators to optimal locations on the system, rather than just accepting it wherever end users install it?

Many distributed generation technologies can supply isolated loads on their own, but massive disconnections from the utility to stand-alone systems are unlikely in the near term. The utility distribution system solves many of the problems with stand-alone generation systems:

- *Reliability* — Generators must be applied in redundant fashion to achieve the reliability of normal distribution service. Together, a utility service along with local generators has much higher than normal reliability.
- *Matching load* — Many small generators are most economical when they run at full power all of the time, but most loads have an average power demand that is well below the peak. A utility source allows the generator to run at full power, and the utility can offset the power

for the load. In addition, much of the renewable generation is intermittent. Without a utility interconnection, renewable sites must have energy storage, which is expensive.

- *Load following* — Many distributed generation designs have limited load-following capability. Step changes in load or motor starting can deeply affect the voltage and frequency.

The utility supply has the benefit of tying thousands of generators together, providing reliability and matching load. In the future, small generators may be applied locally in ways that gain some of these same benefits as the bulk utility system. *Microgrids* tie generators together on a localized scale to gain many of the reliability and load matching advantages of the traditional utility grid (EPRI 1003973, 2001).

Distributed generation technologies continue to advance: cost comes down, performance improves. Projections of penetration of distributed generation into the electrical system vary widely. Since natural gas delivers energy at a cost that is roughly one fourth of the cost of electric energy, if an efficient, low-capital-cost, and low-maintenance distributed generator becomes available, gas energy delivery has an advantage over electric energy delivery. Utilities must prepare for several scenarios and consider distributed generators as another tool for supplying end users with electric power.

14.1 Characteristics of Distributed Generators

14.1.1 Energy Sources

Several energy sources drive distributed generators including

- Reciprocating engines
- Combustion turbines
- Microturbines
- Wind turbines
- Fuel cells
- Photovoltaics

Three power converters convert the power output of the energy source to interface with standard 50- or 60-Hz systems:

- Synchronous generator
- Induction generator
- Inverter

Normally, reciprocating engines and combustion turbines interface through synchronous generators. Microturbines, many wind turbines, fuel cells, and photovoltaics interface through inverters.

Three factors are considerations for the energy source: the stability of the power source, the ability to follow changes in load, and the reliability (availability) of the source.

On reciprocating engines, combustion turbines, and microturbines, the burn rate of the fuel/air mixture controls the power output. The real power output is controllable and responds to load changes fairly quickly. These combustion-based generators are normally very stable sources (unless the fuel supply is poor).

Wind turbines and photovoltaics produce variable output that is normally not controllable. One of the main concerns of wind turbines on distribution circuits is the possibility of voltage flicker due to pulsating output, mainly because the tower shades the blades for part of their rotation. Sunshine variability also causes variation in solar-panel outputs, but the changes are normally slow enough to limit flicker problems. Neither of these can follow load.

Fuel cells have load-following capability, but it is significantly slower than combustion generators. Fuel cells and photovoltaics have no natural provision to provide short-time overloads (some manufacturers include batteries or ultracapacitors to add this capability). This helps control fault currents but limits their ability to supply motor starting current and respond to loads.

The availability and reliability of the energy source plays a role in some applications. Combustion engines and turbines are easily dispatchable. Photovoltaics and wind generation are non-dispatchable; the energy source is not always available. Small hydro units may or may not be dispatchable, depending on the water levels.

14.1.2 Synchronous Generators

The workhorse power generator for over a century, the synchronous generator is the power converter of choice for most distributed power sources. Reciprocating engines and gas turbines almost always interface with a synchronous generator. A synchronous generator can operate independently or in synchronization with the utility. Most distribution-scale generators are salient-pole machines with four or six poles driven by 1800- or 1200-rpm reciprocating engines (with some even slower in larger units). Distribution-scale gas-turbines spin faster but are normally geared down to drive a salient-pole generator at 1800 rpm.

Synchronous generators can operate at leading or lagging power factor and can operate in a voltage control or voltage following mode. Most distributed synchronous generators use a voltage-following control mode. This differs from large generators on the electric system, which have controls to regulate voltage (within the limits of the var capability). Distributed generators normally follow the utility voltage and inject a constant amount of real and reactive

Stator

Rotor

FIGURE 14.1
Synchronous generator. (Courtesy of Caterpillar, Inc.)

power (so the power factor stays constant). Normally, most DG operators run at unity power factor since that produces the most watts for a given kVA rating.

The three-phase ac power system connects to the armature winding on the stator (the outside, nonmoving part) of the synchronous machine (Figure 14.1). The stator wraps around the outside of the generator, so it can couple magnetically with the rotating magnetic field produced on the rotor. The rotor has a dc magnetic field rotating at synchronous speed. The rotor has several poles around the shaft. Each pole is a piece of iron with a coil wrapped around it (the field winding). The field winding forms an electromagnet that creates the dc magnetic field. A separate device, an exciter, generates the dc current for the field winding. The rotating dc magnetic field induces an emf in the stator (armature) winding, which drives power into the ac system. For more detailed descriptions, see one of the many machine books available, such as Fitzgerald et al. (1971).

A voltage source (the emf induced on the armature) behind a reactance models a synchronous machine well. Several impedances come into play depending on the scenario:

- *Subtransient reactance* — X_d'' is the positive-sequence impedance that determines the three-phase fault current during the first few cycles. Normally, X_d'' is between 10 and 30% on the machine rating. Higher values of X_d'' limit the short-circuit current the machine supplies but reduce the capability of the machine to respond quickly to load changes.
- *Negative-sequence reactance* — X_2 impacts the current under voltage unbalance. X_2 is equal to the average of X_d'' and X_q'' (the direct and quadrature impedances) and is typically 10 to 30% on the machine rating.
- *Zero-sequence reactance* — X_0 affects ground-fault currents and the flow of triplen harmonics. The zero-sequence reactance is normally very low, even less than 5% on the machine rating.

In the absence of better information, assume X_d'' = 20%, X_2 = 25%, and X_0 = 5% on the machine rating. In the past, this information was hard to obtain for small machines, but now most generator manufacturers supply them. Table 14.1 shows values of small generator constants from tests, with the characteristics of machines C and D being most representative. Most of the parameters in the table are only needed for very detailed simulations.

For fault calculations, the voltage behind the subtransient reactance is normally in the range of 1.0 to 1.3 per unit with a typical value of 1.1. The power factor ranges from 0.8 lagging to unity. So, with the voltage at 1.1 per unit and X_d'' = 20%, the initial three-phase fault current is 5.5 times the generator rating.

Under a fault, sudden load change, or other transient condition, the subtransient reactance (X_d'' which is often denoted as just X'') dominates. After a time equal to the subtransient time constant (T_d''), which ranges from 1 to 10 cycles, the transient reactance (X_d') more accurately models the generator response.

The direct-axis impedance (indicated by the subscript *d*) is the effective impedance for changes that are in sync with rotation (50- or 60-Hz events). Quadrature-axis impedances (subscript *q*) are the impedances seen with the rotor out of phase with the stator field. For fault calculations we only need to consider the direct-axis impedances. For unbalance or harmonics, the quadrature-axis impedances have impacts.

The *saturated* subtransient reactance (X_{dv}'') is used for short-circuit studies and for harmonics. The subscripted "*v*" means constant voltage; the impedance is the value of *V*/*I* at rated voltage and no load for a short circuit on

TABLE 14.1

Small Synchronous Generator Constants in per Unit on the Machine Base

Machine	A	B	C	D
kVA	69	156	781	1044
kW	55	125	625	835
V	240/480	240/480	240/480	240/480
rpm	1800	1800	1800	1800
H_{gen}	0.26	0.20	0.40	0.43
pf	0.80	0.80	0.80	0.80
X_d	2.02	6.16	2.43	2.38
X_d'	0.171	0.347	0.254	0.264
X_d''	0.087	0.291	0.207	0.201
X_q	1.06	2.49	1.12	1.10
X_q''	0.163	0.503	0.351	0.376
T_{do}' (sec)	0.950	1.87	1.90	2.47
T_d' (sec)	0.080	0.105	0.198	0.273
T_{do}'' (sec)	0.078	0.013	0.024	0.018
T_d'' (sec)	0.004	0.011	0.020	0.014
T_{qo}'' (sec)	0.045	0.020	0.016	0.009
T_q'' (sec)	0.007	0.004	0.005	0.003
r_a	0.011	0.034	0.017	0.013
T_a (sec)	0.014	0.022	0.038	0.032
X_0	0.038	0.054	0.051	0.074
X_2	0.125	0.375	0.279	0.260

Source: Gish, W. B., "Small Induction Generator and Synchronous Generator Constants for DSG Isolation Studies," *IEEE Transactions on Power Systems*, vol. PWRD-1, no. 2, pp. 231–9, April 1986. ©1986 IEEE.

the generator terminals (which is a high-current condition that saturates the magnetic core, which results in a lower impedance). Another version of subtransient reactance, the unsaturated reactance (X_{di}'') is used for stability studies. Usually, if it is not indicated which value it is, it is the saturated value, X_{dv}''.

The negative-sequence reactance (X_2) and the subtransient reactance (X_d'') are determined mainly by the damper windings on the rotor. On a salient-pole machine, damper windings are a shorted winding wound around each pole. Damper windings react to dampen out transients on the machine. The torque produced by the damper windings slows down a machine that is running faster than synchronous speed, and it speeds up a machine that is running slower than synchronous speed. Damper windings are like the rotor on a squirrel-cage induction machine (in fact a squirrel-cage induction machine is a synchronous machine with damper windings and no field on the rotor). At synchronous speed, no current flows in the damper windings since the flux on the rotor is dc, but if transients act to change the frequency relative to the synchronous speed, currents flow in the damper windings and create torque to counteract the disturbance. The damper windings also react to other stator currents that are not 50- or 60-Hz positive-sequence currents (significantly, negative-sequence currents and harmonic currents).

The negative-sequence impedance determines the current flow due to negative-sequence voltages. The impedance to harmonic voltages is also approximately the negative-sequence impedance. The negative-sequence impedance is the same whether the generator is under steady-state, transient, or subtransient conditions.

The zero-sequence impedance of a synchronous machine can have extremely low impedance. It is enough of a problem that many generators are ungrounded or grounded through an impedance to prevent the flow of zero-sequence current. Many generators are not braced to handle the fault current for a line-to-ground fault at the terminals of the machine. Single-phase faults cause more mechanical stress and are higher magnitude. Ground fault currents are 30 to 40% higher than three-phase fault currents (E/X_d'' vs. $3E/(2X_d''+X_0) \approx 1.3$ to $1.4E/X_d''$). The zero-sequence impedance is the same whether it is under steady-state, transient, or subtransient conditions.

The reason that the zero-sequence impedance is so low is that magnetic fields from zero-sequence currents in the stator winding tend to cancel each other. If the fields cancel and couple very little to the rotor, the impedance is very low.

The zero-sequence impedance varies significantly with design. The most prominent difference is due to the pitch of the stator winding. A pole pitch is the number of degrees that the rotor has to move to change from one pole to the other. In a 2-pole machine, one pole pitch is 180°, and in a 4-pole machine, it is 90°. The *pitch factor* (or just the *pitch*) of the stator winding is the portion of the pole pitch that the stator winding spans. A full-pitch stator winding spans the full pitch. A fractional pitch winding spans less than the full pitch. Figure 14.2 shows a comparison of a full-pitch winding and a 2/3-pitch winding. In this figure, each phase has two windings (a_1 and a_2 for example); the current in each winding goes out on one slot conductor (like a_1) and returns in another slot conductor labeled with the prime notation (like a'_1). The 2/3-pitch winding reduces the zero-sequence impedance the most. Because the two conductors in each slot have current in opposite directions, the fields cancel almost completely (since $a = b = c = -a' = -b' = -c'$ for zero-sequence current). Other common pitch factors are 5/8 and 3/4.

Synchronization is important when connecting synchronous generators. If a large machine is out of synch with the utility when it is connected, the sudden torque can damage the generator, and the connecting switch and other devices will have a large voltage disturbance. To avoid these problems, synchronizing relays are required for synchronous generators to ensure that the voltage, frequency, phase angle, and phasing are the same on the utility and generator. Normally, the generator is brought up to speed, and the field current is adjusted to bring the voltage close to the utility system. The frequency is more precisely adjusted to bring it within 0.5 Hz of the power system. Then, the synchronizing relay allows closing when the two voltages are within 10° (CEA 128 D 767, 1994).

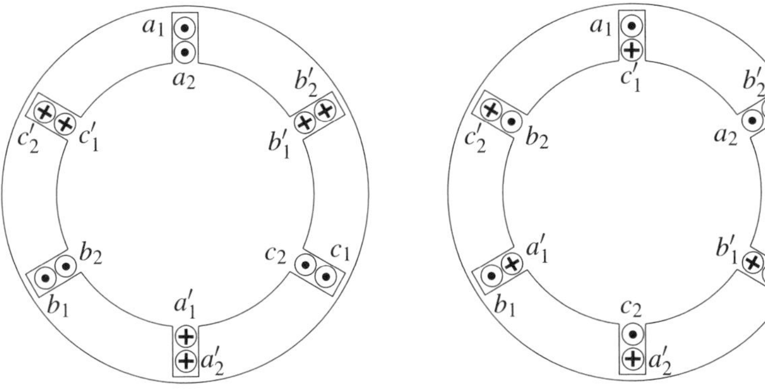

Full-pitch winding Two-thirds-pitch winding

FIGURE 14.2
Comparison of a full-pitch and a 2/3-pitch stator winding for a two-pole machine with a double-layer stator winding with one slot per phase per pole.

14.1.3 Induction Generators

Induction generators are used in a few distributed generation applications, mainly wind-turbine applications where the speed of the prime mover varies. Induction generators are simpler than synchronous generators. They do not have exciters, voltage regulators, governors, or synchronizing equipment.

An induction generator is the same as an induction motor except the prime mover turns the rotor faster than synchronous speed. An induction generator needs supplemental excitation, either supplied by the utility system or local capacitors.

To start some induction generators, the utility may accelerate the prime mover to operational speed, drawing inrush to five or six times the generator rating just like motor starting. A reduced-voltage starter using a reactor or autotransformer reduces the inrush. If possible, the prime mover can accelerate the shaft to near synchronous speed before closing in; this softens the inrush when connecting. This type of connection draws inrush of up to three times the generator rating. Because the induction machine cannot generate voltage on its own, the timing of the connection is unimportant.

Even though the induction generator cannot create voltage on its own, the generator can self-excite with capacitors. Reclosing into a self-excited generator with local capacitors can damage the shaft and other equipment.

Under transient conditions, induction machines respond similarly to synchronous machines. Under fault-type transient conditions, $X_1 = X_2 = 20\%$ (the locked-rotor impedance) and $X_0 = 5\%$. The voltage behind the positive-sequence impedance is also about 1.1 per unit. The main difference between synchronous and induction machines is that the generator driving voltage collapses more quickly in an induction machine.

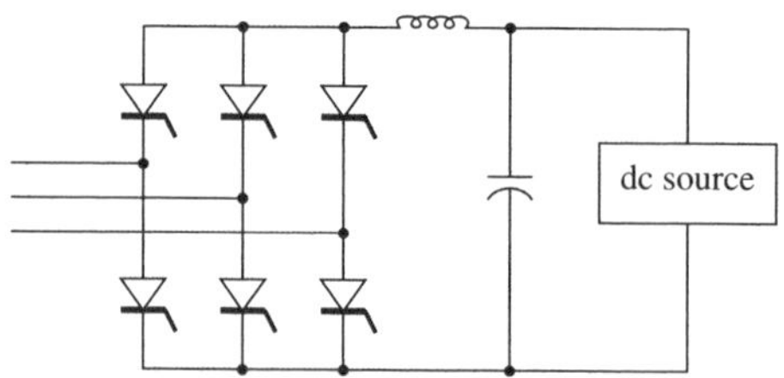

FIGURE 14.3
Common three-phase bridge inverter line-commutated configuration with thyristors.

14.1.4 Inverters

Many types of small generators best interface with utility and utilization power through an inverter. Photovoltaics and fuel cells generate dc. Wind turbines and microturbines generate incompatible frequencies, so they are best applied by rectifying to dc then converting back to ac with an inverter.

Two main classes of inverters are

- Line commutated
- Self commutated

A three-phase line-commutated inverter is the simplest and least expensive converter. Thyristors (also called silicon-controlled rectifiers or SCRs) in a full-wave bridge configuration convert from dc to ac by switching current (see Figure 14.3). Thyristors can switch large amounts of current at low loss. The inverter controls when to turn on the thyristor by applying a voltage pulse to the gate of the thyristor; the thyristor does not turn off until the next current zero.

Since a line-commutated inverter draws currents in square waves, it creates large harmonics, primarily the 5th, 7th, 11th, and 13th. Line-commutated inverters require significant reactive power from the system, between 10 and 40% of the inverter rating. It is possible to use 12-, 18-, or 24-pulse bridge designs with phase-shifting transformers to reduce harmonics.

Line-commutated inverters do not provide significant fault current. The reactor in the inverter limits the fault current. Most designs limit current to 100 to 400% of the inverter rating. Usually inverters have overcurrent protection that removes the firing pulses when current exceeds some threshold. The duration of fault current output also depends on the amount of capacitance on the inverter dc bus. Note that a harmonic filter on the output can increase the momentary fault value.

The self-commutated inverter can generate its own sinusoidal voltage. Although several self-commutated configurations are around, the most common is the fast switching pulse-width modulation (PWM) design using insulated gate bipolar transistors (IGBTs) or other switchable power electronic components. IGBTs can be turned on and off quickly. By controlling the timing and duration of pulses, the controller can shape the current supplied to the system. A common three-phase configuration is a three-phase bridge (the same as Figure 14.3, but with IGBTs instead of thyristors). In a grid-connected application, a PWM inverter can control the power factor; in stand-alone, the inverter can operate alone and control voltage. PWM inverters normally have switching frequencies from 1 to 10 kHz. They produce little low-order harmonics, and the high-frequency harmonics near the switching frequency can be filtered with relatively small filters. Generally, PWM designs have distortion well within IEEE Std. 519-1992.

Self-commutated inverters may be voltage controlled or current controlled. The voltage-controlled inverter is simpler; the controller creates a reference voltage. The current-controlled mode is more suitable for distributed generators; the controller uses the utility voltage as a reference and injects current at the proper angle (usually at unity power factor). Current-controlled inverters produce little fault current. Overcurrent protection removes current almost immediately after an overcurrent (manufacturers implement this to protect their equipment; IGBTs have very limited overcurrent capability). The current pulse before the controller stops firing is typically less than 2 per unit peak and lasts about 200 to 300 μsec.

Self-commutated inverters are used on most smaller inverter-based generators. Very large units still require line-commutated inverters. With continued advances in power electronic switching technologies, larger units will migrate to self-commutated designs.

Many three-phase inverter configurations are ungrounded, like the design in Figure 14.3, which raises interfacing issues on grounded distribution circuits. Often, ungrounded inverter configurations are interfaced with grounded systems through a grounded-wye – delta transformer. This is a grounding transformer; circulating currents in the delta allow the transformer to supply unbalanced load. Some inverter configurations are grounded, such as those in Figure 14.4. Thyristors are shown, but PWM inverters with the same configurations are common. Three single-phase inverters can also supply grounded three-phase systems.

Whether an inverter or a rotating generator, all energy converter technologies have interconnection issues, just different ones. Table 14.2 summarizes some of the most important issues for each major power converter type.

14.1.5 Modeling Small Generators

The simplest load-flow model of distributed generators is as a negative load. The normal load models include constant power, constant impedance, and constant current. More accurate representations depend on the type of generator and its controls.

Since most synchronous DGs operate in a voltage-following mode with a set injection of real and reactive power, the most accurate model is the constant power model. In some programs, the power on each phase must

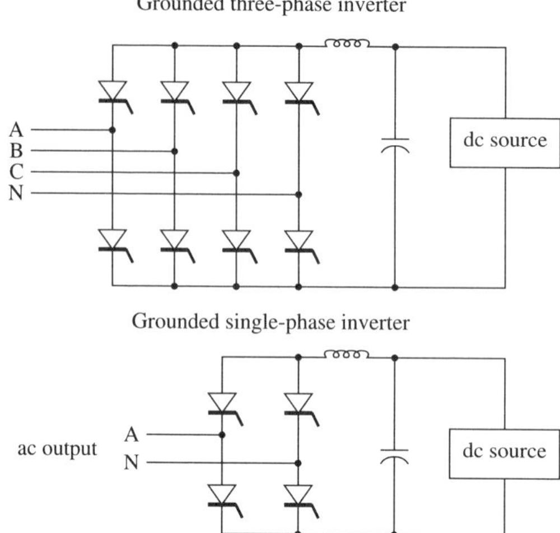

FIGURE 14.4
Grounded inverter configurations.

TABLE 14.2

Power Converter Characteristics from an Interconnection Perspective

Feature	Synchronous Generator	Induction Generator	Inverter
Ability to operate as an isolated (stand-alone) source	Exciter allows it to operate as a stand-alone island	Not intended for stand-alone island but still can accidentally self-excite if isolated with a capacitor	Self-commutated voltage source inverters can function as islands
Synchronism with Utility	Must operate as a synchronous source	Typically operated with several percent slip	Must operate as a synchronous source
Fault contributions	4–8 times rated current for a few cycles	4–8 times rated current for a few cycles	Usually less than 2 times rated current and very short duration
Var injection and control capability	Yes — by adjusting exciter can regulate vars and inject either leading or lagging power into system	No var control — is a lagging source (capacitors needed to correct vars)	Some inverters can vary var output and provide reactive support, most operate near unity power factor
Response speed to load changes and var needs	Exciter and throttle require some time to respond	Induction generator does not have exciter control. Throttle response requires time	Inverter can respond almost instantly to detected conditions to control power and vars if suitably designed with energy storage on bus
Active anti-island	No	No	Yes

Source: EPRI PEAC, "Distributed Generation Course," 2002.

be specified. A somewhat more accurate model is to keep the total power constant while the balance of the three phases is determined by the sequence parameters of the generator (Chen et al., 1991). For induction generators, the reactive power changes with voltage, but the changes are small enough that they can be modeled as constant power devices. See Barker and Johnson (2002) for more on modeling rotating machines for different scenarios.

If a synchronous generator is operated in a voltage-regulating mode, the best load-flow model is as a voltage-regulating, constant real power device. The reactive power limits would also need to be specified.

A constant-power model accurately describes forced-commutated inverters (for microturbines, fuel cells, photovoltaics, etc.), such as those using pulse-width modulation. The inverters have control loops that control both real and reactive power. Line-commutated inverters can also be modeled as a constant power. They are limited in that they can only absorb reactive power and cannot supply vars to the system.

Fault contributions of rotating generators can be modeled with most distribution short-circuit programs using the appropriate subtransient and transient characteristics of the generator. Short-circuit analysis can be done using a standard short-circuit program. Although the addition of distributed resources increases the time-varying nature of the fault current, it is not enough to warrant using a stability or EMTP-type model for fault analysis. Industrial systems with large local generators are still analyzed properly with standard short-circuit analysis programs.

When considering the effects on protection coordination with distributed generators, we can use standard protection programs. With fault contributions from only one source, curves of two overcurrent devices can be plotted on a time–current curve that will demonstrate visually the coordination over a range of currents. With contributions from different sources, it is more difficult to plot curves in a time–current graph. It is more useful to do a table of operating times of different devices for faults at different locations.

More complicated transient modeling is sometimes warranted for large machines or large penetration levels on distribution lines (especially on weak circuits). Transient simulations provide better answers on possible problems related to ferroresonance, voltage flicker, stability, and islanding (EPRI TR-111490, 1998; EPRI TR-112737, 1999).

14.2 Islanding Issues

Islanding is a major interconnection issue. Islanding is a situation where one or more generators and a portion of the utility system operate separately from the rest of the utility system. The formation of an unintentional island is a problem for the utility company. The most important concerns are

- Worker and public safety
- Damage to utility and customer equipment due to out-of-phase reclosure
- Voltage problems
- High overvoltages to utility equipment and customers caused by neutral shifts or ferroresonance

Distributed generators create danger for workers and the public. Line crews might work on a section of line that they thought was de-energized. A distributed generator could energize this line, even though it is disconnected from the utility source. The danger also extends to the general public. Islanded generators may energize downed conductors within public reach (that might have been de-energized by upstream utility switchgear had the island not developed).

Once an island forms, it typically drifts out of phase with the utility system voltage. If the main system is closed into the island, the out-of-phase reclosure may cause damage to the generator, customer loads, and/or utility switchgear as well as being a significant power quality disturbance for customers upstream of the island. An island may also prevent the clearing of fault currents on the system, increasing damage at the location of the fault, and perhaps burning conductors down.

The most common means to prevent islanding is to use voltage and frequency relays on the generator to trip whenever either of these two parameters migrate outside a selected window. This form of islanding protection is known as *passive* protection. It prevents islanding in most cases because when a section of the distribution system and one or more generators separate together, the output of the generator will not match the load on the island. For synchronous or induction generators, the voltage and frequency will drift, which will trip the relays in a short time. Typically the relays are set to a tight frequency range of perhaps ±1 Hz or even ±0.5 Hz. Voltage relays have a bit wider window to allow for typical voltage regulation excursions on the circuit (±5 to 10% is typical). Later in this chapter, specific setting requirements per IEEE standards are discussed.

Synchronous generators, induction generators, self-commutated inverters, and line-commutated inverters — all can island. Synchronous generators and self-commutated inverters are most likely to island because they do not require external excitation. Induction generators and line-commutated inverters can island if they have external excitation, either from capacitor banks or from other generators in the island.

14.2.1 Effect of Transformer Connections on Overvoltages

Generators fed from a transformer with an ungrounded high-side connection can drive a normally grounded utility circuit as an ungrounded island. When a single line-to-ground fault occurs on the distribution system and the sub-

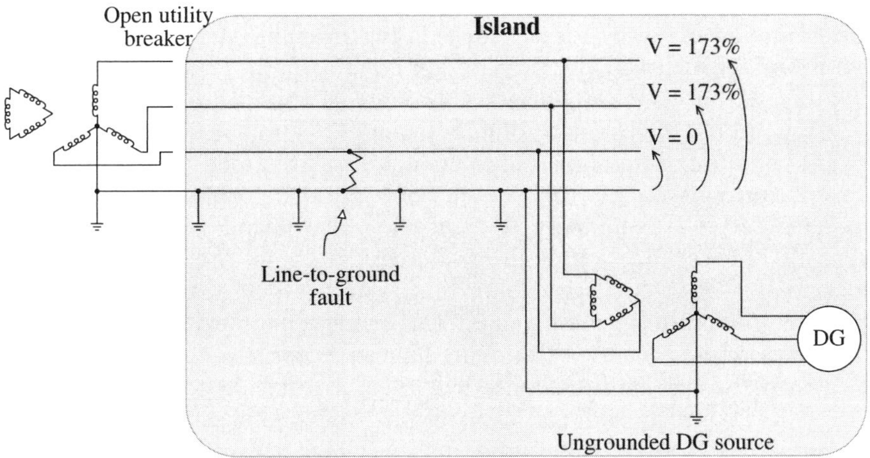

FIGURE 14.5
An ungrounded generator connection driving an island with a line-to-ground fault.

station breaker (or a recloser) opens, the system becomes a three-wire, ungrounded system driven by the ungrounded generator. The fault moves the potential of the ground to the phase potential (see Figure 14.5). That leaves the line-to-neutral voltage on the unfaulted phases equal to the line-to-line voltage ($\sqrt{3}$ = 1.73 times the normal line-to-neutral voltage).

The interconnection transformer configuration and the generator grounding determine if the application is effectively grounded. The most common transformer configurations (the connection notation indicates the primary-side connection, then the secondary-side [generator-side] connection) are

- *Delta – grounded wye* — Leaves the circuit ungrounded.
- *Grounded wye – grounded wye* — Can effectively ground the source *if* the generator is effectively grounded. Is ungrounded if the generator is ungrounded.
- *Grounded wye – delta* — Provides the best way to effectively ground a generator interconnection, but creates a ground source.
- *Grounded wye – delta with a grounding reactor on the high-side wye* — Limits the ground fault source while still maintaining grounding.

Transformer/generator grounding involves a tradeoff. The best grounding arrangements create sources for ground current that may interfere with distribution protection. Limiting ground current creates an interconnection that is not grounded as well (higher overvoltages).

Grounded-wye – grounded-wye transformers are widely used and familiar to utilities. They are often the best connection if the generator is grounded. If the generator neutral connection is not grounded or does not meet effective grounding requirements, then the transformer bank does not effectively ground the source even though the neutral connections to the transformer are grounded on both sides. Some rotating machines are not designed to withstand the forces due to a line-to-ground fault on the generator terminals. For this reason, they are grounded through an impedance or ungrounded (so they are not effectively grounded). Some inverters need isolation from ground and are not designed to operate with a grounded-wye winding on the inverter side of the transformer. For grounded rotating machines, the grounded-wye – grounded-wye connection also supplies significant zero-sequence current (it is a grounding source).

Generator installers often prefer the grounded-wye – delta transformer (with the delta on the generator side). This connection grounds the connection regardless of the generator grounding (the generator may be ungrounded or grounded through an impedance to protect it from line-to-ground faults). Other advantages are reduced harmonics, isolation between the primary and secondary for ground faults, and a smaller voltage sag seen on the generator terminals. The main drawback to using the grounded-wye – delta connection is that it is a grounding bank; it feeds ground faults on the primary. This can interfere with distribution system coordination and desensitize ground relays, improperly blow fuses, or falsely operate sectionalizers, reclosers, or breakers. Large circulating currents in the delta winding on the secondary (usually due to ground faults) can also damage the interconnection transformer. Adding a reactor to the neutral connection of the primary-side wye reduces the grounding transformer effect while still maintaining acceptable grounding.

If generators interface through ungrounded transformer connections, the preferred way to control overvoltages is with a ground overvoltage relay (a 59G, discussed in the next section).

Some large generators are applied right at distribution primary voltage, without an interfacing transformer. This has the same effects as using a grounded-wye – grounded-wye transformer. If the generator is ungrounded, islanded overvoltages are possible. Solidly grounding the generator grounds the source.

Arcing ground faults are another possible overvoltage mode on islands supported by ungrounded generators. On ungrounded systems, intermittent arcing grounds can create switching surge overvoltages from the restriking of the arc (see Chapter 13). A system that is designed to be ungrounded has only line capacitance between the phases and ground. An ungrounded island has the line capacitance and line-to-ground load. The load may help control arcing ground faults. If the island has grounded-wye capacitor banks, arcing ground fault overvoltages may be more likely. This is an area that needs more study.

14.2.1.1 Overvoltage Relays and 59G Ground Fault Detection

Distributed generators usually have over- and undervoltage relays. A typical overvoltage relay uses a type 59T time-delay relay with a pickup of 5

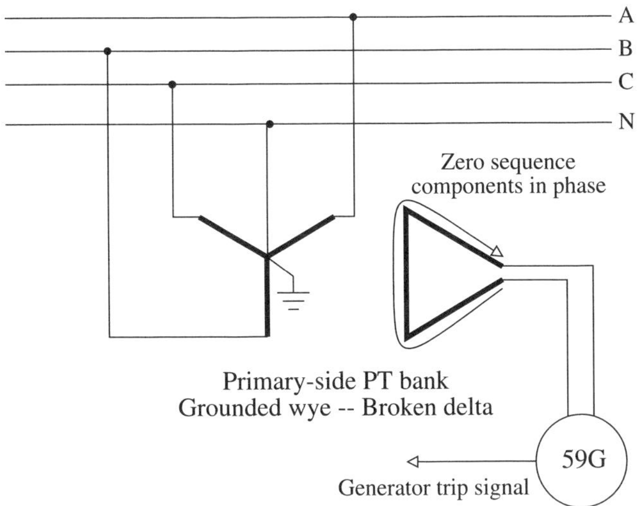

FIGURE 14.6
Ground fault overvoltage detection scheme (59G).

to 10% above nominal voltage. The time setting should be set above the normal clearing time of the feeder relays but less than the substation breaker reclosing time. A reasonable setting is 30 to 60 cycles (ANSI/IEEE Std. 1001-1988). It is best if an instantaneous element (59I) is also used, although a higher setting (such as 40% above nominal) may be needed to prevent excessive nuisance trips. For overvoltages, primary-side measurements are best, especially if the primary winding is ungrounded. Secondary measurements cannot detect the neutral shift on an ungrounded island with a ground fault. To detect and trip on overvoltages, every phase must have a relay, not just on an individual phase and not on an average of the three phases. Undervoltage relays also provide backup for neutral-shift overvoltages by detecting the voltage sag on the phase with the line-to-ground fault.

The 59G relay detects neutral-shift overvoltages with the connection shown in Figure 14.6. The grounded-wye – broken-delta potential transformer arrangement sums the three phase-to-neutral voltages.

The main drawback of relaying schemes is that the overvoltage still occurs; even a short-duration overvoltage may damage some equipment. The overvoltage duration is limited by the speed of the relays and breaker. Depending on the speed of the relaying, these limits may not coordinate with the temporary overvoltage capability of some arresters or sensitive customer equipment. Also, if the relaying fails to operate, the overvoltage could last for some time. Another disadvantage of this approach is that primary-side potential transformers are required. Because of this, many utilities avoid ungrounded transformer connections.

14.2.1.2 Effectively Grounding a Grounded-Wye – Grounded-Wye Transformer Connection

Effectively grounded systems limit the overvoltage caused by a neutral shift to 25% above nominal. In effectively grounded systems the ground fault current is 60% or more of the available three-phase fault current. A grounded-wye – grounded-wye transformer is effectively grounded if the generator applied to it meets effective grounding requirements. The conditions for effective grounding are approximately (ANSI/IEEE C62.92-1987):

$$X_0/X_1 \leq 3 \text{ and } R_0/X_1 \leq 1$$

For effective grounding during islanding for a ground fault on the primary terminals, the generator must be solidly grounded or the impedance must be small enough to meet the criteria above. For a rotating generator with a grounding reactor, the criteria (EPRI 1000419, 2000) is

$$X_{GN} \leq X_{T1} + X_{G1} - X_{T0}/3 - X_{G0}/3$$

where

X_{GN} = generator neutral reactance
X_{T0} = transformer zero-sequence reactance
X_{T1} = transformer positive-sequence reactance
X_{G0} = generator zero-sequence reactance
X_{G1} = generator positive-sequence reactance

14.2.1.3 Sizing a Neutral Grounding Reactor on a Grounded-Wye – Delta Connection to Maintain Effective Grounding

A grounding reactor added to a grounded-wye – delta transformer limits the zero-sequence fault contribution and limits the circulating current in the delta winding. In order to ensure that the connection still maintains an effectively grounded system during a possible island, the grounding reactor cannot be too large. The criteria for effective grounding is: $X_0/X_1 \leq 3$. This depends on the characteristics of the generator and interconnection transformer. For a primary ground fault at the transformer, the sequence impedances are

$$X_0 = X_{T0} + 3X_N$$

$$X_1 = X_{T1} + X_{G1}$$

where

X_N = neutral reactance
X_{T0} = transformer zero-sequence reactance
X_{T1} = transformer positive-sequence reactance
X_{G1} = generator positive-sequence reactance

To keep the interconnection effectively grounded, limit the size of the reactor according to (EPRI 1000419, 2000):

$$X_N \leq X_{T1} + X_{G1} - X_{T0}/3$$

Note that this only guarantees effective grounding for a fault right at the grounding transformer. For faults at other points on the feeder, add the zero and positive-sequence line impedances between the fault location and the generator to X_{T0} and X_{T1} If line impedances are considered, then the zero-sequence line resistance should also be considered. In the equations above, the resistance portion has been ignored because the X/R ratios on generators and transformers are high. Another requirement for effective grounding is that $R \leq 1$. Normally, these are small, so we can ignore them except for a very large generator on a weak system.

14.2.2 Anti-Islanding Protection

At most sites, voltage and frequency relays are the main protection against islanding. When a utility protective device opens and leaves an island, normally, the generation and load will not match. If the generation exceeds the load, the voltage rises and the generators speed up. The opposite occurs if the load exceeds the generation. In either case, when voltage or frequency goes outside of normal limits, voltage or frequency relays should trip the generators.

Over- and undervoltage relays are an important line of defense against islanding. Typically, this consists of

- *Overvoltage time-delay relay (59T)* — A pickup of 5 to 10% above nominal voltage is generally used. The time setting should be set above the normal clearing time of the feeder relays but less than the substation breaker reclosing time. A reasonable setting is 1/2 to 1 sec.
- *Instantaneous overvoltage element (59I)* — An instantaneous overvoltage element is recommended (but not always used). A higher setting (such as 40% above nominal) may be needed to prevent excessive nuisance trips.
- *Undervoltage time-delay relay (27T)* — A pickup of 5 to 10% below nominal voltage is generally used. Similar time settings should be used as the overvoltage time-delay relay.
- *Instantaneous undervoltage relay (27I)* — As with the overvoltage instantaneous element, a loose setting is necessary to prevent nuisance trips (30% below nominal is reasonable).

The time-delay relays are time delays; they are not inverse-time relays. If the voltage on a 59T relay is above the magnitude setting for longer than the time setting, it trips. If the voltage drops back below the magnitude setting, the timer resets after the given reset time. Many generator-protection relays allow multiple pairs of magnitude and time-delay settings to achieve a stair-step voltage-time characteristic.

For overvoltages, it is best if the detecting potential transformers are on the primary side of the generator transformer (especially if the primary winding is ungrounded). For example, the neutral shift on an ungrounded island with a ground fault will not be detected on the secondary side of the transformer. To detect and trip on overvoltages, individual relays should be on each of the three phases (not just on an individual phase and not on an average of the three phases). The undervoltage relays will also provide backup for neutral-shift overvoltages by detecting the voltage sag on the phase with the line-to-ground fault.

The over- and underfrequency relay (81U/O) is another line of defense against islanding. Trip thresholds of ±1% are common. Wide ranges of time delays are used for a 1% setting: 0.1 sec to many seconds. Some utilities specify a wider range for the underfrequency setting to allow the generator to stay in for low-frequency events caused by system-wide disturbances. The low frequency indicates a need for generation, so it helps to have the distributed generation remain connected.

Integrated generator relay packages are available from several manufacturers. These provide voltage and frequency relays along with other interconnection functions such as synchronization (and possibly all of the generator protection functions as well).

EEI (2000) provides guidance for applying anti-islanding relaying and other interconnection equipment for several different types of generators and interconnection transformers.

With voltage and frequency relays, there is still some chance that evenly balanced generation and load can drive islands. Because voltage and frequency relays are not 100% reliable, some large distribution generators have transfer-trip protection schemes. Breakers (or reclosers) upstream of the generator send a trip signal to the generator — if the breaker operates, it trips the generator. This requires a reliable communication line between the breaker and the generator.

14.2.3 Active Anti-Islanding

Inverter-based generators can use more sophisticated ways to detect islands. If the generation and load are almost equal, an island may take some time to drift in voltage and frequency. *Active* anti-islanding inverters attempt to push the voltage and/or frequency out of normal limits. Under normal conditions, the inverter cannot budge the voltage or frequency; with the inverter connected to the utility system, the utility system fixes the voltage and current (assuming the utility system is much bigger than the generation).

The most promising active anti-islanding techniques are the Sandia voltage-shift and frequency-shift methods (Stevens et al., 1999). These use positive feedback to make the voltage or frequency unstable if an island is being driven by generators with this form of anti-islanding. The unstable generator in an island pushes the voltage or frequency out of range, where standard voltage or frequency relays trip the generator. Because it uses positive feedback, the Sandia method allows several of these inverters to work together. Other methods may act to cancel each other; some might try to raise frequency while others try to lower it.

With the frequency-shift method, if the inverter detects that the frequency is increasing (even slightly), the controller injects a current with an even faster frequency. On a weak system, the current controller starts to control the frequency; and in this example, the inverter forces high frequency on an island, which trips the frequency relays.

Fast voltage control devices on distribution circuits may counteract attempts to change the voltage by inverter-based generators. Most regulators and capacitor banks are too slow to interfere in this manner. But, other generators operating in a voltage control mode will hold the island voltage intact. Rotating generators and voltage-source inverters act to stabilize the island's frequency and voltage.

IEEE Std. 929-2000 and UL 1741 (2001) specify an active anti-islanding test for inverter-based relays designed to test any kind of active anti-islanding inverter. Under balanced generation and load and a resonant capacitor-inductor tuned to the fundamental frequency (which tends to hold the frequency), the inverter must disconnect within 2 sec. The test does not verify the performance of multiple inverter-based generators in parallel (with the same type of anti-islanding or with different types of anti-islanding). And, the test does not verify the performance of anti-islanding with rotating generators or motors in the island.

14.2.4 Relaying Issues

Recent industry standards for distributed generator interconnection have specified standard trip thresholds. This is an effort between manufacturers and utilities to ease the complications of interconnection. IEEE 929 and UL 1741 have recommended the thresholds in Table 14.3. The 929 recommenda-

TABLE 14.3

Standard Trip Thresholds for Distributed Generators per the IEEE Std. 929-2000 and UL 1741 (2001)

Voltage Setting						Trip Time	
Volts (120-V Nominal)			Percent			Cycles	Seconds
	V	< 60		V	< 50%	6 cycles	0.1 sec
60 ≤	V	< 106	50% ≤	V	< 88%	120 cycles	2 sec
106 ≤	V	≤ 132	88% ≤	V	≤ 110%	Normal Operation	Normal Operation
132 <	V	< 165	110% <	V	< 137%	120 cycles	2 sec
165 ≤	V		137% ≤	V		2 cycles	0.033 sec

TABLE 14.4

Trip Thresholds for Distributed Generators Based on IEEE P1547-D11-2003

Voltage Setting			Trip Time
	V	< 50%	0.16 sec
50% ≤	V	< 88%	2 sec
110% <	V	< 120%	1 sec
120% ≤	V		0.16 sec

Note: Base voltages are the nominal voltages stated in ANSI C84.1. DR ≤ 30 kW: Maximum Clearing Time. DR > 30 kW: Default Clearing Time or the utility operator may specify different voltage settings or trip times to accommodate utility requirements.

tions are targeted at photovoltaic systems rated at 10 kW or less. UL 1741 applies to all inverter-based generators.

IEEE 929 specifies the voltage measurements at the point of common coupling. UL 1741 specifies a test for an inverter-based DG. No mention is made of the exact measurement location, but it implies measurement at the generator. UL 1741 specifies line-to-ground voltages for three-phase connections, and the DG should cease exporting when any of the three voltages exceeds the limits in Table 14.3. At the time of this writing, IEEE is working on a standard for all generators, and the trip thresholds are similar; Table 14.4 shows the draft settings.

The main concerns with the standard overvoltage trip thresholds are that

- The 110% trip limit on overvoltages allows generators to create voltage in excess of ANSI limits (ANSI C84.1-1995).
- The draft P1547 allows a relatively long time (0.16 sec or 9.6 cycles for 60 Hz) for overvoltages above 120% to exist. IEEE 929 and UL 1741 both require the generator to cease injecting power within 2 cycles if the relay detects an overvoltage greater than or equal to 138%.

Where should relays measure? Right at the generator is easiest, but not always best. To avoid islanding problems with large three-phase generators, measurement on the distribution primary is best. Transformer connections can distort the voltages and make relaying harder; secondary-side measurement on a delta – wye or wye – delta transformer connection cannot detect the overvoltages or undervoltages from a neutral shift on the primary. A second transformer with a delta – wye connection between the utility transformer and the generator interconnection point further distorts the voltages. The distribution primary is usually the most sensitive location for measuring distribution-system overvoltages and undervoltages.

Overall, the best place for relay monitoring is on the primary using line-to-ground voltages. Smaller generator applications (< 1 MW) may use other measuring locations, but consider the transformer connections to ensure effective relay sensitivity. Secondary measurements (either at the generator or at the service entrance) are appropriate for very small generators.

Another one of the questions about relaying set points is whether to use line-to-line voltages, line-to-ground voltages, or both. Generally, line-to-ground voltages are more sensitive, so they should be used, but it is not always straightforward. Figure 14.7 shows where a neutral shift has occurred on the distribution system to a line-to-ground fault. Phase A has a voltage sag, and phases B and C have voltage swells due to the neutral shift. This

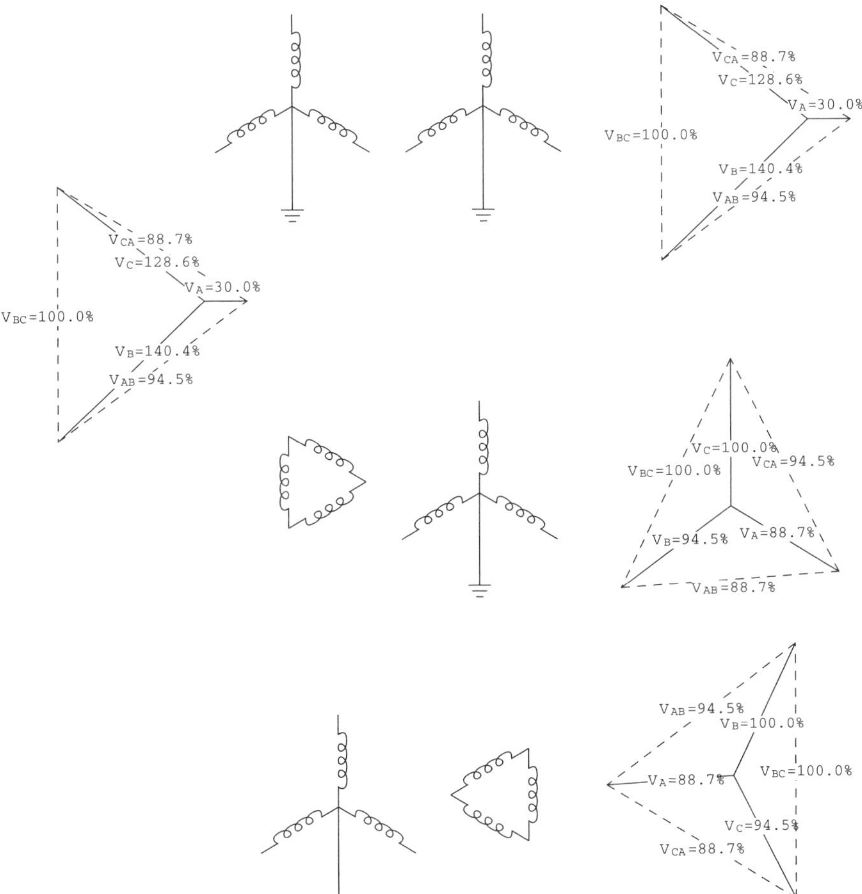

FIGURE 14.7
Line-to-line and line-to-ground voltages on the primary and secondary of several transformer connections for an overvoltage due to a neutral shift for a fault on primary phase A. (Copyright © 2001. Electric Power Research Institute. 1005917. *Distributed Generation Relaying Impacts on Power Quality.* Reprinted with permission.)

is a more severe neutral shift than is usual on a four-wire multigrounded distribution circuit; it could represent an island being driven by distributed generators. If a grounded-wye – grounded-wye transformer is used, the line-to-ground voltages on the secondary are most sensitive. If a delta – grounded-wye transformer is used, the line-to-line voltages on the secondary are slightly more sensitive (although, overall the ability to sense primary line-to-ground overvoltages is reduced by the delta – wye connection). A grounded-wye – delta connection has line-to-ground voltages more sensitive to the overvoltage (again, the ability to sense primary line-to-ground overvoltages is reduced because of the transformer connection). The voltages shown assume that the DG does not change the voltage on the primary or secondary, when in reality the DG will normally be changing the voltage on the secondary and maybe even the primary.

One important point is that all three voltages must be measured (whether they are line-to-line or line-to-ground voltages). If voltages are assumed balanced and only one of the three voltages is used, sustained overvoltages could occur on the unmonitored phases.

Power quality disturbances also impact generators because the generator must trip for voltage excursions. An analysis of EPRI's Distribution Power Quality (DPQ) data showed that the number of generator trips at most sites is not particularly severe. Generators at most sites should have less than five trips per year when using the UL 1741/IEEE 929 or IEEE P1547 settings (see Figure 14.8). Trips become excessive only if the generator undervoltage relaying is set at a very sensitive level.

Distributed generators based on inverters can do much of the interconnection relaying in the inverter controller. Utilities worry about inverter-based relays, mainly because

- Inverter-based relays are not utility-grade.
- Testing in the field is very difficult. (They do not have external test points for relay test sets.)

Inverter-based relays do have the advantage that they can perform anti-islanding relaying as discussed in the previous section.

Many utilities require utility-grade relays for small installations, and essentially all require them for large installations. But, what is a utility-grade relay? There is no fixed definition, but normally, they must pass industry standards (ANSI/IEEE C37.90-1989; ANSI/IEEE C37.90.1-1989; IEEE Std. C37.90.2-1995). Utility-grade relays are normally more robust than industrial relays and have test ports for external testing.

Inverter-based generators do not have external test ports for testing. Separation of the relaying component from the other controller functionalities is difficult to do because the inverter controller uses the voltage and current sensing for other control aspects.

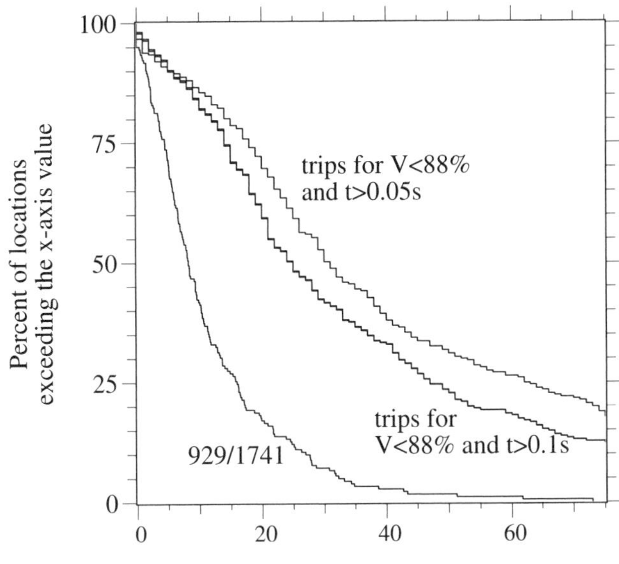

FIGURE 14.8
Comparison of the number of generator trips with the IEEE 929/UL 1741 voltage/time thresholds to very sensitive trip settings (less than 88% voltage for 0.1 or 0.05 sec). (Copyright © 2001. Electric Power Research Institute. 1005917. *Distributed Generation Relaying Impacts on Power Quality.* Reprinted with permission.)

14.2.5 Self-Excitation

During islanding, the generator can resonate with system capacitor banks (see Figure 14.9). This series resonance can happen with any type of interconnection transformer configuration, with or without ground faults on the system. The simplest resonance is a series resonance between the generator subtransient impedance and system capacitors. Load on the island helps hold the voltage down, so the highest voltages occur under light load. Transformer saturation is not enough to significantly reduce this type of

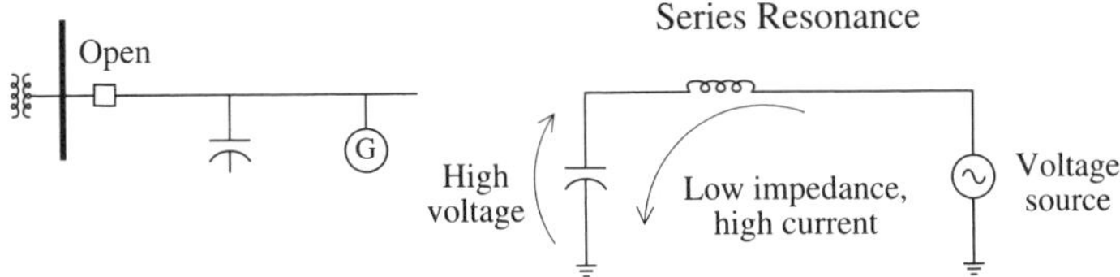

FIGURE 14.9
Series resonance between a generator and capacitor bank during islanding.

overvoltage. This resonance is also referred to as self-excitation, the same phenomenon seen on induction motors with capacitor banks. It is also similar to a resonance seen on distribution feeders that feed low-voltage secondary networks and have capacitor banks. After a fault, the backfeed through the network transformers can resonate.

The circuit resonates when the power-frequency impedance of the capacitor equals the impedance of the inductance. And, overvoltages still occur even if the impedances do not exactly match. A criteria to determine if self-excitation may occur is that the ratio of capacitance to inductance is within a ratio of one to two (in either direction):

$$0.5 < X_C/X_L < 2$$

where

X_C = impedance of the capacitor banks = $(\text{kV}_{\text{LL}})^2/(\text{Mvar})$

X_L = impedance of the system inductance, which includes the generator impedance, the transformer impedance, and the line impedance between the generator and the capacitor bank

Load on the circuit helps hold down the overvoltage. For example, if the load is entirely resistive (R), the magnitude of the per unit overvoltage at the capacitor is

$$V_C = \frac{RX_C}{\left|RX_L - X_C(R + X_L)\right|}$$

For cases where problems might occur, some options are

- *Reactance* — Change the reactive impedance (generator size or transformer impedance).
- *Capacitors* — Remove or change the size of the capacitor banks.
- *Relaying* — Rely on the overvoltage relaying to remove the generator during an overvoltage condition. Use an instantaneous element (59I), apply relays on each of the three phases, and use potential transformers for the relays on the primary side of the interconnection transformer.

During a line-to-ground fault with an ungrounded transformer connection, a different resonant mode can occur (Dugan and Rizy, 1984). The circuit resonates with a different combination of impedances. The zero-sequence capacitance resonates against a combination of the positive- and negative-sequence impedances and capacitances (see Figure 14.10). This overvoltage adds to any overvoltage due to a neutral shift.

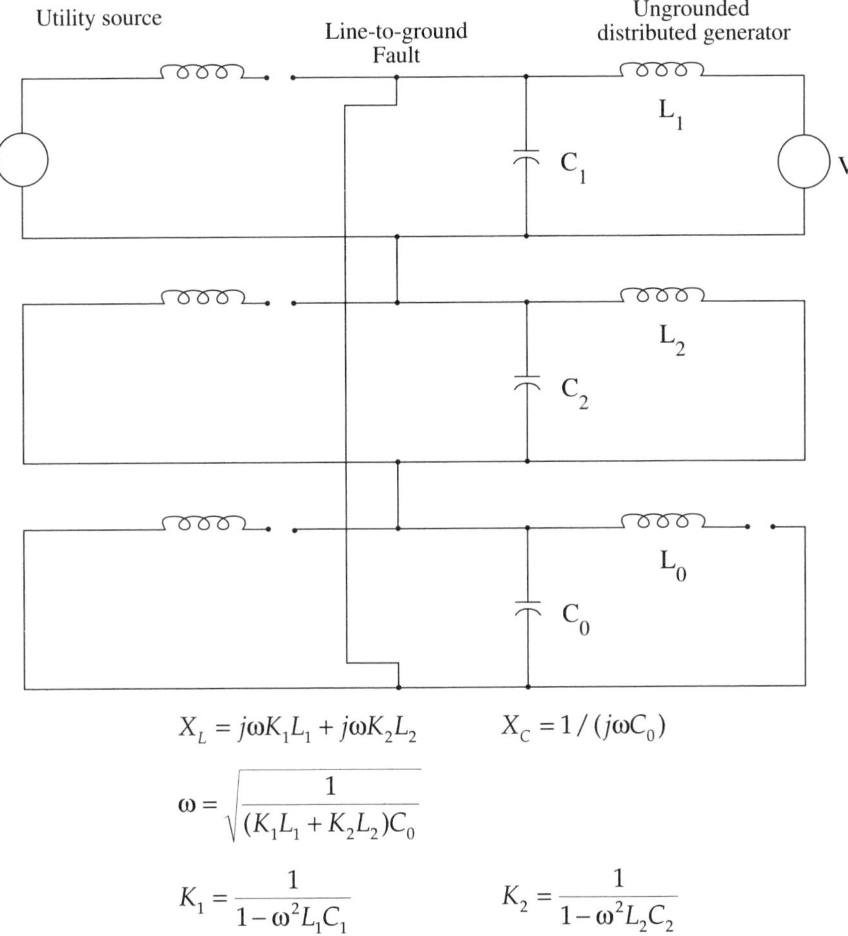

FIGURE 14.10
Series resonance between a generator and capacitor banks during a line-to-ground fault on an ungrounded islanding.

The zero-sequence capacitance is most likely from grounded capacitor banks, but long cable runs may have enough capacitance. The solutions given above for positive-sequence resonances also help with this zero-sequence resonance. Another option for this zero-sequence resonance is to unground capacitor banks (to eliminate the zero-sequence capacitance).

14.2.6 Ferroresonance

A generator in an island can drive the circuit into ferroresonance (Feero and Gish, 1986; Wagner et al., 1989). The peak voltage during this ferroresonance can reach three per unit. Both induction and synchronous generators can create ferroresonance, and it can occur with all three phases connected (single phasing is normally involved with ferroresonance). The resonance can occur no matter how the generator interconnection transformer is configured (although the overvoltage is worse if the ferroresonance occurs simultaneously with a neutral shift on an ungrounded island). Four conditions are necessary for DG islanding ferroresonance to occur:

1. The island driven by the generator must be isolated from the utility.
2. The generator must supply more power than there is load on the island.
3. The isolated circuit must have enough capacitance to resonate (30 to 400% of the generator rating). This can be due to utility capacitor banks or from capacitor banks at the generator.
4. A transformer group must be present in the island.

The ferroresonant circuit is very similar to the self-excited case in Figure 14.9, with the only addition being the distribution transformer group. When the generator and the capacitance resonate, the overvoltage on the distribution circuit causes the transformer to saturate. The saturated transformer has a low impedance. Consider it like a switch: when it is saturated, it is a short circuit; when it is not saturated, it is an open circuit. The switching action of the saturating transformer causes the chaotic ferroresonant type waveform. Ferroresonance tends to widen the range of capacitance over which the generator can resonate with the capacitor bank.

Solution options for this type of ferroresonance include

- *Reconfiguration* — Changing the distribution system characteristics to change the criteria given above (limit or expand the area that could island or remove or change the size of the capacitor bank).
- *Relaying* — Rely on the overvoltage relaying to remove the generator during an overvoltage condition. Use an instantaneous element (59I), apply relays on each of the three phases, and use potential transformers for the relays on the primary side of the interconnection transformer.

14.2.7 Backfeed to a Downed Conductor and Backfeed Voltages

During a line-to-ground fault where one phase opens (either due to a fuse operation or a broken conductor), backfeed through a three-phase generator can cause issues on distribution systems. While this type of backfeed occurs without distributed generators (see Chapter 4), distributed generators make backfeed more likely.

With a distributed generator on the load side of the transformer, the backfeed current and voltage (EPRI 1000419, 2000) is

$$I_F = \frac{(A - 3Z_0Z_2)V + 3Z_0Z_2V_G}{3Z_0Z_1Z_2 + R_FA}$$

$$V_F = R_FI_F$$

where

A = $Z_0Z_1 + Z_1Z_2 + Z_0Z_2$
Z_1 = positive-sequence impedance of the generator, Ω
Z_2 = negative-sequence impedance of the generator, Ω
Z_0 = zero-sequence impedance of the generator and load, Ω
R_F = fault resistance, Ω
V = line-to-neutral voltage, V
V_G = line-to-neutral voltage equivalent of the generator, V

Under an open circuit ($R_F = \infty$), the backfeed voltage is

$$V_F = \frac{(A - 3Z_0Z_2)V + 3Z_0Z_2V_G}{A}$$

The presence of the distributed generator strengthens the backfeed current and voltage. The generator provides a source of voltage, and the impedance of a rotating generator (especially the low zero-sequence and negative-sequence impedances) increases the backfeed. An induction generator does not require supplementary excitation (capacitance) to backfeed under these circumstances.

Having a 59G ground overvoltage trip relay, a directional (or even nondirectional) ground relay, and phase undervoltage relays should trip the circuit in most cases of backfeed. It is best if relaying trips the primary-side protective device (usually this is a breaker) rather than just tripping the generator, since backfeeds can still occur through the transformer even if the generator is taken offline.

14.3 Protection Issues

Figure 14.11 shows a scenario where a distributed generator may falsely operate an upstream breaker, recloser, fuse, or sectionalizer for a fault just upstream of the protective device.

The fault current from the generator (and through the upstream protective device) for a three-phase fault can be found by the following equation:

$$I_{3p} = V/Z_1$$

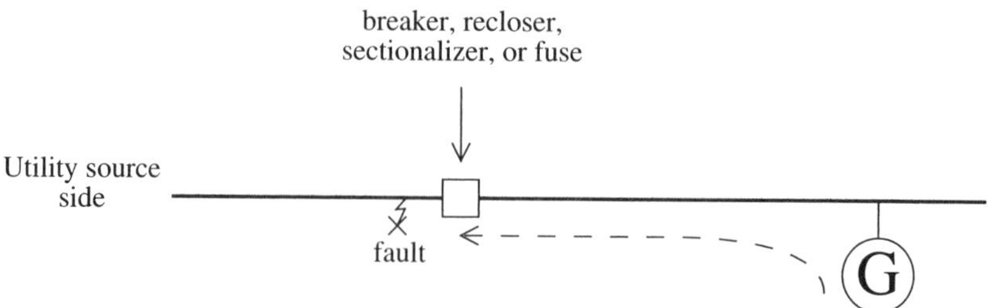

FIGURE 14.11
Scenario where a distributed generator trips an upstream device. (Copyright © 2000. Electric Power Research Institute. *Engineering Guide for Integration of Distributed Generation and Storage Into Power Distribution Systems*. Reprinted with permission.)

where

Z_1 = positive-sequence impedance between the generator and upstream device. This includes the equivalent generator impedance reflected to the primary-side voltage, the transformer impedance, and the line impedance from the generator to the upstream device.

V = line-to-ground voltage

This assumes that the generator voltage source is the same as the utility voltage source. A more conservative and appropriate choice is

$$I_{3p} = 1.1\ V/Z_1$$

since the internal voltage behind the generator is normally higher than one. For synchronous generators, use the subtransient reactance, $Z_1 = jX_d''$.

For a line-to-ground fault, it is more complicated because current can "route through" the generator impedances. The sequence diagram for this situation is shown in Figure 14.12. The sequence currents from the generator are

$$I_{G1} = \frac{Z_{S1}}{Z_{S1} + Z_1}\left(\frac{V}{Z'}\right)$$

$$I_{G2} = \frac{Z_{S2}}{Z_{S2} + Z_2}\left(\frac{V}{Z'}\right)$$

$$I_{G0} = \frac{Z_{S0}}{Z_{S0} + Z_0}\left(\frac{V}{Z'}\right)$$

where

$Z' = Z_{S1} \,||\, Z_1 + Z_{S2} \,||\, Z_2 + Z_{S0} \,||\, Z_0$

Z_{S0}, Z_{S1}, Z_{S2} = zero, positive, and negative-sequence impedances upstream of the fault. This includes the line impedance, the substation

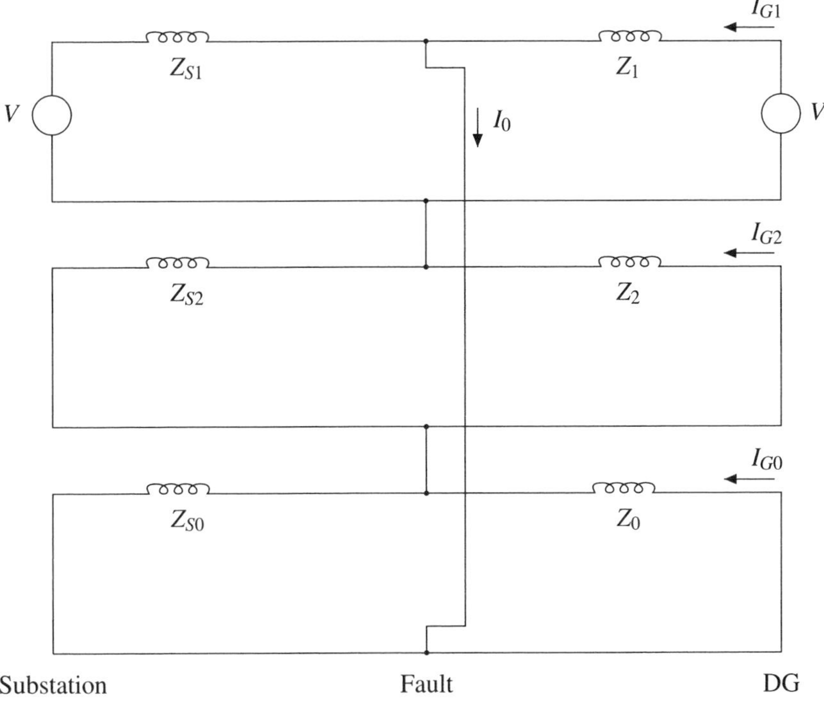

FIGURE 14.12
Sequence diagram of a line-to-ground fault with a downstream generator feed. (Copyright © 2000. Electric Power Research Institute. 1000419. *Engineering Guide for Integration of Distributed Generation and Storage Into Power Distribution Systems.* Reprinted with permission.)

transformer impedance, and the transmission equivalent impedance

Z_0, Z_1, Z_2 = zero, positive, and negative-sequence impedance between the generator and fault. This includes the equivalent generator impedance reflected to the primary-side voltage, the transformer impedance, and the line impedance from the generator to the upstream device

The phase relay on the faulted phase sees the following current:

$$I_{faulted\ phase} = I_{G1} + I_{G2} + I_{G0}$$

The ground relay operates on

$$I_{ground\ relay} = 3I_{G0}$$

While the line-to-ground fault current is normally smaller than the three-phase fault current, it is important because of

- *Relay sensitivity* — The upstream protective device (breaker, recloser, or sectionalizer) may have a ground relay that is more sensitive than the phase relay setting. Even if the ground-fault current is lower, increased ground-relay sensitivity may cause miscoordination.
- *Grounding sources* — Two common generator transformer configurations supply significantly more ground fault current; their zero-sequence impedance is low (see Figure 14.13):
    - Grounded-wye – delta transformer connection — This grounding transformer presents a zero-sequence impedance equal to the transformer impedance (about 5% of its rating).
    - Grounded-wye – grounded-wye transformer connection with a rotating generator solidly grounded — A rotating generator has low zero-sequence impedance (5% of its rating). The total zero-sequence impedance is the transformer impedance plus the gen-

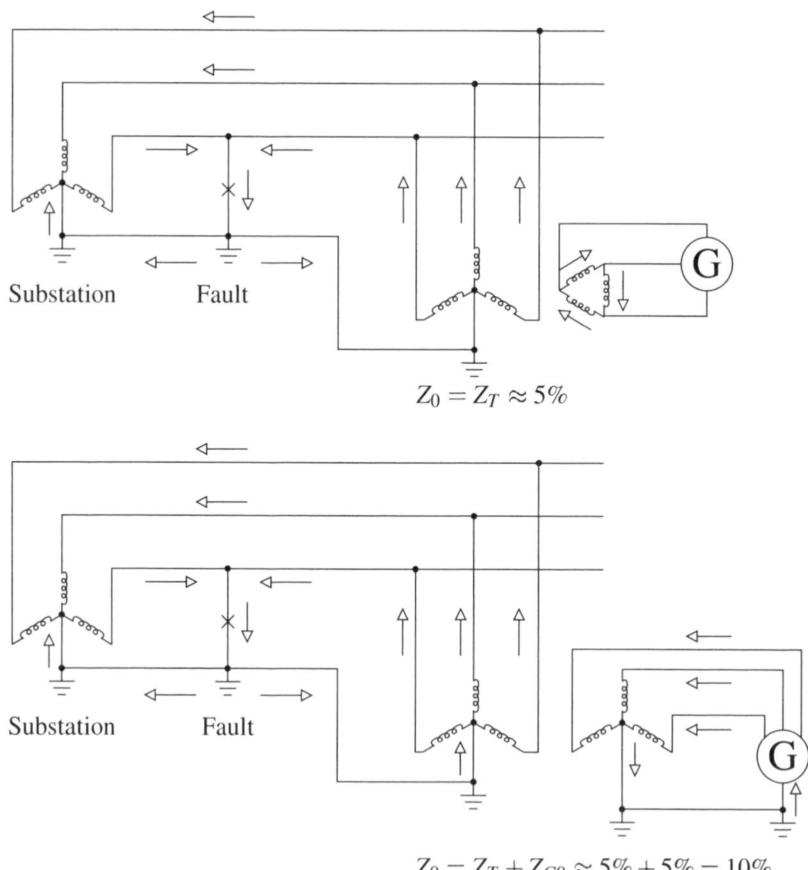

FIGURE 14.13
DG grounding source connections feeding single-phase faults.

erator impedance ($Z_T + Z_{G0} \approx 5\% + 5\% = 10\%$, assuming both the transformer and generator have the same kVA rating).

To determine whether the upstream device will misoperate, compare the fault current with the pickup setting of the device. If it exceeds the pickup, then examine the time-current characteristics of the protective device. If it is an instantaneous element, then assume it miscoordinates. If the device has a time-delay characteristic, then compare the operating time for the given fault current to the operating time of the next upstream device.

If false tripping of utility protective devices is found, there are several possible remedies. On the generator side, the corrective options are

- *Size* — Limit the size of the generator to the point where coordination is achieved.
- *Impedance* — Increase the impedance of the generator transformer or generator.
- *Ground faults* — If miscoordination occurs from ground faults, add a grounding reactor to the generator (for a grounded-wye – grounded-wye interconnection).

On the utility side, the options are

- *Settings* — Use a higher pickup setting on the upstream device or use a slower characteristic that coordinates with the generator contribution. Of course, this requires checking to make sure that the distribution circuit is still adequately protected and that the new setting coordinates with other protective devices on the distribution circuit.
- *Instantaneous* — Disable the instantaneous element or add a time delay to the instantaneous element of sufficient length to coordinate.
- *Fuses* — Use a larger and/or slower fuse.
- *Ground faults* — If it is a ground-protection issue with a grounded-wye – delta interconnection transformer, add a primary neutral reactor.

If the generator is a large rotating unit that is close to the nearby protective device, then there may be a significant offset to the fault waveshape. This may affect certain protective devices. If the device is close to miscoordinating given the equations above, then use a more advanced approach with a short-circuit program that accounts for the offset.

14.3.1 Tradeoff Between Overvoltages and Ground Fault Current

The transformer/generator grounding plays important roles in supplying ground fault currents and overvoltages that may occur during islanding.

Better grounding limits the overvoltage, but the generator supplies more ground fault current, which may affect coordination.

At one extreme are the sources that form grounding transformers that can provide significant fault current:

- *Grounded wye – delta* — The zero-sequence impedance of the interconnection is the transformer impedance, normally about 5% on its rating.
- *Grounded wye – grounded wye with the generator solidly grounded* — The zero-sequence impedance is about 10% (5% for the transformer and 5% for the generator).

At the other extreme are ungrounded connections, which provide no ground fault current but cause an overvoltage of 1.73 per unit on the unfaulted phases during a ground fault (note that these can feed ground faults, just not with zero sequence):

- Delta – delta
- Delta – grounded wye
- Grounded wye – grounded wye with the generator ungrounded

Where ground-fault coordination is a problem, in-between approaches are often best. Adding a grounding reactor to a grounded wye – delta or adding a grounding reactor to a generator limits the ground-fault current.

It is normally possible to do both — have effective grounding and reasonably limit the zero-sequence fault contribution.

Figure 14.14 shows the tradeoffs with overvoltages and grounding for an example case. The case has a generator with a grounded-wye – delta transformer connection on a 12.47-kV circuit, 4 mi (6.4 km) from the substation. The transformer has a reactor in the grounded-wye primary. The example assumes a 5% transformer impedance, 20% positive and negative-sequence reactances for the generator, and a 1.1 per-unit voltage behind the generator impedance. Reactor size determines the grounding effectiveness. The reactor impedance is shown as per-unit multiples of the transformer impedance. Except for very large generators relative to the distribution system capability, using a reactor can effectively ground the circuit (overvoltages less than 1.25 per unit) and still reasonably limit the zero-sequence current.

Note that a grounded-wye – grounded-wye transformer with the generator solidly grounded has similar results to the case with a reactor of 0.3 per unit in Figure 14.14. In this solidly grounded scenario, the generator supplies significant fault current.

14.3.2 Fuse Saving Coordination

The majority of faults on overhead distribution lines are temporary (meaning the fault will be cleared if power is interrupted and restored). Temporary

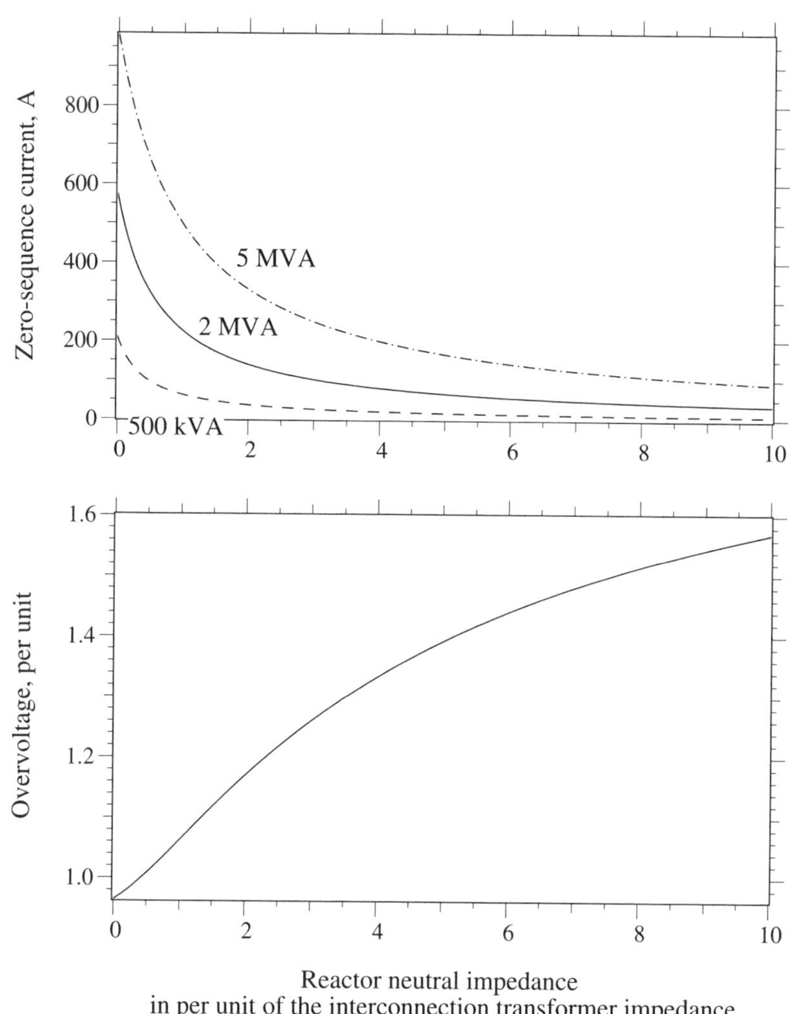

FIGURE 14.14
Overvoltages and zero-sequence fault contribution for a grounded-wye – delta interconnection transformer with a neutral reactor for different size generators (the overvoltage remains the same with size).

faults on lateral taps can be cleared by the feeder breaker before the lateral fuse blows. This is usually done with the instantaneous element of the breaker relay or recloser in the substation. This practice is known as fault selective feeder relaying or simply as "fuse saving." Distributed generators increase the fault current through the fuse, which may blow the fuse before the breaker opens.

Locations with lower fault currents are where generators are most likely to blow fuses. At locations where fault currents are already high, fuses will not coordinate with a station breaker (so generators will not make the situ-

TABLE 14.5

The Smallest K-Link Fuse That Coordinates with a 5-Cycle Breaker with a Fuse-Saving Scheme with Fault Current Contributions from the Substation and Distributed Generator(s)

Generator Fault Current, A	Fault Current Contribution from the Substation, A								Generator kW[a]
	500	1000	1500	2000	2500	3000	3500	4000	
0	40K	80K	100K	140K	140K	200K	200K	200K	0
100	50K	80K	140K	140K	140K	200K	200K	—	432
200	50K	100K	140K	140K	200K	200K	200K	—	864
300	65K	100K	140K	140K	200K	200K	200K	—	1296
400	65K	100K	140K	140K	200K	200K	200K	—	1728
500	80K	100K	140K	140K	200K	200K	200K	—	2160
600	80K	140K	140K	140K	200K	200K	—	—	2592
700	100K	140K	140K	200K	200K	200K	—	—	3024
800	100K	140K	140K	200K	200K	200K	—	—	3456
900	100K	140K	140K	200K	200K	200K	—	—	3888
1000	100K	140K	140K	200K	200K	200K	—	—	4320

Note: No line impedance between the generator and the fault.

[a] Assuming V_{LL} = 12.47 kV and generator X_d'' = 20%.

Source: Copyright © 2000. Electric Power Research Institute. 1000419. *Engineering Guide for Integration of Distributed Generation and Storage Into Power Distribution Systems*. Reprinted with permission.

ation worse). Table 14.5 shows this for the commonly used K links. The smallest fuses that can coordinate during fuse saving are shown given the fault current from the substation and a distributed generator. The generator size in the last column is specifically for a 12.47-kV system with a synchronous generator with X_d'' = 0.2; the generator fault current contributions in the first column of the table can be applied for other applications.

14.4 Power Quality Impacts

14.4.1 Voltage Regulation

Distributed generation often helps voltage on a distribution circuit because it offsets some of the voltage drop caused by loads. But, distributed generation can cause voltages outside of ANSI C84.1 limits under the following scenarios:

- *Voltage boost* — Distributed generators can cause high voltages because they inject real power back upstream into the system, causing a voltage rise.
- *Regulator interaction* — Distributed generators can interact with line regulators, which can cause low or high voltages on a circuit depending on the scenario.

Distributed generators can influence distribution voltage regulation. Even if the generator is in a "voltage following" mode, it still influences voltage by injecting power. In many cases this is beneficial but can cause low or high voltages, depending on the scenario. Comfort et al. (2001) showed an example where a 5-MW combustion turbine raised the voltage to 130 V (1.086%) 3 mi from the substation on a 12.5-kV circuit.

There are also concerns with generators causing high voltages on distribution circuits because of reverse power flow. It is possible to estimate the effect of a generator by using the standard voltage drop equations with reverse power flow. The voltage drop along a feeder due to a load is approximately equal to

$$V_{drop} \approx I_R \cdot R + I_X \cdot X$$

where

V_{drop} = voltage drop along the feeder, V

R = line resistance, Ω

X = line reactance, Ω

I_R = line current due to real power flow (negative for a generator injecting power), A

I_X = line current due to reactive power flow (negative for a capacitor), A

Distributed generators raise voltage the most where X/R ratios are low. The real power portion causes the largest voltage rise when the line resistance is high. If the generator injects vars like a capacitor or has local capacitors, the voltage rise is worse.

Note that this approximation is no substitute for a proper load flow. It does not fully model the response of the load to the change in voltage, and it does not consider regulator response. It is useful for a first cut at estimating whether the voltage rise due to the generator might be a problem.

If voltage rise is a problem, there are several options:

- *Size* — Limit the size of the generator to below the level necessary to cause problems.
- *Relocation* — Relocate the generator to a more suitable location on the distribution circuit.
- *Vars* — Have the generator absorb more reactive power (by removing local capacitors or operating a synchronous generator or self-commutated inverter at reduced power factor). This is the opposite of what is normally done. Reducing the power factor of the generator causes voltage drop to counter the real-power voltage rise.
- *Operation time* — Limit the operation of the generator to peak periods of the day (when the voltage drop along the circuit tends to be highest). This is not an exact solution since the highest voltages at a particular point on a feeder do not always coincide with light load.

High voltages can occur at heavy load (especially close to a voltage regulator or capacitor bank on the distribution feeder).

- *Relaying* — Rely on the generator overvoltage relay to remove the generator during high voltages. This may work out fine if the high voltage conditions are infrequent.
- *Impedance* — A utility option would be to reduce the resistance of the lines and transformers from the substation bus to the generator. This would be done by using larger conductors on lines and cables and specifying lower copper losses on transformers. Another option is to build an express feeder to the generator.
- *Regulation equipment* — Another utility-side option is to add regulation equipment (capacitors or regulators) to counteract the voltage rise from the distributed generator.

Normally, the utility only needs to consider the service voltage, but utility engineers may become involved with cases of high voltage within the facility. ANSI C84.1 voltage standards account for voltage drop within a facility in the utilization voltage standard — they do not account for a voltage rise that generators cause.

Figure 14.15 shows a scenario where a generator just downstream of a regulator can cause low voltage on the end of the circuit due to the regulator's line drop compensation controller. The power injected by the generator fools the regulator into not raising voltage as much as it should.

Many regulators have a special concern — with reverse power, they ratchet to the extreme tap, causing high or low voltage (EPRI TR-105589, 1995). Most distribution regulators are bi-directional — they measure voltage on both sides of the regulator, and power flow determines which side the regulator uses for compensation. Bi-directional regulators are meant for distribution system locations where the source may change directions, commonly normally-open loops that can be reconfigured. The regulator always tries to regulate the voltage on the downstream side, determined by the direction of power flow. If a downstream distributed generator injects enough real power to reverse power flow on the regulator, the regulator thinks the source has moved to the other side. It tries to regulate voltage on the upstream side of the regulator (V_1 in Figure 14.16). If the measured voltage is higher than the setpoint, the regulator changes its tap to try to lower the voltage. Nothing happens to the voltage upstream of the generator (V_1) since the utility source holds the voltage constant. The regulator keeps trying to lower the voltage, moving from tap to tap until the regulator ratchets all the way to the limit, trying to lower voltage as much as possible.

What the regulator did was *raise* the voltage downstream of the regulator. It raised the voltage when the feeder voltage was already high — to the maximum amount. If the feeder voltage is lower than the regulator setpoint, the opposite occurs: the regulator ratchets to the extreme tap setting that

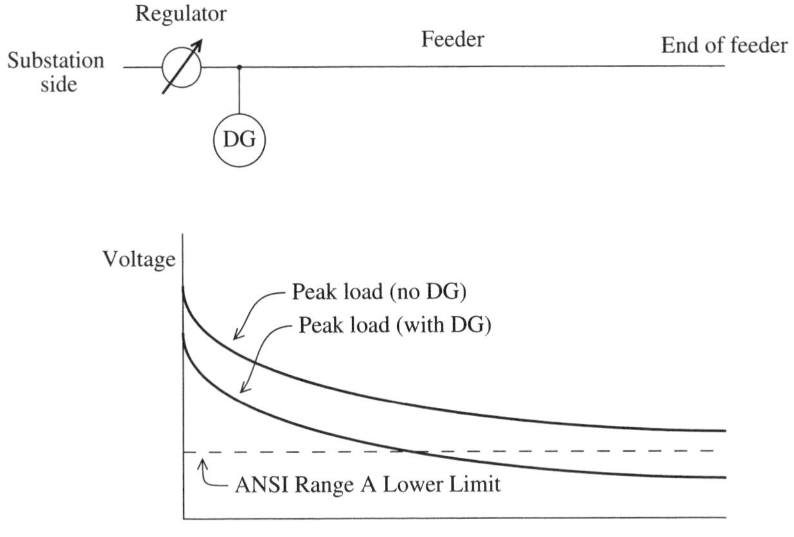

FIGURE 14.15
A generator just downstream of a regulator that leads to low voltage at the end of the feeder. (Copyright © 2000. Electric Power Research Institute. 1000419. *Engineering Guide for Integration of Distributed Generation and Storage Into Power Distribution Systems.* Reprinted with permission.)

causes even lower voltage downstream of the regulator. These scenarios may lead to voltage outside of ANSI C84.1 limits.

If the size of the generator or generators is less than 10% of the load at the regulator, problems are unlikely. If the size exceeds this simple screening criteria and the generator is close to the regulator, then more detailed study is warranted. If problems are found, options include the following:

- *Remove LDC* — Remove the line drop compensation on the regulator. Generally, the voltage setpoint must be raised to provide a good voltage profile on the circuit (if so, check that the end of the circuit does not have high voltages due to the generator).
- *Settings* — Reduce the line drop compensator settings and raise the voltage setpoint a smaller amount. This moves the regulator constant voltage point closer to the regulator. The key is the location of the constant voltage point (also called the load center — the fictitious point on the feeder where voltage is held constant). Generators upstream of this point lower the voltage downstream of this point. Generators downstream of the constant voltage point are okay.
- *Advanced controls* — If problems are due to reverse power flow (on bi-directional regulators), advanced regulator controllers are available to change operating mode during reverse power flow to prevent the regulator from dropping to the lowest tap. If power

Regulator fed from source 1: regulates V_2

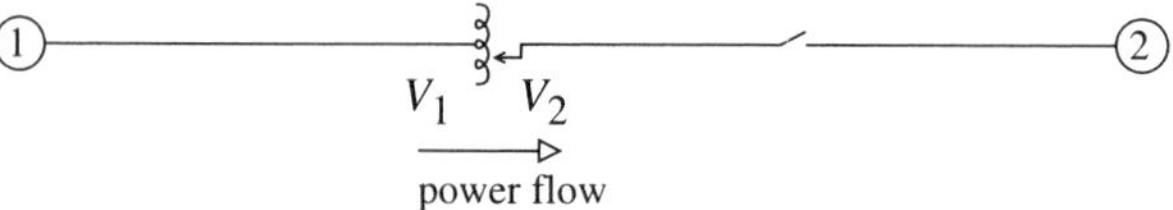

Regulator fed from source 2: regulates V_1

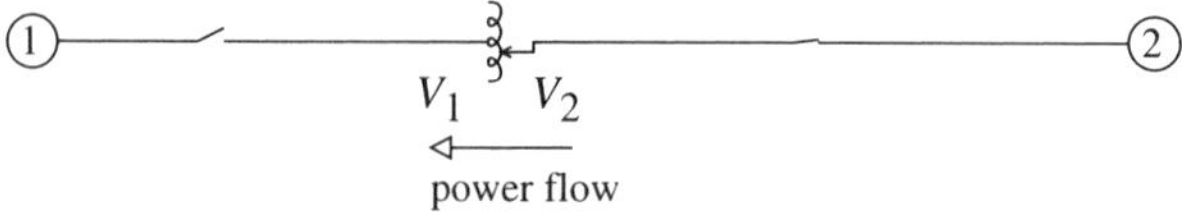

Generator downstream of a regulator: tries to regulate V_1

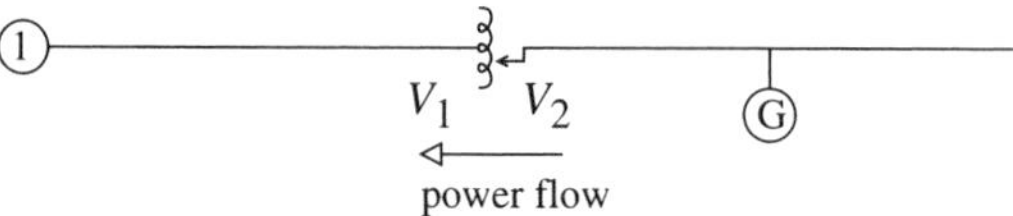

If V_1 is higher than the setpoint: it ratchets to the highest boost

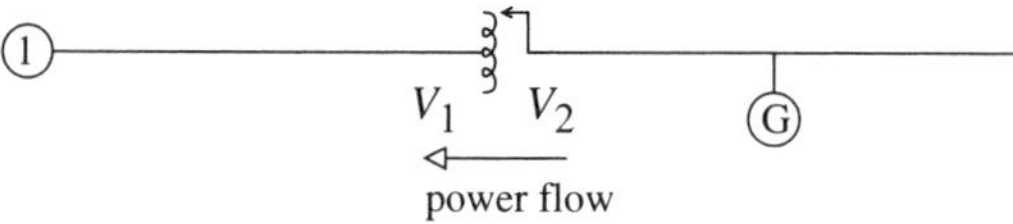

FIGURE 14.16
Misoperation of a bi-directional regulator due to reverse power flow.

reverses, some controllers can block tap changes or move to the neutral position.

- *Relocation* — Relocate the generator to upstream of the regulator, or relocate the regulator to downstream of the generator.
- *Equipment* — Add additional regulators or switched capacitors downstream of the generator.
- *Automation* — Use an automated distribution scheme to control the regulator based on data from other locations on the circuit.

Another consideration with regulators (and switched capacitor banks) is unwanted interaction with generators (especially fluctuating sources). Widening the regulator bandwidth setting helps avoid this. Increasing the time delay may also help prevent excessive regulator tap changes.

In addition to upsetting regulation, distributed generators have potential for significantly improving voltage profiles and voltage regulation. Generators inject real power, and many types can vary reactive power (capacitively or reactively). Distributed generators could be operated in a voltage regulation mode where voltage is regulated at the point of common coupling (or

TABLE 14.6

IEEE Harmonic Current Distortion Limits for Generation in Percent (IEEE P1547-D11-2003 and IEEE Std. 519-1992)

Individual Harmonics					Total Demand Distortion
n < 11	11 ≤ n < 17	17 ≤ n < 23	23 ≤ n < 35	35 ≤ n	
4.0	2.0	1.5	0.6	0.3	5.0

Note: Even harmonics are limited to 25% of the odd harmonic limits.

at some other point). This is a complex application that requires significant analysis to avoid problems.

14.4.2 Harmonics

Distributed generators must adhere to strict industry standards for harmonic injection. Table 14.6 shows the industry limits on harmonics, the same limits imposed on the largest loads. PWM-based inverters and rotating generators can meet these limits, so harmonic problems should be rare. Line-commutated inverters normally need filtering to meet these limits. For these limits, IEEE P1547-D11-2003 defines the current as the greater of the local system maximum current without the generator (15 or 30 min demand) and the generator rated current (transformed to the primary if needed).

Even though most generators do not produce significant harmonics, they can still be involved with harmonic problems. Generators have higher impedances to harmonics than the normal utility connection. Synchronous generators have about a 20 to 25% impedance to harmonics on its full-load rating. Compare that with a utility-source of the same size with an impedance of about 5% (approximately the service transformer impedance). If a high-harmonic load switches from the utility supply to a generator supply, the voltage distortion is worse, because the generator supply is not as stiff as the utility supply. The most prominent problem this causes is for load served by UPSs. High voltage distortion causes the UPSs to switch to battery, which drains the battery, possibly causing an interruption to the sensitive load on the UPS.

Calculate the voltage at the *n*th frequency with

$$V_n = n \cdot X \cdot I_n$$

where $X = X_2$ for positive or negative-sequence harmonics, $X = X_0$ for zero-sequence harmonics, and I_n is the harmonic current generated at the *n*th harmonic. Use superposition to calculate the harmonic voltage distortion due to multiple loads.

The classic model for the impedance to harmonics is the negative-sequence impedance. It is used for positive- and negative-sequence harmonics. Arseneau et al. (1979) showed that a constant impedance that linearly increases with frequency is relatively accurate. The negative-sequence impedance is $(X_d'' + X_q'')/2$. The direct-axis subtransient reactance is also a common approximation. Normally, this is accurate since X_d'' and X_q'' are usually close to each other. If no other information is available, use $X_2 = 25\%$.

EPRI PEAC tests of generators with different X_d'' and X_q'' showed that X_d'' is a better approximation for generators supplying single-phase power-supply load (EPRI 1005922, 2001; Langley et al., 2002). Single-phase power-supply load draws a burst of current near the voltage crest; from the generator's point of view, it is not a harmonic but more like a fault every half cycle. So, only the direct-axis impedance plays a role. Now, if a resonance occurs, the quadrature axis is expected to play more of a role, making voltage distortion worse since $(X_d'' + X_q'')/2$ is larger than X_d''.

Generators have a low impedance to the third harmonic and other zero-sequence harmonics. It is well known that balanced in-phase third-harmonic values are zero sequence (so $V_a=V_b=V_c$ for voltages and $I_a=I_b=I_c$ for currents). All harmonics that are multiples of three are zero sequence; 1, 4, 7 ... are positive sequence, and 2, 5, 8 ... are negative sequence.

A low zero-sequence harmonic impedance is usually an advantage in that it keeps the voltage distortion low for third-harmonic load currents. One problem is occasionally seen when two generators in parallel feed load and are separated from the utility. Generators also create harmonics. If the generators have different characteristics or are loaded differently, a third-harmonic current can circulate on the neutral between the generators. This can trip one or both generators on high neutral current. The circulating current is pronounced when one generator has a 3/4-pitch stator winding (which acts as a harmonic voltage source that is relatively high in third harmonics) and the other generator has a 2/3-pitch stator winding (which has a low impedance to harmonics).

14.4.3 Flicker

Large generators on weak distribution systems can generate voltage flicker. If it is large enough, any generator can cause a noticeable voltage change at starting or stopping. Voltage flicker is worse with generators that have varying power outputs, such as

- *Reciprocating engines* — Poor fuel or maintenance may result in misfiring cylinders. Misfiring leads to a fluctuating power output. Very large, low-rpm engines have a slow enough beat that the power output fluctuates under normal operations.
- *Wind turbines* — Tower shading causes flicker. As a blade passes the tower holding the turbine, the torque produced by the blade drops, so the power output drops.

Starting of induction generators also causes voltage flicker, just like starting an induction motor. Slower power changes normally do not cause flicker problems. For example, although photovoltaic power output varies significantly as clouds cover and then reveal the sun, the sunlight intensity change is slow enough that it does not cause severe flicker.

The flicker from reciprocating engines depends on the rotational speed of the generator and the quality of the fuel supply. Anderson and Mirheydar (1995) showed a case where misfiring cylinders caused voltage flicker on a 12-kV circuit and analyzed this problem in detail. Poor fuel was the cause — methane recovered from a former landfill.

The frequency of oscillations determines how severe the flicker is, with oscillations in the range of 5 to 10 Hz being most noticeable. The frequency of oscillation of the power stroke of a reciprocating engine operating in hertz is

$$f_F = \frac{N \cdot rpm}{30k}$$

(for 60 Hz only; for 50 Hz, divide by 25 instead of 30)
where

N = number of cylinders
rpm = rotational speed of the generator, rev/min

$$k = \begin{cases} 2 \text{ for two-cycle engine} \\ 4 \text{ for a four-cycle engine} \end{cases}$$

For all but the very slowest generators with few cylinders, the frequency of this power stroke beat is fast enough; it cannot cause flicker. But, if cylinders misfire, the frequency of power oscillation with the misfire in hertz is

$$f_{PS} = \frac{rpm}{30k}$$

For many common generators, the power oscillation frequency with misfiring cylinders lies right in the range where it is very noticeable. A 4-cycle 900-rpm engine produces flicker at 7.5 Hz, at the most sensitive part of the voltage flicker curve. An 1800-rpm engine is better (and fortunately more standard); these produce flicker at 15 Hz; although this can still cause problems, it is on the less sensitive part of the voltage flicker curve.

The main way to prevent reciprocating engine flicker is controlling the fuel quality and performing proper maintenance to minimize misfiring.

Most of the experience with flicker and wind turbines comes from Europe where wind turbines have higher penetration levels (and the higher-voltage lamp filament makes flicker worse). Gerdes and Santjer (1996) found that flicker is not a problem on most wind turbines if

$$I_{SC}/I_L > 100$$

where

I_L = wind turbine rated load current

I_{SC} = three-phase short-circuit current at the point of common coupling

Their measurements found most problems where the short-circuit ratio was less than 40, but there were large differences between wind turbines. Some were fine even at weak locations with a short-circuit ratio as low as 10, but the worst performers caused flicker even with a short-circuit ratio of over 90. Variations in performance were mainly due to aerodynamic and mechanical differences.

There are several possible generator-side remedies for reducing the flicker caused by generators. These are not normally done, but some options include

- *Controls* — Use internal controls to regulate real power output to slow down the rate of output change. For example, many inverters and rotational generators include soft-start functions that allow them to slowly ramp power up over many seconds rather than step the power on.
- *Storage* — Use a battery or other energy-storage device to help reduce power output swings.
- *Vars* — Use changes in reactive power to offset the changes in real power. PWM inverters can change the reactive power output, so the generator could operate lagging when the unit is producing full real power and change to a leading power factor (to increase voltage) when the unit is not producing any real power. Synchronous generators with fast exciters may also be able to correct for flicker.
- *Size* — Limit the size of the generator to prevent flicker.

Solutions based on individual types of generators include

- *Wind turbines* — Choose a design that reduces the effect of tower shading.
- *Reciprocating engines* — Use proper fuel grades, and keep the engine tuned up.
- *Induction generators* — Use a reduced-voltage starter that uses a method such as an autotransformer, extra impedance, solid-state electronic voltage reduction, or transformer reconnection from wye during starting to delta during normal operation. Alternately, use a self-starter to get the generator up to speed before connecting to the grid.

Other solutions for flicker include any of the standard techniques discussed in Chapter 11 for reducing voltage flicker from fluctuating loads.

A final consideration for voltage flicker is that a large generator located on a weak system may improve the voltage flicker caused by adjacent fluctuating loads. The generator control algorithm determines how effectively it can correct voltage. A constant-voltage control with a fast response (such as a synchronous generator with a fast exciter regulating voltage or a PWM inverter) is most likely to respond to flicker. Of course, a large generator located on a weak system is more likely to have other issues, such as its own flicker, voltage-regulation problems, and islanding problems.

14.4.4 Other Impacts on Power Quality

Distribution systems normally do not have stability issues. That may change with large penetration of generators. The largest concern is with faults that cause the distributed generators to trip (since they have sensitive voltage sag settings). Miller et al. (2001) found that with a 40% penetration of generation on a circuit, motor load on the circuit could stall, causing a voltage collapse on the circuit.

Large penetration of distributed generators may also impact bulk system stability. Miller (2002) performed simulations of the U.S. western grid and found that a 20% penetration of distributed generation seriously erodes stability when the distributed generators trip off quickly because of voltage or frequency relays. Another concern with the bulk system stability is with active anti-islanding inverters. Significant penetration of these will degrade the overall system stability.

Large synchronous generators on a distribution system might require changes to reclosing practices. Because of the danger of reclosing into an island that is out of phase, the reclosing delay might have to be extended. Many utilities use an immediate reclose attempt, which means there is no intentional delay. Circuit breakers typically reclose in about one-third to one-half of a second. This may not give the generator time to clear, especially in a near-island situation. Extending the reclose time on utility breakers and reclosers gives the generator more time to clear.

As an alternative to extending the time before reclose, a voltage-supervising relay can block reclosing on a circuit breaker until the line voltage is dead.

14.4.5 High Quality Power Configurations

To enhance reliability, most configurations of generators only help with long-duration interruptions, not with voltage sags, not with momentary interruptions. Figure 14.17 shows two configurations that use continuously running generators to supply critical loads with higher-quality power. The basic idea is to use a static switch for fast transfers between the utility and the generator. The basic configurations are

- *On-line configuration:* The generator supplies some or all of the load. If a sag or interruption hits the utility supply, a static switch opens up, and the generator supplies all of the load.
- *Off-line configuration:* The load normally runs off of the generator. If the generator fails, a static transfer switch operates and switches the load to the utility input.

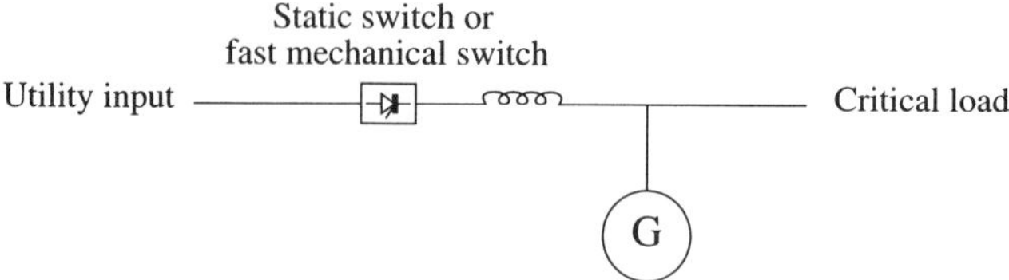

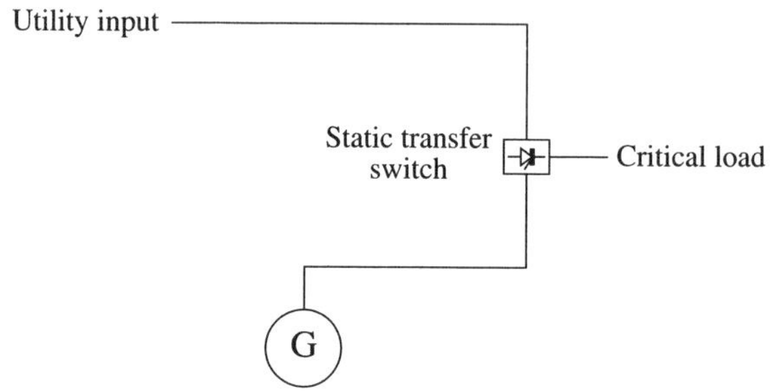

FIGURE 14.17
Generator configurations supplying protection against short-duration as well as long-duration interruptions. (Copyright © 2002. Electric Power Research Institute. 1007281. *Analysis of Extremely Reliable Power Delivery Systems.* Reprinted with permission.)

In both of these cases, the generator must match the load quickly. This is difficult, as many generation technologies have limited load-following capability. Some of the ways of matching generation to the load are as follows:

- Shed load by tripping load breakers to match the load to the generator.
- Use some sort of power electronics with short-term energy storage to dynamically correct for mismatches to give the generator time to match the load.

Also consider that the generator must supply motor starting current if the critical load has motors.

14.5 Generator Reliability

Generator reliability is important when the distributed generator is critical: stand-alone customer systems, micro-grids, generation applied to relieve peak loading (it does not help if it is offline). Figure 14.18 shows that indi-

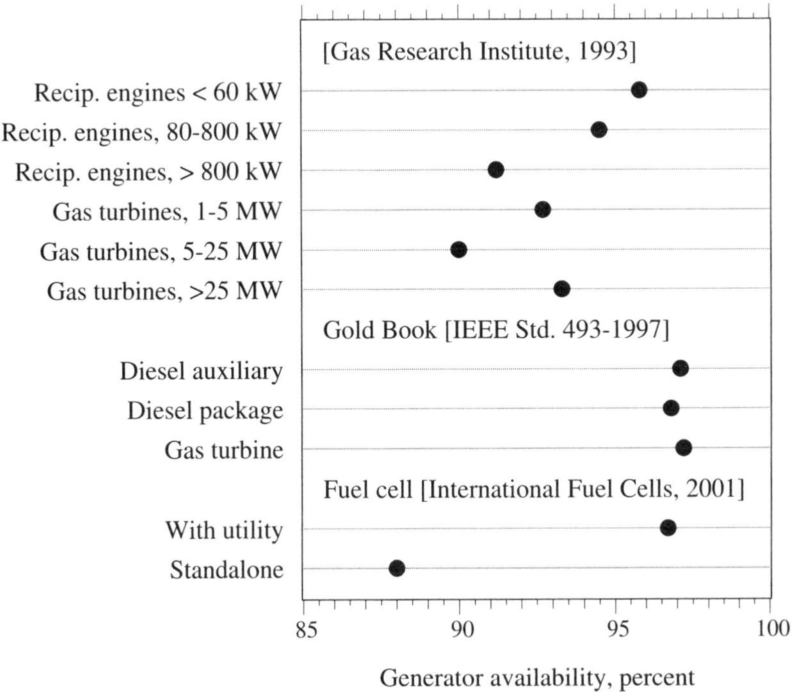

FIGURE 14.18
Generator availability. (Data from [GRI, 1993; IEEE Std. 493-1997; International Fuel Cells, 2001]. Figure copyright © 2002. Electric Power Research Institute. 1007281. *Analysis of Extremely Reliable Power Delivery Systems.* Reprinted with permission.)

vidual generators typically have availabilities ranging from 90 to 97%. This is much lower than the availability of typical electric utility supplies. A 97% availability is 11 days interrupted per year. The *Gold Book* (IEEE Std. 493-1997 based on Smith et al., 1990) cites failure rates of 32.9, 11.8, and 15 failures per year for diesels with auxiliary support systems, diesels packaged as units, and gas turbines.

As a supplement to a utility connection, a generator can greatly increase the reliability to the end user. Consider a utility supply with an availability of 99.966% (3.5 h of down time annually). If a generator with an availability of 97% (11 days of unavailability) is paralleled, the total availability of the parallel sources improves to 99.9988% (1–[1–0.97][1–0.99966]) or an average of 6 min of down time per year. Without the utility, designing a reliable supply becomes more difficult.

Without a utility supply, any critical application needs redundant generators. Assuming that an individual generator has 97% availability, Table 14.7 shows availabilities for several combinations of generators. This data is repeated as a graph in Figure 14.19 showing the number of nines of availability (a 99.9% availability is three nines, a 99.99% availability is four nines) against the generation capacity and notations showing the configuration.

TABLE 14.7

Configurations of Redundant Generators

Number of Units Needed	Total Number of Units	Notation	Capacity	System Availability
1	1	N	100%	0.97
1	2	N+1	200%	0.9991
1	3	N+2	300%	0.999973
1	4	N+3	400%	0.99999919
2	2	N	100%	0.941
2	3	N+1	150%	0.9974
2	4	N+2	200%	0.99989
2	5	N+3	250%	0.9999960
2	6	N+4	300%	0.99999986
3	3	N	100%	0.912
3	4	N+1	133%	0.9948
3	5	N+2	167%	0.99974
3	6	N+3	200%	0.999988
4	4	N	100%	0.885
4	5	N+1	125%	0.9915
4	6	N+2	150%	0.99950
4	7	N+3	175%	0.999974
4	8	N+4	200%	0.9999987
5	5	N	100%	0.858
5	6	N+1	120%	0.988
5	7	N+2	140%	0.99914
5	8	N+3	160%	0.999949
5	9	N+4	180%	0.9999972
5	10	N+5	200%	0.99999986

The binomial distribution shows the failure probabilities for r out of n generators. The probability of exactly r successes in n trials with a probability of success p (Billinton and Allan, 1984) is

$$P_r = {}_nC_r p^r (1-p)^{n-r} = \frac{n!}{r!(n-r)!} p^r (1-p)^{n-r}$$

So, for an example with $p = 0.97$, and three total generators, the probabilities of a given numbers of generators being out of service are

Units Out of Service	Probability of the Given Number of Units Out of Service
0	$0.97^3 = 0.912673$
1 ($r = 2, n = 3$)	$3\ (0.97)^2\ (1–0.97) = 0.084681$
2 ($r = 1, n = 3$)	$3\ (0.97)\ (1–0.97)^2 = 0.002619$
3	$(1–0.97)^3 = 0.000027$
	Sum = 1.0

Maintenance plays a key role in the reliability of generators. Not counting down time for maintenance, the availability of generators jumps from 97%

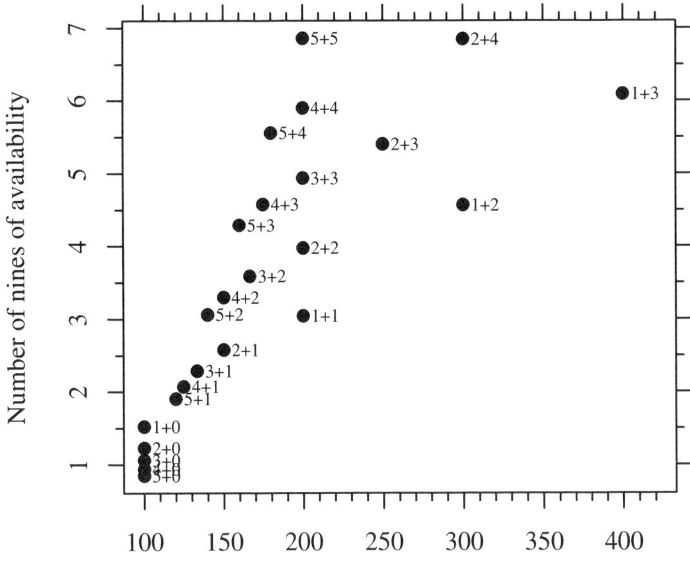

FIGURE 14.19
Theoretical availability with different generation configurations. The graph assumes an individual unit is 97% available. The $k + n$ notation means k generators are needed to meet the peak load, and the site has n extra generators. (Copyright © 2002. Electric Power Research Institute. 1007281. *Analysis of Extremely Reliable Power Delivery Systems.* Reprinted with permission.)

TABLE 14.8

Generator Availabilities, Percent

	Including Maintenance	Excluding Maintenance
Diesel auxiliary	97.1	98.9
Diesel package	96.8	99.1
Gas turbine	97.2	99.4

Source: IEEE Std. 493-1997, *IEEE Recommended Practice for the Design of Reliable Industrial and Commercial Power Systems (Gold Book)*; Smith, C. A., Donovan, M. D., and Bartos, M. J., "Reliability Survey of 600 to 1800 kW Diesel and Gas-Turbine Generating Units," *IEEE Transactions on Industry Applications*, vol. 26, no. 4, pp. 741–55, July/August 1990.

to 99% (see Table 14.8). If maintenance can be performed at non-critical times, the reliability of the installation improves greatly.

The availabilities in Figure 14.19 are theoretical and do not include overlapping or common-mode failures. For generators, several factors contribute to common-mode failures:

- *Fuel supply* — Continuously run generators likely have a common supply of natural gas. Backup generators are more likely to have independent tanks for supply. Units with dual-fuel capability can reduce the likelihood of this failure mode.
- *Controls* — Generators may have common controls for starting.
- *Space* — If generators are in close proximity, failure of one may require adjacent units to be shut down.
- *Design flaws* — Common design flaws may fail all units under the right conditions.

Good design, especially with redundant supporting systems, helps minimize the probability of common-mode failures, and periodic testing helps minimize hidden failures. And because maintenance is such a large component of down time, coordinating maintenance among parallel units and scheduling for non-critical times helps reduce the chance of failures.

References

Anderson, P. M. and Mirheydar, M., "Analysis of a Diesel-Engine Driven Generating Unit and the Possibility for Voltage Flicker," *IEEE Transactions on Energy Conversion*, vol. 10, no. 1, pp. 37–47, March 1995.

ANSI C84.1-1995, *American National Standards for Electric Power Systems and Equipment — Voltage Ratings (60 Hz).*

ANSI/IEEE C37.90-1989, *IEEE Standard for Relays and Relay Systems Associated with Electric Power Apparatus.*

ANSI/IEEE C37.90.1-1989, *IEEE Standard Surge Withstand Capability (SWC) Tests for Protective Relays and Relay Systems.*

ANSI/IEEE C62.92-1987, *IEEE Guide for the Application of Neutral Grounding in Electrical Utility Systems. Part I — Introduction.*

ANSI/IEEE Std. 1001-1988, *IEEE Guide for Interfacing Dispersed Storage and Generation Facilities with Electric Utility Systems.*

Arseneau, R., Hill, E. F., Szabados, R., and Ferry, D., "Synchronous Motor Impedances Measured at Harmonic Frequencies," IEEE/PES Summer Power Meeting, Vancouver, Canada, 1979.

Barker, P. P. and De Mello, R. W., "Determining the Impact of Distributed Generation on Power Systems. I. Radial Distribution Systems," IEEE Power Engineering Society Summer Meeting, 2000.

Barker, P. P. and Johnson, B. K., "Power System Modeling Requirements for Rotating Machine Interfaced Distributed Resources," IEEE Power Engineering Society Summer Meeting, 2002.

Billinton, R. and Allan, R. N., *Reliability Evaluation of Power Systems*, Pitman Advanced Publishing Program, 1984.

CEA 128 D 767, *Connecting Small Generators to Utility Distribution Systems*, Canadian Electrical Association, 1994.

Chen, T. H., Chen, M. S., Inoue, T., Kotas, P., and Chebli, E. A., "Three-Phase Cogenerator and Transformer Models for Distribution System Analysis," *IEEE Transactions on Power Delivery*, vol. 6, no. 4, pp. 1671–81, October 1991.

Comfort, R., Gonzalez, M., Mansoor, A., Barker, P. P., Short, T. A., and Sundaram, A., "Power Quality Impact of Distributed Generation: Effect on Steady State Voltage Regulation," PQA Conference, 2001.

Dugan, R. C. and Rizy, D. T., "Electric Distribution Protection Problems Associated with the Interconnection of Small, Dispersed Generation Devices," *IEEE Transactions on Power Apparatus and Systems*, vol. PAS-103, no. 6, pp. 1121–7, June 1984.

EEI, *Distributed Resources Task Force Interconnection Study*, Edison Electric Institute, 2000. Chair: Murray Davis of Detroit Edison.

EPRI 1007281, *Analysis of Extremely Reliable Power Delivery Systems: A Proposal for Development and Application of Security, Quality, Reliability, and Availability (SQRA) Modeling for Optimizing Power System Configurations for the Digital Economy*, Electric Power Research Institute, Palo Alto, CA, 2002.

EPRI 1000419, *Engineering Guide for Integration of Distributed Generation and Storage Into Power Distribution Systems*, Electric Power Research Institute, Palo Alto, CA, 2000.

EPRI 1003973, *Investigation of the Technical and Economic Feasibility of Micro-grid Based Power Systems*, Electric Power Research Institute, Palo Alto, CA, 2001.

EPRI 1005917, *Distributed Generation Relaying Impacts on Power Quality*, Electric Power Research Institute, Palo Alto, CA, 2001.

EPRI 1005922, *Distributed Generation Source Stiffness and its Impact on Voltage Distortion: Laboratory Test Results and Analytical Model Verification*, Electric Power Research Institute, Palo Alto, CA, 2001.

EPRI PEAC, "Distributed Generation Course," 2002.

EPRI TR-105589, *Engineering Handbook for Dispersed Energy Systems on Utility Distribution Systems*, Electric Power Research Institute, Palo Alto, CA, 1995.

EPRI TR-111489, *Integration of Distributed Resources in Electric Utility Systems: Current Interconnection Practice and Unified Approach*, Electric Power Research Institute, Palo Alto, CA, 1998.

EPRI TR-111490, *Integration of Distributed Resources in Electric Utility Distribution Systems: Distribution System Behavior Analysis for Suburban Feeder*, Electric Power Research Institute, Palo Alto, CA, 1998.

EPRI TR-112737, *Integration of Distributed Resources in Electric Utility Distribution Systems: Distribution System Behavior Analysis for Urban and Rural Feeders*, Electric Power Research Institute, Palo Alto, CA, 1999.

Feero, W. E. and Gish, W. B., "Overvoltages Caused by DSG Operation: Synchronous and Induction Generators," *IEEE Transactions on Power Delivery*, vol. PWRD-1, no. 1, pp. 258–64, January 1986.

Fitzgerald, A. E., Kingsley Jr., C., and Kusko, A., *Electric Machinery*, McGraw-Hill, New York, 1971.

Gerdes, G. J. and Santjer, F., "Results from Power Quality Measurements in Germany," American Wind Energy Association Windpower '96, Denver, CO, June 1996.

GRI, "Reliability of Natural Gas Power Generation Systems," Gas Research Institute (GRI) White Paper, 1993. Citing results from: Brown, H. W. and Stuber, F. S., "Reliability of Natural Gas Cogeneration Systems," Final Report, GRI-93/0020, September 1993.

IEEE P1547-D11-2003, *Draft Standard for Distributed Resources Interconnected with Electric Power Systems.*

IEEE Std. 493-1997, *IEEE Recommended Practice for the Design of Reliable Industrial and Commercial Power Systems (Gold Book).*

IEEE Std. 519-1992, *IEEE Recommended Practices and Requirements for Harmonic Control in Electrical Power Systems.*

IEEE Std. 929-2000, *IEEE Recommended Practice for Utility Interface of Photovoltaic (PV) Systems.*

IEEE Std. C37.90.2-1995, *IEEE Standard for Withstand Capability of Relay Systems to Radiated Electromagnetic Interference from Transceivers.*

International Fuel Cells, "Fuel Cell Reliability for the 12-Month Period Ending Nov. 1, 2001," 2001. As cited at http://www.ifc.com.

Langley, R., Short, T. A., and Mansoor, A., "Impact of Non-Linear Loads on M-G Sets," PQA Conference, Portland, OR, 2002.

Miller, N. W., Presentation at the IEEE/PES Winter Power Meeting, 2002.

Miller, N. W., Walling, R. A., and Achilles, A. S., "Impact of Distributed Resources on System Dynamic Performance," IEEE/PES Transmission and Distribution Conference and Exposition, 2001.

Smith, C. A., Donovan, M. D., and Bartos, M. J., "Reliability Survey of 600 to 1800 kW Diesel and Gas-Turbine Generating Units," *IEEE Transactions on Industry Applications*, vol. 26, no. 4, pp. 741–55, July/August 1990.

Stevens, J., Bonn, R., Ginn, J., Gonzalez, S., and Kern, G., "Development and Testing of an Approach to Anti-Islanding In Utility-Interconnected Photovoltaic Systems," Sandia National Laboratories, 1999. SAND98-1684.

UL 1741, *Inverters, Converters, and Controllers for Use in Independent Power Systems*, Underwriters Laboratories, Inc., Northbrook, IL, 2001.

Wagner, C. L., Feero, W. E., Gish, W. B., and Jones, R. H., "Relay Performance in DSG Islands," *IEEE Transactions on Power Delivery*, vol. 4, no. 1, pp. 122–31, January 1989.

Tree Trimmer Crew Leader Dies When He Contacts Energized Power Line in Puerto Rico

A tree trimming crew was en route to a job site when they noticed that trees had fallen over a 4,000-volt power line in front of a radio tower. The damage had been caused by Hurricane Hugo. Since all the power lines in the area had been de-energized, the crew leader decided to clear the area in front of the radio tower. The crew cut and pruned the trees to clear the area, and upon finishing, returned to the truck while the crew leader made a final inspection of the work area. Crew members heard their leader cry out and ran to the work site, where they found him lying on his back. No vital signs could be detected. Investigation revealed that although the power lines had been de-energized, a gas station was using a portable gas-powered electric generator to supply electrical power to the gas pumps. Since the main circuit breaker at the gas station had not been opened, electrical current from the generator flowed back through the transformer and energized the power line at the work area.

U.S. National Institute for Occupational Safety and Health

Index

A

B

C

D

E

F

G

H

I

L

M

N

O

P

Q

R

S

U

V

W

X

Z